Physics, the Human Adventure

Physics, the Human Adventure

From Copernicus to Einstein and Beyond

Gerald Holton and Stephen G. Brush
Harvard University *University of Maryland*

RUTGERS UNIVERSITY PRESS
New Brunswick, New Jersey and London

Library of Congress Cataloging-in-Publication Data

Holton, Gerald James.
Physics, the human adventure : from Copernicus to Einstein and beyond / Gerald Holton and Stephen G. Brush. — 3rd ed.
p. cm.
Rev. ed. of: Introduction to concepts and theories in physical science. 1952.
Includes bibliographical references and index.
ISBN 0-8135-2907-7 (alk. paper) — ISBN 0-8135-2908-5 (pbk. : alk. paper)
1. Physical sciences. I. Brush, Stephen G. II. Holton, Gerlad James. Introduction to concecpts and theories in physical science. III. Title.

Q160 .H654 2001
500—dc21 00-062534

British Cataloging-in-Publication data for this book is available from the British Library

First edition published by Addison-Wesley 1952 under the title
Introduction to Concepts and Theories in Physical Science

Second Addison-Wesley edition, 1972

Paperback of second edition published by Princeton University Press, 1985

Manufactured in the United States of America

To Nina and Phyllis

And to the countless men and women worldwide whose painstaking experiments and leaps of scientific imagination revealed our universe and invigorated our civilization

Contents

Preface xiii

PART A
The Origins of Scientific Cosmology

CHAPTER 1
The Astronomy of Ancient Greece 3

1.1 The Motions of Stars, Suns, and Planets 3
1.2 Plato's Problem 5
1.3 The Aristotelian System 6
1.4 How Big Is the Earth? 8
1.5 The Heliocentric Theory 10
1.6 Modified Geocentric Theories 11
1.7 The Success of the Ptolemaic System 14

CHAPTER 2
Copernicus' Heliocentric Theory 17

2.1 Europe Reborn 17
2.2 The Copernican System 17
2.3 Bracing the System 22
2.4 The Opposition to Copernicus's Theory 23
2.5 Historic Consequences 25

CHAPTER 3
On the Nature of Scientific Theory 27

3.1 The Purpose of Theories 27
3.2 The Problem of Change: Atomism 30
3.3 Theories of Vision 31
3.4 Criteria for a Good Theory in Physical Science 35

CHAPTER 4
Kepler's Laws 40

4.1 The Life of Johannes Kepler 40
4.2 Kepler's First Law 41
4.3 Kepler's Second Law 43
4.4 Kepler's Third Law 45
4.5 Kepler's Theory of Vision 46
4.5 The New Concept of Physical Law 47

CHAPTER 5
Galileo and the New Astronomy 50

5.1 The Life of Galileo 50
5.2 The Telescopic Evidences for the Copernican System 52
5.3 Toward a Physical Basis for the Heliocentric System 54
5.4 Science and Freedom 58

PART B
The Study of Motion

CHAPTER 6
Mathematics and the Description of Motion 63

6.1 René Descartes 63
6.2 Constant Velocity 65
6.3 The Concept of Average Speed 67
6.4 Instantaneous Speed 68
6.5 Acceleration 70
6.6 Oresme's Graphical Proof of the Mean-speed Theorem 72
6.7 Equations of Motion for Constant Acceleration 73

CHAPTER 7
Galileo and the Kinematics of Free Fall 77

7.1 Introduction 77
7.2 Aristotelian Physics 78
7.3 Galileo's *Two New Sciences* 80
7.4 Galileo's Study of Accelerated Motion 83

CHAPTER 8
Projectile Motion 88

8.1 Projectile with Initial Horizontal Motion 88
8.2 Introduction to Vectors 91
8.3 The General Case of Projectile Motion 93
8.4 Applications of the Law of Projectile Motion 96
8.5 Galileo's Conclusions 97
8.6 Summary 99

PART C
Newton's Laws and His System of the World

CHAPTER 9
Newton's Laws of Motion 103

9.1 Science in the Seventeenth Century 103
9.2 A Short Sketch of Newton's Life 104
9.3 Newton's *Principia* 105
9.4 Newton's First Law of Motion 108
9.5 Newton's Second Law of Motion 109
9.6 Standard of Mass 111
9.7 Weight 112
9.8 The Equal-Arm Balance 114
9.9 Inertial and Gravitational Mass 115
9.10 Examples and Applications of Newton's Second Law of Motion 116
9.11 Newton's Third Law of Motion 118
9.12 Examples and Applications of Newton's Third Law 119

CHAPTER 10
Rotational Motion 123

10.1 Kinematics of Uniform Circular Motion 123
10.2 Centripetal Acceleration 125
10.3 Derivation of the Formula for Centripetal Acceleration and Force 127
10.4 The Earth's Centripetal Acceleration and Absolute Distances in the Solar System 128

CHAPTER 11
Newton's Law of Universal Gravitation 131

11.1 Derivation of the Law of Universal Gravitation 131
11.2 Gravitating Planets and Kepler's Third Law 135
11.3 The Cavendish Experiment: The Constant of Gravitation 136
11.4 The Masses of the Earth, Sun, and Planets 138
11.5 Some Influences on Newton's Work 139
11.6 Some Consequences of the Law of Universal Gravitation 140
11.7 The Discovery of New Planets Using Newton's Theory of Gravity 144
11.8 Bode's Law: An Apparent Regularity in the Positions of the Planets 146
11.9 Gravity and the Galaxies 149
11.10 "I Do Not Feign Hypotheses" 151
11.11 Newton's Place in Modern Science 153

PART D
Structure and Method in Physical Science

CHAPTER 12
On the Nature of Concepts 157

12.1 Introduction: The Search for Constancies in Change 157
12.2 Science and Nonscience 158
12.3 The Lack of a Single Method 159
12.4 Physical Concepts: Measurement and Definition 161
12.5 Physically Meaningless Concepts and Statements 163
12.6 Primary and Secondary Qualities 164
12.7 Mathematical Law and Abstraction 165
12.8 Explanation 167

CHAPTER 13
On the Duality and Growth of Science 170
13.1 The Free License of Creativity 170
13.2 "Private" Science and "Public" Science 171
13.3 The Natural Selection of Physical Concepts 172
13.4 Motivation 174
13.5 Objectivity 176
13.6 Fact and Interpretation 177
13.7 How Science Grows 178
13.8 Consequences of the Model 180

CHAPTER 14
On the Discovery of Laws 187
14.1 Opinions on Scientific Procedure 187
14.2 A Sequence of Elements in Formulations of Laws 191
14.3 The Limitations of Physical Law 195
14.4 The Content of Science: Summary 197

PART E
The Laws of Conservation

CHAPTER 15
The Law of Conservation of Mass 203
15.1 Prelude to the Conservation Law 203
15.2 Steps Toward a Formulation 203
15.3 Lavoisier's Experimental Proof 204
15.4 Is Mass Really Conserved? 206

CHAPTER 16
The Law of Conservation of Momentum 209
16.1 Introduction 209
16.2 Definition of Momentum 210
16.3 Momentum and Newton's Laws of Motion 212
16.4 Examples Involving Collisions 213
16.5 Examples Involving Explosions 215
16.6 Further Examples 215
16.7 Does Light Have Momentum? 216
16.8 Angular Momentum 217

CHAPTER 17
The Law of Conservation of Energy 219
17.1 Christiaan Huygens and the Kinetic Energy (*Vis Viva*) Concept 219
17.2 Preliminary Questions: The Pile Driver 222
17.3 The Concept of Work 223
17.4 Various Forms of Energy 224
17.5 The Conservation Law: First Form and Applications 226
17.6 Extensions of the Conservation Law 229
17.7 Historical Background of the Generalized Law of Conservation of Energy: The Nature of Heat 234
17.8 Mayer's Discovery of Energy Conservation 239
17.9 Joule's Experiments on Energy Conservation 242
17.10 General Illustration of the Law of Conservation of Energy 245
17.11 Conservation Laws and Symmetry 247

CHAPTER 18
The Law of Dissipation of Energy 251
18.1 Newton's Rejection of the "Newtonian World Machine" 251
18.2 The Problem of the Cooling of the Earth 253
18.3 The Second Law of Thermodynamics and the Dissipation of Energy 256
18.4 Entropy and the Heat Death 259

PART F
Origins of the Atomic Theory in Physics and Chemistry

CHAPTER 19
The Physics of Gases 265
19.1 The Nature of Gases—Early Concepts 265
19.2 Air Pressure 267
19.3 The General Gas Law 270
19.4 Two Gas Models 272

CHAPTER 20
The Atomic Theory of Chemistry 275
20.1 Chemical Elements and Atoms 275
20.2 Dalton's Model of Gases 276
20.3 Properties of Dalton's Chemical Atom 278
20.4 Dalton's Symbols for Representing Atoms 279
20.5 The Law of Definite Proportions 280
20.6 Dalton's Rule of Simplicity 281
20.7 The Early Achievements of Dalton's Theory 282
20.8 Gay-Lussac's Law of Combining Volumes of Reacting Gases 284
20.9 Avogadro's Model of Gases 285
20.10 An Evaluation of Avogadro's Theory 288
20.11 Chemistry After Avogadro: The Concept of Valence 289
20.12 Molecular Weights 292

CHAPTER 21
The Periodic Table of Elements 296
21.1 The Search for Regularity in the List of Elements 296
21.2 The Early Periodic Table of Elements 297
21.3 Consequences of the Periodic Law 301
21.4 The Modern Periodic Table 303

CHAPTER 22
The Kinetic-Molecular Theory of Gases 308
22.1 Introduction 308
22.2 Some Qualitative Successes of the Kinetic- Molecular Theory 310
22.3 Model of a Gas and Assumptions in the Kinetic Theory 311
22.4 The Derivation of the Pressure Formula 315
22.5 Consequences and Verification of the Kinetic Theory 318
22.6 The Distribution of Molecular Velocities 322
22.7 Additional Results and Verifications of the Kinetic Theory 327
22.8 Specific Heats of Gases 329
22.9 The Problem of Irreversibility in the Kinetic Theory: Maxwell's Demon 333
22.10 The Recurrence Paradox 336

PART G
Light and Electromagnetism

CHAPTER 23
The Wave Theory of Light 341
23.1 Theories of Refraction and the Speed of Light 341
23.2 The Propagation of Periodic Waves 344
23.3 The Wave Theory of Young and Fresnel 347
23.4 Color 350

CHAPTER 24
Electrostatics 352
24.1 Introduction 352
24.2 Electrification by Friction 352
24.3 Law of Conservation of Charge 353
24.4 A Modern Model for Electrification 353
24.5 Insulators and Conductors 354
24.6 The Electroscope 356
24.7 Coulomb's Law of Electrostatics 357
24.8 The Electrostatic Field 359
24.9 Lines of Force 361
24.10 Electric Potential Difference—Qualitative Discussion 362
24.11 Potential Difference—Quantitative Discussion 363
24.12 Uses of the Concept of Potential 364
24.13 Electrochemistry 365
24.14 Atomicity of Charge 366

CHAPTER 25
Electromagnetism, X-Rays, and Electrons 369
25.1 Introduction 369
25.2 Currents and Magnets 369

25.3 Electromagnetic Waves and Ether 374
25.4 Hertz's Experiments 377
25.5 Cathode Rays 379
25.6 X-rays and the Turn of the Century 382
25.7 The "Discovery of the Electron" 385

CHAPTER 26
The Quantum Theory of Light 388
26.1 Continuous Emission Spectra 388
26.2 Planck's Empirical Emission Formula 391
26.3 The Quantum Hypothesis 392
26.4 The Photoelectric Effect 396
26.5 Einstein's Photon Theory 398
26.6 The Photon-Wave Dilemma 400
26.7 Applications of the Photon Concept 402
26.8 Quantization in Science 403

PART H
The Atom and the Universe in Modern Physics

CHAPTER 27
Radioactivity and the Nuclear Atom 409
27.1 Early Research on Radioactivity and Isotopes 409
27.2 Radioactive Half-Life 413
27.3 Radioactive Series 415
27.4 Rutherford's Nuclear Model 417
27.5 Moseley's X-Ray Spectra 422
27.6 Further Concepts of Nuclear Structure 424

CHAPTER 28
Bohr's Model of the Atom 427
28.1 Line Emission Spectra 427
28.2 Absorption Line Spectra 428
28.3 Balmer's Formula 432
28.4 Niels Bohr and the Problem of Atomic Structure 434
28.5 Energy Levels in Hydrogen Atoms 435
28.6 Further Developments 441

CHAPTER 29
Quantum Mechanics 446
29.1 Recasting the Foundations of Physics Once More 446
29.2 The Wave Nature of Matter 447
29.3 Knowledge and Reality in Quantum Mechanics 451
29.4 Systems of Identical Particles 456

CHAPTER 30
Einstein's Theory of Relativity 462
30.1 Biographical Sketch of Albert Einstein 462
30.2 The FitzGerald-Lorentz Contraction 464
30.3 Einstein's Formulation (1905) 467
30.4 Galilean Transformation Equations 468
30.5 The Relativity of Simultaneity 470
30.6 The Relativistic (Lorentz) Transformation Equations 472
30.7 Consequences and Examples 474
30.8 The Equivalence of Mass and Energy 474
30.9 Relativistic Quantum Mechanics 477
30.10 The General Theory of Relativity 480

CHAPTER 31
The Origin of the Solar System and the Expanding Universe 487
31.1 The Nebular Hypothesis 487
31.2 Planetesimal and Tidal Theories 489
31.3 Revival of Monistic Theories After 1940 491
31.4 Nebulae and Galaxies 494
31.5 The Expanding Universe 495
31.6 Lemaître's Primeval Atom 496

CHAPTER 32
Construction of the Elements and the Universe 499
32.1 Nuclear Physics in the 1930s 499
32.2 Formation of the Elements in Stars 503
32.3 Fission and the Atomic Bomb 506
32.4 Big Bang or Steady State? 509
32.5 Discovery of the Cosmic Microwave Radiation 512
32.6 Beyond the Big Bang 513

CHAPTER 33
Thematic Elements and Styles in Science 517

33.1 The Thematic Element in Science 517
33.2 Themata in the History of Science 520
33.3 Styles of Thought in Science and Culture 522
33.4 Epilogue 525

APPENDIXES

APPENDIX I
Abbreviations and Symbols 531

APPENDIX II
Metric System Prefixes, Greek Alphabet, Roman Numerals 535

APPENDIX III
Defined Values, Fundamental Constants and Astronomical Data 537

APPENDIX IV
Conversion Factors 539

APPENDIX V
Systems of Units 541

APPENDIX VI
Alphabetic List of the Elements 543

APPENDIX VII
Periodic Table of Elements 545

APPENDIX VIII
Summary of Some Trigonometric Relations 547

APPENDIX IX
Vector Algebra 551

General Bibliography 555

Credits 559

Index 561

Refer to the website for this book, www.ipst.umd.edu/Faculty/brush/physicsbibliography.htm, for Sources, Interpretations, and Reference Works and for Answers to Selected Numerical Problems.

Preface

For well over a century, the lives of most people in the developed countries were changed—more often than not, improved—by the application of discoveries in physics. Radio and television, computers and the Internet, x-rays, and lasers are only a few well-known examples. Some physicists became celebrities because of their astonishing discoveries and theories: Marie Curie and Ernest Rutherford, for deciphering the mysterious phenomena of radioactivity and transmutation, and Albert Einstein and Stephen Hawking, for overturning common-sense notions of space, time, matter, and energy. Yet physics itself, which should be a major part of the intellectual heritage of all properly educated men and women, is often regarded—wrongly—as an esoteric subject, understood (at least in the United States) by only a small elite. The sincere and heroic efforts of educators and writers have produced progress in teaching the public and its political leaders what physics is all about, but much more will be needed in the twenty-first century.

The authors believe that physics can be effectively taught by showing *how* its basic principles and results were established by Galileo Galilei, Isaac Newton, James Clerk Maxwell, Albert Einstein, and others, and how that science is a true part of our developing cultural heritage. That means not just presenting the theories we now consider correct, but trying to demonstrate how those theories replaced others that seemed, at the time, more plausible. One cannot appreciate Galileo's achievement without first understanding the widely accepted Aristotelian doctrines he struggled to defeat; one cannot appreciate Einstein's achievement without first understanding the well-founded and thoroughly tested Newtonian theories he had to revise. If Galileo and Einstein themselves gave lucid explanations and justifications of their own ideas, why not use their own words?

Moreover, "physics" must be interpreted in a very broad sense, as inseparable from astronomy, chemistry, and other sciences. Galileo's and Newton's research on motion and forces was stimulated by, and tested on, the astronomical problems arising from the battle between the Ptolemaic and Copernican views of the world. Modern physics owes much to the chemical investigations of Antoine Lavoisier, John Dalton, and Dmitrii Mendeléeff, and repays the debt by providing useful theories of the properties of atoms and molecules. Ernest Rutherford's theory of radioactive decay acquires some of its significance from its application to estimates of the ages of rocks and of the earth, important not only to geology but also to the credibility of Charles Darwin's theory of biological evolution.

A more indirect contribution to other sciences came from calculations by astronomer Cecilia Payne, based on atomic physics, which suggested that the sun's atmosphere is mostly hydrogen; other astronomers confirmed and generalized her result, concluding that hydrogen is the most abundant element in stars and in the entire universe. Starting from that conclusion, physicists Hans Bethe and William Fowler, with astronomer Fred Hoyle, showed how the other elements could be constructed from hydrogen, by a process that also enables the sun and other stars to shine for billions of years—thereby providing conditions that allow the evolution of living organisms.

And of course the most spectacular consequences of Einstein's general theory of relativity are found in astronomy and cosmology. A person ignorant of all these connections to other sciences can hardly claim to know the true significance of physics.

Perhaps the most unusual characteristic of our book, compared to other physics textbooks, is the emphasis it puts on the nature of discovery, reasoning, concept-formation and theory-testing in science as a fascinating topic in its own right. This means that the historical and philosophical aspects of the exposition are not merely sugar-coating to enable the reader to swallow the material as easily as possible, but are presented for their own inherent interest.

Indeed, this is not intended to be an "easy" book; that is, it does not pretend that physics can be understood without any mathematics or without logical reasoning. At the same time, it is not assumed that there is any obligation to "cover" a standard amount of material. We do not apologize for omitting a few "fundamental" equations and even entire subfields of physical science that are conventionally included in textbooks, nor for including topics such as early theories of vision that are rarely if ever mentioned in physics or optics textbooks. The selection of topics (and depth of treatment of each topic) is to a considerable extent governed by our goal of presenting a comprehensible account—a continuous story line, as it were—of how science evolves through the interactions of theories, experiments, and actual scientists. We hope the reader will thereby get to understand the scientific worldview. And equally important, by following the steps in key arguments and in the derivation of fundamental equations, the readers will learn *how scientists think*.

Like physics itself, this book has a history. In 1945, a committee at Harvard University published a report on *General Education in a Free Society* that asserted:

> From the viewpoint of general education the principal criticism to be leveled at much of present college instruction in science is that it consists of courses in special fields, directed toward training the future specialist and making few concessions to the general student. . . . Comparatively little serious attention is given to the examination of basic concepts, the nature of the scientific enterprise, the historical development of the subject, its great literature, or its interrelationships with other areas of interest and activity. What such courses frequently supply are only the bricks of the scientific structure. The student who goes on into more advanced work can build something from them. The general student is more likely to be left simply with bricks.

In 1952, Gerald Holton published *Introduction to Concepts and Theories in Physical Science,* intended as a text for general education courses of the kind proposed by the Harvard committee and for the liberal-arts physics courses that were introduced in many colleges and universities in the 1950s and 1960s. The book was successful in reaching that audience; in fact it eventually had to compete with several imitators, including those for the more superficial "physics for poets" type of course.

In 1973, a second edition of *Introduction,* revised and with new material by Stephen G. Brush, was published and stayed in print until 1997. The present book, which may be considered a third edition of *Introduction,* has been extensively revised. All the chapters in the second edition have been reworked to further clarify the physics concepts, updated to take account of recent physical advances and historical research, and modified so as to be consistent with SI units. New topics include theories of vision (Chapters 3 and 5), estimates of distances in the solar system (Chapter 10), the eighteenth-century prediction of the return of Halley's comet and analysis of deviations from Kepler's laws (Chapter 11), angular momentum conservation and Laplace's nebular hypothesis (Chapter 16), relation between symmetries and conservation laws (Chapter 17), first estimates of atomic sizes (Chapter 22), research on cathode rays in relation to the discovery of x-rays and of the electron (Chapter 25), Marie Curie's discoveries in radioactivity (Chapter 27), Pauli's exclusion principle and electron spin (Chapter 28), applications of quantum mechanics to many-particle systems (Chapter 29), and Dirac's relativistic quantum equation leading to the prediction of antiparticles (Chapter 30). New chapters discuss theories of the origin of the solar system and the expanding universe (Chapter 31); fission, fusion, and the big bang-steady state controversy (Chapter 32), and thematic elements and styles in scientific thought (Chapter 33).

The book is intended for a year course (two semesters or three quarters) in a general education or core program, taken primarily by nonscience majors who have an adequate background in mathematics (up to but not including calculus). Like the previous edition, it may well be used in parallel with another text in a course for science majors. (The use of the now-standard SI units should make it possible to coordinate this book with any other modern text.) In a shorter survey course, or in a course on the history and philosophy of science, some of the more technical sections in Chapters 6, 8, 9, 10, 11, 16, 17, 22, 24, 26, and 30 could be omitted, and a few of the Recommended Readings could be assigned.

We also anticipate that this book could be used for self-study by more mature students or by those preparing to teach physics in college or high school. Finally, we hope that advanced students in the history and philosophy of science will profit from our detailed treatment of the work of Galileo, Newton, and Einstein, and from the extensive bibliographies, keyed to individual chapters and sections of the text, that we provide on the website for this book at www.ipst.umd.edu/Faculty/brush/physicsbibliography.htm. Answers to selected numerical problems may also be found there.

* * *

S. G. B. thanks the directors (G. Holton, F. J. Rutherford, and F. Watson) and staff (especially A. Ahlgren, A. Bork, B. Hoffmann, and J. Rigden) of Harvard Project Physics, for many discussions and suggestions about how to use the historical approach in teaching physics. He is also indebted to S. J. Gates, Jr., at the University of Maryland, for useful information on the relation between symmetry and conservation laws. New diagrams were prepared by Eugene Kim, and new photographs were supplied by the Emilio Segrè Visual Archives at the American Institute of Physics (see the credits at the end of book). Nancy Hall provided invaluable assistance in obtaining permissions to use other new illustrations.

PART A

The Origins of Scientific Cosmology

1 The Astronomy of Ancient Greece

2 Copernicus' Heliocentric Theory

3 On the Nature of Scientific Theory

4 Kepler's Laws

5 Galileo and the New Astronomy

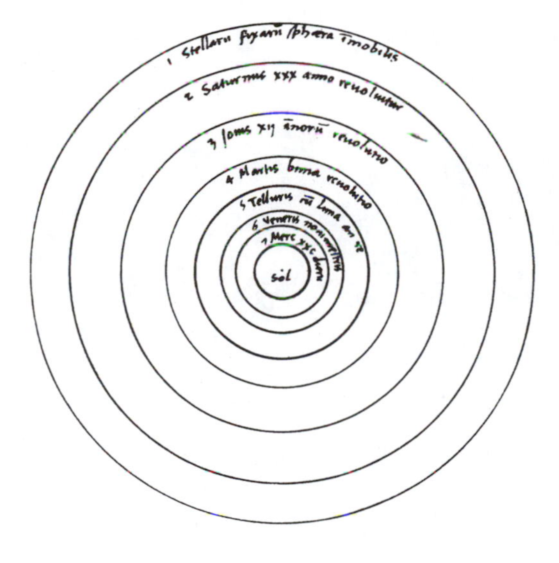

The heliocentric system as depicted in Copernicus' own manuscript.

Our purpose in this book is to tell the story of the major ideas that have led to our current understanding of how the physical universe functions. At the same time we also aim to show that the growth of science was part of the general development of our civilization, as it is to this day. We think that you will find it an amazing, even inspiring, story of the power of the ideas of men and women who dedicated their lives to the search for truths.

Although most of the book deals with physics, we will begin with astronomy, for several reasons. The methods of science were perhaps first introduced when people attempted to reduce the apparently chaotic yet recurring motions of the stars, sun, moon, and planets to an orderly system.

Second, many of our guiding ideas in science, not only in methodology but also in such specific concepts as time and space, force, and motion are derived from early astronomical schemes. We shall see that a continuum of ideas leads from the early naive speculations about stellar motion right to twenty-first century science, paralleling the expansion of our universe from the idea of a near sky to that of the rotating celestial sphere, to the recognition of the solar system as part of a huge galaxy, and finally to the universe of innumerable galaxies scattered through unbounded space.

Third, the development of the world system involves many of the scientists responsible for the strength and content of present physical theory. Their place in science and the ambience of their times are also part of our story.

Fourth, the work to be discussed—the planetary systems, the law of universal gravitation—had a most profound effect on Western culture. For example, without an understanding of the Newtonian revolution and its background, one cannot hope to comprehend fully the upheavals in eighteenth-century history, or the development of modern theories of economics and government, or the philosophies of Locke, Berkeley, and Hume.

Finally, we may use the opportunity now to watch the rise and fall of physical theories, to analyze their structure, and to sharpen our faculties of discrimination between theories that are useful and those that are misleading. For example, one misleading feature of the ancient geocentric system in astronomy was the way it conflated the human perspective on the world—the fact that we are observing the heavenly bodies from one particular planet—with the assumption that we are at the center of the universe and everything moves around us. For comparison we also discuss a rather different problem in which a similar mistake occurred: the nature of human vision. Here the most natural assumption seemed to be that the eye *sends out* rays *to* the objects it sees, and this assumption seemed to account for optical phenomena; it took many centuries of debate before scientists could convince themselves that the eye *receives* rays *from* objects. That theory, like the heliocentric theory of the universe, was resisted because it seemed to demote humans from their exalted position as Masters of the Universe. We may not always like the knowledge science brings us, but an essential part of growing up, as a human being and as a civilization, is to abandon the infantile assumption that the world revolves around us and is there to be controlled by our actions and desires.

CHAPTER 1

The Astronomy of Ancient Greece

1.1 The Motions of Stars, Suns, and Planets

Although there is now ample evidence that civilizations in earlier centuries and in other parts of the world had already gathered significant pieces of astronomical and mathematical knowledge, we can trace the beginnings of science as we know it to the imaginative minds of the great Greek thinkers. In order to have a definite starting place, let us try to imagine ourselves in the position of the leading scientific community of the ancient world around 400 B.C., in Athens.

Although the use of precision optical instruments was still 2000 years away, simple observation of the night sky had by about 400 B.C. established enough data and interpretations concerning the motions of the heavens to nourish several world theories. The stars and the Milky Way seem to move throughout the night as if they were rigidly attached to an invisible bowl that rotates around a fixed center in the sky (now called the *North Celestial Pole*). From observations made from different points on the earth's surface, it could be inferred that this bowl is really more like a very large sphere surrounding the earth, and that the earth itself is also a sphere. Parmenides of Elea, in the sixth century B.C., was one of the first scientists to reach the first conclusion.

As we will see in Section 1.4, it was possible by the third century B.C. to obtain a fairly accurate estimate of the size of the earth; but it was not until the nineteenth century A.D. that scientists could determine even approximately the distance of some of the stars from the earth.

The Greeks were quite familiar with the fact that the hypothetical "celestial sphere" containing the stars appears to rotate uniformly from east to west, returning to its starting point every day. We will refer to this motion as the *diurnal* (daily) rotation. (If we define the "day" by the rising and setting of the sun, then the stars appear to go around in a period about 4 minutes shorter than a day.)

Of course we now know that it is the earth rather than the stars that rotates on its axis every day, and that the appearance of stars arranged on a big sphere is an illusion. But that is not something you should simply accept "on authority." Instead you should consider carefully which of the following observations can be explained equally well by assuming either that the stars move or that the earth moves.

There is now a particular star, called the Pole Star or Polaris, that happens to be very close to the North Celestial Pole (the apparent center of rotation of the stars in the northern hemisphere). However, in 400 B.C., Polaris was several degrees away from the North Celestial Pole; there was no obvious Pole Star at that time. The Greeks knew that the North Celestial Pole does move very slowly with respect to the stars, a phenomenon associated with the so-called *precession of the equinoxes* (see below); but the available observations did not extend over a long enough period of time for them to understand the precise nature of this motion. It is now known that the North Celestial Pole itself moves in a small circle, returning to its original position after about 26,000 years.

It was also known to the Greeks (though, it appears, not so well known to people today) that while the sun shares the diurnal motion of the stars it does not quite keep up with them. By observing the stars just before sunrise and just after sunset, one can discover that the sun slowly changes its position relative to the stars every day. In fact it follows a path from west to east through the stars, called the *ecliptic,* going through the "signs of the Zodiac," and returning to its starting point after about 365¼ days.

To describe the motion of the sun, we define "noon" as the time each day when the sun reaches its highest point in the sky, before starting to set. This time is easily found by looking at the shadow of a vertical stick (called a

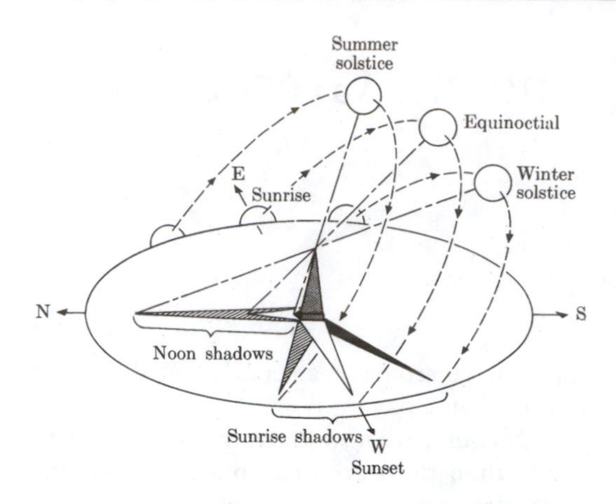

Fig. 1.1. Motion of the sun as seen by an observer in the northern hemisphere at four different times of the year. The gnomon's shadow at noon is longest when the sun is lowest in the sky (winter solstice), shortest when the sun is highest, and intermediate at the two equinoxes. The changing positions of the sunrise shadows show the motion of the sun from south (winter) to north (summer).

gnomon): the length of the shadow is shortest when the sun is at its highest point. This shortest length itself varies throughout the year, reaching a minimum at the *summer solstice* when the sun is most nearly directly overhead at noon and a maximum at the *winter solstice* when the sun does not rise very high in the sky. These seasonal variations in the sun's height at noon correspond to variations in the length of time the sun is above the horizon—longest in summer, shortest in winter—as we will discuss below.

These motions can be more easily comprehended if we recognize that the sun does not simply move from east to west each day, but also has a north-south motion throughout the year (Fig. 1.1). Moreover, the sun's motions are different when seen from different places on the earth's surface; these variations are best understood by using a three-dimensional classroom model. On about March 21 (vernal equinox) the sun is directly overhead at noon for an observer at the equator, and it then moves northward every day until about June 21 (summer solstice) when it is directly overhead at noon for places 23½° north of the equator (Tropic of Cancer). The sun then moves southward, so that about September 23 (autumnal equinox) it is directly overhead at noon on the equator again, and about December 22 (winter solstice) it is directly overhead at noon for places 23½° south of the equator (Tropic of Capricorn). The sun then moves northward again and the cycle repeats itself.

Notice that the seasons are not equally long. The period from vernal equinox to autumnal equinox (spring and summer) is a few days longer than the period from autumnal equinox to the next vernal equinox (fall and winter).

The north-south motion of the sun is of course the major factor that determines temperatures on the surface of the earth. Between March 21 and September 23 the day will be more than 12 hours long in the northern hemisphere, and the sun will rise relatively high in the sky (depending on latitude). Between September 23 and March 21 the day will be less than 12 hours long in the northern hemisphere, and the sun will not rise very high. (These circumstances will simply be reversed for the southern hemisphere.) On March 21 and September 23, both day and night will be 12 hours long everywhere; hence the term *equinox* (Latin for *equal night*).

The correlation between the seasons and the sun's motion through the stars was a scientific finding of vital importance to the ancient agricultural civilizations. By setting up a calendar of 365 days, the ancient astronomers could predict the coming of spring and thus tell the farmer when to plant his crops. Eventually it was found that such a calendar would become more and more inaccurate unless occasional extra days were added to take account of the awkward fact that the year (as measured by the time from the sun's leaving a spot among the stars to its return to that spot) is about a quarter of a day longer than 365 days.

Another difficulty was that the position of the sun in the Zodiac at the vernal equinox gradually changed over the centuries. For more than a thousand years before the birth of Christ, the sun, on March 21, was among the stars in the constellation Aries, but by and by it moved into Pisces, through which it is still moving each year on that date. In a few more centuries it will be in Aquarius. (Hence the 1960's slogan, derived from popular astrology, proclaiming the "Coming of the Age of Aquarius.") This phenomenon is called the *precession of the equinoxes* and will later on be seen to be simply another aspect of the gradual motion of the North Celestial Pole mentioned above.

Finally, the ancients noticed that certain special celestial objects that look like stars did not stay in fixed positions on the celestial sphere, but wandered around in a complicated but regular manner (always staying close to the ecliptic). These objects became known as planets (from a Greek word meaning wanderer), and the study of their motions was one of the chief occupations of astronomers up to the seventeenth century A.D.

The ancient Greek theories of planetary motion were not narrowly technical attempts to correlate a particular set of observations in the fashion that might nowadays be called "scientific." The Greek philosophers favored a broader approach. On the basis of preliminary observations, an astronomer might formulate a scheme of planetary motion that could account for the data available at the time; but for such a scheme to have lasting value it did not so much need to be in accord with all subsequent observations, but instead should dovetail with other, more strictly philosophical or theological ideas. The followers of Pythagoras, for example, conceived that the relative sizes of the orbits of planets were proportional to the lengths of the successive strings on a harmoniously tuned stringed instrument. This was thought to assure a "harmony of the spheres," more satisfying and important to the general philosophical requirements of the Pythagoreans than would be the relatively narrower demands of quantitative predictability of physical events made by the modern astronomer.

Or, to cite an example involving an entirely different school in Greek science, Aristotle found in qualitative observations of the free fall of large and small bodies enough evidence to integrate satisfactorily these physical phenomena with his philosophically more wide-reaching, and therefore to him more satisfying, scheme of elements and of natural place.

1.2 Plato's Problem

Nothing is easier and more erroneous than to underestimate the point of view of the early Greeks. Not only did the mechanics of the Greeks "work" within the limits of their current interest, but in some schools of thought it was also intellectually a majestic and deeply meaningful construct. Above all, it presents, as it were, the childhood of science, not to be judged in terms of the maturer point of view of contemporary knowledge. It may be true that eventually it outlived its usefulness, but it gave moderns the direction along which they could develop fruitfully. The influence of Greek thought is present in any contemporary activity, in science no less than in art or law, government or education.

Astronomy is a case in point. The great philosopher Plato (427–347 B.C.) set a problem for his students along these lines: The stars—considered to be eternal, divine, unchanging beings—appear to move around the earth, in that eminently perfect path, the circle. But a few "stars" appear to wander rather disturbingly across the sky, tracing perplexingly irregular figures in their yearly paths. These are the planets. Surely they too must *really* move in uniform, ordered circles, or rather, in their case, combinations of circles. Now, how can we account for the observations of planetary motion, and so "save the appearances" (or "fit the data," as we would now say)?

Plato's important question may be paraphrased as follows: "Determine what combination of uniform and ordered motions must be assumed for each of the planets to account for the apparent, more irregular movements." The very formulation of this historic problem strikingly summarizes for us the two main contributions of Greek philosophers to the topic of our discussion:

a) Physical theory (for example, the theory of planetary motions) is intelligible only in the context of a prior metaphysical[1] assumption (for example, the assumption that heavenly bodies must execute "perfect," circular motions). This proved to be an unfruitful doctrine on which to base science. By the sixteenth and seventeenth centuries, after the great struggle that we will discuss in later chapters, it began to be abandoned in favor of the experimental sciences.

b) Physical theory is built on observable and measurable phenomena (for example, the apparent motion of the planets), concerns itself with uniformities of behavior underlying the apparent irregularities, and expresses itself in the language of number and geometry. This guiding idea, which Plato did not extend beyond astronomy and which was derived in part from the Pythagoreans, was a treasured hint that

[1]The term *metaphysics* is used in a specific sense: the discipline which studies principles of knowledge or of being in terms of intuitive, self-evident concepts, concepts of direct "everyday" experience, and analogies.

reappeared when Kepler and Galileo fashioned their experimental science. But Plato held that physical laws can be found from directly intuited principles, the aim being to explain specific phenomena in the context of a philosophic system. The truth of a principle was not measured, as it is today, by its usefulness in every conceivable known or predicted physical situation. Consider this sentence from Plato's *Phaedon:* "This was the method I adopted: I first assumed some principle, which I judged to be the strongest, and then I affirmed as true whatever seemed to agree with this, whether relating to the cause or to anything else; and that which disagreed I regarded as untrue."

Plato's specific astronomical question, which he did not seriously attempt to answer himself, became the prime concern of astronomers to the time of Galileo. Let us see how they tried to construct a system of the world in accordance with the axiom of "uniform and ordered motion," that is, a system that allows the celestial objects to move only in uniform (constant) circular motion or in combinations of several such motions. We shall discover that ultimately all such systems proved unsatisfactory; but modern science was born out of this failure.

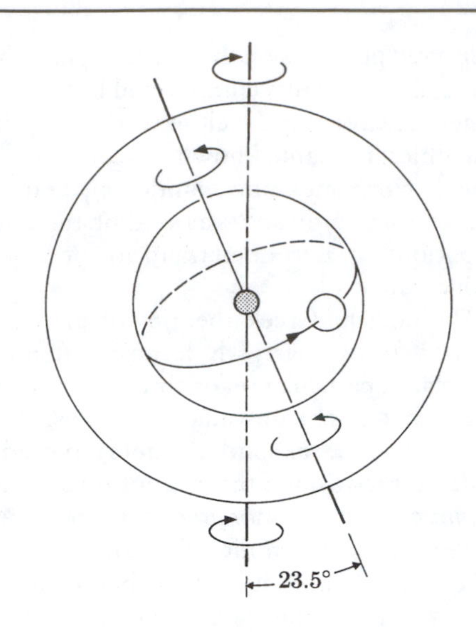

Fig. 1.2. The annual north-south (seasonal) motion of the sun (empty circle) around the earth (shaded circle) was explained by assuming it to be fixed on a sphere whose axis was tilted 23½° from the axis of the stars. The outer circle represents the sphere of the stars, which rotates around the vertical axis.

1.3 The Aristotelian System

Obviously, the first guess to investigate in rationalizing the motion of celestial objects is the most simple model, the *geocentric system,* which places the earth at the center of the celestial sphere carrying the fixed stars. (*Gē* is the Greek word for earth.) The observed daily motion of those fixed stars will result from our model at once if we require the large celestial sphere to rotate uniformly on a north-south axis once a day. We may then attempt to explain the apparent motion of the sun (Fig. 1.2), the moon, and the five visible planets about the fixed earth by letting each be carried on a transparent sphere of its own, one within the other, all seven enclosed by the sphere of fixed stars, and the earth at the center of the whole scheme (Fig. 1.3).

But since those seven celestial bodies do not rise and set always at the same point of the horizon, since in fact the planets travel on most complicated paths against the fixed stars, sometimes even briefly reversing the direction of motion, we must give each of their spheres a whole set of simultaneous rotations about different axes, each rotation of the proper speed and direction, each axis of the proper inclination, to duplicate the actually observed paths.

Here the mathematical genius of the Greek was indeed put to the test! The sun and the moon were relatively simple to discuss, but some of the planets offered much difficulty. Plato's pupil Eudoxus thought that 26 simultaneous uniform motions would do for all seven celestial bodies. Others proposed to abandon the assumption that the sun and planets are fixed on celestial spheres—which means that they must always remain at the same distance from the earth—and developed the more complicated combinations of circles that we shall briefly summarize in Sections 1.6 and 1.7.

But is the task of the astronomer (or any scientist) strictly limited to constructing mathematical systems, no matter how artificial, that will agree with observations? Is the job done when one can "save the appearances"—or should one also look for a physically plausible explanation of how nature works?

A clear answer to this question was given by Aristotle (384–322 B.C.), Plato's successor and the greatest philosopher of antiquity (some would say the greatest of all time). Aristotle's writings on cosmology were closely integrated with his overall philosophy; he united in one conceptual scheme elements that we now separate into separate components—scientific, poetic, theological, ethical. It was precisely because his theory of the universe concentrated on physical understanding rather than mathematical computation that it was so widely adopted, especially in the medieval period just before the birth of modern science.

What does the universe consist of? Proceeding from a then current notion, Aristotle postulated that all matter within our physical reach is a mixture of four elements: Earth, Water, Air, and Fire. Actually one would never see the pure elements; it was understood that a lump of earth or a stone might contain mostly the element Earth, mixed with smaller amounts of the other three elements. A vessel full of the purest water that anyone can obtain would be expected to contain beside the element Water also some earthy substance. Indeed, on boiling away the liquid, one does find a solid residue. (So convincing was this experiment and this interpretation that when the great chemist Lavoisier by ingenious experimental demonstration first proved in 1770 that most of the residue came from the walls of the vessel itself, he also reported that what was "thought by the old philosophers, still is thought by some chemists of the day.")

A second postulate prescribed that each of these four elements has a tendency or desire to reach its "natural place" of repose: Earth at the bottom (or center of the universe), next Water, then Air, and finally Fire at the top.

A third postulate directed that the actual motion of an object is determined by the tendency of the element most abundantly present. So the behavior of steam rising from a boiling vessel was explained as the upward motion owing to the introduction of the element Fire into the heated water; when, on cooling, the steam gave up its Fire, the then predominant element, Water, could assert itself again, and the condensed moisture could precipitate down to its natural place below.

One consequence of this view was that the motion of an object up or down to its natural place, the so-called "natural motion," was so governed by the balance of its constituent elements that its speed of motion must be proportional to the amount of the predominant element. A big stone, containing evidently more Earth than a small one, consequently was expected to descend that much faster when allowed to execute its natural motion of free fall. We will return to this point in Chapter 7, where we will see that this prediction about the speed of falling objects provided one of the handles by which Galileo could overturn the Aristotelian system.

Aristotle also postulated that these four elements are found only in the terrestrial or "sublunar" (below the moon) domain. Beyond that, there is an entirely different kind of element, the Ether,[2] from which the heavens are constructed. Whereas the four terrestrial elements are always involved in processes of change—"generation and corruption," in Aristotle's vivid phrase—the Ether is by definition pure and immutable. It has its own natural motion, appropriate to its nature, a motion that has no beginning and no end and that always keeps it in its natural place: *circular motion*. Thus the use of circles to explain the motions of heavenly bodies, as suggested by Plato, is not merely a mathematical convenience: it became a philosophical necessity in Aristotle's system.

The earth, however, cannot have a natural motion of rotation—that would be inconsistent with its nature. Our planet is (by definition) composed primarily of the element Earth, whose natural motion—as one can easily see by dropping a stone—is downward to its resting place, not circular.

Aristotle conceived of the system of celestial spheres in a mechanical way that is not entirely alien to the physics of the seventeenth, eighteenth, and nineteenth centuries. Outside the largest sphere (in which the stars are embedded) is the divine "prime mover" (labeled as *primum mobile* in Fig. 1.3). The prime mover turns the starry sphere at a regular rate; this motion is transferred, by friction with some loss, to the spheres of the outer planets, and thence to the spheres of the sun and inner planets. While Aristotle accepted the 26 spheres of Eudoxus mentioned above, he prescribed additional ones, partly to account for some of the most obvious discrepancies between Eudoxus' system and the observed paths of the planets, and partly as intermediate spheres needed to "neutralize" the reversed motion of some of the spheres.

Aristotle easily proved that the rotating starry sphere must be finite, not infinite. Parts of

[2]Also known as the "quintessence," from the Latin term for "fifth element." Ether, sometimes spelled Aether, comes from a Greek word that means to burn or glow.

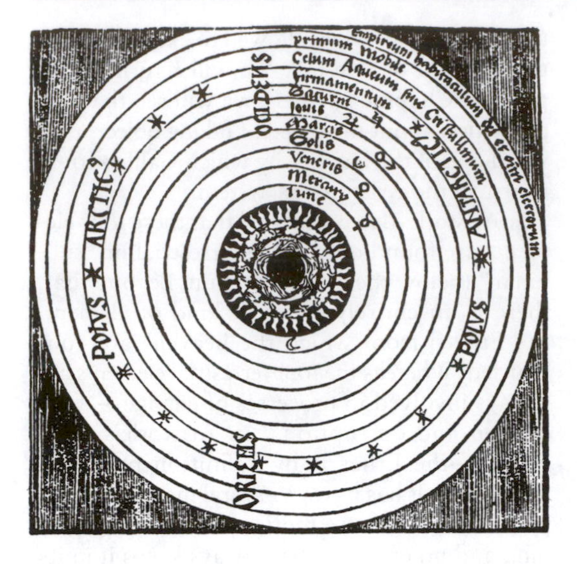

Fig. 1.3. A late medieval conception of the world. The sphere of the moon (lune) separates the terrestrial region (composed of concentric shells of the four elements Earth, Water, Air, and Fire) from the celestial region. Beyond the moon are the concentric spheres carrying Mercury, Venus, Sun, Mars, Jupiter, Saturn, and the fixed stars.

the sphere farther from the center of rotation must move faster in order to complete a rotation in the same time as parts closer to the center; a point at distance R from the center must travel the distance $2\pi R$ in 24 hours. But this is impossible if R is infinite, for that would imply an infinite speed—which is (according to Aristotle) inconceivable.

And yet, there remained easily observed features of the sky that were unexplained by Aristotle's system—notably the fact that the sun, the moon, Venus, Mars, and Jupiter at times seemed brighter or nearer and at other times farther away from the earth. A set of uniform rotations of the celestial bodies on spheres concentric with the earth could never allow them to change their distance from the earth.

Aristotle was aware of this anomaly, but discounted the importance of the simple, though eventually fatal, argument against the fundamental hypothesis of his system. He did not take seriously the consequences that to us seem deceptively obvious, mainly because it was then, as it is now, incomparably easier to leave in doubt the testimony of a few contrary observations than to give up a whole world scheme that seemed conclusive and necessary from the point of view both of his philosophy and his mechanics.

Of course, this does not mean that he proposed a theory he knew to be false. The original Aristotelian science was not simply bad modern science but an activity fundamentally different from it. Perhaps we may also see here an example of an important human trait, which colors all scientific work and which even the greatest thinkers can never hope to overcome entirely: We all tend to deny the importance of facts or observations not in accord with our convictions and preconceptions, so that sometimes we ignore them altogether, even though, from another point of view, they would stand before our very eyes. Moreover, even the most general and modern scientific theory does not hope or even seriously attempt to accommodate every single detail of every specific case. One is always forced to idealize the observations before attempting a match between "facts" and theory—not only because there are usually unavoidable experimental uncertainties in observation, but because conceptual schemes are consciously designed to apply to selected observations rather than to the totality of raw experience. As a consequence, the history of science is studded with cases in which it turned out later that the neglected part of a phenomenon was actually its most significant aspect. But on the other hand, if there had been no tentative, half-true, or even plainly wrong theories in science, we should probably never see a correct one evolving at all. Since the task cannot humanly be done in one jump, we must be satisfied with successive approximations.

1.4 How Big Is the Earth?

Alexander the Great (356–323 B.C.), who as a youth had been tutored by Aristotle, conquered most of the world known to the Greeks during his short life. In 331 B.C. he added Egypt to his empire and founded there the city of Alexandria. During the following decades the center of Greek civilization shifted from Athens to Alexandria. Although Alexander's own empire broke up after his death, the kingdom of the Ptolemies in Egypt maintained Greek civilization at a high level for the next two centuries. Alexandria itself became a cosmopolitan mixture of races and creeds, the population being mostly Egyptian with a Greek-Macedonian upper class, many Jews, Africans, Arabs, Syrians, Hindus, etc. The Belgian-American historian of science George Sarton, who wrote extensively about Greek and

Alexandrian science, compared the relation between Alexandria and Athens to that between New York and London.

Two of the most famous cultural institutions in the ancient world were the Museum and the Library at Alexandria. The Museum was the center for scientific and mathematical research, conducted by Euclid, Apollonius, Eratosthenes, Hipparchus, perhaps Aristarchus, and (much later) the astronomer Ptolemy (not to be confused with the kings named Ptolemy). Some of these names will appear again later in this chapter.

Eratosthenes (273–192 B.C.) was educated in Athens but spent more than half his life in Alexandria. At a time when many scientists were already becoming overspecialized in mathematics or astronomy, Eratosthenes, like Aristotle, dared to be a generalist. As chief librarian of the Library at Alexandria, he was responsible not only for collecting, arranging, and preserving several thousand rolls of papyrus manuscripts, but also for knowing what was in them. He applied mathematics and his knowledge of the world to found the science of geography. His contemporaries gave him the deprecatory nicknames "beta" (meaning second-rate) and "pentathlos" (the modern equivalent would be a five-letter man in athletics). They assumed that anyone who was jack of all trades must be master of none. Nevertheless Eratosthenes accomplished one outstanding feat: He measured the size of the earth!

A brief sketch of Eratosthenes' calculation will give some indication of the level of sophistication that Greek science had reached by the third century B.C. Eratosthenes made the following assumptions:

1. The earth is spherical.
2. The sun's rays are parallel when they strike the earth (which means that the sun must be very far away from the earth).
3. The two places on the earth's surface where he made his observations, Alexandria and Syene (now Aswan, Egypt), lie on the same north-south line.
4. Syene is exactly on the Tropic of Cancer (latitude 23½° N), so that the sun is directly overhead at noon on the day of the summer solstice.

None of these assumptions is exactly correct, but they are so nearly correct that Eratosthenes was able to obtain a surprisingly accurate result.

The calculation was based on two measurements: the angle that the sun's rays make with

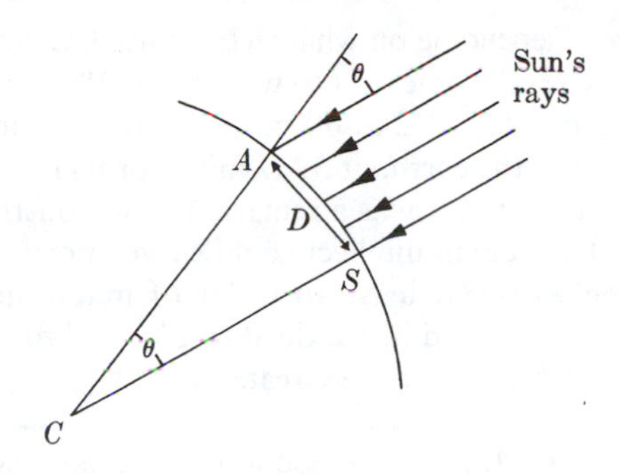

Fig. 1.4. How Eratosthenes estimated the size of the earth.

a vertical stick at noon at the summer solstice at Alexandria (an easy measurement); and the linear distance (D) between Alexandria (A) and Syene (S) (a difficult one, done by timing the march from A to S of a well-trained group of soldiers whose rate of marching was known and fairly constant). According to the fourth assumption, the angle at A must be the same angle θ subtended at the earth's center, C, by the arc AS; in other words, it is the difference in latitude between Alexandria and Syene. Eratosthenes could then set up a proportion between the angle θ and the distance D: the ratio of θ to a complete circle (360°) is the same as the ratio of D to the circumference of the circle, c.

$$D/c = \theta/360$$

where θ = angle ACS, c = circumference of earth, and D = length of arc AS.

According to Eratosthenes, the angle θ was, by measurement, exactly the one-fiftieth part of a whole circle:

$$\theta = 360/50 = 7.2°$$

and the measured distance was

$$D = 5000 \text{ stadia}.$$

Hence, solving the above equation for the circumference c, he found

$$c = 360D/\theta = 250{,}000 \text{ stadia}.$$

To compare this result with modern values, all we need to know is the conversion factor between the Greek unit of distance, the stadium, and some modern unit (mile or kilometer). Unfortunately there is some doubt about this conversion factor; the best guess is that a stadium was about one-tenth of a mile, or perhaps a little longer. Depending on what factor is used, Eratosthenes' result comes out to within 1 or 2% of the modern value of 24,860 miles for the circumference of the earth, or 3957 miles for its radius R. In any case, it was a remarkable demonstration that human intellect could survey portions of the world at least an order of magnitude larger than could be subdued by physical force, even by Alexander the Great.

Problem 1.1. The data reported by Eratosthenes look as if they had been rounded off. Assuming that the value of the angle θ is really between 360/49 degrees and 360/51 degrees, and that the distance D is really between 4900 and 5100 stadia, find the upper and lower limit for the measurement of the circumference and radius of the earth in stadia. Convert these values to miles, assuming 1 stadium = 1/10 mile.

There is a curious postscript to this story. Many people today believe that when Christopher Columbus set out on his first voyage across the Atlantic to seek an ocean route to Asia, other Europeans still believed that the earth is flat. Supposedly Columbus, by discovering the continent now called America, proved that the earth is round. But, although it is true that a few writers in the centuries before 1492 described a flat earth and this may well have been the view of uneducated people at the time of Columbus, scientists since Greek antiquity have believed it to be spherical.

Columbus himself did not doubt that the earth is round, but his estimate of its size was much too small. Ignoring the estimates of Eratosthenes and others, which we now know to be fairly accurate, he asserted that the distance he would have to travel from the Canary Islands, off the west coast of Europe, to the east coast of Japan, was about 2500 miles—approximately one-fourth of the actual distance (at 28° north latitude). The error, perhaps motivated by wishful thinking or the desire to make his project look realistic, indirectly helped him to get financial support for his voyage. Had he used the results of Eratosthenes, the goal of reaching Asia by sailing west might well have seemed unattainable.

1.5 The Heliocentric Theory

Of course, the problem of planetary motion persisted. There were two different major types of attack after Aristotle—the *heliocentric theory* (Greek *helios,* sun) and the modified geocentric theory. Let us now discuss the first of these. Aristarchus of Samos (third century B.C.), perhaps influenced by the work of Heraclides of Pontus (fourth century B.C.), suggested that a simple world system would result if the sun were put at the center of the universe and if the moon, the earth, and the five then-known planets revolved around the sun in orbits of different sizes and speeds. We do not know many details; his work on this subject is known to us only through references in other old writings. But evidently he assumed that the earth has a daily rotation on its north-south axis, as well as a yearly revolution in the orbit around the sun, and he placed the whole system in the sphere of fixed stars, which thereby could be considered at rest with respect to the center of the universe.

This heliocentric hypothesis has one immediate advantage. It explains the bothersome observation that the planets are at times nearer to the earth and at other times farther away. But the ancient world saw three very serious flaws in Aristarchus' suggestion:

First, it did violence to philosophical doctrines (for example, that the earth, by its very "immobility" and position is differentiated from the "celestial bodies," and that the natural "place" of the earth is the center of the universe). In fact, his contemporaries considered Aristarchus impious "for putting in motion the hearth of the Universe." Also, the new picture of the solar system contradicted common sense and everyday observation; the very words used in astronomy (sunrise, progression of planets, etc.) reflect the intuitive certainty that the earth must be at rest.

Problem 1.2. List the common-sense observations concerning sun, stars, and planets from which the geocentric theory sprang, apparently so convincingly. Be careful not to *assume* the geocentric or any other theory in your description.

Second, Aristarchus does not seem to have fortified his system with detailed calculations and quantitative predictions of planetary paths; by our present standards these are obvious conditions for assuring recognition in the physical sciences. This work seems to have been purely qualitative, although in some other accomplishments he is said to have shown considerable mathematical powers.

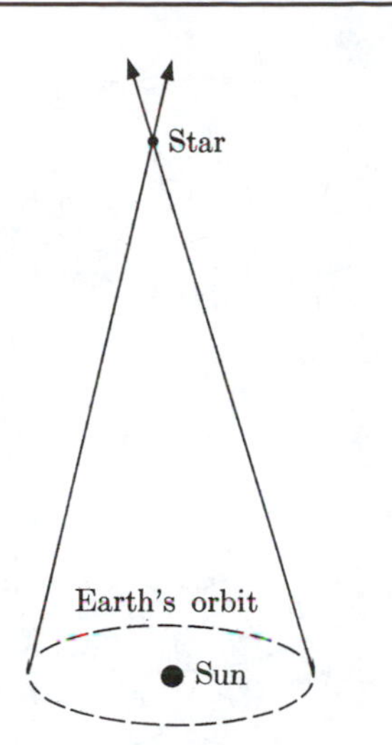

Fig. 1.5. Parallax of a star as seen from the earth. The diagram is not to scale; the star is much farther away, relative to the size of the earth's orbit.

Third, the Greek thinkers offered an ingenious bit of reasoning to refute Aristarchus. If the earth is to move around the sun, its large orbit will carry it sometimes near to a given fixed star on the celestial sphere and sometimes farther away from it. Thus the angle at which we have to look for this star will be different as seen from the various points in the earth's annual sweep (Fig. 1.5). This phenomenon, called the annual parallax of the fixed stars, should occur on the basis of Aristarchus' heliocentric hypothesis; but it was not observed by the Greek astronomers.

In explanation we may say either: (a) that the stellar parallax is so small as to be unobservable with the naked eye—which in turn requires the fixed stars to be unimaginably distant compared with the diameter of the yearly orbit of the earth, or (b) that Aristarchus was wrong and the earth does not move around within the celestial sphere. It seems natural that the ancients, predisposed to reject the heliocentric system, and also loath to consider a seemingly infinite universe, chose the second of these alternatives. The first, however, proved eventually to be the correct one. The parallax was indeed present, although so small that even telescopic measurements did not reveal it until 1838.

Problem 1.3. The annual parallax of a given star may be crudely defined as half the angle between the two lines of sight drawn from the center of the earth to the star from opposite ends of a diameter across the earth's orbit. Since the Greeks could not detect the parallax, and since their accuracy of measurement was often only about ½°, what must be the least distance from the earth's orbit to the nearest fixed star? (Express in astronomical units, or AU, 1 AU being the sun's mean distance from us, about 93×10^6 mi or 15×10^7 km.) F. W. Bessel in 1838 first observed the annual parallax, which for the nearest bright star, α Centauri, is of the order of ¼ second of arc. What is its approximate distance from us? (Modern telescopes can measure parallaxes of about 0.001 second, and therefore can directly determine distances of stars about 1000 times as far away as α Centauri.)

The heliocentric theories of Aristarchus and others seem to have been so uninfluential in Greek thinking that we should ordinarily spend no time on them. But these speculations stimulated the crucial work of Copernicus 18 centuries later. Fruitful ideas, it is evident, are not bound by time or space, and can never be evaluated with final certainty.

1.6 Modified Geocentric Theories

We now turn to the other and more vigorous offshoot of the early astronomical schemes. This will be an example of how science progresses, despite what we now regard as a seriously mistaken basic assumption (that the universe revolves around the earth).

To allow the planets to have variable distances from the earth while still keeping to the old belief of an immovable earth, the system of concentric spheres was modified in ingenious ways, principally by Apollonius,[3] Hipparchus,[4] and finally the influential astronomer and

[3]Apollonius (third century B.C.) suggested the epicycle; he also worked out, without any reference to astronomy, the properties of conic sections such as the parabola and the ellipse, which became important later in the Kepler-Newton theory of the seventeenth century (see Chapters 4 and 11).

[4]Hipparchus (second century B.C.) invented a method of calculation equivalent to what is now called trigonometry. His work on the theory of the motions of the sun, moon, and planets was incorporated into Ptolemy'system; since most of what is known of Hipparchus' work comes from ambiguous references by Ptolemy, it is difficult to know how much credit each should get for creating the "Ptolemaic system." Hipparchus is also credited with having discovered the precession of the equinoxes, although something was probably known about this phenomenon long before him.

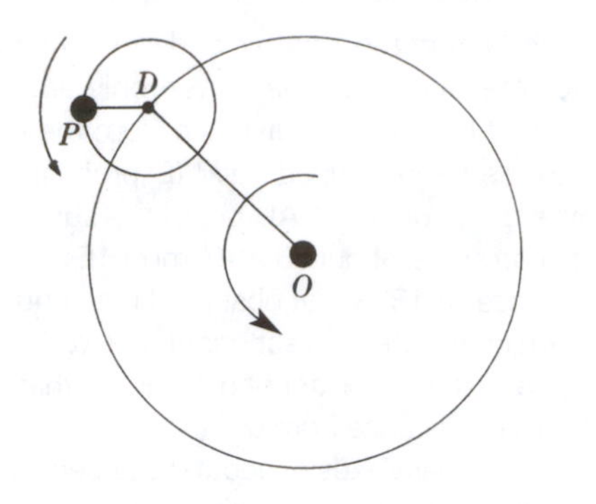

Fig. 1.6. An example of epicyclic motion of a planet *P.*

geographer, Claudius Ptolemy (second century A.D.). (Note the enormous time interval between Hipparchus and Ptolemy: We have no record of any major contributions to astronomy during the period that saw the rise of the Roman Empire and of Christianity.)

a) *Eccentric motion.* If the stationary earth were not exactly at the center of rotation of a uniformly moving celestial object, the latter would move along an eccentric path as seen from the earth, and vary in distance at different times. This scheme fits the apparent yearly motion of the sun fairly well, since it appears larger (and consequently nearer) at noon in our winter compared with our summer. Note that by admitting eccentric motions the astronomers really were dodging the old principle that required the planet motions to be circular around the center of the earth.

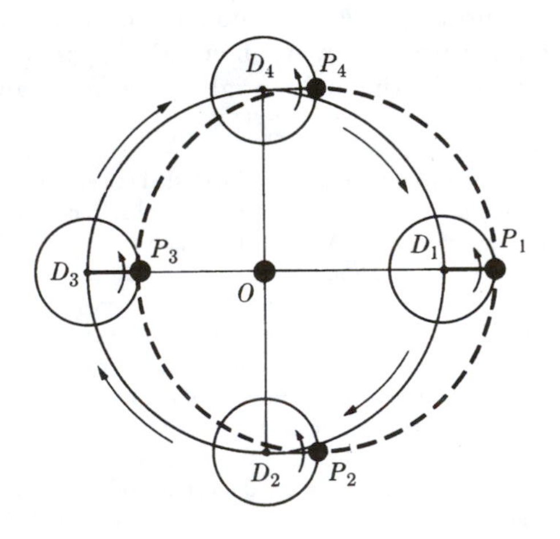

Fig. 1.7. Eccentric path represented by epicyclic motion (no retrograde motion in this case).

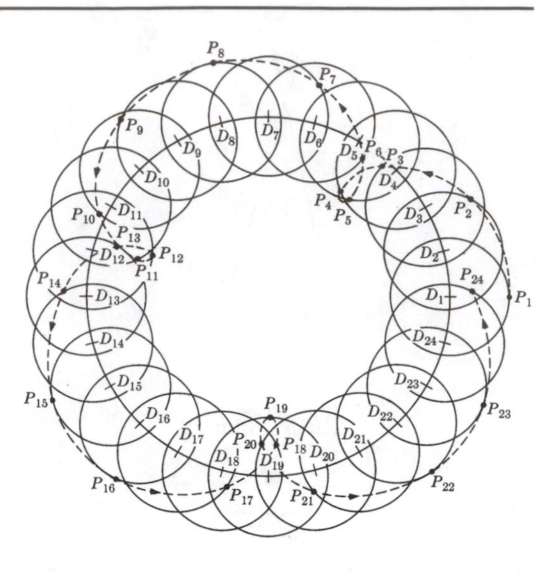

Fig. 1.8. Epicyclic motion of *P,* showing three temporary reversals of direction of motion at P_4, P_{11}, and P_{18}.

b) *Epicyclic motion.* Figure 1.6 represents an object *P* (such as the sun or a planet) having two simultaneous uniform rotary motions, one a rotation of *P* about a point *D* (in space) with radius *PD,* and the other a rotation of the line *OD* about point *O* (the position of the earth). The small circle is called an epicycle and the large circle the deferent. The two motions may have entirely independent speeds, directions, and radii. Figure 1.7 indicates the special case in which the epicycle scheme yields an eccentric path (dashed line), and Fig. 1.8 shows, by the dashed line connecting 24 successive positions of *P,* the complicated motion resulting when *P* rotates about *D* several times while *D* moves once around *O*.

Problem 1.4. What is the ratio of the speeds of revolution along the epicycle and the deferent in Fig. 1.8? What might be the path if the ratio were exactly 3:1?

This last type of motion does, in fact, exhibit most of the observed complications in the paths of planets as seen from the earth against the background of the fixed stars; note particularly the reversal in the direction of motion, called *retrograde motion,* at positions P_4, P_{11}, and P_{18}. Jupiter's path, one with 11 such loops, is covered in approximately 12 years per complete revolution around the earth.

Notice also that the planet should appear *brighter* during retrograde motion because it is closer to earth, and this is (qualitatively) what is observed. Thus the epicycle model is an improvement over the Aristotelian system, which placed each planet on a sphere that kept it always the same distance from the earth.

By a proper choice of radii, speeds, and directions, an eccentric path can be traced just as well by an epicyclic motion. Ptolemy used one or the other, and sometimes mixtures of both types, depending on the problem. We must also note that the device of epicyclic motion suffers from the same philosophical difficulties as the eccentric. The rotation of *P* about *D* is, if we wish to be strictly logical, just as much in conflict with old axioms as was the circular motion of *P* about a center other than the earth.

Problem 1.5. Construct as well as you can with compass and ruler the path of *P* (similar to Fig. 1.7), if *P* revolves around *D* twice while *D* goes around *O* once.

Problem 1.6. From observations we may form the following hypotheses: The sun moves around the earth from east to west once a day, just as the celestial sphere does, but about 4 min per day less quickly. The sun also approaches the earth during our winter (i.e., for observers in the northern hemisphere) and recedes in summer, and in addition it slowly travels north of the equator in summer and south of the equator in winter. Which of these features can be incorporated into the spherical model shown in Fig. 1.2, and which cannot? Suggest and draw the outlines of a qualitative geocentric model for these motions of the sun about the earth, using the devices described in this section.

Now we can summarize what in the Ptolemaic system has been retained from Plato and Aristotle, and what has been amended. Use is still made of *uniform circular motions* and of a *stationary earth*. Vanished is the scheme of spheres all concentric at the earth, and with it the need to have all rotations exactly earth-centered. This is even more strikingly evident when we discover that Ptolemy found it necessary to add still another device to the list in order to represent more faithfully some features of celestial motion. This device is the *equant*.

c) *The equant.* The sun moves halfway around its path through the stars (that is, 180° out of a full cycle of 360°) between the vernal equinox, March 21, and the autumnal equinox, September 23. It then completes the other half of the cycle between September 23 and March 21. Thus the sun appears to move a little more slowly during the summer, and a little faster during the winter; it takes almost 6 days more to cover the same angular distance (measured against the celestial sphere) in the summer than in the winter. This fact could not be explained by any combination of rotating spheres or epicycles, since it was always assumed that the rate of circular rotation is constant. Similar but less obvious difficulties arose in accounting for planetary motions.

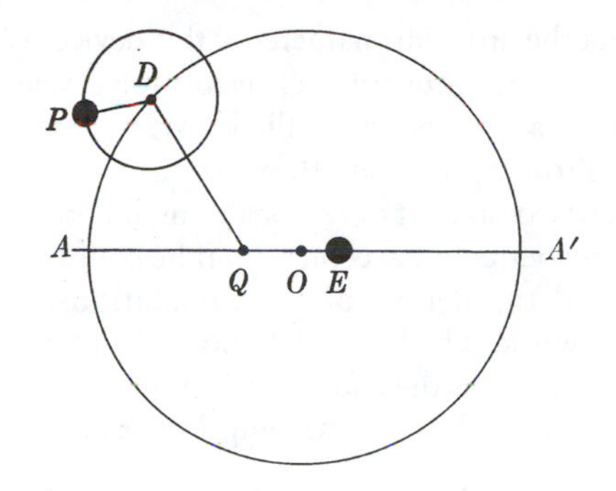

Fig. 1.9. Motion with respect to equant at *Q*.

Ptolemy's equant device is shown in Fig. 1.9. An object *P* is in cyclic motion around *D*, which in turn moves on a circle whose center is at O. The earth may be at O or, if this is a mixture of epicyclic and eccentric motion, the earth may be anywhere along the line *AA*', say at position *E*. So far the motion of *D* has been specified as uniform with respect to O, but to represent the irregular motions of the sun and planets mentioned above, Ptolemy proposed to let *D* revolve uniformly with respect to *Q*, a point called the equant. That is, the angle *DQA* changes at a constant rate while *D* executes its circular sweep. Now *D* is no longer in strictly uniform circular motion, although its motion is still uniform (seen from *Q*) and circular (seen from O).

To those astronomers whose only concern was to "save the appearances," that is, to establish a system that would provide accurate predictions of all celestial phenomena, the equant was a brilliant success. But those who wanted a system that would be compatible with general philosophical principles, such as the necessity of uniform circular motion, would be unhappy about the artificial character of this device. Which is the more important criterion for a scientific theory: accuracy or intelligibility?

Ptolemy himself thought that his system was based on a set of reasonable assumptions that any reader could accept, even if he could not follow all the details of the calculations. In his great work, which became known through Arab translations as the *Almagest* (literally, "the greatest"), he stated these assumptions as follows.

1. That the heaven is spherical in form and rotates as a sphere.
2. That the earth, too, viewed as a complete whole, is spherical in that form.
3. That it is situated in the middle of the whole heaven, like a center.
4. That by reason of its size and its distance from the sphere of fixed stars, the earth bears to this sphere the relation of a point.
5. That the earth does not move in any way.

Implicit in his work is also the old, now somewhat distorted doctrine of uniform circular motion as the only behavior thinkable for celestial objects.

Problem 1.7. What is the significance of the fourth point in Ptolemy's preliminary assumptions?

1.7 The Success of the Ptolemaic System

By adjusting the respective axes, directions of motions, rates and radii of rotations, and number and size of epicycles, eccentrics, and equants; by fitting his devices to the observed paths in a cut-and-try fashion; by utilizing partial results of generations of previous astronomers—so did Ptolemy manage to assemble the apparatus that proved still useful to astronomers, navigators, and astrologers more than 14 centuries later. It was an answer to Plato's original question, and on the whole a magnificent piece of work.

One pleasing feature was that centers of the epicycles of the moon, Mercury, and Venus fell on the same line between earth and sun (see Fig. 1.10)—though it may have seemed curious that the positions of two particular planets, Mercury and Venus, were linked to that of the sun while the other planets seemed to move more freely.

But, whereas the Aristotelian system could not account for the variations in brightness of

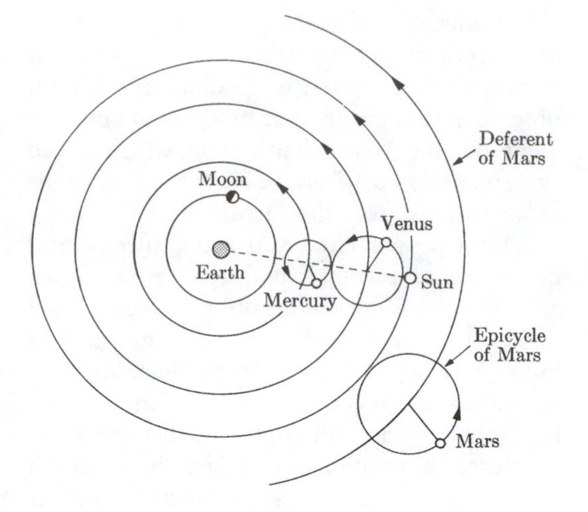

Fig. 1.10. Partial and schematic diagram of Ptolemaic system of planetary motion.

planets during retrograde motion, the Ptolemaic system failed in the opposite direction by attributing to the moon, for example, a very great variation in distance, inconsistent with the rather small variations in brightness actually observed.

But the complexity and inaccuracy of detail in the whole scheme was considerable, and subsequent observations required amendment of some features of the model from time to time, for example, the change of an assigned equant or the addition of another epicycle. By the time of Copernicus, some geocentric systems then in use required more than 70 simultaneous motions for the seven celestial bodies.

Against this defect were five powerful reasons why the system was generally accepted wherever it became known.

a) It gave an accurate enough description of what could be observed with the instruments of the times.

b) It predicted the future paths of planets well enough for the purposes of the times, if only after cumbersome calculations; and when a serious discrepancy between predictions and observations occurred, it could be resolved by tampering a little with the gears of the flexible apparatus. Today, even as at that time, the geocentric system is still preferred for calculations in navigation on earth—and in astrology!

c) It explains naturally why the fixed stars show no annual parallax.

d) In most details it coincided with Greek philosophical and physical doctrine concerning the nature of the earth and the celestial bodies. Later, when reintroduced to Europe through the writings of Arab scholars, the Ptolemaic system[5] was given theological significance by the scholastics. Furthermore, it was also in line with contemporary physics (e.g., projectile motion), based on the same philosophical doctrine of natural motion, natural place, and so forth (see Chapter 7).

e) It had common-sense appeal. It is almost unavoidable to hold the opinion that we actually can "see" the sun and stars moving around us, and it is comforting to think of ourselves on a stable, immovable earth.

But for all that, Ptolemy's theory was eventually displaced by a heliocentric one. Why did this happen? What were the most significant deficiencies in Ptolemy's picture? More generally, when is a scientific theory, from our present point of view, successful or unsuccessful? We shall want to answer these questions in detail after we have looked at an outline of the great rival scheme of planetary motion.

RECOMENDED FOR FURTHER READING

Note: The books and articles mentioned here and at the end of the book are intended to guide you to stimulating sources if you wish to carry this subject further. Those in the "Recommended for Further Reading" category at the end of each chapter are generally written at a technical level no higher than that of the text itself. "Sources, Interpretations, and Reference Works," listed in a website for this book at www.ipst.umd.edu/Faculty/brush/physicsbibliography.htm are intended primarily for the use of more advanced students and for instructors. Answers to selected numerical problems will also be found in the website along with the bibliography of sources. When only the author and a short title are given, the complete reference can be found in the Bibliography at the end of the book (this includes books cited in more than one chapter and those of general interest). The excerpts from the original writings are especially recommended, for no historian's summary can really convey the impact and nuances of the scientist's own words.

Marshall Clagett, *Greek Science in Antiquity,* Princeton Junction, NJ: Scholar's Bookshelf, 1988, Chapters 2, 3, 7

Michael J. Crowe, *Theories of the World,* Chapters 1-4, with excerpts from Ptolemy's *Almagest;* Appendix on archaeoastronomy

C. C. Gillispie (editor), *Dictionary of Scientific Biography,* articles by G. E. L. Owen and L. Minio-Paluello on Aristotle, Vol. 1, pages 250–258, 267–281; by W. H. Stahl on Aristarchus of Samos, Vol. 1, pages 246–250; by D. R. Dicks on Eratosthenes, Vol. 4, pages 388–393

Norriss S. Hetherington (editor), *Cosmology,* articles by Hetherington, "To Save the Phenomena," "Plato's Cosmology, and "Aristotle's Cosmology," pages 69–103; by James Evans, "Ptolemy," pages 105–145

Thomas S. Kuhn, *The Copernican Revolution,* Chapters 1, 2, 3

David Lindberg, *The Beginnings of Western Science,* Chapters 1–5.

A. G. Molland, "Aristotelian science," in *Companion* (edited by Olby), pages 557–567

John North, *Norton History of Astronomy,* Chapter 4

S. Sambursky, *Physical Thought,* Introduction and selections from Presocratics, Pythagoreans, Plato and Aristotle, pages 1–48, 62–83, 104–112

G. Sarton, *Ancient Science and Modern Civilization,* Lincoln, Neb.: University of Nebraska Press, 1954. Three essays dealing with the world of Hellenism as personified by Euclid and Ptolemy, and with the decline of Greek science and culture. For a more detailed account see his *Hellenistic Science and Culture in the Last Three Centuries* B.C., New York, Dover, 1993.

C. C. W. Taylor, R. M. Hare, and Jonathan Barnes, *Greek Philosophers,* New York: Oxford University Press, 1999, on Socrates, Plato, and Aristotle

GENERAL BOOKS ON THE HISTORY OF SCIENCE AND THE HISTORY OF PHYSICS

(These books deal primarily with the pre-twentieth-century period. See the end of Chapter 26 for a list of books on the history of modern physical science. Complete references are given in the Bibliography at the end of this book.)

Anthony M. Alioto, *History of Western Science.*

Bernadette Bensaude-Vincent and Isabelle Stengers, *History of Chemistry*

Michel Blay, *Reasoning with the Infinite. From the Closed World to the Mathematical Universe*

[5]In connection with our study of the Ptolemaic system there is surely no more thrilling experience than to read that famous excursion through the medieval universe, the *Paradiso* of Dante.

H. A. Boorse and L. Motz (editors), *World of the Atom*
S. G. Brush (editor), *Resources for the History of Physics*
S. G. Brush (editor), *History of Physics: Selected Reprints*
Herbert Butterfield, *Origins of Modern Science*
I. B. Cohen, *The Birth of a New Physics.* A brief, readable account covering most of the same topics as Parts A, B, and C of this book.
R. G. Collingwood, *The Idea of Nature.* A stimulating survey of the metaphysical ideas of ancient and early modern scientists.
A. C. Crombie, *Science, Art and Nature in Medieval and Modern Thought*
Michael J. Crowe, *Theories of the World from Antiquity to the Copernican Revolution*
Peter Dear (editor), *The Scientific Enterprise in Early Modern Europe*
René Dugas, *History of Mechanics*
Elizabeth Garber, *The Language of Physics*
C. C. Gillispie (editor), *Dictionary of Scientific Biography*
Edward Grant, *The Foundations of Modern Science in the Middle Ages*
A. Rupert Hall and Marie Boas Hall, *A Brief History of Science*
J. L. Heilbron, *Elements of Early Modern Physics*
John Henry, *The Scientific Revolution*
Norriss S. Hetherington, *Cosmology: Historical, Literary, Philosophical, Religious, and Scientific Perspectives*
Michael Hoskin (editor), *The Cambridge Illustrated History of Astronomy*
James R. Jacob, *The Scientific Revolution: Aspirations and Achievements, 1500-1700*
Christa Jungnickel and Russell McCormmach, *Intellectual Mastery of Nature,* Vol. I, *The Torch of Mathematics 1800–1870*
David Knight, *Ideas in Chemistry*
Alexandre Koyré, *From the Closed World to the Infinite Universe*
John Lankford (editor), *History of Astronomy*
David Lindberg, *The Beginnings of Western Science*
Stephen Mason, *A History of the Sciences*
James E. McClellan III and Harold Dorn, *Science and Technology in World History*
A. E. E. McKenzie, *The Major Achievements of Science*
John North, *Norton History of Astronomy and Cosmology*
R. C. Olby *et al.* (editors), *Companion to the History of Modern Science*
Lewis Pyenson and Susan Sheets-Pyenson, *Servants of Nature*
Colin A. Ronan, *Science: Its History and Development among the World's Cultures*
S. Sambursky, *Physical Thought* (anthology of sources)
Richard H. Schlagel, *From Myth to Modern Mind*
Cecil J. Schneer, *The Evolution of Physical Science*
Cecil J. Schneer, *Mind and Matter*
Ronald A. Schorn, *Planetary Astronomy*
Stephen Toulmin and June Goodfield, *The Architecture of Matter*
Stephen Toulmin and June Goodfield, *The Fabric of the Heavens*
Albert Van Helden, *Measuring the Universe*
J. H. Weaver, *The World of Physics*
E. T. Whittaker, *History of the Theories of Aether and Electricity,* Vol. I, *The Classical Theories*

CHAPTER 2

Copernicus' Heliocentric Theory

2.1 Europe Reborn

Our story of the rise of modern science takes us now to Renaissance Europe of the years around 1500 A.D. As the word "renaissance" suggests, this was a time of "rebirth"—of the long-dormant arts and sciences and of Western civilization in general. Astronomical theory had not progressed significantly since Ptolemy. St. Thomas Aquinas (1225–1274) had joined the Aristotelian ideas of celestial motions with Christian theology. Thus the geocentric theory had received new meaning in terms of current philosophical doctrine; to question one was to attack the other, and so far there had appeared no one who thought it necessary or had the daring to battle seriously against such formidable allies.

But the spirit of the times was changing; the Renaissance movements spread out from Italy to sweep the Western world, and within a few generations there arose a new ideal of man, the confident individual full of curiosity and *joie de vivre*. (A similar ideal of woman had to wait until the twentieth century to be generally accepted.) Concern with spiritual abstractions was balanced by a new enthusiasm for exploring the natural world, both among scientists and among artists (and especially when combined in one person who was both, such as Leonardo da Vinci). The invention of printing revolutionized the intellectual world by assuring that the works of a writer could be rapidly distributed to an ever-growing audience, rather than restricted to the few lucky enough to obtain copies laboriously transcribed by hand. Even the works of the ancient Greek scientists and philosophers were more widely available than they had been at the time they were first written; but their authority was soon to be undermined by new ideas and discoveries.

In the topography of world history this great period of summing up old wisdom and of forming new attitudes is analogous to a watershed in which great rivers have their sources. It was the time in which lived as contemporaries or within a generation of one another most of the men whose work heralded the new age: Gutenberg and da Vinci, Columbus and Vasco da Gama, Michelangelo and Dürer, Erasmus, Vesalius, and Agricola, Luther and Henry VIII. One cannot help comparing this short time interval of decisive change with a similar break in the traditions of Western thought four hundred years later—the period between the publication of Darwin's *The Origin of Species* (1859) and the first large-scale release of atomic energy, which bracketed such names as Mendel and Pasteur, Planck and Einstein, Rutherford and Bohr, and also such new forces as Marx and Lenin, Freud and Pareto, Picasso and Stravinsky, Shaw and Joyce.

2.2 The Copernican System

In the year that the New World across the Atlantic Ocean was discovered, Nicolaus Copernicus (1473–1543) was a young student in Poland. During his life he watched in his old world the gathering of great cultural changes. And it is said that on the very day he died he saw the first copy of his great book, the *Revolutions,* which gave us a whole new universe.

The full title of Copernicus' main work, *Six Books Concerning the Revolutions of the Heavenly Spheres,* startles us at the very outset with the implication of an Aristotelian notion of concentric spheres. He was indeed concerned with the old problem of Plato, the construction of a planetary system by combination of the fewest possible uniform circular motions. Far from being a revolutionary who wanted to replace traditional theories by a radical new one, he wanted to eliminate some of the innovations Ptolemy had introduced—in particular, the equant, described above in Section 1.6—and return to the principles established by earlier Greek astronomers. He was careful to point out

Fig. 2.1. Nicolaus Copernicus (1473–1543), Polish astronomer who founded the modern heliocentric theory.

that his proposal of a heliocentric system—letting the earth as well as the other planets revolve around a fixed sun—was not a new idea, but was sanctioned by some of the authorities of antiquity.

The early ideas of Copernicus can be traced in a brief sketch, called *Commentariolus,* which he wrote several years before composing the *Revolutions.*[1] Over the years a misleading myth has arisen about why Copernicus proposed the heliocentric system. It is said that as the astronomers before Copernicus made more accurate observations, they found it necessary to revise the Ptolemaic system, adding additional "epicycles on epicycles" to fit the data. Finally the system became so implausibly bizarre that it fell of its own weight, creating a "crisis" to which Copernicus responded with a radical new idea. This myth is so firmly embedded in modern discourse that the phrase "epicycles on epicycles" is used to characterize any theory that has become too complicated to be credible.

But, according to historian Owen Gingerich, there was no such crisis at the time of Copernicus. There was no flood of new data that had to be accommodated, and none of the most popular versions of the Ptolemaic system used more than one epicycle per planet. In fact, Copernicus himself tells us that the problem with Ptolemy's theory was not its difficulty in fitting observations but the unsatisfactory way it did so:

> . . . the planetary theories of Ptolemy and most other astronomers, although consistent with the numerical data, seemed . . . to present no small difficulty. For these theories were not adequate unless certain equants were also conceived; it then appeared that a planet moved with uniform velocity neither on its deferent nor about the center of its epicycle. Hence a system of this sort seemed neither sufficiently absolute nor sufficiently pleasing to the mind.
>
> Having become aware of these defects, I often considered whether there could perhaps be found a more reasonable arrangement of circles, from which every apparent inequality would be derived and in which everything would move uniformly about its proper center, as the rule of absolute motion requires.

To Copernicus, any type of celestial motion other than uniform circular motion was "obviously" impossible: " . . . the intellect recoils with horror" from any other suggestion; " . . . it would be unworthy to suppose such a thing in a Creation constituted in the best possible way." These arguments are of the same type as those of his scholastic opponents, except that to them the immobility of the earth was equally "obvious."

Copernicus then argued that a more reasonable arrangement of circles could be constructed by putting the sun rather than the earth at the center. He reported later in the *Revolutions* that he found this idea by reading the classics:

> . . . according to Cicero, Nicetas had thought the earth moved, . . . according to Plutarch certain others [including Aristarchus] had held the same opinion . . . when from this, therefore, I had conceived its possibility, I myself also began to meditate upon the mobility of the earth. And although it seemed an absurd opinion, yet, because I knew that others before me had been granted the liberty of supposing

[1]According to Edward Rosen, who has published an English translation of the *Commentariolus,* it was not printed during the author's lifetime, but a number of handwritten copies circulated among astronomers, and then disappeared for three centuries. It was first published in 1878.

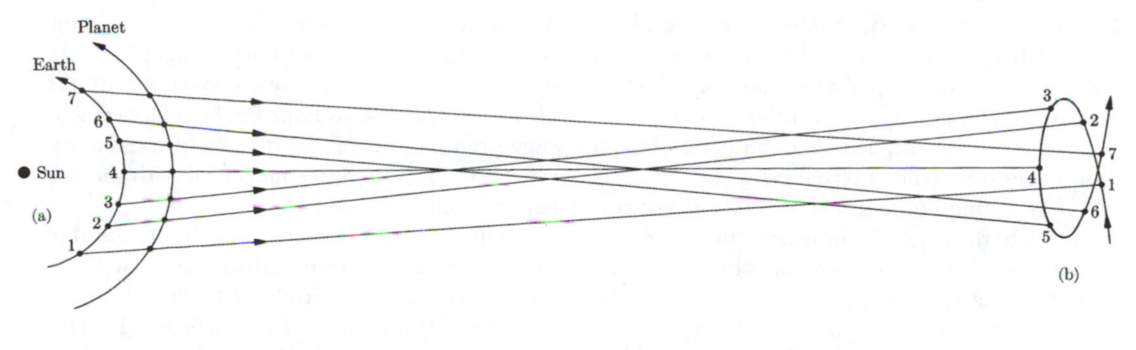

Fig. 2.2. (a). Actual configurations of sun, earth, and planet, and changing line of sight during the planet's apparently retrograde motion. (b) Apparent path of the planet against the background stars at far right, as seen from the earth.

> whatever circles they chose in order to demonstrate the observations concerning the celestial bodies, I considered that I too might well be allowed to try whether sounder demonstrations of the revolutions of the heavenly orbs might be discovered by supposing some motion of the earth. . . . I found after much and long observation, that if the motions of the other planets were added to the motions [daily rotation, and yearly revolution about the sun] of the earth, . . . not only did the apparent behavior of the others follow from this, but the system so connects the orders and sizes of the planets and their orbits, and of the whole heaven, that no single feature can be altered without confusion among the other parts and in all the Universe. For this reason, therefore, . . . have I followed this system.

Thus, in essence, Copernicus proposed a change of viewpoint for the representation of celestial motions along the lines of Aristarchus' heliocentric system. By highly gifted calculations, using some observations of his own, Copernicus then did what Aristarchus had failed to do: he proved that the motion of celestial objects as then known could indeed be represented by a combination of a few uniform circular motions in a sun-centered system. All planets, the earth included, could be imagined as moving on concentric spheres, with relatively few and small epicycles and eccentrics still needed to account for the finer details of motion. Moreover, the same direction of motion along almost all deferents and epicycles could be assumed, which had not been true for the geocentric model. But above all, the odious equant could be discarded. All motions were now truly circular and uniform with respect to their own centers. Plato's question was answered again, in an alternative way.

The most obvious advantage of the heliocentric system is that it gives a much more natural explanation of the retrograde motion of the planets. This explanation can be visualized by imagining first that you are riding in a fast car, which is passing a slower-moving car on a highway. Forgetting about the motion of your own car, observe the changing position of the slower car as seen against the background of distant trees and buildings (these correspond to the fixed stars). Initially, when it is still far away, the other car appears to be moving forward, but as your car passes it, it appears to be moving backward for a short time, simply because of the rotation of the line of sight. This phenomenon is precisely analogous to the retrograde motion of a planet, e.g., Mars, as seen from the earth (see Fig. 2.2). As Copernicus explained it in the *Commentariolus,*

> This happens by reason of the motion, not of the planet, but of the earth changing its position in the great circle [its orbit around the sun]. For since the earth moves more rapidly than the planet, the line of sight directed toward the firmament regresses, and the earth more than neutralizes the motion of the planet. This regression is most notable when the earth is nearest to the planet.

But the earth has two distinct motions in Copernicus' system: it rotates around its own axis, and it revolves around the sun. Not only does this simplify the explanation of the planetary motions, it allows the outer sphere of fixed stars to be at rest. Copernicus did not foresee that this apparently trivial change could lead to enormous consequences: if the stars no longer have to rotate together, they need not all be placed at the same distance from the earth; perhaps they are scattered through space out to infinity.

Copernicus argued that the precession of the equinoxes could also be more easily explained by attributing the motions of rotation and revolution to the earth. In the geocentric system this phenomenon could be explained by assuming a gradual change in the relative rotations of the celestial spheres carrying the sun and stars, a change that could be rationalized only by adding yet another celestial sphere to the system. Copernicus wrote:

> Hence it is the common opinion that the firmament has several motions in conformity with a law not sufficiently understood. But the motion of the earth can explain all these changes in a less surprising way.[2]

It is worth noting that the precession problem had theological as well as astronomical aspects, and that Copernicus was a canon of the Roman Catholic Church. At the Council of Nicaea in 325 A.D., the Church had defined Easter as the first Sunday after the first full moon on or after the vernal equinox. The Julian calendar, established by Julius Caesar, added one day to the normal 365 days every four years, on the assumption that the actual length of the year is 365¼ years. But in fact the year is slightly shorter (365.2422 days). The vernal equinox had been originally set at March 21, but because of the inaccuracy of the Julian calendar the date of the vernal equinox had actually slipped back to March 11 by the thirteenth century. Furthermore, the computation of full moons was somewhat inaccurate.

With the growth of commerce and communications in Europe in the late Middle Ages, the problem of calendar reform was becoming increasingly urgent, and in 1514 Pope Leo X asked Copernicus and other astronomers for their assistance. Copernicus declined the invitation to come to Rome at that time, on the grounds that he must first complete his work on the motions of the sun and moon before he could help straighten out the calendar. Later, in dedicating his book, *The Revolutions of the Heavenly Spheres,* to Pope Paul III, Copernicus suggested that his results may be of some assistance to the Church in connection with the calendar problem.

The Copernican system was in fact used in the astronomical computations on which the Gregorian calendar (introduced by Pope Gregory XIII in 1582) was based. This calendar, which we now use, modifies the Julian by not adding a leap day to years divisible by 100 unless they are also divisible by 400 (thus 2000 was a leap year but 1900 was not). But that did not necessarily mean that the Church had accepted the physical assumptions—in particular the motion of the earth—from which the Copernican system had been developed; it was simply admitted to be a more efficient system, or "hypothesis," for doing calculations.

Copernicus himself emphasized that the heliocentric system provided a unique pattern into which all the planets must fit in a definite way, in contrast to the Ptolemaic system, which required a special and separate construction for each planet. For example, Copernicus found that his system provided a definite ordering of the planets in terms of their orbital period, which agreed with the order of orbital sizes (we shall shortly say how relative orbital sizes were obtained):

> Saturn, first of the planets, which accomplishes its revolution in thirty years, is nearest to the first sphere [the immobile sphere of the fixed stars]. Jupiter, making its revolution in twelve years, is next. Then comes Mars, revolving once in two years. The fourth place in the series is occupied by the sphere which contains the earth and the sphere of the moon, and which performs an annual revolution. The fifth place is that of Venus, revolving in nine months. Finally, the sixth place is occupied by Mercury, revolving in eighty days.

In the Ptolemaic system, on the other hand, Mercury, Venus, and the sun all had the same period (one year) so that there was some disagreement as to how to order their spheres and no unique way to determine relative orbital sizes. Moreover, in the geocentric systems there might be two or more equally good ways of "saving the appearances"—the observed motions of the planets might be explained by eccentrics, by epicycles,

---

[2]Although Copernicus did not in fact accomplish any significant improvement over the geocentric system in his theory of the precession of the equinoxes, one historian of science has argued that it was his attempt to deal with this particular problem that led him to work out the details of the heliocentric system. See J. Ravetz, "The origins of the Copernican revolution."

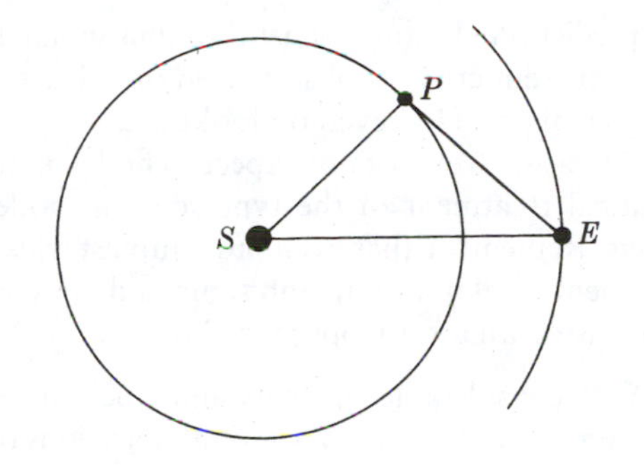

Fig. 2.3. Method for computing relative distances between an inferior planet *P* and the sun *S*, observed from earth *E*. (The planet may be Mercury or Venus.)

by equants, or by various different combinations of these devices. Hence it was difficult to believe in the physical reality of any one particular combination. Copernicus claimed that his own system "binds together so closely the order and magnitudes of all the planets and of their spheres or orbital circles and the heavens themselves that nothing can be shifted around in any part of them without disrupting the remaining parts and the universe as a whole."

Actually this last claim is a little exaggerated: The Copernican system in its full elaboration suffers from some of the same ambiguities as the Ptolemaic, while requiring nearly as many circles. But it does at least provide a good method for determining the relative sizes of the planetary orbits. The method is most simply illustrated for the inferior planets (those closer to the sun than is the earth, i.e., Venus and Mercury).

Let ∠*SEP* be the angle of maximum elongation of the planet away from the sun, as viewed from the earth. Then the angle ∠*SPE* must be a right angle; hence the ratio *SP*/*SE* must be equal to the sine of the angle ∠*SEP*. A direct observation of the maximum elongation of Mercury or Venus can therefore be used to compute the ratio of its distance from the sun to that of the earth from the sun. The relative distances of the outer planets from the sun can be found by a somewhat more complicated method.[3] The results are shown in Table 2.1.

Table 2.1. Relative Distances of Planets from the Sun

	Copernicus	*Modern Value*
Mercury	0.3763	0.3871
Venus	0.7193	0.7233
Earth	1.0000	1.0000 (by definition)
Mars	1.5198	1.5237
Jupiter	5.2192	5.2028
Saturn	9.1742	9.5389

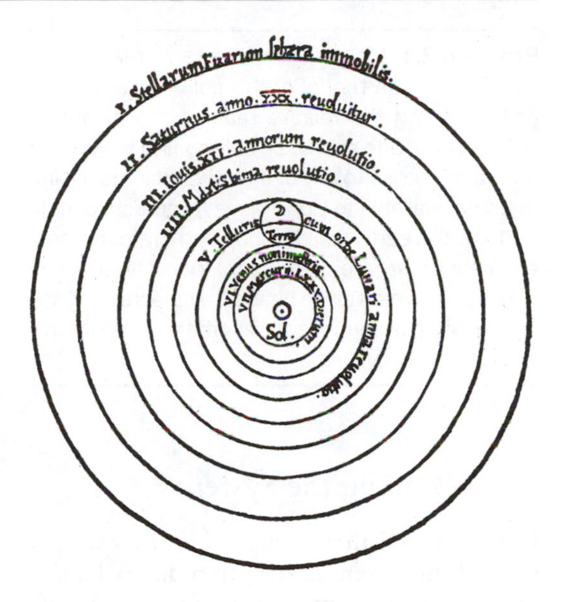

Fig. 2.4. The heliocentric system as shown in Copernicus' book *The Revolutions of the Heavenly Spheres* (1543). The spheres around the sun (Sol) are identified as follows: (I) immobile sphere of the fixed stars; (II) Saturn, revolves in 30 years; (III) Jupiter, revolves in 12 years; (IV) Mars, revolves in 2 years; (V) earth, revolves in 1 year, with moon's orbit; (VI) Venus, revolves in 9 months; (VII) Mercury, revolves in 80 days.

There was still no accurate method for estimating the *absolute* distances, but at least one now had, for the first time, an idea of the relative distances. Copernicus did not try to improve on the value of the absolute distance of the earth as estimated (on poor hypotheses) by the Greek astronomer Hipparchus for the earth-sun distance, namely 1142 times the radius of the earth. The modern value, first roughly estimated by Richer and Cassini in the 1670s (see Section 10.4), is about 93 million miles, or 23,000 times the radius of the earth.

---

[3]See, e.g., T. S. Kuhn, *The Copernican Revolution*, pages 175–176.

Problem 2.1. One of the impressive features of the Copernican model of the solar system is its ability to find the relative radii of the planetary orbits. Thus, one should be able to find how close the earth comes to any of the other planets. Using the data available to Copernicus (Hipparchus' estimate of the earth-sun distance and Eratosthenes' estimate of the size of the earth), calculate the minimum possible distance between the earth and any other planet, assuming circular orbits. Compare this to the modern value.

2.3 Bracing the System

Knowing well that to many his work would seem absurd, even contrary to ordinary human understanding, Copernicus attempted to fortify it against anticipated criticism in four ways.

a) He tried to make plausible that his assumptions coincided with Platonic principles at least as well as Ptolemy's. He has many passages (we have seen samples) on the deficiencies of the Ptolemaic system, on how harmonious and orderly his own seems, and how pleasingly and evidently it reflects the mind of the Divine Architect. To Copernicus, as to everyone of his period (and to many of our own), the observable world was but a symbol of the working of God's mind; to find symmetry and order in the apparent chaos of sense data was to him an act of reverence, a renewed proof of the Deity. He would have been horrified if he had known that his theory was ultimately responsible for the sharp clash, in Galileo's time, between science and religion. Copernicus, we must not forget, was a highly placed and honored church dignitary. In matters of doctrine he might have regarded himself as a conservative who was objecting to current scholastic teaching only to the extent of wishing to make it more consistent with Aristotle's principles of mechanics. (Compare the forms of the systems in Figs. 1.3 and 2.4.)

b) Copernicus prepared enough quantitative material to put his book mathematically on the same footing as Ptolemy's; that is, he calculated the relative radii and speeds of the elements in his system so that tables of planetary motion could be made. (Note how this compares with Aristarchus' more qualitative work.) The two theories were thus equally correct with respect to prediction of future planet positions within the then current error of observation of at least 1/6 degree of arc. However, on looking at Copernicus' treatise, we must not expect to find a mathematical treatment of the type seen in modern texts. Remember that even our simplest trick of mathematical notation, and signs, did not come into use until after Copernicus' death.

c) With considerable ingenuity and much success, Copernicus tried to answer several objections that were as certain to be raised against his heliocentric system as they had been, long ago, against Aristarchus'. To the argument that his earth, rotating so rapidly about its own axis, would surely burst like a flywheel driven too fast, he countered, "Why does the defender of the geocentric theory not fear the same fate for his rotating celestial sphere—so much faster because so much larger?" Aristotle had argued that the celestial sphere must be finite to avoid the consequence that its most distant parts would have to move infinitely fast (which is impossible); Copernicus inverted this argument, saying it is safer to ascribe the rotation to the earth, which we know to be finite.

To the argument that birds in flight and the like should be left behind by the rapidly rotating and revolving earth,[4] he answered that the atmosphere is dragged along with the earth.

As to the old question of the absence of parallax among the fixed stars, Copernicus could give only the same answer as Aristarchus (a good answer even though the Greeks did not accept it):

> . . . the dimensions of the world are so vast that though the distance from the sun to the earth appears very large compared with the size of the orbs of some planets, yet compared with the dimensions of the sphere of fixed stars, it is as nothing.

This distance to the fixed stars, he says elsewhere, is "so immense as to render imperceptible to us even their apparent annual motion. . . ." He cites several other good arguments (and a few erroneous ones) to support his system, but these will sufficiently illustrate his point.

d) To us the most impressive success of Copernicus is the simplification of the geometrical rep-

[4]The earth's orbital speed around the sun is about 70,000 mi/hr, and the speed of west-east rotation alone of a point on the equator is more than 1000 mi/hr.

resentations needed in his system and the consequent greater ease with which his apparatus can be used by the astronomer for the solution of practical problems. Most striking is the explanation of retrograde motion in terms of the changing apparent position of a planet as seen from the moving earth (Fig. 2.2). By eliminating the artificial epicycles in this case, Copernicus seemed to have found a simpler system. But merely moving the origin of the coordinate system from the earth to the sun, while keeping enough circular motions to fit the same observations to the same degree of accuracy, could not yield a substantially simpler system. That had to wait for Kepler's brilliant (but puzzling) innovation: replacing circles by ellipses.

Problem 2.2. Make a list of the assumptions underlying Copernicus' theory, analogous to this list for Ptolemy's (see Section 1.6).

Problem 2.3. Copernicus postulated that the atmosphere is dragged along by the rotating earth, presumably being unaware of any observable effects of rotation on the atmosphere. Find out and explain one such effect that is now considered well established.

2.4 The Opposition to Copernicus' Theory

But Copernicus' hope was not quickly fulfilled. It took more than a century for the heliocentric system to be generally accepted by astronomers; in the meantime the theory and its few champions were faced with powerful opposition, some of it the very same that antagonists had used against the heliocentric ideas of the Greek philosophers.

a) First came the argument from dogma concerning the immobility and central position of the earth. For all his efforts, Copernicus was in general unsuccessful in persuading his readers that the heliocentric system was at least as close as the geocentric one to the mind and intent of the Deity. All religious faiths in Europe, including the newly emerged Protestants, found enough Biblical quotations (e.g., *Joshua* 10:13) to assert that the Divine Architect had worked from a Ptolemaic blueprint.

Problem 2.4. Read *Joshua* 10:13, and interpret the astronomical event referred to in terms of the two rival planetary systems.

The Roman Catholic Church eventually put the *Revolutions* on the Index of forbidden books as "false and altogether opposed to Holy Scriptures," withdrawing its approval of an earlier outline of Copernicus' work. Some Jewish communities forbade the teaching of the heliocentric theory. It was as if man's egocentric philosophy demanded the middle of the stage for his earth, the scene both of his daily life and prayer in a world created especially for his use, and of the drama of salvation with the expected eventual coming of the Savior or Messiah.

Although it was allowed that for mathematical calculations the Copernican scheme might indeed be useful, and although even St. Thomas Aquinas had grave misgivings about the Ptolemaic system because it did not stick to strictly uniform circular motions as interpreted by Aristotle, yet it seemed "philosophically false and absurd," dangerous, and fantastic to abandon the geocentric hypothesis. And in all truth, what other reaction could one have expected? We must try to appreciate the frame of reference of the time. In general, Europe then recognized as its two supreme sources of authority the Bible (either the literal text or its interpretation by Rome) and Aristotle (as interpreted by his followers). Physical science as we now think of it was only gradually evolving; that was still the period in which scientists like Kepler and Paracelsus could also practice the pseudosciences of astrology and alchemy.

b) A further explanation for the resistance that the Copernican theory had to overcome is to be sought in the damage it brought even to contemporary physics. This point was well discussed by the historian Herbert Butterfield:

> . . . at least some of the economy of the Copernican system is rather an optical illusion of more recent centuries. We nowadays may say that it requires smaller effort to move the earth round upon its axis than to swing the whole universe in a twenty-four-hour revolution about the earth; but in the Aristotelian physics it required something colossal to shift the heavy and sluggish earth, while all the skies were made of a subtle substance that was supposed to have no weight, and they were

comparatively easy to turn, since turning was concordant with their nature. Above all, if you grant Copernicus a certain advantage in respect of geometrical simplicity, the sacrifice that had to be made for the sake of this was nothing less than tremendous. You lost the whole cosmology associated with Aristotelianism—the whole intricately dovetailed system in which the nobility of the various elements and the hierarchical arrangement of these had been so beautifully interlocked. In fact, you had to throw overboard the very framework of existing science, and it was here that Copernicus clearly failed to discover a satisfactory alternative. He provided a neater geometry of the heavens, but it was one that made nonsense of the reasons and explanations that had previously been given to account for the movements in the sky.

(*Origins of Modern Science*)

Problem 2.5. Read Milton's *Paradise Lost,* Book VIII, lines 1–202, in which Adam and Raphael discuss the two systems from a classic point of view, which may well have mirrored the general opinion of educated persons in England until the full impact of Newton's work was established.

c) Third, argument arose from the oft-mentioned lack of observable parallax among fixed stars. Copernicus' only possible (and correct) reply was still unacceptable because it involved expanding the celestial sphere to practically an infinite distance away from the earth. This is to us no longer intellectually difficult, but it was a manifestly absurd notion at the time. Among the many reasons for this, perhaps the intensely religious mind of the period could not be comfortable with a threatening Hell so close below the ground and a saving Heaven so infinitely far above.

If we consider for a moment, it is evident that as a consequence of this attitude an actual observation of annual parallax at that time would not necessarily have settled the dispute in favor of Copernicus, for an obvious and simple Ptolemaic assumption of an additional yearly epicyclic motion superposed on the great daily rotation of the fixed stars would explain away such an observation in full accord with the geocentric system.

d) A fourth important argument against the Copernican astronomy at the time was that apart from its powerful simplicity it offered to contemporary astronomers no overruling scientific advantages over their geocentric astronomy, i.e., there was then no important observation that was explainable only by one and not by the other, no experiment to pit one against the other in a clear-cut decision. Copernicus introduced no fundamentally new experimental facts into his work, nor was the accuracy of his final predictions significantly better than previous estimates.

To us it is clear, although it did not enter the argument then, that the scientific content of both theories—the power of prediction of planetary motion—was about the same at that time. As the English philosopher Francis Bacon wrote in the early seventeenth century: "Now it is easy to see that both they who think the earth revolves and they who hold the *primum mobile* and the old construction, are about equally and indifferently supported by the phenomena." In our modern terminology we would say (although this is not what Bacon had in mind) that the rival systems differed mainly in the choice of the coordinate system used to describe the observed movements. As measured with respect to the earth, the sun and stars do move, and the earth, of course, is at rest. On the other hand, measured with respect to the sun, the earth is not at rest. Any other system of reference, for instance, one placing the moon at the origin of the coordinate system and referring all motions to that new center of the universe, is quite conceivable and equally "correct," although, of course, evidently most complex.

Bacon's empirical relativism, which led him to downgrade the Copernican system, is refuted by certain observations made later. Consider the pattern of "trade winds" in the earth's atmosphere: near the equator they tend to blow from east to west. At higher latitudes (north or south of the equator) they tend to blow from west to east. (Thus weather conditions in the United States are propagated from west to east, as can easily be seen from the weather maps shown on television.) This simple fact was known to Columbus; he took advantage of the pattern in planning his voyages across the Atlantic Ocean. Instead of sailing directly west from Spain he went south to the Canaries to pick up the strong trade winds blowing toward the west; after exploring the West Indies he headed north until he reached the region of strong winds blowing toward the east, and used them to expedite his return home. Columbus did not know about the rotation of the earth, but scientists in the eighteenth century pointed out that the pattern is just what one would expect: The linear speed of rotation is greatest at the equator, so if the atmosphere is not completely dragged along by the rotating earth, it will (like the sun) move westward relative to the earth. But the slower linear speed at higher latitudes can set up a circular motion, so that the wind blows "back"—that is, eastward. The earth's rotation in space is made more plausible still by the famous pendulum experiment first performed by the French physicist J.B.L. Foucault in the middle of the nineteenth century, and now repeated daily in science museums all over the world.

But at the beginning of the seventeenth century every newly discovered peculiarity of planetary motion could equally easily have been accommodated in some geocentric representation as well as in a heliocentric one. In fact, to many people it seemed that the compromise system proposed by the Danish astronomer Tycho Brahe (1546–1601) combined the best features of both: All the planets except the earth revolved around the sun, and the sun then revolved around the earth, which remained at rest.

2.5 Historic Consequences

Eventually the vision of Copernicus did triumph; in a moment we shall follow the work leading to the universal acceptance of the heliocentric theory, even though his specific system of uniform circular motions will have to be sacrificed along the way. We shall see that the real scientific significance and the reason for the eventual admiration of Copernicus' work lie in a circumstance he would never have known or understood—in the fact that only a heliocentric formulation opens the way for an integration of the problem of planetary motion with the simple laws of "ordinary" (terrestrial) mechanics as developed during the following 150 years—a synthesis of two sciences, even of two methods, which was achieved through Newton's Theory of Universal Gravitation. Consequently, it became possible to explain, on the basis of the supposition that the earth turns and travels, such diverse phenomena as the diurnal and the annual apparent motion of the stars; the flattening of the earth; the behavior of cyclones, tradewinds, and gyroscopes; and much else that could not be bound together into so simple a scheme in a geocentric physics.

But apart from this historic triumph, the memory of Copernicus is hallowed for two more reasons. First, he was one of those giants of the fifteenth and sixteenth centuries who challenged the contemporary world picture and thereby gave life to that new and strange idea that later was to grow into science as we now know it. Secondly, his theory proved to be a main force in the gathering intellectual revolution, which shook man out of his self-centered preoccupations. Out with the Ptolemaic system has to go also the self-confident certainty that a symbolic uniqueness about our position among planets shows that man is the sum and summit, the chief beneficiary of Creation, to whom the whole is given as if it were his toy or his conquest. Wherever this certainty still exists, at least it cannot claim that science backs it up.

Copernicus thus seems in retrospect to have played a role that he never foresaw or intended: instigator of the "Scientific Revolution." This revolution, which created modern science during the period from about 1550 to 1800, also gave science the right to claim authoritative knowledge. As historian Herbert Butterfield wrote:

> Since [the Scientific Revolution] overturned the authority in science not only of the middle ages but of the ancient world—since it ended not only in the eclipse of scholastic philosophy but in the destruction of Aristotelian physics—it outshines everything since the rise of Christianity and reduces the Renaissance and Reformation to the rank of mere episodes, mere internal displacements, within the system of medieval Christendom. Since it changed the character of men's habitual mental operations even in the conduct of the non-material sciences, while transforming the whole diagram of the physical universe and the very texture of human life itself, it looms so large as the real origin both of the modern world and of the modern mentality that our customary periodisation of European history has become an anachronism and an encumbrance. There can hardly be a field in which it is of greater moment to us to see at somewhat closer range the precise operations that underlay a particular historical transition, a particular chapter of intellectual development.
>
> (*Origins of Modern Science*)

An important purpose of our story of planetary motion to this point has been preparation for the following digression—whose relevance is highlighted by Butterfield's comment—before returning to our main topic. We may now ask briefly: By what standards do we judge a scientific theory?

RECOMMENDED FOR FURTHER READING

Marie Boas, *The Scientific Renaissance 1450–1630*

Herbert Butterfield, *The Origins of Modern Science,* Chapter 2

Michael J. Crowe, *Theories of the World,* Chapters 5 and 6 (with excerpts from Copernicus' *On the Revolutions*)

C. C. Gillispie, *Dictionary of Scientific Biography,* article by E. Rosen on Copernicus, Vol. 3, pages 401–411 (1971)

Owen Gingerich, "Copernicus: A modern reappraisal," in *Man's Place in the Universe: Changing Concepts* (edited by D. W. Corson), pages 27–49, Tucson: University of Arizona, College of Liberal Arts, 1977

Edward Grant, "Celestial motions in the late Middle Ages," *Early Science and Medicine,* Vol. 2, pages 129–148 (1997)

Norriss S. Hetherington, *Cosmology,* articles by M.-P. Lerner and J.-P. Verdet, "Copernicus," pages 147–173, by Edward Grant, "Medieval cosmology," pages 181–199, by Alison Cornish, "Dante's moral cosmology," pages 201–215, and by J. D. North, "Chaucer," pages 217–224

T. S. Kuhn, *The Copernican Revolution,* Chapters 4 and 5

John North, *Norton History,* Chapter 11

J. Ravetz, "The origins of the Copernican Revolution," *Scientific American,* October 1966, pages 88–98; "The Copernican Revolution," in *Companion* (edited by Olby), pages 201–216

S. Sambursky, *Physical Thought,* excerpts from Al-Biruni, Averroes, Moses Maimonides, Oresme, Nicolas Cusanus, Osiander and Copernicus, pages 133–135, 142–147, 161–188

For additional sources, interpretations, and reference works see the website www.ipst.umd.edu/Faculty/brush/physicsbibliography.htm

CHAPTER 3

On the Nature of Scientific Theory

As we now turn to the standards by which to judge scientific theories, it is obvious that we cannot pretend to lay down criteria by which working scientists should check their own progress during the construction of a theory. What we shall do here is something else, namely, to discuss how scientists have evaluated theories. As examples we will consider the rival theories of the universe, discussed in the previous two chapters, and a very different type of theory: atomism and its application to explaining vision.

3.1 The Purpose of Theories

We have argued that a main task of science, as of all thought, is to penetrate beyond the immediate and visible to the unseen, and thereby to place the visible into a new, larger context. For like a distant floating iceberg whose bulk is largely hidden under the sea, only the smallest part of reality impresses itself upon us directly. To help us grasp the whole picture is the supreme function of theory. On a simple level, a theory helps us to interpret the unknown in terms of the known. It is a conceptual scheme we initially invent or postulate in order to explain to ourselves, and to others, observed phenomena and the relationships among them, thereby bringing together into one structure the concepts, laws, principles, hypotheses, and observations from often very widely different fields. These functions may equally well be claimed for the hypothesis itself. In truth, we need not lay down a precise dividing line, but might regard theory and hypothesis as differing in degree of generality. Therefore at one extreme we might find the limited working hypothesis by which we guide our way through a specific experiment, placing at the other end of the spectrum the general theory, which guides the design and interpretation of all experiments in that field of study.

Examples of general theories suggest themselves readily, even if we decide, perhaps rather arbitrarily, to use the word "theory" only for those few historic and general schemes of thought, such as the theories of planetary motion, of universal gravitation, of vision, and the like. Galileo's theory of projectile motion welded together the laws of uniformly accelerated motion, including free fall and the principle of superposition of velocities, to produce one conceptual scheme, one overall method of attacking, predicting, and interpreting every conceivable problem involving bodies moving under the influence of a constant force. Similarly, when Charles Darwin had pondered the results of such fields as paleontology and comparative anatomy, he could find an inspired connection in widely different types of observations, explaining previously separate truths in different disciplines as part of one grand scheme of thought—his theory of evolution.

Problem 3.1. Examine some one theory discussed in other courses with which you are thoroughly familiar—perhaps Adam Smith's theory of economics, Gregor Mendel's theory of heredity, the Marxian theory of society, Sigmund Freud's theory of the unconscious, or a theory of "business cycles." Differentiate clearly in each case between the *data* available to the theorizer on one hand and the *hypotheses* used by him to explain the data on the other.

Problem 3.2. Do the same for Ptolemy's and Copernicus' theory of planetary motion.

If we inquire more deeply into the main purposes of theories, we find that there are, on the whole, three functions. Let us summarize them here briefly:

1) A theory correlates many separate facts in a logical, easily grasped structure of thought. By correlating, by putting in juxtaposition and order previously unrelated observations, we can understand them; we always explain by pointing to a

relationship. The Platonists explained the motion of planets by relating planets to certain "necessary" attributes of divine beings. Ptolemy and Copernicus explained it by relating the planetary paths to the mathematical combination of certain more or less simple geometrical figures. And not only will a fruitful theory explain the laws that it relates within its framework, it also will show where and why these laws in practice may not hold precisely, as, e.g., Galileo's specific use of his theory of projectile motion to explain why projectiles are not expected actually to follow parabolas for very long paths.

It is a basic fact about human beings that when they look at the moving parts of some intriguing new toy or gadget, they usually try to visualize a mechanical model of the hidden mechanism, to explain the separate motions observed by one theoretical system. This type of activity seems to be a necessity for the human mind. It is as if the intellect is restless until it can subordinate individual events to more inclusive schemes. It may be a matter of economy of thought, for a good theory allows us to grasp, remember, and deduce a large number of otherwise elusive facts. (Think of the wealth of facts summarized by the Copernican theory.)

And simple theories in physics are often based on mechanical models. We recall Lord Kelvin's famous remark, "I never satisfy myself until I can make a mechanical model of a thing. If I can make a mechanical model I can understand it" (*Lectures on Molecular Dynamics*). This is, of course, not true for all scientists or all branches of science. Conceptual schemes can certainly not always be phrased in mechanical models, and there are several examples in the history of physics that prove that at times progress may be seriously delayed by a too strong belief in a mechanical model. Particularly, the last 150 years have shown that such models, like all analogies, while often helpful as guides to the imagination, can cause it to fall into traps. For example, it undoubtedly was—and is—easier to think of light as a vibration moving through a material ether than as energy propagated through a void; and yet the ether picture carried with it some eventually inescapable fallacies. Similarly, we owe modern chemistry in large part to the simple, almost naive pictorial representation that Dalton used to order his thoughts about atoms and molecules; but, again, much early trouble in his atomic theory can be traced to his too concrete and prematurely complete mental picture, as we shall see in the section on the chemical atom (Chapter 20).

In fact, it has sometimes been suggested that the science of the ancients, in relying frequently on analogies involving the behavior and drives of living beings, was organismic, whereas science from Newton on became mechanistic, to turn finally in our century more and more to abstract, mathematical sets of mental images. But neither is the mathematical type of model free from the dangers inherent in a mechanistic science. Yet our thoughts proceed, as it were, on crutches, and so depend on these schemes and pictures, no matter how incomplete.

2) Whether general or limited, theories and hypotheses are expected to suggest new relations, to start the imagination along hitherto unsuspected connecting paths between old and new facts. In this sense, even the theories that were eventually abandoned or found altogether faulty—such as the Ptolemaic theory of planetary motion, the phlogiston theory of chemistry, or the caloric theory of heat—played a vital and positive role in the life of science, for they tended to propose problems, focus the attention, and integrate the effort of many scientists along the same lines rather than allowing it to be completely dispersed in random directions. Even a fallacious theory, if actively and widely pursued, can soon lead to the key observations needed for a better theory; for, as Francis Bacon remarked, truth arises more easily from error than from confusion.

Problem 3.3. The following table indicates the period of revolution of the planets about the sun as measured relative to the stars (the so-called *sidereal period*) in the Copernican system. Instead of Copernicus' values, we shall list approximate modern results. (a) Examine these values and the order of spheres in the two systems (consult the diagrams) and then explain why Copernicus felt that there was exhibited order and harmony in his scheme, "a precise relation between the motion and sizes of the orbs." (b) If a new planet were discovered with a sidereal period of about 1830 days, what would the theory suggest about the limits of its orbit? If it should turn out that the actual orbit is larger than Jupiter's, is the whole theory wrong? Explain.

	Sidereal Period (days)
Mercury	88
Venus	225
Earth	365¼
Mars	687
Jupiter	7,330
Saturn	10,760

Problem 3.4. Compare the orbit of Venus in the two rival theories by referring to Figs. 1.10 and 2.4. If it were found that Venus was not self-luminous but obtained its light from the sun, what would these two theories suggest concerning the nature of the shadow on Venus that might be seen at different times through a telescope from the earth? [Hint: Where would the sun have to be with respect to Venus and the earth when we see Venus fully illuminated? Make sketches of planetary position.]

3) The foregoing paragraphs hint at a further purpose of theories: the *prediction* of specific new observable phenomena and the solution of practical problems. In Shakespeare's great tragedy, Hamlet's theory of his uncle's guilt in the death of the old king yields him a prediction of how the suspect will react to the little play-within-the-play; conversely, the expected reaction proves Hamlet's theory to be very likely correct, and so gives him the desired solution to the problem of the old king's death.

In astronomy and physics, where the problems and predictions needed are usually numerical ones, the theories are, as a rule quantitative. The problems to be solved may be quite abstract, e.g., why the planets sometimes seem near, at other times far, or may be very practical ones. For example, just as some of the earliest speculations on celestial motions seem to have been prompted by the need to make calendars and to predict eclipses, so one of the problems Copernicus' theory helped to solve was the exact length of the year and of the lunar month—data needed for the revision of the calendar.

Karl Popper, an influential twentieth-century philosopher, argued that prediction is such an important function of theories that it can be used to distinguish between science and non-science or pseudoscience. He pointed out that, strictly speaking, a theory can never be proved *correct* by its plausible explanations of known facts or its successful predictions of new facts, unless one can also prove (which seems impossible) that *no other theory could offer the same explanations or predictions.* But, he claimed, a theory can be proved *wrong* if one of its predictions can be refuted by experiment.

Thus the way to make progress in science, according to Popper, is to propose testable theories and try to refute or "falsify" them. A theory that has survived several attempts to falsify it is not necessarily correct but is preferable to one that has been refuted. Even worse than a refuted theory is one that pretends to be scientific but is so flexible it can explain all the known facts, yet never makes any predictions that *could* be falsified by a conceivable experiment. Examples of such *unfalsifiable* theories, according to Popper, are psychoanalysis and Marxism: they cannot be said to be wrong, they are just not *scientific* theories because they can't be tested.

Popper's Principle: We give the name "Popper's Principle" to the thesis that science progresses by "conjectures and refutations," to use the title of one of his books. The principle assumes that scientists actually do give up their theories when experiments refute their predictions, and are willing to accept, provisionally, a new theory that has survived all tests so far. Moreover, it assumes that any scientific theory can be expected to predict new facts, not just explain old ones.

We do *not* assert that Popper's Principle is correct. Rather, we consider it to be an interesting hypothesis *about* science, which should itself be tested by looking at the past and present behavior of scientists. If we find a theory that is generally accepted by scientists but has not made successful predictions, we would have to say that Popper's Principle has been violated *in this case.* It could still be valid in other cases; as we will argue later (Section 12.3), there is no reason to believe that all scientists must follow a single method.

Problem 3.5. Read Copernicus' treatise in extensive extract (see the sources listed at the end of the preceding chapter) and determine to what extent his work seems to exhibit these three purposes of theory.

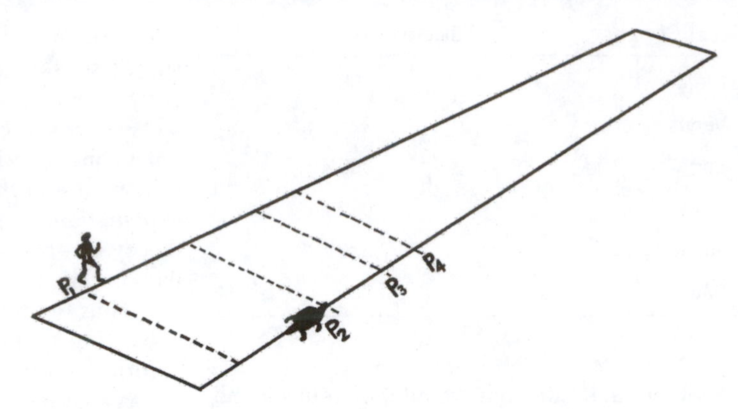

Fig. 3.1. Zeno's paradox: a handicap race between Achilles and the tortoise. Though Achilles runs much faster he can never catch up because he would have to cover an infinite number of space intervals (each smaller than the previous one) in a finite time.

3.2 The Problem of Change: Atomism

Parmenides of Elea, in addition to being an early advocate of the doctrine that the earth is a sphere at the center of the universe (Section 1.1), raised an important philosophical objection to the use of scientific reasoning to describe processes in the natural world. He started from the axiom: "Only that can really exist which can also be thought." This might be called a "principle of intelligibility": humans have the ability to understand the nature of the world; if a concept or proposition is not intelligible to us, it cannot be true or logically coherent.

The principle of intelligibility was superficially attractive but turned out to be rather dangerous to the scientific enterprise, when employed by a philosopher of skeptical temper. Parmenides used it to "prove" that *change is impossible.*[1]

His pupil Zeno of Elea drove home this conclusion by proposing a famous set of paradoxes. He used them to prove that the simplest kind of change, *motion,* is inconceivable (or at least cannot be consistently described by the mathematical tools available to the Greeks).

The first of Zeno's paradoxes, called "the dichotomy," states that whatever moves from point A to point B must reach the middle of its path (halfway between A and B) before it reaches the end. But before it reaches the middle it must reach the one-quarter mark, and before that it must reach the one-eighth mark, and so on indefinitely. Hence the motion can never even start!

The second paradox is called "Achilles and the Tortoise." Imagine a handicap race between Achilles, the legendary hero who fought in the Trojan War, starting at point P_1, and the tortoise starting at point P_2 farther along toward the finish line. Even though Achilles can run much faster than the tortoise he can never pass him, because by the time he has reached P_2 the tortoise will have advanced to P_3; when Achilles gets to P_3 the tortoise will have gone a little way further to P_4, and so on (Fig. 3.1).[2]

In ancient Greece, the views of Parmenides and Zeno were opposed by the atomists. The Greek philosopher Democritus accepted Parmenides' Principle of Intelligibility but wanted to show that change *is* intelligible, hence can really exist. Contrary to Zeno, he asserted that *motion* is an intelligible form of change. He proposed that all changes that we observe in the world are not

[1]He began by asserting that "not-being" does not exist; the only thing that exists is "Being" (apparently he meant some universal substance). Moreover, Being must be *one,* for if it were many, there must be something that divides it. But there is nothing that is *not* Being, hence what could divide it? He then analyzed "change." There are only two conceivable kinds: (a) from one kind of Being to another, and (b) from non-being to Being. But (a) is impossible since there is only one Being, and (b) is impossible since non-being does not exist so it isn't available to become anything. Hence Being is unchangeable.

[2]These two paradoxes depend on the assumptions that (1) both space and time are infinitely divisible, and that (2) you cannot take an infinite number of steps in a finite time. But suppose you try to escape them by assuming that space and time are made of discrete parts. Then Zeno has two more paradoxes to confound you. The simpler one is called "The Arrow." Call the smallest possible time interval an "instant." At any given instant an arrow is either at rest or moving. But if it were moving then that instant would thereby have to be divided into smaller parts (in the first part the arrow is here, in the next part it is there, etc.) contrary to your assumption that the instant is indivisible. So the arrow cannot be moving at any instant of time and therefore it can never move at all. So motion is impossible.

really changes in the *nature* of things but are due to the movement from one place to another of atoms, which retain their nature. (The Greek word *atomos* means "indivisible.")

Contrary to Parmenides, Democritus believed that "being" (all the matter in the world) is *not* one but is divided into many Beings—the "atoms." But, in agreement with Parmenides, *each* Being is unchangeable (aside from its motion) and indivisible. Moreover, "non-Being"—empty space—does exist, it separates the atoms and gives them room to move. Change, for Democritus, can also include union or separation of atoms, but they retain their separate unchangeable identities while joined together. The atoms are infinite in number; they are all made of the same substance, but differ in size and shape. All differences that we perceive in things are supposed to be the result of differences in the size and shape of the atoms, and their arrangement in space. Change is to be explained as the result of changes in the positions of atoms, along with possible unions and separations.

Democritus made one hypothesis that proved nearly fatal for his theory: the soul consists of "round atoms" that permeate the human body. This, together with his general policy of substituting atomistic explanations for traditional beliefs that attributed many phenomena to divine action, led other philosophers to label the atomic theory as atheistic. As a result it was not considered an acceptable scientific theory until the seventeenth century, and survived only because it was presented in a work of literature, the great poem *On the Nature of the Universe* (often known by its Latin title *De Rerum Natura*), written by the Roman philosopher Lucretius in the first century B.C.

3.3 Theories of Vision

How do we see an object? Does the eye send out something to the object, or does the object send something to the eye? Before reading any further, write down your answer to this question, and at least one good reason to support it. How could you test the correctness of your answer?

We introduce the problem of how vision works, and the story of how that problem was solved, for two reasons. First, it shows how physics is linked to an important aspect of human experience, and thereby to problems in biology and the psychology of sense perception. Second, it illustrates how science often grew in a roundabout fashion: some very clever people, in the absence of experiments, stumbled along plausible, mathematically sophisticated but (from our modern perspective) wrong routes, while other people were, for centuries, less successful in producing a satisfactory quantitative theory but eventually managed to develop what we now consider the correct solution. The first theory, going back to the Pythagoreans, is called *extramission:* the eye sends something *out* to explore the world and perceive objects. But according to Epicurus, one of the Greek atomists, objects emit "images"—particles that have the same appearance as the objects. When received by the eye, these images create the perception of the object. This became known as the *intromission theory of vision* (the eye brings something *in*).

There is one obvious difficulty with the intromission theory: shouldn't the object be getting smaller as a result of this process? The theory apparently makes a prediction that is easily refuted by observation. Epicurus has a ready answer: "particles are continually streaming off from the surface of bodies, although no diminution of the bodies is observed, because other particles take their place."

Plato, like the Pythagoreans, taught that vision depends on something emitted from the eye, a kind of fire that illuminates the object to be seen. But, like the atomists, he stated that vision also depends on something emitted from the object (intromission hypothesis). Neither, he pointed out, is sufficient: we also need daylight to see. So, he proposed, the visual fire emitted by the eye coalesces with daylight to form a single homogeneous body extending from the eye to the object; this body passes on the image from the object to the eye.

Aristotle was not an atomist but, like the atomists, he had to face Parmenides' challenge by offering an intelligible explanation of change. To explain change in the terrestrial domain Aristotle introduced the concept of *potential* being. An object may have a potential property that becomes actual: thus an acorn has the potential to become an oak tree.

In his discussion of vision, Aristotle criticized extramission theories: it is absurd to think that the human eye can send out rays of visual fire reaching the Sun and all the stars. Clearly, he says, the object must send something to the eye, but—contrary to the atomists—he argues that the medium between object and eye must play an active role. He points out first that if you put a colored object in contact with the eye you don't see the color at all. You need a transparent medium, such as air, water, or glass, to transmit the image. *Transparency,* however, is only potential unless there is a luminous body present to make it actual, producing light. When that happens the entire medium becomes transparent all at once. Thus there is no process by which the image moves from object to eye in time; we see the object instantly.

To us it makes sense to say that vision depends on something coming from the visible object into our eyes, whether or not we agree with Aristotle's views on the role of the intervening medium. But supporters of the intromission hypothesis were not able to develop a quantitative theory that could be compared with experiment. That was first done by extramissionists.

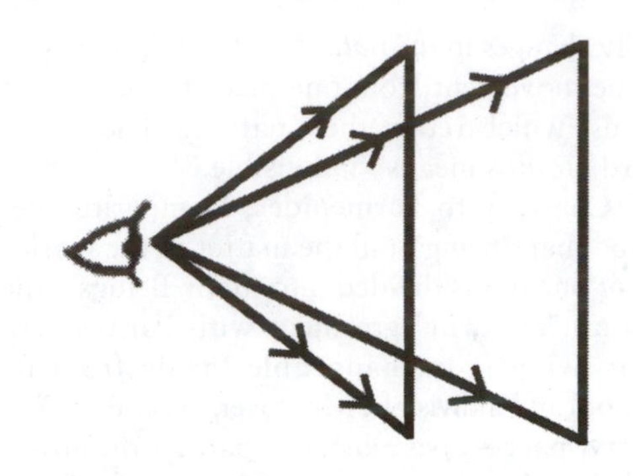

Fig. 3.2. Euclid proposed that the eye sends out rays to the object seen; the apparent size of the object is the angle formed at the eye by the rays, and thus depends on its distance.

Euclid's Geometrical Ray Theory

About 300 B.C. the famous Greek mathematician Euclid wrote a book on optics, treated as a branch of geometry. He postulated that the eye sends out a set of visual rays—straight lines that form a cone, with its apex inside the eye and its base on the surface of the objects seen (Fig. 3.2). Using rays, he was able to explain several phenomena, such as the formation of images by reflection.

What difference does it make whether the rays go out from the eye (as shown in Fig. 3.2) or come in? Euclid reminded his readers that you might have difficulty finding a needle even though it is right in front of you; you must look directly at the needle in order to see it. This common experience would be hard to understand if the needle is sending rays to your eye all the time; you should be able to see it if your eyes are open regardless of where you "look." In fact you must *actively* send rays from your eye to the needle in order to see it, rather than wait *passively* for the needle to send its images to you.

But now the extramission theory encounters a puzzle, similar to the problem highlighted by Zeno's paradoxes: Is there a finite number of rays with gradually widening spaces between them, or do the rays in the visual cone completely fill it? In the first case, we would expect objects that happen to be placed in between two rays cannot be seen—very small nearby objects and larger objects at greater distances. Isn't this contrary to common experience, or does it simply support Euclid's needle argument (Fig. 3.3)? In the second case, in order to have an infinite number of rays at the base of the cone, we must start with an infinite number emerging from a single point inside the eye, which seems impossible. (We could see the needle no matter how far away it is.)

The astronomer Ptolemy generally followed Euclid's geometrical treatment of optics, but

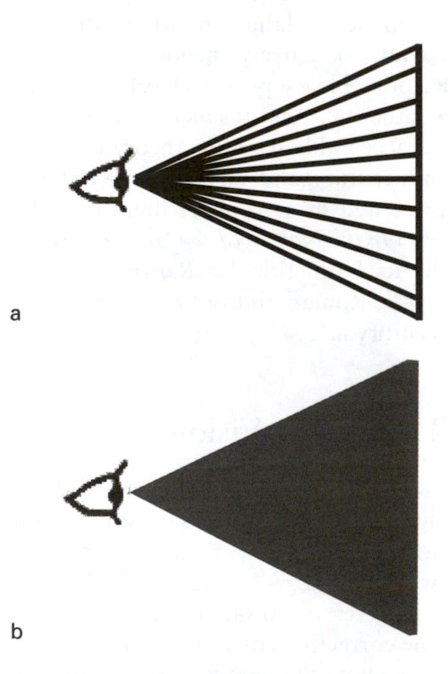

Fig. 3.3. Are the rays from the eye discrete (a) or do they fill space continuously (b)? If the rays are discrete, objects located entirely between two rays could not be seen without moving the eyeball slightly so the rays strike different points. Does this situation agree with experience?

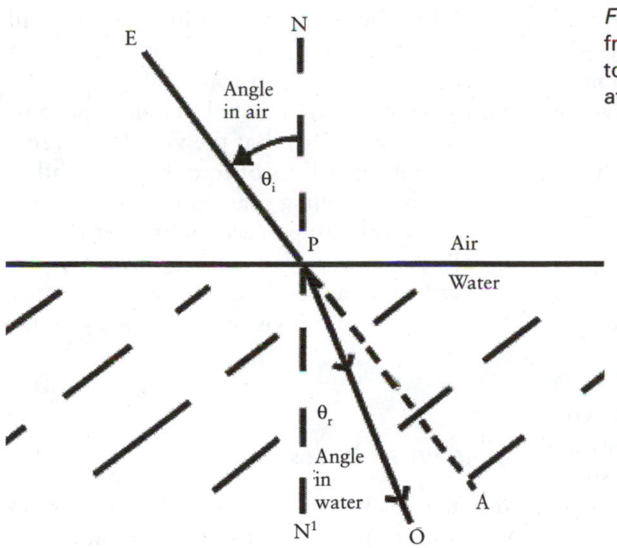

Fig. 3.4. Refraction of a ray of light coming from the eye (*E*) as it passes from air into water toward an object *O*. The object appears to be at *A*.

since the first part of his treatise has been lost we cannot be sure about his physical explanation of vision. He supported the extramission theory but, unlike Euclid, he assumed that the rays continuously filled space within the visual cone. This would avoid the peculiar feature of Euclid's theory—small objects at a distance would be alternately visible and invisible as you turn your head slowly.

Ptolemy reported systematic observations of the *refraction* of rays of light at an interface between air and water. This is the phenomenon that makes a spoon in a half-filled glass of water appear to be bent. Thus if a ray comes from the eye (*E*) and strikes the air-water surface at the point *P* making an angle θ_i (called the *angle of incidence*) with the line *NP* perpendicular to the surface (called the *normal*), it will change direction so that it moves toward the object *O*. The bent ray *PO* makes an angle θ_r (called the *angle of refraction*) with the extension of the normal into the water, *N′P*. The apparent position of the object is the point *A*, along the extension of the original ray *EP*.

Ptolemy's numerical values for the angles of refraction corresponding to different angles of incidence are somewhat questionable (see Problem 3.10). Nevertheless he opened up a new field of research that would later provide an important test for theories about the nature of light (Section 23.1).

We have no surviving records of any further progress on theories of vision until the 9th century A.D., when Arab scientists took up the problem. Abu Yusuf Ya'qub ibn Ishaq al-Kindi, in Baghdad, reviewed the intromission and extramission theories presented by the Greeks, and presented four arguments against the former.

1. A flat circular disk should, according to the intromission theory, look the same regardless of its orientation, since it simply sends out little circular images to the eye; but in reality if you hold the disk edgewise you just see a line segment, not a circle, as predicted by Euclid's extramission theory. (Try this with a dime.) Note that al-Kindi means by "image" the three-dimensional shape of the object, not just the surface facing the observer. Note also that in the early intromission theory we do not have rays emitted separately from each point of the object (as in later theories), just coherent images of the object as a whole.
2. Everyone agreed that hearing works by intromission: Sounds come into the ear from a noisy object, not the reverse. If vision worked by intromission as hearing does, then the eye would be shaped like the ear to collect the air that carries sound. But in fact the eye is a movable sphere, which can be directed toward the object to which it sends its rays.
3. If vision worked by intromission, then we would see objects at the edge of the visual field as clearly as those at the center, since once the images have entered the eye it doesn't matter from which direction they

have come. But in reality we see objects in the center of the visual field more clearly. This fact can be explained with the extramission theory if we assume that visual rays are emitted from all points on the open part of the surface of the eyeball; al-Kindi shows geometrically that in this case objects at the center of the visual field would then receive more rays than those at the edge. (But this implies that there is no longer just one visual cone, as Euclid and Ptolemy assumed, but rather one for each point on the open surface.)

4. When you look at a page of a book you don't read the entire page simultaneously (as the intromission theory predicts), instead you move your eye along to select the particular letters you want to read, in a particular sequence (a fact easily explained by extramission).

While al-Kindi supported the extramission theory of Euclid and Ptolemy, he did not explain *what* is being extramitted—the physical nature of the visual ray—in a way that is comprehensible to modern readers. The "visual fire" is not a stream of particles, rather it is some kind of force that acts on the medium between the eye and the object to facilitate the transmission of visual information. Nevertheless, thanks to al-Kindi the extramission theory was widely accepted by Arab scientists from the eleventh through the thirteenth century, although challenged by the revival of Aristotelian philosophy.

The Aristotelian theory of vision was advocated, and the extramission theory sharply criticized, by another Arab scientist, Abu Ali al-Husain ibn Abdullah ibn Sina (980–1037), known to Europeans by his Latinized name Avicenna. Like al-Kindi, he argued by showing that the opposing theory led to consequences contrary to common experience. Thus:

1. If it were true, as al-Kindi proposed, that the eye acts on the medium to facilitate the transmission of images, then you would expect that two or several people with weak vision would be able to see better if they stand close together so as to reinforce each other's action on the medium. But we know that a partly blind person can't improve his vision by getting close to another partly blind person, even if they try to see the same object.
2. It is absurd to think that your eye can produce a substance or power great enough to fill half the entire universe in order to see all the stars above the horizon—and do it again every time you open your eye!
3. If you try to avoid that objection by proposing (with Euclid) that the visual rays are a set of discrete lines of zero thickness rather than a cone filling space continuously, then since a ray can perceive only what it actually encounters, you will see only a few tiny parts of distant bodies, missing most of their surfaces. But our vision is not spotty in this way.

Alhazen's Optics

Although it's not too hard to think of decisive objections to the extramission theory, it is very difficult to show that the intromission theory, based on the assumption that a coherent miniature image of an entire object is received by the eye, can explain the observed facts about vision, especially the quantitative-geometrized aspects of perspective. The first scientist to propose the intromission theory in its modern form was Abu 'Ali al-Hasan ibn al-Hasan ibn al'Haytham (965–1039), known to Europeans as Alhazen. Although Alhazen lived at about the same time as Avicenna, he spent most of his life in Cairo (Egypt), while Avicenna lived in the Eastern region of the Islamic world near Buhkhara (now in Uzbekistan). There is no evidence that they ever met or read each other's work.

Alhazen presented a few of the objections to the extramission theory advanced by earlier writers, but his primary criticism was that whether or not visual rays are really emitted from the eye, they cannot explain vision. This is because the perception of objects ultimately has to occur in the eye, not at the object itself, and even the extramission theories, if they were to be at all plausible, had to assume that the visual ray doesn't just go out to the object; it has to *bring back* (or make the object send back) some kind of image to the eye. This return trip is all we really need; the hypothetical outgoing ray is superfluous.

As often happens in science, it is difficult to get people to give up an established theory no matter how defective it may be, unless a better one is available to replace it. (This is one reason why Popper's Principle doesn't always hold.) Alhazen proposed a new version of the intromission theory: Rather than assuming that the entire object sends out tiny copies of itself, he postulated that each point or small region on the surface of the body radiates in all directions.

It might seem that this makes no sense at all, for every point of the eye's open surface would receive an incoherent jumble of images from all the points of all the objects in its visual field. But Alhazen argued that of all the rays received at the center of the eye's open surface from the object, only one—the ray coming from the point directly in front—is perpendicular to the surface of the eye and is not *refracted.* All other rays will be refracted to some extent and thus (he claimed) weaker.

Alhazen then asserted (without a satisfactory justification) that the refracted rays are either not perceived at all or, if they are, are interpreted by the eye as having come from the same place as the central nonrefracted ray. This was a weakness in his theory (later rectified by Kepler), but if one is willing to accept his assertion, then the mathematical analysis of ray optics developed by Euclid and Ptolemy for the extramission theory can be applied to the intromission theory by simply reversing the directions of the rays.

Alhazen's book, *De Aspectibus,* integrated the physical, mathematical, and physiological aspects of vision. It strongly influenced the development of optics in Europe from the thirteenth to seventeenth centuries, culminating in the definitive work of Johannes Kepler (Section 4.6).

3.4 Criteria for a Good Theory in Physical Science

When we examine why scientists as a whole, in the historical development of science, have favored or rejected a given theory, we may discern a few criteria that seemed to have implicitly or unconsciously dominated the slowly evolving process of decision. (But by no means must we suppose that a theory is necessarily ever rejected solely because it fails to answer perfectly one or another of the questions in the listing that follows.)

Three qualifications have already been cited:

1) A fruitful theory correlates many separate facts—particularly the important prior observations—in a logical, preferably easily grasped structure of thought.

2) In the course of continued use it suggests new relations and stimulates directed research.

3) The theory permits us to deduce predictions that actually check with experience by test, and it is also useful for solving long-standing puzzles and quantitative problems.

To illustrate: In his work on general relativity, Einstein first shows that those observations explained by contemporary physics are not in conflict with his brief set of assumptions, but that a large variety of facts from previously separate fields can all be deduced from these assumptions, thereby gathering them into one structure. Second, he demonstrates that a few hitherto unlooked-for effects should exist (such as the deflection of light beams passing near large masses) and he stimulates the experimental astrophysicists to search for these phenomena. The sensational confirmation of his light-bending prediction in 1919 helped make Einstein the world's most famous scientist. But other physicists and especially astronomers were equally impressed by his success in explaining some old, previously unexplainable observations—especially a slow rotation of the orbit of the planet Mercury (called "advance of the perihelion" by astronomers).

Note that scientists often use the word "prediction" to mean the deduction from theory of an observable result, whether or not that result is already known. Thus both light-bending and the change in Mercury's orbit are called "predictions," even though one effect was new, the other old.

The history of science has shown us that prosperous theories frequently have some of the following additional properties:

4) When the smoke of initial battle has lifted, the successful theory often appears to have simple and few hypotheses—to the extent that the words "simple" and "few" have any meaning at all in science. The test of a theory comes through its use; hence the survival value of condensing, in an economical way, much information. A theory that needs a separate mechanism for each fact it wishes to explain is nothing but an elaborate and sterile tautology.

But simplicity is not always evident at first. The intromission theory of vision, despite its plausibility, seemed inferior to the extramission theory for several centuries because it could not easily be formulated in terms of geometrical rays. Alhazen's theory seemed incomplete and arbitrary until Kepler simplified it by introducing a new model for the structure of the eye, as we will show in Section 4.5.

5) Ideally, the assumptions should be plausible—to contemporary scientists, of course—even if they are not immediately subject to test; and the whole tenor of the theory should not be in conflict with current ideas. Where this could not be arranged, the theory faced an often stormy and hostile reception and had to submit to long and careful scrutiny before its general acceptance. Therefore it did not always grow as quickly and widely as one might have hoped. On the other hand, this cannot be helped, if nature disagrees with current common-sense preconceptions; in fact, precisely these preconceptions may block the road to the core of the difficult problems.

We saw that in ancient Greece Parmenides insisted a concept must be *intelligible*—but then convinced himself and his students that the very idea of *change* is unintelligible. This conclusion may seem absurd; but it is not so easy to dismiss Zeno's clever paradoxes, which seem to prove the impossibility of one important kind of change: motion.

As the work of Copernicus and later Galileo shows us (a strange truth, which will become even more evident in subsequent discussions), major advances in scientific theories often depend on the daring and persistence of one investigator who questions the obvious and stubbornly upholds the unbelievable.

This problem of the plausibility of new theories is difficult and fascinating. As the English philosopher Alfred North Whitehead reminds us, all truly great ideas seemed somewhat absurd when first proposed. Perhaps we call them great ideas just because it takes an unusual mind to break through the pattern of contemporary thought and to discern the truth in implausible, "absurd" forms.

To Copernicus the geocentric theory accepted by his predecessors was ugly and arbitrary: "neither sufficiently absolute nor sufficiently pleasing to the mind." They might have said the same about his heliocentric theory—but he and his followers, like composers of "modern" music that sounds dissonant on first hearing, taught their audiences to enjoy a more sophisticated harmony.

Why then do we not drop the tentative requirement of conformity to accepted ideas? If almost all great innovators in science, from Copernicus to Einstein, initially met with skepticism or worse from most of their colleagues, should we not simply conclude that radical new ideas should always be welcome, and that opposition to such ideas is just a symptom of unreasoning conservatism on the part of the scientific fraternity?

Not at all. In fact, the person who proclaims: "they thought Galileo was crazy but he turned out to be right; they think *I*'m crazy, therefore *I* will turn out to be right!" is using bad logic to justify what is almost certainly a worthless theory.

In this collection of selected cases, we are discussing only theories considered revolutionary at the time, but we must not allow this to distort our perceptions: Great revolutionary ideas arise only very rarely, compared with the large number of workable or even great ideas conceived within the traditional setting. Therefore individual scientists are naturally predisposed to favor the traditional type of advance, which they know and believe in from personal experience. They quite rightly must defend their fundamental concepts of nature against large-scale destruction, particularly at the earlier stages, where the innovators cannot present very many results and confirmations of their new ideas. Indeed, sometimes the discoverers themselves are so strongly committed to the established ideas, so startled by the implications of their own work, that they predict the storm of condemnation or even simply fail to draw the final conclusions.

A now famous instance of this sort appears in the publication (1939) by Otto Hahn and Fritz Strassmann, who originally (but unfairly) received all the credit for the experimental discovery of nuclear fission. At that time it was still axiomatic in the thinking of many scientists that nonradioactive atomic nuclei are stable, that bombardment with small particles (e.g., neutrons) may at best dislodge an alpha particle or two. But these two men, after firing neutrons into pure uranium, were left with material that by chemical test was proved to contain barium (Ba) and other elements whose atoms are about half as large as uranium atoms—"evidently" as we would say now, the result of splitting uranium atoms. But in an almost agonized expression of the labor pains attending the birth of all great recognitions, Hahn and Strassmann could not dare to accept publicly the evidence of their own chemical experiments, and so they wrote:

> As 'nuclear chemists,' in many ways closely associated with physics, we cannot yet bring

ourselves to make this leap in contradiction to all previous lessons of nuclear physics. Perhaps, after all, our results have been rendered deceptive by some chain of strange accidents.

("Concerning the existence . . . ," *Naturwissenschaften*)

A note on the human dimension behind scientific publications: It was their colleague Lise Meitner, forced to leave Germany by Nazi assaults on Jews, who first recognized that the Hahn-Strassman observations implied nuclear fission, and with her nephew Otto Frisch worked out the first rough theory of the process. Yet Meitner was denied a share of the Nobel Prize for the discovery, and she was only much later given some recognition for it.

Later, after having developed our view that science has its laws of evolution analogous to those governing the evolution of species, we shall appreciate more fully the significance of the bitter and long struggles that new theories may create within science. The fitness of truths is most advantageously shaped and most convincingly demonstrated in vigorous contest. The situation is, after all, not so very different in other fields. The predominant religious and social concepts of our time have not all developed quietly or been accepted spontaneously. If the physical sciences sometimes seem to have progressed so very much faster, if the recent struggles have been relatively short, we might, to a great extent, credit this to the emergence, from the seventeenth century onward, of a more or less tacit agreement among scientists on standards for judging a new conceptual scheme.

Planck's Principle: Even so, scientists are, above all, human; a really new and startling idea may in retrospect clearly have fulfilled all these requirements without being widely accepted. Max Planck, with perhaps only a little too much bitterness about his own early struggles, says, "An important scientific innovation rarely makes its way by gradually winning over and converting its opponents: it rarely happens that Saul becomes Paul. What does happen is that its opponents gradually die out, and that the growing generation is familiarized with the ideas from the beginning: another instance of the fact that the future lies with youth."(*Philosophy of Physics*) This view, which may apply to only a small fraction of cases, has become known as "Planck's Principle" in the sociology of science. It represents the opposite extreme of Popper's Principle, and may be no more accurate as a general description of how science works.

6) History points out another feature of successful theory: It is flexible enough to grow, and to undergo minor modifications where necessary. But if, after a full life, it eventually dies, it dies gracefully, leaving a minimum of wreckage—and a descendant.

Let us consider these interesting points. In any field a prediction is no more than a guess unless it follows from some theory.[3] And just as the meaning of a prediction depends on a theory, so does the validity of a theory depend on the correctness of the predictions. Like the very laws it correlates, a theory must be based on a finite, though perhaps small, number of observations; yet to be useful it may have to predict correctly a very large number of future observations. Think again of the Ptolemaic theory; it was built on many separate observations of planetary positions, but it yielded enough reasonably correct predictions of positions to still be useful more than 1400 years later.

Yet we know from the history of science that physical theories are not likely to last forever. Sooner or later there will appear observations to contradict the predictions, probably in a region for which the conditions had to be extrapolated from the original range of the theory (e.g., the region of very small or very large quantities, or of measurements much more precise than served initially as a basis for the theory), or perhaps as a result of a general widening of the scientific horizon. With ingenuity one can usually find that the old theory can be modified to apply to the new phenomena as well. Like an apple tree in an orchard, we keep a theory for the sake of its fruits; when the apple crop becomes poor, we then try to save the tree by judicious pruning and the like. In the endangered theory, perhaps a reexamination of some assumptions or concepts will save the scheme. Otherwise, the only alternatives are to retain the old theory as is, with the clear understanding of the restricted range to

[3]In the 1960s, the followers of the psychoanalyst Immanuel Velikovsky argued that scientists should accept his bizarre theory of ancient planetary encounters because a few of its many predictions, such as the emission of radio noise from Jupiter, were later confirmed. Yet those few predictions did not follow logically from any theory compatible with the well-established laws of physics (see H. H. Bauer, *Beyond Velikovsky*).

which it still applies, and to develop a new structure to take care of the new range; or, if the flaws appeared within the proper jurisdiction of the old theory and the assumptions stand challenged beyond repair, to abandon it as soon as a new, better one can be contrived.

Note the last phrase: a defective theory will be abandoned "as soon as a new, better one can be contrived." Scientists sometimes prefer an inadequate theory to none at all. Hence the frustration of critics who laboriously "refute" an accepted theory but see it survive because they have nothing better to offer in its place. The heliocentric theory was apparently refuted by the failure of astronomers to see the predicted stellar parallax effect (Section 2.4), yet after the work of Galileo and Newton it was impossible to go back to the geocentric theory. When the parallaxes of several stars were finally measured by Bessel and other astronomers in the nineteenth century (see Problem 1.3), they were now merely yet another kind of expected effect, in accord with the already-universal acceptance of the earth's motion around the sun. Later we will see that the kinetic theory of gases was retained despite its failure to account quantitatively for the specific heats of diatomic molecules (Section 22.8). In this case the anomaly was eventually explained by quantum theory, but in the meantime it would have been foolish to throw out the kinetic theory, which was quite successful in explaining and predicting other properties of gases.

These three choices—expansion, restriction, or death—appear in the history of theories and of laws generally. We tend to remember and to honor most those schemes that indicated the solution in a period of perplexity or left some hint of a new start at their demise. In the words of Niels Bohr, "The utmost any theory can do [is] to be instrumental in suggesting and guiding new developments beyond its original scope."

But finally, lest our brief paragraphs on these criteria of good theories be mistaken for scientific dogma, we should echo the very humble opinion of Einstein on this same topic. He distinguishes between two main criteria: (a) the external confirmation of a theory, which informs us in experimental checks of the correctness of the theory, and (b) the inner perfection of a theory, which judges its "logical simplicity" or "naturalness." He then qualifies these remarks as follows: "The meager precision of the assertions [(a) and (b) above] . . . I shall not attempt to excuse by lack of sufficient printing space at my disposal, but confess herewith that I am not, without more ado, and perhaps not at all, capable to replace these hints by more precise definitions" ("Autobiographical Notes").

Additional Problems

Problem 3.6. Investigate the death of a theory from another field with which you are well acquainted (such as the phlogiston theory in chemistry, or the theory of spontaneous generation in biology) in terms of our six criteria. What were the predictions that failed to come true, or the phenomena that contradicted the assumptions of the theory? Could the theory have been modified to serve for a time? Did a new theory rise directly from the old?

Problem 3.7. P. W. Bridgman wrote in *Reflections of a Physicist* that in the current flux in ideologies, moral ideas, and other social concepts the intelligent scientist sees ". . . an exemplification of what his physical experience has taught him; namely, ideas are to be inspected and re-examined when the domain of application is extended beyond that in which they arose. In one very important respect he recognizes that the present epoch differs from former epochs in that the enormous increase in invention, bringing peoples nearer together and increasing their command over forces more advantageous to man, effectively provides just that extension in the domain of application of social concepts which he is prepared to expect would demand fundamental revision." Write a short essay on your opinion about this quotation, and also examine what other guiding ideas you think may be properly transferred to the field of social studies from the study of physical theories.

Problem 3.8. Was the geocentric theory as set forth by Ptolemy false? Was it true?

Problem 3.9. Discuss the relative advantages of the Ptolemaic and Copernican theories of celestial motion, comparing them according to each of the six criteria developed in Section 3.2.

Problem 3.10. Ptolemy presented the following data for the angles of incidence and refraction when a ray of light is refracted at the surface of air and water (Fig. 3.4, page 33):

Angle in Air	*Angle in Water*
10°	8°
20°	15½°
30°	22½°
40°	29°
50°	35°
60°	40½°
70°	45°
80°	50°

Do you think he really observed these numbers? Do they seem plausible (including the implication that he could measure angles to within half a degree)? *Hint:* Compute the differences of successive values for "angle in water" and then the differences of those differences.

Problem 3.11. Discuss the relative validity of Popper's and Planck's principles on the basis of two or three examples about which you have some knowledge.

RECOMMENDED FOR FURTHER READING

Note: For further reading on theories and the growth of science, consult the bibliography at the end of Chapter 14. We list here only the works pertaining to atomism and optics.

C. C. Gillispie, *Dictionary of Scientific Biography,* articles by A. I. Sabra on Alhazen [Ibn al-Haytham], Vol. 6, pages 189–210 (1972); G. B. Kerferd on Democritus, Vol. 4, pages 30–35 (1971); Leonardo Tarán on Parmenides, Vol. 10, pages 324–325 (1974); Kurt von Fritz on Zeno of Elea, Vol. 14, pages 607–612

William L. McLaughlin, "Resolving Zeno's paradoxes," *Scientific American,* Vol. 271, no. 5, pages 84–89 (1994)

S. Sambursky, *Physical Thought,* excerpts from Zeno, Leucippus, Democritus, Plato, Epicurus, Lucretius, Al-Hazen, and Moses Maimonides, pages 50, 55–61, 83–92, 135–139, 147–150

J. H. Weaver, *World of Physics,* excerpts from Aristotle, Epicurus, Lucretius, von Weizsäcker, and Bridgman, Vol. I, pages 291–306, 309–322, 329–347; Vol. III, pages 103–119, 837–850

CHAPTER 4

Kepler's Laws

4.1 The Life of Johannes Kepler

As we follow the development of the theory of planetary motion, the pace now gathers momentum. We have reached the years around 1600 A.D. The Renaissance and Reformation are passing. Copernicus has been read by a few astronomers who see the computational advantages of his system but will not take seriously its physical and philosophical implications. Through this silence we hear one striking voice sounding the first cries of a coming battle. The antiorthodox pantheist and evangelizing Copernican, Giordano Bruno, is traveling through Europe, announcing that the boundaries of the universe are infinitely far away and that our solar system is one of infinitely many. For his several outspoken heresies he is tried by the Inquisition and burned at the stake in 1600.

But the seeds of a new science are sprouting vigorously here and there. In England there are Francis Bacon (1561–1626) and William Gilbert (1540–1603); in Italy, Galileo Galilei (1564–1642). And in Copenhagen, Tycho Brahe (1546–1601), the first person since the Greeks to bring real improvements into observational astronomy, spends nearly a lifetime in patient recording of planetary motion with unheard-of precision. His data are often accurate to less than half a minute of arc, more than twenty times better than those of Copernicus, even though the telescope has not yet been invented.

After Tycho's death, his German assistant Johannes Kepler (Fig. 4.1) continued the observations and, above all, the theoretical interpretation of the voluminous data. Whereas Tycho had developed a planetary system of his own (mentioned at the end of Section 2.4), which became widely accepted for a time, Kepler was a Copernican. The announced purpose of his work was the construction of better tables of planetary motion than were then available on the basis of the more dubious data of Copernicus' own period. But Kepler's motivation and main preoccupation was the perfection of the heliocentric theory, whose harmony and simplicity he contemplated "with incredible and ravishing delight." Throughout his long labors he was strongly influenced by the metaphysical viewpoint associated with the Pythagorean and "neoPlatonic" tradition; this tradition had revived in the Renaissance as one of the challenges to the hegemony of Aristotle and his followers.

To Kepler even more than to Copernicus, the clue to God's mind was geometric order and numerical relation, expressed in the features of the simple heliocentric scheme. Among his earliest works we find, typically, an enthusiastic attempt to link the six then-known planets and their relative distances from the sun (see Section 2.2) with the relationships among the five regular solids of geometry. The best result of this work was that it served to bring Kepler to the attention of Tycho and Galileo.

In attempting to fit the accurate new data on Mars' orbit to a Copernican system of simple uniform motion (even if equants were used), Kepler was completely unsuccessful. After four years of calculations he still found that the new data placed the orbit just eight minutes of arc outside the scheme of Copernican devices. This is many minutes less than the limit of precision that Copernicus knew applied to his own data; therefore Copernicus would not have worried about such a deficiency. But Kepler knew that Tycho Brahe, with his unfailing eye and superb instruments, had recorded planetary positions with uncertainty much smaller than the eight minutes of arc. With an integrity that has become a characteristic attitude of scientists in the face of quantitative fact, he would not allow himself to hide this fatal discrepancy behind some convenient assumptions. To Kepler, these eight minutes meant simply that the Copernican scheme of concentric spheres and epicycles failed to explain the actual motion of Mars when the observations were made accurately enough.

Fig. 4.1. Johannes Kepler (1571–1630), German astronomer who discovered three empirical laws of planetary motion and proposed a theory of retinal vision.

4.2 Kepler's First Law

Kepler must have been stunned; after all, he was a convinced Copernican. After years of continuous labor he discovered at length how the Copernican theory could be amended to make it applicable to both the old and the new observations—by dropping the one assumption that bound it most explicitly to the doctrines derived from ancient Greece. When he examined the paths of planets on the basis of a heliocentric representation, he found that they corresponded to a simple type of figure, an ellipse, whose properties had been known to mathematicians since the second century B.C. (It is ironic that Apollonius, who proposed the epicycle device discussed in Section 1.6, also developed the theory of ellipses without any thought for their possible application in astronomy.)

If, therefore, the ellipse is recognized to be the natural path of celestial bodies, then we can use it to build a geometrically truly simple world scheme. *Planets move in elliptical paths, with the sun at one focus of the ellipse.* This is the "Law of Elliptical Paths," the first of Kepler's three great laws.

Kepler's first law, by amending the Copernican heliocentric theory, gives us a wonderfully simple mental picture of the solar system. Gone are all the epicycles and eccentrics; the orbits are clean and elliptical. Figure 4.2 shows a schematic representation of the present conception of the solar system—in essence Kepler's, but with the addition of the planets Uranus, Neptune, and Pluto, discovered much later. An attempt has been made to represent on the figure the relative sizes of the planets. The approximate relative sizes of the orbits, to a different scale, are also indicated, but since most of these ellipses are quite nearly circular, they are shown as circles where possible. All orbits are nearly in the same plane, except for the more pronounced tilt of Pluto's (which means that Neptune's and Pluto's paths do not cross at any point in space, although Fig. 4.2 may give that impression).

Problem 4.1. (The important properties of the ellipse; refer to Fig. 4.3.) The equation for an ellipse in rectangular coordinates is

$$\frac{x^2}{a^2} + \frac{y^2}{b^2} = 1$$

where a and b are respectively half the major and half the minor axis. On graph paper or a computer screen, construct a large ellipse by deciding on some constant values for a and b and by then solving for a few positive and negative y values for assumed positive and negative x values. (Note that if $a = b$, the figure is a circle.) The distance c from the origin of the coordinates to one or the other focus (F_1 or F_2) is given by

$$c = \sqrt{a^2 - b^2}.$$

Enter the foci on your ellipse.

Now we introduce the term "eccentricity" (e) for an ellipse—not to be confused with the word "eccentric" as used in Section 1.6. The eccentricity is a quantity indicating by how much an ellipse differs from a circle. By definition,

$$e = \sqrt{1 - \frac{b^2}{a^2}}.$$

Find e for the ellipse you have drawn. Test by measurements that for a point P *anywhere* on the ellipse, $PF_1 + PF_2$ = a constant. Lastly, note that for a given set of coordinates an ellipse in a given plane is completely defined (i.e., can be correctly drawn) if just its values for a and b are given; these two data suffice (Fig. 4.3).

Problem 4.2. Determine the approximate value of e for Pluto from measurements on the projection

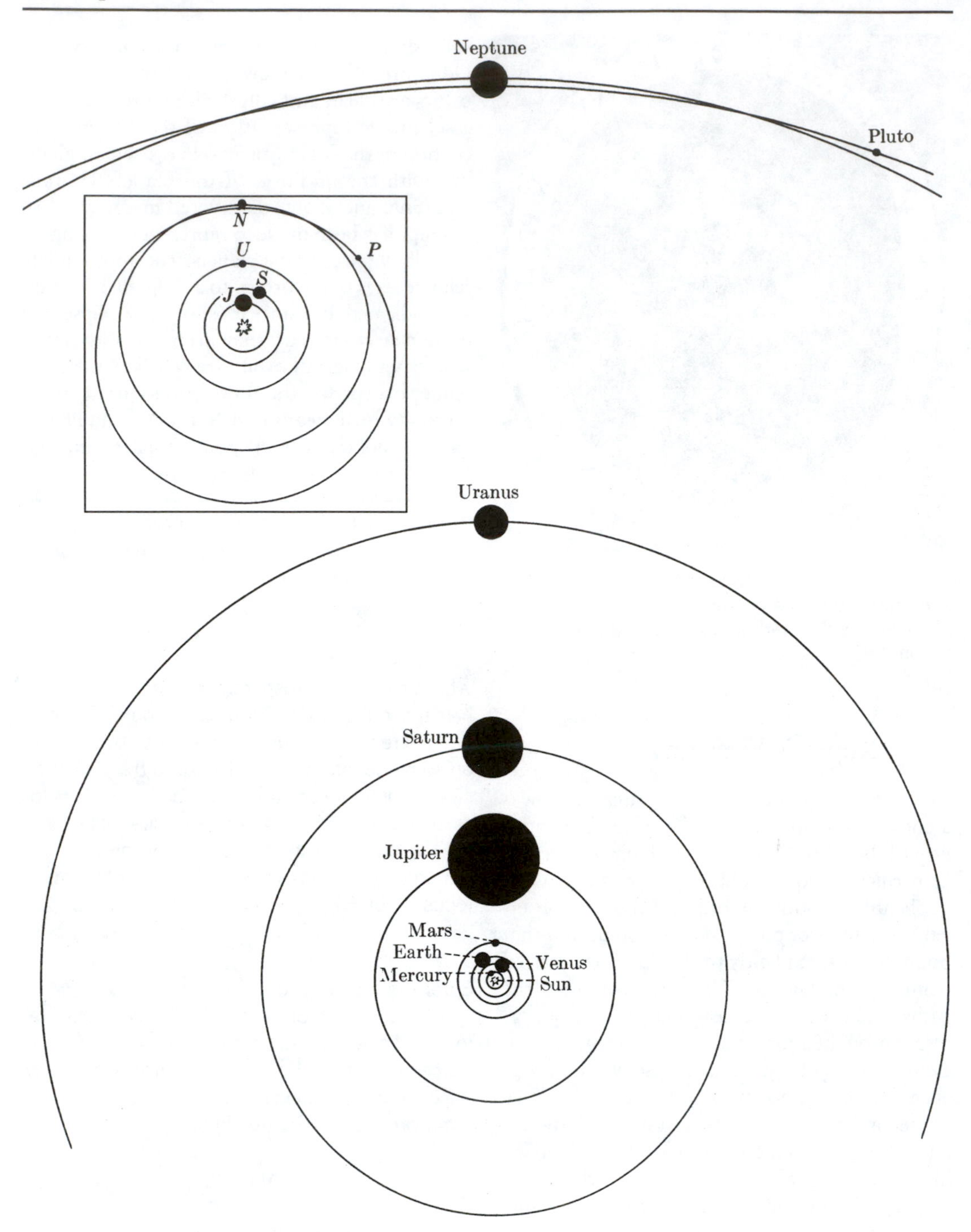

Fig. 4.2. Schematic outline of the solar system, showing relative sizes of the orbits and, to a different scale, the relative sizes of the planets. The inset traces Pluto's full orbit.

of its orbit shown in Fig. 4.2. (Note: e for Mercury is about 0.2, and for the remaining planets even less.) What is the shape of the ellipse if $e = 0$? If $e = \infty$?

Although Kepler was delighted to find that he could replace the complicated combinations of epicycles and eccentrics previously used to describe a planet's orbit by a simple ellipse, he may have asked himself the obvious question, "Is

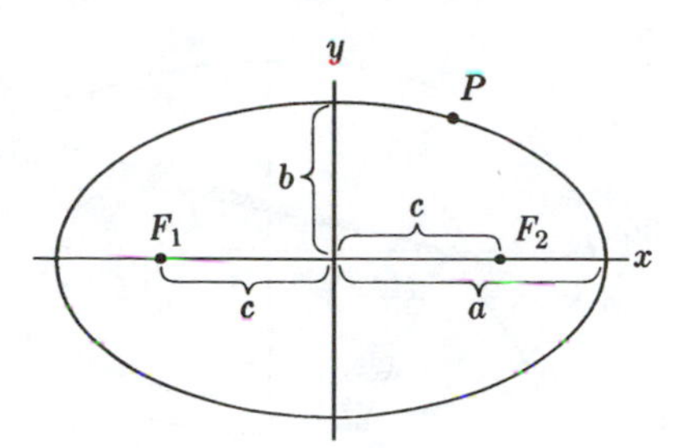

Fig. 4.3. Ellipse, with foci at F_1 and F_2.

it not rather mysterious that of all possible types of paths the planets all select just the ellipse?" We could perhaps understand Plato's predisposition for uniform circular motions, but we cannot readily comprehend nature's insistence on the ellipse! It is no longer a man-made mystery, but therefore all the darker. In fact, there was no rational answer until a reticent genius in England almost 80 years later showed the law of the ellipse to be one of the many surprising consequences of a much more far-reaching law of nature. But we are not yet ready to take up his reasoning, though we shall in Chapter 11.

If for the present we accept Kepler's first law purely as a summary of observed facts—an *empirical* law—we note that by describing the paths as elliptical, the law gives us all the possible locations of a given planet, but it does not tell us when it will be at any of these positions; it talks about the shape of the orbit, but not the changing speed along the orbit. This makes the law rather inadequate for astronomers who wish to know where they should expect to look for a planet at a given time, or even for the layperson who already knows (as we noted in Section 1.6 in connection with the equant) that the sun appears to move faster through the stars in December than in June. Of course Kepler was well aware of all this, and in fact even before finding what we now call his "first" law he had already established another law governing changes in the speed of a planet.

4.3 Kepler's Second Law

Kepler knew that he needed to find the mathematical relation between the speed of a planet in any position of its orbit and the speed in any other position. If such a relation could be found, the motion of any one planet would be specifiable by just a few separate figures: two data to specify the ellipse (i.e., the major and minor axes); one more to give the speed at a certain portion of the path (e.g., when the planet is nearest to the sun, at the "perihelion"); and enough additional data to place this ellipse at the correct tilt with respect to those for the other planets. Thus we could hope to summarize all features of the motion of every planet in the solar system in a compact, elegant way.

But does the hoped-for relation between speed and path actually exist? No necessary reason that it should has appeared so far. Kepler is said to have been in ecstasy when by sheer hard work and ingenuity he finally did read the needed, "second" law out of his own voluminous data. Well he might have been; his whole labor would have been of little use without this discovery.

Kepler's route to the second law was an amazing performance in which the correct result emerged as a deduction from several then-reasonable but incorrect assumptions. It will give us a glimpse how, to this day, the truth emerges more likely from "error than from confusion," as Francis Bacon declared.

First, Kepler assumed that planets are driven around their orbits by a force emanating from the sun, with the "rays" of force acting like paddles, pushing the planets at an angle to the length of the paddle.

Second, he assumed that the strength of this force is inversely proportional to the planet's distance from the sun. (The reason for this, in Kepler's thinking, and using his imagery, was that the force at any given distance r must be uniformly spread over the circumference of a circle in the orbital plane; at a greater distance, say $2r$, the same total force must be spread over a circle whose circumference is twice as great, so the strength of the force at any given point on that circle can only be half as much.[1])

Third, he then assumed that the speed of the planet must be proportional to the force pushing it, and hence inversely proportional to the distance:

[1]He knew that if the force were spread out in all directions in three-dimensional space it would decrease in strength, like the intensity of light, as the inverse *square* of the distance. But such a force, he thought, would be wasteful since the planets all move very nearly in the same plane; why would a force extend to regions where there is nothing to be moved? Kepler had not freed himself from the teleological thinking of Aristotelian philosophy.

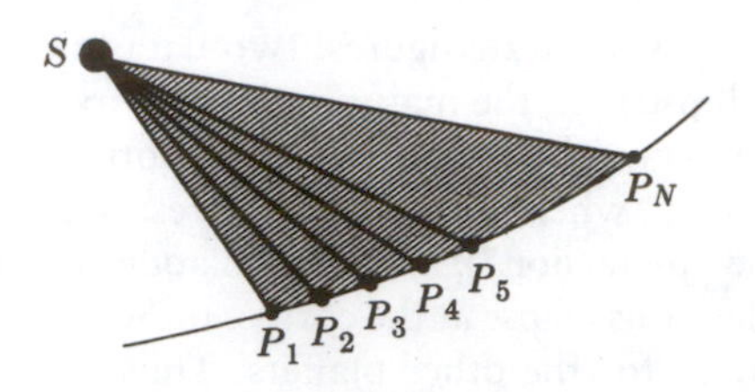

Fig. 4.4. Kepler assumed that the sum of all the lines drawn from the sun to the planet as it moves along its path, $SP_1 + SP_2 + SP_3 + \ldots + SP_N$, could be approximated by the shaded area, segment SP_1P_N (here the eccentricity of the orbit is exaggerated).

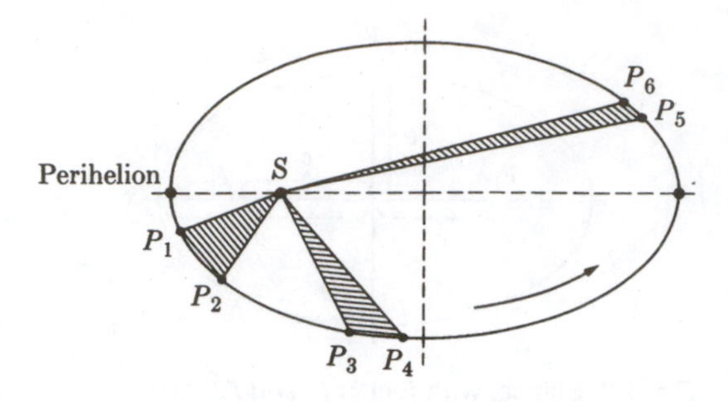

Fig. 4.5. Elliptical path of planets around the sun S (at left focus), illustrating Kepler's Second Law (eccentricity much exaggerated).

$$v \propto 1/r. \tag{4.1}$$

The assumption that speed is proportional to the net force is of course incompatible with modern principles of physics, as we shall see in Chapters 7 and 9; it was simply one of the Aristotelian or common-sense ideas that Kepler shared with all his contemporaries.

According to Kepler's first assumption, the time it takes a planet to go a short distance along its path would be proportional to its distance from the sun. This is roughly correct, and happens to be exactly correct at certain special points in the orbit. Kepler then proposed to compute the time it takes the planet to cover a long segment of the path (during which the distance from the sun will be changing) by adding up the planet-sun distances for each of the small arcs composing this long segment.

Fourth, he assumes that the sum of these distances will be equal to the area swept out by the straight line drawn from sun to planet (see Fig. 4.4):

$$t \propto \text{area swept out by planet-sun line.} \tag{4.2}$$

This happens to be a good approximation for the case of the actual orbits Kepler was dealing with; but the mathematics needed to make it exact (the "calculus" of Newton and Leibniz) was not to be invented for another half century.

Now Kepler introduced a fifth assumption, that the orbit is a circle. This is again only an approximation, which happens to be fairly accurate for almost all planetary orbits (Kepler had not yet established his "first" law, which required the orbits to be elliptical); but in fact it is not even necessary to make such an approximation.

Kepler's second law, which he stumbled on by a line of reasoning that is unconvincing to a modern reader, is already contained in Eq. (4.2) above: the area swept out by the planet-sun line is proportional to the time elapsed. Or, to rephrase it in the way that has now become standard: *During a given time interval a line drawn from the planet to the sun sweeps out an equal area anywhere along its path.* This is also referred to as the "Law of Equal Areas." In spite of the inaccuracy of the assumptions used in its original derivation, the law itself describes exactly the motion of any planet around the sun.[2] It also applies to the motion of the moon around the earth, or of a satellite around any planet.

Figure 4.5 illustrates the meaning of the second law. We are looking down on a greatly exaggerated representation of a planetary path around the sun S at one focus of the ellipse. The planet may be our earth; it is found to move from position P_1 to P_2 (say during a week in our winter). During an equal time interval in the spring the journey proceeds from P_3 to P_4, and from P_5 to P_6 in summer. The implication from the drawing is that the planet's speed along the orbit is greatest when nearest the sun and less when more distant. This is qualitatively in agreement with the first assumption, $v \propto 1/r$, which Kepler used to derive the law.

The fact that the earth does move faster (or that the sun as seen from the earth's northern hemisphere appears to move faster against the background of stars) in the winter than in the summer had been well known to astronomers long before Kepler; it was an effect that could be

[2]This statement must be qualified slightly, for reasons that will become apparent when we discuss Newton's derivation in Chapter 9: the law is accurate provided the influence of any third body can be neglected.

explained by the introduction of the "equant" device in the geocentric system (see Section 1.6), and it was one reason why the Copernican system without equants was not quite adequate to represent the details of planetary motion. Kepler's second law serves the same purpose as the equant, but in a way that ultimately turns out to be much more satisfactory. However, in Kepler's own work it is only an empirical rule that, though accurate, does not yet have any theoretical explanation.

Problem 4.3. As the text mentioned, the earth is actually closer to the sun in winter (for residents of the northern hemisphere) than in summer. Explain this apparent paradox.

4.4 Kepler's Third Law

Kepler's first and second laws were published together in 1609 in his *New Astronomy.* But he was still unsatisfied with one aspect of his achievement: It had not revealed any connection between the motions of the different planets. So far, each planet seemed to have its own elliptical orbit and speeds, but there appeared to be no overall pattern for all planets. Nor was there any good reason why one should expect such a relationship.

Kepler, however, was convinced that on investigating different possibilities he might discover one simple rule linking the whole solar system. He looked for this rule even in the domain of musical theory, hoping like the Pythagoreans (Section 1.1) to find a connection between planetary orbits and musical notes; his last great book (1619) is entitled *The Harmonies of the World.*

This conviction that a simple rule exists, so strong that it seems to us like an obsession, was partly a remnant of his earlier numerological preoccupations, and partly the good instinct of genius for the right thing to work on. But it also indicates a deep undercurrent running through the whole history of science: the belief in the simplicity and uniformity of nature. It has always been a source of inspiration, helping scientists over the inevitable unforeseen obstacles, and sustaining their spirit during periods of long and fruitless labor. For Kepler it made bearable a life of heartbreaking personal misfortunes, so that he could write triumphantly, on finally finding the great third law:

> . . . after I had by unceasing toil through a long period of time, using the observations of Brahe, discovered the true distances of the orbits, at last, at last, the true relation . . . overcame by storm the shadows of my mind, with such fullness of agreement between my seventeen years' labor on the observations of Brahe and this present study of mine that I at first believed that I was dreaming . . . .
>
> (*Harmonies of the World*)

The law itself, in modern terminology, states that if T is the sidereal period of any chosen planet (i.e., the time for one complete orbital revolution by the planet in its path around the sun), and R_m is the mean radius of the orbit of that planet,[3] then

$$T^2 \propto (R_m)^3$$

or equally

$$T^2 = K(R_m)^3,$$

where K is a constant having the *same value* for all planets. (This equation is written in the modern form, not the way Kepler wrote it.) That fact binds all planets together in one system. But what is the value of K?

If $T^2/(R_m)^3$ is the same for all planets, we can calculate its numerical value for *one* planet for which we know T and R_m well. For our earth: $T_E = 1$ year, $(R_m)_E \approx 9.3 \times 10^7$ miles. Knowing now K, we are always able to compute for any other planet its T if its (R_m) is given, and vice versa.

Problem 4.4. Using the third law and the data for T_E and $(R_m)_E$ for the earth, compute values of R_m for a few other planets, taking the corresponding values of T from the quotation from Copernicus' *Revolutions* in Section 2.2.

Appropriately, Kepler's third law is often called the *Harmonic Law* because it establishes one beautifully simple relationship among all the planets. From this peak we may survey our progress so far. Starting from the disconnected multitude of Ptolemaic gears we have reached a heliocentric formulation that views the solar

[3]The value of R_m for an elliptical path is half the length of a straight line from perihelion to aphelion (see Fig. 4.5); conveniently, most planetary paths are nearly enough circular so that R_m is then simply the radius of the orbit, taken to be circular.

system as a simple, logically connected unit. Our mind's eye grasps the Keplerian universe at one glance, and recognizes the main motions in it as the expression of simple laws expressed in a mathematical form. This has ever since been a model of highest achievement of empirical science.

4.5 Kepler's Theory of Vision

Sixteenth-century astronomers found that it was convenient to make measurements during a solar eclipse by letting the sunlight pass through a small opening and form an image on the opposite wall; this was the principle of the "pinhole camera" or "camera obscura" (Fig. 4.6). But Tycho Brahe found that some kind of correction had to be made for the effects of the finite size of the opening itself, otherwise the apparent sizes of the sun and moon would be incorrect. Kepler decided it was necessary to understand how radiation goes through apertures in order to make accurate astronomical observations.

Starting from the medieval optical tradition going back to Alhazen (Section 3.3), Kepler realized that it was necessary to establish a one-to-one correspondence between each point in the visual field and a point in the eye. His predecessors had failed to do this satisfactorily because they had to assume that rays striking the eye exactly perpendicular to its surface were the only ones seen, while those failing (even by a very small amount) to be perpendicular contributed nothing at all to vision.

Kepler solved the problem in 1604 by using the rudimentary knowledge of the eye's anatomy available in his time. He postulated that the aqueous and crystalline humors together act like a spherical lens. This means that rays from a given point of the visual field, which diverge and strike different points on the cornea at the front of the eye, will be refracted in such a way that they converge and strike a single point on the retina at the back of the eye. (Kepler did not give a diagram to show this, but his theory is illustrated in a diagram published by Descartes, see Fig. 4.7). This establishes the required one-to-one correspondence.

Yet, Kepler's theory of the retinal image raises one difficulty he was unable to solve. The image formed on the retina by this process will be reversed left-to-right and upside down, as one can already see from the ray diagram for the pinhole camera (Fig. 4.6). How do we manage to see objects as they actually are? Kepler avoided the difficulty by asserting that it is not a problem in optics, because the path from the retina to the brain goes through opaque substances, so the image obviously has to be converted from light to some other kind of signal. Someone else would have to explain that part of vision.

We cannot criticize Kepler for failing to solve the entire problem of vision; instead we have to applaud his success in separating out one part of the problem that *can* be solved by optical theory, independently of the rest of the problem. That is how science often progresses: find a meaningful problem you can solve rather than waste your time trying to explain everything at once—and end up explaining nothing.

The modern view is that the "rest of the problem" is actually a pseudoproblem. There is no physiological or biochemical process that somehow inverts the image as it travels from the retina to the brain. Instead, as the British philosopher George Berkeley pointed out in 1709, we

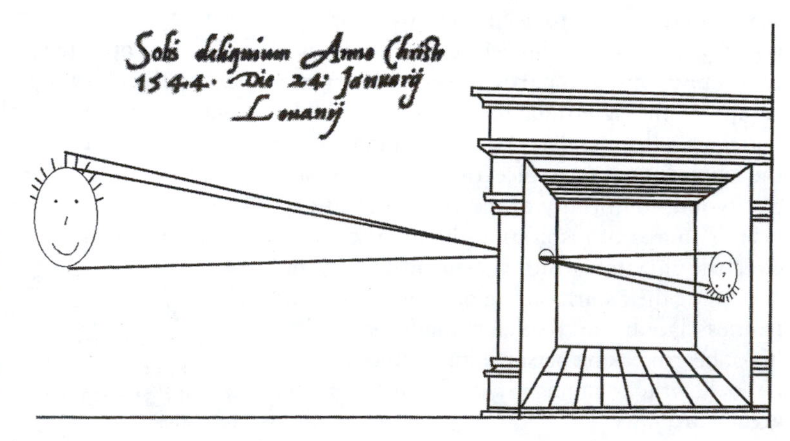

Fig. 4.6. Diagram by Gemma Frisius (1545) showing how a pinhole camera (camera obscura) can be used to study a solar eclipse. Note that the image will be upside-down and reversed left-to-right.

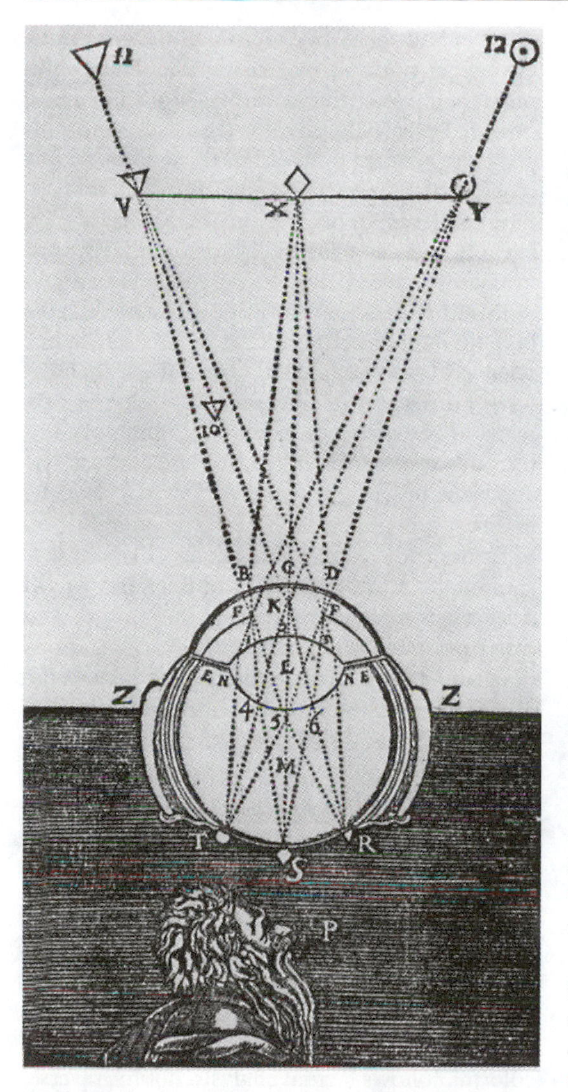

Fig. 4.7. Kepler's theory of vision, as illustrated by Descartes (*Optics,* 1637). Rays from each point on the object (V, X, Y) are refracted through the cornea (BCD) and lens (L) to foci (RST) on the retina, where they form an inverted image of the object.

react to objects, not to retinal images, and these reactions involve coordination between the eye and the tactual sensations of our bodies. In other words, we see "up" and "down," "right" and "left" by making a correlation between what the eye perceives and what the hand touches. All we need from the retina is an image that provides the correct *relative* positions of objects in the visual field; the brain learns to do the rest.

Berkeley's view was confirmed in a famous experiment by the American psychologist George M. Stratton in 1896. Stratton used a combination of lenses, strapped to his head, to invert the images coming to his eyes, so that he would see everything upside-down and left-right reversed. After a couple of days of confusion he found that he could adapt to these images so that things looked normal almost all the time, and he could move around safely and effectively in his environment. Then, after a week, he removed the lenses. Again there was a period of confusion but his perception of the world finally returned to normal again. In other words, he had established, in this elegant experiment, that the brain copes with the inverted image by a *psychological process* that is obviously learned and can be unlearned.

4.6 The New Concept of Physical Law

With Tycho's work, his own observations, and his powerful three laws, Kepler was able to construct the long-needed accurate tables of planetary motion, which remained useful for a century. We honor him for all these great accomplishments in astronomy (and these are only a few of the achievements of this prodigious man), but we must not forget to single out two features of his work that had a profound effect on the development of all physical sciences.

One, which has been discussed, is the rather new attitude toward observed facts. We noted the change in Kepler's work from his earlier insistence on a geometric model and its form as the main tool of explanation to the discussion of the movement itself and of the numerical relations underlying it. The other is his successful attempt to formulate physical laws in mathematical form, in the language of geometry and algebra. From then on, after the time of Kepler, the equation develops naturally as the prototype of most laws in physical science. (We shall see some of this development when we come to the work of Galileo and Descartes in the following chapters.)

In this sense, Kepler's science was truly modern. He, more than anyone before him, bowed to the relentless and supreme arbiter of physical theory, namely, the evidence of precise and quantitative observation. But another step was needed to reach full modernity. True, in the Keplerian system the planets no longer are considered to move in their orbits because of their divine nature or the continual effort of angels, as in scholastic teaching, nor because their spherical shapes themselves were self-evident explanation for their circular motion, as in Copernicus' thought. So we are left without any physical agency to "explain" or give plausibility to the planetary motion so well described in these three laws.

Kepler himself felt a need to back up his mathematical descriptions with physical mechanisms. In one of his later books he tells how his own views have changed:

> Once I firmly believed that the motive force of a planet was a soul. . . . Yet as I reflected that this cause of motion diminishes in proportion to distance, just as the light of the sun diminishes in proportion to distance from the sun, I came to the conclusion that this force must be substantial—'substantial' not in the literal sense but . . . in the same manner as we say that light is something substantial, meaning by this an unsubstantial entity emanating from a substantial body.
>
> (*Mysterium Cosmographicum*, second edition)

Although it was left to Newton to work out the theory of gravitational forces, and thereby to tie Kepler's three laws together with the heliocentric conception and the laws of terrestrial mechanics in a monumental synthesis, Kepler entertained a then quite promising hypothesis: The recently published work of the Englishman William Gilbert (1544–1603) on magnetism had greatly intrigued him, and his rich imagination could picture magnetic forces emanating from the sun to "drive" the planets in their orbits.

Magnetism in fact does not explain Kepler's laws. Newton later felt it necessary to prove in some detail that this hypothetical agency would not account for the quantitative observations. But in a more general sense Kepler did anticipate the *type* of explanation that Newton was to establish. As he wrote to a friend in 1605:

> My aim is to show that the heavenly machine is not a kind of divine, live being, but a kind of clockwork (and he who believes that a clock has a soul, attributes the maker's glory to the work), insofar as nearly all the manifold motions are caused by a most simple, magnetic, and material force, just as all motions of the clock are caused by a simple weight. And I also show how these physical causes are to be given geometrical expression.
>
> (*Johannes Kepler in seinen Briefen*)

We have here an example of the enormous change in outlook in Europe that had begun more than two centuries before. More and more, observed events ceased to be regarded as *symbols* and were allowed to stand for themselves. People ceased to be preoccupied with anthropomorphic riddles in an organismic world, and slowly became factual observers and theorizers in a mechanistic world.

Without this new attitude, there could have been no modern science, for if we are to start our science from experimental observables, we must have faith that we are dealing with the raw material of experience, not with symbols of complex mysteries. We had to become enthusiastic about the observable world for its own sake, we had to attain a tacit faith in the meaningfulness of nature and its direct accessibility to our understanding, before generations of scientists would arise to devote themselves to the minute and often most tedious quantitative investigations of nature. In this sense Kepler's work heralds the change toward the modern scientific attitude—to regard a wide variety of phenomena as explained when they can all be described by one simple, preferably mathematical, pattern of behavior.

It seems at first astonishing that Kepler should be the one to follow this road. He had begun his career as a symbol-seeking mystic; but now we may discern a reflection of the great change in his complex soul: he fashions his physical laws—and looks for his symbolism afterward. Philosophical speculation, still often rather colorful, follows factual analysis, not the reverse; and to this day many scientists have found it quite possible to reconcile their physics and their personal philosophy on the basis of this sequence.

Additional Problems

Problem 4.5. Compare and evaluate Kepler's two theories—of planetary motion and of vision—in terms of our criteria for a good theory.

Problem 4.6. Trace the changes and evolution of the Copernican heliocentric theory in the hands of Kepler, and then compare this with the development and expansion of some important theory from another field familiar to you (e.g., Darwin's theory of evolution, from its exposition in 1859 to the "neo-Darwinian synthesis" of the 1950s).

Problem 4.7. Does it seem possible that a revision of the geocentric theory can be made, doing away with equants, eccentric motions, etc., and instead describing the planetary paths *about the earth* directly by simple geometric figures? Is it likely that laws could be found for this simplified geocentric system that relate the paths, speeds, and periods of revolution of the different planets?

RECOMMENDED FOR FURTHER READING

M. Boas Hall, *Scientific Renaissance*, Chapter X.
Max Caspar, *Kepler*, New York: Dover, 1993
M. J. Crowe, *Theories of the World*, Chapters 7 and 8
W. H. Donahue, "Kepler," in *Cosmology* (edited by N. S. Hetherington), pages 239–262
Owen Gingerich, article on Kepler in *Dictionary of Scientific Biography* (edited by C. C. Gillispie), Vol. 7, pages 289–312
Gerald Holton, "Johannes Kepler's universe: its physics and metaphysics," in Holton, *Thematic Origins*, pages 53–74
Job Khozamthadan, "Kepler and the origin of modern science," *Indian Journal of History of Science*, Vol. 33, pages 63–86 (1998)
John North, *Norton History of Astronomy*, Chapter 12
S. Sambursky, *Physical Thought*, excerpts from Kepler, pages 206–215
Curtis Wilson, "How did Kepler discover his first two laws?" *Scientific American*, Vol. 226, No. 3, pages 92–106 (March 1972)
Arthur Zajonc, *Catching the Light*, Chapter 2

CHAPTER 5

Galileo and the New Astronomy

"There are more things in heaven and earth,
Horatio,
Than are dreamt of in your philosophy. . . ."

Shakespeare, *Hamlet,* Act I

One of the friends and fellow scientists with whom Kepler corresponded and exchanged news of the latest findings was Galileo Galilei (1564–1642). Although the Italian's *scientific* contribution to planetary theory is not so substantial as that of his correspondent across the Alps, he nevertheless made himself an unforgettable key figure in this subject. In a sense, Kepler and Galileo complemented each other in preparing the world for the eventual acceptance of the heliocentric theory—the former laying the scientific foundation with his astronomical work, and the latter fighting the dogmatic objections and, in his work on mechanics, helping to overturn the whole structure of scholastic physics with which the old cosmology was entwined.

For it was Galileo more than anyone else who challenged the fruitfulness of the ancient interpretation of experience, and focused the attention of physical science on the productive concepts—time and distance, velocity and acceleration, force and matter—and not on qualities or essences, ultimate causes or harmonies, which were still the motivation of a Copernicus and at times the ecstasy of a Kepler. Galileo's insistence, so clearly expressed in his work on freely falling bodies (see Chapter 7), on fitting the concepts and conclusions to observable facts, on expressing his results in terms of mathematics, are now accepted as central achievements, reinforcing the same traits then emerging from parts of the work of Kepler.

But perhaps the greatest difference between the work of Galileo and that of his Aristotelian contemporaries was his orientation, his viewpoint, the kind of question he considered important. To most of his opponents, Galileo's specific, science-centered problems were not general enough, since he excluded the orthodox philosophical problems. Then, too, his procedure for discovering truth seemed to them fantastic, his conclusions preposterous, haughty, often impious. There exists an uncanny parallel between these objections to Galileo's point of view and the outraged derision and even violence initially heaped on the discoverers of new ways of viewing the world in art, e.g., the French painter Edouard Manet and his fellow impressionists in the late nineteenth century.

5.1 The Life of Galileo

Galileo Galilei, one of our chief symbols of the great struggle out of which modern science arose, was born in Pisa, Italy, in 1564, the year of Shakespeare's birth and Michelangelo's death. His father was a poor but noble Florentine from whom he acquired an active and competent interest in music, poetry, and the classics; some of Vincenzo Galilei's compositions are now available in recorded modern performances.

Galileo was sent to the University to study medicine, but preferred mathematics instead. According to one of the many stories (or legends) about him, his first scientific discovery was made in 1583 while he was sitting in the cathedral at Pisa. The sermon was rather boring that day, and he started watching a lamp, suspended from the ceiling, swinging back and forth. He noticed that even though the swings of the lamp got shorter and shorter, the *time* required for each swing remained the same. This (true for relatively small arcs of swing) is the fundamental law of the *pendulum,* which he discussed in quantitative detail in his book on mechanics in 1638.

Nor did medicine miss out completely on the fruits of Galileo's genius: He invented a simple pendulum-type timing device for the accurate measurement of pulse rates. But even at that time he made himself conspicuous by constantly challenging the authoritative opinions of his elders.

Fig. 5.1. Galileo Galilei (1564–1642), Italian astronomer and physicist who supported the heliocentric system and established some of the basic properties of motion (Chapters 7 and 8).

Lured from medicine to physical sciences by reading Euclid and Archimedes, he quickly became known for his unusual intellect. Although he had dropped out of the University of Pisa without taking his degree, he soon acquired such a reputation that he was appointed professor of mathematics at the same university at the age of 26. His characteristic traits now showed themselves: an independence of spirit and an inquisitive and forthright intellect unmellowed by tact or patience. He protested against the traditional attire of the faculty by circulating a poetic lampoon, "Against wearing the toga." He attacked the scientific views of his colleagues, who were of course almost all dogmatic Aristotelians and, as disciples too often are, were rabid defenders of their often erroneous interpretations of their master's works.

It was around 1590, while Galileo was at Pisa, that he may have made a public experiment on the speeds of unequal weights dropped from the famous Campanile of Pisa, though it is more likely that the story is only a legend. Because the incident is so widely known and is popularly supposed to have been a crucial experiment in the history of physics, we may, for what it is worth, look at a source of the story, the biographical notes on Galileo by one of his latest and closest students, Vincenzo Viviani—an account whose accuracy has often been challenged:

> As it seemed to him [Galileo] that a true knowledge of the nature of motion was required for the investigation of the natural effects, he gave himself completely to the contemplation of that [motion]: and then to the great confusion of all the philosophers, very many conclusions of Aristotle himself about the nature of motion, which had been theretofore held as most clear and indubitable, were convicted of falseness, by means of experiments and by sound demonstrations and discourses; as, among others, that the velocity of moving bodies of the same composition, unequal in weight, moving through the same medium, do not attain the proportion of their weights, as Aristotle assigned it to them, but rather that they move with equal velocity, proving this by repeated experiments performed from the summit of the Campanile of Pisa, in the presence of all other teachers and philosophers and of all students.
>
> (*Racconto istorico;* translation by Arnolfo Ferruolo)

It should be noted that the speeds of bodies of the *same composition* are being compared. Apparently in 1590 Galileo believed that bodies of the same density would fall at the same rate, but that the rate of fall might still depend on the difference in density between the object and the medium through which it falls. Galileo's writings on mechanics during this period indicate that he had not yet developed the theory presented in his definitive work on mechanics published in 1638, according to which all bodies of whatever composition must fall at the same rate (if air resistance can be neglected, as in a vacuum). So Galileo's interpretation of the famous Leaning Tower experiment, if in fact it was actually performed at that time, would not have been quite the same as the modern one.

In 1591, Galileo's father died, leaving a large family to be supported. Galileo's salary at Pisa was inadequate, and moreover his appointment was not likely to be renewed when the three-year contract expired, because he had made too many enemies by his blunt attacks on the

Aristotelians. Fortunately he was able to obtain an appointment at Padua, where he spent the next 18 years in more congenial surroundings and with a larger salary. Marina Gamba bore him three children; but they separated in 1610, when Galileo returned to Florence in his native Tuscany.

At Padua, Galileo had begun his work in astronomy. The first evidence that he had accepted the Copernican system is found in two letters written in 1597, one of them to Kepler in response to the latter's 1596 book, *Mysterium Cosmographicum.* Galileo told Kepler that he had been a Copernican for several years, and had found several physical arguments in favor of the earth's motion. (Probably these arguments were based on the periodicity of ocean tides.) But Galileo paid little attention to the details of Kepler's work on planetary orbits; he never adopted Kepler's ellipse in place of the traditional circle.

In 1609, Galileo learned that someone in Holland had discovered how to magnify the appearance of distant objects by putting together two glass lenses. On the basis of this report Galileo constructed his own "optical tube," or telescope, and pointed it toward the heavens, with the extraordinary results that we shall describe in the next section. He also demonstrated the device to the leaders of the Venetian republic, showing how it could be used to detect ships approaching the city long before they were visible to the naked eye. They were sufficiently impressed to give Galileo a lifetime contract as professor, at a considerable increase in salary.

Yet Galileo was not satisfied with his position. He dedicated his book, *The Sidereal Messenger,* an account of his first discoveries with the telescope, to Cosimo de Medici, Grand Duke of Tuscany, and even called the newly discovered satellites of Jupiter the "Medicean stars." The reason was obvious: He hoped to flatter the duke into giving him a job, or rather, into giving him financial support for his research without time-consuming obligations. As he explained his reasons for leaving the position he had just been given at Padua in the republic of Venice:

> It is impossible to obtain wages from a republic, however splendid and generous it may be, without having duties attached. For to have anything from the public one must satisfy the public and not any one individual; and so long as I am capable of lecturing and serving, no one in the republic can exempt me from duty while I receive pay. In brief, I can hope to enjoy these benefits only from an absolute ruler.
>
> (*Discoveries and Opinions of Galileo*)

But when Galileo left the republic of Venice to accept the patronage of the authoritarian ruler of Florence, he was making himself much more vulnerable to the attacks of his future enemies in the Church. Venice (so we are told by historians) would never have surrendered Galileo to the Inquisition.

At any rate, in 1610 Galileo became Court Mathematician and Philosopher (a title he chose himself) to the Grand Duke of Tuscany. From then, until his death at 78 in 1642, his life, despite recurring illness and family and money troubles, was filled with continuous excellent work, with teaching and writing, and of course with the relentless and ultimately tragic fight with his enemies that is to be described in this chapter.

5.2 The Telescopic Evidences for the Copernican System

In 1610, Galileo published a little booklet entitled *Sidereus Nuncius,* which may be translated *The Sidereal Messenger* (or "Message from the Stars"). In this book Galileo announced the discoveries he had made with his telescope:

1. The planet Jupiter has four smaller planets revolving around it. These were later called *satellites* by Kepler and other astronomers, and Jupiter is now known to have at least 16 of them. But the existence of even one such satellite was a blow to traditional ideas, for two reasons. First, some philosophers had convinced themselves that there *must* be exactly seven heavenly bodies (not counting the stars); hence the discovery of any more would be a metaphysical impossibility.[1]

---

[1] The Florentine astronomer Francesco Sizzi argued in 1611 that there could not be any satellites around Jupiter for the following reasons: "There are seven windows in the head, two nostrils, two ears, two eyes and a mouth; so in the heavens there are two favorable stars, two unpropitious, two luminaries, and Mercury alone undecided and indifferent. From which and many other similar phenomena of nature such as the seven metals, etc., which it were tedious to enumerate, we gather that the number of planets is necessarily seven. . . . Besides, the Jews and other ancient nations, as well as modern Europeans, have adopted the division of the week into seven days, and have named them from the seven planets: now if we increase the number of plan-

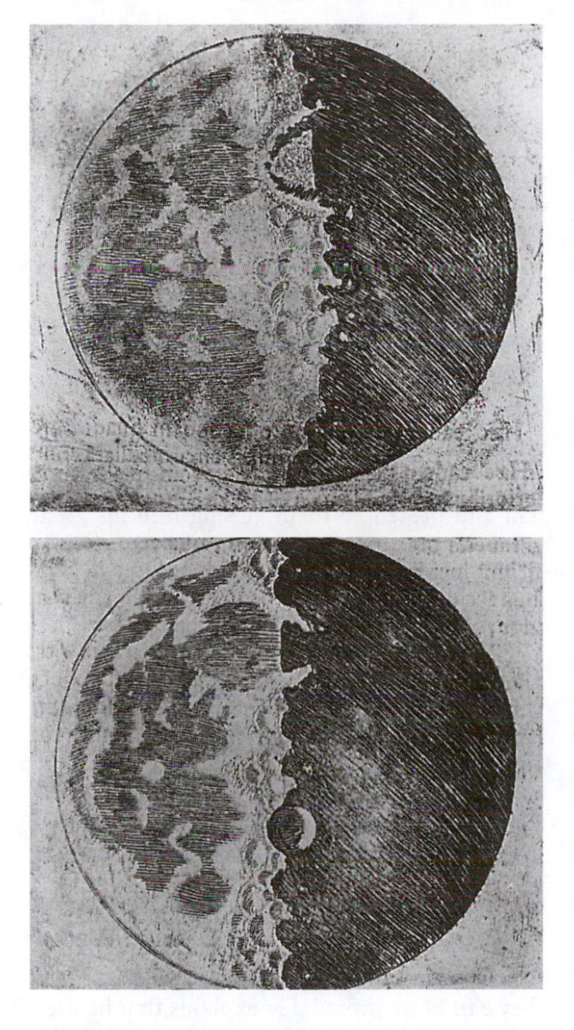

Fig. 5.2. Galileo's drawings of the moon, from *The Sidereal Messenger.*

Second, whereas all the other celestial objects *appeared* to revolve around the earth with a motion that could only be explained or understood by rather sophisticated theories, the satellites of Jupiter slowly but obviously revolved around Jupiter; hence the earth cannot be the center of rotation for all objects in the universe.

2. By telescopic observation, we see "the surface of the moon to be not smooth, even and perfectly spherical, as the great crowd of philosophers have believed about this and other heavenly bodies, but on the contrary, to be uneven, rough, and crowded with depressions and bulges. And it is like the face of the earth itself, which is marked here and there with chains of mountains and depths of valleys." Some of the mountains are as high as four miles. Galileo alludes here to the fact that the same set of doctrines that then still generally upheld the geocentric system as the only possible choice also required that the celestial objects be "perfect"—spherical and unblemished—yet he could see not only mountains on the moon but also spots on the sun.

3. The fixed stars do not look much larger when viewed through a telescope; in fact, almost all of them are still just pinpoints of light. Galileo concludes that their apparent size as seen by the naked eye must be misleadingly large. We can imagine them as being extremely far away without having to attribute an incredibly immense size to them, and thus put aside for the moment the vexing problem of the absence of parallax (see Sections 1.5 and 2.4).

4. The Milky Way, which appears as a continuous region of light to the naked eye, is resolved by the telescope into many individual stars. Indeed, with the early telescopes hundreds if not thousands of stars could now be seen that were not visible to the naked eye. Such facts could not easily be explained by those who believed that the whole universe had been created solely for the benefit and contemplation of mankind. Why would God put so many invisible things in the sky?

Many of Galileo's contemporaries refused to accept the scientific validity of his telescopic discoveries. After all, it was known that all kinds of visual tricks could be played with lenses. The only scientist who publicly supported Galileo at the time was Kepler, who wrote a pamphlet titled *Conversation with the Starry Messenger,* pointing out that the new discoveries were consistent with his own Copernican-based theories. Having gained the endorsement of the leading astronomer of Europe, Galileo could not be ignored; yet even Kepler and Galileo together could not immediately convert the rest of the world.

A year after his discoveries, Galileo complained in a letter to Kepler:

> You are the first and almost the only person who, after a cursory investigation, has given entire credit to my statements. . . . What do you

---

ets, this whole system falls to the ground. . . . Moreover, the satellites are invisible to the naked eye and therefore can have no influence on the earth and therefore would be useless and therefore do not exist." (*Dianoia Astronomica*) say of the leading philosophers here to whom I have offered a thousand times of my own accord to show my studies, but who, with the lazy obstinacy of a serpent who has eaten his fill, have never consented to look at the planets, or moon, or telescope?"

In his characteristic enthusiasm, Galileo had thought that through his telescopic discoveries everyone would see, as with his own eyes, the absurdity of the assumptions that prevented a general acceptance of the Copernican system. But people can believe only what they are ready to believe. In their fight against the new Copernicans the scholastics were convinced that they were surely "sticking to facts" and that the heliocentric theory was obviously false and in contradiction with both sense observation and common sense, not to speak of the theological heresies implied in the heliocentric view. They had made Aristotelian science their exclusive tool for understanding observations, just as today most nonscientists make their understanding of physical theory depend on their ability to visualize it in terms of simple mechanical models obeying Newtonian laws.

But at the root of the tragic position of the Aristotelians was, in part, the fact that an acceptance of the Copernican theory as even a possible theory would have had to be preceded by a most far-reaching reexamination and reevaluation of their personal beliefs. It would have required them to do the humanly almost impossible—to discard their common-sense ideas, to seek new bases for their old moral and theological doctrines, and to learn their science anew (which was of course what Galileo himself did to an amazing degree, for which his contemporaries called him fool, or worse, and for which we call him genius). Being satisfied with their system, the Aristotelians were, of course, unaware that history would soon prove their point of view to be far less effective in the quest to understand nature.

Problem 5.1. Explain carefully the implications of Galileo's discovery that the stars do not appear larger when viewed through a telescope.

Problem 5.2. Another of Galileo's telescopic discoveries was that the planet Venus sometimes seemed to be fully illuminated by the sun and at other times not at all; it has phases, like the moon. Referring to your answers to Problem 3.4, explain the significance of this discovery. How would you account for the fact that Venus, as seen by the naked eye, shows little change in brightness throughout the year, even though it goes through moonlike phases?

5.3 Toward a Physical Basis for the Heliocentric System

During the next two decades Galileo developed further his arguments for the Copernican system and presented them at length in his great work, *Dialogue Concerning the Two Chief World Systems,* published in 1632. Here he stressed the rational arguments that seemed to him to be conclusive, apart from observational evidence of the type presented in *The Sidereal Messenger.* Observations alone, says Galileo, do not decide uniquely between a heliocentric and a geocentric hypothesis, since most phenomena can be explained by either. Galileo nevertheless thinks of the earth's motion as "real," because it seems more reasonable and simplifies the picture.

The dialogue format allows Galileo to recognize many of the objections raised by opponents of the Copernican system and show how they can be answered. He explains that he used to have lengthy discussions in Venice with "Giovanni Francesco Sagredo, a man of noble extraction and trenchant wit," and, from Florence, "Filippo Salviati, the least of whose glories were the eminence of his blood and the magnificence of his fortune. His was a sublime intellect . . ." The other member of the group was an Aristotelian philosopher who admired the commentaries of the medieval writer Simplicius. Galileo honors his deceased friends Salviati and Sagredo by giving the name of the former to his own spokesman, and calling the nonpartisan chairman, who facilitates the discussion, Sagredo. Since the Aristotelian proponent was apparently still alive when Galileo wrote the *Dialogue,* he does not mention his real name but gives him the pseudonym Simplicio. This was perhaps to spare him embarrassment since, following the plan of the "Socratic dialogues," Simplicio was to play the role of the honest but ignorant person who is led by clever questioning to a conclusion contrary to the one he started with.[2]

Galileo/Salviati first presents several arguments for the heliocentric system, largely following Copernicus:

> Let us consider only the immense bulk of the starry sphere in contrast with the smallness of the terrestrial globe . . . Now if we think of the velocity of motion required to make a complete rotation in a single day and night, I cannot persuade myself that anyone could be found who would think it the more reasonable and credible thing that it is the celestial sphere which did the turning, and the terrestrial globe which remained fixed.

As a second point, Galileo reminds his readers that in the geocentric model it is necessary to ascribe to the planets a motion mostly opposite to that of the celestial sphere (why?), again an unreasonable, we might almost say unharmonious or unesthetic, assumption.

Third, Jupiter's four moons had shown him that there, too, existed the rule that the larger the orbit of the rotating body as reckoned from the center of rotation, the longer the period of rotation (qualitatively, Kepler's third law). Copernicus had pointed out long before that the same relation exists for the case of the planets themselves, and it held true even in the Ptolemaic system—but with this disharmony: In the Ptolemaic system the characteristic periods of revolution around the earth increase from the short one (273 days) for the moon to the very large one (30 years) for Saturn and then suddenly drop back to 24 hours for the celestial sphere.

In the Copernican system, however,

> by giving mobility to the earth, order becomes very well observed among the periods; from the very slow sphere of Saturn one passes on to the entirely immovable fixed stars, and manages to escape a fourth difficulty necessitated by supposing the stellar sphere to be movable. The difficulty is the immense disparity between the motions of the stars, some of which would be moving very rapidly in vast circles, and others very slowly in little tiny circles, according as they are located farther from or closer to the poles.

Fifth, owing to the slightly changing tilt of the earth's axis (as we would now put it), the apparent paths of the stars on the celestial sphere change slowly over the centuries (this refers to the precession of the equinoxes, mentioned in Sections 1.2 and 2.2), again an improbable or at any rate an unreasonable feature of a geocentric theory that claimed to be based on the immutable, ideal, eternal characteristics of the heavenly bodies.

Next, Galileo found it impossible to conceive in what manner the stars could be fixed in the celestial sphere to allow them rapid rotation and even motion as a group (cf. last point), while also having them preserve their relative distances so perfectly:

> It seems to me that it is as much more effective and convenient to make them immovable than to have them roam around, as it is easier to count the myriad tiles set in a courtyard than to number the troop of children running around on them.

Lastly, Galileo declares it to be quite implausible that the earth could be stationery while the rest of the universe rushes around it:

> I do not understand why the earth, a suspended body balanced on its center and indifferent to motion or to rest, placed in and surrounded by an enclosing fluid, should not give in to such force and be carried around too. We encounter no such objections if we give the motion to the earth, a small and trifling body in comparison with the universe, and hence unable to do it any violence.

In the text, Galileo's hypothetical opponent Simplicio answers these points:

> It seems to me that you base your case throughout upon the greater ease and simplicity of producing the same effects. As to their causation, you consider the moving of the earth along equal to the moving of all the rest of the universe, while from the standpoint of action, you consider the former much easier than the latter. To this I answer that it seems that way to me also when I consider *my own* powers, which are not finite merely, but very feeble. But with respect to the power of *the Mover* [God],

[2]The suspicion that Simplicio's name is used because it means "simpleton" is only partly allayed by his statement that it comes from the name of the Aristotelian commentator. Moreover, some readers may have realized that Simplicio's arguments were similar to those used by the Pope.

> which is infinite, it is just as easy to move the universe as the earth.

Analyze carefully this part of Galileo's modern-sounding and conclusive reply (and incidentally notice the sly final sentence):

> If I had ever said that the universe does not move because of any lack of power in the Mover, I should have been mistaken . . . But what I have been saying was with regard not to the Mover, but only the movables [bodies being moved] . . .
>
> Giving our attention, then, to the movable bodies, and not questioning that it is a shorter and readier operation to move the earth than the universe, and paying attention to the many other simplifications and conveniences that flow from merely this one, it is much more probable that the diurnal motion belongs to the earth alone than to the rest of the universe *excepting* the earth. This is supported by a very true maxim of Aristotle's which teaches that, "It is vain to expend many means where a few are sufficient'.

But no matter how simple it might be to assume that the earth moves, Galileo realized that common sense must rebel at the idea, for we do not observe with our eyes the phenomena that we would expect to occur if the surface of the earth were in fact rotating at several hundred miles per hour. Salviati, Galileo's spokesman in the *Dialogue,* recalls that according to Aristotle, the "strongest reason" for the earth's immobility is the motion of heavy bodies,

> . . . which, falling down from on high, go by a straight and vertical line to the surface of the earth. This is considered an irrefutable argument for the earth being motionless. For if it made the diurnal rotation, a tower from whose top a rock was let fall, being carried by the whirling of the earth, would travel many hundreds of yards to the east in the time the rock would consume in its fall, and the rock ought to strike the earth that distance away from the base of the tower.

Salviati/Galileo then attacks Aristotle's reasoning in an indirect way by reconstructing the description that Aristotle would have to give of a stone's motion when dropped from a tower, supposing the earth did move. In order to account for the actual observation that the stone does fall along the side of the tower, Aristotle would have to compound this motion out of the stone's natural motion downward toward the center of the earth and a circular motion around the center; Aristotle would then have to claim that such a combination of two different natural motions in the same object is impossible.

Simplicio/Aristotle now argues that the impossibility of such motion is confirmed by the behavior of a stone dropped from the top of the mast of a boat, which falls to close to the foot of the mast if the ship is at rest " . . . but falls as far from that same point when the ship is sailing as the ship is perceived to have advanced during the time of the fall, this being several yards when the ships course is rapid."

Salviati and Sagredo (the latter playing the role of the "neutral observer") help Simplicio to reconstruct the Aristotelian argument, showing that although a rock dropped from a tower on land might conceivably share the "natural" circular motion of the tower if the earth did rotate, the stone dropped from the ship could not do so because the ship's motion is not natural. The supposed falling behind in the latter case is thus set up as the crucial proof of the earth's immobility.

Now we might expect that Galileo will turn to the test of experiment and portray his opponent's credulous reliance on authority. And this seems to be the case, at first:

> *Salviati:* . . . Now, have you ever made this experiment of the ship?
>
> *Simplicio:* I have never made it, but I certainly believe that the authorities who adduced it had carefully observed it. Besides, the cause of the difference is so exactly known that there is no room for doubt.
>
> *Salviati:* You yourself are sufficient evidence that those authorities may have offered it without having performed it, for you take it as certain without having done it, and commit yourself to the good faith of their dictum. Similarly it not only may be, but must be that they did the same thing too—I mean, put faith in their predecessors, right on back without ever arriving at anyone who had performed it. For anyone who does it will find that the experiment shows exactly the opposite of what is written; that is, it will show that the stone always falls in the same place on the ship, whether the ship is standing still or moving with any speed you please. Therefore, the same cause holding good on the earth as on the ship, nothing can be inferred about the earth's

> motion or rest from the stone falling always perpendicularly to the foot of the tower.
>
> *Simplicio:* If you had referred me to any other agency than experiment, I think that our dispute would not soon come to an end; for this appears to me to be a thing so remote from human reason that there is no place in it for credulity or probability.

But suddenly Simplicio realizes that Salviati has not actually done the experiment either!

> *Simplicio:* So you have not made a hundred tests, or even one? And yet you so freely declare it to be certain? I shall retain my incredulity, and my own confidence that the experiment has been made by the most important authors who make use of it, and that it shows what they say it does.
>
> *Salviati:* Without experiment, I am sure that the effect will happen as I tell you, because it must happen that way; and I might add that you yourself also know that it cannot happen otherwise, no matter how you may pretend not to know it—or give that impression. But I am so handy at picking people's brains that I shall make you confess this in spite of yourself.

Have we, then, come up to the great turning point that divides ancient, scholastic, verbalistic, qualitative, a-prioristic Aristotelian science from modern, experimental, inductive science—only to find that the transition from Aristotelian to (eventually) Newtonian dynamics is effected by the methods of the Socratic dialogue without actually *doing* any experiment at all? Should Galileo's *Dialogue* be banned for its heresy against what even today is so commonly thought to be the "scientific method"?

Galileo's proof that the stone would fall at the foot of the mast of a moving ship (experimentally verified a few years later by Pierre Gassendi) introduces what we would now call the *law of inertia* or *Newton's first law of motion.* The proof is based on what the modern physicist calls a *thought experiment,* a device often used by Albert Einstein and others in discussing fundamental problems. Rather than either performing an actual experiment or reasoning syllogistically from general principles (in the Aristotelian fashion), Galileo asks us to consider what would happen in certain specified situations, and then shows that the conclusion of his opponent cannot be right because it leads to self-contradictory statements, or else that the conclusion Galileo is after is inherent in, and a necessary consequence of, sound convictions about those situations, convictions that we formed on the basis of ideas already tested (e.g., by daily experience).

> *Salviati:* . . . Suppose you have a plane surface as smooth as a mirror and made of some hard material like steel. This is not parallel to the horizon, but somewhat inclined, and upon it you have placed a ball which is perfectly spherical and of some hard and heavy material like bronze. What do you believe this ball will do when released?"

Simplicio agrees that the ball will roll down the plane spontaneously, continuously accelerating; to keep it at rest requires the use of force; and if all air resistance and other obstacles were removed, it would keep moving indefinitely "as far as the slope of the surface extended." Already the suggestion of motion to infinity in a vacuum has started to sneak in, but without arousing Simplicio's Aristotelian suspicions.

Now Salviati asks whether the ball would also roll upward on the slope, and of course Simplicio replies that this could happen only if a force were impressed on it; and moreover, "The motion would constantly slow down and be retarded, being contrary to nature." He also agrees that (in accord with Aristotelian physics) the greater the downward slope, the greater the speed.

> *Salviati:* Now tell me what would happen to the same movable body placed upon a surface with no slope upward or downward."

In this case, says Simplicio, the body will not move at all if it is simply set down at rest on the plane; but if it is given any initial impetus in a particular direction, there is no cause for either acceleration or deceleration.

> *Salviati:* Exactly so. But if there is no cause for the ball's retardation, there ought to be still less for its coming to rest; so how far would you have the ball continue to move?
>
> *Simplicio:* As far as the extension of the surface continued without rising or falling.
>
> *Salviati:* Then if such a space were unbounded, the motion on it would likewise be boundless? That is, perpetual?
>
> *Simplicio:* It seems so to me . . .

So, starting out from the Aristotelian postulate that a force is always needed to sustain an unnatural motion, Simplicio has been forced to concede that a motion that he considers not natural can nevertheless continue indefinitely, without a force acting on the body!

In other words, Galileo has found a loophole in the apparently impregnable structure of Aristotelian physics—even if it seems not a very large loophole. He concentrates on the single point on the boundary between natural motions downward and unnatural motions upward. Nevertheless, this point will prove to be a hole big enough for Galileo to sneak through, carrying with him the stone dropped from the mast of the ship, the moving earth, and one of the fundamental principles of Newtonian mechanics.

Continuing the discussion, Galileo seems to abandon his infinite horizontal plane, which we all know does not exist in nature anyway, and returns to local motion on the surface of the earth. Now such motion is, in this context, equivalent to motion on the infinite horizontal plane, since motion on the surface of a truly spherical earth is neither upward nor downward. Hence, he argues, the stone must continue to move along with the ship even when it is no longer physically connected with it—we could now say, continue to have also the forward momentum it had when it was still part of the ship—and will therefore fall directly at the foot of the mast, whether the ship itself is moving or not.

Galileo is thus able to complete the original argument about the stone dropped from the mast of a moving ship and thereby refute Aristotle's major objection to the rotation of the earth; he has shown that no conclusions about the motion of the earth can be drawn directly from observations of falling objects.[3] Crucial to this proof is his principle of inertia, which he states again in his later book *Two New Sciences* (1638):

> Furthermore we may remark that any velocity once imparted to a moving body will be rigidly maintained as long as the external causes of acceleration or retardation are removed, a condition which is found only on horizontal planes; for in the case of planes which slope downward there is already present a cause of acceleration, while on planes sloping upward there is retardation; from this it follows that motion along a horizontal plane is perpetual; for, if the velocity is uniform, it cannot be diminished or slackened, much less destroyed.

Since the law of inertia was to become a cornerstone of Newton's theory of mechanics and of planetary motion, one might ask why Galileo himself applied it only to local motion at the surface of the earth, and not to the motion of the earth and planets around the sun. Perhaps he would have done so; but before writing his *Two New Sciences* he had had to solemnly swear not to speak of such matters.

Problem 5.3. The experiment described by Galileo, of dropping an object from the mast of a moving ship, was first performed in 1641 (by the French philosopher Gassendi). Within the error of observation, the object did land directly at the bottom of the mast, just as if the ship were not moving. What does this result by itself, apart from Galileo's arguments presented in the *Dialogue,* prove about the motion of the earth?

5.4 Science and Freedom

The tragedy that descended on Galileo is described in many places, and it is impossible to do justice to the whole story without referring to the details. Briefly, he was warned in 1616 by the Inquisition to cease teaching the Copernican theory, for it was now held "contrary to Holy Scripture." At the same time Copernicus' book itself was placed on the *Index Expurgatorius* and was suspended "until corrected." But Galileo could not suppress what he felt deeply to be the truth. Whereas Copernicus had still invoked Aristotelian doctrine to make his theory plausible, Galileo had reached the new point of view in which he urged acceptance of the heliocentric system on its own merits of simplicity and usefulness, and apart from such questions as those of faith and salvation. This was the great break.

In 1623, Cardinal Barberini, formerly his friend, was elevated to the papal throne, and Galileo seems to have considered it safe enough to write again on the controversial topic. In

---

[3]In fact he has proved a little too much. Strictly speaking it is *not* true that a stone dropped from the top of a tower on land, for example, will land precisely at the bottom if the earth is rotating. There is a very small displacement, too small to measure until the end of the eighteenth century. The question is taken up again in Problem 10.4 in this book.

1632, after making some required changes, Galileo obtained the Church's necessary consent to publish the work *Dialogue Concerning the Two Chief World Systems* (from which the previous arguments for the Copernican theory were drawn), setting forth most persuasively the Copernican view in a thinly disguised discussion of the relative merits of the Ptolemaic and Copernican systems. After publication it was realized by the authorities or his enemies among the clergy that he may have tried to fool an unsophisticated censor and circumvent the 1616 warning. Furthermore, Galileo's forthright and tactless behavior, and the Inquisition's need to demonstrate its power over suspected heretics, conspired to mark him for punishment.

Among the many other factors in this complex story a prominent role is to be assigned to Galileo's religious attitude—which he himself believed to be devoutly faithful, but which had come under suspicion by the Inquisition. Galileo's 1615 letter to the Grand Duchess Christina showed he held that God's mind contains all the natural laws, and that the occasional glimpses of these laws that the human investigator may laboriously achieve are proofs and direct revelations of the Deity, quite as valid and grandiose as those recorded in the Bible: "The holy Bible and the phenomena of nature proceed alike from the Divine Word . . . nor is God any less excellently revealed in Nature's actions than in the sacred statements of the Bible." These opinions—held, incidentally, by many present-day scientists—can, however, be taken for symptoms of pantheism, one of the heresies for which Galileo's contemporary, Giordano Bruno, had been burned at the stake in 1600. Nor did Galileo help his case by such phrases as his quotation of Cardinal Baronius' saying: "The intention of the Holy Ghost is to teach us how one goes to heaven, not how heaven goes."

Now old and ailing, Galileo was called to Rome and confined for a few months. From the partly still secret proceedings we gather that he was tried (in absentia), threatened with torture, induced to give an oath that formally renounced the Copernican theory, and finally sentenced to perpetual confinement (house arrest).

None of his friends in Italy dared to defend Galileo publicly. His book was placed on the *Index* (where it remained along with Copernicus' and one of Kepler's until 1835). In short, and this is the only point of interest to us here, he was set up at the time as a warning for all, that the demand for spiritual and ideological obedience indisputably entails intellectual obedience also. His famous Abjuration, later ordered to be read from the pulpits throughout Italy and posted as a warning, has an ominously modern sound for those who remember the "confessions" elicited from dissidents by totalitarian regimes in the twentieth century.

After his notorious trial, it was rumored that despite having publicly renounced the Copernican mobility of the earth, Galileo muttered under his breath "*Eppur si muove*"—"and yet, it moves." This phrase has become a classic expression of the idea that you can't make someone change their mind by force, though you may compel them to *say* they have changed it.

But without freedom to debate and publish new ideas, science will not flourish for long. It is perhaps not simply a coincidence that after Galileo, Italy—the mother of outstanding thinkers till then—produced for the next 200 years only a few great physicists, while elsewhere in Europe they arose in great numbers. To scientists today this famous facet in the story of planetary theories is not just an episode in passing. Not a few teachers and scientists in our time have had to face powerful enemies of open-minded inquiry and of free teaching, and again had to stand up before those rulers who fear the strength of unindoctrinated intellect.

Even Plato knew that an authoritarian state is threatened by intellectual nonconformists, and he recommended for them the now time-honored treatment: "reeducation," prison, or death. In the late unlamented Soviet Union, geneticists at one time were expected to reject well-established theories, not on grounds of conclusive new *scientific* evidence, but because of doctrinal conflicts. This same struggle explains the banishment of discussion of at least the origins of relativity theory from Nazi Germany's textbooks in the 1930s, because, according to racist metaphysics, Einstein's Jewish origins invalidated his work for Germans.

In the United States, too, there has not always been freedom to express scientific views. One of the most notorious examples was the 1925 "Monkey Trial" in Tennessee, in which the teaching of Darwin's theory of evolution was punished because it conflicted with certain types of Bible interpretations; the teaching of biology in American high schools suffered for decades afterward, and in some places suffers still.

The warfare of authoritarianism against science, like the warfare of ignorance against knowledge, has not diminished since Galileo's day. Scientists take what comfort they can from the verdict of history. The theory that got Galileo into trouble has since become quite acceptable to most religious authorities. Although the Vatican did not announce that it had reversed its 1633 condemnation of Galileo until 1992, the scientific usefulness of his work was not so long delayed. Less than 50 years after Galileo's death, Newton's great book, the *Principia,* had appeared, integrating the work of Copernicus, Kepler, and Galileo so brilliantly with the principles of mechanics that the triumph of their ideas was irrevocable—and as we shall soon see, more significant than they themselves might have hoped.

Problem 5.4. In J. B. Conant's book, *On Understanding Science* are summarized "Certain Principles of the Tactics and Strategy of Science" and "The Interaction of Science and Society." The following main points are made: (a) New concepts evolve from experiments or observations and are fruitful of new experiments and observations. (b) Significant observations are the result of "controlled" experiments or observations; the difficulties of experimentation must not be overlooked. (c) New techniques arise as a result of experimentation (and invention) and influence further experimentation. (d) An important aspect of scientific work is the interaction between science and society. Examine the development of planetary theory in terms of each of the four points.

Problem 5.5. Between 1600 and 1700, most European astronomers rejected the geocentric theory and adopted some form of heliocentric theory. Compare the validity of Planck's principle and Popper's principle (see Chapter 3) as applied to this historical event. Do both principles leave out some essential factors?

RECOMMENDED FOR FURTHER READING

Herbert Butterfield, *The Origins of Modern Science,* Chapter IV

Alan Cowell, "After 350 years, Vatican says Galileo was right after all," *New York Times,* 31 October 1992, pages 1, 4

M. J. Crowe, *Theories of the World,* Chapter 9 and Epilogue (with quotations showing the cultural impact of the Copernican Revolution)

Stillman Drake, "Galileo," in *Dictionary of Scientific Biography* (edited by Gillispie), Vol. 5, pages 237–249 (1972)

Galileo, *Sidereus Nuncius, or The Sidereal Messenger,* translated by Albert van Helden, Chicago: University of Chicago Press, 1989.

Owen Gingerich, "How Galileo changed the rules of science," *Sky & Telescope,* Vol. 85, no. 3, pages 32–36 (1993).

Marie Boas Hall, *The Scientific Renaissance 1450–1630,* Chapter XI

N. S. Hetherington, *Cosmology,* articles by Hetherington, S. Drake, and E. McMullin, pages 227–238, 575–606

T. S. Kuhn, *The Copernican Revolution,* Chapter 6

S. Sambursky, *Physical Thought,* excerpts from Galileo and Gassendi, pages 216–225, 255–25

Darva Sobel, *Galileo's Daughter: A Historical Memoir of Science, Faith, and Love,* New York: Walker, 2000

For further discussion of the fascinating subject of this chapter, see the "Sources" at the website www.ipst.umd.edu/Faculty/brush/physics bibliography.htm

PART B

The Study of Motion

6 Mathematics and the Description of Motion

7 Galileo and the Kinematics of Free Fall

8 Projectile Motion

Historically as well as logically mechanics represents the foundation of physics and the prototype for the study of other physical sciences. The concepts developed in this field will appear again and again in this book. Moreover, mechanics is to physics what the skeleton is to the human figure. At first glance it may appear stiff, cold, and somewhat ghastly, but after even a brief study of its functions one experiences with mounting excitement the discovery of an astonishingly successful design, of a structure that is ingeniously complex, yet so simple as to be almost inevitable.

We shall begin with the key topic in mechanics, the laws governing some of the simpler motions of objects—movement along a straight line and a curved trajectory. For even in Galileo's time there was this axiom: ignorato motu, ignoratur natura ("not to know motion is not to know nature").

CHAPTER 6

Mathematics and the Description of Motion

6.1 René Descartes

The Aristotelian philosophy was rapidly fading away under the attacks of such men as Galileo in Italy and Francis Bacon (1561–1626) in England; but what was to replace it? A collection of experimental facts and mathematical theories, no matter how precisely they fit together, does not constitute a satisfying philosophical system for interpreting the cosmos and our place in it.

The cosmology that was ultimately to be associated with the name of Isaac Newton owes much of its general aim and character to a French philosopher, René Descartes (1596–1650), even though his detailed scientific theories were overthrown by Newton. It was Descartes who first attempted to provide a general philosophical framework for the new science of the seventeenth century, defined some of the basic problems that scientists were concerned with, and suggested methods for solving them. Yet his name does not often appear in physics books, because most of his detailed solutions were rejected by later generations. What remained, aside from a general attitude toward the physical world, were the beginnings of a powerful mathematical technique for representing geometric forms and physical processes in symbolic language that greatly facilitated logical deductions: Descartes' *analytic geometry,* later to be wedded to the *differential and integral calculus* of Newton and Leibniz.

In this chapter, we will summarize only a few of the more elementary concepts of the mathematical description of motion: not in their historical sequence, but from a viewpoint that now seems best adapted to physical applications. But first let us glance briefly at the career of Descartes himself.

Descartes was educated in a French Jesuit school from the age of 8 to 16, but his health was poor, so he was allowed to stay in bed late in the morning. Perhaps this helped him to develop his lifelong habits of prolonged meditation. At 17 he went to Paris, but it is said that he "had sufficient strength of character to hold himself aloof from the distractions of the capital." He traveled for several years, serving as an officer in various armies. In 1619 he conceived his reform of philosophy, taking mathematics as a model since that discipline alone seems to produce certain results. Although some of his discoveries in optics and mathematics were first worked out during this period, it was not until 1628 that he settled in Holland and devoted himself wholeheartedly to science.

In 1633, Descartes was about to publish a book on his view of the system of the world when he heard of Galileo's condemnation by the Church in Rome. Descartes himself was a good Catholic and did not wish to come in conflict with the Church, so he suspended the publication of his book even though it would not have been officially censored or suppressed in Protestant Holland.

Four years later he published his first work, actually a collection of four treatises: The best known of these, the *Discourse on Method,* was intended as a preface to the other three. The second part, *Optics,* includes the first publication and first proof of "Snell's law" of refraction, together with a description of how the eye works and instructions for making better telescope lenses. We shall discuss Descartes' *Optics* later in connection with Newton's theory of light (Section 23.1). The third part, on *Meteorology,* discusses weather, clouds, snow, etc., and in particular gives a very good theory of the rainbow. The fourth part, *Geometry,* lays the foundations for analytical geometry, one of the essential mathematical tools of modern physical science.

In 1644, Descartes published his major work, *Principles of Philosophy.* Here he made a magnificent attempt to construct a complete theory of the world out of nothing but the concepts of matter and motion. It was a failure, but it left an indelible imprint on subsequent attempts to think about the nature of the physical

Fig. 6.1. René Descartes (1596–1650), French philosopher and mathematician. The Cartesian mechanistic philosophy was popular in the seventeenth century and afterward. His (incorrect) principle of "conservation of motion" was a precursor of the modern law of conservation of momentum (Chapter 16), and his ideas about the propagation of light influenced both the particle and wave theories (Chapter 23). Descartes' method of representing curves by algebraic formulas ("analytic geometry") is an essential tool for physical scientists.

world. Even after the triumph of Newtonian physics, many scientists still shared Descartes' preference for avoiding the concept of force—that is, of "action at a distance"—and for postulating instead that space is filled with pieces of matter that can interact only when they touch. What appear in Newton's theory to be long-range forces acting across empty space, such as gravity, would in Descartes' view be explained by the propagation of impulses through an invisible ethereal matter that is imagined to fill the intervening space.

For Descartes, all motion is relative: One piece of matter can be said to be moving only with respect to other pieces of matter in its vicinity. This allows him to assert that the earth is "at rest" without abandoning the Copernican system! Descartes considers the heavens to be undergoing a continuous circular motion with respect to the sun; the heavens carry along with them all the bodies that they contain, including the stars, planets, and earth, each whirling in its own vortex. Each body is thus at rest with respect to its own local vortex, while the vortex moves around the sun. Perhaps, with this doctrine of relativity of motion, he hoped to avoid conflict with his Church, whose theologians sometimes asserted that science should concern itself only with describing the "appearances" and leave to theology the task of determining "reality."

Descartes' cosmology, though permeated with references to God, nevertheless represents a major step toward the elimination of supernatural agents from any active role in moving the celestial objects or other matter. For example, he asserts that God always conserves the same amount of "movement" in the world; this means in effect that while each piece of matter may have been created with any arbitrary movement, the transfer of movement from one piece to another by collisions is governed by a deterministic rule.

From a scientific viewpoint, this means that Descartes not only asserts the principle of inertia—a body remains in the same state of motion in a straight line or of rest until acted on by another body—but that he also has some inkling of the principle of conservation of momentum (see Chapter 16).

From a philosophical viewpoint, it means that the Cartesian world is very much like a machine. Indeed, Descartes admits that he made use of the example of machines in constructing his model of how the world acts, and he asserts that there is no real difference between machines made by men and objects found in nature, except that in the former all the moving parts must be large enough for us to see them. Even animals may be like machines.

On the other hand, mechanistic theories cannot explain *human* behavior because, according to Descartes, the world of spirit and thought is completely different and separate from the world of matter, and is not governed by the same laws. With this *dualistic* approach, he attempted to mark off a realm of natural phenomena that could be studied by the scientist without interference from the theologian, while leaving the theologian supreme in the realm of the spirit.

In 1649, Descartes was summoned to Sweden to serve as tutor to Queen Christina. The Swedish climate was too much for him. Within a few months he caught a chill, suffered inflammation of the lungs or perhaps pneumonia, and died.

6.2 Constant Velocity

According to Galileo and Descartes, the natural state of a physical object is rest or *motion along a straight line, at constant speed.* Let us see how the physicist describes such motion in a way that can easily be generalized to other kinds of motion.

Modern physicists use the *metric system* of units, which has been adopted in all countries for scientific work. It may be helpful to remember that 1 meter (1 m) = 39.37 inches, or a little over 1 yard; 1 centimeter (1/100 of a meter) is about 0.4 inch. See Appendix IV for other conversion factors.

Suppose we are watching the motion of a car traveling along a perfectly straight and smooth road, and that we wish to find the exact relationship between the *distance* covered and the *time* required to cover it. The progress of this car may be timed at stations regularly spaced at, say, 30 meters all along the line. The car has reached full speed before it goes past the first station, and all our time measurements will be expressed in terms of the time elapsed since the first station was reached, the moment at which our experiment begins. If the motion has been uniform (constant speed), a tabular arrangement of the time schedule for the first five stations might read as in Table 6.1.

A more useful way of representing the same information is a time-vs.-distance graph (Fig. 6.2). Along the horizontal axis we plot the time elapsed in seconds, and along the vertical axis the distance covered in meters.[1] On this graph the five known data are entered, and since they fall on a straight line, we draw that line as shown to indicate our not-unreasonable belief that any additional data taken would also have fallen along that line. For example, we are fairly confident that at a point 45 meters from station A the car would have passed by 3 seconds after the beginning of our series of observations. This conclusion is an *interpolation,* and is most conveniently done on a graph of this type.

On the other hand, we are also fairly sure that after 9 seconds, the vehicle, if it continued to move at the same speed, reached a position 135 meters beyond station A. This is an *extrapolation.* Obviously, if it is not already known that the graph describes a uniform motion, one must not place too much confidence in results obtained by interpolation or extrapolation, particularly when the basic data are few or widely spaced.

Table 6.1.

Station	*Distance from station A in meters*	*Time of travel from station A in seconds*
A	0	0
B	30	2
C	60	4
D	90	6
E	120	8

From Table 6.1 it should already be evident that the car travels twice the distance in twice the time, three times the distance in three times the time, and so forth; in other words, the distance covered is directly proportional to the time elapsed. The straight line in the graph is another way of representing the same thing. Descartes' system of analytic geometry enables us to translate the straight line into an algebraic equation: Let the time elapsed since station A was passed be represented by the letter t, and let the distance covered during that time interval be called s; then for the motion here described we can write

$$s \propto t,$$

where the symbol $\propto$ means "is proportional to." But this is equivalent to the equation

$$s = kt$$

where k is a constant that depends on neither s nor t.

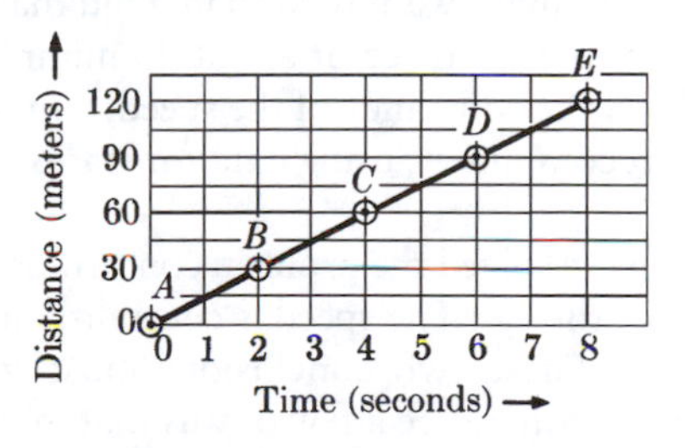

Fig. 6.2. Graph for uniform speed.

[1]The interval between the origin and a point measured along the horizontal axis is sometimes called the *abscissa,* and the interval along the vertical axis is called the *ordinate.*

In analytic geometry, the equation $s = kt$ represents a straight line passing through the origin of coordinates. The constant k is called the *slope.* It is also the *tangent* of the angle between the line and the horizontal axis. In this case, its value is simply the ratio of distance covered to time elapsed,

$$k = s/t.$$

This ratio of distance covered to time elapsed is, by definition, the *speed;* the fact that it is equal to a constant brings us back to (and is a consequence of) our original assumption that the car travels with constant speed.

Consider for a moment what an amazing thing has actually happened. First we watched the actual motion of a car along a straight road. Then, from the multitude of ever-changing impressions—the blur, the noise, the turning of wheels, the whole chaos of events progressing in time and space—we have rescued two measurable quantities, s and t, both of which take on different values every instant, and we have found that their ratio is a constant, an unchanging theme underlying the flux of otherwise meaningless, unrelated data. We have defined a concept, speed, and so have been led to discover a simple feature in an otherwise complex situation. Perhaps familiarity with the *concept of speed* prevents you from appreciating this experience of *creating order* from a *chaos* of sense impressions by abstracting from it some measurable data and by perceiving or inventing or intuiting a suitable concept to describe that portion of the total phenomenon; however, we shall return to this method, the very heart of scientific procedure, again and again.

The exact numerical value of the speed is to be found by substituting corresponding values for s and t. For example, from the data taken at station B, $s = 30$ m and $t = 2$ sec; from this, it follows that the speed, s/t, is 15 m/sec. (If you have not yet become accustomed to using the metric system, you might want to keep in mind that 13 m/sec is about 44 ft/sec or about 30 mi/hr.) As a check on the constancy of the speed, you may wish to recompute it, using data for stations C, D, or E.

We have solved the problem concerning the motion of the car. The speed is constant, and is 15 m/sec. These two conclusions must seem obvious, but there are a few details that may be worth an extra word. To compute the speed we do not really need to rely on the raw data in Table 6.1. We can use the time-vs.-distance graph directly if we have any confidence in it at all. Consider, e.g, the uniform motion indicated by the line *OQPU* in Fig. 6.3. From the previous paragraphs we know that for such a case the speed is given by s_1/t_1 or by s_2/t_2 (since $s_1/t_1 = s_2/t_2$ by similar triangles), but it is given equally well by $(s_2 - s_1)/(t_2 - t_1)$, i.e., by the ratio of any small distance along the line of motion to the time needed to pass through that distance. This statement may seem intuitively acceptable, but we should check it rigorously.

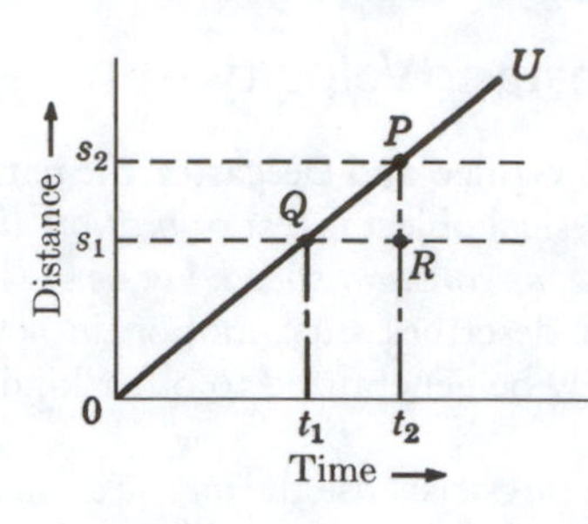

Fig. 6.3. Distance-time graph showing equality of slopes for different intervals.

Problem 6.1. Prove that $(s_2 - s_1)/(t_2 - t_1) = s_1/t_1$.
Solution: Note that in Fig. 6.3 the distance $QR = t_2 - t_1$, and $PR = s_2 - s_1$; also that the following three are similar triangles: Os_1Q, Os_2P, and QRP. Therefore corresponding distances have equal ratios, that is,

$$s_1/t_1 = s_2/t_2 = (s_2 - s_1)/(t_2 - t_1) \qquad \text{Q.E.D.}$$

(The initials "Q.E.D." stand for the Latin phrase *quod erat demonstrandum,* meaning "which was to be proved.")

Before you leave these relatively harmless findings, be sure that you can answer to your own full satisfaction the following questions:

- What is the definition of constant speed (use algebraic notation)?
- What experiment must I make to find out whether a certain body moves with constant speed?
- How do I find the value of the speed once I know that it is constant?
- How short a distance $(s_2 - s_1)$ may I choose in order to calculate the speed?

After that, try to tackle the superficially deceptive questions upon which, as we shall find, depends the understanding of a large section of our physics:

- For what reason do we here accept these arguments from Euclidean geometry or algebra?
- What does mathematics tell us about real physical events, such as the motion of material bodies?

Another problem is raised by a second critical look at our concept of constant speed along a straight line. Table 6.1 presented raw experimental data in tabular form and was our first step in formulating the idea of constant speed. Figure 6.1 represented these data in the form of a graph and expanded their usefulness by permitting convenient interpolation and extrapolation. The proportion $s \propto t$ abstracted from this graph a concise statement that we expressed by the equation s/t = a constant (15 m/sec in this specific case).

What has happened here is that in four stages we have proceeded from a bulky table of raw data to a generalization (s/t = 15 m/sec). If on repeated observations of this car we always arrive at this same equation, in other words, if the car always moves at a constant speed of 15 m/sec, we would regard the equation as a *law* applicable to this restricted type of observation. As it is, our equation arose from a single experiment and is simply a general description of this one motion.

At first glance all four stages (the table, the graph, the proportion, and the equation) seem to contain the same information, with the last stage appearing vastly preferable because of its more economical demands on paper and on mental effort. However, we must not let our innate love for conciseness misguide us; the equation s/t = 15 m/sec does in fact include everything that the table offers, but it also tells indiscriminately a great deal more that may actually be quite unwarranted. For example, unless you had seen the tabulated data, you might believe on the basis of the equation alone that surely s = 15 mm if t = 0.001 sec, or that $s = 4.7 \times 10^8$ m if t = 1 yr.

Both statements are correct only if the motion under investigation is truly uniform within the limits of such very exact measurements as 0.001 sec and for such a span of time as 1 yr, but in our actual, limited experience neither of these conditions is likely to hold. Even less do we concern ourselves with the difficulties that would arise if we tried to imagine time and space divided into infinitesimal intervals (as in Zeno's paradoxes, discussed in Section 3.2) or to contemplate the possible rotation of an infinitely large sphere (the argument Aristotle used to prove that the universe is finite, see Section 1.3).

The lesson to be drawn here is that equations in physics, other than a few purely axiomatic definitions, must be thought to have attached an unseen "text," a statement that describes the actual limitations and other implied assumptions under which the equation may be applied. These conditions, this text, may simply be of the following form. "The equation s/t = 15 m/sec describes the result of a certain experiment in which t did not exceed 8 sec, and in which only a few measurements were made, each with such and such accuracy." Without the clear understanding that equations in physics almost always carry hidden limitations, we cannot expect to understand a single physical law; we would make unwarranted extrapolations and interpolations. We would be in the catastrophic position of a navigator who has to negotiate a rocky channel without having any idea of the length, width, and draft of his ship.

Although the concept of motion at constant speed is commonplace today, when anyone driving a car can verify it by glancing at the speedometer, it was not widely familiar or even understood in previous centuries, when most motions were experienced as rough and jerky. For followers of Aristotelian philosophy, it was not even meaningful to talk about the ratio of two different physical quantities like distance and time; the idea would have to be expressed by a cumbersome statement such as "the ratio of s_2 to s_1 is the same as the ratio of t_2 to t_1."

6.3 The Concept of Average Speed

Prolonged motion with constant speed occurs so rarely in nature (despite its importance as an "ideal" motion in the theories of Galileo and Descartes) that, as you may have noticed, we have denied it the honor of a special letter symbol of its own. (We did not introduce the letter v, which you might have expected to be used for constant velocity or constant speed; this will instead be used for instantaneous speed, in the next section.)

We turn now to motions with nonuniform speed (accelerated or decelerated motions) but still taking place only along a straight line. A general

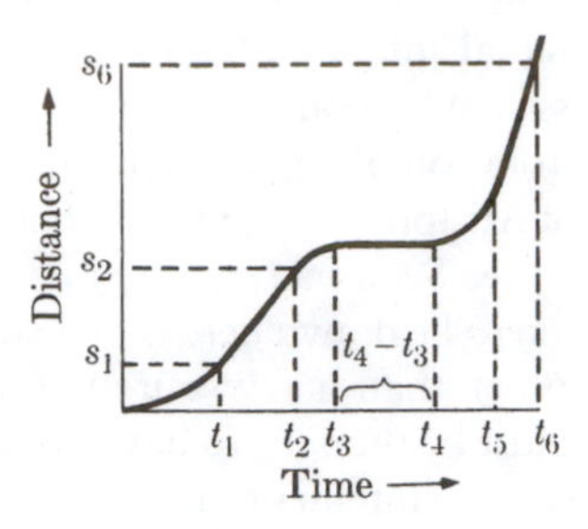

Fig. 6.4. Graph for nonuniform speed.

case of this type is shown in Fig. 6.4, where a car starts from rest, moves in a straight line gaining speed until time t_1, and proceeds with constant speed until time t_2 (as indicated by the straight-line portion between t_1 and t_2), after which it slows down to rest (t_3), waits without moving until time t_4, and then again accelerates to a new, higher constant speed. To deal with such variations in speed, it is necessary to introduce the familiar term, average speed or average velocity, symbolized by $\bar{v}$ or $\langle v \rangle_{av}$. (The latter is sometimes more convenient, and calls attention to the fact that an average has to be computed.)

DEFINITION
The average speed $\bar{v}$ of an object during a time interval t is the distance covered during this time interval divided by t.

Note that for any motion other than prolonged constant speed, the value of $\bar{v}$ depends on the length of the time interval chosen. For example, a glance at Fig. 6.4 suggests that since the curve is becoming steeper between t_1 and t_2 (its slope is increasing), $\bar{v}$ for the first t_1 seconds is smaller than $\bar{v}$ computed for the first t_2 seconds.

On the other hand, during the interval $(t_4 - t_3)$ the average speed is zero. Although the speed from t_1 to t_2 might be 30 m/sec and between t_5 and t_6 50 m/sec, the average speed $\bar{v}$ for the entire time to t_6, which is calculated by $\bar{v} = (s_6/t_6)$, is perhaps 20 m/sec. We recognize that the meaning of $\bar{v}$ is this: No matter how irregular the motion actually is during a time interval t, if the object had traveled instead with a constant speed of magnitude $\bar{v}$ it would have traveled the same distance during those t seconds.

Instead of writing out "the interval from t_1 to t_2" we will often use the abbreviation Δt. The Greek letter Δ, delta, means "a small change in (whatever quantity follows)." Similarly for distance intervals: Δs does not mean "delta times s" but reads "delta s," i.e., a small increment in distance, a small displacement covered during the corresponding short time interval Δt. Therefore we can write the definition of average speed as

$$\bar{v} = \Delta s/\Delta t$$

6.4 Instantaneous Speed

All our speed measurements so far have involved more or less extended time measurements. However, we certainly also want to have a concept that tells us the speed of a moving object at one instant, the type of information that we read off a speedometer. For example, we may ask, "What is the speed at time t_a for the motion pictured in Fig. 6.5?" Now we *do* know how to find the average speed $\bar{v}$ during a time interval from t_1 to t_2, which includes the instant t_a, i.e.,

$$\bar{v} = (s_2 - s_1)/(t_2 - t_1).$$

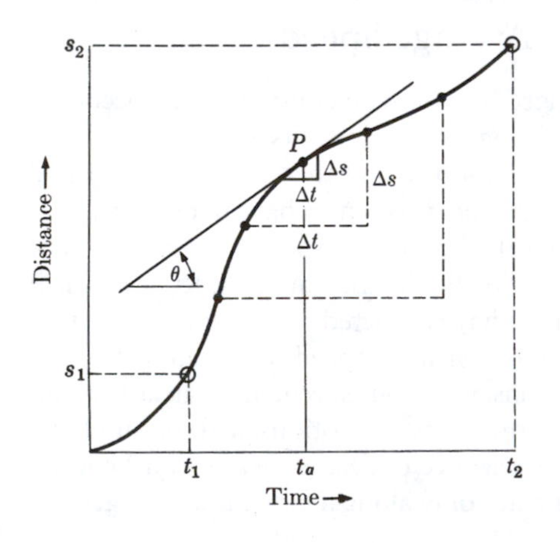

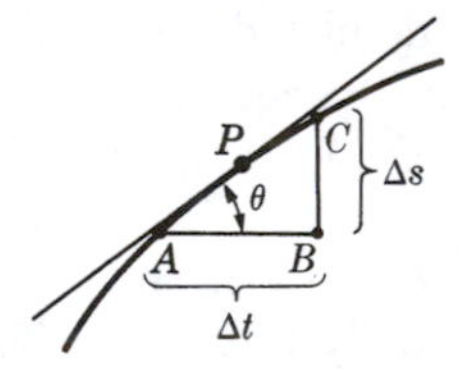

Fig. 6.5. Determination of instantaneous speed.

Table 6.2. Data for Estimation of Instantaneous Speed at the Point P ($t = 7$ sec, $s = 20$ m)

t_B	t_A	s_B	s_A	$\Delta t = t_B - t_A$	$\Delta s = s_B - s_A$	$\bar{v}$ (m/sec)
12	2	30	5	10	25	2.5
7.5	6.5	21	19	1.0	2.0	2.0
7.05	6.95	20.10	19.92	0.10	0.18	1.8
7.005	6.995	20.010	19.993	0.010	0.017	1.7

However, since the speed is not constant from t_1 to t_2, we certainly cannot identify the speed at the instant t_a with the average speed $\bar{v}$ *during* $(t_2 - t_1)$. Therefore we must resort to what undoubtedly will look like a trick: we calculate $\bar{v}$ for a very short time interval Δt encompassing the instant t_a—so short an interval that the value of $\bar{v}$ would not change materially if it were made even shorter. The instantaneous speed (symbolized simply by v) at t_a is therefore calculated by successive approximations from the average speed $\bar{v}$, and may be defined by

$$v = \lim_{\Delta t \to 0} \bar{v} = \lim_{\Delta t \to 0} \left(\frac{\Delta s}{\Delta t}\right)$$

Putting this equation into words, the instantaneous speed is the value the average speed ($\Delta s/\Delta t$) attains when the time interval (Δt) is chosen sufficiently small, approaching zero in the limiting case.

A concrete example will help here. From the data on which Fig. 6.5 is based we might construct Table 6.2 for space and time intervals including point P on the curve. The intervals are nested inside $t_2 - t_1$ and $s_2 - s_1$, converging on the point P as they become smaller and smaller. For each pair Δs and Δt we can compute the quotient, the average speed $\bar{v} = \Delta s/\Delta t$ for that interval. Although both Δs and Δt become indefinitely small, $\bar{v}$ approaches a definite limiting value as long as the curve is smooth. In the table the subscripts A and B are used to identify the variable lower and upper limits of the intervals.

As you perhaps remember from mathematics courses, what we have just done is the algebraic equivalent of finding the slope of the straight line tangent to the curve at the point P in Fig. 6.5. As Δt and Δs become smaller and smaller, the section of the curve that they include is more nearly a straight line, playing the role of the hypotenuse in the right triangle ABC (see inset in Fig. 6.5). But by trigonometry, $\Delta s/\Delta t = \tan\theta$, where θ is the angle made by the hypotenuse with the horizontal; so that ultimately, when Δt is sufficiently small, $\Delta s/\Delta t$ becomes simultaneously both the numerical value of the instantaneous speed v at point P and the value of $\tan\theta$ of the line tangent at point P, i.e., the slope of that line.[2]

We gratefully accept this unexpected result because it cuts short the labor involved in finding instantaneous speeds for nonuniform motions. In the future we need not construct such tables as Table 6.2 in order to find the instantaneous speed at a given point during the motion of an object. We now simply refer to the distance-vs.-time graph, find the line that is tangent to the curve at the point in question, and calculate its slope.

In fact, this last full sentence may stand as the definition of the concept of instantaneous speed in terms of actual operations needed for its determination. If ascertaining speeds by looking at a speedometer seems far simpler, realize that in the initial construction and calibration of the speedometer we must still take good care that it does automatically perform tasks equivalent to all those outlined in our definition of instantaneous speed.

It is this concept, instantaneous speed, that we shall find most useful from now on, and we shall discuss typical problems in which the (instantaneous) speed of a body changes from some initial value v_0 to some other value v during a specified time interval t. We shall sometimes make plots of (instantaneous) speed vs. time. For the case shown in Fig. 6.2, such a plot would evidently be simply a straight horizontal line, but for Fig. 6.4 it is more complex.

[2]Strictly speaking the equation $\Delta s/\Delta t = \tan\theta$ is incorrect since the left-hand side must have units (or "dimensions") m/sec, while the right-hand side is a pure number having no units. This is why we have to specify "the numerical value" of $\Delta s/\Delta t$, as distinct from the physical quantity.

Problem 6.2. Copy Fig. 6.4 on a large scale and superpose on it a plot of the speed vs. time.

6.5 Acceleration

Of all motions, the one most important for our purposes is motion with uniformly changing velocity, i.e., motion with constant acceleration or constant deceleration. To a fair approximation this is the characteristic behavior of ordinary freely falling bodies, objects sliding down a smooth inclined plane, carts coasting to a stop on level ground, and generally of all bodies that experience a constant force.

At the time of Galileo it was not obvious how "constant acceleration" should be defined. Does it mean that velocity changes by equal amounts in equal *distances* ("the farther you go the faster you go") or in equal *times* ("the longer you travel the faster you go")? After thinking about this question for some time, he eventually decided that the first definition is unsatisfactory, so he settled on the second. He might have reasoned as follows: If constant acceleration means Δv is proportional to Δs, then if you are at rest ($v = 0$), Δs is stuck at zero, so you can never start moving.[3]

Having eliminated the only plausible alternative, Galileo concluded that *constant acceleration must be defined as a state in which the change in velocity is proportional to the time interval.*

Graphically, there are three simple cases involving uniformly changing velocity. In Fig.6.6a the speed at the start is zero ($v_0 = 0$) and reaches a value v after t sec; in Fig. 6.6b v_0 is not zero, and the speed increases uniformly; and in Fig. 6.6c the speed decreases from a large initial value v_0 to nearly zero during time t. Concrete experimental situations corresponding to each drawing will suggest themselves readily.

For all such cases it is easy to define the concept of acceleration, for which we shall use the letter *a:* Acceleration is the ratio of the change of speed ($v - v_0$) to the time (t) during which this change occurs, or

$$a = (v - v_0)/t. \qquad (6.1)$$

As we have agreed to restrict our attention for the time being to motion with constant acceleration, we need not be concerned with such refinements as average acceleration and instantaneous acceleration, for, as in the case of constant speed, the average and instantaneous values are numerically equal, with that restriction.

To illustrate this new concept let us consider some very straightforward cases.

EXAMPLE 1

A skier starts down an incline and gains speed with uniform acceleration, reaching a speed of 10 m/sec after 5 sec. What is her acceleration?

Solution: As in every problem, we must first *translate* each separate phrase into mathematical symbols, or else extract from each some clues as to which physical concepts, laws, and limitations are here applicable. This is not a trivial process; most students find that

[3] Galileo himself gave what he called a "very clear proof" for the case of an accelerated falling body: "When speeds have the same ratio as the spaces passed or to be passed, those spaces come to be passed in equal times; if therefore the speeds with which the falling body passed the space of four braccia were the doubles of the speeds with which it passed the first two braccia, as one space is double the other space, then the times of those passages are equal; but for the same moveable [moving body] to pass the four braccia and the two in the same time cannot take place except in instantaneous motion. But we will see that the falling heavy body makes its motion in time, and passes the two braccia in less [time] that the four; therefore it is false that its speed increases as the space.' (*Two New Sciences,* Third Day, Drake translation). To the modern reader this proof is not satisfactory since it seems to confuse the instantaneous speed with the average speed during an interval, and the mean-speed theorem (Section 6.6) cannot be used to calculate the latter if speed increases uniformly with distance rather than time. The situation is somewhat like Zeno's "dichotomy" paradox (Section 3.2), as Galileo seems to have realized when he discussed the problem in the *Dialogue* (First Day). It may be difficult to understand, says Salviati, how a body can pass "through the infinite gradations of slowness in approaching the velocity acquired during the given time." The answer is that the body "does pass through these gradations, but without pausing in any of them. So that even if the passage requires but a single instant of time, still, since each small time interval contains infinitely many instants, we shall not lack a sufficiency of them to assign to each its own part of the infinite degrees of slowness, though the time be as short as you please." Galileo displays here an intuitive understanding of the mathematics of infinity, good enough to give satisfactory definitions of physical concepts like speed and acceleration; mathematicians two centuries later found more consistent ways to express his ideas.

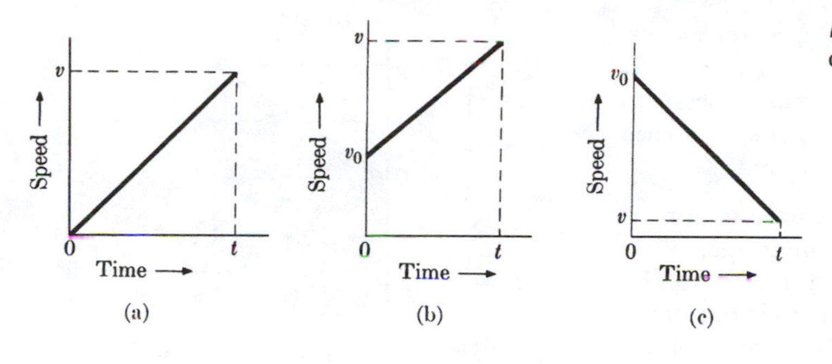

Fig. 6.6. Three cases of uniformly changing velocity.

a kind of intuition gained from experience in working out such problems is more important than facility in solving mathematical equations. The phrase "a skier starts down an incline" is translated to $v_0 = 0$; "gains speed with uniform acceleration" means a is constant, therefore Eq. (6.1) will hold for this case; "reaching a speed of 10 m/sec after 5 sec" can be simply restated as $v = 10$ m/sec at $t = 5$ sec. The final phrase does not mean "what is the cause or physical nature of his acceleration" (that will come later), but simply asks for a numerical value, $a = ?$

Now we put all our symbolic statements side by side and read:

$$v_0 = 0; \quad v = 10 \text{ m/sec}; \quad t = 5 \text{ sec};$$

$$a = (v - v_0)/t \text{ holds here}; \quad a = ?$$

Evidently, by simple substitution,

$$a = (10 \text{ m/sec} - 0)/(5 \text{ sec}) = 2 \text{ (m/sec)/sec}.$$

The result may be read as "2 meters per second, per second," and may be written more concisely 2 m·sec^2. Although this abbreviation can be read "2 meters per second squared," remember that the unusual concept of "square seconds" arises merely from the fact that we have defined acceleration as a change in speed (units: m/sec) divided by a time interval (units: sec).

Although this example must still appear exceedingly simple, you will find that the steps taken here are the same as in every single problem, no matter how sophisticated. If you train yourself from the very start to *translate each phrase separately,* you will have avoided half the initial troubles with physics problems.

EXAMPLE 2

A car has been traveling on a level road at a speed of 70 km/hr when suddenly the motor fails and the car coasts along, gradually slowing down. During the first minute it loses 30 km/hr. What was the acceleration during that time?

Solution: Translating, we read $v_0 = 70$ km/hr, $v = 40$ km/hr at $t = 1$ min $= 60$ sec; $a = ?$ (Evidently it will be a negative acceleration—sometimes called deceleration—but that is a superfluous term.)

It is very tempting to plunge ahead and say that

$$a = (40 \text{ km/hr} - 70 \text{ km/hr})/(60 \text{ sec}) = -\tfrac{1}{2} \text{(km/hr)/sec}.$$

However, this is true only if we assume that the car coasted with uniform acceleration—a guess, but not necessarily the only correct one. Therefore we must say "$a = -\frac{1}{2}$ (km/hr)/sec *if* the acceleration was uniform." Notice that in this case *the speed is positive but the acceleration is negative.*

EXAMPLE 3

A car on a roller coaster with many curves changed its speed from 5 km/hr to 25 km/hr in 20 sec. What was the acceleration?

Translation: $v_0 = 5$ km/hr, $v = 25$ km/hr, $t = 20$ sec, $a = ?$ However, since there is little likelihood that the acceleration was constant, we must simply confess that we cannot solve this problem until further data are given from which the type of motion may be deduced. It would be a mistake to regard such a statement as an admission of dishonorable defeat; on the contrary, in science, as in every other field, the possibility of solving a problem offhand is the exception rather than the rule, and it is a good enough preliminary victory to know what additional information is needed for an eventual solution.

The units for acceleration have been variously given as (m/sec)/sec or (km/hr)/sec; others are evidently also possible; for example, (m/sec)/min. (If you happen to live in one of the few countries in the world that has not yet adopted the metric system for daily use, you may have to use the table of conversion factors in Appendix IV in order to translate these units into others that are more familiar to you.) In all these cases, the units correspond to the dimension of (length/time)/time, or in a symbolic shorthand frequently encountered, L/T^2.

6.6 Oresme's Graphical Proof of the Mean-Speed Theorem

In the last four sections we have temporarily set aside our historical account in order to introduce some basic concepts in modern notation. Many of the same problems of accelerated motion were discussed as long ago as the thirteenth century by mathematicians at Oxford and Paris, though many of their arguments, expressed verbally without the aid of algebra and analytic geometry, are hard for modern readers to follow. One particular result of their work that is of great importance for physics is the *mean-speed theorem,* sometimes called the Merton theorem, after Merton College, Oxford, where it was first developed.

In modern notation, the mean-speed theorem refers to a motion that is uniformly accelerated with initial speed v_0 and final speed v, and occupies a time interval t. The theorem states that the distance traveled is the same as that traveled in another motion that takes place at the mean speed (that is, at a constant speed equal to the average of v_0 and v) during the same time interval t. In other words, if the entire motion had taken place at a constant speed v_0 for a time t, the distance would be v_0t; if it had taken place at a constant speed v for a time t it would be vt; but in the case in which the speed changes uniformly from v_0 to v, the distance is the same as if the speed had been equal to a constant speed of amount $\frac{1}{2}(v_0 + v)$, that is, the distance would be $\frac{1}{2}(v_0 + v)t$.

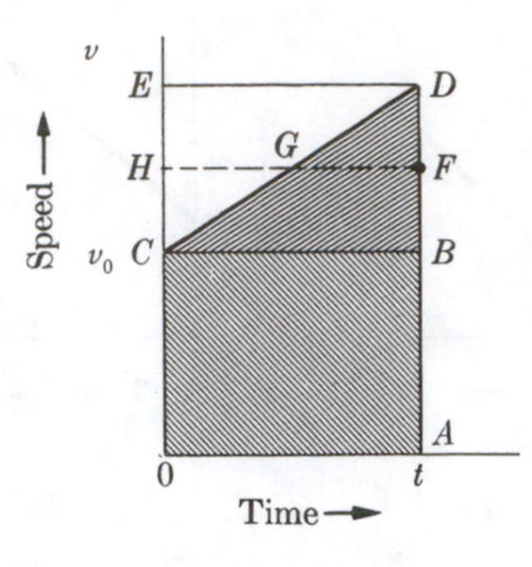

Fig. 6.7. Oresme's graphical representation of the relation between speed, time, and distance.

This result may seem uninteresting, even trivial. It can easily be derived from the algebraic equations to be discussed in the next section. More significant, however, is a particular *method* used to derive it by Nicolas Oresme at the University of Paris, early in the fourteenth century. Oresme (*circa* 1320–1382), chaplain to King Charles V and later Bishop of Lisieux, has attracted the attention of historians of science because of his ideas on kinematics and astronomy, which anticipated to some extent those developed by Galileo in the seventeenth century.

Oresme realized that since a quantity like v_0t is a product of two numbers, it can be represented as the area of a rectangle whose sides are v_0 and t. This would be the rectangle *OABC* in Fig. 6.7. Similarly, vt is the area of the rectangle *OADE*. Therefore, Oresme argued, the actual distance traveled when the velocity is changing uniformly from v_0 to v must be represented by the area of the shaded part of the figure, namely the rectangle *OABC* plus the triangle *CBD*. (If this is not obvious, after a little study of the figure it should seem at least reasonably plausible.)

Once it is accepted that the distance covered in the accelerated motion is represented by the shaded area *OABDC,* the proof of the mean-speed theorem is simple enough. If to this area we add the triangle *CGH,* and subtract the triangle *GFD,* we obtain the rectangle *OAFH.* Now if the line *HGF* is parallel to *ED* and *CB* and is exactly halfway between them, then the point *G* will be exactly halfway between *C* and *D,* and the line segment *CH* will be equal to *DF.* Also, *HG* = *GF* and ∠*DGF* = ∠*HGC*. Therefore the triangles *HCG* and *GDF* are congruent and have equal area, so that the area added is equal to the area subtracted. So the area of the rectangle

OAFH, which represents the distance traveled at a constant speed halfway between v_0 and v, is equal to the area of the original figure (rectangle *OABC* plus triangle *CBD*) representing the area covered in uniformly accelerated motion with speed increasing constantly from v_0 to v.

The mathematical reasoning presented above is about as difficult as any we shall need in this book, aside from the solution of a few simple algebraic equations. In fact, the first step of Oresme's proof hints at the fundamental theorem of the calculus of Newton and Leibniz: The area under the speed-time curve is equal to the distance traveled. Since the velocity itself is defined as the rate of change of distance with time, we could generalize this theorem to obtain a general relation between any quantity (such as distance) that varies with changes in another quantity (such as time), and the rate of change of the first quantity with respect to the second; this general relation would involve finding the area under the curve of "rate of change" vs. the second quantity, and may be valid even when the change of the rate of change (in this case the acceleration) is not uniform.

Although the calculus based on this theorem is an extremely powerful tool, used throughout all of theoretical physics since the seventeenth century, we shall not find it necessary for our basic explanations of physical concepts and theories.

6.7 Equations of Motion for Constant Acceleration

The reason for our present preoccupation with uniformly accelerated motion is not only its occasional appearance in nature but also, and primarily, the *ease* with which we can formulate some simple equations of motions, i.e., equations that relate the five important variables:

distance covered (s)
time taken (t)
initial velocity (v_0)
final velocity (v)
acceleration (a).

Equation (6.1) is a good example. We may rewrite it as

$$v = v_0 + at \qquad (6.2)$$

or, in words, the velocity at the end of a time interval t is the sum of the initial velocity and the change in velocity, which is the constant acceleration times the time elapsed.

There are other similar equations relating the five variables. Let us now construct three of them. We can begin with the property of speed-vs.-time graphs mentioned in the previous section: The area under the curve is numerically equal to the distance (s) covered. Thus, in Fig. 6.7, the shaded area under the curve is that of a rectangle of sides v_0 and t plus a triangle of base t and height $(v - v_0)$:

$$s = v_0t + (v - v_0)t. \qquad (6.3)$$

We could interpret this as the distance v_0t covered during time t if the motion had proceeded with constant speed v_0, plus the distance $\frac{1}{2}(v - v_0)t$ added by virtue of the acceleration of the motion.

If we did not originally know the final velocity v, but were given the initial speed v_0 and the acceleration a, we could use Eq. (6.1) to eliminate v from the above equation. The quickest way to do this is to notice that the expression $(v - v_0)$ would be the same as the right-hand side of Eq. (6.1) if it were divided by t. So let us multiply and divide the second term on the right side of Eq. (6.3) by t in order to put it in this form:

$$s = v_0t + \frac{1}{2}\left[\frac{v - v_0}{t}\right]t^2 .$$

We can now use Eq. (6.1) to replace the quantity in brackets [ ] by a:

$$s = v_0t + \tfrac{1}{2}at^2. \qquad (6.4)$$

Equation (6.4) is a very useful equation. It lets us compute the distance covered in time t, starting from initial speed v_0, whenever the motion is uniformly accelerated with acceleration a. It is often used in the special case in which the initial speed is zero; then:

$$s = \tfrac{1}{2}at^2 \text{ (when } v_0 = 0). \qquad (6.5)$$

On the other hand, sometimes a somewhat more general form of (6.4) is needed to cover situations in which the distance variable s does not begin at a zero mark when the time variable t is zero. This is best illustrated by an example.

EXAMPLE 4

An automobile was 10 km west of Boston, going west on a straight east-west turnpike at a speed of 50 km/hr, when a uniform acceleration of 1 km/hr^2 was applied for an hour. How far from Boston was the car at the end of that hour?

Translation: We must define the initial distance $s_0 = 10$ km; we are given $v_0 = 50$ km/hr, $a = 1$ km/hr^2, $t = 1$ hr. If we use the letter s to represent total distance from Boston, then the distance covered in this time interval t will be $s - s_0$; it is this quantity rather than s that must be put on the left-hand side of Eq. (6.4):

$$s - s_0 = v_0 t + \tfrac{1}{2}at^2;$$

hence we have

$$s = s_0 + v_0 t + \tfrac{1}{2}at^2. \quad (6.6)$$

Substituting our numerical values, we find

$$\begin{aligned} s &= 10 \text{ km} + (50 \text{ km/hr}) \times (1 \text{ hr}) + \tfrac{1}{2}(1 \text{ km/hr}^2) \times (1 \text{ hr})^2 \\ &= (10 + 50 + \tfrac{1}{2}) \text{ km} = 60.5 \text{ km}. \end{aligned}$$

Equation (6.6) is the most general equation of this type we shall need, as long as we restrict ourselves to cases with uniform acceleration.

So far we have been assuming that the motion proceeds for a given time interval t, and we have derived equations for the distance and speed. But we might also want to deal with problems in which the distance is given and we are required to find the time or speed. In the simplest such case, if the initial velocity and initial distance are both known to be zero, we can use Eq. (6.5) and simply solve for t in terms of s:

$$t = \sqrt{(2s/a)}.$$

In more general cases we might have to solve a quadratic equation such as Eq. (6.6), with given values of s, s_0, and v_0.

A problem that will turn out to have considerable importance in physics is: Given the initial speed, the distance covered, and the acceleration, find the final speed. Now that you have seen the general procedure, you should try to derive an expression to find the answer for this yourself.

Problem 6.3. Prove that if $s_0 = 0$,

$$v^2 = v_0^2 + 2as. \quad (6.7)$$

[*Hint:* According to the mean-speed theorem, $s = \frac{1}{2}(v_0 + v)t$. Combine this with Eq. (6.1).]

Now let us collect together the main results of this section, four equations relating the six variables s_0, s, t, v_0, v, and a, starting with the mean-speed theorem:

$$s = \tfrac{1}{2}(v_0 + v)t \quad \text{(I)}$$

$$s = s_0 + v_0 t + \tfrac{1}{2}at^2 \quad \text{(II)}$$

$$v = v_0 + at \quad \text{(III)}$$

$$v^2 = v_0^2 + 2as \quad \text{(IV)}$$

(Unless otherwise stated, it can be assumed that $s_0 = 0$.)

Now it is possible to step back, so to speak, and to study the overall pattern of our work so far. We are now convinced that the Eqs. (I) through (IV) contain all the important and interesting relations between distances traveled, time needed, and so forth, for those motions that specifically are uniformly accelerated. Whenever we encounter this type of motion, be it executed by a falling stone or whatever, we shall at once know how to predict any two of the five variables when given the remaining three. For example, if it is known that a stone dropped from the Leaning Tower of Pisa strikes the ground 60 m below with a measured speed of 36 m/sec, we may at once calculate its time of fall and its acceleration. Translating the data to $v_0 = 0$, $v =$ 36 m/sec, $s = 60$ m, $t = ?$, $a = ?$, we scan our four trusted equations and note that t can be found from Eq. (I) [$t = (2s)/(v_0 + v) = 3.33$ sec] and the acceleration can be found either from (IV) [$a = (v^2 - v_0^2)/2s$] or, now that t is known, from (II) or (III).[4]

The fact is that Eqs. (I) through (IV) become our *tools* for dealing with this type of problem, and now that we have derived and established them, we shall simply remember and apply them as occasion arises, in much the same manner as a craftsman, once having carefully selected a proper tool kit for his work, ceases to reinvent or justify his tools at every turn, but is content

[4]The assumption that a is indeed constant for free fall of real bodies at Pisa must be checked first if reason for doubt exists. If the motion should in fact turn out not to be uniformly accelerated, we should investigate how good an approximation our idealizations make toward the true case of free fall. This will be discussed in the next chapter.

to reach into his bag for the proper instrument demanded by the situation at hand.

Before our tool kit begins to fill up, we must take the important step of gaining complete mastery over and confidence in each new concept or law as it is introduced. We can then apply it when called upon without feeling a need to re-examine it in detail. For example, when later we are to find v from v_0, a, and s for a uniformly accelerating body, we can use equation IV (that is, $v^2 = v_0^2 + 2as$, therefore $v = \sqrt{(v_0^2 + 2as)}$ even though undoubtedly one can have only little intuitive feeling about this relation. If you have initially derived Eq. (IV) to your satisfaction, you must then believe in it.

This is, of course, not at all the same thing as the self-defeating practice of just learning equations by heart or of looking them up in the notes to apply them when called upon. Such a ruse cannot work, because one cannot apply an equation properly without knowing all its intimate strengths and weaknesses—the implied "text."

EXAMPLE 5

A baseball is thrown straight up with an initial speed of 19.6 m/sec. Assuming that it has a constant acceleration downward (due to gravity) of 9.8 m/sec^2, and that air resistance can be ignored, how many seconds does it take to get to a height at which it stops going up and starts coming down? What is that height?

Translation: We define $s = 0$, $t = 0$ at the moment the ball was thrown, and use positive numbers to indicate upward motion, so v_0 = +19.6 m/sec. Moreover, the acceleration *downward* should be a *negative* number. We want to know the value of t when $v = 0$.

According to Eq. (III), setting $v = 0$ we have:

$$0 = v_0 + at, \text{ or } at = -v_0 = -19.6 \text{ m/sec}$$

$$t = -v_0/a = -(19.6 \text{ m/sec})/(-9.8 \text{ m/sec}^2)$$
$$= +2 \text{ sec.}$$

We now use Eq. (II) to calculate the height (s) at this instant of time:

$$s = s_0 + v_0t + \tfrac{1}{2}at^2 = 0 + (19.6 \text{ m/sec})(2 \text{ sec})$$
$$+ \tfrac{1}{2}(-9.8 \text{ m/sec}^2)(2 \text{ sec})^2$$
$$= 39.2 \text{ m} - 4.9 \times 4 \text{ m} = 19.6 \text{ m.}$$

The baseball attains a height of 19.6 m, after 2 sec, then falls down. (How long does it take to fall back to the same height from which it was thrown?)

This example illustrates an important fact: Just as in Example 2, the speed can be *positive* while the acceleration is *negative* (as the ball is going up); the speed can also be *negative* while the acceleration is *negative* (as the ball is falling down); but, even more remarkable, an object can be *accelerated* at an instant when its speed is *zero* (at $t = 2$).

To give a final example on this belabored but important point, let us imagine that we are asked to find the time of fall of a brick from the top of the Empire State building. Given are v_0, a, and s. The temptation to "plug in" the readily memorized Eq. (II) must here yield to the knowledge that springs from closer study that the four equations undoubtedly do not apply in this case; for such large a distance of fall, the air friction will prevent constancy of acceleration. As everyone knows by observation of falling leaves, snow, rain, etc., each body sooner or later may reach a constant terminal velocity of fall. This velocity, beyond which there is no further acceleration at all, depends on the surface area of the body, its weight, and the properties of the medium (air, etc.) through which it falls.

This completes our formal study of simple motion, the basis for our coming discussion of the laws of projectile motion.

Additional Problems

Note: As a rule, do not spend more than about half an hour at the most on any problem at one session. If you are unable to solve it in that time, plan to come back to it later, or else finish it by stating what you foresee is likely to be the subsequent procedure and final solution. Note also that not all problems have simple numerical answers, even if the statement of the problem is numerical.

Problem 6.4. (a) Find by graphical interpolation the approximate position of the car described in Fig. 6.2 corresponding to a time of travel of 5 sec, 0.5 sec, 7.005 sec. Extrapolate to find the position corresponding to 10 sec, 50 sec, –2 sec (do this work on graph paper or computer screen with corresponding lines, and providing a printout). (b) Discuss the limitations of the process of interpolation and of extrapolation.

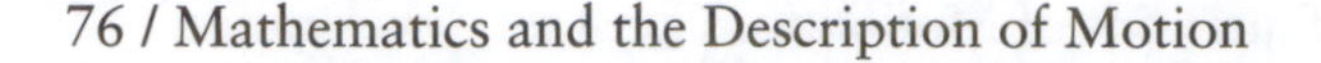

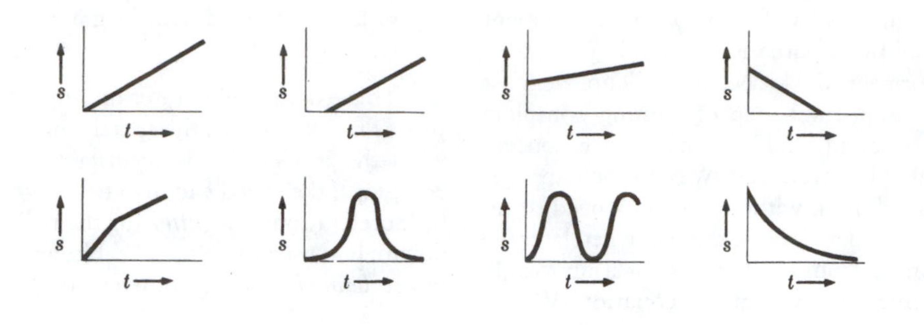

Fig. 6.8.

Problem 6.5. On graph paper or computer screen, plot a graph corresponding to Fig. 6.2, but use data taken at stations A, B, C, D, and E when a particle with a constant velocity of 50 mi/hr (= 22.2 m/sec) was passing by. You must first construct a table corresponding to Table 6.1. (Keep your eyes open; the numerous calculations can be simplified.)

Problem 6.6. Describe the possible experimental situations that gave rise to each of the eight graphs of distance vs. time (Fig. 6.8) for motion along one line. Then try to express each of four or five of these graphs in terms of an equation relating *s* and *t* in algebraic fashion.

Problem 6.7. A leisurely vacation trip from Washington, D.C., to San Francisco, California took 15 days, including several long stops en route. Calculate the average speed for this trip in miles per hour. [*Note:* Whenever the data given seem insufficient, you will find the necessary additional information in pertinent books in your library. If you cannot find such books, use an intelligent guess.]

Problem 6.8. Make some reasonable assumption about the speed that a car can pick up during the first 10 sec after starting from rest, then calculate the acceleration. [*Note:* In all future problems, assume that the acceleration of freely falling bodies at the surface of the earth is 9.80 m/sec^2. This happens to be true, within fairly narrow limits, over the whole earth. The symbol employed for this "constant" (the acceleration due to gravity) is the letter *g* instead of *a*. It is italicized, like other letters representing physical quantities, and should not be confused with "g" for gram.

Problem 6.9. Five observation posts are spaced 20 m apart along a straight vertical line. A freely falling body is released at the topmost station and is timed as it passes each of the others. Compute a table of the data expected, and represent this information in a graph (use graph paper).

Problem 6.10. Make a table of the value of *s*, *v*, and $\overline{v}$ for a freely falling body if $v_0 = 0$ and *t* is, in turn, 1 sec, 2 sec, 3 sec, 5 sec, and 10 sec. Represent this information in three graphs.

Problem 6.11. A boy violently throws a stone vertically downward from a building 50 m high, imparting to it a substantial initial speed. If the stone takes 1.8 sec to hit the pavement below, what was its initial speed?

Problem 6.12. Make a list of all the relevant limitations, conditions, and idealizations that were implied in your solution of Problem 6.11.

Problem 6.13. The nucleus of a helium atom is to travel along the inside of a straight hollow tube 2 m long, which forms a section of a particle accelerator. If the nucleus is to enter at 1000 m/sec and leave at the other end at 10,000 m/sec, how long is it in the tube? State your assumption.

RECOMMENDED FOR FURTHER READING

Marie Boas, *The Scientific Renaissance 1450–1630,* Chapter VII

I. B. Cohen, *Birth of a New Physics,* Chapter 5

C. C. Gillispie, *Dictionary of Scientific Biography,* articles by Theodore M. Brown on Descartes, Vol. 4, pages 51–65, and by Marshall Clagett on Oresme, Vol. 10, pages 223–230

Geneviève Rodis-Lewis, *Descartes: His Life and Thought,* translated by Jane Marie Todd, Ithaca: Cornell University Press, 1998

S. Sambursky, *Physical Thought,* excerpts from Descartes, pages 238–251

CHAPTER 7

Galileo and the Kinematics of Free Fall

7.1 Introduction

The material in the preceding chapter on velocity and acceleration for simple types of motion, which is the logical and historical groundwork of physics, is to a large extent the work of Galileo Galilei. It is a small but critical part of the voluminous work of this remarkable person, who is often called the father of modern science. Now we shall examine in some detail Galileo's contribution to the problem of freely falling bodies as a historical, scientific, and methodological achievement.

Galileo is important and interesting to us for several reasons, quite apart from his dramatic role in the history of astronomy described in Chapter 5. First, if we understand what he had to struggle against, what contemporary attitudes he had to discard in founding the new "Mechanical Philosophy," then perhaps we shall be clearer about the nature of our present-day structure of science and our own attitudes and preconceptions. To cite an analogy, we certainly must not fail to study the period of the American Revolution if we wish to understand the present democratic institutions that grew out of it in the United States.

Second, after the historical setting of Galileo's work has been clarified, he will speak to us as a scientist at his best, as an investigator whose virtuosity in argument and discovery illustrates clearly the profound excitement of scientific research. He will be for us an example of a physicist at work, and although in his case circumstances were extraordinary, that fact does not make him unrepresentative: If he had been born in the twentieth century, we would find it not at all hard to imagine him making major new discoveries and playing a leading part in many branches of modern science.

Third, Galileo's approach to the problem of motion of freely falling bodies will provide us, both in this chapter and later, with an opportunity for a short introspective discussion on procedure in science. We need this. Within recent years, many people have displayed misunderstanding and even hostility toward science and technology, while at the same time workers in the life sciences and social sciences have debated vigorously whether the methods of the physical sciences are applicable to their own fields. Since the physical sciences have in fact been spectacularly successful in discovering the laws of nature (whether or not one approves of all the use that society has made of those discoveries), a clear understanding of how the physicist works must be central to the larger question of the nature of science. Without an awareness of the place of definition, hypothesis, experimentation, and intuition in physics, without a knowledge of the vital processes by which this field functions and grows, a science must seem a dry formalism, to be memorized without understanding, reproduced on examinations without enthusiasm, and perhaps applied to practical problems without due consideration of its limitations.

To begin with, let us consider what ideas about motion Galileo might have learned at his university and what type of physics he eventually had to reject. We will be discussing what is now called *kinematics* (from the Greek *kinema,* motion), the branch of physics that deals with the description of motion, excluding the study of the specific forces responsible for the motion (the latter topic is called *dynamics*). It was, of course, the science of the scholastics, the heirs and often misinterpreters of that greatest of scientist-philosophers of antiquity, Aristotle.

But at this point, a warning: The history of science is a sophisticated field in its own right. By the very nature of its subject matter, there is a good deal of inconclusiveness and uncertainty at important points, and we shall in this book neither attempt nor presume to evaluate such delicate questions as, for example, to what extent Galileo shared many opinions and methods with the Aristotelians or drew on the work of other investigators—Archimedes, the whole tradition of scholars at the venerable University of Paris, Tartaglia, Benedetti, or Stevinus. For science, like any other intellectual enterprise, cannot start from scratch without any roots in some tradition. Galileo, too, had the often unacknowledged encouragement of many enlightened bits of contemporary physics in his struggle to shape a consistent whole. But his fight, judging by his own words, does appear to have been against a group of exponents of the stricter contemporary scholastic view. In short, the statement that Galileo started modern science is so abrupt and leaves out so much that it can justly be called wrong, but no more wrong than the statement that the American Republic dates from June 1788, when the Constitution went into effect.

7.2 Aristotelian Physics

Despite the widespread story of Galileo's experiment from the Leaning Tower of Pisa, most people believe (erroneously) that heavy bodies gain speed far more quickly when falling than do lighter ones. This was the view recorded by Aristotle (Section 1.3), much of whose works, through the mediation of St. Thomas Aquinas and his followers, became the basis of Christian thinking in Europe from the thirteenth century on. However, Aristotle's view on the motion of falling bodies was not simply an error or the result of miscarried experimentation, but a characteristic consequence of his acceptance of a great overall scheme; indeed it was a rather minor member of an imposing structure that he erected as a result of his search for a unity of all thought and experience. His system embraced at the same time elements that are now separated into component parts: scientific, poetic, theological, and ethical.

According to Aristotle's system, described in Section 1.3, the natural motion of an object must depend on how much it contains of each of the elements; objects consisting mostly of Water and Earth must seek their natural place at the center of the universe, which is also the center of the earth. Thus a big stone would be expected to fall faster than a small one, and this expectation seems to be borne out by common experience, at least if we compare extremes (a boulder and a leaf).

A modern scientist would immediately object that Aristotle's theory must fail any quantitative test. In his book *On the Heavens*, Aristotle states that the speed of fall is proportional to the weight of the object:

> A given weight moves a given distance in a given time; a weight which is heavier moves the same distance in less time, the time being inversely proportional to the weights. For instance, if one weight is twice another, it will take half as long over a given distance.

Note that Aristotle does not write out equations—nor did most physicists until after the time of Galileo.

Casual or careful observation (which Aristotle did indeed use well) does prepare us to note differences in speed of fall. But one would think that even with a crude method for actually measuring times of fall, anyone could discover that a stone twice as heavy as another one does not fall twice as fast, and certainly that a stone ten times as heavy as another one does not fall ten times as fast. But that is the wisdom of hindsight. It was then not natural to think in terms of such actual tests on earth.

Aristotle took into account not only the weight of the falling object but also the resistance offered by the medium through which it falls. Thus a stone falls faster in air than in water. It was perhaps natural to suggest, therefore, that the speed is inversely proportional to the resisting force. One could think of the weight of the object and the resistance of the medium as two opposing forces: There is motion only if the weight exceeds the resistance (a stone cannot fall through solid earth), and the greater the weight in proportion to the resistance, the greater the speed. (Aristotle apparently thought of v as remaining constant once the motion begins, unless there is some change in the resistance or other forces acting on the object.)

Stated verbally, this sounds like a plausible rule, although it too may not stand up to experimental test. But as soon as we try to express it in mathematical terms, unexpected consequences emerge. If speed v is directly proportional to weight W and inversely proportional to resistance R, we can write

$$v \propto W/R \quad \text{or} \quad v = k(W/R),$$

where k is an unspecified proportionality constant. We also have the proviso that

$$v = 0 \quad \text{unless} \quad W > R$$

(no motion unless the weight exceeds the resisting force).

It was not until the fourteenth century that Thomas Bradwardine, at Merton College, Oxford, realized that these two equations do not go together harmoniously. He considered the following imaginary (or thought) experiment: Start with a given weight and a very small resisting force, and gradually let the resistance R increase until it finally exceeds the weight W. From the qualitative description of motion according to Aristotle, one would expect that the speed v would gradually decrease to zero in this experiment. But the equations tell us that v gradually decreases to a value slightly greater than k, as R approaches W, but then v suddenly drops to zero when R is equal to W. This discontinuous change in speed is not at all consistent with the basic viewpoint of Aristotle's physics, and, therefore, according to Bradwardine, the relation $v \propto W/R$ must be rejected.

It is actually possible to find a mathematical function of W and R that has the desired property—decreasing continuously to zero as R, initially less than W, approaches it.[1] But that was not a fruitful approach, and so we shall not pursue it. Perhaps it is enough to point out the useful role of mathematics in the development of physical theories: Forcing physicists to make their ideas precise sometimes exposes their inconsistencies and leads to the rejection of hypotheses without the need for any experimental test. Yet in the search for valid hypotheses, mathematics cannot be a substitute for *all* experiments; a growing theory must somehow be fertilized by contact with the real world.

Another consequence of Aristotle's relation, assuming we can express it by the equation $v \propto W/R$, is that motion in a vacuum, with no resistance at all ($R = 0$), would have to take place with infinite speed, no matter what the value of the weight W. This absurdity was cited by Aristotle and his followers as a reason why a vacuum itself is inconceivable! It was unthinkable to them that the paradox might be attributed to some inadequacy in the original postulate.

By the end of the sixteenth century, Aristotle's theory of motion had been strongly criticized by a number of eminent scientists, and it was perfectly clear that it could not hope to provide quantitative agreement with the results of the simplest experiments. Yet the theory was still firmly entrenched in the learned world. According to modern tenets of scientific method, this should have been a scandal: Scientists are not supposed to adhere dogmatically to a theory that does not measure up in experimental tests. But before we rush to condemn our ancestors (for behavior that is in truth not so much different from that observed in many fields even in our own time), let us inquire into the reasons for the continuing attractiveness of the Aristotelian philosophy.

According to Aristotle, a theory that merely describes and predicts the facts of observation, no matter how accurately, is not good enough: It must also give them meaning in a wider sense, by showing that the facts are in accord with the general postulates of the whole philosophical system. To the Aristotelian of the late sixteenth century, the postulates concerning elements and tendencies (like the postulates about circular motions in astronomy), and all conclusions from them, were true ("philosophically true"), clear, and certain, because they were also part of a larger, important, and satisfying scheme, reaching into theology and other fields. The postulates could not be given up just because the detailed motion of two weights dropped from a tower contradicted one of their many conclusions. He would give "philosophic truth" precedence over contradictory "scientific truth"—although the latter did not yet exist as a distinct concept. His physics was the investigation of the nature of things in a context so large that the preoccupation of later science—for example, with detailed measurement of falling bodies—would seem to him like a small, artificial, incomplete subject with little significance by itself.

As the historian F. S. Taylor put it:

> The science of Aristotle and of the scholastics had been primarily a science of purpose. It was transparently clear to the latter that the world and all that's in it was created for the service of man, and that man had been created for the service of God. That was a perfectly intelligible scheme of the world. The sun was there to give us light and to tell the time and mark out the calendar by his motions. The stars and planets were a means of distributing beneficent or maleficent influences to the things on earth to which they were sympathetically linked; plants and animals were there to give

---

[1]Bradwardine himself suggested a logarithmic function: $v = \log W/R$ (recall that $\log 1 = 0$). For an account of Bradwardine's life and work, see the article on him by J. E. Murdoch in the *Dictionary of Scientific Biography*. Another possibility was suggested by John Philoponus of Alexandria in the sixth century: $v = W - R$.

> us food and pleasure, and we were there to please God by doing His will. Modern science does not produce any evidence that all this is not perfectly true, but it does not regard it as a scientific explanation; it is not what science wants to know about the world.
>
> (*A Short History of Science*)

Conversely, in general (though there are notable exceptions), students of nature before Galileo's time would necessarily have found it unreasonable or even inconceivable to be seriously interested only in the type of questions our science asks. That such questions are indeed interesting or important, or that the methods employed in our scientific investigations can give results at all, could not have been apparent before the results themselves began to pour forth in the seventeenth century. But so overwhelming was the unsuspected, ever-growing stream that we, who are caught in the flood, can hardly think in prescientific terms. We must deplore this unresourceful imagination of ours. If each of us even for a moment could transcend his or her own pattern to appreciate clearly the point of view from which an Aristotelian would have proceeded, we should be better equipped for dealing with the everyday conflicts of differing temperaments and opinions.

It was not easy for people to give up the Aristotelian view of the world, in which everything had its own purpose and place, carefully planned in relation to the whole. To many devout and sensitive intellects of the early seventeenth century, it must have been extremely disturbing to learn of new discoveries of stars randomly distributed through the sky with no evident relation to the earth, which itself no longer had a privileged position, and to read of theories that, like the old atheistic doctrines of Lucretius and Epicurus, seemed to describe the world as a meaningless, fortuitous concourse of atoms moving through empty space. The poet John Donne lamented the changing times in some famous lines, written in 1611:

> And new Philosophy calls all in doubt,
> The Element of fire is quite put out;
> The Sun is lost, and th'earth, and no man's wit
> Can well direct him where to looke for it.
> And freely men confesse that this world's spent,
> When in the Planets, and the Firmament
> They seeke so many new; then see that this
> Is crumbled out again to his Atomies.
> 'Tis all in peeces, all coherence gone;
> All just supply, and all Relation . . .

7.3 Galileo's *Two New Sciences*

Galileo's early writings on mechanics were in the tradition of the standard medieval critics of Aristotle, and, as we mentioned in Section 5.1, at the time when the legendary Tower of Pisa experiment might have been performed, he had not yet developed the general theory for which he is now best known. During his mature years, his chief interest was in astronomy, and it was in connection with the theoretical effects of the earth's rotation that he proposed his principle of inertia (Section 5.3). When his *Dialogues Concerning the Two Chief World Systems* (1632) was condemned by the Roman Catholic Inquisition and he was forbidden to teach the "new" astronomy, Galileo decided to concentrate on mechanics. This work led to his book *Discourses and Mathematical Demonstrations Concerning Two New Sciences Pertaining to Mechanics and Local Motion,* usually referred to as the *Two New Sciences.* It was written while he was technically a prisoner of the Inquisition, and was published surreptitiously in Holland in 1638, for in his own Italy his books were then being suppressed. The Church authorities could not have known that this book would prove to be as dangerous as the *Dialogues:* It was to provide the foundation for Newton's work, which drove the last nail into the coffin of Aristotelian philosophy.

Galileo was old, sick, and nearly blind at the time he wrote *Two New Sciences,* yet his style in it is sprightly and delightful. As in his early work on astronomy, he uses the dialogue form to allow a lively conversation among three speakers: Simplicio, who represents the Aristotelian view; Salviati, who presents the new views of Galileo; and Sagredo, the uncommitted man of good will and open mind, eager to learn. But something else has been added: Galileo, in his own voice, as the author of a technical Latin treatise being discussed by the three speakers.

Let us listen to Galileo's characters as they discuss the problem of free fall:[2]

---

[2]We have largely used Stillman Drake's translation, occasionally adding italics for emphasis.

Salviati: I seriously doubt that Aristotle ever tested whether it is true that two stones, one ten times as heavy as the other, both released at the same instant to fall from a height of, say, one hundred braccia, differed so much in their speed that upon the arrival of the larger stone upon the ground, the other would be found to have descended no more than ten braccia. *(Note that one braccio is about 23 inches; 100 braccia happens to be nearly the height of the Tower of Pisa. The Italian word braccio comes from the word for "arm.")*

Simplicio: But it is seen from his words that he appears to have tested this, for he says, "We see the heavier. . . ." Now this "We see" suggests that he had made the experiment.

Sagredo: But I, Simplicio, who have made the test, assure you that a cannonball that weighs one hundred pounds (or two hundred, or even more) does not anticipate by even one span the arrival on the ground of a musket ball of no more than half [an ounce], both coming from a height of two hundred braccia.

In the above passage Sagredo uses the unit of length: 1 span = 9 inches (distance from tip of thumb to tip of little finger when extended).

At this point one might well have expected to find a detailed report on an experiment done by Galileo or one of his colleagues. Instead, Galileo presents us with a thought experiment—an analysis of what would happen in an imaginary experiment. He also ironically uses Aristotle's own method of logical reasoning to attack Aristotle's theory of motion (just as he had done in his proof of the law of inertia, discussed in Section 5.3):

Salviati: But without other experiences, by a short and conclusive demonstration, we can prove clearly that it is not true that a heavier moveable [body] is moved more swiftly than another, less heavy, these being of the same material, and in a word, those of which Aristotle speaks. Tell me, Simplicio, whether you assume that for every heavy falling body there is a speed determined by nature such that this cannot be increased or diminished except by using force or opposing some impediment to it.

Simplicio: There can be no doubt that a given moveable in a given medium has an established speed determined by nature, which cannot be increased except by conferring on it some new impetus, nor diminished except by some impediment that retards it.

Salviati: Then if we had two moveables whose natural speeds were unequal, it is evident that were we to connect the slower to the faster, the latter would be partly retarded by the slower, and this would be partly speeded up by the faster. Do you not agree with me in this opinion?

Simplicio: It seems to me that this would undoubtedly follow.

Salviati: But if this is so, and if it is also true that a large stone is moved with eight degrees of speed, for example, and a smaller one with four [degrees], then joining both together, their composite will be moved with a speed less than eight degrees. But the two stones joined together make a larger stone than the first one which was moved with eight degrees of speed; therefore this greater stone is moved less swiftly than the lesser one. But this is contrary to your assumption. So you see how, from the supposition that the heavier body is moved *more* swiftly than the less heavy, I conclude that the heavier moves *less* swiftly.

Simplicio: I find myself in a tangle . . . that indeed is beyond my comprehension.

As Simplicio/Aristotle retreats in confusion, Salviati/Galileo presses on with the argument, showing that it is self-contradictory to assume that an object would fall faster if its weight were increased by a small amount. Simplicio cannot refute Salviati's logic. But, on the other hand, both Aristotle's book and his own casual observations tell him that a heavy object does, at least to some extent, fall faster than a light object:

Simplicio: Truly, your reasoning goes along very smoothly, yet I find it hard to believe that a birdshot must move as swiftly as a cannon ball.

Salviati: You should say "a grain of sand as [fast as] a millstone." But I don't want you, Simplicio, to do what many others do, and divert the argument from its principal purpose, attacking something I said that departs by a hair from the truth, and then trying to hide under this hair another's fault that is as big as a ship's hawser. Aristotle says, "A hundred-pound iron ball falling from the height of a hundred braccia hits the ground before one of just one pound has descended a single braccio." I say that they arrive at the same time. You find, on making the experiment, that the larger anticipates the smaller by two inches—And now you want to hide, behind these two inches, the ninety-nine braccia of Aristotle, and speaking only of my tiny error, remain silent about his enormous one.

This is as clear a statement as we would wish of the idea that a naive first glance at natural events is not at all a sufficient basis for a physical theory; different freely falling bodies in air indeed do *not* arrive at the very same instant at the foot of the tower; but this, on further thought, turns out to be far less significant than the fact that they do *almost* arrive at the same instant. In a vacuum the fall would be equal for both objects. Attention to this last point, a point that is actually far removed from the immediate sense impressions gained in this experiment, is suggestive and fruitful, because it regards the failure of equal arrivals in fall through air as a minor discrepancy instead of a major truth, as an experimental circumstance explainable by air friction. And indeed, when the air pump was invented soon after the time of Galileo, this hypothesis was confirmed by the observation of the truly simultaneous fall of a light feather and a heavy coin inside an evacuated glass tube.

But Galileo's invocation of a thought experiment rather than a real one in *Two New Sciences* created the impression that he had never done the Leaning Tower experiment at all, and was relying entirely on logical reasoning for his belief that objects of different weight fall together. The distinguished French philosopher-historian Alexandre Koyré, surveying Galileo's published works, suggested that he was really a Platonist rather than an experimental physicist and that the results of several of his thought experiments did not agree with the actual behavior of objects in the physical world.

The American historian of science Stillman Drake, on the basis of close scrutiny of Galileo's unpublished manuscripts, concluded that Galileo actually did do experiments even though he didn't publish his numerical results. Other scholars such as Thomas Settle repeated the experiments themselves with apparatus similar to that described by Galileo, and found that one could obtain similar results—contrary to Koyré, who apparently didn't try to replicate the experiments himself.

For example, Galileo's manuscript notes state that when one holds a light ball in one hand and a heavy ball in the other and tries to drop them simultaneously from a height, sometimes "the lighter body will, at the beginning of the motion, move ahead of the heavier" although the heavy ball overtakes it, so that they usually land nearly at the same time. When Settle repeated this experiment, photographing a person who tries to release the two balls simultaneously, he found that the hand holding the heavy ball nevertheless opens slightly later, apparently because of differential muscular fatigue. Historian I. B. Cohen concludes: "The discovery that in this case, as in others, the results of experiment accord with Galileo's reports gives us confidence in Galileo as a gifted experimenter, who recorded and reported exactly what he observed" (*Birth of a New Physics*). At the same time Galileo did not allow puzzling small deviations from the expected result to obscure his more important conclusion: The fall of heavy and light bodies is basically similar, and shows nothing like the large variation claimed by the Aristotelians.

As Galileo reveals his attitude toward experiments, we are reminded of the old quip that "science has grown almost more by what it has learned to ignore than by what it has had to take into account." Here again we meet a recurring theme in science: Regard observable events with a penetrating eye to search behind the immediate confusion of appearances for an underlying simplicity and mathematical lawfulness. In this particular case, it was easier first to sweep away a contrary hypothesis (Aristotle's postulate that speed of fall depends on weight) by reasoning in a thought experiment than to discuss extensive experiments that could never be completely convincing. To arrive at the correct result, it would have been futile to rely on the observational methods available at the time; instead, everything depended on being able to "think away" the air and its effect on free fall.

This is easy enough for us, who know of pumps and the properties of gases, but it was at that time conceptually almost impossible for several reasons—some evident, but some quite subtle. For example, Aristotle, who had examined the question of the existence of a vacuum, had denied its possibility as contrary not only to experience but to reason. One of his arguments was: Where there is a vacuum, there is *nothing*. If nothing remains, he continued, then also space itself and the physical laws can no longer exist in this vacuum, for how could a stone, for example, know which way is up or down except in relation to and through the influence of a medium?

The answer depended on dissociating the words *vacuum* and *nothingness,* and so preventing space from becoming annihilated merely because air has been withdrawn. But although Galileo found it possible to do this in thought, Descartes and his followers continued to reject the possible existence of a vacuum, and we may believe that other scientists eventually accepted it mainly on the basis of subsequent *experimental demonstrations* that bodies could fall, and that heat, light, magnetism, etc., could be propagated through vessels from which air had been withdrawn.

7.4 Galileo's Study of Accelerated Motion

Others had known before Galileo that the Aristotelians were wrong about free fall, but it is to his credit that he proceeded to discover the details of the correct description of this motion and to make it part of a more general system of mechanics. The task he set himself in the *Two New Sciences* was to invent concepts, methods of calculation and measurement, and so forth, to arrive at a description of the motion of objects in rigorous mathematical form.

In the following excerpts we must not lose sight of the main plan. First, Galileo discusses the mathematics of a *possible,* simple type of motion, defined as uniformly accelerated. Then he proposes the hypothesis that the motion of free fall occurring in nature is actually uniformly accelerated. However, it is not possible to test this hypothesis directly by experiment with the instruments available in Galileo's time. He therefore argues that the hypothesis will be confirmed if it is successful in describing another kind of motion, one closely related to free fall: the rolling of a ball down an inclined plane. (In modern terms, the same force of gravity will act on the ball whether it is dropped vertically downward or allowed to roll down an inclined plane; but in the second case the full pull of gravity cannot act and the actual motion is slowed down, therefore it can be more easily measured.) Finally, he describes an experimental test of quantitative conclusions derived from his theory of accelerated motion, as applied to the particular experimental situation.

Galileo introduces the discussion with these remarks: "*Salviati* (reading from Galileo's Latin treatise): We bring forward a brand new science concerning dealing with a very old subject":

> There is perhaps nothing in nature older than MOTION, about which volumes neither few nor small have been written by philosophers; yet I find many essentials of it that are worth knowing which have not even been remarked. Certain commonplaces have been noted, as, for example, that in natural motion, heavy falling things continually accelerate; but the proportion according to which this acceleration takes place has not yet been set forth . . . It has been observed that missiles or projectiles trace out a line somehow curved, but no one has brought out that this is a parabola. That it is, and other things neither few nor less worthy [than this] of being known, will be demonstrated by me; and (what is in my opinion more worthwhile) there will be opened a gateway and a road to a large and excellent science, of which these labors of ours shall be the elements, [a science] into which minds more piercing than mine shall penetrate to recesses still deeper.

There follows first a very thorough and searching discussion of uniform (unaccelerated) motion, along the lines indicated in our previous chapter. Then begins the section on naturally accelerated motion. Salviati again reads from the treatise:

> Those things that happen which relate to equable motion have been considered in the preceding book; next, accelerated motion is to be treated of. And first, it is appropriate to seek out and clarify the definition that best agrees with that [accelerated motion] which nature employs. Not that there is anything wrong with inventing at our pleasure some kind of motion and theorizing about its consequent properties, in the way that some men have derived spiral and conchoidal lines from certain motions, though nature makes no use of these paths; and by pretending these, men have laudably demonstrated their essentials from assumptions [*ex suppositione*]. But since nature does employ a certain kind of acceleration for descending heavy things, we decided to look into their properties so that we might be sure that the definition of accelerated motion which we are about to adduce agrees with the essence of naturally accelerated motion. And at length, after continual agitation of mind, we

are confident that this has been found, chiefly for the very powerful reason that the essentials successively demonstrated by us correspond to, and are seen to be in agreement with, that which physical experiments show forth to the senses. Further, it is as though we have been led by the hand to the investigation of naturally accelerated motion by consideration of the custom and procedure of nature herself in all her other works, in the performance of which she habitually employs the first, simplest, and easiest means. . . .

Thus when I consider that a stone, falling from rest at some height, successively acquires new increments of speed, why should I not believe that these additions are made by the simplest and most evident rule? For if we look into this attentively, we can discover no simpler addition and increase than that which is added on always in the same way. We easily understand that the closest affinity holds between time and motion, and thus equable and uniform motion is defined through uniformities of times and spaces; and indeed, we call movement equable when in equal times equal spaces are traversed. And by the same equality of parts of time, we can perceive the increase of swiftness to be made simply, conceiving mentally that this motion is uniformly and continually accelerated in the same way whenever, in any equal times, equal additions of swiftness are added on. . . . the definition . . . may be put thus: "*I say that that motion is equably or uniformly accelerated which, abandoning rest, adds on to itself equal momenta of swiftness in equal times.*"

In the preceding passage Galileo has been speaking for himself, as it were, in the role of an "Academician" who has written a treatise on mechanics, which the characters in the dialogue are discussing. He has introduced what is in fact the modern definition of uniformly accelerated motion (and of acceleration itself): Just as velocity was defined as the change in distance divided by change in time,

$$v = \frac{\Delta s}{\Delta t},$$

so will acceleration be defined as change in velocity divided by change in time,

$$a = \frac{\Delta v}{\Delta t},$$

and a uniformly accelerated motion is simply one for which a is constant for any time interval.

But from his own earlier struggles with the concept of acceleration, Galileo knows that it is far from obvious that this is the only possible definition (Section 6.5). He therefore inserts at this point a critical discussion, which gives us considerable insight into the problem of constructing a reasonable mathematical description of nature:

Sagredo: Just as it would be unreasonable for me to oppose this, or any other definition whatever assigned by any author, all [definitions] being arbitrary, so I may, without offence, doubt whether this definition, conceived and assumed in the abstract, is adapted to, suitable for, and verified in the kind of accelerated motion that heavy bodies in fact employ in falling naturally.

The discussion now turns toward the only correct answer to this excellent question, namely, that Galileo's "arbitrary" definition of acceleration happens to be most useful to describe the experimental facts of real, observable motion. But there is first a significant little excursion when Sagredo proposes:

Sagredo: From this reasoning, it seems to me that a very appropriate answer can be deduced for the question agitated among philosophers as to the possible *cause* of acceleration of the natural motion of heavy bodies.

Salviati sternly turns away this persistent preoccupation of the previous two millennia with a simple and modern type of statement, which, roughly speaking, directs us to ask first about the "how" instead of the "why" of motion, not to waste energy on a theory of motion while the *descriptive law* of motion is yet unknown:

Salviati: The present does not seem to be an opportune time to enter into the investigation of the cause of the acceleration of natural motion, concerning which philosophers have produced various opinions, some of them reducing this approach to the center [gravitational attraction], others to the presence of successively fewer parts of the medium [remaining] to be divided, and others to a certain extrusion by the surrounding medium which, in rejoining itself behind the moveable, goes pressing on and thus continually pushing it out. Such fantasies, and others like them, would have to be examined and resolved, with little

> gain. For the present, it suffices our Author [Galileo] that we understand him to want us to investigate and demonstrate some attributes of a motion so accelerated (whatever the cause of its acceleration may be).

Galileo then scrutinizes his decision that for uniformly accelerated motion he should regard the increase in velocity proportional to time elapsed rather than distance covered, a proposition that equally well fulfills his requirement of simplicity of relationships. He admits having at one time entertained the latter alternative, but shows by a set of plausible experiments, which he asks us to imagine (in a set of thought experiments, one of the very fruitful devices in the scientist's kit of methods) that such a hypothetical motion does not correspond to actual free fall. In modern terms, a motion for which $\Delta v \propto \Delta s$ may arbitrarily be *called* uniformly accelerated, but such a definition is useless in describing motion in the physical world; for example, as we noted in the previous chapter, it implies that a uniformly accelerated object, if at rest, could never start moving.

Having disposed of this alternative, Galileo might now have produced a set of experimental data to show that in free fall, say for a stone dropped from a high tower, the measured quantity $\Delta v/\Delta t$ does indeed remain constant for various time intervals Δt during the descent. This would have settled everything. But consider the experimental troubles involved in such a direct attack! Today we might use high-speed motion picture equipment to record the successive positions of the falling stone, from which a graph of distance vs. time could be constructed. From this graph one could calculate the necessary data for an instantaneous velocity-vs.-time graph, and we could then check the hypothesis that $\Delta v/\Delta t$ is constant throughout the motion. But in Galileo's time, not even a good clock for timing rapid motions was available.

Experiments on fast free fall being impossible, he characteristically turned to a more easily testable consequence of his hypothesis, "the truth of which will be established when we find that the inferences from it correspond to and agree exactly with experiment."

First he convinced himself analytically that a ball rolling down a smooth incline obeys the same rules that apply to motion in free fall, that it is in short a "diluted" or slowed-down case of that motion, whatever its law. If the ball is found to move with constant acceleration, so must also a freely falling body (though the numerical value of the acceleration will of course be different).

His attention therefore turns finally to a simple experiment. (Thomas Settle replicated it in 1961, using equipment similar to that employed by Galileo; this experiment is often done, with modern apparatus, in introductory physics courses, so you may have an opportunity to judge for yourself whether Galileo could have attained the accuracy he claimed.)

> *Salviati:* . . . In a wooden beam or rafter about twelve braccia long, half a braccio wide, and three inches thick, a channel was rabbeted [cut] in along the narrowest dimension, a little over an inch wide and made very straight; so that this would be clean and smooth, there was glued within it a piece of vellum, as much smoothed and cleaned as possible. In this there was made to descend a very hard bronze ball, well rounded and polished, the beam having been tilted by elevating one end of it above the horizontal plane from one to two braccia at will. As I said, the ball was allowed to descend along the said groove, and we noted (in the manner I shall presently tell you), the time that it consumed in running all the way, repeating the same process many times, in order to be quite sure as to the amount of time, in which we never found a difference of even the tenth part of a pulse-beat.
>
> This operation being precisely established, we made the same ball descend only one-quarter the length of this channel, and the time of its descent being measured, this was found always to be precisely one-half the other. Next making the experiment for other lengths, examining now the time for the whole length [in comparison] with the time of one-half, or with that of two-thirds, or of three-quarters, and finally with any other division, by experiments repeated a full hundred times, the spaces were always found to be to one another as the squares of the times. And this [held] for all inclinations of the plane; that is, of the channel in which the ball was made to descend . . .
>
> As to the measure of time, we had a large pail filled with water and fastened from above, which had a slender tube affixed to its bottom, through which a narrow thread of water ran; this was received in a little beaker during the entire time that the ball descended along the channel or parts of it. The small amounts of

> water collected in this way were weighed from time to time on a delicate balance, the differences and ratios of the weights giving us the differences and ratios of the times, and with such precision that, as I have said, these operations repeated time and again never differed by any notable amount.

Do not fail to notice the further ingenious twist in the argument. Even in these inclined-plane experiments it was not possible to test directly whether $\Delta v/\Delta t$ is constant, because this would have demanded direct measurements of instantaneous velocities, whereas Galileo could measure directly only distances and time intervals. It is for just this reason that Galileo devoted a considerable portion of the *Two New Sciences* to deriving various relations between distance, time, speed, and acceleration, equivalent to the equations we have already presented in Section 6.7. In particular, he was able to use a theorem stating that: *in uniformly accelerated motion starting from rest, the distance covered is proportional to the square of the time of descent.* That theorem is just our (Eq. 6.5),

$$s = at^2 \text{ (when } v_0 = 0)$$

where s is the distance covered and t is the time elapsed. This was precisely the result Galileo claimed to have found in his inclined-plane experiment, and therefore he concluded that the motion is indeed uniformly accelerated.

It is easy to criticize Galileo's account of his experiment, on the grounds that the results may not be applicable to free fall. For example, if the plane (or beam) is tilted more steeply toward the vertical, the ball starts to slip as well as roll; and it is not evident that one can extrapolate the values of the acceleration found for various angles of inclination to the free-fall value for vertical planes. For that matter, Galileo does not even report a numerical value for acceleration, either for the rolling ball or for free fall. (A value roughly equivalent to the modern value of $g = 9.8$ m/sec^2 for free fall was first obtained by Christiaan Huygens.)

Although the complete experimental verification of Galileo's view of the way bodies move would have to come later, the important thing was that it had been clearly stated, bolstered by plausible reasoning and demonstrations, and put into forms that could be quantitatively tested. Let us now recapitulate these results:

1. Any body set into motion on a *horizontal* frictionless plane will continue to move indefinitely at the same speed (law of inertia
2. In free fall through a vacuum, all objects—of whatever weight, size, or constitution—will fall a given distance in the same time.
3. The motion of an object in free fall, or while rolling down an inclined plane, is uniformly accelerated, i.e., it gains equal increments of speed in equal times.

Note that the third result also applies to objects that started with some initial velocity, a baseball thrown straight up in the air (Example 5 in Section 6.7). In this case the acceleration is in the direction opposite to the initial speed, so the baseball *loses* that speed by equal increments in equal times until its speed is reduced to zero. But even then it continues to be accelerated: Its *downward* speed now starts from zero and gains equal increments in equal times until it hits an obstacle.

These findings are not enough by themselves to constitute a complete science of motion. So far, Galileo and his colleagues have not come to the real secret power of physics—that it finds *laws,* relatively few, and from these one can deduce an infinity of possible special cases. We are heading for them. But in the meantime, some people call the equation for free fall, in its generalized form as Eq. (6.6),

$$s = s_0 + v_0 t + \tfrac{1}{2}at^2$$

"Galileo's Law of Free Fall," with the acceleration a now set equal to g, the acceleration due to gravity. Because of its power to predict an infinite number of cases, it is not too wrong to call it that (though Galileo did not use algebraic formulas). Yet, we shall see it is just one aspect of a truly majestic law, Newton's Law of Gravitation, combined with his Laws of Motion.

Perhaps Galileo had some inkling of this future progress as he concluded the Third Day of his *Two New Sciences:*

> *Sagredo:* . . . These [properties of motion] which have been produced and demonstrated in this brief treatise, when they have passed into the hands of others of a speculative turn of mind, will become the path to many others, still more marvelous. . . .
>
> This has been a long and laborious day, in which I have enjoyed the bare propositions more than their demonstrations, many of which I believe are such that it would take me more than an hour to understand a single one

of them. That study I reserve to carry out in quiet, if you will leave the book in my hands after we have seen this part that remains, which concerns the motion of projectiles. This will be done tomorrow, if that suits you.

Salviati: I shall not fail to be with you.

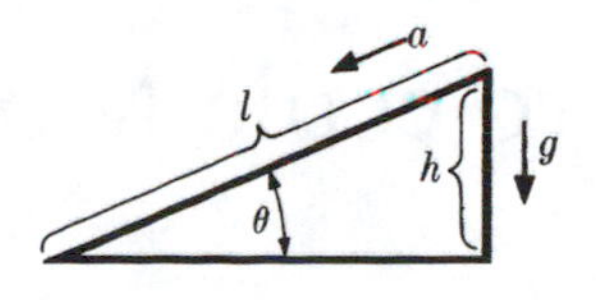

Fig. 7.1.

Problems

Problem 7.1. Read through Galileo's Third Day in the *Two New Sciences* in order to find out what his arguments were for thinking that the laws of motion on an inclined plane are the same as for free fall. (The book is available in a modern English translation by Stillman Drake.)

Problem 7.2. List the steps outlined in this chapter by which Galileo progressed from his first definition of what he means by uniformly accelerated motion to his final confirmation that his definition coincides with the actually observable motion of free fall. What limitations and idealizations entered into the argument?

Problem 7.3. Read through Galileo's Third Day to Theorem IV. Then restate as concisely as possible the argument that the acceleration a of an object sliding on a smooth inclined plane is to the acceleration in free fall g as the vertical height of the plane h is to the length of the plane l, or, in short, that $a = g \sin \theta$ (Fig. 7.1).

Problem 7.4. Summarize what specific methods of attack and of solution you may have discerned in this discussion of Galileo's work on freely falling bodies.

RECOMMENDED FOR FURTHER READING

S. G. Brush (editor), *History of Physics,* articles by W. A. Wallace, "Mechanics from Bradwardine to Galileo," and by S. Drake, "Galileo's discovery of the law of free fall," pages 10–32

H. Butterfield, *Origins of Modern Science,* Chapter 5

I. Bernard Cohen, *The Birth of a New Physics,* Chapter 5 and Supplements 3–7

Stillman Drake, *Galileo*

Galileo Galilei, "Notes on Motion," on the World Wide Web at www.mpiwg-berlin.mpg.de/Galileo_Prototype/index.htm [announced in *Science,* Vol. 280, page 1663 (1998)].

Ludovico Geymonat, *Galileo Galilei*

James McLachlan, *Galileo Galilei, First Physicist,* New York: Oxford University Press, 1999

John Murdoch, article on Bradwardine in *Dictionary of Scientific Biography* (edited by C. C. Gillispie), Vol. 2, pages 390–397

S. Sambursky, *Physical Thought,* pages 226–237

For further discussion of the subject of this chapter, see the "Sources" at the website www.ipst.umd.edu/Faculty/brush/physicsbibliography.htm

CHAPTER 8

Projectile Motion

Turning now to the more general motion of projectiles, we leave the relatively simple case of movements along a *straight line* only and expand our methods to deal with motion *in a plane*. Our whole understanding of this field, a typical and historically important one, will hinge on a far-reaching discovery: The observed motion of a projectile may be thought of as the result of two *separate* motions, combined and followed *simultaneously* by the projectile, the one component of motion being an unchanging, unaccelerated horizontal translation, the other component being a vertical, accelerating motion obeying the laws of free fall. Furthermore, these two components do not impede or interfere with each other; on the contrary, the resultant at any moment is the simple effect of a superposition of the two individual components.

Again it was Galileo who perfected this subject, in the section The Fourth Day immediately following the excerpts from the *Two New Sciences* that were presented in our last chapter. Let us refashion his original arguments to yield somewhat more general conclusions and to illustrate by one painstaking and searching inquiry how an understanding of a complex problem may grow. In later chapters it will usually be necessary for you to supply some of the intermediate steps of a problem, so this effort now will be a good investment. Thus, if Chapter 7 represented a historically inclined chapter, this one may be called a factually directed one.

8.1 Projectile with Initial Horizontal Motion

We may start with two simple but perhaps surprising experiments:

a) Consider an airplane in steady horizontal flight that drops a small, heavy Red Cross parcel as part of a rescue effort. Watching this, we realize that (except for air friction) the parcel *remains directly below the plane* while, of course, dropping closer and closer to the ground. This is represented in Fig. 8.1. If we watched this event from a balloon, very high up and directly above this region, we should, of course, see only the horizontal part (component) of the motion; and if from there we now could see the parcel at all, we should think it traveled directly along with the plane, instead of falling away from it. The clear implication is that the horizontal component of the motion of the parcel remains what it was at the moment of release from the plane (unchanged, as suggested by the law of inertia), even though there is superposed on it the other, ever-increasing component of vertical velocity.

b) In the second experiment we shall place two similar small spheres at the edge of a table (cf. Fig. 8.2). At the very moment that one is given a strong horizontal push, so that it flies off rapidly and makes a wide, curved trajectory, the other is touched only very gently and thus drops straight down to the floor. (By "trajectory" we mean here the path followed by a projectile under the influence of gravity.) Which will land first?

The experimental fact is that they both land at the same moment, no matter what the initial horizontal speed of the pushed ball; furthermore, the balls remain at equal levels throughout their fall, though of course they separate in the horizontal dimension, one having a large (and by the previous experiment, presumably undiminishing) horizontal velocity and the other having none. The conclusion here is that the vertical component of the motion is quite independent of any additional horizontal movement.

The joint result of both experiments is that motion in two dimensions can be resolved into two components, the horizontal and vertical, and in the case of projectiles (parcel, spheres, etc.) moving near the earth, these two components of motion are independent. This allows us to answer some inquiries into the motion of our projectiles,

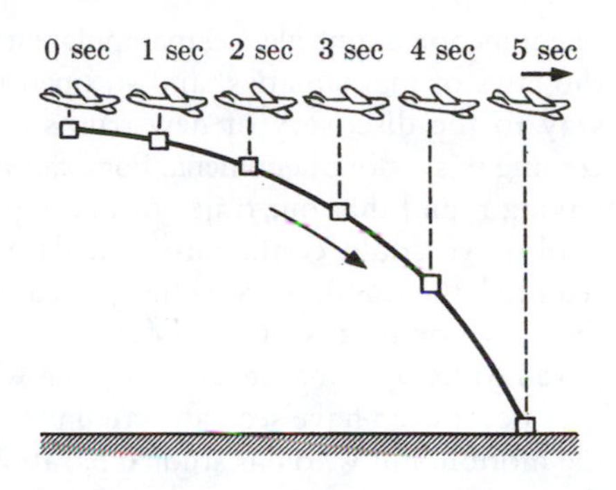

Fig. 8.1. Successive positions of a parcel dropped from a plane.

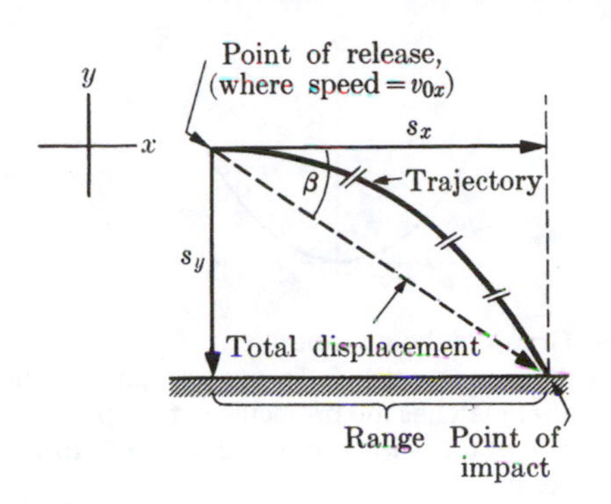

Fig. 8.3. Trajectory and total displacement of a projectile.

the simplest being, "How far away from the starting point does the parcel or the second sphere land?" During the time taken for the fall or flight, say t seconds, the horizontal, unchanging component of motion will transport the projectile a distance equal to the initial velocity imparted in the forward direction at the moment of release multiplied by the time t; simultaneously the growing vertical component of motion contributes a downward displacement by an amount corresponding to the distance of free fall during that time t under the influence of the gravitational acceleration g. If we agree to give the subscript x to all horizontal components, and y to all vertical components, as indicated in Fig. 8.3, we may then rewrite the previous sentence more concisely:

$$s_x = v_{0x}t \quad \text{or} \quad s_x = v_x t, \tag{8.1}$$

because $v_{0x} = v_x$, the horizontal motion being by assumption unaccelerated; and second,

$$s_y = \tfrac{1}{2}gt^2, \tag{8.2}$$

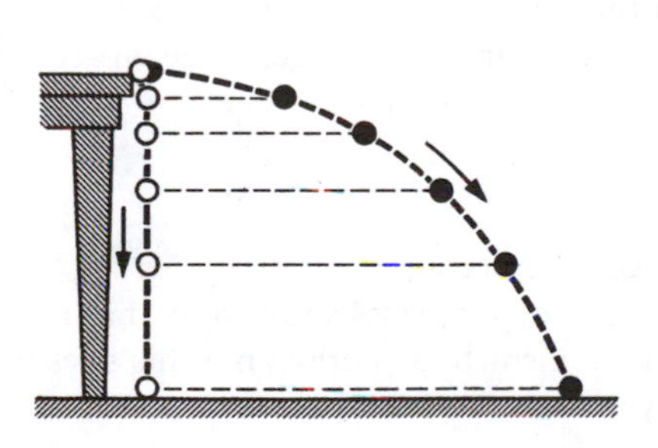

Fig. 8.2. Illustrating that the horizontal and vertical components of motion are independent.

because v_{0y} was here assumed to be zero, the initial motion at the time of release having been altogether in the horizontal direction.

If t is the actual total time of flight, s_x so calculated is the so-called "range" of the projectile and s_y is the distance of drop, but our question above inquired about neither s_x nor s_y. If we still wish to find the total displacement from the starting point (a quantity we might call s), a glance at Fig. 8.3 will at once solve our problem. By the Pythagorean theorem, for displacement in a plane,

$$s = \sqrt{s_x^2 + s_y^2}\,. \tag{8.3}$$

We may now substitute, and find that

$$s = \sqrt{(v_{0x}t)^2 + (\tfrac{1}{2}gt^2)^2}\,.$$

In order to produce a numerical answer, we must therefore be given the initial velocity and the time of flight, quite apart from the value of g, which in all our work will be assumed to be numerically equal to 9.8 m/sec^2 or 32.2 ft/sec^2.

Furthermore, the angle that direction s makes with the horizontal, β, is clearly given[1] by $\tan \beta = s_y/s_x$. It is at this angle that a "sight" will have to be set to ascertain the proper instant of release of this projectile from the plane.

It would be a more difficult problem, and one we shall not try to solve, to discover not the total *displacement* but the length of the actual path along the curved trajectory. It happens to be a fairly unimportant question to us, the range s_x and the distance s_y being the more sought-after

[1]See Appendix VIII for a summary of trigonometric relations.

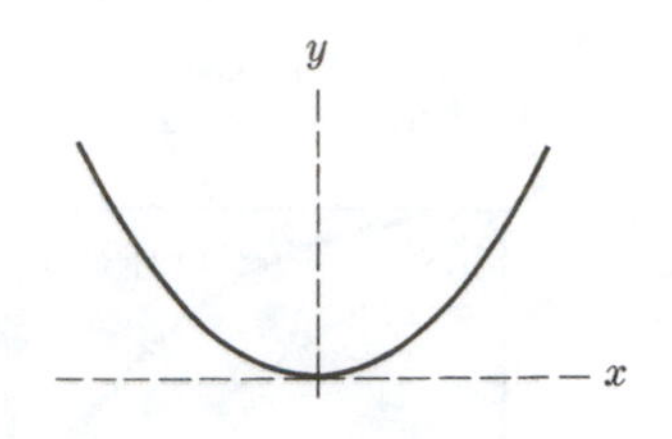

Fig. 8.4. Graph of the equation y = (positive constant) x^2. To see the part of the parabola that matches the projectile path in previous figures, look at the mirror image of Fig. 8.4 (turn it upside down).

quantities; yet we may deduce an interesting result about the shape of the path. From Eqs. (8.1) and (8.2),

$$s_x = v_{0x}t, \therefore t = s_x/v_{0x}, \quad \text{and} \quad t^2 = s_x^2/v_{0x}^2.$$

But

$$s_y = \tfrac{1}{2}gt^2 = \tfrac{1}{2}g(s_x^2/v_{0x}^2),$$

which may be written

$$s_y = \left(\frac{g}{2v_{0x}^2}\right)s_x^2. \tag{8.4}$$

Since the quantities in parentheses are constant throughout a particular projectile motion, it appears that s_y is proportional to s_x^2; hence $s_y = ks_x^2$, where $k = (g/2v_{0x}^2)$. But this equation, according to the analytic geometry of Descartes, corresponds simply to a parabola, as shown in Fig. 8.4, and that is (ideally) the shape of the path.

Problem 8.1. Plot on graph paper or computer screen the equation $y = (-\frac{1}{3})x^2$. (a) Give x only + values; (b) give x both + and – values. Recall the useful convention that to distances above the origin or to the right of the origin, positive values are assigned; for distances below or to the left of the origin, negative values.

After Galileo had deduced the parabolic nature of the trajectory (by an original and similar argument), projectile motion immediately became much simpler to understand; for the geometrical properties of the parabola itself had long been established by the mathematicians in their disinterested pursuit of abstract geometrical worlds. We find here a first clue to three important facts of life in science:

1. If we can express phenomena quantitatively and cast the relation between observables into equation form, then we can grasp the phenomenon at one glance, manipulate it by the laws of mathematics, and so open the way to the discovery of new truths concerning this set of phenomena. For example, having found that our trajectories are parabolic, we could confidently calculate, if required, the length of the actual path along the curve by means of some formula proposed to us by a mathematician, one who may never even have seen an actual projectile motion, but who has studied parabolas thoroughly.
2. Consequently there is always an imperative need for a well-developed system of pure mathematics from which the physicist may draw.
3. We can see why the physical scientist always tries to cast a problem into such a form that established approaches, procedures, or tricks from another branch of science or from mathematics will aid in the solution.

As an example of the last point, just as Galileo applied the mathematics of parabolas to actual projectile motions, so do modern acoustical engineers solve their problems by means of the mathematical schemes developed quite independently by electrical engineers for their own very complex field. Whatever the methods of science may be, they have shown themselves to be transferable from one specialty to the other in a remarkable and fruitful way.

Yet another question to be asked about our projectile motion is this: What is the actual velocity of the moving object at some time t after release? We may be prompted by our two initial experiments and by the discussion leading to Eq. (8.3) to represent the two observed components of the actual velocity at several portions along the path, as has been done in Fig. 8.5a. At each of the three selected points the horizontal velocity component $v_{0x} = v_x$ = constant, but the vertical component grows linearly from $v_{0y} = 0$ to a larger and larger value, in accordance with the law for falling bodies,

$$v_y = v_{0y} + gt,$$

which here becomes simply $v_y = gt$.

Our experiments showed us that v_x and v_y will not disturb each other, but this does not tell us yet what the actual velocity v is at any chosen point on the trajectory. Now it was Galileo's very important discovery that the total effect of both velocity components, i.e., the actual veloc-

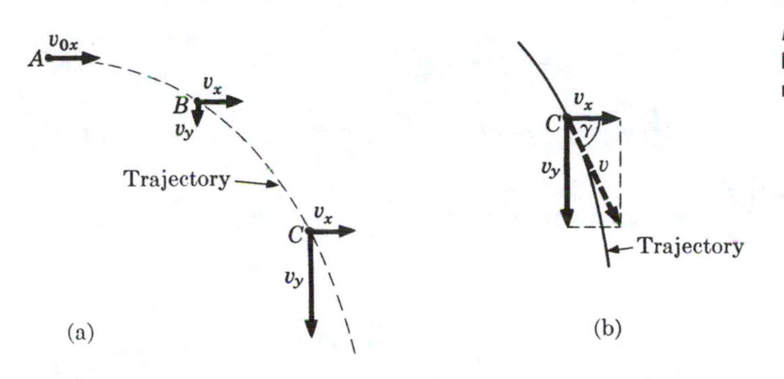

Fig. 8.5. The actual velocity v can be compounded from its components v_x and v_y.

ity v, could be compounded from the two components v_x and v_y by the very same simple scheme that enabled us to find s when s_x and s_y were known.

In Fig. 8.5b, the velocity at point C is constructed by completing the parallelogram (in this case simply a rectangle, since v_x and v_y are at 90°) formed by the arrows corresponding to v_x and v_y; the total, the "real" velocity, is then indicated by the diagonal v, the line being momentarily tangent at every point along the path of the projectileand having the value

$$v = \sqrt{v_x^2 + v_y^2} \qquad (8.5)$$

Consequently, the momentary direction of motion at point C can be specified in terms of the angle γ between v and the horizontal direction, and is given by

$$\tan \gamma = v_y/v_x \qquad (8.6)$$

This, to repeat, is a postulate that is *checked by experiment*. We may cast it into formal language: At any instant the instantaneous velocity of a projectile, and its direction of motion, may be obtained by *superposing the two independent components v_x and v_y in a simple parallelogram construction,* where v, the resultant, is given in both direction and magnitude by the diagonal, as shown in Fig. 8.5b.

This *principle of the superposition of velocity components* is so simple as to seem now self-evident; but that is deceptive. The *only* justification for postulating the crucial Eqs. (8.5) and (8.6) is that they are essentially experimentally confirmed discoveries. In one form or another the superposition principles are such simplifying devices that they might seem to have been too much to hope for; yet natural phenomena graciously do exhibit such features at times, and it is practically unthinkable that physics could have embarked on its great journey of vigorous discovery from Galileo on without his having intuited many such "simple" fundamentals.

EXAMPLE 1

A stone is dropped from a plane flying parallel to the ground at a speed of 200 m/sec. If the stone needs 5 sec to reach the ground, how high did the plane fly, what is the range of this projectile, what is its total displacement, and what is its final speed just before striking?

As we now translate phrase by phrase, we write: $v_{0y} = 0$, $v_{0x} = 200$ m/sec, $t = 5$ sec, $s_y = ?$, $s_x = ?$, $s = ?$, $v = ?$ We usually assume that $g = 9.8$ m/sec^2, but for this example 10 m/sec^2 is accurate enough; also, we assume that air friction on the projectile is so minimal that its horizontal motion is not affected and its vertical motion remains uniformly accelerated. Both these assumptions are, incidentally, experimentally found to hold quite well for the first few seconds of motion, but increasingly less well thereafter.

For a solution we find $s_y = \frac{1}{2}gt^2 = 125$ m, $s_x = v_{0x}t = 1000$ m, $s = \sqrt{125^2 + 1000^2}$ m, and so forth.

8.2 Introduction to Vectors

The foregoing discussion has exhausted every worthwhile question we may put at present about projectile motion, *provided the motion started with a straight horizontal velocity* and no vertical component of velocity at all. Now we shall go on to the general case, i.e., to the motion of projectiles of all sorts (not least those in sports such as baseball, soccer, and tennis) and with all

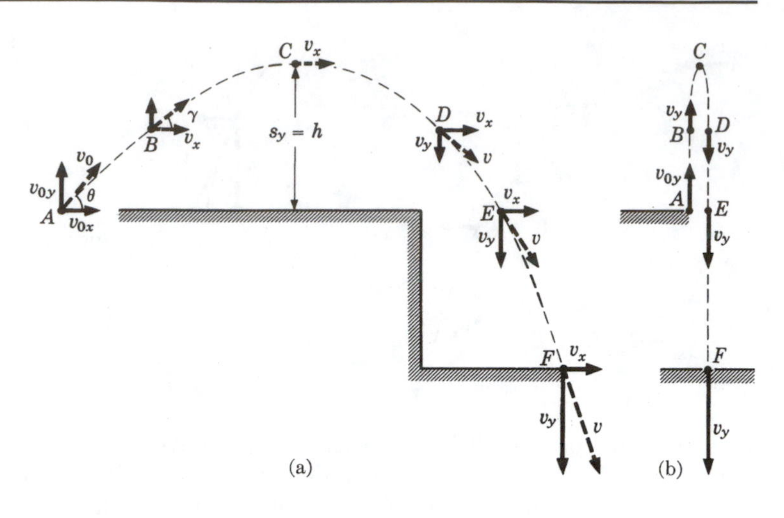

Fig. 8.6. Generalized projectile motion.

conceivable initial conditions of velocity, not neglecting projectiles shot from a gun pointing at any angle above or below the horizontal line. Leaving aside the sad abuses made of the findings of scientists, as in the aggressive uses of guns, our interest here is that we can regard this exposition as a typical formulation of a large and successful *general scheme to handle all possible specific instances* of a problem. Furthermore, in this lengthy effort we shall of necessity have to develop or invent some new physical concepts that will be useful in a wider context later.

Imagine now the progress of a projectile that leaves the muzzle of a gun with an initial velocity v_0 and at an angle θ with the horizontal (Fig. 8.6a), rises to some height *h*, and then drops into a valley below the horizon. The horizontal and vertical components of the initial velocity, v_{0x} and v_{0y}, can be computed by *resolution* from v_0 and the angle θ by means of the same trigonometric approach that leads us to the converse proportion, the *composition* of v from v_x and v_y:

$$\text{Horizontal component of } v_0 = v_{0x} = v_0 \cos\theta, \tag{8.7}$$

$$\text{Vertical component of } v_0 = v_{0y} = v_0 \sin\theta$$

As before, v_{0x} remains also the value of the horizontal component v_x at all later points, including the six positions shown. (The point *F* in Fig. 8.6 refers to the state of affairs just *before* the projectile hits.) On the other hand, v_{0y} at point *A* is subsequently diminished as the projectile rises against the opposing, downward acceleration of gravity, until, at point C, $v_y = 0$. From then on, v_y no longer points upward, but reverses and becomes larger and larger in the previously examined manner. *So far as this vertical component alone is concerned,* its history would have been exactly the same if v_{0x} had been zero, i.e., if we had been thinking about a stone thrown vertically upward at the edge of a cliff with that same initial value of v_{0y} (Fig. 8.6b). (Recall Example 5 in Section 6.7.)

The question now arises as to how to calculate the total actual velocity v of the projectile in Fig. 8.6a at some point *B*. We still have faith that at *B*, as elsewhere, $v = \sqrt{v_x^2 + v_y^2}$ (and in the absence of faith an experiment would prove the point). v_x at *B* is still, in magnitude, equal to v_{0x} and therefore equal to $v_0 \cos\theta$. But when we now turn to the vertical component, we find that its initial magnitude has been decreasing between points *A* and *B* through constant deceleration due to the gravitational pull in a direction opposite to the motion. We may regard v_y at *B* as a constant speed component v_{0y} diminished by a contrary, growing component of magnitude *gt*, where *t* is the time needed to reach point *B* (see

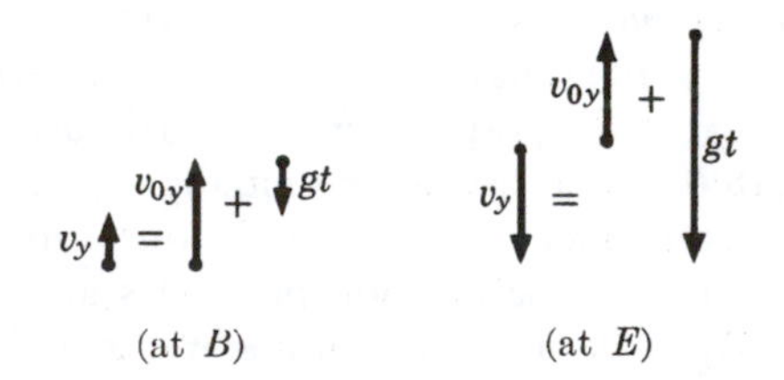

Fig. 8.7. Calculation of v_y at two points along the trajectory.

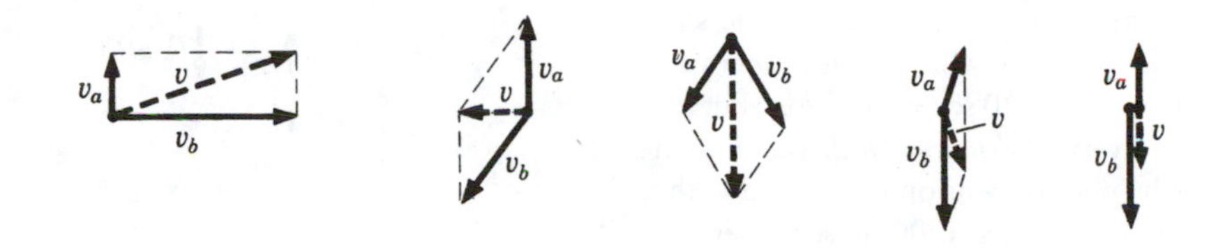

Fig. 8.8. Graphical method for adding two velocity vectors.

Fig. 8.7, "at *B*"). At point *C*, evidently $v_y = 0$, since there the term *gt* has become exactly as large as v_{0y}. Beyond that point we may become somewhat uncertain as to how to calculate v_y from v_{0y}, *t*, and *g*, and here we must have recourse to a geometric scheme.

Consider Fig. 8.6b. Each arrow representing v_y at any one point may itself be thought of as the result of simple addition or subtraction of the length of two other arrows, representing, respectively, the constant velocity v_{0y} (upward), and the cumulative effect of gravity, *gt*, downward (Fig. 8.7). But this is just a special case of a general scheme of additions of quantities such as displacement (for which we used it in Section 8.1) and of velocities, the scheme due to Galileo that we previously employed to compose v from v_x and v_y: No matter whether two velocities simultaneously at work on one body are active at right angles to each other (as before), or along the same line (as now while we consider only the *vertical* component), or make any other angle with each other, we can find the resulting total velocity by the so-called parallelogram method of construction. In this construction, the arrows representing the two velocity components are drawn to some scale (for example, 1 cm = 10 m/sec) and placed on paper in correct relation end to end; the parallelogram is completed; and the diagonal is drawn, which, pointing in the direction away from the meeting point of the component arrows, gives to the same previous scale the magnitude and direction of the resultant total velocity.

Figure 8.8 shows several such hypothetical cases, v being the resultant of two components v_a and v_b. The first of these examples is familiar to us. The last is a special case of the one preceding; the two components lie in opposite directions along one line and the drawing of a clear parallelogram becomes practically impossible, although the result is obviously that shown. Similarly, if the angle between v_a and v_b in the third sketch were to shrink to zero, the resultant v would increase to the plain numerical sum of the components.

Quantities that permit such a graphic method of addition are called *vectors;* besides displacement and velocity we shall find many other important vector quantities, notably force. In each case the method for vector addition (that is, for finding the resultant of two separate components) is this parallelogram method, based on a simple superposition law. (For further discussion of vectors, see Appendix IX.)

8.3 The General Case of Projectile Motion

We shall soon have ample opportunity to make use of vector addition in other contexts, but for the moment it is enough to consider the scheme in Fig. 8.7 as a restatement of the special case of vector addition shown in the last drawing of Fig. 8.8. Now a simple convention suggests itself that circumvents the unavoidably inaccurate and somewhat messy scheme of drawing these arrows to scale to find the resultant in each case. If we agree to the convention to give positive values to quantities represented by an upright arrow, and negative values to those oppositely directed, then we can employ the simple equation

$$v_y = v_{0y} + gt \tag{8.8}$$

in its original form to compute v_y at any time *t* after the projectile motion starts. A glance at Fig. 8.7 will show qualitatively that this equation, together with that convention, assures the correct calculation of the numerical values for v.

EXAMPLE 2

A projectile is shot from a gun tilted at 60° above the horizontal; the muzzle velocity v_0 is 1000 m/sec: (a) What is v_y after 20 sec of flight, (b) what is the total velocity v of the projectile after 20 sec of flight, (c) at what angle with the horizontal does it then travel?

Solution: (a) $v_y = v_{0y} + gt$, where $v_{0y} = v_0 \sin \theta$ = 870 m/sec (positive value); gt is numerically about 10 (m/sec^2) × 20 sec = 200 m/sec, but representing always a *downward directed* quantity, it is by our convention to be used with a negative sign, that is, as (200) m/sec. Therefore v_y = 870 m/sec + (200) m/sec = 670 m/sec. (The fact that our answer has a positive sign itself means that the velocity is still directed upward; this means we might be now at some position on the trajectory like point *B* in Fig. 8.6a.) (b) To compute v by $\sqrt{v_x^2 + v_y^2}$ and then the angle from $\tan \gamma = v_y/v_x$, we now find v_x from $v_x = v_{0x} = v_0 \cos \theta$. (Continue as an exercise.)

EXAMPLE 3

How long will the above-mentioned projectile take to reach the highest point? The translation of this question depends on the realization that $v_y = 0$ at the top point of the trajectory, i.e., at point C in Fig. 8a. Then $gt = -v_y$ by Eq. (3.8), or $(-10 \times t)$ m/sec = 870 m/sec, and consequently $t = -870/(-10)$ sec ≈ 87 sec. (We note at once our suspicion that for such conditions the physical realities of air friction undoubtedly make our calculations quite academic. But in the "real world" the effect of air resistance can of course be computed separately and adjustments made.)

EXAMPLE 4

What is v_y, the vertical component of velocity of this projectile, after 174 sec?

Solution: From Eq. (8.8) we find v_y = 870 m/sec + (−10 × 174) m/sec = 870 m/sec − 1740 m/sec = −870 m/sec (negative = downward). In short, after a time as long again as it took to reach the top of the trajectory, the velocity component downward is numerically equal to but oppositely directed to the initial vertical component. This is pictured at point *E* in Fig. 8.6.

The last two examples make it clear that it might be wise to assign the minus sign permanently to the value for g, to write $g = -9.8$ m/sec^2 once and for all, and so banish the temptation to change the plus sign in Eq. (8.8) at any time. We shall do this from now on.

But you may have wondered why we did not adopt these conventions before, in Section 8.1, or, since we failed to do so at that time, whether conclusions obtained there are not now erroneous. The answer is that our convention of + and − signs is arbitrary and could be turned around completely, as long as we somehow distinguish in our numerical work between "up" and "down" components (g is constant, even at "the top"). In Section 8.1, where v_{0y} was kept jealously at zero value, only components of one variety, namely "downward," appeared in any problem; hence there was no need to introduce a differentiating sign convention then.

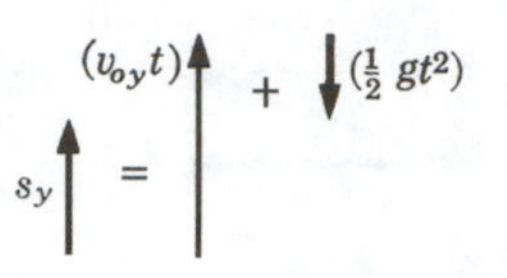

Fig. 8.9. Calculation of displacement vector from velocity, acceleration, and time.

With this treatment of one type of vector components as a background, we need perhaps only point out that the same sign convention, if applied to *displacement* components, will allow us to solve in a similar manner all problems involving the height of a trajectory, the range, etc. After all, displacements also are vectors in the sense that they can be represented by arrows and can be added (composed) or resolved according to the same simple parallelogram scheme that we used for velocities.

For example, our use of the Pythagorean theorem in Eq. (8.3) shows this vector aspect of displacements clearly. Another example might be the graphical derivation of s_y at, say, point *B* in Fig. 8.6, as shown in Fig. 8.9.

Therefore s_y will also attain positive values if represented by an arrow pointing upward, i.e., when the displacement is to a point *above the level of release* of the projectile; and s_y will assume negative values for displacements measured in the other direction. Then we can use in their original form the pertinent equations involving displacement, such as

$$s_y = v_{0y}t + \tfrac{1}{2}gt^2, \qquad (8.9)$$

provided again that we substitute in all our calculations for g the value −9.80 m/sec^2.

EXAMPLE 5

The projectile mentioned in the previous examples falls into a valley 300 m below the level of the gun. How long was it in the air, and what was the range? Translating, we note that the old data are $v_0 = 1000$ m/sec at 60°, therefore $v_{0y} = 870$ m/sec, $v_{0x} \approx 500$ m/sec; $g \approx 10$ m/sec^2. To this we add $s_y = 300$ m (a negative sign to indicate that the vertical displacement is to a point *below* the level of release), $t = ?$, $s_x = ?$

Equation (8.9) can be written: $\frac{1}{2}gt^2 + v_{0y}t - s_y = 0$, a quadratic equation easily solved for t:

$$t = \frac{-v_{0y} \pm \sqrt{(v_{0y})^2 - 4(\frac{1}{2}g)(-s_y)}}{2(\frac{1}{2}g)} .$$

So far, we have not called in our sign convention. Now we may substitute our values with their proper signs to obtain a numerical answer. (Do this.) Having calculated t, we can solve for s_x, since it is still true that $s_x = v_{0x}t$.

In completing the last example you had to recall that of two possible solutions of such quadratics, one may be physically irrelevant or even absurd (for example, if t comes out zero or negative or if the quantity under the square root is negative). What does this reveal about the role of mathematics in physical science?

To summarize, we now feel justified in using the following equations of motion in projectile problems, where all symbols have their discussed meaning and the new convention is used to substitute numerical values with correct signs.

$$s_y = \tfrac{1}{2}(v_{0y} + v_y)t \qquad \text{(I)}$$

$$s_y = v_{0y}t + \tfrac{1}{2}gt^2 \qquad \text{(II)}$$

$$v_y = v_{0y} + gt \qquad \text{(III)}$$

$$v_y^2 = v_{0y}^2 + 2gs_y \qquad \text{(IV)}$$

$$s_x = v_{0x}t = v_xt \qquad \text{(V)}$$

$$v_{0x} = v_x = v_0 \cos\theta \qquad \text{(VI)}$$

$$v_{0y} = v_0 \sin\theta \qquad \text{(VII)}$$

$$v = \sqrt{v_x^2 + v_y^2}, \qquad \tan\gamma = v_y/v_x \qquad \text{(VIII)}$$

$$s = \sqrt{s_x^2 + s_y^2}, \qquad \tan\beta = s_y/s_x \qquad \text{(IX)}$$

Conventions regarding sign:

- Vector components in the $+y$ direction are given + values.
- Vector components in the $-y$ direction are given – values.

Consequently, s_y has + values if the displacement is above the level of release; s_y has – values if the displacement is below the level of release; and v_{0y} and v_y have + values while the corresponding motion is upward, whereas v_{0y} and v_y have – values while the corresponding motion is downward. The value of g, the acceleration of gravity on earth, is taken to be about –9.80 m/sec^2.

This set is not quite as formidable as it might seem at first glance. We recall that Eq. (I) follows from the definition of average velocity for uniformly accelerated motion, that is, $\bar{v} = (v_{0y} + v_y)/2 = s/t$. Equation (II) is an experimental law (from free-fall experiments), and therefore, so to speak, on a different plane or a different kind of knowledge. Equations (III) and (IV) can be derived from (I) and (II). Equation (V) is a second experimental law (that is, the horizontal component of motion v_x is unaccelerated). The remaining equations (VI) to (IX) simply refer to the mathematical properties of vectors: Eq. (VI) includes the rule defining how to find the x component of vector v_0, while Eq. (VII) provides for the y component. Equations (VIII) and (IX) give the rule for vector addition (magnitude and direction of resultant, first for the total actual velocity and second for the total actual displacement). Finally, the sign conventions, which quickly become second nature in use, reflect only a common-sensical way of differentiating between addition and subtraction of quantities.

When practicing physicists have to solve projectile problems, they do not really need to have before them all these defining equations and the conventions. They understand these tacitly; what they will regard as important and essential about ideal projectile motion are Eqs. (II) and (IV), modified as follows:

$$s_y = (v_0 \sin\theta)t + \tfrac{1}{2}gt^2;$$

$$s_x = (v_0 \cos\theta)t.$$

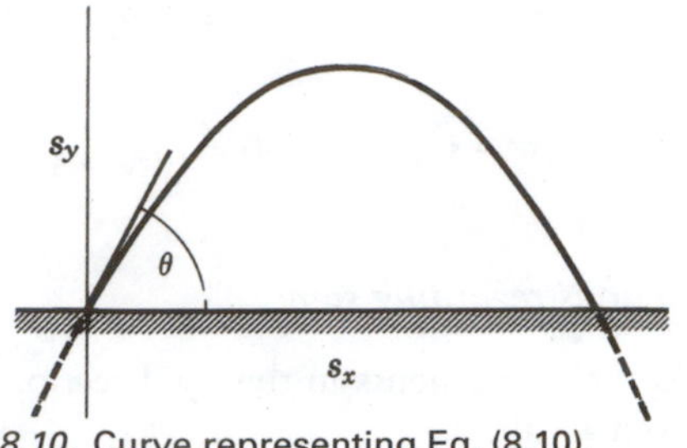

Fig. 8.10. Curve representing Eq. (8.10).

They will ordinarily combine these equations by eliminating t between them (that is, in the left equation, replacing t by $s_x/v_0 \cos \theta$), thus obtaining

$$s_y = (\tan \theta)s_x + \tfrac{1}{2}g\left(\frac{s_x^2}{v_0^2 \cos^2 \theta}\right). \qquad (8.10)$$

This last equation they will consider to be *the* equation of the trajectory *in vacuo,* and they can attack most problems with this formula (see Fig. 8.10). For us it suffices to realize that the whole truth about projectile motion, through mathematics, can be so economically expressed and so conveniently handled. Also, we should realize that Eq. (8.10) is of the following form:

$$s_y = (\text{a constant}) \times s_x + (\text{another constant}) \times s_x^2.$$

This once more implies that the *trajectory is parabolic* [no longer simply the right half of the parabola in Fig. 8.4, which applied where $\theta = 0$, whence $\tan \theta = 0$ and $s_y = (\text{constant}) \times s_x^2$ only].

Problem 8.2. The two curves of Fig. 8.11 differ only in the placement of the origin of the coordinates, so that in (a) $\theta = 0$, in (b) $\theta = 60°$. Copy the two parabolas on a sheet of graph paper and show by computation for several points that the equation for (a) is $Y = (\text{a constant}) \times X^2$, and for (b) $Y = (\text{a constant}) \times X + (\text{another constant}) \times X^2$. Once more we see that a knowledge of parabolas is all we should really need to solve practical projectile problems. Nevertheless, at our level it will prove easier for us if we continue to solve our problems by means of Eqs. (I)–(IX) rather than, say, by Eq. (8.10) plus the theory of parabolas.

8.4 Applications of the Law of Projectile Motion

Let us review what we have done. First, starting from some experimental observations, we have found the separate equations for the horizontal and the vertical components of a general projectile motion. In particular, the horizontal component is found to be unaccelerated, and the vertical one obeys the laws of motion of uniformly accelerated objects. Second, the total actual motion (displacement s and speed v) is obtained in direction and magnitude by a vector addition of the components, following an empirically justified procedure of adding by the parallelogram method. Now comes the reward. We see what a wide variety of projectile problems we can solve with this set of equations (which we may regard as the algebraic expression of *the general law of projectile motion*).

EXAMPLE 6

A gun pointed at an angle θ above the horizon shoots a projectile with muzzle velocity v_0. What is the time t needed for the projectile to return to level ground? That is, given v_0, θ, and, of course, g, find t if $s_y = 0$ (Fig. 8.12).

Solution: Since

$$s_y = v_{0y}t + \tfrac{1}{2}gt^2 \qquad (\text{II})$$

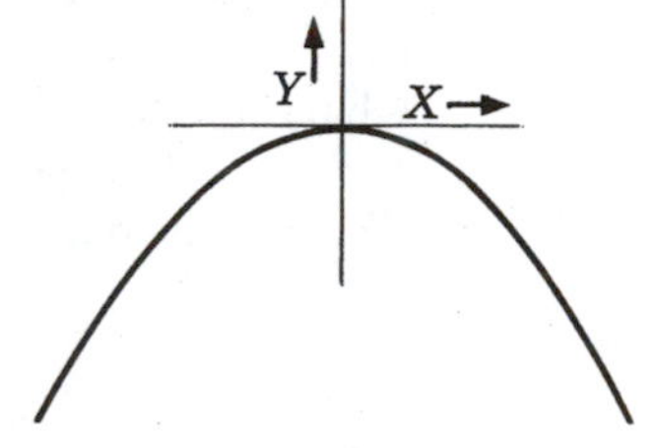

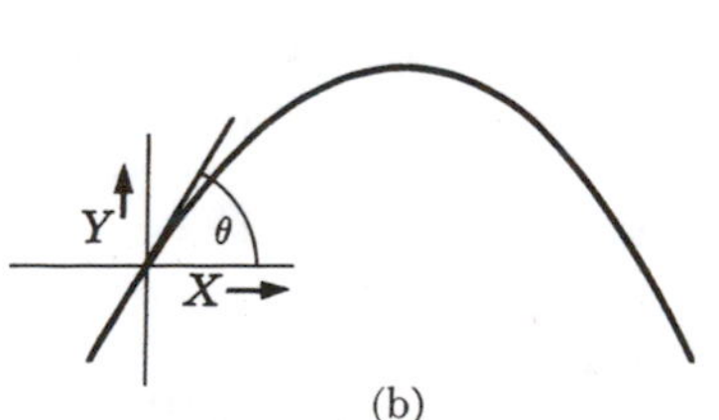

(b)

Fig. 8.11. Curves for Problem 8.2.

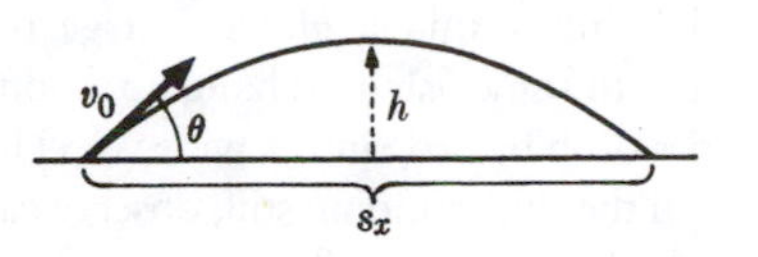

Fig. 8.12. Trajectory of a projectile shot from a gun (see Example 6).

$$0 = (v_0 \sin \theta)t + \tfrac{1}{2}gt^2$$

and

$$t = (-2v_0 \sin \theta)/g.$$

Since g takes on a negative value, t will come out positive, as it surely must.

EXAMPLE 7

In the previous example, what is the maximum height reached? That is, given v_0, θ, g, find s_y ($\equiv h$) if $v_y = 0$.

Solution: By Eq. (IV), $0 = (v_0 \sin \theta)^2 + 2gs_y$, and

$$s_y\ (\equiv h) = -(v_0 \sin\theta)^2/2g.$$

Again s_y is positive, since g has a negative value.

EXAMPLE 8

In the first example, what is the range, that is, if v_0, θ, and g are given, and $s_y = 0$, what is s_x?

Solution: $s_x = v_{0x}t = (v_0 \cos \theta)t$. However, since t is not explicitly known, it must first be calculated, and this can be done as shown in Example 1. (This may be called a two-step problem.) By direct substitution, it follows that

$$s_x = \frac{-v_0^2 2 \cos \theta \sin \theta}{g} = \frac{-v_0^2 (\sin 2\theta)}{g}$$

(using a formula from Appendix VIII-6). It should be emphasized again that these results would hold strictly only for projectiles moving in a vacuum above a flat earth. The actual effects of air resistance alone for modern high-speed projectiles may be found by computation to decrease the effective range by 50%. Evidently, for practical applications, we should have to investigate how such effects modify these results through theoretical and empirical corrections.

8.5 Galileo's Conclusions

Galileo himself carried his work to the point of computing fine tables of ranges and heights of trajectories over level ground for different angles θ. These calculations proved that for a given initial velocity the range was a maximum if θ = 45°, as we can ascertain merely by inspection of the general equation $s_x = [\,v_0^2 (\sin 2\theta)]/g$; for $\sin (2\theta)$ takes on its maximum value, 1, if θ = 45°. (In punting a football to the longest distance, athletes know intuitively to launch the ball at some angle near that[2]—and not at, say, 20° or 80°) Galileo also used this result to show that s_x for a given type of projectile is equally large for any two values of 0 differing from 45° by an equal amount in each direction (for example, 52° and 38°). Convince yourself of this unexpected fact by substitution, if no simpler way suggests itself.

Galileo very penetratingly remarks at this point:

> *Sagredo:* The force of rigorous demonstrations such as occur only by use of mathematics fills me with wonder and delight. From accounts given by gunners, I was already aware of the fact that in the use of cannon and mortars, the maximum range, that is, the one in which the shot goes farthest, is obtained when the elevation is 45° . . . ; but to understand *why* this happens far outweighs the mere information obtained by the testimony of others or even by repeated experiment. . . .
>
> *Salviati:* The knowledge of a single effect apprehended through its cause opens the mind to understand and ascertain other facts without need of recourse to experiment, precisely as in the present case, where, having won by demonstration the certainty that the maximum range occurs when the elevation [θ] is 45°, the Author [Galileo] demonstrates what has perhaps never been observed in practice, namely, that for elevations which exceed or fall short of 45° by equal amounts, the ranges are equal. . . .
>
> (*Two New Sciences*)

[2]The 45° result cannot be applied directly in this case because of the large effect of air resistance on a football. Even for the relatively compact baseball, air resistance changes the optimum angle (for greatest total distance) to about 35°. For a discussion of these and other points see Robert Adair's fascinating book, *The Physics of Baseball* (written for "those interested in baseball, not in the simple principles of physics").

Note the pleasure that he finds in discovering a mathematical law, and that he regards it as the way "to understand *why*" projectiles move as they are observed to do. Note also especially the phrase "without need of recourse to experiment"; Galileo, traditionally regarded as the patron saint of experimental science, clearly warns us that the continual experimental verification of a prediction from a law is unnecessary once the law is sufficiently well established. After the initial mandatory doubts are satisfied, one must sufficiently believe in the law in order to obtain any benefit from it at all.

A second important conclusion from the work on projectiles may seem almost trivial at first glance. In support of the Copernican thesis that the earth moves about its axis and around the sun, Galileo offered an answer to critics who argued that if the earth moved, a stone dropped from a tower would be left behind while falling through the air, and consequently would not land directly at the foot of the tower as observed, but much beyond. Galileo assumes that during the *short* time of fall, the earth, and the top and the foot of the tower may be thought to move forward equally far with uniform velocity. If, then, the whole tower moves with the same speed v_{0x}, the dropped stone must fall along the tower, because it will retain this "initial" horizontal component of speed, as truly as a parcel dropped from a moving plane lands below the plane or, in Galileo's analogy, as an object falling from the mast of a moving ship lands at the foot of the mast.

From this and equivalent observations concerning the other laws of mechanics has been developed a most valuable generalization, usually called the *Galilean relativity principle:* Any mechanical experiment, such as on the fall of bodies, done in a stationary "laboratory" (for example, on a stationary ship) will come out precisely the same way when repeated in a second "laboratory" (say, on a smoothly moving ship) as long as the second place of experimentation moves with constant velocity as measured from the first.

Galileo suggested that the same principle could be demonstrated more conveniently, without having to climb to the top of the mast of a moving ship:

> Shut yourself up with some friend in the main cabin below decks on some large ship, and have with you there some flies, butterflies, and other small flying animals. Have a large bowl of water with some fish in it; hang up a bottle that empties drop by drop into a wide vessel beneath it. With the ship standing still, observe carefully how the little animals fly with equal speed to all sides of the cabin. The fish swim indifferently in all directions; the drops fall into the vessel beneath; and, in throwing something to your friend, you need throw it no more strongly in one direction than another, the distances being equal; jumping with your feet together, you pass equal spaces in every direction.
>
> When you have observed all these things carefully (though there is no doubt that when the ship is standing still everything must happen in this way), have the ship proceed with any speed you like, so long as the motion is uniform and not fluctuating this way and that. You will discover not the least change in all the effects named, nor could you tell from any of them whether the ship was moving or standing still. In jumping, you will pass on the floor the same spaces as before, nor will you make larger jumps toward the stern than toward the prow even though the ship is moving quite rapidly, despite the fact that during the time you are in the air the floor under you will be going in a direction opposite to your jump. . . . the butterflies and flies will continue their flights indifferently toward every side, nor will it ever happen than they are concentrated toward the stern, as if tired out from keeping up with the course of the ship, from which they will have been separated by long intervals by keeping themselves in the air
>
> . . . The cause of all these correspondences of effects is the fact that the ship's motion is common to all the things contained in it, and to the air also.
>
> (*Dialogue Concerning the Two Chief World Systems,* Second Day)

To express Galileo's principle more concisely, let us use the words "coordinate system" instead of "ship" or "laboratory," since all that counts in the description of experiments is the system of reference used for measurements. Some corner of the rectangular laboratory table might be the origin for all our measurements of distance, and the x, y, and z directions may correspond to the directions of the three edges. Then we may restate the *Galilean relativity principle:* "All laws of mechanics observed in one coordinate system are equally valid in any other coordinate system moving with a constant velocity relative to the first." If you drop a suitcase in a compartment in a train, be it at rest or speeding at a constant 40 or 80 km/hr across level terrain, the case will always hit the floor below its point of release, and will fall downward with the usual acceleration as measured in the car. Consequently, from the motion of a projectile in a car you cannot decide whether the car itself is moving. If the principle is true in its general form, it follows that no mechanical experiment made on our earth can decide such intriguing questions as whether the solar system as a whole proceeds with some steady forward motion through the space about us.

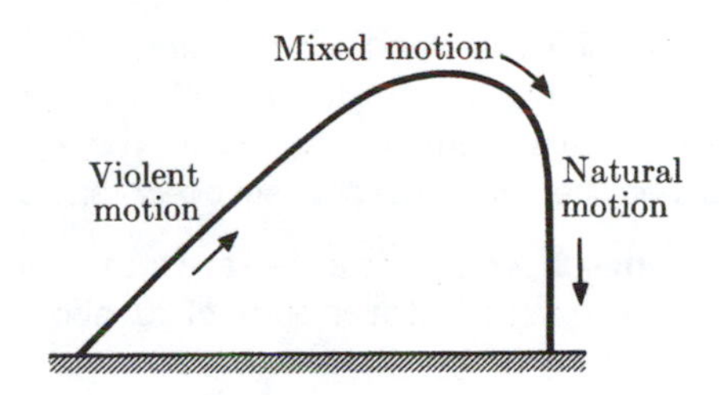

Fig. 8.13. An early sixteenth-century conception of projectile motion using the Aristotelian terminology. ("Violent" motion was considered to be that produced by a propelling force acting on the object.)

8.6 Summary

Evidently, this chapter has carried us far from the starting point of our discussion. Recall that we began with the simple case of projectile motion in a plane, θ being 0° at first. Then we deduced the superposition principles for velocities, learned how to add vector components by the simple parallelogram method, and extended our treatment to the general case, obtaining the law of projectile motion without any restrictions on the initial direction of motion. Finally we saw that even more general principles were induced from these laws.

We also noted in passing the power and meaning of the mathematical formulation of physical events, and the multifold predictions flowing from such mathematical laws—whether theoretical, as the determination of a maximum range for $\theta = 45°$, or practical, as the construction of gunnery tables. And from this example we can already perceive the moral problem that faces the modern scientist: the obvious possibility that his discoveries may be applied by others in warfare or for other purposes, whether or not he likes it.

With these examples of the development and power of physical law in mind, we shall return later to a discussion of "scientific methods." But throughout the rest of this study we shall encounter influences of Galileo's work from time to time, as one might find an ancestral trait running through generations of a family.

For the moment, an urgent task remains. In this subdivision, Part B of the text, we have studied motions of various sorts, but without regard to the forces that cause them. To kinematics, we must now add the concept of force to achieve a *dynamics*. These are as two sides of a coin; without both, we cannot purchase the full understanding of any mechanical phenomenon.

Additional Problems

Problem 8.3. What is meant by saying that the paths of projectiles *in vacuo* are parabolic? When we find by experiment that the paths are parabolic, what help is that in solving projectile problems? *Why* should trajectories follow parabolic paths instead of perhaps that shown in Fig. 8.13?

Problem 8.4. A stone is thrown virtually vertically upward with an initial speed v_{0y} = 10 m/sec (Fig. 8.14). Find (at the end of the first, third, and seventh second), (a) the displacement s_y of the stone, (b) the instantaneous velocity. Use your data to plot a graph of v vs. t (on graph paper), and find from it (*graphically*) (c) when the stone will have a speed of 20 m/sec, and (numerically) (d) what the acceleration is at the instant that the stone has reached its topmost position and just begins to turn downward.

Fig. 8.14. Trajectory of a stone thrown nearly upward from a small platform.

Problem 8.5. How long will the stone in Problem 8.4 take to return to the level of release? Obtain this result first from your graph for Problem 8.4, and then check it by independent direct calculation.

Problem 8.6. A gun fires a shell at a muzzle velocity of 1000 m/sec and at an angle of 30° above the horizon. How long will it take to return to earth, how high will the projectile rise, and what is its range on level ground? (Neglect air friction.)

Problem 8.7. A baseball is thrown by a pitcher horizontally with a speed of 40 m/sec from a vertical height of 2 m above home plate. The batter is 20 m away from the pitcher. Will this pitch be called a strike if the batter does not swing?

Problem 8.8. The famous gun known as "Big Bertha" in World War I had a maximum range on flat territory of about 100 km. What would have been the muzzle velocity of the projectile if it had exhibited such a range *in vacuo?* (The actual muzzle velocity under ordinary circumstances would be somewhat larger.)

Problem 8.9. A baseball is thrown with v_0 = 15 m/sec at an elevation (θ) of 60°. Show that after 2 sec it has risen 6.6 m, has traveled 15 m horizontally, and that its direction of motion is now tilted downward, making an angle of about –40.9° (i.e., 40.9° below the horizontal).

Problem 8.10. Modern cosmic ray research sometimes employs rockets to sample automatically the radiation at high altitudes. For the sake of simple calculation, consider that a self-propelled rocket constantly increases its speed on a straight vertical path so that at an altitude of 25 km, when the fuel is exhausted, it has attained a speed of 1500 m/sec. At this point the stabilizers automatically turn the rocket at an angle of 60°, after which the missile continues as an ordinary projectile. What time interval (in seconds) is available for measurements in the relatively air-free region above 25 km? What height does the rocket reach? What is the total time from the firing of the rocket to the instant it hits the ground? What assumptions have you introduced?

Problem 8.11. A hunter aims his gun barrel directly at a monkey in a distant palm tree. Where will the bullet go? If the animal, startled by the flash, drops out of the branches at the very instant of firing, will it then be hit by the bullet? Explain. Would this last answer hold if the acceleration of gravity were not 9.8 m/sec^2 but only 1/6 as great, as on the moon?

Problem 8.12. For each of the following cases sketch two graphs, one of vertical displacement vs. time elapsed and the other of vertical velocity vs. time elapsed. Carefully use the convention concerning positive and negative values. (a) A parachutist falls from a plane, after a while opens the `chute, and floats to the ground. (b) A marble drops from your hand and bounces three times before coming to rest on the floor. (c) A shell is fired at 60° elevation and falls into a deep valley.

Problem 8.13. Read through the Fourth Day (Motion of Projectiles) in Galileo's *Two New Sciences,* and on this basis briefly discuss the following: (a) Galileo's examination of the role of air resistance in projectile motion; (b) Galileo's use of experiments to further his argument; (c) Galileo's interest in practical applications of his work.

RECOMMENDED FOR FURTHER READING

Robert K. Adair, *The Physics of Baseball,* second edition, New York: HarperCollins, 1994

I. B. Cohen, *The Birth of New Physics,* pages 109–117, 212–213.

Stillman Drake, *Galileo*

A. R. Hall, *From Galileo to Newton 1630–1720,* New York: Harper, 1963; Chapters I–V. Covers several aspects of the development of seventeenth-century science not mentioned in this book.

PART C

Newton's Laws and His System of the World

9 Newton's Laws of Motion

10 Rotational Motion

11 Newton's Law of Universal Gravitation

This subject of the formation of the three laws of motion and of the law of gravitation deserves critical attention. The whole development of thought occupied exactly two generations. It commenced with Galileo and ended with Newton's *Principia;* and Newton was born in the year that Galileo died. Also the lives of Descartes and Huygens fall within the period occupied by these great terminal figures. The issue of the combined labors of these four men has some right to be considered as the greatest single intellectual success which mankind has achieved.

A. N. Whitehead, *Science and the Modern World*

As we look into history, it seems that sometimes progress in a field of learning depended on one person's incisive formulation of the right problem at the right time. This is so with the section of mechanics called *dynamics,* the science that deals with the effects of forces on moving bodies.

The person was Isaac Newton, and the formulation was that of the concepts *force* and *mass,* expounded in three interconnected statements that have come to be called *Newton's three laws of motion.* These, together with an introduction to rotational motion, complement and extend the study of pure kinematics in Part B. We can then not only understand a great many simple but important mechanical phenomena, but also, with the help of Newton's law of universal gravitation, we can solve some of the outstanding problems in astronomy introduced in Part A. Once we have successfully dealt with such problems, Newtonian dynamics becomes a powerful tool for advancing to the next level of understanding the physical sciences.

Newton's Laws of Motion

9.1 Science in the Seventeenth Century

Between the time of Galileo's death and the publication of Isaac Newton's *Philosophia Naturalis Principia Mathematica* (1687) lie barely 44 years, yet an amazing change in the intellectual climate of science has taken place in that brief interval. On the one side, the "New Philosophy" of experimental science is becoming a respected and respectable tool in the hands of vigorous and inventive investigators; and, on the other, this attitude is responsible for a gathering storm of inventions, discoveries, and theories. Even a very abbreviated list of these, covering less than half of the seventeenth century and only the physical sciences, will show the justification for the term "the century of genius": the work on vacuums and pneumatics by Torricelli, Pascal, von Guericke, Boyle, and Mariotte; Descartes' great studies on analytical geometry and on optics; Huygens' work in astronomy and on centripetal force, his perfection of the pendulum clock, and his book on light; the establishment of the laws of collisions by Wallis, Wren, and Huygens; Newton's work on optics, including the interpretation of the solar spectrum, and his invention of calculus almost simultaneously with Leibniz; the opening of the famous Greenwich observatory; and Hooke's work, including that on elasticity.

Scientists are no longer isolated. Although we must not suppose that suddenly all scientists are of a modern bent of mind, the adherents of the New Philosophy are many and growing in numbers. They have formed several scientific societies (in Italy, England, and France), of which the most famous is the Royal Society of London (as of 1662). They meet and communicate with one another freely now, they cooperate and debate, write copiously, and sometimes quarrel vigorously. As a group, they solicit support for their work, combat attacks by antagonists, and have a widely read scientific journal. Science is clearly becoming well defined, strong, and international.

What is behind this sudden blossoming? Even a partial answer would lead us far away from science itself, to an examination of the whole picture of quickening cultural, political, and economic changes in the sixteenth and seventeenth centuries. One aspect is that both craftsmen and men of leisure and money begin to turn to science, the one group for improvement of methods and products and the other for a new and exciting hobby—as *amateurs* (in the original sense, as lovers of the subject). But availability of money and time, the need for science, and the presence of interest and organizations do not alone explain or sustain such a thriving enterprise. Even more important ingredients are able, well-educated persons, well-formulated problems, and good mathematical and experimental tools.

Able scientists—these were indeed at hand. Some, like Newton, found freedom to work even in universities still dominated by medieval scholasticism. Others, like Wren and Hooke, found time to pursue scientific research in the midst of busy careers in other fields such as architecture, if they did not, like Robert Boyle, possess the initial advantages of inherited wealth.

Well-formulated problems—these had recently become clear in the writings of Galileo and others. It was Galileo, above all, who had directed attention to mathematics as the fruitful language of science and had presented a new way of viewing the world of fact and experiment. While Francis Bacon's emphasis on observation and induction was acclaimed by many, especially in the Royal Society, Galileo had also demonstrated the fruitfulness of bold hypotheses combined with mathematical deduction. His scorn of sterile introspection and blind subservience to ancient dogma echoed through all fields of science. Plato's old question, "By the assumption of what uniform and ordered motions can the apparent motions of the planets be accounted for?" had lost its original meaning in the new science. The new preoccupation is illustrated by what may be called the two most critical problems in seventeenth-century science: "What forces act on the planets to explain their actually observed paths?" and "What accounts for the observed facts of terrestrial motions now that Aristotelian doctrines have failed us?"

Good mathematical and experimental tools were also being created. With mathematics finding expression in physics, the two cross-fertilized and gave rich harvests, as might be expected when the same people (Descartes, Newton, and Leibniz) were making major discoveries in both fields. Analytic geometry and the calculus are parts of the rich legacy of the seventeenth century; they were still useful in science even after many of the scientific theories of that century were ultimately superseded. The telescope, the microscope, and the vacuum pump opened up vast new realms of phenomena for scientists to study, and the need for more accurate measurement of these phenomena stimulated the invention of other ingenious devices, thus initiating the collaboration between scientist and instrument maker that has become typical of modern science.

But in trying to solve the riddle of the spectacular rise of science after Galileo, perhaps we must simply realize that science has an explosive potential for growth once the necessary conditions are in place.[1] Around the time of Galileo the necessary conditions for growth were established: At last there were many men with similar attitudes working in the same fields; they could communicate more freely with one another and had access to the accumulated achievements of the past (partly through the relatively young art of printing); they had become impatient with qualitative reasoning and began to be more and more intrigued with the quantitative approach. To use a modern analogy, a critical mass was reached, and a chain reaction could proceed.

This, from the point of view of science, was the new age in which Newton lived. But a brief word of caution before we follow his work. The history of ideas, in science as in any other field, is not just an account of visits with the most important names. The work of each genius is made possible, stabilized, sometimes even provoked, but always connected with the whole structure of science only through the labors of lesser-known persons, just as a brick is surrounded by fortifying mortar. A house is not just a heap of bricks—and science cannot be made by giants only. Therefore, properly, we should trace in each person's contribution the heritage of the past, the influence of contemporaries, and the meaning for successors. It would be rewarding, but here we can touch on only the barest outlines.

9.2 A Short Sketch of Newton's Life

Isaac Newton was born on Christmas day in 1642, in the small village of Woolsthorpe in Lincolnshire, England. He was a quiet farm boy who, like young Galileo, loved to build and tinker with mechanical gadgets and seemed to have a secret liking for mathematics. Through the fortunate intervention of an uncle he was allowed to go to Trinity College, Cambridge University, in 1661 (where he appears to have initially enrolled in the study of mathematics as applied to astrology!). He proved an eager and excellent student. By 1666, at age 24, he had quietly made spectacular discoveries in mathematics (binomial theorem, differential calculus), optics (theory of colors), and mechanics. Referring to this period, Newton once wrote:

> And the same year I began to think of gravity extending to the orb of the Moon, and . . . from Kepler's Rule [third law] . . . I deduced that the forces which keep the Planets in their orbs must [be] reciprocally as the squares of their distances from the centers about which they revolve: and thereby compared the force requisite to keep the Moon in her orb with the force of gravity at the surface of the earth, and found them to answer pretty nearly. All this was in the two plague years of 1665 and 1666, for in those days I was in the prime of my age for invention, and minded Mathematicks and Philosophy more than at any time since.
>
> (Quoted in Westfall, *Never at Rest*)

[1]Historians in recent decades have been debating the role of religious factors, in particular the rise of Puritanism in England, in creating a favorable social climate for science. Some articles summarizing this controversy are cited in the website bibliography for this book. Another interpretation, from the viewpoint of the discipline of sociology of science, is well presented in Joseph Ben-David's book, *The Scientist's Role in Society.*

From his descriptions we may conclude that during those years of the plague, having left Cambridge for the time to study in isolation at his home in Woolsthorpe, Newton had developed a clear idea of the first two laws of motion and of the formula for centripetal acceleration, although he did not announce the latter until many years after Huygens' equivalent statement.[2]

After his return to Cambridge, he did such creditable work that he followed his teacher as professor of mathematics. He lectured and contributed papers to the Royal Society, at first particularly on optics. His *Theory of Light and Colors,* when finally published, involved him in so long and bitter a controversy with rivals that the shy and introspective man even resolved not to publish anything else. As Bertrand Russell pointed out, "If he had encountered the sort of opposition with which Galileo had to contend, it is probable that he would never have published a line."

Newton now concentrated most on an extension of his early efforts in celestial mechanics—the study of planetary motions as a problem of physics. In 1684 the devoted friend Halley came to ask his advice in a dispute with Wren and Hooke as to the force that would have to act on a body executing motion along an ellipse in accord with Kepler's laws; Newton had some time before found the rigorous solution to this problem ("and much other matter"). Halley persuaded his reluctant friend to publish the work, which touched on one of the most debated and intriguing questions of the time.

In less than two years of incredible labors the *Principia* was ready for the printer; the publication of the volume (divided into three "Books") in 1687 established Newton almost at once as one of the greatest thinkers in history.

A few years afterward, Newton, who had always been in delicate health, appears to have had what we would now call a nervous breakdown. On recovering, and for the next 35 years until his death in 1727, he made no major new discoveries, but rounded out earlier studies on heat and optics, and turned more and more to writing on theology. During those years he received honors in abundance. In 1699 he was appointed Master of the Mint, partly because of his great interest in, and competent knowledge of, matters concerning the chemistry of metals, and he seems to have helped in reestablishing the debased currency of the country. In 1689 and 1701 he represented his university in Parliament. He was knighted in 1705. From 1703 to his death he was president of the Royal Society. He was buried in Westminster Abbey.

9.3 Newton's *Principia*

In the original preface to Newton's work—probably, with Charles Darwin's later work on evolution, one of the greatest volumes in the history of science—we find a clear outline:

> Since the ancients (according to Pappus) considered *mechanics* to be of the greatest importance in the investigation of nature and science, and since the moderns—rejecting substantial forms and occult qualities—have undertaken to reduce the phenomena of nature to mathematical laws, it has seemed best in this treatise to concentrate on *mathematics* as it relates to natural philosophy [we would say "physical science"] . . . For the basic problem of philosophy seems to be to discover the forces of nature from the phenomena of motions and then demonstrate the other phenomena from these forces. It is to these ends that the general propositions in Books 1 and 2 are directed, while in Book 3 our explanation of the system of the world illustrates these propositions. For in Book 3, by means of propositions demonstrated mathematically in Books 1 and 2, we derive from celestial phenomena the gravitational forces by which bodies tend toward the sun and toward the individual planets. Then the motions of the planets, the comets, the moon, and the sea [tides] are deduced from these forces by propositions that are also mathematical.

What a prospect! But—analogous to Euclid's *Geometry,* the work begins with a set of definitions: in this case, mass, momentum, inertia, force, and centripetal force. Then follows a section on *absolute* and *relative* space, time, and motion. Since critiques of these definitions have

[2]This too must have been the time of the famous fall of the apple. One of the better authorities for this story is a biography of Newton written by his friend Stukely in 1752, where we read that on one occasion Stukely was having tea with Newton in a garden under some apple trees, when Newton told him he realized "he was just in the same situation, as when formerly, the notion of gravitation came to his mind. It was occasioned by the fall of an apple, as he sat in a contemplative mood."

been of some importance in the development of modern physics, some of them are worth noting here.

Mass. Newton was very successful in using the concept of mass, but less so in clarifying its meaning. He states that mass or "quantity of matter" is "the product of density and bulk" (bulk = volume). But what is density? Later on in the *Principia,* he defines density as the ratio of "inertia" to bulk; yet at the beginning, he had defined inertia as proportional to mass. So the definition appears to be circular, as Ernst Mach pointed out toward the end of the nineteenth century.

There have been various attempts to justify or replace Newton's definition of mass, but they are unnecessary, since the logical structure of the *Principia* does not really depend on this definition. In fact, it is now recognized by most scientists that *any* new theory is likely to postulate a certain number of concepts whose meaning must first of all be grasped intuitively (though later they may be defined operationally).

Thus a modern text or encyclopedia typically states that mass is simply an undefined concept that cannot be defined in terms of anything more fundamental, but must be understood operationally as the entity that relates two observable quantities, force and acceleration. Newton's second law, to be introduced below, can be regarded as a definition of mass.

What is more important is that Newton clearly established the modern distinction between *mass* and *weight*—the former being an inherent property of a body, whereas the latter depends on the acceleration due to gravity at a particular location (see next section).

Time. Newton writes:

> Absolute, true, and mathematical time, in and of itself and of its own nature, without relation to anything external, flows uniformly, and by another name is called duration. Relative, apparent, and common time, is any sensible and external measure (precise or imprecise) of duration by means of motion such a measure—for example, an hour, a day, a month, a year—is commonly used instead of true time.

No one dreamed of disputing this twofold definition until Mach complained that absolute time is "an idle metaphysical conception," and at first few took that criticism seriously. Albert Einstein reported that he was deeply influenced by reading Mach's book *The Science of Mechanics* (first published in German in 1883), and his theory of relativity, published in 1905, is in agreement with Mach at least on the point of rejecting absolutes such as time. Still we shall accept Newton's definition provisionally until Chapter 30, where, following Einstein, we shall argue that in a strict sense there cannot be any such thing as absolute time in physical theory, because it has no meaning in terms of operations.

Space. Newton continues:

> Absolute space, of its own nature without reference to anything external, always remains homogeneous and immovable. Relative space is any movable measure or dimension of this absolute space; such a measure or dimension is determined by our senses from the situation of the space with respect to bodies . . . Since these parts of space cannot be seen and cannot be distinguished from one another by our senses, we use sensible [perceptible by the senses] measures in their stead . . . instead of absolute places and motions we use relative ones, which is not inappropriate in ordinary human affairs, although in philosophy, abstraction from the senses is required. For it is possible that there is no body truly at rest to which places and motions may be referred.

From the last sentence it might seem that Newton was not sure if there is such a thing as absolute space. That would be a mistake. He thought he had proved by experiment that rotation is a kind of motion in absolute space.[3] The well-known Foucault pendulum experiment, first done in the middle of the nineteenth century, seems like another demonstration that the earth moves with respect to something fixed. At least we should recognize that Newton's belief in absolute space was plausible; at any rate it was not damaging to the structure of his system,

---

[3]A bucket, half filled with water, is suspended by a long rope, which is twisted so much that, when it is released, the bucket starts spinning rapidly. A few seconds later the water begins to follow the motion of the bucket, and also rises up on the sides, thus showing that the water is rotating in space, though at rest relative to the bucket with which it is in contact. Mach later countered that this is no proof that the water is rising owing to forces brought about while rotating in absolute space. He speculated, for example, that the same observation would be made if the bucket of water were kept still and the "fixed stars" were set in rotation with respect to the bucket.

since he hardly made explicit use of it anywhere (or, for that matter, of absolute time).

Immediately thereafter, still in the introductory section of the *Principia,* Newton stated his famous three laws of motion (which we shall discuss in the following section) and the principles of composition of vectors (for example, of forces and of velocities). Book 1, titled "The Motion of Bodies," applies these laws to problems of interest in theoretical astronomy: the determination of the orbit described by one body around another, when they are assumed to interact according to various force laws, and mathematical theorems concerning the summation of gravitational forces exerted by different parts of the same body on another body inside or outside the first one.

Another kind of application is relevant to the particle model of light: Newton computes the action of surfaces on small particles reflected or refracted by them. Book 2, on the motion of bodies in resisting mediums, seems to have as its main purpose the proof that Descartes' vortex model of ether and matter cannot adequately account for the observed motions of the planets; but along with this there are a number of theorems and conjectures about the properties of fluids.

Book 3, "The System of the World," is the culmination: It makes use of the general results derived in Book 1 to explain the motions of the planets and other gravitational phenomena such as the tides. It begins with a remarkable passage on "Rules of Reasoning in Philosophy." The four rules, reflecting the profound faith in the uniformity of all nature, are intended to guide scientists in making hypotheses, and in that function they are still up to date. The first has been called a *principle of parsimony,* and the second and third, *principles of unity.* The fourth is a faith without which we could not use the processes of logic. Because of their signal importance in the history of science we quote these rules almost in full:

RULE 1

No more causes of natural things should be admitted than are both true and sufficient to explain their phenomena.

As the philosophers say: 'Nature does nothing in vain, and more causes are in vain when fewer suffice. For nature is simple and does not indulge in the luxury of superfluous causes.'

RULE 2

Therefore, the causes assigned to natural effects of the same kind must be, so far as possible, the same.

Examples are the cause of respiration in man and beast, or the falling of stones in Europe and in America, or of the light of a kitchen fire and the sun, or of the reflection of light on the earth and the planets.

RULE 3

Those qualities of bodies that cannot be intended and remitted [that is, qualities that cannot be increased and diminished][4] *and that belong to all bodies on which experiments can be made should be taken as qualities of all bodies universally.*

For since the qualities of bodies can be known only through experiments; and therefore qualities that square with experiments universally are to be regarded as universal qualities; and qualities that cannot be diminished cannot be taken away from bodies. Certainly idle fancies ought not to be fabricated recklessly against the evidence of experiments, nor should we depart from the analogy of nature, since nature is always simple and ever consonant with itself. . . . We find those bodies that we handle to be impenetrable, and hence we conclude that impenetrability is a property of all bodies universally . . .

RULE 4

In experimental philosophy, propositions gathered from phenomena by induction should be considered exactly or very nearly true notwithstanding any contrary hypotheses, until yet other phenomena make such propositions either more exact or liable to exceptions.

"This rule should be followed so that the arguments based on induction may not be nullified by hypotheses.

At the end of Book 3 is a famous "General Scholium," which has had as much influence on subsequent discussions of scientific method as have the "Rules of Reasoning," because of one phrase which occurs in it. Having concluded that a satisfactory system of the world can be based on the postulate that there exists a universal force of gravitation between all pieces of matter in the universe, Newton confesses himself unable to discover any deeper explanation of the cause of this force which stands up to rigorous scrutiny—despite much effort he seems to have

[4]Newton means qualitative properties such as impenetrability, as distinguished from quantitative properties such as temperature.

expended on finding such a cause. Unwilling to put forward an artificial hypothesis unsupported by sufficient evidence, he simply states: "*Hypotheses non fingo*"—"I do not feign[5] hypotheses."

Following Book 3 in most editions of the *Principia* is an essay on "The System of the World," which provides a nonmathematical summary of the principal results of the third book. Now, as in the past, few readers ever get all the way through Books 1, 2, and 3; so we recommend that in order to get an overview of the main results of the work from its author's viewpoint, this concluding section be read immediately after his Introduction.

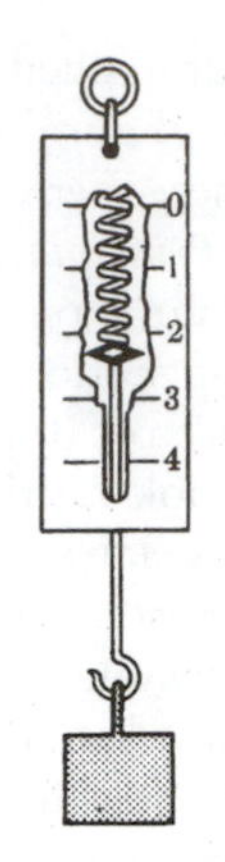

Fig. 9.1. Cutaway view of a spring balance.

9.4 Newton's First Law of Motion

Since we have already discussed the historical background of this law (Sections 5.3 and 6.1) we can proceed directly to a fairly didactic treatment from the modern point of view.

We may phrase Newton's first law of motion, or law of inertia, as follows: *Every material body persists in its state of rest or of uniform, unaccelerated motion in a straight line, if and only if it is not acted upon by a net (that is, unbalanced external) force.*

The essence is this: If you see a moving body deviating from a straight line, or accelerating in any way, then you must assume that a net force (of whatever kind) is acting on the body; here is the criterion for recognizing qualitatively the presence of an unbalanced force. But note well that this law does not help you to discover either how large the force is or its origin. There is implied only the definition of force as the "cause" of change of velocity, a definition we already alluded to in our discussion of Galileo's work (see Section 7.3). We recall that the Aristotelian scholastics had a rather different view in this matter; they held that force was also the cause of uniform (unaccelerated) motion.

As an example, consider the two opposing opinions that would exist when watching a cart pulled by a horse across a plane in steady motion. The Newtonian commentator would hold that the horse's efforts serve solely to equal and cancel the force of friction on the wheels, so that the net force on the cart, that is, the pull of the animal on it minus the force of friction between wheels and ground, is zero, the cart being consequently in a state of equilibrium, a state that by definition includes uniform unaccelerated motion as well as rest. On the other hand, the Aristotelian scholar would explain that since the natural state of the cart is to be at rest, a force has to be supplied to the cart by the horse to keep it in uniform motion—even in the absence of friction under the cart, if such a circumstance could be approached. We have here not so much a dispute about experimentally observable facts as a difference in viewpoint, complicated by the use of the same word, *force,* in two such disparate meanings.

As a matter of fact, most people who have not studied a little physics are intuitively Aristotelians, not Newtonians. The Aristotelian view, here as in some other contexts, is the one that is closer to contemporary common-sense opinion. Since friction is in actuality never absent and is often a very real hindrance in moving an object, it is natural that one develops the intuitive idea that a force is necessary to "keep things moving," and, as a next step, to define force as "the cause of continued motion." The Newtonian turns from this anthropomorphic position to consider the net forces on the moving body, and even dares to cast his definition of force in terms of the vector sum (that is, net force) without having first identified the individual components separately. Only the great practical success that followed it

[5]The Latin word *fingo* is sometimes translated "frame," but this obscures the connotation that Newton seems to have intended: To feign a hypothesis is to suggest it without really believing in it. We shall discuss the significance of Newton's assertion in Section 11.10.

could have justified a step that logically is so precarious.

Other points are also raised by the first law of motion. It has been made plain that a net force must be supplied to change a body's state from rest to motion or from motion to rest, from one speed to another or from one direction of motion to another even at the same speed. In somewhat unrigorous language still, we may say that material bodies are endowed with a sluggishness, a laziness toward change of motion, an inertia. There is very little profit in trying to account for this experimental fact in terms of some physical model or picture—let us simply accept that physical bodies are characterized by inertia just as they are by such other fundamental attributes as volume, chemical constitution, and so forth. The reason for the existence of mass (inertia) is still a frontier problem of physics.

If the change of direction of motion implies the action of a net force, we cannot think—as the ancients did—that no force need act on the planets in their orbits to keep their movement circular; on the contrary, we must consider that they are subjected to forces that draw them continually from a straight-line motion.

Problem 9.1. Explain in terms of Newton's first law of motion the common experience of lurching forward when a moving train suddenly decelerates and stops. Explain what happens to the passengers in a car that makes a sharp and quick turn. A penny is put on a record, and the motor is started; when the record has reached a certain speed of rotation, the coin flies off. Why?

Problem 9.2. Assume that a horizontal, perfectly frictionless table is set up in your laboratory. A block is placed on it and given a small push. What will be the motion of the block? How will this be affected by a straight-line motion of the whole laboratory during the experiment? What if the block moves in an arc sideways? How could you decide whether or not it is your laboratory that is moving in a curve?

Problem 9.3. An object weighing *X* units hangs on a spring balance (Fig. 9.1) in an elevator. Decide in each of the following cases whether the balance will read more than *X* or less, and why: the elevator accelerates upward; accelerates downward; moves without acceleration down; decelerates on the way down.

9.5 Newton's Second Law of Motion

So far, we have only a qualitative notion of the concept of force and, incidentally, of the property of bodies called *inertia*. But to develop a science of dynamics we must be able to determine forces and inertias quantitatively, instead of merely calling a body's inertia large because, in a qualitative way, a large force is needed to change its state of motion.

In Newton's own formulation of the second law, he states that the force acting on a body is equal to the change of its quantity of motion, where "quantity of motion" (later called "momentum") is defined as the product of mass and velocity. This is a simple generalization arising naturally from observations on collisions, in which a sudden blow produces a finite change of motion in a short period of time. But for continuously acting forces, such as gravity, it has been far more convenient to define force differently, that is, to use the *rate of change* of motion, which thus brings in Galileo's concept of acceleration; it is this version of the second law, formalized by the Swiss mathematician Leonhard Euler in 1750, that was eventually adopted in physics, and that we shall be using.[6]

It is customary to begin by stating the second law of motion in the form: *The net (external unbalanced) force acting on a material body is directly proportional to, and in the same direction as, its acceleration.*

Here the point is this: Once a net force is qualitatively detected by noticing changes of velocity (first law), let us define the net force precisely by the exact rate of change of velocity. If we adopt the symbols F_{net} and a, we can put the sentence above in the following form:

$$F_{net} \propto a \text{ (for a given body),}$$

or

$$F_{net}/a = \text{a constant for a given body.} \quad (9.1)$$

---

[6]It might appear that there would be a conflict between these two versions, yet Newton was able to use both of them in appropriate situations. He glossed over the difference in the definitions implied in the two equations $F = \Delta mv$ and $F = \Delta mv/\Delta t$. He did this either by taking $\Delta t = 1$ or by implicitly absorbing the factor Δt into the definition of F in the first case. Although Newton's second version is equivalent to the modern equation $F = ma$ discussed below, that equation itself never appeared in the *Principia*.

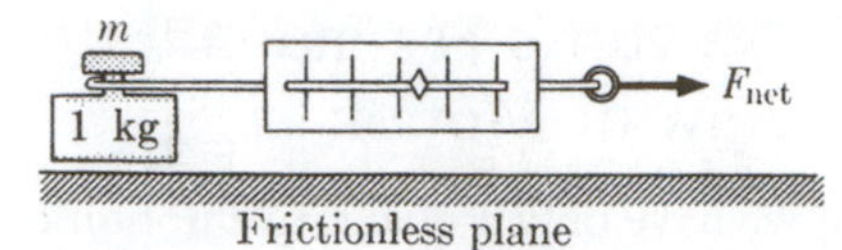

Fig. 9.2. Thought experiment to define the unit of force.

The constant so defined is a measure of the body's inertia, for clearly a large ratio of F_{net} to a means that a large force is needed to produce a desired acceleration, which is just what we expect to find true for large, bulky objects, to which we intuitively assign a larger inertia than to small, light objects.

If we now symbolize the constant in Eq. (9.1), the measure of inertia, by a letter of its own, *m,* and give it the alternative name *mass,* we can write the second law of motion:

$$F_{net}/a = m \quad \text{or } F_{net} = m \times a. \qquad (9.2)$$

Note that *m,* italicized, should not be confused with m, the abbreviation for "meter." (If you wonder why we make some letters do double or triple duty in physical science, take a look at Appendix I: There are many more than 26 concepts that need abbreviations!)

Equation (9.2) enables us in principle to assign a numerical value to a net force if we measure the acceleration it produces on a body of known mass, or conversely we could in principle obtain a numerical value for the mass in terms of the acceleration and the net force. But of course you notice the vicious circle: Unhappily we find that inertia and force, the two concepts we are trying to quantify rigorously, are mutually interconnected. To find the one we seemingly would have to know the other in advance.

There is one straightforward solution for this apparent dilemma. Before we try to measure any masses or forces, we first choose some one unique object, perhaps a certain piece of polished rock or metal, as the universal *standard of mass,* and regard it arbitrarily as having an inertia of unity, or *unit mass;* for example, name it the *standard mass of 1 kilogram* (abbreviated 1 kg). (It corresponds to about 2.2 pounds.) Then, by Eq. (9.2), any net force can be evaluated by means of the observable acceleration it produces on our standard object.

This important point will be clearer if we resort to an imaginary, idealized experiment ("thought experiment"): Place the standard 1-kg object on a smooth horizontal plane where friction is so negligible that we can speak of it as a frictionless plane. Now attach to the standard an uncalibrated spring balance (of negligible mass) and pull on it horizontally, as shown in Fig. 9.2. Here the applied force is indeed the net force. If this net force is constant (as seen by the pointer of the spring balance remaining at the same place throughout), the acceleration of the object is also constant, and therefore ascertainable without fundamental difficulties. Let us say that it comes by measurement to 4 m/sec^2 for a given pull in this concrete example. But that result tells us at once the magnitude of the pull also, for by Eq. (9.2), $F_{net} = m \times a$; thus

$$F_{net} = 1 \text{ kg} \times 4 \text{ m/sec}^2 = 4 \text{ kg}\cdot\text{m/sec}^2.$$

The pointer reading corresponding to that pull may now be marked[7] on the spring balance as 4 kg·m/sec^2; or, since the unit kg·m/sec^2 has traditionally been given the shorter name "newton" (abbreviated N, not italicized), the marking should be 4 newtons.

A single point will, of course, not serve to fix the calibration of a spring balance. Ideally, therefore, we repeat the whole procedure several times with different applied forces, by observing in each case the corresponding acceleration of the 1-kg object. Eventually we shall thus develop a fully calibrated balance; with its aid we may not only measure other forces directly, but—more important—may reverse the argument to determine quickly the unknown mass m_x of any other object. For example, the now calibrated balance is applied to m_x and drawn out to read 15 newtons while accelerating m_x on the frictionless plane. If the acceleration, measured simultaneously, is 6.2 m/sec^2, then, by Eq. (9.2),

$$m_x = F_{net}/a = 15 \text{ N}/(6.2 \text{ m/sec}^2)$$
$$= (15 \text{ kg}\cdot\text{m/sec}^2)/(6.2 \text{ m/sec}^2) = 2.4 \text{ kg}.$$

In summary: Newton's second law, in conjunction with the essentially arbitrary choice of

[7]That the force was constant can, strictly speaking, be judged only from the constancy of the observed acceleration. However, happily, there is a separate law of nature governing the behavior of elastic bodies like springs in spring balances, by which the displacement of the spring, and therefore the position of the pointer, is steady and unique for constant forces.

one standard of mass, conveniently fixes the unit of force, permits the calibration of balances, and gives us an operational determination of the mass of all other bodies.

Problem 9.4. Recount in detail what steps you must take (in idealized experimentation) to determine the unknown mass m_x (in kilograms) of a certain object if you are given nothing but a frictionless horizontal plane, a 1-kg standard, an uncalibrated spring balance, a meter stick, and a stop watch. [Note: Your answer represents a first approach toward an *operational definition* of mass, and will prove a great aid in comprehension of dynamical problems.]

Problem 9.5. What is the mass m_x if the observed acceleration of the object on the frictionless horizontal plane is 0.85 m/sec^2, while the balance reads off a net accelerating force of 22 N? What acceleration will a net force of 1.0 N produce on a 3.8-kg mass?

Problem 9.6. On a certain horizontal plane the friction under a moving body is not zero but 7.4 N, while the balance pulls horizontally with 13.6 N. If the observed acceleration of the body is 2.2 cm/sec^2, what is its mass?

Problem 9.7. The mass of a student is about 75 kg. As she sits in a car and starts from rest, her acceleration may be about 2 m/sec^2. What net force must the car seat exert to accelerate her, and in which direction? If the car comes from 60 km/hr to a sudden stop (in about 1 second) during an accident, what force is exerted on her and what applies the force?

9.6 Standard of Mass

It will readily be appreciated that the Standard of Mass that represents 1 kilogram, though essentially arbitrary, has been chosen with care. For scientific work, 1/1000 of 1 kilogram, equal to 1 gram, was originally defined as the mass of a quantity of 1 cubic centimeter (1 cm^3) of distilled water at 4°C. This decision, dating from the late eighteenth century, is rather inconvenient in practice. Although standardizing on the basis of a certain amount of water has the important advantage that it permits cheap and easy reproduction of the standard anywhere on earth, there are obvious experimental difficulties owing to the effects of evaporation, the additional inertia of the necessary containers, relatively poor accuracy for measuring volumes, and so on.

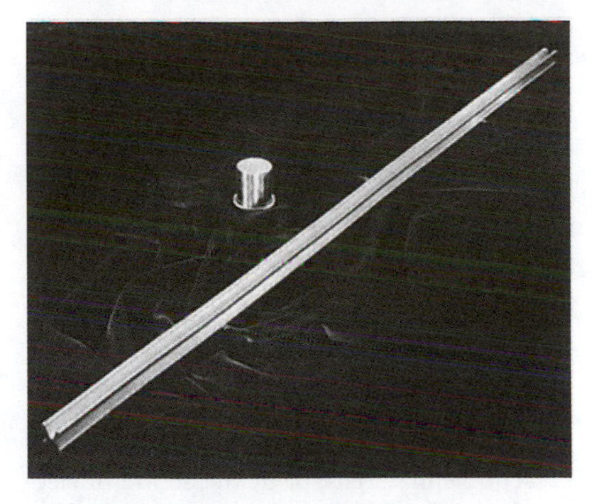

Fig. 9.3. **Standard kilogram, a platinum-iridium cylinder constructed in 1878, held at Sèvres, France, together with a standard meter.**

Therefore, it became accepted custom to use as the standard of mass a certain piece of precious metal, a cylinder of platinum alloy kept under careful guard at the Bureau Internationale des Poids et Mesures at Sèvres, a suburb of Paris (alongside what for a long time was defined as the standard of length, a metal bar regarded as representing the distance of 1 meter[8]). For use elsewhere, accurate replicas of this international standard of mass have been deposited at the various bureaus of standards throughout the world; and from these, in turn, auxiliary replicas are made for distribution to manufacturers, laboratories, etc.

Again, it was a matter of convenience and accuracy to make this metal block have an inertia 1000 times that of a 1-g mass; and thus the standard object (see Fig. 9.3) is a cylinder about 1 in. high of mass 1000 g, called 1 kg. When we think of 1 g of mass we can think of the 1/1000 part of that large kilogram standard. For all our discussions and thought experiments, we shall speak and think of our standard as being indeed a metal block having the mass of 1 kg. The weight of this block is about 2.2 times as great as 1 lb, the unit of weight generally used in the United States for commercial purposes. (More

[8]As indicated in footnote 1 to Chapter 23 and in Appendix III, this bar is no longer the primary standard for the meter.

strictly, 1 kg corresponds to 2.204622 lb avoirdupois, or 1 lb corresponds to 0.4535924 kg. But the easily remembered relation 1 lb $\leftrightarrow$ 0.454 kg is sufficient for us. For further discussion and tables of conversion factors, see the appendixes.)

9.7 Weight

Objects can be acted on by all kinds of forces; by a push from the hand; or by the pull on a string or spring balance attached to the object; or by a collision with another object; or by a magnetic attraction if the object is made of iron or other susceptible materials; or by the action of electric charges; or by the gravitational attraction that the earth exerts on bodies. But no matter what the origin or cause of the force, and no matter where in the universe it happens, its effect is always given by the same equation, $F_{net} = ma$. Newton's second law is so powerful precisely because it is so general, and because we can apply it even though at this stage we may be completely at a loss to understand exactly why and how a particular force (like magnetism or gravity) should act on a body. We glimpse here again the awesome beauty and generality of physics.

If the net force is, in fact, of magnetic origin, we might write $F_{mag} = ma$; if electric, $F_{el} = ma$; and so forth. Along the same line, we shall use the symbol F_{grav} when the particular force involved is the gravitational pull of the earth. Because this case is so frequently considered, a special name for F_{grav}, namely *weight,* or a special symbol, W, is generally used.

Of all the forces a body may experience, the gravitational force is the most remarkable. Let us discuss it in detail. First of all, we are struck by the fact that gravitational forces arise between two bodies (such as a stone and the earth below) without any need to prepare them for it—simply by placing them near each other. Before two pieces of steel attract each other by magnetism, we must specially treat at least one of them to magnetize it, and before electric forces can act between two spheres, at least one has to be sprayed with charges. But gravitational pulls are inherently present between *any* two bodies having mass, although ordinarily the mutual attraction is too small to become noticeable between objects lying on a table. Also, on the moon, sun, or other bodies the local acceleration of gravity is different from g on earth. At any rate, we shall concentrate at first on the more spectacular case of the *earth*'s pull on bodies at or near its surface. There, F_{grav} acts whether we wish it or no, whereas we can apply or remove at will mechanical, magnetic, and electric influences.

Our second point is this: It is easy to change the mechanical and electric forces on a material particle—by pulling harder on the string to which it is tied, or by depositing more charges on it, etc.—but it is not at all like that with its weight. No matter what we do to it, a body's weight, F_{grav}, at a given locality is essentially unaffected. To be sure, the farther the two bodies are separated, the less the gravitational attraction between them, as borne out by experiment, so that the weight of an object disappears completely as we remove it to a spot extremely far from all other objects, for example to interstellar space.[9] But if at a certain spot on this globe the weight of a block of metal is 1 newton, nothing will change this result short of destroying the identity of the object by cutting something off or by adding to it. Even if we interpose screens of all kinds between block and earth, the gravitational pull of the earth remains unchanged, as was proved long ago by careful experimentation.

Presupposed throughout the previous paragraphs was some accurate method for measuring F_{grav}. We might simply drop the object, allowing F_{grav} to pull on the object and to accelerate it in free fall, and then find the magnitude of F_{grav} by the relation

$$F_{grav} = \text{mass } (m) \times \text{acceleration owing to gravity } (g).$$

Knowing m by our previous type of experiment (Section 9.5), and observing experimentally the acceleration g in free fall, we can at once calculate F_{grav}. (Since this equation contains a mass, we again caution the reader not to confuse g, in italics, with g, the abbreviation for gram.)

Happily there is another method that is easier and more direct. We need only our previously calibrated spring balance; from it we hang the body for which F_{grav} is to be determined, and then wait until equilibrium is established. Now we do not allow F_{grav} downward to be the only force on the body, but instead we balance it out by the pull upward exerted by the spring balance. When

[9] The same result, though for an entirely different reason, is obtained by placing the object—in thought—at the center of our planet, where the gravitational pulls from all sides will fairly well cancel.

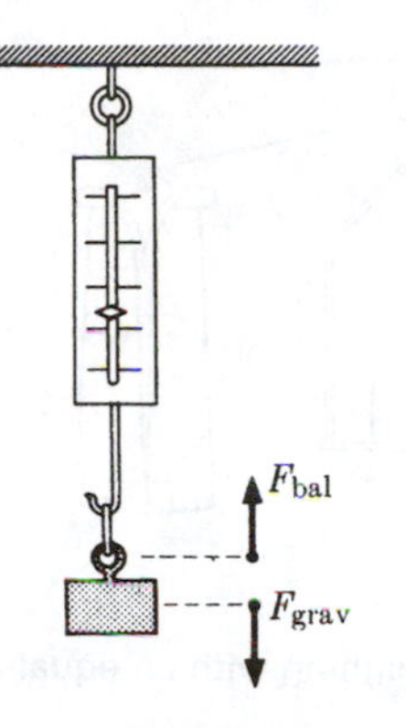

Fig. 9.4. Weighing with a spring balance.

the pointer comes to rest—say on the 5-newton reading—then we know (by Newton's first law) that the upward pull of the spring, F_{bal}, just counterbalances the downward pull of F_{grav} on the object. The net force on the body is zero. While oppositely directed, these two forces on the same object are numerically equal, and therefore F_{grav}, the weight in question, must also be 6 newtons (Fig. 9.4).

In passing we note the curious fact that in "weighing" we did not "read off" the weight, but read rather the magnitude of an oppositely directed balancing force. This might be called a *static* determination of weight, whereas the previous one was *dynamic*. As a matter of experimental fact, both procedures, though so different, give equal values at the same locality.

Yet another remarkable attribute of the gravitational force is revealed by Galileo's famous observation that (again at a given locality and neglecting air resistance) all objects fall with the same gravitational acceleration g. This can mean only that the gravitational attraction F_{grav} is simply proportional to the mass of the attracted object, that its composition, shape, etc. do not influence F_{grav} at all.

On reflection, this is an astonishing discovery. The magnitude of other forces (for example, electric and magnetic ones) is not at all simply proportional to the mass of the affected object. On the contrary; two spheres of equal mass but of different material generally behave entirely differently when in the presence of the same magnet or electric charge.

That there should exist a universal and strictly linear proportionality between weight and mass is logically as unexpected as if we had found that the weights of objects are proportional exactly to the square root of their volumes, or, to use an analogy, if it were discovered that on some far-off planet a man's wealth is proportional to the size of his body. Certainly, wealth and size would not thereby become identical concepts, but such a simple empirical relationship between them should allow us quickly to judge a man's bank account by surveying his bulk, and vice versa.

Similarly, although F_{grav}, the weight, and m, the mass, of an object refer to entirely different concepts, the simple relationship between the corresponding values soon puts us in the habit of judging the relative weight from an estimate of the mass, and vice versa.

In summary, the dynamic method of measuring weights by $F_{grav} = m \times g$ involves a prior determination of mass m and also a measurement of g. Now, while g is constant for all types of objects at a given locality and may for most purposes be taken as 9.80 m/sec^2 or 32.2 ft/sec^2 on the surface of the earth, the exact value is measurably different at different localities. Even the composition of the earth mass nearby affects it a little. As Table 9.1 shows, the variation is from about 9.7804 m/sec^2 at the equator to 9.8322 m/sec^2 at the North Pole—fully ½% of difference. A 1-kg mass will weigh about 9.81 newtons in Maine and 9.79 newtons in Florida. And at the same latitude, g will decrease with increasing height from sea level. Although the effect is

Table 9.1. Gravitational Acceleration

Location	*Value of g (m/sec²)*
Boston	9.804
Chicago	9.803
Denver	9.796
Key West	9.790
San Francisco	9.800
Washington, D.C.	9.801
Paris	9.809
Latitude 0°, sea level	9.78039
Latitude 90°, sea level	9.83217
Sun's surface	274.40
Moon's surface	1.67

generally quite feeble, so accurate are the methods of measurement that the minute change in g from one floor of a laboratory to the next is actually discernible!

Contemplation of the above table, which indicates that an object weighs only 1/6 as much on the moon as it does on earth, may remind you of the term "weightlessness" used to characterize the experience of astronauts. What does this mean? It could simply mean that if you are out in space, far from any object that exerts a gravitational force, g is very small so your weight mg is also very small. But the phenomenon is more general than that. In fact there are two important situations in which an object (or a person) is weightless.

First, consider the following thought experiment. As you stand on a scale to "weigh" yourself, the floor (which, while sagging slightly, has been pushing up on the scale) suddenly gives way. You and the scale drop into a deep well in "free fall." At every instant, your falling speed and the scale's falling speed will be equal, since you fall with the same acceleration (as Galileo proved). Your feet now touch the scale only barely (if at all). You look at the dial and see that the scale registers zero (the reason is explained in Example 1 of Section 9.10, below). This does not mean you have lost your weight—that could happen only if the earth suddenly disappeared or if you were suddenly removed to deep space. F_{grav} still acts on you as before, accelerating you downward. But since the scale is accelerating *with* you, you are no longer pushing down on it, nor is it pushing up on you. (You can get a slight hint of this experience in an elevator that rapidly accelerates downward.)

The second situation, which will be explained more fully in Chapter 11, is the weightlessness experienced by an astronaut in orbit around the earth. In this case the gravitational force *toward* the earth is balanced by the "centrifugal" force *away* from the earth produced by the astronaut's motion around it. (As we will see, the term "centrifugal force" is not correct, though commonly used.)

Problem 9.8. You are given a frictionless horizontal plane, a 1-kg standard mass, an uncalibrated spring balance, a meter stick, stop watches, and assistants. Without further assumptions or aids, and calling only upon Newton's laws of motion, recount in complete detail what steps you should take in idealized experimentation to determine the unknown weight F_{grav} (in newtons) of an object in two essentially different ways.

Problem 9.9. A standard mass is taken from Paris, France, to our National Institute of Standards and Technology outside Washington, D.C. How much is the mass near Washington if it was 1 kg in Paris? What is its weight in Paris and near Washington? What is the percentage change in its weight? (Consult Table 9.1 for necessary data.)

Problem 9.10. At the earth's equator (latitude 0°), the gravitational acceleration g is 9.78039 m/sec^2 at sea level and 9.77630 m/sec^2 at a height 1000 m above sea level. If these results were to be confirmed by comparing the weight of a 500-g object at these two levels on the same spring balance, what must be the magnitude of the smallest division on the scale that one can accurately estimate?

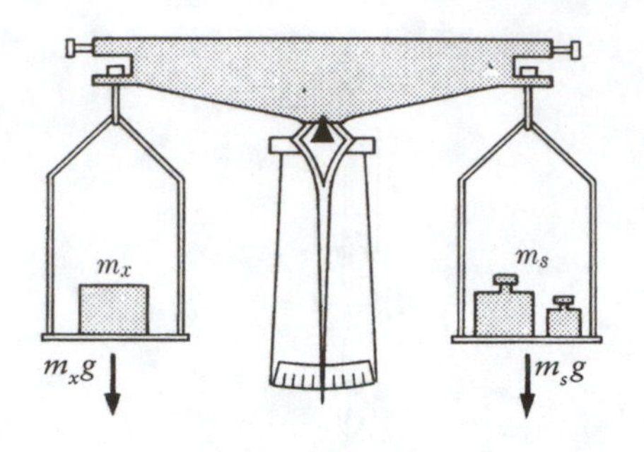

Fig. 9.5. Weighing with an equal-arm balance.

9.8 The Equal-Arm Balance

Before we leave the crucial—and initially perhaps troublesome—concept of mass, we must mention a third way of measuring the mass of objects, in everyday practice by far the most favored and accurate method.

By way of review, recall first that we need the essentially arbitrary standard. Once this has been generally agreed on, we can calibrate a spring balance by using it to give the standard measurable accelerations on a smooth horizontal plane. Then the calibrated balance can serve to determine other, unknown masses, either (a) by a new observation of pull and resulting acceleration on the horizontal plane, or (b) by measuring on the balance F_{grav} for the object in question and dividing this value by the measured gravitational acceleration g at that locality.

To methods (a) and (b) we now add method (c), known of course as "weighing" on an equal-arm balance—at first glance seemingly simple and straightforward, but in fact conceptually most deceptive. We place the unknown mass m_x on one pan (Fig. 9.5) and add a sufficient quantity of calibrated and marked auxiliary-standard masses on the other side to keep the beam horizontal. When this point is reached, the gravitational pull on the unknown mass, namely $m_x g$, is counterbalanced by the weight of the standards, $m_s g$. (Note that we assume here the sufficiently verified experimental result concerning the equality of gravitational accelerations for all masses at one locality.) But if $m_x g = m_s g$, then $m_x = m_s$. Counting up the value of all standard masses in one pan tells us directly the value of the mass on the other side.

The whole procedure is much to be preferred over methods (a) and (b), as the range, the accuracy of calibration, and the consistency of a spring balance are all quite limited, whereas in method (c) the equal-arm balance can handle accurately a very large range of masses, and the accuracy of individual measurements (which can be as good as 10^{-7} gram with special care) even without refinements is easily better than 1/1000 of a gram. And the results are repeatable and consistent, there being only these three sound and verifiable assumptions implied: that the arms of the balance are straight and equally long, that the value of g is the same under each pan, and that the initial marking of the auxiliary standard masses is permanently correct.

9.9 Inertial and Gravitational Mass

We probably take for granted that a result obtained by method (a), described in the previous section, is always equal to the result, for the same object, involving methods (b) or (c). And yet, a moment's thought should convince us that this identity is completely unforeseeable—either an incredible coincidence or a symptom of a profound new law of nature.

To begin with, consider that the mass was obtained in case (a) in a dynamic measurement: The object actually is allowed to "give in" to the force, is actually accelerated by the pull of the horizontally moving spring balance, and the mass so determined is truly a measure of the inertia of the object. Gravitational forces do not enter at all into the picture; in fact, the experimental measurement could be carried out just as well or better in gravitation-free space. On the other hand, methods (b) and (c), the static procedures of weighing on a vertically held spring balance or on an equal-arm balance, could not be invoked at all in the absence of gravity. Here the whole measurement depends on one property of material bodies—that they are attracted to other objects such as the earth—and inertia plays no role whatsoever.

We see that case (a) on one hand and, (b) and (c) on the other measure two entirely different attributes of matter, to which we may assign the terms *inertial mass* and *gravitational mass*, respectively. For practical purposes we shall make little distinction between the two types of mass. But in order to remind ourselves how essential and rewarding it may be to keep a clear notion of the operational significance of scientific concepts, and that, historically, considerable consequences may follow from a reconsideration of long-established facts in a new light, there are these words from Albert Einstein's and Leopold Infeld's book *The Evolution of Physics:*

> Is this identity of the two kinds of mass purely accidental, or does it have a deeper significance? The answer, from the point of view of classical physics, is: the identity of the two masses is accidental and no deeper significance should be attached to it. The answer of modern physics is just the opposite: the identity of the two masses is fundamental and forms a new and essential clue leading to a more profound understanding. This was, in fact, one of the most important clues from which the so-called general theory of relativity was developed.
>
> A mystery story seems inferior if it explains strange events as accidents. It is certainly more satisfying to have the story follow a rational pattern. In exactly the same way a theory which offers an explanation for the identity of gravitational and inertial mass is superior to the one which interprets their identity as accidental, provided, of course, that the two theories are equally consistent with observed facts.

Problem 9.11. Imagine that, in another universe, some bodies have only gravitational mass and others only inertial mass. Describe how some everyday objects would behave if made only of one or only of the other type of material.

Problem 9.12. An astronaut, standing on the ground in San Francisco and wearing a heavy space suit, throws a ball straight up in the air. It reaches a maximum height of about 11 m before it starts to fall back down. Then she does the same thing while standing on the surface of the Moon, throwing the ball with the same initial speed. How high does it go?

9.10 Examples and Applications of Newton's Second Law of Motion

Newton's second law, on which were built the previous paragraphs, is so central, so indispensable in physical science, that it will be well worth our looking at a few examples of its application to simple mechanical problems.

EXAMPLE 1

An object of mass m (measured in kg) hangs from a calibrated spring balance in an elevator (Fig. 9.6). The whole assembly moves upward with a known acceleration a (measured in m/sec^2). What is the reading on the balance?

Solution: As long as the elevator is stationary, the upward pull of the balance, and hence its reading F_1, will be equal in magnitude to the weight mg of the object. The same is true if the elevator is moving *up* or *down* with constant speed, which is also a condition of equilibrium and of cancellation of all forces acting on m. However, in order to accelerate upward with a m/sec^2, m must be acted on by a net force of ma newtons in that direction. In symbol form, $F_{net} = ma$; but

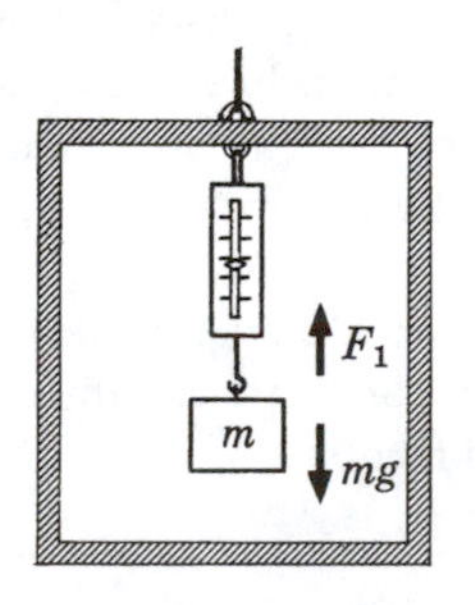

Fig. 9.6. Weight of an object suspended in an elevator.

$$F_{net} = F_1 - mg,$$

so

$$F_1 = ma + mg = m(a + g).$$

The reading F_1 will be larger than before. Conversely, on decelerating by a (in m/sec^2) while on the way up or accelerating on the way down, the spring balance will relax a little so that there remains on m some net force ($mg - F_1$) in the downward direction and, of course, of magnitude ma. Hence the reading $F_1 = mg - ma$. Finally, if the elevator were to fall freely ($a = g$), the net force on all objects within must be simply their weight. That is, the last equation becomes $F_1 = mg - mg$, that is, the spring balance reads zero. The object has only one force acting on it: its own weight F_{grav}. And a person standing on a balance while the elevator is falling will see his "weight" go to zero—the balance is falling away at the same rate under him. In this way one can experience "weightlessness" without changing either F_{grav} or g.

Once more we are impressed with the need to speak clearly: A spring balance evidently does not always simply register the weight of an object hanging from it. This would be true only if the whole system were in equilibrium, that is, at rest or in uniform, unaccelerated motion; in other cases, the balance shows a larger or smaller reading depending on the accelerating forces it must supply to the object.

EXAMPLE 2

A string, thrown over a very light and frictionless pulley, has attached at its ends the

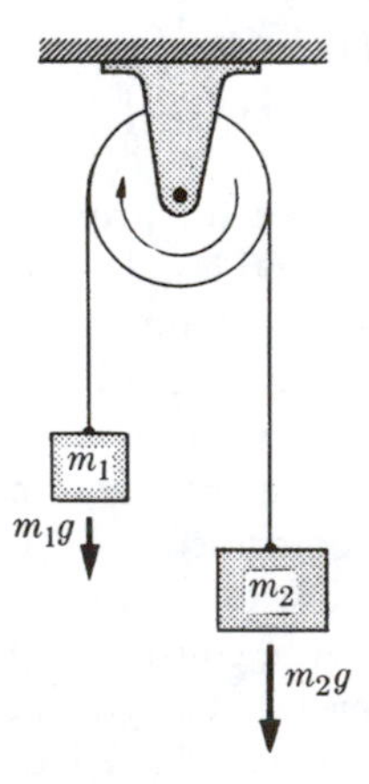

Fig. 9.7. Atwood's machine.

two known masses m_1 and m_2, as shown in Fig. 9.7. Find the magnitude of the acceleration of the masses.

Solution: In this arrangement, called Atwood's machine, after the eighteenth-century British physicist who originated it, the net external force on the system of bodies is $m_2g - m_1g$ (again assuming m_2 to be larger than m_1). The total mass being accelerated is $m_1 + m_2$, and, consequently,

$$a = \frac{(m_2 - m_1)}{(m_1 + m_2)} g, \tag{9.3}$$

which we may solve with the known values of m_1, m_2, and g for that locality. Conversely, if m_1, m_2, and a are experimentally determined with accuracy, and if corrections are made for the effects of string, pulley, and air resistance, a good value for g is obtainable.

EXAMPLE 3

In the Atwood's machine problem above, Example 2, we may wish to inquire what the tension is in the string while the masses accelerate, a problem not unlike that of Example 1. First of all we must define clearly the meaning of "tension in a string": for our purposes it will simply be the force reading if a spring balance were cut into the string or rope (Fig. 9.8a). For a rope of negligible mass (the usual specification in our problem), this force is the same at each end.

To obtain the tension T in terms of the other data or observables, consider Fig. 9.8b closely. If we fasten our attention on one or the other of the two masses only, say m_2, to the exclusion of the rest of the system, then it is evident that m_2 is acted on by a net force $(m_2g - T)$; therefore $(m_2g - T) = m_2a$, for m_2 is the only mass acted on by that net force. Rewriting, we obtain

$$T = m_2g - m_2a = m_2(g - a). \tag{9.4}$$

On the other hand, exclusive concentration on m_1 gives $F_{net} = T - m_1g$, or

$$T = m_1a + m_1g = m_1(g + a). \tag{9.5}$$

In answering the question as to the tension in the rope, either of these two results is usable. Incidentally, juxtaposing the two expressions for T, we obtain $m_2(g - a) = m_1(g + a)$, which yields again

$$a = \frac{(m_2 - m_1)}{(m_1 + m_2)} g.$$

This we had already derived on the basis of more direct arguments, which gives us confidence in the method of "isolation" adopted here. Conversely, the last equation may be used to determine the magnitude of the mass on one side of the pulley (say m_1) exclusively by observations made on the other side (that is, the magnitude of m_2 and a). For example,

$$m_1 = m_2 \frac{(g - a)}{(g + a)}.$$

In a casual way this last example has hinted at a most powerful method of argument in physical science, namely, *the isolation of a specific and small part of the total situation.* As Fig. 9.8b symbolized, T (and from it, m_1) could be found by observations made solely on one small portion of the assembly, the rest being perhaps hidden in the inaccessible darkness of a black box. We

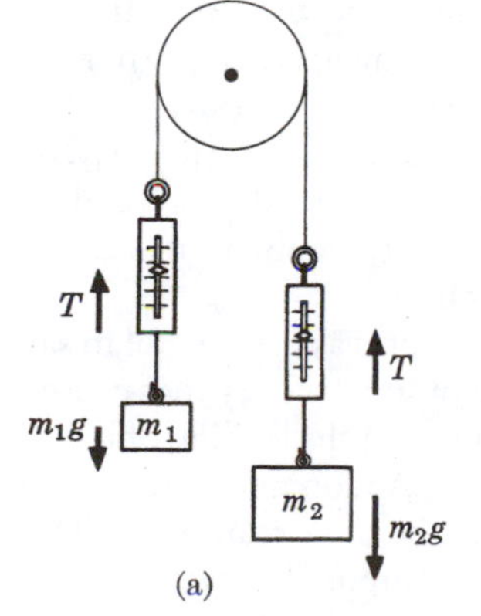

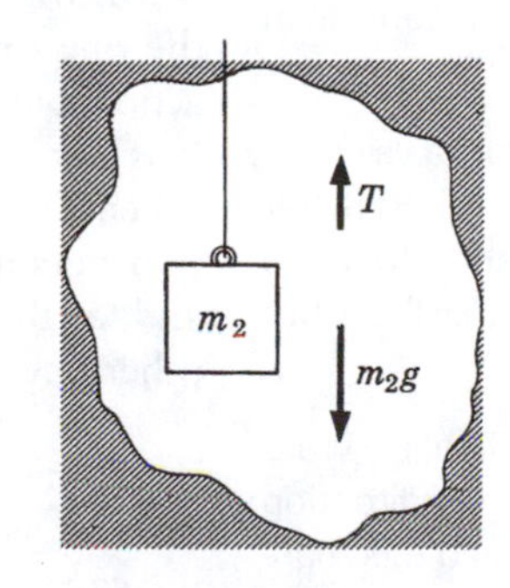

Fig. 9.8. Atwood's machine with spring balances.

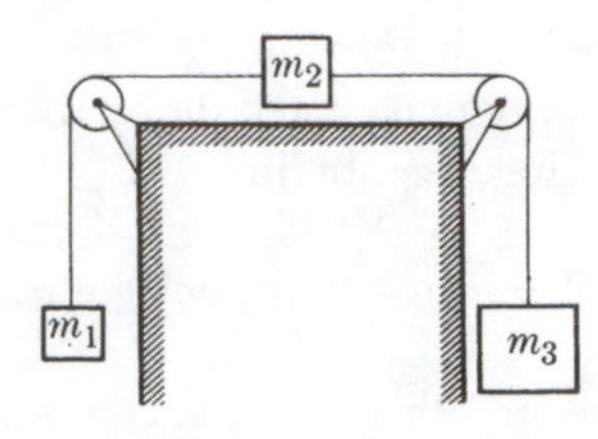

Fig. 9.9. Arrangement of masses for Problem 9.15.

can thereby tell something about the "inside" of the box though we must remain "outside."

In a sense, nature always confronts scientists with a "black-box problem": They are given some general laws of nature and a few observable manifestations of a complex scheme or hidden mechanism, and their task is to know as much as possible about the scheme or mechanism despite the necessarily scant and incomplete direct information. The problem may be as humble as this one—finding m_1 by observation made on m_2—or it may be as general as deducing the temperature, motion, size, mass, and composition of a distant star by observations made on a thin beam of its weak light.

Problem 9.13. A lamp hangs vertically from a cord in a descending elevator. The elevator decelerates by 300 cm/sec^2 before coming to a stop. If during this event the tension in the cord is 20 N, what is the mass of the lamp?

Problem 9.14. A 160-lb man stands on a platform (spring) balance in a moving elevator. Under what conditions will the balance read ¼ more than when at rest?

Problem 9.15. On a frictionless horizontal table rests a 15-kg mass m_2 connected as in Fig. 9.9 to hanging masses m_1 = 10 kg and m_3 = 25 kg. Find the acceleration of the system, and the tension in the rope above m_1.

Problem 9.16. A woman tries to obtain a value for *g* with an Atwood's machine but, without her knowing it, she and her whole laboratory are freely falling through space. What will be her result for *g*, and why?

Problem 9.17. If tomorrow the whole earth were to move with large acceleration in a direction toward the North Pole Star, how would this manifest itself mechanically in different parts of the world?

9.11 Newton's Third Law of Motion

Newton's first law defined the force concept qualitatively, and the second law quantified the concept while at the same time providing a meaning for the idea of mass. To these, Newton added another highly original and important law of motion, the third law, which completes the general characterization of the concept of force by explaining, in essence, that *each existing force on one body has its mirror-image twin on another body.* In Newton's words,

> *To any action there is always an opposite and equal reaction; in other words, the actions of two bodies upon each other are always equal and always opposite in direction:* Whatever presses or draws something else is pressed or drawn just as much by it. If anyone presses a stone with a finger, the finger is also pressed by the stone. If a horse draws a stone tied to a rope, the horse will (so to speak) also be drawn back equally towards the stone, for the rope, stretched out at both ends, will urge the horse toward the stone and the stone toward the horse by one and the same endeavor to go slack and will impede the forward motion of the one as much as it promotes the forward motion of the other.
>
> (*Principia*)

The point is rather startling: A single particle, existing all by itself, can neither exert nor experience any force at all. Forces spring up only as the result of interaction of two entities, and then the one pushes or pulls the other just as much as it is being pushed or pulled in return. The earth is attracted upward to a falling apple as much as the falling apple is attracted downward toward the earth. We may choose to call the one force the action and the other the reaction, but the order of naming is really arbitrary. It is not that one of the two forces *causes* the other. They cause each other simultaneously, as two persons, caught in a crowd, may collide and, at the same moment, complain it was the other one who had pushed.

In the same manner, it is as meaningful to say it was the borrowing of some money that caused the debt as it is to reply that the lending of it was the cause of the credit. Action and reaction do stand to each other in the same relation as debit to credit: The one is impossible without the other, they are equally large but in an opposite sense, the causal connection is introduced only artificially, and, most important, they happen respectively to two different bodies.

To emphasize all these points, we might rephrase the third law of motion as follows:

Whenever two bodies A and B interact so that body A experiences a force (whether by contact, by gravity, by magnetic interaction, or whatever), then body B experiences simultaneously an equally large and oppositely directed force.

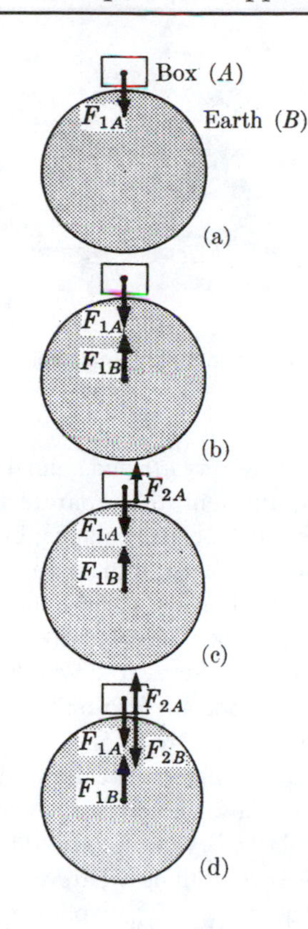

Fig. 9.10. Forces on box and earth in contact.

9.12 Examples and Applications of Newton's Third Law

It is easiest to understand the third law in terms of worked-out examples. The first two involve bodies in equilibrium, the last two accelerating systems.

EXAMPLE 1

The simplest case concerns a box (body A) standing on the earth (body B). Let us identify the forces that act on each. Probably the one that comes to mind first is the weight of the box, F_{grav}. We name it here F_{1A} and enter it as a vertically downward arrow, "anchored" to the box A at its center of gravity (see Fig. 9.10a). The reaction that must exist simultaneously with this pull of the earth on the box is the pull of the box on the earth, equally large (by the third law) and entered as a vertical, upward arrow, F_{1B} at the center of the earth in Fig. 9.10b. This completely fulfills the third law.

However, as the second law informs us, if this were the complete scheme of force, the box should fall down while the earth accelerates up. This is indeed what can and does happen while the box drops to the earth, settles in the sand, or compresses the stones beneath it. In short, the two bodies do move toward each other until enough mutual elastic forces are built up to balance the previous set. Specifically, the earth provides an upward push on the box at the surface of contact, as shown in Fig. 9.10c by F_{2A} an upward arrow "attached" to the box, while, by the law now under discussion, there exists also an equal, oppositely directed force on the ground, indicated by F_{2B} in Fig. 9.10d.

There are now two forces on each body. Equilibrium is achieved by the equality in magnitude of F_{1A} and F_{2A} on A, and by F_{1B} and F_{2B} on B. But beware! F_{1A} and F_{2A} are not to be interpreted as action and reaction, nor are F_{1B} and F_{2B}. The reaction to F_{1A} is F_{1B}, and the reaction to F_{2A} is F_{2B}. Furthermore, F_1 and F_2 are by nature entirely different sets of forces—the one gravitational and the other elastic. In short, F_{1A} and F_{1B} are equally large by Newton's third law, but F_{1A} is as large as F_{2A} (and F_{1B} as large as F_{2B}) by the condition of equilibrium, derived from the second law.

EXAMPLE 2

The sketch in Fig. 9.11 involves a system of four elements—a horizontal stretch of earth E on which a recalcitrant beast of burden B is being pulled by its owner M by means of a rope R. Follow these four force couples: F_{1E} is the push experienced by the earth, communicated to it by the man's heels (essentially by static friction). The reaction to F_{1E} is the equally large force F_{1M} exerted on the man by the earth. The man pulls on the rope to the left with a force F_{2R} and the reaction to this is the force F_{2M} with which the rope pulls on the man to the right. A third set is F_{3B} and F_{3R} acting respectively on the donkey and on the rope. Finally, the

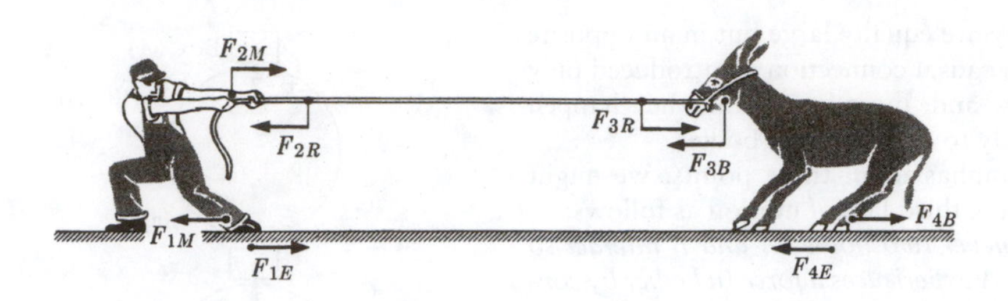

Fig. 9.11. Man and donkey pulling each other.

interaction between earth and animal is F_{4E} and F_{4B}. In equilibrium, the separate forces on each of the four objects balance; but if equilibrium does not exist, if the man succeeds in increasing the donkey's speed to the left, then $F_{3B} - F_{4B} = m_{\text{beast}} \times a$, and similarly for the other members of the system. And whether there is equilibrium or not, any "action" force is equal and opposite to its "reaction."

The whole point may be phrased this way: By the third law, the forces F_{1E} and F_{1M} are equal; similarly, F_{2M} and F_{2R} are equal. But the third law says nothing whatever about the relationship of F_{1M} to F_{2M}, two forces arranged to act on the same body by virtue of the man's decision to pull on a rope, not by any necessity or law of physics. If there happens to be equilibrium, then F_{1M} will be as large as F_{2M} by Newton's second law (or, if you prefer, by definition of "equilibrium").

Incidentally, to the owner of the animal the difference between zero velocity and a larger, though constant, velocity is of course all that matters; and so he would naturally consider Newtonian physics contrary to common sense if he heard that we lump both these so differently prized states, rest and uniform motion, under one name equilibrium—and treat them as equivalent. The donkey driver's attitude is certainly understandable, but surely we must avoid it if, as happens again and again, science forces us to break with commonsense preconceptions. The reward is always greater insight into a whole class of natural phenomena.

EXAMPLE 3

Consider an automobile accelerating on a straight road. The force that propels it is essentially the (static) friction between the tires and road at their surfaces of momentary contact, and this force is, by the third law, equal to the contrary force that the earth experiences. Consequently an observer on a fixed outpost in the universe, on a star perhaps, should ideally be in a position to observe the earth moving backward under the car while the latter moves forward over the earth.

But evidently these equal forces, if they were the only ones acting, would produce very unequal accelerations. The inertia of the earth being effectively infinitely large compared with that of the car, the former will experience only an infinitely small acceleration.

Nevertheless, an important conceptual clarification is implied: Strictly speaking, the a in $F_{\text{net}} = ma$, the acceleration produced by forces on the earth, should be measured not with respect to the earth itself, but rather with respect to some fixed point in space—a "fixed star" would serve.

In practice, we do allow ourselves to use the earth as a frame of reference for the measurement of accelerations because the earth's large inertia ensures that it will respond only negligibly to the forces of reaction communicated to it by mutual gravitation, friction, elastic interaction, and so on.

EXAMPLE 4

We come now to a humble-looking experiment with a lot of meaning. Consider two blocks at rest on a horizontal, frictionless track (or fitted as cars with light, friction-free wheels). They are connected to each other by a tightly compressed spring (Fig. 9.12). Of course, the

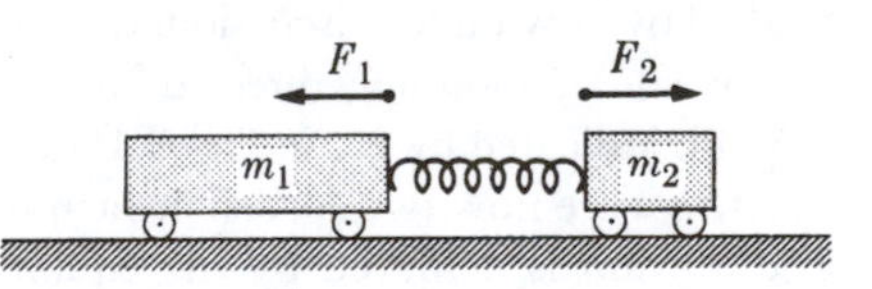

Fig. 9.12. The reaction-car experiment.

forces F_1, and F_2 on the two blocks owing to their mutual pressure are equal though oppositely directed, and only the balancing forces exerted by a string stretched from one block to the other prevents them from shooting off in opposite directions. Now the string is cut or burned off, and the freed spring propels the cars to the right and the left, respectively. At any instant, the opposing forces on the two cars are still equally large; we express this by an equation,

$$F_1 = -F_2.$$

As the spring quickly relaxes more and more, the forces at its two ends diminish in magnitude until finally the mutual forces on the cars are zero as they separate completely and the spring drops out entirely. Taking an average force $\bar{F}$ for the whole performance, we have

$$\bar{F}_1 = -\bar{F}_2.$$

However, in the absence of friction, $\bar{F} = m\bar{a}$ (where $\bar{a}$ stands for average acceleration), or

$$m_1\bar{a}_1 = -m_2\bar{a}_2. \quad (9.6)$$

The average acceleration is given by $\bar{a} = (v - v_0)/t$. For both cars, $v_0 = 0$ (start from rest), and t, the time of interaction, is of course the same on each. Thus we rewrite Eq. (9.6):

$$m_1v_1 = -m_2v_2. \quad (9.7)$$

Considering simply the *numerical magnitudes* of the velocities, that is, neglecting the minus sign (which only serves to remind us that the velocities are toward opposite directions), we note that

$$m_1/m_2 = v_2/v_1, \quad \text{or}$$
$$m_1 = m_2(v_2/v_1). \quad (9.8)$$

In words, these equations reveal that the velocities achieved by two objects by mutually interacting forces are inversely proportional to their masses.

Equation (9.8) has special further importance. If one of these objects, say, m_2, is chosen to be a standard of mass, then the inertia of any other object m_1 can be directly compared by means of the easily found ratio of measured velocities v_2 and v_1. Specifically, if in some special case m_2 = 1 kg (by definition), and v_2 = 20 cm/sec and v_1 = 5 cm/sec by observation, then m_1 = 4 kg.

This so-called *reaction-car experiment* immediately offers us a fourth and last method, which we shall call (d) in analogy with the three methods discussed in Section 9.8, for the determination of the mass of an object—namely, by using a standard as mass m_2 and finding the unknown mass from Eq. (9.8). As in method (a), we find here its inertial mass relative to a previously chosen standard; but this new case has the profound advantage that nowhere is there a calibration of spring balances or even a *measurement* of forces! As long as we know, by ample experimental verification of Newton's third law, that the mutual forces between these two cars are equally large and oppositely directed, there is no further need to investigate their actual magnitudes.

Problem 9.18. A $3\frac{1}{3}$-ton truck starts from rest and reaches a speed of 40 km/hr in half a minute. What will be the change of speed of the earth during that same time? (Mass of earth = 6.6×10^{21} tons or about 6×10^{24} kg.) Is the effect on the earth different if the truck takes three times as long to reach the same final speed?

Problem 9.19. The thought experiment involving the reaction cars for the determination of relative masses was first proposed by Ernst Mach, the nineteenth-century Austrian physicist, in one of the early self-conscious examinations of the foundations of mechanics. (His works later stimulated a good deal of modern thought in physics and the philosophy of science.) Describe a detailed step-by-step procedure for using his method to find the inertial mass of an object (in kilograms). Then describe how you would go on to prove experimentally that the gravitational mass of the object (in kilograms) has an identical value; and finally, show how you would find the accurate weight of the object (in newtons) at that locality.

RECOMMENDED FOR FURTHER READING

Gale E. Christianson, *Isaac Newton and the Scientific Revolution,* New York: Oxford University Press, 1996

I. Bernard Cohen, *The Birth of a New Physics,* revised edition, Chapter VII; "Newton's second law and the concept of force in the *Principia,* in

History of Physics (edited by S. G. Brush), pages 33–63

I. Bernard Cohen and Richard S. Westfall (editors), *Newton: Texts, Backgrounds, Commentaries,* New York: Norton, 1995; selections from the *Principia,* pages 116–118, 219–238, and commentary by A. Koyré, pages 58–72

A. R. Hall, *From Galileo to Newton 1630–1720,* New York: Harper, 1963; Chapters VI, VII, and IX

John Henry, *The Scientific Revolution,* Chapter 4

James R. Jacob, *The Scientific Revolution, Chapter 5*

R. S. Westfall, "The career of Isaac Newton: A scientific life in the seventeenth century," *American Scholar,* Vol 50, pages 341–353 (1981); *The Life of Isaac Newton,* New York: Cambridge University Press, 1993

CHAPTER 10

Rotational Motion

At this point in our study, we have at our command enough knowledge and tools to deal with a large variety of problems concerning motions and forces. The basic structure has been set up for an understanding of the type of concepts, questions, and methods of answer in the repertoire of the physical scientist. But there is still one dangerous gap to be filled, one main supporting pillar to be moved into place before the next level can be constructed.

In the previous chapters we first acquainted ourselves with the description of uniformly accelerated motions along a straight line, and in particular with that historically so important case of free fall. Next came general projectile motion, an example of motion in a plane considered as the superposition of two simple motions. Then we turned to a consideration of the *forces* needed to accelerate bodies along a straight line. But there exists in nature another type of behavior, not amenable to discussion in the terms that we have used so far, and that is *rotational motion,* the motion of an object in a plane and around a center, acted on by a force that continually changes its direction of action. This topic subsumes the movement of planets, flywheels, and elementary particles in cyclotrons.

We shall follow the same pattern as before: concentrating on a simple case of this type, namely, circular motion. We shall first discuss the kinematics of rotation without regard to the forces involved, and finally study the dynamics of rotation and its close ally, vibration.

10.1 Kinematics of Uniform Circular Motion

Consider a point whirling with constant speed in a circular path about a center *O;* the point may be a spot on a record turntable, or a place on our rotating globe, or, to a good approximation, the planet Venus in its path around the sun.

Before we can investigate this motion, we must be able to describe it. How shall we do so with economy and precision? Some simple new concepts are needed:

a) The *frequency* of rotation is the number of revolutions made per second (letter symbol n), expressed in 1/sec (or sec^{-1}). A wheel that revolves 10 times per second therefore has a frequency of $n = 10\ \text{sec}^{-1}$. While useful and necessary, the phrase "number of revolutions" does not belong among such fundamental physical quantities as mass, length, and time.

b) Next, we define the concept *period of rotation* (symbol T) as the number of seconds needed per complete revolution, exactly the reciprocal of n, and expressed in units of seconds.

$$T = \frac{1}{n} \qquad (10.1)$$

The wheel would consequently have a period of rotation of 0.1 sec.

c) An angular measure is required. The angle θ swept through by a point going from P_1 to P_2 (Fig. 10.1) can, of course, be measured in degrees, but it is more convenient in many problems to express θ by the defining equation

$$\theta = \frac{s}{r} \qquad (10.2)$$

where s is the length of the arc and r is the radius of the circle. This ratio of arc to radius is a dimensionless quantity; however, the name *radians* (abbreviation: rad) is nevertheless attached to this measure of angle, partly to distinguish it from degrees of arc.

To fix the rate of exchange between these two types of angular measure, consider the special case of $\theta = 360°$. For that angle, $s = 2\pi r$, namely, the circumference of the circle, and therefore $\theta = 2\pi r/r = 2\pi$ rad. If $360° = 2\pi$ rad, $1° = 0.0175$ rad and 1 rad $= 57.3°$.

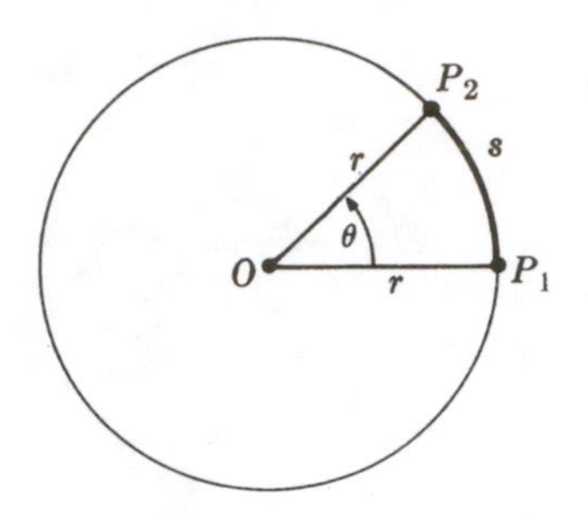

Fig. 10.1. Definition of an angle in radian measure: $\theta = s/r$.

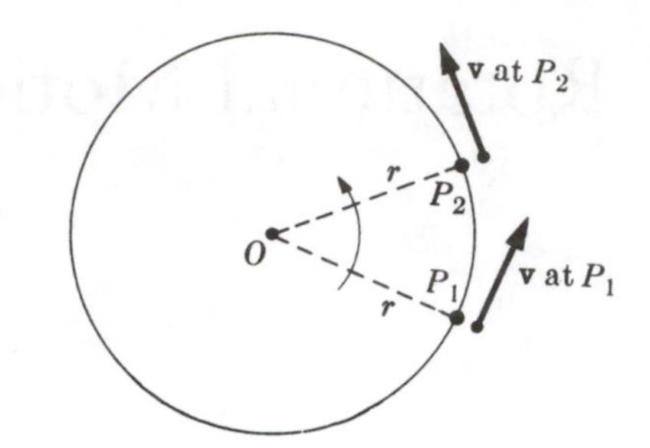

Fig. 10.2.

d) We now inquire into the velocity of a particle executing uniform circular motion. The word "uniform" means, of course, that the rate of spin (the speed s/t) does not change. Nevertheless, for future reference, it is well to remember that the velocity vector associated with the rotating point does change in *direction* from one instant to the next, although its magnitude, represented by the length of the arrows in Fig. 10.2, is here constant.

Let us now concentrate entirely on the magnitude of the velocity, the speed, given again by the ratio of distance covered to time taken; if we know the period or frequency of the motion and the distance r from the spot to the center of the circle, v is directly found (usually in cm/sec) by realizing that $s = 2\pi r$ if $t = T$, that is,

$$v = 2\pi r/T = 2\pi nr. \tag{10.3}$$

e) The quantity v defined in this last equation refers to the magnitude of the tangential or linear velocity, i.e., to the velocity of the point along the direction of its path. Analogous to this time rate of change of distance stands the powerful concept *angular velocity* [symbolized by the Greek letter ω (omega)], which is the time rate of change of angle. By definition, for the type of motion we are here considering,

$$\omega = \theta/t \tag{10.4}$$

a quantity to which we may assign units of sec^{-1} or, more colloquially, radians/sec. If we happen to know n or T, we can find the magnitude of the angular velocity from the fact that $\theta = 2\pi$ if $t = T$, or

$$\omega = 2\pi/T = 2\pi n. \tag{10.5}$$

The formal relation between ω and v is evident if we compare Eqs. (10.3) and (10.5):

$$v = \omega r. \tag{10.6}$$

But direct inspection of the defining equation (10.4) for ω reveals a significant difference between v and ω. Every point on a rigid rotating body (for example, a disk), no matter where it is placed on that body, has the *same* ω at the same instant, whereas those different points take on very different values for v, depending on their respective distances r from the axis of rotation. The concept of angular velocity ω therefore introduces into problems of rotational motion the possibility of constancies not attainable directly in terms of v alone. There is no better reason for coining a new concept in any field of science.

Problem 10.1. Before about 1970, most American homes owned a "phonograph" or "record-player" designed to reproduce music and speech from large plastic disk-shaped "records" (these were the "analog" precursors of the digital "compact discs" popular in the 1990s). These records would be placed on a motor-driven "turntable" that could be set to rotate at one or more of three speeds: 33, 45, or 78 revolutions per minute (rpm). The music was represented by variations in the size and shape of a continuous spiral "groove," starting near the outer rim of the record. A needle, attached to a metal arm pivoting from the side of the turntable, transmitted these variations to an electromechanical device that re-created the original sound. Some of the important features of rotational motion could be easily observed by watching the record go around. How great an angle (in radians and in degrees) is covered by a spot on a 78-rpm record in 5 min of constant rotation? What are the linear speeds of two points, respectively 3 and 12 cm from the center of rotation? What are the corresponding angular speeds?

Problem 10.2. How long is the groove that travels under the phonograph needle described in Problem 10.1 while playing the whole of one side of a 12-in 33-rpm record? Assume that each side contains about 30 min of music.

Problem 10.3. Find the frequency of revolution, the linear velocity, and the angular velocity for a spot on the equator (earth's radius = 6380 km; period of rotation about its own axis = 23 hr 56 min per rotation).

Problem 10.4. Suppose a high tower stands at a point on the earth's equator. Using the data given in Problem 10.3, calculate the linear velocity of an object at the top of the tower. Is an object at the foot of the tower going with exactly the same linear velocity? If the object at the top were dropped, would you expect it to land exactly at the base of the tower (neglecting air resistance, and assuming an absolutely vertical tower)? If not, in which direction would you expect a deviation? (You are not expected to calculate the actual amount of deviation, but you should be able to explain qualitatively the reason for a deviation and evaluate Galileo's argument given in a footnote [number 3] near the end of Section 5.3.)

Problem 10.5. Find the approximate linear speed of our planet in its yearly path around the sun (assuming it to be nearly enough a circular orbit with radius 1.5×10^{11} m).

10.2 Centripetal Acceleration

It was noted in the previous section that motion with constant speed around a circle implies that the velocity vector is continually changing in direction though not in magnitude. According to Newton's laws of motion, a force must be acting on a body whose velocity vector is changing in any way, since if there were no net force it would continue to move at constant velocity in a straight line. And if there is a force, there must be an acceleration. So when in circular motion with constant speed, a body is in the seemingly paradoxical situation of being accelerated but never going any faster (or slower)!

To clarify this situation we have to emphasize the distinction that the physicist makes between velocity and speed. As we saw in Section 8.2, velocity is a *vector* quantity, having both magnitude and direction; speed is a *scalar* quantity, having magnitude only (see Appendix IX). We will use boldface type for vectors when they are likely to be confused with the scalar magnitude of the same quantity. Any change in the velocity is an acceleration, whether or not there is a change in the speed.

In the case of circular motion, the change in direction of the velocity vector was shown in Fig. 10.2; we now need to analyze this change in a little more detail (Fig. 10.3). The vector labeled "$\boldsymbol{v}$ at P_2" is the resultant of two other vectors, which must be added together: the vector "$\boldsymbol{v}$ at P_1" and the vector "$\Delta\boldsymbol{v}$" which represents the change in velocity that occurs during the time interval Δt as the body moves along the circle from P_1 to P_2.

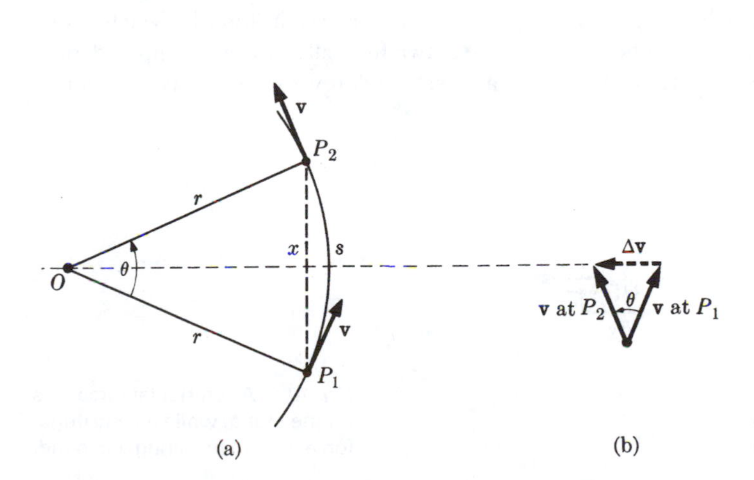

Fig. 10.3. How the velocity vector changes in circular motion at constant speed.

As can be seen from the diagram, the vector $\Delta \boldsymbol{v}$ is directed *toward the center of the circle.* (A mathematical proof of this fact could be given, but it seems unnecessary.) The acceleration is defined as the change in velocity divided by the time interval, in the limit as the time interval becomes very small. In symbols:

$$\boldsymbol{a} = \lim_{\Delta t \to 0} \left(\frac{\Delta \boldsymbol{v}}{\Delta t} \right).$$

Acceleration is also a vector, directed toward the center of the circle; hence it is called the *centripetal acceleration* ("centripetal" means "seeking the center"). The corresponding force that must be acting on the body with mass m to produce the acceleration is a vector in the same direction, $\boldsymbol{F} = m\boldsymbol{a}$ according to Newton's second law; in this case, $\boldsymbol{F}$ is called the *centripetal force.*

Unfortunately the ordinary language that we use to describe motion causes confusion at this point. We are accustomed to hearing talk about "centrifugal" force, a force that is said to act on a whirling body in a direction away from the center. We shall not use this term, because there is no such force, despite the illusion. If you tie a weight on a string and twirl it around your head (Fig. 10.4), you think you can feel such a force, but that is a force acting on you at the center, not a force acting on the whirling body; it is in fact the third-law reaction to the centripetal force acting on the whirling weight. Or, if the string breaks, you see the object flying away along the tangent of the circular path it was following just before the break. But no force pulls it in flight (not counting gravity, downward) and it does not move off along a radius either.

Isaac Newton was one of the first to recognize that all these phenomena are due to the natural tendency—*inertia*—of any body to keep on moving in the same direction if it is not constrained to do otherwise. If the string breaks when the weight is at the point P_1 (Fig. 10.3), it will "fly off on a tangent" (not along the radius)—that is, it will continue to move in the direction indicated by the arrow of the velocity vector at P_1. While the object is still attached to the string and moving in a circle, you have to provide a force on it toward yourself—the centripetal force—to prevent it from flying off. And since you are exerting a force on the string, the string must also, by Newton's third law, exert an equal and opposite force on your hand. The outward force that *you* feel is the reaction to the force you apply.

What is the magnitude of the centripetal acceleration? In other words, how does it depend on the speed of rotation and the size of the circular orbit? The answer to this question was crucial to Newton's theory of the solar system; as we shall see in the next chapter, it was needed as a step to the law of gravitation. Newton worked out the theory of centripetal acceleration during his burst of creative activity in the Plague years 1665–1666, although his publication of it, in 1687, was 14 years later than the published derivation by his great contemporary, the Dutch physicist Christiaan Huygens. The result for the magnitude of the centripetal acceleration is very simple:

$$a = v^2/r \qquad (10.7)$$

Thus the centripetal acceleration increases with the square of the linear speed of the moving body in its orbit, but decreases inversely as the radius of the circle.

Before turning to the derivation of this result, let us summarize its physical meaning by quoting Newton in the *Principia* of 1687, a very clear discussion and one that indicates how Newton was able to make connections between a great variety of situations:

> Centripetal force is the force by which bodies are drawn from all sides, or are impelled, or in any way tend, toward a point as to a center.

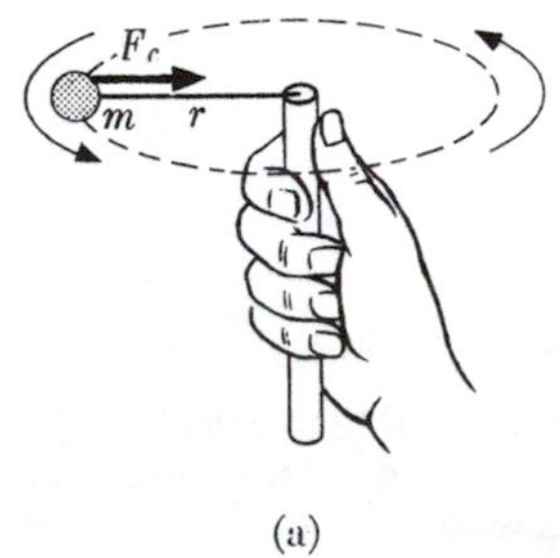

(a)

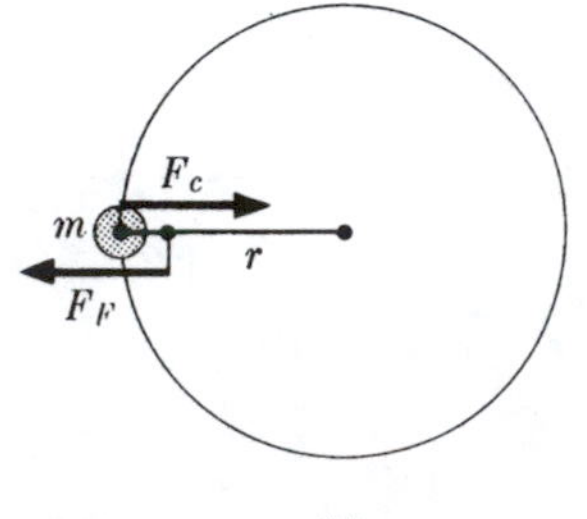

(b)

Fig. 10.4. A centripetal force acts on the stone, while a centrifugal force acts on the string and hand.

One force of this kind is gravity, by which bodies tend toward the center of the earth; another is magnetic force, by which iron seeks a lodestone; and yet another is that force, whatever it may be, by which the planets are continually drawn aside from rectilinear motions and compelled to revolve in curved lines orbits.

A stone whirled in a sling endeavors to leave the hand that is whirling it, and by its endeavor it stretches the sling, doing so the more strongly the more swiftly it revolves; and as soon as it is released, it flies away. The force opposed to that endeavor, that is, the force by which the sling continually draws the stone back toward the hand and keeps it in an orbit, I call centripetal, since it is directed toward the hand as toward the center of an orbit.

And the same applies to all bodies that are made to move in orbits. They all endeavor to recede from the centers of their orbits, and unless some endeavor is present, restraining them and keeping them in orbits and hence called by me centripetal, they will go off in straight lines with a uniform motion. . . .

So too the moon, whether by the force of gravity—if it has gravity—or by another other force by which it may be urged toward the earth, can always be drawn back toward the earth from a rectilinear course and deflected into its orbit; and without such a force the moon cannot be kept in its orbit. If this force were too small, it would not deflect the moon sufficiently from a rectilinear course; if it were too great, it would deflect the moon excessively and draw it down from its orbit towards the earth. In fact, it must be of just the right magnitude, and mathematicians have the task of finding the force by which a body can be kept exactly in any given orbit into which a body leaving any given place with a given velocity is deflected by a given force.

10.3 Derivation of the Formula for Centripetal Acceleration and Centripetal Force

In Fig. 10.3, a point moves uniformly through an angle θ from P_1 to P_2 along the arc s. The speeds at P_1 and P_2 are equal, but the direction of the velocity vectors changes through angle θ between P_1 and P_2. The change of velocity, Δv, is obtained graphically in the usual manner in Fig. 10.3b. Now note that the triangle there, and the one contained by P_1OP_2 in Fig. 10.3a, are similar isosceles triangles. Therefore $\Delta v/v = x/r$ and $\Delta v = vx/r$. On dividing both sides by Δt, the time interval needed for this motion, we obtain $\Delta v/\Delta t = vx/r\,\Delta t$. The left side represents the average acceleration $\bar{a}$ during Δt; and if, as in Chapter 6, we restrict Δt to smaller and smaller values, in the limit as Δt approaches 0, the average acceleration becomes equal to the instantaneous acceleration a:

$$a = \lim_{\Delta t \to 0} \left(\frac{\Delta v}{\Delta t}\right) = \lim_{\Delta t \to 0} \left(\frac{vx}{r\Delta t}\right).$$

At the same time, however, as Δt (and with it, θ) decreases, the line x becomes more and more nearly equal to the arc s, so that in the limit, $x = s$, and we write

$$a = \lim_{\Delta t \to 0} \left(\frac{vs}{r\Delta t}\right)$$

We note finally that as $\Delta t \to 0$, $s/\Delta t \to v$, the instantaneous velocity of the moving point, so that the formula for centripetal acceleration becomes

$$a = v^2/r.$$

Earlier we found $v = \omega r$; thus a is also expressed by

$$a = \omega^2 r.$$

As can be seen from Fig. 10.3, the direction of Δv, and hence the direction of a, is perpendicular to the chord marked x. In the limit as P_1 and P_2 coalesce, it will therefore be perpendicular to the instantaneous velocity vector, which points along the tangent to the curve. Since the acceleration is directed toward the center, the term "centripetal" can be used, and we shall henceforth attach a subscript c. Similarly, the centripetal force will be denoted by F_c; it is, by the second law that always applies, equal to ma_c.

The physical causes and means for providing a rotating object with the necessary centripetal force vary greatly. A rotating flywheel is held together by the strength of the material itself, which can supply the stress, at least up to a point. A car rounding a curve on a level road is kept from flying off tangentially in a straight line by the force of friction, which supplies the centripetal force by acting (sideways) on the tires. The moon, as Newton first explained, is kept captive in its orbit about the earth by the gravitational pull continually experienced by it. The electron circulating around an atomic nucleus in Niels Bohr's model (Chapter 28) does so by virtue of electric attraction to the center. But in each case, insofar as the motion is circular and uniform, $F_c = ma_c$, or, to clothe this expression in all its alternative forms, the centripetal acceleration is

$$F_c = ma_c = mv^2/r = m\omega^2 r$$
$$= m(4\pi^2/T^2)r = m(4\pi^2)n^2 r. \quad (10.8)$$

10.4 The Earth's Centripetal Acceleration and Absolute Distances in the Solar System

In 1672–1673, astronomers of the Paris Observatory carried out a remarkable project that provided two important kinds of data: the variation in the (apparent) gravitational acceleration g of objects at the earth's surface, subsequently attributed to the effects of the earth's rotation, and the sizes of planetary orbits in the solar system. The project was carried out by Jean Richer (d.1696) in collaboration with Giovanni Domenico Cassini (1625–1712), director of the observatory. The first kind of data depended on comparing results determined at Paris with those at a significantly different *latitude,* for example near the equator; the second kind was facilitated by making measurements at a greatly different *longitude.*

These criteria were both satisfied by Cayenne, about 5° north of the equator on the coast of South America, now part of French Guiana and known for its pepper. The French were attempting to establish a profitable colony there, and since the government sponsored the Academy's projects, travel to Cayenne was relatively convenient. Richer arrived in 1672 to make a number of scientific measurements, including the change in its length required to make a pendulum swing with a period of exactly one second, and the position of Mars at one of its closest approaches to earth (opposition near perihelion).

a) *Variation of g.* The period T of a pendulum of length L_0 is, as Galileo noticed, independent of the length of its swing, at least for relatively small arcs; according to Newtonian mechanics, it depends on L_0 and g_0 according to the formula

$$T = 2\pi\sqrt{(L_0/g_0)} \quad (10.9)$$

where g_0 is the local acceleration owing to gravity. A "seconds" pendulum is one whose length has been adjusted so that T is equal to exactly 1 sec.

If a seconds pendulum, adjusted at Paris, is taken to another place where g is different, the period will no longer be 1 sec. (The period can be determined, for example, by counting the number of swings during 24 hr from noon to noon, using the shortest-shadow definition of noon given in Section 1.1.) If the pendulum's length is changed from L_0 to a new length L_1 so that the period is again exactly 1 sec, then we can infer that g_0 has changed to

$$g_1 = g_0 L_1/L_0$$

Richer found that his pendulum, which executed a certain number of swings in 24 hr at Paris, took only 23 hr, 57 min, and 32 sec at Cayenne. He therefore had to shorten its length somewhat in order to make it swing in exactly 1 sec. We can now interpret this result by saying that g_1, the acceleration of gravity at Cayenne, is smaller than g_0, the acceleration of gravity at Paris.

What makes it smaller? The most obvious explanation, in the light of this chapter, is that the centripetal acceleration a_c associated with the earth's rotation at the equator is greater than at the higher latitude of Paris, because a point on the surface of a rotating sphere has to move a greater distance to make one complete daily trip at the equator. This extra centripetal acceleration has to be *subtracted* from the acceleration that would be produced by gravity if the earth were not rotating.

One of the old objections to the earth's rotation was the assertion that if the earth were really spinning fast enough to go around on its axis every day, it would fly apart. While this does not happen for the present rate of rotation since a_c is much less than the acceleration due to gravity, the effect would indeed occur if the earth were spinning several times as fast. Huygens pointed out that if the rotation rate were increased by a factor of about 17, centripetal acceleration would just cancel gravitational acceleration at the equator (check this). If you jumped straight up in the air you might not come down again!

There is also an indirect effect of the earth's rotation that was not clearly understood until Newton developed his theory of gravity (see Section 11.6), though Huygens suspected it. If the earth was originally a spinning fluid sphere, the greater centripetal acceleration at the equator compared to other parts of its surface would cause it to bulge at the equator and flatten at the poles. Points on the equator would then be farther from the earth's center than points closer to the poles, and the force of gravity there would be weaker. But the amount of this effect could not be estimated until the shape of the earth had been determined.

b) *Sizes of planetary orbits.* Copernicus was able to calculate the *relative* sizes of all the planetary orbits (Table 2.1) but had no way to determine precisely the absolute distances of any of them. That would require at least one direct measurement.

One way to measure the distance of an object C too far away to reach physically is *triangulation:* observe it from two places A and B and measure the angles $\angle CAB$ and $\angle CBA$. If the distance of the baseline AB is known, then the other sides of the triangle CA and CB can be found by trigonometry. If C is very far away compared to AB, these angles will be nearly the same and a small error in measuring the angles will produce large errors in CA and CB. This is why attempts by the ancient Greeks to measure the distances of the sun and moon failed to give accurate values.

Richer and Cassini used the Paris-Cayenne distance as the baseline for measuring the distance of Mars at a close opposition by triangulation. They could then estimate the earth-sun distance, knowing that the relative distance is about 2½ times as great.

They had two advantages over the Greeks. First, they could measure angles more accurately, thanks to the invention of the telescope and other instruments in the seventeenth century. Second, they could use a much larger baseline BC by making observations on different continents separated by the Atlantic Ocean. These observations could supplement those made by a single observer at different times during the day, using the observer's own motion (due to the earth's rotation) to provide the baseline (see Fig. 1.5 for the analogous method of measuring stellar parallax using the earth's revolution around the sun.) The single-observer method was also used at the same time by the astronomer John Flamsteed in England to estimate the distances of Mars and the sun.

The result of the Richer-Cassini triangulation of Mars was an estimate of about 87,000,000 miles for the average earth-sun distance. Although this happens to be reasonably close to the modern value, the expected errors in their measurements were so large that no one could place much confidence in the accuracy of their results. But Flamsteed reported a very similar result at the same time, so there was some reason to believe that the *order of magnitude* of the earth-sun distance was about 100 million miles (rather than 10 million or 1000 million). This was about 15 times as great as a rough guess by Copernicus.

Although the precise value of the earth-sun distance remained in doubt for another century, the measurements by Richer, Cassini, and Flamsteed constituted a tremendous advance in astronomical knowledge. Among other things, they made it possible for Huygens to make the first reasonable estimate of the speed of light (Section 23.1).

EXAMPLE 1

Find the centripetal acceleration of an object on the equator.

Solution: This object follows a circular path, covering an angle of 2π rad per day ($= 8.6 \times 10^4$ sec); $\omega = (2\pi/8.6 \times 10^4)$ rad/sec $= 7.3 \times 10^{-5}$ rad/sec. The radius of the earth is about 6.4×10^6 m. Therefore $a_c = \omega^2 r \approx 0.034$ m/sec^2.

Problem 10.6. Find the centripetal acceleration owing to the earth's rotation at a place with latitude 30°. Find the centripetal acceleration for the whole earth, owing to its yearly swing around the sun. (Approximate distance $= 1.5 \times 10^{11}$ m; the assumption of a circular orbit is necessary here, though not entirely correct.)

Problem 10.7. Derive, step by step, from first principles, the relation $F_c = m(4\pi^2/T^2)r$, and explain at each point what law of physics, concept, approximation, etc., has been introduced.

Problem 10.8. The rim of a flywheel 3 m in height is a metal band of 1000 kg mass. The spokes can support a tension of up to 10^6 newtons. At what speed of rotation will the wheel burst? Make a drawing to show how the pieces will fly away.

Problem 10.9. Mention several other examples of rotational motion, and explain in each case how centripetal force is supplied.

Problem 10.10. Find the distance from Paris to Cayenne AB from a modern reference work. Using the value 25″ for the parallax of Mars (see Problem 1.3 for definition of parallax) compute the earth-Mars distance at opposition. Then assuming the planetary orbits are circular and using the Copernican estimates of the relative sizes of the planetary orbits (Table 2.1), calculate the earth-sun distance. For this purpose you may ignore the curvature of the earth and treat *AB* as a straight line.

RECOMMENDED FOR FURTHER READING

C. C. Gillispie, *Dictionary of Scientific Biography*, articles by E. Rosen on Richer, Vol. 11, pages 423–425, and by R. Taton on Cassini, Vol. 3, pages 100–104

J. Herivel, "Newton's discovery of the law of centrifugal force," *Isis*, Vol. 51, pages 546–553 (1960)

CHAPTER 11

Newton's Law of Universal Gravitation

In the second half of the seventeenth century, as science was beginning to take on a recognizably modern form, one long-standing problem was ripe for solution: the structure and dynamics of the solar system. Most of the preceding chapters in this book have been concerned with assembling the pieces of the puzzle—not because the major goal of physics is to understand the solar system, but because the successful solution of that problem provided at the same time a powerful method for attacking the whole range of problems in terrestrial physics and even a model for inquiring into how the universe as a whole is constructed.

In this chapter we will see how Newton, in his *Principia,* was able to apply his theory of forces and motion to the astronomical system developed by Copernicus, Kepler, and Galileo. In this application the formula for centripetal acceleration derived in Chapter 10 will play an essential part. After showing how Newton's law of universal gravitation provides a convincing explanation for some of the phenomena known in the seventeenth century, we shall briefly survey its role in later developments in astronomy. Finally, we shall attempt to assess Newton's place in the overall development of science in the light of his own theories, as well as their impact on the modern view of the universe.

11.1 Derivation of the Law of Universal Gravitation

The three main sections of the *Principia* contain an overwhelming wealth of mathematical and physical discoveries; included are the proofs leading to the law of universal gravitation, proofs so rigidly patterned in the Euclidean style that it seems better to present here a derivation of this historic law in another plausible sequence. The arguments are sometimes rather subtle, and thus they afford a splendid illustration of the interplay among established laws, new hypotheses, experimental observations, and theoretical deductions in physical theory. The purpose of the following pages is to gain an understanding of the process, rather than to invite the memorizing of the individual steps.

a) Planets and satellites are not in equilibrium. An unbalanced (net) force acts on them. If they were in equilibrium, if no net force were acting on them, they would be moving in straight lines instead of elliptical paths, in accord with Newton's first law of motion.

b) Whatever the nature or the magnitude of the net force acting on a planet or on a satellite, its direction must be toward the center of motion at every instant. Newton derived this conclusion directly from Kepler's second law (*Principia,* Book I, Propositions I and II), and we may paraphrase his own argument as follows:

A body moves along a straight line at constant velocity; at equal time intervals Δt it traverses equal spaces, i.e., in Fig. 11.1a, the length of the line segment $PQ = QR = RS$, etc. With respect to any fixed point O, the line from O to the moving body sweeps out equal areas during equal time intervals, for the triangles PQO, QRO, RSO, etc., are all of equal area, having equal bases and equal heights. Now imagine that this body experiences a brief and sudden blow at Q, a force directed exactly along QO. Of course, the direction of motion is modified. A velocity component has been added which, during time Δt, by itself would move the body from Q to Q' (Fig. 11.1b), but which, in conjunction with the original velocity toward R, adds up to a total displacement from Q to R'. But the area swept out during time Δt is *not* affected by this reorientation! The area of OQR, which would have been covered in time Δt if no blow had been applied, is the same as the area of OQR' that actually was covered. (Proof: RR' is parallel to QQ', thus triangles OQR and OQR' have the same base and equal heights.) Therefore, the area of OQR' is also equal to the area of OPQ.

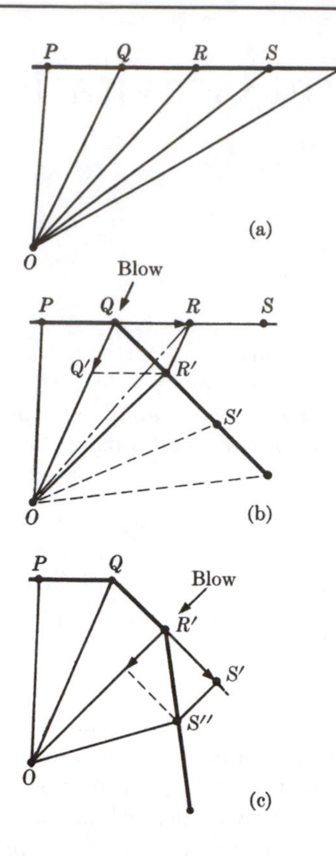

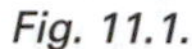
Fig. 11.1.

If no further force acts, the motion during equal time intervals Δt then proceeds from Q to R', from R' to S', etc. But a second blow at R', again directed exactly to O, modifies the motion once more (Fig. 11.1c). By the very same argument as before, we know that area $OR'S'' = OR'S'$. In general, we conclude that *centrally directed* forces applied at equal time intervals do not affect the areas swept out per unit time.

There is no reason why those time intervals should not be chosen as we please, so that in the limit as Δt approaches zero, the centrally directed force becomes a continuously acting centripetal force, and the broken curve melts into a smooth one. Finally, we turn the argument around and, with Newton, say that since planets by Kepler's empirical second law do indeed sweep out equal areas per unit time, the force acting on each must be a continuous, centrally directed force. In the case of the ellipse, this center of force is one of the foci; for the circle it is the center of the figure.

Problem 11.1. Strictly speaking, we should also prove that blows *not* all directed to the same center O, but instead to any other points, will fail to yield an equal area law. Attempt such a proof.

---

c) Now that we accept that the force must be centrally directed, a *centripetal* force, there appears the following crucial problem: "If a body revolves in an ellipse (including the special case of a circle), it is required to find the law of the centripetal force tending to the focus of the ellipse." Newton's mathematics proved for the first time that for paths along conic sections such as hyperbolas, parabolas, ellipses, and circles, the centripetal force at any instant must be proportional to the inverse square of the distance of the body to the focus. In short, any body obeying Kepler's first law of elliptical paths must be acted on by a force following the law $F = C/R^2$, where C is a constant for that body and R is measured from the center of the body to the center of forces, the focus of the ellipse.

We cannot at this point follow the general proof, but we shall show that if for a planet in a circular path the centripetal force is granted to be equal to $F = C/R^2$, there follows by derivation, without further assumptions, that the celestial body should also obey the law

$$T^2 = KR^3.$$

Conversely, since we can observe the latter to be actually true—it is Kepler's third law—we judge that the hypothesis $F = C/R^2$ is well founded for the case of planetary motion.

The derivation proceeds as follows: The centripetal force F_c (or now simply F), on the planet that, as assumed, is given by C/R^2, is also by Newton's second law equal to $m_p a_c$, where m_p is the mass of the planet and a_c is the centripetal acceleration. For circular paths around the sun, as are actually followed by almost all planets with but little deviation,

$$a_c = \frac{v^2}{R}$$

v being the speed of the planet in its orbit. But

$$a_c = \frac{v^2}{R} = \frac{4\pi^2 R^2}{T^2 R} = \frac{4\pi R}{T^2},$$

where T is the period of orbital revolution of the planet. It follows [see Eq. (10.8)] that

$$F = m_p a_c = m_p \frac{4\pi^2 R}{T^2} \qquad (11.1)$$

Combining the last result with our assumed value for F, we have

$$\frac{C}{R^2} = m_p \frac{4\pi^2 R}{T^2}, \text{ or } T^2 = \left[m_p \frac{4\pi^2}{C}\right] R^3. \qquad (11.2)$$

Because m_p and C are constant, at least for a given planetary orbit, the bracketed term in Eq. (11.2) is constant for a given planet, no matter what the size of its orbit. T^2 here is proportional to R^3. This is reminiscent of the form of Kepler's third law, but unless we can prove (as we have not yet done above) that the bracketed term in Eq. (11.2) is really the same constant for *all* planets, the quantity T^2/R^3 from Eq. (11.2) may have a different value for different planets.

But we must recognize for this reason also that this was no proof of the inverse square law for the centripetal force, either. Kepler's law requires that

$$m_p(4\pi^2/C) = K,$$

where K is, for our solar system, the same constant for all planets in all their orbits around the sun. Not until we discover what C contains can we know whether the bracketed term actually yields the same constant for all planets. We shall strengthen and complete the above proof as soon as we can, for we have decided not to bypass the difficult points. Let us note in passing that use has been made of Newton's second law of motion and of the equation for centripetal acceleration.

Historically, Newton's demonstration that elliptical planetary paths imply an inverse square law of the force came at a time when the idea of such a law was generally "in the air." In fact, Halley had come to Newton in 1684 just in order to ask whether he could supply the proof that others were looking for in vain!

d) The *origin* of the centripetal force needed to keep the planets in their orbits has not been touched on so far. We recall that Kepler speculated that some driving magnetic force reached out from the sun to move the planets. He was wrong, but at least he was the first to regard the sun as the controlling mechanical agency behind planetary motion. Another picture had been given by the French philosopher René Descartes (Section 6.1), who proposed that all space was filled with a subtle invisible fluid of contiguous material corpuscles; the planets of the solar system were supposed to be caught in a huge vortexlike motion of this fluid about the sun. This idea was attractive to the minds of the day, and consequently was widely accepted for a time, but Newton proved that this mechanism could not account for the quantitative observations on planetary motion as summarized, for example, in Kepler's laws. The problem remained.

At this point Newton proposed a dramatic solution: The centripetal force on the planets is nothing but a gravitational attraction of the sun, and the centripetal force on satellites revolving around planets is also completely given by the gravitational pull by the planets. (Less than a century earlier it would have been impious or "foolish" to suggest that terrestrial laws and forces regulated the whole universe, but now, after Kepler and Galileo had unified the physics of heaven and earth, it had become a natural thing to suspect.) If the earth attracts the moon with the same type of force with which it attracts a falling apple or a projectile, and if the sun attracts the earth, the moon, and all the other celestial bodies with the same type of force, then there is no need for any additional cosmic force or prime mover and gravity then becomes a universal, unifying principle, which, while in fundamental contradiction to the axioms of the scholastics, would have gladdened the heart of Kepler.

But we must follow the proof. First paralleling young Newton's thoughts, let us see whether the centripetal force F needed to keep the moon in its (nearly enough) circular orbit about the earth *can* be identified with terrestrial gravity. The *direction* of F is by definition toward the center of the earth, and this checks with the direction of the force of gravity. But what about the *magnitude* of F? We apply the equation for centripetal force to this case, and find

$$F = m_m(4\pi^2 R/T^2), \qquad (11.3)$$

where m_m is the mass of the moon, R is now its distance from the center of rotation about the earth, and T is its period of revolution.

Does this value for F actually coincide with the gravitational pull that the earth exerts on our satellite, as Newton proposed? That depends on the nature of the gravitational force. If gravity is propagated undiminished through all space, the weight of the moon will be simply $m_m \times g$, the same as a stone of the moon's mass when placed on giant scales somewhere on the surface of the earth. But not only does it seem unlikely that the gravitational acceleration stays the same no matter how far away from the earth we go, we also recall that in part (c) above we found evidence that the centripetal force (whatever its final nature) must fall off as the square of the increasing distance. If gravity is to account completely for the centripetal force, it too will have to follow an inverse square law. Let us therefore assume that the weight of an object falls off according to such a law, and consider now whether the gravitational pull of the earth on the moon just equals the centripetal force in Eq. (11.3).

This is our argument: An object with the same mass as the moon, m_m, has the weight $m_m g$ when weighed at the surface of the earth, i.e., at a distance r (the radius of the earth) from the earth's center. That same object, when removed to a great distance R from the earth's center, will assume a smaller weight, which we shall call W_R and which must fulfill the following proportion if the inverse square law is to be obeyed:

$$\frac{m_m g}{W_R} = \frac{(1/r^2)}{(1/R^2)} \quad \text{or} \quad W_R = m_m g \frac{r^2}{R^2}. \tag{11.4}$$

If the centripetal force F acting on the mass m_m rotating about the earth at distance R with a period T is really equivalent to the gravitational force W_R at that distance, the terms on the right side of Eqs. (11.3) and (11.4) should also be equivalent:

$$m_m \frac{4\pi^2 R}{T^2} = m_m g \frac{r^2}{R^2}$$

or

$$T^2 = \frac{4\pi^2}{gr^2} R^3. \tag{11.5}$$

Conversely, if we substitute observed values of T, g, r, and R in Eq. (11.5) and find that the equation yields equal numbers on each side, then we are justified in regarding our hypotheses as valid; then the gravitational force *does* fall off as the square of the increasing distance, and does fully account for the needed centripetal force. [Incidentally, we note with satisfaction that Kepler's third law is implied in our result, Eq. (11.5).]

Problem 11.2. Substitute the needed values in consistent units and check to what extent Eq. (11.5) holds. The period T of the moon is 27 d, 7 hr, 43 min; $g = 9.80$ m/sec^2; $r = 6380$ km; $R = 380{,}000$ km.

This was the calculation that Newton, with contemporary data, found to "answer pretty nearly," probably within a few percent. The assumption of a strictly circular path and somewhat inaccurate values available to him for r and g made it clear from the start that no *perfect* agreement could have been expected.

It has been a source of much speculation why Newton did not tell anyone of this remarkable result when he first conceived it, or for nearly 20 years thereafter. Apart from his reticence and his fear of litigations with jealous men, he seems to have been unable at the time to account clearly for one implied assumption in the argument, namely, that the earth's gravitational force acts as if originating at the very center of the globe, and that consequently the measurement of distances must be made, not to the globe's surface, but to its center. He understood this requirement later when writing the *Principia,* where he proved in general that two homogeneous spheres attract each other as though their masses were concentrated at their centers. Another reason for the delay may have been that Newton did not know or properly appreciate the significance of Kepler's law of areas until about 1684; before that time his manuscripts show that he was trying to base his planetary theory on the equant device, applied to elliptical orbits.

Another, more skeptical, explanation is that in telling, in later years, the story of his youthful achievements, Newton exaggerated his progress toward the mature theory of gravity, in order to retain for himself the entire credit for discovering that theory—against the claims of Robert Hooke who demanded recognition for his own unproven conjecture, announced in a letter to Newton around 1680, that gravity follows an inverse square law.

As a summary for part (d) we may use Newton's own statement that "*the moon gravitates toward the earth and by the force of gravity is always drawn back from rectilinear motion and kept in its orbit.*" Apart from the assumptions and approximations alluded to, we have made use of only the following new arguments: that the gravitational force falls off as the square of the distance, and that by Rules I and II of Section 9.3 we can identify terrestrial gravity with the centripetal force the earth exerts on the moon.

11.2 Gravitating Planets and Kepler's Third Law

The previous paragraphs have not involved the forces between the sun and the planets, but we should again suspect that these ideas may be extended to the whole solar system. In Newton's words:

> Hitherto we have called "centripetal" that force by which celestial bodies are kept in their orbits. It is now established that this force is gravity, and therefore we shall call it gravity from now on. For the cause of the centripetal force by which the moon is kept in its orbit ought to be extended to all the planets, by rules 1, 2, and 4.

But such rules (quoted in Section 9.3) are just guiding, not prescriptive ones. They suggest, but do not prove; this labor remains to be done.

We remember that the centripetal force on planets in circular orbits around the sun was given by Eq. (11.1) as

$$F = m_p(4\pi^2 R/T^2).$$

Does this actually also correspond in each case to the gravitational attraction of the sun on the particular planet? If we knew the value of g on the surface of the sun we could fashion the arguments as in the previous section; but of course we do not know it at this point, and so we have recourse to another argument. Perhaps we can decide on a theoretical formula for the gravitational force of any one body on another. With the confidence gained by our success in the previous discussion of the earth and moon, we now boldly suggest that the gravitational force F_{grav} between *any* two spherically symmetric bodies is proportional to the inverse square of the distance between the two centers, everything else being kept constant:

$$F_{grav} \propto 1/R^2.$$

Next, let us consider two specific solid bodies, quite isolated from the rest of the universe, say a stone (m_1) and the earth (m_2) at a distance R between their centers. The pull of gravity, the weight of m_1 at the given distance R, is F_{grav}. But by Newton's law of action and reaction, the pull of the earth (m_2) on m_1 is just as great as the pull of m_1 on m_2; the weight of a stone F_{grav} as measured by its attraction toward the earth *is equal* to the "weight of the earth" as measured by its attraction to the stone, strange though this may sound at first, and *either one may be called* F_{grav} (Fig. 11.2).

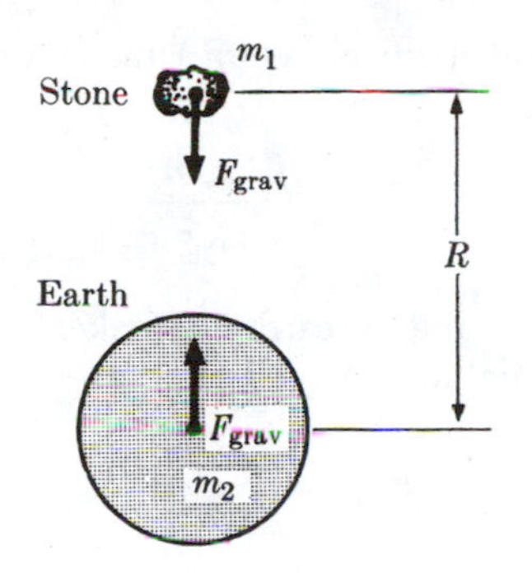

Fig. 11.2. The force of attraction of the earth on a stone is equal in magnitude to the attraction of the stone on the earth.

However, we know by experiment that at a given locality the weight of the stone grows directly with the mass of the stone, or $F_{grav} \propto m_1$. On the other hand, if the mass of the planet changed, the weight of a given stone would change also (in fact, you would weigh much less on our little moon than on earth). In short, if experiments prove that $F_{grav} \propto m_1$ at a constant distance R, then we must also accept $F_{grav} \propto m_2$, for otherwise we should have to make the assumption that the mutual gravitational pull F_{grav} depends on something besides the magnitude of the masses and the distances between them.

Combining these three proportionalities,

$$F_{grav} \propto 1/R^2, F_{grav} \propto m_1, F_{grav} \propto m_2,$$

we have

$$F_{grav} \propto m_1 m_2/R^2 \text{ or } F_{grav} = Gm_1 m_2/R^2, \quad (11.6)$$

where R is the center distance between the two bodies (let us assume them to be homogeneous spheres) and G is a constant of proportionality.

If we have confidence that Eq. (11.6) gives, in fact, the correct attraction that two masses exert on each other, and if one were the sun

(m_s), the other any planet (m_p), then the sun's pull would be

$$F_{\text{grav}} = G\frac{m_p m_s}{R_{ps}^2}.$$

If the moon and the earth were taken separately, their mutual force would be

$$F_{\text{grav}} = G\frac{m_m m_e}{R_{me}^2}.$$

We seem to have in Eq. (11.6) a *law of universal gravitation.*

But now we begin to be assailed by doubts. If we are right, this newly found law must certainly be compatible with all three of Kepler's laws. There is no trouble about the first two—elliptical orbits and the equal area relations must indeed result from this type of gravitational force, since it is centrally directed and is proportional to the inverse square of the distances in accordance with just those requirements, laid down in parts (b) and (c). But what of Kepler's third law? We remember that our previous argument, centering on Eq. (11.2), was a little shaky. We had not really proved, with the force assumed there, that T^2/R^3 had to be a truly universal constant, with the same value for all planets in our universe. Now that we know more about the force acting in the solar system, we can try more rigorously to obtain such a result; for if we do *not* obtain it, gravity cannot be the sole force acting among the celestial bodies.

Let us test our law of universal gravitation against the nearly circular orbital motion of our own planet. The centripetal force that must exist on our mass m_p at distance R_{ps} from the sun is given by

$$m_p\frac{4\pi^2 R_{ps}}{T^2}.$$

The gravitational force supposedly available is

$$G\frac{m_p m_s}{R_{ps}^2}.$$

If the two coincide, if the gravitational force "supplies" and so "explains" the centripetal force, then

$$m_p\frac{4\pi^2 R_{ps}}{T^2} = G\frac{m_p m_s}{R_{ps}^2}, \quad \text{or}$$

$$T^2 = \left[\frac{4\pi^2}{G m_s}\right] R_{ps}^3. \qquad (11.7)$$

Now we really can see whether $T^2/(R_{ps})^3$ is a constant for all planets, as required by Kepler's third law. The expression in brackets in Eq. (11.7), which is to be identified with that constant, contains only the constant of proportionality G, the mass of the sun, and a numerical factor; none of these is variable from planet to planet. Therefore $T^2/(R_{ps})^3$ is truly constant, Kepler's law is fulfilled, and the assumption of universal gravitation is vindicated!

One immediate consequence of the universality of Newtonian gravity is that any object in orbit around the earth is *falling* toward the earth. Newton, in the famous story, started his path toward the *Principia* with the insight that the moon is falling toward the earth for the same reason that the apple is falling to the ground. Both are in "free fall," meaning that they are "weightless." So an astronaut in a space ship orbiting the earth (or moon) is "weightless," just like the person in an elevator that drops into a well, discussed in Section 9.7. The force of gravity, which is responsible for the weight of the person standing on the surface of the earth, is still acting, and is directed toward the earth's center, but now serves as the centripetal force that keeps the person in orbital motion and not flying off in a straight line into space.

11.3 The Cavendish Experiment: The Constant of Gravitation

There remains yet another doubt. Is it proper to have assumed just now so glibly that G does in fact have the same value for all planets? (If it does not, the bracketed term above is, after all, not constant for all bodies.) Here we turn to experiment. We can first measure G only for various materials on earth; using the equation

$$F_{\text{grav}} = G(m_1 m_2/R^2),$$

we get

$$G = F_{\text{grav}}\,(R^2/m_1 m_2).$$

We may propose to measure F_{grav} with spring balances, neglecting all other disturbing attractions except that between two known masses m_1 and m_2 at a measured distance R from each other (see Fig. 11.3). It turns out that for any reasonably small masses in our laboratories, F_{grav} is so exceedingly small that special techniques and precautions have to be adopted to measure any attraction between two such masses; for example, two 1-kg masses separated by 10-cm pull on each other with a force smaller than 10^{-8} newton!

The most serious technical problems of measurement were solved by Henry Cavendish more than 100 years after the publication of the *Principia;* his value, in modern MKS units, was 6.754×10^{-11} N·m²/kg². The best present value for G is about 6.674×10^{-11} N·m²/kg² for all substances (not as precise as one might have expected for the beginning of the high-tech twenty-first century). Cavendish's measurements were made with a delicate torsion balance, a method still used though it must compete with other techniques.

Problem 11.3. If a remeasurement of G were attempted with the arrangement shown in Fig. 11.3, and if the error inherent in our measuring devices is about 0.0002 N, what is the smallest usable radius for each of two similar gold spheres that almost touch? (Density of gold = 19.3 g/cm³. This is a fairly representative example of what one has to go through before designing apparatus for experimentation.)

Problem 11.4. If by chance our standard of mass were not 1 kg, but some other unit 1.225×10^5 times larger (call it 1 tm), and if all other standards (of force, distance, etc.) stayed the same, what would be the value of G in that other system of units? Does this result seem to teach any particular lesson?

Cavendish's results, and all the more accurate ones since, have shown that G has indeed the same value, no matter what the composition of m_1 and m_2, even for meteorite material. Being disposed to apply terrestrial laws to celestial bodies, we extend our findings, and maintain that in the absence of evidence to the contrary all materials in the world, including sun, planets, and satellites, obey the same law of gravitation.

Although Newton knew in principle how to go about measuring G accurately, he lacked the precise instruments; nevertheless he devised an ingenious proof for the constancy of G, which is the main point at issue here. Consider a mass

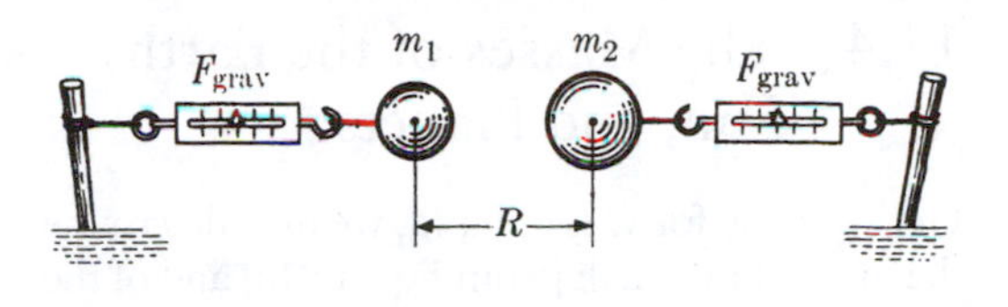

Fig. 11.3. $F_{grav} = G(m_1 m_2/R^2)$

m_1 on the surface of our earth (of mass m_e), that is, at a distance r from the center of our globe. Its weight, which we called F_{grav}, is of course also given by $m_1 g$; therefore, by our new law,

$$m_1 g = G(m_1 m_e)/r^2 \text{ or } G = (r^2/m_e)g. \quad (11.8)$$

At a given locality, r^2/m_e, is, of course, a constant, regardless of its numerical value. If at that locality all substances show precisely the same value for g (the gravitational acceleration), then we have established that G, too, is constant, regardless of chemical composition, texture, shape, etc. This is just what Newton showed experimentally. His measurements of g were made not by just dropping small and large bodies, from which even Galileo had concluded that g cannot vary significantly, but by the much more accurate method of timing pendulums of equal lengths but of different materials. It was known by then that for a given length the period T of a simple pendulum is proportional only to $1/\sqrt{g}$. After exhaustive experiments, all pointing to the constancy of G, he could write: "This is the quality of all bodies within the reach of our experiments; and therefore (by rule 3) to be affirmed of all bodies whatsoever."

Thus G attains the status of the universal constant of gravitation (one of the very few truly universal constants in nature), and the proposed law of universal gravitation in Eq. (11.6) can be applied on a cosmic scale. And quite incidentally, Eq. (11.8) clears up a question about which Aristotelians and many present-day students harbor the strongest doubts, namely, the reason why at a given locality the acceleration of gravity is constant for all bodies. The answer is that g depends only on G, m, and r^2, and all these parameters are constant at one locality.

11.4 The Masses of the Earth, Sun, and Planets

Once a value for G is at hand, we may determine the mass of the earth [from Eq. (11.8)] and of the sun [from Eq. (11.7)]. Moreover, Eq. (11.7) should apply in the same form for a satellite moving around a planet of mass m_p, with a period T, and a radius (or a half-major axis of the elliptical orbit) $R_{p\text{-sat}}$:

$$T^2_{\text{sat}} = \frac{4\pi^2}{Gm_p} R^3_{p\text{-sat}}. \qquad (11.9)$$

This yields the mass of any planet whose satellites we can observe.

Problem 11.5. Determine the mass of the earth and of the sun as indicated. (Answer: $m_{\text{earth}} = 5.98 \times 10^{24}$ kg, $m_{\text{sun}} = 333{,}000 \times m_{\text{earth}}$.)

Problem 11.6. The innermost of Saturn's satellites, Mimas, has a fairly circular orbit of 187,000-km radius and a period of revolution of about 23 hr. Find the mass of Saturn.

For the masses of the satellites themselves (including our own moon), and of the planets that boast of no moons (Mercury, Venus), no such simple calculations exist. There we may use the relatively minor and elaborate interactions among the celestial bodies, which we have conveniently neglected until now. Some of these are called *perturbations,* i.e., slight digressions from the regular path of a body because of the pull of other celestial objects. Another detail that may be used to derive the relative mass of a satellite is the fact that Kepler's laws hold, strictly speaking, only if the center of mass of the system is considered to be at the focus of the ellipse. Furthermore, it is not the earth that moves in accordance with the first law in an elliptical orbit around the sun, but the center of mass of the earth and moon together (a point about 1000 miles below the surface of the earth and on a line between the centers of those two spheres).

Except for such mass determinations (or for making accurate astronomical and nautical tables), we usually forget about the complications. But one of the results obtained is that the moon's mass is about 1/81 of the earth's, and that the sun's mass, when determined in this indirect fashion, has the same value as followed from Eq. (11.7).

Owing to perturbations, Kepler's laws cannot be expected to hold *exactly,* except hypothetically for a single planet whose mass is infinitesimal compared with that of the sun, and which is not affected by the gravitational force of any other body. (In fact, astronomers before Newton had begun to worry more and more about these discrepancies, particularly for Saturn and for the moon.) Once more we find a set of laws breaking down when the precision of observations is increased beyond the original range. But this breakdown is not serious; it does not invalidate the overall outlines of the theory of planetary motion, particularly now that it is possible in principle to account in detail for each discrepancy by means of the law of gravitation.

We are reminded of the 8 min of arc that Copernicus did not know of (Section 4.1), and are tempted to marvel again at how significant it is that these secondary effects were too small to appear earlier and confuse the original investigators. But in this case there seems to be an extra twist; if the perturbations were any larger, we might well not have a solar system to contemplate, because it would hardly be stable enough to last these billions of years without catastrophic collisions.

As a summary for the preceding paragraphs, we may state that the law of gravitation, which we formulated originally by arguments for terrestrial bodies, was found to hold among the planets as well, because by its assumption Newton could predict relationships (such as Kepler's third law), already well established by observation. Furthermore, the law opened up ways of calculating the mass of the sun, the planets, and the satellites. Table 11.1 gives some approximate relative masses.

Problem 11.7. With the data given in Table 11.1, compare the gravitational pulls of sun and earth on the moon. Why does the earth not lose the moon to the sun? Can the path of the moon ever be convex as seen from the sun? [For hints you may wish to read F. L. Whipple, *Earth, Moon and Planets*]

Problem 11.8. Equation (11.7) can be written as

$$K = \frac{4\pi^2}{Gm_s},$$

where K is the constant in Kepler's third law as obtained by observation of T and R. You have

Table 11.1. Mass Relative to Earth (1.00 = 5.97×10^{24} kg).

Sun	332,900	Jupiter	317.8
Moon	0.0123	Saturn	95.2
Mercury	0.0553	Uranus	14.5
Venus	0.815	Neptune	17.2
Mars	0.1074	Pluto	0.0025

Source: Astronomical Almanac.

taken this equation to find m_s, using also the measured value of the universal constant G. The value of m_s so obtained actually checks against independent calculations of m_s using perturbation effects. Exactly in what way is our whole argument strengthened by this gratifying confirmation?

Problem 11.9. (a) Using the values of the masses of the earth, moon, and sun given in Table 11.1, the distances of sun and moon from earth (Problems 10.6 and 11.2), and the fact that the angle from our eye to either edge of both sun and moon is about ½°, find the density of the earth, moon, and sun. (The density of most rocks is about 3×10^3 kg/m^3.) (b) What do these results suggest about the internal structure of the sun, earth, and moon? (c) From Eq. (11.8), find the local acceleration of gravity experienced by someone (i) on the moon, (ii) near the sun.

Problem 11.10. An artificial satellite is to be sent up to revolve around the earth like a little moon. What must be its period and speed of revolution around the center of the earth if it is to stay at a height of 1000 miles above the surface? Outline qualitatively the requirements for the initial velocity if this satellite is to be shot from a big gun on the earth; are these requirements reasonable?

We have extended our law of universal gravitation in *one* direction—to all planets and satellites; it seems logical to extend it in the other direction—to all *parts* of every body. Newton writes in Book III of the *Principia:*

> Proposition VII. Theorem VII: That there is a power of gravity tending to all bodies proportional to the several quantities of matter [i.e., the product of the masses] they contain. That all the planets mutually gravitate one towards another, we have proved before; . . . [we now introduce the idea that] the force of gravity towards any whole planet arises from, and is compounded of, the forces of gravity toward all its parts. . . . If it be objected, that, according to this law, all bodies with us must mutually gravitate one toward another, whereas no such gravitation anywhere appears, I answer, that . . . the gravitation toward them must be far less than to fall under the observation of our senses.

With the aid of his own calculus Newton proceeds to show that assuming the same universal law of gravitation for each smallest particle of a body (say of a sphere), we obtain in sum a resultant force of gravity for the whole body of such characteristics as are actually observable.

There offers itself now an overall view of the enormous range and sweep of the simple formula for the gravitational attraction of two bodies. *Postulating* the law just for each of the parts of a body, it gives us the attraction of the whole for some outside object. For the special case of spherical celestial bodies, it provides a force necessary and sufficient for the description of all observed motions, for the derivation of all three laws of Kepler, and for the small, long-observed deviations therefrom. This was the vindication of Copernicus, Kepler, and Galileo—in the context of the greater, more far-reaching theory of universal gravitation. The whole structure is often referred to as the *Newtonian Synthesis*. No wonder it was thought, by scientists as well as by poets, to be an almost superhuman achievement.

Problem 11.11. List the fundamental hypotheses of the theory of universal gravitation and the experimental consequences. Which hypothesis is not *directly* confirmable by experiment (although its consequences are)?

11.5 Some Influences on Newton's Work

What were the main intellectual tools, concepts, and attitudes with which Newton worked? If we wish to claim any insight into his time and his work we must at least summarize these facts now; most of them have already appeared scattered here and there through the discussion.

Newton was not a "hard-boiled" scientist typical of the current type; rather he shared many characteristics of the late-Renaissance scholar. He was not free from traces of what we would regard as pseudoscientific beliefs; apart from some early interest in astrology, he seems to have spent much time in his "elaboratory," cooking potions that to us would smell more of alchemy than of chemistry—though the aim there, as in all his activities, seems to have been the search for underlying general principles rather than quick practical gains. In his essay "Newton, the Man," the great economist J. M. Keynes presented an intriguing picture, based on Newton's own papers, of Newton as a Faust, a searcher for the key to all knowledge in science, theology, and magic.

By our present standards we might also say that his belief in absolutes and his anthropomorphic conception of a Creator carried very far into some of his scientific writings. But here we touch on the very secret of motivation for his scientific work; though we cannot enter this field here, such a study is fascinating in its own right, as a glance at the biographies by Frank Manuel and R. S. Westfall will show.

In the category of more strictly scientific factors, we should first of all consider the clear Galilean influence on Newton's formulation of the physical concepts of mass and force. The decisive attitude throughout his work is that celestial phenomena are explainable by quantitative terrestrial laws, and that these laws have legitimate general meaning and are not just mathematical conveniences covering unattainable "true" laws. Second, recent historical studies have demonstrated Newton's intellectual debt to Descartes, a debt that Newton did not publicly acknowledge because after long struggles he finally found a solution radically different from the one proposed by the French philosopher-scientist to the same problem: the dynamics of the solar system.

b) To Newton's fundamental faith in the proximity and accessibility of natural law, we must add a word about his methodology. His debt to the pioneers of the new experimental science is clear (for example, he made ingenious pieces of equipment and performed good experiments when the theory needed verification), but he also successfully combined this principally inductive approach with the deductive method then most prominently displayed in Descartes' work. With his mathematical powers enriching the experimental attitude, he set a clear, straight course for the methods of physical science.

c) Not just Newton's attitude toward concepts, but many of the concepts themselves, came to him from Galileo and his followers—above all, force, acceleration, the addition of vector quantities, and the first law of motion. Newton also drew, of course, from Kepler; and through their books and reports to the Royal Society he kept in touch with the work of such contemporaries as Huygens and Hooke.

d) Apart from his own experiments, he took his data from a great variety of sources; for instance, Tycho Brahe was one of several astronomers, old and new, whose observations of the moon's motion he consulted. When he could not carry out his own measurements he knew whom to ask, and we find him corresponding widely with men like Flamsteed and Halley, both Royal Astronomers. There is evidence that he searched the literature very carefully when he was in need of exact data, for example, on the radius of the earth and the distance to the moon.

e) Lastly, we must not fail to consider how fruitfully and exhaustively his own specific contributions were used repeatedly throughout his work. The laws of motion and the mathematical inventions appear again and again, at every turn in the development. But he was modest about this part of his achievement, and once said (echoing what had become a well-known saying even in the seventeenth century) that if he sees further than others, "*it is by standing upon the shoulders of Giants.*"[1]

11.6 Some Consequences of the Law of Universal Gravitation

What amazed Newton's contemporaries and increases our own admiration for him was not only the range and genius of his work on mechanics, not only the originality and elegance of his proofs, but also the detail with which he developed each idea to its fullest fruition. It took almost a century for science to fully comprehend, verify, and round out his work, and at the end of a second century an important scientist

[1]Newton actually used this phrase in connection with his theory of light in a letter to Robert Hooke in 1676. Its earlier history has been traced by Robert K. Merton, in his book *On the Shoulders of Giants.*

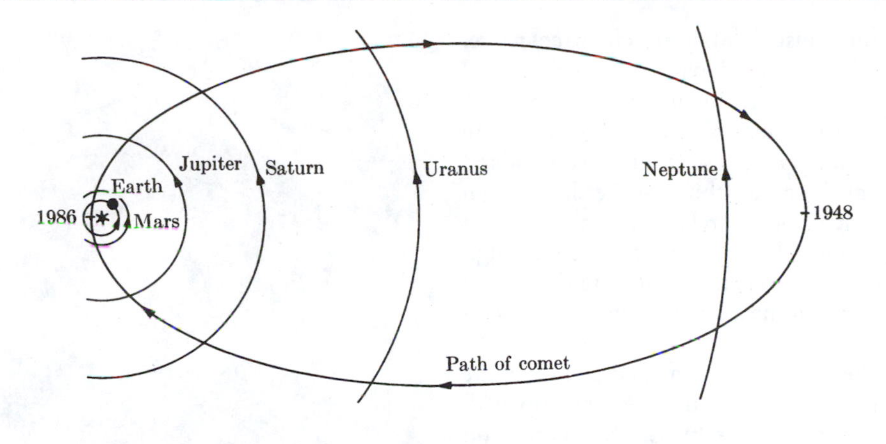

Fig. 11.4. Path of Halley's comet.

and philosopher still had to confess: "Since his time no essentially new principle [in mechanics] has been stated. All that has been accomplished in mechanics since his day has been a deductive, formal, and mathematical development of mechanics on the basis of Newton's laws."

The technical difficulties involved in settling some of the gravitational problems first attacked by Newton were severe enough to tax the brightest mathematicians of the eighteenth century. Many of them were French, by birth or adoption, including Alexis Claude Clairaut (1713–1765), Joseph Louis Lagrange (1736–1813, born in Italy with the name Giuseppe Lodovico Lagrangia), and Pierre Simon de Laplace (1749–1827). No less influential in converting the French intellectuals of the Enlightenment from Cartesian to Newtonian science were popularizers like François Marie Arouet de Voltaire (1694–1778) and his friend Gabrielle-Emilie du Châtelet (1706–1749), she being the first French translator of Newton's *Principia*.

a) An example of Newton's thorough mastery was his treatment of the moon's perturbations. The influence on its path of the gravitational forces from almost all other celestial bodies was considered, and the repeated use of the gravitational law yielded an astonishingly close approximation to every small complexity of motion. (A full list of influential factors in tabular form would now fill hundreds of pages.) For one minor variation of the path, however, Newton's published theoretical result was only half the observed value. He did not camouflage this defect, but clearly stated that his calculations gave the wrong figure. Consequently a loud, long battle arose among some scientists, several of whom called for complete abandonment of the whole theory of gravitation on account of this discrepancy. Eventually the mathematician Alexis Clairaut noted a small error in the long calculations, clearing up the difficulty. Later still, some unpublished notes of Newton were found, showing that he, too, 50 years before Clairaut, had discovered and corrected his own mistake!

b) Comets, whose dreaded appearance had been interpreted as sure signs of disaster throughout antiquity and the middle ages, were now shown to be nothing but passing masses of material obeying the laws of gravitation, made visible by reflected light while near the sun. The great comet of 1682, whose path Edmond Halley watched carefully, indicated to him a period of recurrence of approximately every 75 years if the comet, although an eccentric member of the solar family, nevertheless were to obey the usual laws of mechanics, including Kepler's laws. Its returns in 1759 and three times since, after covering a wide ellipse carrying it far beyond the last planet, were heralded as significant symbols of the triumph of Newtonian science (see Fig. 11.4). The 1759 return was especially dramatic because Clairaut was able to predict the exact date of its appearance, not just "about 75 years" after the previous one but within a 30-day margin of error, by taking account of the perturbations on the comet caused by Jupiter and Saturn.

c) Newton reasoned that the shape of planets and satellites might be explained as the result of the mutual gravity of the separate parts, which would pull a large quantity of initially liquid material (or a dust cloud) into a compact sphere. Further, although a motionless body might indeed form a perfect sphere, a planet rotating about an axis ought to assume the shape of an oblate spheroid, i.e., it ought to bulge at the equator and be flatter at the poles (think of the shape of an onion). In fact, from the relative magnitude of the equatorial bulge, say, of Jupiter as seen through a good telescope, one may calculate the period of rotation about the axis (the length of a day on Jupiter).

For the earth, Newton predicted that the same effect would be found. As it happened, geodetic measurements by G. D. Cassini and his son Jacques Cassini II for a limited region of the earth's surface suggested the opposite conclusion: The earth is a prolate spheroid, its polar axis being longer than its equatorial axis (think of a lemon). The disagreement stimulated French scientists in the mid-eighteenth century to investigate the shape of the earth as a crucial test for Newton's gravitational theory. The Paris Academy of Science sponsored expeditions to the polar and equatorial regions to make precise measurements of the curvature in each region.

Fig. 11.5. Alexis Claude Clairaut (1713–1765), French mathematician who worked out several successful predictions from Newton's gravitational theory.

The results of this project, perhaps the first-ever large cooperative government-funded effort to resolve a scientific question, provided a decisive confirmation of Newton's theory. The diameter of the earth from pole to pole was found to be a few dozen miles less than that across the equatorial plane; the modern value is about 27 miles less. Again it was Clairaut who extended Newton's work by developing a general theory of the shape of planets and stars, taking account of possible variations of their internal density.

One consequence of the oblate shape is that the gravitational force on an object at the pole is somewhat greater than at the equator, owing to the shorter distance from the center of gravity of the earth to the pole. This effect is relatively small, but another cause joins to diminish the acceleration of a freely falling body at the equator—the effect observed by Jean Richer at Cayenne in 1672 (Section 10.4). At that region the surface velocity of a point on earth is about 1600 km/hr, because of the rotation of the globe about its axis. Since the centripetal acceleration a_c is given by v^2/r, a_c for any object at the equator is about 0.03 m/sec^2, as we calculated at the end of Chapter 10. This much acceleration is required just to keep the body from flying off along a tangent; consequently the acceleration of free fall (and therefore the apparent weight of the body) is proportionately less than it would be if our globe were not rotating.

Measured values of free-fall acceleration, all at sea level, range from about 9.83 m/sec^2 near the poles to 9.78 m/sec^2 near the equator (Table 9.1). These observed values are, of course, the true acceleration due to gravity *minus* the centripetal acceleration at that latitude. But we shall be content with the convention that the letter g may stand for those *observed* values and yet usually be called "acceleration due to gravity."

Problem 11.12. Because of the bulging of the earth near the equator, the source of the Mississippi River, although high above sea level, is nearer to the center of the earth than is its mouth. How can the river flow "uphill"? [*Hint:* The word "uphill" needs contemplation.]

Problem 11.13. According to a speculative theory going back to George Howard Darwin (son of the biologist Charles Darwin), our moon might have been formed from material of the earth's crust flung off by the rotating earth. How fast would the earth have had to rotate at that time to make the latter picture plausible?

Problem 11.14. What approximate percentage error does the quoted value for *g* in Boston imply? How might such an accurate experiment be performed?

At present, methods of measurement have been refined to such an extent that small local fluctuations in *g* can be detected and used to give clues of possible mineral deposits underground. But once one goes below the surface into a mine, *g* is found to change, becoming zero as the center of the earth is approached. The law of gravitation in its simple form holds only if the two bodies whose mutual attraction is to be measured are separated and not entwined or intergrown. Newton predicted this excellently, and a moment's thought will persuade us that at the center of the earth, at any rate, we could have no weight because the forces of the earth's attraction from all sides would cancel.

d) The phenomenon of the tides, so important to navigators, tradesmen, and explorers through the ages, had remained a mystery despite the efforts of such scientists as Galileo, but at least the main features, the diurnal high tide and the semimonthly spring tide (maximum high tide), were explained by Newton through application of the law of universal gravitation. He recognized that the moon (and to a lesser extent the other celestial bodies) would pull on the nearest part of the ocean, and so tend to "heap up" the waters. We do not want to take time for the many details involved, except to look at the conclusions. The bulge of water, which actually is raised at the same time on both sides of the globe, does not stay directly under the moon but is displaced to a position a little ahead by the rotation of the earth. Thus a high tide at a point in the ocean is encountered some time after seeing the moon pass overhead, and again about 12 hours later (Fig. 11.6).

The sun has a similar, though smaller, effect, and since the sun, the moon, and the earth are in line twice a month (at the new and the full moon), the two tidal forces coincide to produce semimonthly extremes of tide level. Between "spring tides," when the maximum difference between high and low tides occurs, the sun and moon may partially cancel each other's effects, leading to a minimum difference ("neap tides").

The direct pull of the sun on the earth is about 175 times that of the moon, but tides are generated by the difference of pull on the waters on both sides of the globe; and that is larger for the moon. (Why?)

The exact features of this complex phenomenon depend to a great extent on the topography of shore line and ocean floor. Long series of actual observations help to predict individual features of the tides at each locality. In passing, we should notice that the "solid" earth has similarly explained tides, which must be allowed for in such exact experiments as refined astronomical observations, and that it is just such a tidal effect that offers one of the methods for the approximate calculation of the mass of the moon.

e) Astronomers in the eighteenth century realized that there are several small but persistent discrepancies between Kepler's laws and the observed motions of the moon and the planets. The most important of these discrepancies, called *inequalities,* were:

1. *The secular acceleration of the moon:* Edmund Halley, using eclipse records going back to antiquity, found that the moon seems to go around the earth a little faster every century.
2. *The long inequality of Jupiter and Saturn:* Over the past few centuries Jupiter seems to be moving more rapidly in its orbit around the sun, while Saturn is moving more slowly.
3. *The decrease in the obliquity of the ecliptic:.* The 23½° angle between the axis of the sun's sphere and that of the stars' sphere (Fig. 1.2), or (in the heliocentric system) the angle between the earth's rotation axis and the axis

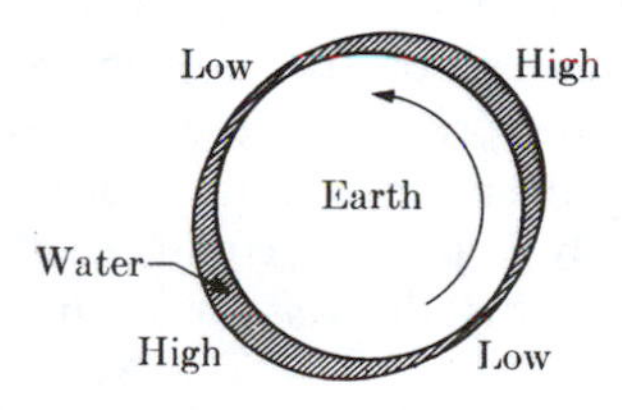

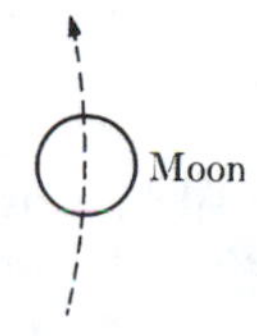

Fig. 11.6. Tides caused by the moon. The earth is to be imagined as rotating in this (much exaggerated) envelope of water. The high tide stays relatively fixed with respect to the moon.

of its plane of revolution around the sun is not constant but has been gradually decreasing.

If these changes continue indefinitely in the same direction—if they are *secular* inequalities—the ultimate effects on all human life could be severe. Recall Kepler's Third Law, $T^2 \propto R^3$. If the period T decreases, the orbital radius R must also decrease. So, over millennia, the moon would gradually get closer to the earth, raising enormous tides and eventually crashing into the surface, destroying most life on earth. (The now-famous asteroid impact 65 million years ago that is believed to have annihilated the dinosaurs was a gentle nudge by comparison.) Even more ominous, if Jupiter's period and orbital radius became small enough, this giant planet would wipe out the entire earth. At the same time Saturn would move outward—destroying Uranus, Neptune, and Pluto—eventually escaping the solar system.

If the obliquity of the ecliptic simply decreased to zero and did not change after that, we would no longer have seasonal variations of temperature. The increased intensity of the sun's radiation due to the tilt of one hemisphere toward the sun, alternating with the decreased intensity when it tilts away 6 months later, would be replaced by a nearly constant intensity. (There would be a small residual variation because of the eccentricity of the earth's orbit.)

Those changes would not be noticeable for centuries or millennia, but the recognition that they are consequences of Newtonian gravitational theory was to have an immediate effect on the status of the "clockwork universe" view of the world. That view, articulated by Kepler, Descartes, and Robert Boyle, assumes there are *no* such secular inequalities. Instead, all changes must be *cyclic*.

So it was a remarkable reinforcement of the clockwork universe theory when Lagrange and Laplace proved that these three inequalities are cyclic: the deviations of the motions of the moon, Jupiter, and Saturn from Kepler's laws will eventually reverse themselves, and the obliquity of the ecliptic will oscillate between finite limits. (At least the proofs seemed satisfactory at the beginning of the nineteenth century, though more accurate calculations later forced revisions of these simple conclusions; on a scale of 10^{18} y (a quintillion years) the motion of the outer planets is chaotic and Uranus may eventually be ejected from the solar system as a result of its interactions with Jupiter and Saturn.)

11.7 The Discovery of New Planets Using Newton's Theory of Gravity

One of the most intriguing aspects of Newton's work still remains to be told: how, more than 100 years after his death, Newton's law of universal gravitation helped astronomers to discover new planets. The British physicist Oliver Lodge refers to it as follows:

> The explanation by Newton of the observed facts of the motions of the moon, the way he accounted for precession and nutation and for the tides, the way in which Laplace [whose mathematical work extended Newton's calculations] explained every detail of the planetary motions—these achievements may seem to the professional astronomer equally, if not more, striking and wonderful; . . . But to predict in the solitude of the study, with no weapons other than pen, ink, and paper, an unknown and enormously distant world, to calculate its orbit when as yet it had never been seen, and to be able to say to a practical astronomer, "Point your telescope in such a direction at such a time, and you will see a new planet hitherto unknown to man"—this must always appeal to the imagination with dramatic intensity. . . .
>
> (*Pioneers of Science*)

One night in 1781 William Herschel of Bath, England, an extraordinarily energetic mixture of professional musician and gifted amateur astronomer, was searching the sky with his homemade, 10-ft telescope. For years he had patiently looked at and reexamined every corner of the heavens, finding new stars, nebulae, and comets, and he was becoming well known among astronomers. On that particular night he again found a celestial object, hitherto uncatalogued and of such "uncommon appearance" that he suspected it to be a new comet. Through the Royal Society the news spread. As observations continued night after night and other astronomers, especially A. J. Lexell, computed its orbit, it became evident that this was not a comet but a newly discovered planet, more than 100 times larger than the earth and nearly twice as far from the sun as Saturn, until then the outermost member of the solar family. So *Uranus* was discovered—an unsuspected and sensational widening of the ancient horizon, a planet just barely visible to the naked eye, but previously mistaken for a star when occasionally observed.

By that time it was known how to compute the elliptic orbit of a planet from a few widely separated observations of its varying positions. Also, the expected small deviations from the true ellipse owing to the perturbing force of the other planets were accurately predictable on the basis of Newton's law of gravitation. Uranus' 84-year orbit was thus mapped out for it by calculation, and all went well for many years. But by 1830 it became more and more evident that Uranus was misbehaving, and that the assumptions on which its schedule had been worked out were in need of revision.

Some thought that Newton's theory might not apply accurately, after all, over such immense distances, but they had nothing better to offer (and as has been well pointed out, theories are in general not overthrown by isolated contradictory facts but only by other, better theories). Others suggested that a hitherto unsuspected comet or a more distant planet might introduce additional perturbations into Uranus' path; but they, too, were merely guessing and made no concrete quantitative predictions.

But the idea of an undiscovered planet beyond Uranus intrigued two young men, who independently undertook the immensely difficult mathematical task of locating the positions of this suspected perturbing body solely from the observed motions of Uranus, using throughout the law of gravitation in unmodified form. John Couch Adams started his work as an undergraduate at Cambridge University and had his mathematical result of the problem two years after graduation For the necessary confirmation he wrote to the Royal Observatory at Greenwich, asking that their powerful telescope search for the hypothetical new planet at a predicted location beyond Uranus. Since Adams was an unknown junior mathematician, however, he was not taken seriously enough to interrupt the current work there with what would have been a tedious search.

A few months later, U. J. J. Leverrier in France published the result of similar, independent calculations, giving very nearly the same position for the suspected planet as Adams had. While finally some observations were slowly being undertaken in England to verify Adams' theoretical conclusions, Leverrier sent his own prediction to J. G. Galle at the Berlin observatory, who, happily having at hand a new star map to aid in the search, on the very evening of the letter's arrival, searched for and recognized the planet at very nearly its predicted position. Thus *Neptune* was added to the solar system in 1846; and here was indeed a triumph of the law of universal gravitation!

Neptune, in turn, was closely watched. The orbital radius is about 30 times the earth's, and its period is therefore given by Kepler's law (correctly) as 164.8 years. But in time the perturbations of this planet and those of Uranus again became larger than could be accounted for in terms of known forces, and the hypothesis of still another planet was naturally made. An arduous 25-year-long search yielded the discovery of Pluto in 1930, announced on the double anniversary of Herschel's discovery of Uranus and of the birthday of Percival Lowell, the American astronomer whose calculations had led to the search and who had founded the observatory in Arizona at which the discovery was made by C. W. Tombaugh. The name appropriately follows the tradition of using classical mythology (Pluto was the God of the underworld) but the astronomical symbol ♇ also incorporates the initials of Percival Lowell.

Another astronomer, W. H. Pickering, had made independent calculations and predictions of Pluto's position as far back as 1909, and had initiated a telescopic search for the planet at the Mount Wilson Observatory in California. Nothing was found then, but after the discovery at the Lowell Observatory in 1930, the old Mount Wilson photographs were reexamined and showed that Pluto could have been found in 1919 if its image had not fallen directly on a small flaw in the photographic emulsion!

This story dramatizes the frequently forgotten possibility that for every overpublicized discovery that is made by "accident," without elaborate preparations, there may well be an equally important discovery that, also by accident, *failed* to be made despite careful research.

11.8 Bode's Law: An Apparent Regularity in the Positions of the Planets

a) This chapter would be incomplete without a short discussion of the influence of a strange and simple rule on the discovery of Neptune. This rule is known as *Bode's law,* after Johann Elbert Bode (1747–1826), director of the astronomical observatory at Berlin, who published it in 1772. (It is also sometimes called the *Titius-Bode rule,* since it had been developed earlier by Johann Daniel Titius, a professor in Wittenberg, Germany.)

At the start of his long search for regularity in the solar system, young Kepler had discovered a crude relationship between the sizes of the various planetary orbits, but it was a rather fruitless jumble of numerological statements. The best that can be said of it was that its mathematical ingenuity first brought him to the attention of Brahe and Galileo. Later, the three Keplerian laws, unified by Newton's work, showed that there existed a simple universal relation between each planet's speed and orbit, but it left unanswered the question of why a given planet did not move in some other possible orbit with a correspondingly different speed. Was there, for example, any specific reason why the earth could not pursue a course nearer to Mars', although necessarily with a relatively longer period of revolution? Nothing in Kepler's or Newton's work showed that the chosen orbit was in any way unique or necessary.

Since the whole effort of science is bent on finding simple regularities, it is no wonder that scientists continued to look for clues. Bode's law, obtained no doubt, like Kepler's laws, by incessant mathematical play with the data to be explained, established just such a clue, which, in the absence of physical reasons, must certainly look like a strange coincidence.

If we number the successively more distant planets from the sun with $n = 1$ (Mercury), 2 (Venus), 3 (Earth), . . . , then the radius of the orbit of any planet (in astronomical units) is given by the *empirical* formula

$$R = 0.3 \times 2^{(n-2)} + 0.4 \text{ (in AU).}$$

But this formula works only if two exceptions are made. One is that a zero must be used instead of the factors $0.3 \times 2^{(n-2)}$ for the calculation of Mercury's orbital radius. Furthermore, although Mars can be assigned $n = 4$, Jupiter, the next planet, has to be assigned $n = 6$; Saturn, the next and outermost planet then known, has to have $n = 7$. The number $n = 5$ could not be given to any existing planet. This at once suggested to Bode that the disproportionately large space that actually extends between Mars and Jupiter harbored a yet-unnoticed planet for which $n = 5$ and which, *according to this rule,* could be expected to have an orbital radius of 2.8 AU. (Derive this predicted radius from the last formula.)

Here was a chance to confirm whether the rule was anything more than coincidence and fantasy! Bode was convinced that a search would reveal the suspected planet, and he wrote, in a manner that reminds us again of Copernicus' and Kepler's motivations, "From Mars outward there follows a space . . . in which, up to now, no planet has been seen. Can we believe that the Creator of the world has left this space empty? Certainly Not!"

At first no one discovered anything in that seemingly God-forsaken gap, and interest in the law must have lagged (one might say it did not fulfill the criterion of stimulation of further discoveries). But then, 9 years after Bode's announcement, came Herschel's discovery of distant Uranus. With $n = 8$ as a logical number to assign, the law predicted an orbital radius of 19.6 AU, only about 2% larger than the actual value derived by subsequent observations.

This agreement directed attention to the other predictions inherent in Bode's law, and the search for the "missing" planet was renewed with vigor. In time, the first result came, but, as so often in science, only as the by-product of research not directly concerned with this problem. In 1801, the Sicilian astronomer Giuseppe Piazzi, while compiling a new star catalog, chanced on a new celestial object that was moving as rapidly as might a comet or a planet. In the exciting months that followed, astronomers and mathematicians cooperated to derive the orbit of the foundling, now christened Ceres. It was disappointingly puny—less than 500 miles in diameter—but the orbital radius came to 2.77 AU, only about 1% less than Bode's law predicted for the missing planet in that region between Mars and Jupiter!

Even while Ceres was being hailed as the long-sought sister planet, another, smaller body was discovered with about the same orbit, and then others still, all smaller than Ceres. Today we know of several thousand such objects, some with very eccentric paths, but all evidently springing from one family with a common history—perhaps fragments that were forming into a planet when its evolution was somehow interrupted or perhaps the shattered remnants of a larger planet in the predicted orbit. It is possible to speculate that there were two (or more) planets trying to establish orbits near each other, but that they eventually collided "because" Bode's law would permit only one orbit to be there. This line of reasoning would put Bode's law into a new light; we might be tempted to search later for other clues that the law expresses simply the series of dynamically possible, stable orbits for single planets.

The discovery of these *asteroids* or *planetoids* bore out Bode's law so strikingly that it was natural to apply it when the suspicion of a planet beyond Uranus first arose. The first available value for n being 9, the corresponding orbital radius would be $(0.3 \times 2^{(9-2)} + 0.4)$ AU, that is, 38.8 AU.

Table 11.2. Bode's Law

n	*R* (theoretical)	*Planet*	*R* (observed)
1	0.4	Mercury	0.39
2	0.7	Venus	0.73
3	1.0	Earth	1.00
4	1.6	Mars	1.53
5	2.8	(Asteroids)	(2.3 to 3.3)
6	5.2	Jupiter	5.22
7	10.0	Saturn	9.6
8	19.6	Uranus	19.3
		Neptune	30.2
9	38.8	Pluto	39.5

The theoretical values are computed from the formula $R = 0.3 \times 2^{n-2} + 0.4$ (in AU), except that $R = 0.4$ for $n = 1$.

Adams and Leverrier based their calculations on the predictions of Bode's law to find the probable location, mass, and orbit of their hypothetical planet. When Neptune was afterward seen and its course plotted, the actual orbit was found to be about 20% smaller than Bode's law had predicted—the first failure of the law, but by luck one not so serious as to invalidate the calculations that led to Neptune's discovery. Later, when Pluto was found, its actual orbital radius of 39.5 AU much more seriously contradicted the value of 77.2 AU that would follow from Bode's law for $n = 10$, although it is very close to the place predicted for Neptune, with $n = 9$.

The irregularities of Neptune and Pluto with respect to Bode's law exemplify in a very simple way a general problem that is frequently encountered in testing scientific theories. If a theory is successful in explaining all known observations and in predicting some results that were not known at the time it was first proposed, must we reject it when it fails to account for a further observation? Or is it legitimate to introduce, at least provisionally, an *ad hoc* hypothesis to explain the exception?

For example, we might conjecture that the original solar system contained nine planets, with orbits corresponding to the integers $n = 1$ through 9 in the formula, including Pluto as number 9 rather than Neptune. Then, just as we have suggested that planet number 5 broke up, leaving the belt of asteroids to occupy the "empty" orbit, we might propose that Neptune was captured into the solar system after its original formation, or perhaps that it was originally in orbit number 10 but suffered a near-collision with a comet that pushed it into its present orbit between numbers 8 and 9. (Current views are actually quite different: It is Pluto whose credentials as a "real planet" rather than an asteroid or escaped satellite are now challenged.)

Some writers on scientific method would object that a theory loses its scientific value if it can be too easily modified to explain away all apparent exceptions: If such a procedure were permitted, how would we ever be able to dispose of false theories?

On the other hand, if we were too strict in requiring that a theory be abandoned as soon as it is contradicted by new data, we might lose all hope of finding rational explanations of complex phenomena, especially those that depend primarily on one or two physical factors, but weakly or occasionally on others. If a hypothesis, or even a numerical rule, is helpful in the search for new knowledge, why reject it simply because it is not yet a completely comprehensive and accurate theory? Astronomers would be extremely surprised if a modern theory of the origin of the solar system led *exactly* to a simple formula such as Bode's law; yet this or a similar algebraic formula might well be an adequate first approximation, just as the circle is a fairly good approximation for the shape of the orbit of most planets.

Until we have a more fundamental theory that explains why Bode's law is approximately correct for the planets up to and including Uranus, we cannot know why it breaks down beyond Uranus. The exceptional case of Neptune, which might lead us to examine carefully whether that planet is different in some way from all the others, may yet prove to be the most interesting episode in the entire history of Bode's law; for without such a law, not only would it have been much more difficult to discover Neptune, but there would have been little reason to suspect that there is anything exceptional about it!

b) This may be a good place to review briefly the *different types of laws* we have encountered here. At the first level of sophistication there is the type of law usually called *empirical:* It seems to summarize simply some fairly directly observed regularity, without attempting to provide a theoretical explanation for it. As it happens, the empirical laws we have discussed up to now all pertain to planetary orbits (Kepler's laws and Bode's law); later on we shall encounter Boyle's law of gas pressure, Gay-Lussac's law of combining volumes for chemical reactions, Mendeléeff's periodic law of the elements, and Balmer's formula in spectroscopy. Since in most cases we now have a satisfactory theoretical understanding of the regularities expressed by these laws, we no longer call them merely empirical laws. But as we have already seen in our discussion of Kepler's work, even a regularity appearing rather obvious to us now may have been an insight that presupposed a difficult reorientation within science and stimulated important further developments. At any rate, we must allow these empirical rules a vital place in science.

Next in order, we may perhaps place those laws that represent an induction of one regulating principle from a variety of apparently very different phenomena. For example, Galileo's law of projectile motion and the fundamental equation $F = ma$ of Newtonian physics were perceptions not obvious from direct observation. In fact, in these cases the formulation of the law was preceded by the formulation of new concepts: in Galileo's case, acceleration and the independence of horizontal and vertical velocity components; in Newton's case, mass and to some extent force. Because these laws usually carry with them the definition of fundamentally important concepts involved therein, we might call them *definitional* laws. Examples of this type might also include the powerful laws of conservation of energy and of momentum, which we shall discuss shortly.

Third, the remaining group is formed by those laws that represent a general conclusion derived *from some postulate or theory,* whether new or old. For example, the lens laws, which make possible the design of optical instruments, can be derived from a theory of the propagation of light. The law of pendulum motion can be deduced from the conceptual scheme of mechanics built around Newton's three laws of motion. And the law of universal gravitation certainly was not an empirical rule, nor did it serve to exhibit a new fundamental concept (all the factors in $F = Gm_1m_2/R^2$ were previously defined concepts, G being essentially a proportionality constant). Instead, as we have seen, it was a generalization derived from the conceptual scheme featuring Kepler's rules, Newton's own laws of motion, and the hypothesis of long-range attraction among the planets and the sun.

This type of law, which we may perhaps name *derivative,* often seems the most satisfying of the three because, since it has been derived from some underlying theory as well as based on empirical facts, we are tempted to feel that it is also somehow more fully "explained." Indeed, scientists continually seek to reduce purely empirical laws to derivative ones, thereby giving the empirical law a measure of additional "meaning,"[2] while at the same time extending and fortifying the theory itself.

There is a kind of symbiotic relationship here between law and theory. A theory becomes more and more respected and powerful the more phenomena that can be derived from it, and the law describing these phenomena becomes more meaningful and useful if it can be made part of a theory. Thus Newton's theory of universal gravitation gained greatly in stature because it enabled one to derive the laws that govern the moon's motion, known by empirical rules since

[2]This feeling appears somewhat fallacious if subjected to rigorous analysis. After all, the theory itself is acceptable only if its derived laws can be checked by experiments, and may in fact be incorrect *even if* the derived laws turn out to be correct. We shall see such an example when we come to deal with the theory of light. Nevertheless, great theories are relatively stable for long periods, and laws derived from them are as secure as we have any right to expect.

the days of the Babylonian observers. These rules in turn, now that they were "understandable" in terms of the theory, could be reformulated more accurately; in fact, the theory even enriched them, for example, by calling attention to hitherto unnoticed peculiarities of the moon's motion.

Another important example of this kind will be presented in the section on the nuclear atom: That theory has been so enormously important in modern science because it swallowed up whole textbooks full of isolated empirical and definitional laws from all branches of physics and chemistry, and reissued them as derivative laws springing from the unifying concept of the atom.

From this discussion we may perhaps draw some general conclusions that are valid beyond the physical sciences. Empirical rules may be valid enough in a limited field, but they do not contain enough information to warn the naive user when some particular new case is really outside the field of applicability of the empirical rule. Not until the damage is done is it clear that the rule did not apply.

Most of our own everyday actions tend to be governed by this type of uncomprehended rule and plausible generalization. This is doubly unfortunate as these rules, unlike those in science, are usually not even based on accurate clear-headed observation. While our life thereby may become simpler, it also is made more brutal and senseless, and we open ourselves up to an invasion of pseudoscience, superstition, and prejudice, even in matter subject to test and measurement.

Problem 11.15. Would the credibility of Bode's law be strengthened or weakened if a series of small planets was discovered between Venus and Mercury, with orbits whose radii are 0.55 A U, 0.475 AU, 0.4375 AU, . . . ?

Problem 11.16. Can we conclude that an empirical law represents absolute truth if it agrees with all known facts (within limits of observational error) at a certain time? How would your answer to this question change if the number of independent known facts were 3? 10? 1000?

Problem 11.17. If an empirical law works satisfactorily for five cases, but is definitely wrong for the sixth, are we justified in proposing an *ad hoc* hypothesis to account for the single discrepancy? How would your answer change if the number of successful cases was fifty instead of five?

Problem 11.18. Assume that in a planetary system other than our own, the distance from the central star to each of the four inner planets is given by the following table:

Sequence number of planet (*n*):	1	2	3	4
Radius of orbit, in 10^6 mi (*R*):	3	5	9	17

Obtain from these data a "law," a formula that will relate *R* to *n* for each planet. At what distance do you expect the fifth planet to be?

11.9 Gravity and the Galaxies

Even in Newton's work we find further consequences of the law of universal gravitation. For example, he gave a good semiquantitative explanation for the long-standing mystery of the precession of the equinoxes, i.e., a very slow rotation of the earth's axis, much like the wobbling motion of a rapidly spinning top. But now we turn to a more ambitious question. Do Newton's laws, so serviceable and fruitful within the solar system, continue to apply beyond, among the "fixed" stars?

To Copernicus or Galileo this question would have been meaningless, for it was not until Newton's time that European astronomers (beginning with Halley) noticed relative motions in the celestial sphere.[3] In fact, our whole solar system was found to move with respect to those distant stars. William Herschel, in 1803, also discovered that some star neighbors rotated about each other (double stars), and his son John Herschel showed that their motions are compatible with the assumption of central forces between them, such as exist in the solar system.

This new picture of our living, moving universe, which the prophetic philosopher Giordano Bruno had sensed long ago, places the solar system in the uncrowded company of many billions of other suns and their possible attendants (of which fewer than 6000 suns are visible without a telescope). The whole cloud of stars forms

[3]In China, where science was vigorously pursued during the centuries before its revival in Europe, the proper motion of a few stars was estimated by a Buddhist monk, I-Hsing (682–727). He was able to compare his measurements of their positions with those recorded by others much earlier.

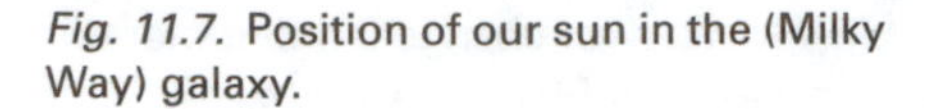

Fig. 11.7. Position of our sun in the (Milky Way) galaxy.

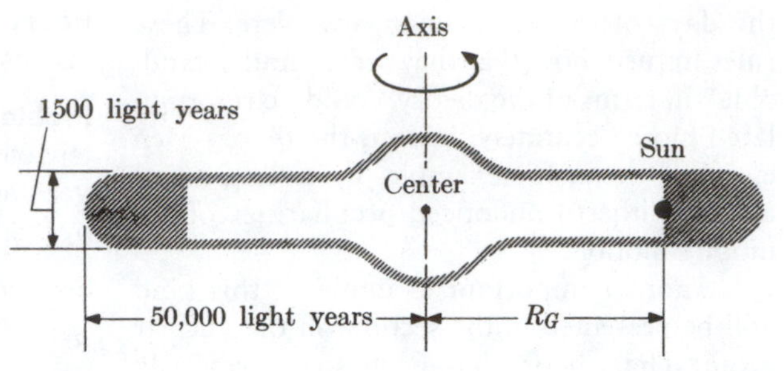

our galaxy, a vaguely disk-shaped region over 100,000 light-years across and about 1000 light-years thick. At a distance of about 26,000 light-years from the galactic center (not even at the center of all solar systems!) our planetary system seems rather lost in this structure.

We use the familiar term *light-year,* defined as the *distance* traveled by light in one year (in a vacuum), equal to 9.46×10^{13} km or 5.88×10^{12} mi. Astronomers also measure distance by parallax: the angle at the object subtended by the radius of the earth's orbit around the sun (see Problem 1.3). Thus a parallax of 1 sec, or "parsec" for short, is 206,265 AU = 3.084×10^{16} m.

It was William Herschel who popularized the notion that when we look far out into space with powerful telescopes such as he constructed, we are also looking far back in time. So when we say that we see a star a million light-years away from us, we mean that we see it now as it was a million years ago. Such statements obviously contradicted, and helped to discredit, the seventeenth-century idea (held even by Newton) that the universe was created only a few thousand years ago.

We might wonder whether the mutual gravitational attraction of the stars should make the galaxy slowly gather up and coalesce into one solid mass at the center. Since this does not happen, motion of the stars being not all toward the center, there are then two explanations. Either the law of gravitation does not extend itself to the whole universe or, in analogy with the solar system itself, the whole galaxy is spinning around its axis, "distracting" the threatening gravitational attractions by making them do the chores of centripetal forces.

On examination, the second choice actually seems to hold. Our own solar system revolves about the distant center of the galaxy at a speed of about 250 km/sec. According to our present view, the edge of the whole galaxy, whose framework we see as the Milky Way, completes one revolution in perhaps a quarter-billion years. (The word "galaxy" is derived from the Greek word *gala,* milk.)

Now let us make a bold little computation. If Newton's laws hold, we should be able to compute the approximate mass of the whole galaxy from the rate of (circular) revolution of our own sun about the galactic center.

If we neglect that part of our galaxy lying in the shaded portion in Fig. 11.7, we may say that the centripetal force on our sun is the gravitational pull exerted on it by the mass of the galaxy (acting approximately as if it were all located at the center). Thus we expect that the following equation holds:

$$m_S \frac{4\pi^2 R_G}{T^2} = G \frac{m_S m_G}{R_G^2},$$

where T stands for the period of revolution of the sun (about 2×10^8 years), m_G for the mass of the galaxy, m_S for the mass of the sun. The only unknown in this equation is m_G, and on solving for it, $m_G = 2 \times 10^{11}$ times the mass of the sun (check this).

But when we add up the mass of all the stars that our galaxy is likely to hold, even in the most generous estimate, we can account for fewer than half of those 200 billion units. Either our assumptions are wrong, or we can find some galactic mass other than that contained in the stars of our galaxy. The assumptions were, of course, rather crude, but short of abandoning the law of gravitation, no tinkering with those admittedly qualitative details will explain the calculated overabundance of mass. So we turn to the alternative—and we certainly *can* find evidences that our galaxy contains matter not congregated into suns and solar systems, namely, nebulae, interstellar dust, and most rarefied gases.

There are many such clues to the presence of this tenuous matter that would rescue the

Fig. 11.8. Spiral nebula, a galaxy about a million light-years from our own.

Newtonian laws, mainly its effect on starlight. The density is inconceivably low, less than for the best vacuums in the laboratory, perhaps only a few atoms per cubic centimeter. But this is spread out over equally inconceivable distances, and there are several alternative estimates of the total distributed material that could supplement the known star masses by about the missing amount. Thus our confidence in the law of gravitation is not destroyed, but we see that it opens up a new avenue of research.

Beyond our "local" world, starting at a distance of about thirty times our own galactic diameter, we find other galaxies scattered through space, as far as our biggest telescopes can reach. These island universes do not seem to differ greatly in size from ours and often appear like a spiral or pinwheel (see Fig. 11.8), which is probably the shape of our own galaxy as well. (It is clear that if Kepler's third law holds for a whole galaxy, the outlying material will revolve more slowly than the masses near the center; hence the possibility of the spiral shape. On the other hand, if the whole system were to rotate like a solid disk, each part having the same period of revolution and angular velocity, then the centripetal force holding the system together could not obey an inverse square law but would have to be linearly proportional to the distance from the center.)

Even as these distant galaxies seem to recede from us and from one another at enormous speeds, they rotate at the same time at rates close to our own; hence it may be assumed that they contain about the same mass and obey the same laws of mechanics. And with this extension of the law of gravitation to the "expanding universe" (to be discussed in Chapters 31 and 32), we find ourselves at the limit of our imagination and at the height of admiration for so universal a principle. There have been few products of human genius to match its ambitious promise and its exciting fulfillment.

11.10 "I Do Not Feign Hypotheses"

The *theory* of gravitational forces, whose main hypothesis is the attraction of all particles of matter for one another, yields the derived *law* of universal gravitation, which in turn explains, as we have seen, Kepler's empirical laws and a wealth of other phenomena. Since one purpose of any theory is this type of explanation and summary, Newton's theory strikes us as eminently satisfactory. But there remained one feature that gravely bothered Newton, his contemporaries, and, indeed, students to our day. How could one account for gravity itself? What is it that causes the attraction of one body for another? Is there not some intervening medium (such as the frequently postulated *ether*) that somehow transmits the pull in a mechanical fashion?

The very statement of these questions reflects how firmly the mind is committed to physical models and to mechanical explanations and how unsatisfied a mathematical, abstract argument leaves our emotions initially. Rather than accept *action at a distance,* i.e., the idea that bodies can exert forces on one another without the agency of a medium between them, most scientists preferred to think of all space as filled with some kind of omnipresent fluid, which, apart from being otherwise so tenuous by itself as to be undetectable by any experiment, still had to be strong and versatile enough to communicate all gravitational forces (and incidentally perhaps also serve to propagate light and electric and magnetic effects).

The search for some such medium was long and the arguments loud. At the end of Book III of the *Principia,* Newton put his often-misinterpreted remarks:

> I have not as yet been able to deduce from phenomena the reason for these properties of gravity, and I do not feign hypotheses. . . . It is enough that gravity really exists, and acts according to the laws that we have set forth and is sufficient to explain all the motions of the heavenly bodies and of our sea.

This is a famous statement, clearly an echo of Galileo's admonition in very similar circumstances (see Section 7.4). With the words, "I do not feign hypotheses," Newton exempts himself from the obligation to account—at least in public—for the observed consequences of the gravitational theory by additional hypotheses (for example, of a mechanical ether) beyond those needed to derive the laws and observations.[4] The mathematical law of gravity explains a wide range of observations; that is enough justification for accepting it. It is easily conceivable that one might eventually explain the law itself in terms of something still more fundamental, but that, Newton suggests, he is at the time unprepared to do. Nor does he feel that his theory suffers by this inability, for the purpose of scientific theory is not to find final causes and ultimate explanations, but to explain observables by observables and by mathematical argument; that he has done.

On the other hand, his refusal, in the absence of experimental clues, to propose a mechanism by which the effect of gravity may pass from one body to another does not mean he will permit the opposite error, namely, dismissal of the question by the invention of some principle of gravity innate in matter, such as might well have satisfied a medieval scholar. This is brought out in a letter to the theologian Richard Bentley, who was preparing a popular lecture on Newton's theories and asked for clarification of their implications. Newton wrote:

> It is inconceivable, that inanimate brute matter should, without the mediation of something else, which is not material, operate upon, and affect other matter without mutual contact; as it must do, if gravitation, in the sense of Epicurus [an ancient Greek atomist philosopher] be essential and inherent in it. And this is one reason, why I desired you would not ascribe innate gravity to me. That gravity should be innate, inherent, and essential to matter, so that one body may act upon another at a distance through a *vacuum,* without the mediation of anything else, by and through which their action and force may be conveyed from one to another, is to me so great an absurdity, that I believe no man who has in philosophical matters a competent faculty of thinking, can ever fall into it. Gravity must be caused by an agent acting constantly according to certain laws; but whether this agent be material or immaterial, I have left to the consideration of my readers.
>
> (25 February 1693, reprinted in *Newton's Philosophy of Nature,* edited by Thayer.)

The question of how one body can be acted on by another across empty space has an obstinate appeal for our picture-seeking mind. Part of the reason for the great popularity of Descartes' vortex theory, the scheme generally accepted during most of Newton's lifetime, was that it provided a plausible picture of a universe completely filled with a whirlpool of material corpuscles whose action on one another and on the planets was simple physical contact. There we have an example of a plausible, easily grasped conceptual scheme that had the fault, fully appreciated by Newton but not by the majority of his contemporaries, of failing to agree quantitatively with observations, specifically, of failing to allow the derivation of Kepler's laws.

But by contrast with the most successful feature of Descartes' picture, its inherent appeal to the common sense of the muscles, Newton's lack of an intuitive "reason" for planetary motion stood out that much more glaringly. Newton had to spend much time in his later life to uphold his less intuitively appealing system against the Cartesians, and at the same time to refute insistent attempts within his own camp to introduce metaphysical implications into the law of universal gravitation. He declared again and again that he was neither able nor willing "to adjudge natural causes" of gravitation.

This was all the more remarkable (and a big step toward the modern conception of what is required of physical theory) since Newton—

[4]For a discussion of the significance of the word "feign" (rather than "frame" as the Latin *fingo* is sometimes translated), see A. Koyré, *Newtonian Studies* (1965), page 35.

who, we must remember, later regarded himself as much a theologian as a scientist—might have introduced an explicit theistic assumption at that point. As has been hinted at before, the existence of a Creator is a persistent implicit hypothesis in his work, and at times it does come out into the open, rarely in the *Principia* but more often in his other writings. Thus he could say in the *Opticks* in explanation of the term "absolute space," "God moves the bodies within his boundless uniform sensorium."

Privately, Newton, like his contemporaries, did in fact believe that some material agency would eventually be found to explain the apparent action at a distance involved in his law of gravitation. There are several strong hints in his writings that he entertained the idea of an all-pervading ether, for example, *Opticks,* Query 18, where he asks whether heat is not "conveyed through the vacuum by the vibration of a much subtler medium than air, which, after the air was drawn out [as by evacuating a vessel with a pump], remained in the vacuum"; this medium he imagined as "expanded through the heavens." At the time, this was a most reasonable speculation, proceeding as it did from some simple observations on the transmission of (radiant) heat to a thermometer suspended in an evacuated vessel, and anticipating the *wave theory of heat* popular in the early nineteenth century (see Section 17.7). It is the more interesting that this tempting but vague picture was not permitted in the *Principia.*

11.11 Newton's Place in Modern Science

So impressive were the victories of Newtonian mechanics in the early eighteenth century that, through the customary extrapolation from science to philosophy, there spread a mechanistic world view that asserted that man's confident intellect could eventually reduce *all* phenomena and problems to the level of mechanical interpretation. In economics, philosophy, religion, the "science of man"—everywhere the success of the work of Newton and the Newtonians was one of the strongest influences to encourage the rising "Age of Reason."

One of the consequences of the mechanistic attitude, lingering on to the present day, was a widespread belief that with the knowledge of Newton's laws (and those of electrodynamics later) one could predict the whole future of the whole universe and each of its parts if only one were given the several positions, velocities, and accelerations of all particles at any one instant. This view became known as *Laplacean determinism* as it found its most famous expression in statements by the French astronomer P. S. de Laplace around 1800. It was a veiled way of saying that everything worth knowing was understandable in terms of physics, and that all of physics was essentially known.

Today we honor Newtonian mechanics for less all-inclusive but more valid reasons. The factual content of the *Principia* historically formed the basis of most of our physics and technology, and the success of Newton's approach to his problems revealed the most fruitful method to guide the work of the subsequent two centuries in the physical sciences.

We also recognize now that despite its breathtaking scope of applicability, Newton's mechanics holds in its stated form only within a definable region of our science. For example, although the forces within each galaxy may be Newtonian, one can speculate that forces of repulsion instead of attraction operate between each galaxy and its neighbors (one interpretation of the greater and apparently increasing speed with which more distant systems recede). And at the other end of the scale, among atoms and subatomic particles, an entirely non-Newtonian set of concepts has to be presented to account for the behavior of those small-scale worlds, as we shall see in Part H.

Even within the solar system, there are a few small residual discrepancies between predictions and facts. The most famous is the erroneous value of the slow change in the position of closest approach to the sun by the planet Mercury; calculations and observations differ by some 43 sec of arc per 100 years. A similar small failure of Newton's laws of motion and of gravitation exists for the path of a recurring comet that comes close to the sun and, to a lesser degree, for the rest of the members of the solar system.

But these difficulties cannot be traced to a small inaccuracy in the law of gravitation (for example, to a slight uncertainty in the exponent in the expression Gm_1m_2/R^2). On the contrary, as in the case of the failure of the Copernican system to account accurately for the detail and "fine structure" of planetary motion, we are here again confronted with the necessity of scrutinizing our assumptions. Out of this study has come the realization that no single revision can account for all these deviations from classical predictions, so that at present Newtonian science is joined at one end with relativity theory, which, among many other achievements, covers phenomena involving bodies moving at high speeds and/or the close approach of large masses, and at the other end borders on quantum mechanics, which gives us a physics of atoms and molecules. As for the middle ground, Newtonian mechanics still describes the world of ordinary experience and of "classical" physics as accurately and satisfyingly as it always did.

Additional Problems

Problem 11.19. Would Newton's laws of motion, as usually stated, hold in a geocentric system like Ptolemy's? [*Hint:* The term "motion along a straight line" in Newton's first law refers to a line in space such as might be drawn from the sun to a relatively "fixed" star.] If they do not, explain why not, and whether it is plausible that equivalent laws might be formulated to describe motions in a geocentric system.

Problem 11.20. Turn to Section 3.4 and use those paragraphs to rate the Newtonian theory of universal gravitation point by point. (Do not assume that such a conscious rating is ever seriously undertaken in scientific work.)

Problem 11.21. Newton proved in the *Principia* that the gravitational force on an object at a point inside a uniform sphere, at a distance R from the center, would be the same as that exerted by the part of the sphere situated closer to the center (i.e., by a sphere of radius R, ignoring the outer spherical shell). Using this fact, show that the gravitational force inside a sphere is directly proportional to R.

RECOMMENDED FOR FURTHER READING

Herbert Butterfield, *The Origins of Modern Science,* Chapters 8–10

B. E. Clotfelter, "The Cavendish experiment as Cavendish knew it," in *Physics History II* (edited by French and Greenslade), pages 104–107

I. Bernard Cohen, "Newton's Discovery of Gravity," *Scientific American,* Vol. 244, no. 3, pages 167–179 (1981)

I. Bernard Cohen and Richard S. Westfall, *Newton,* extracts from *Principia* and essays by Cohen, Colin Maclaurin, and David Kubrin, pages 116–119, 126–143, 238–248, 253–296, 339–342

Betty Jo Teeter Dobbs and Margaret C. Jacob, *Newton and the Culture of Newtonianism*

Stillman Drake, "Newton's apple and Galileo's *Dialogue,*" *Scientific American,* Vol. 243, no. 2, pages 151–156 (1980). Galileo's *Dialogue* contains a diagram that may have influenced Newton in linking the falling apple with the moon in orbit.

Richard P. Feynman, *Character of Physical Law,* Chapters 1 and 2

C. C. Gillispie (editor), *Dictionary of Scientific Biography,* articles by Jean Itard on Clairaut and by Colin Ronan on Halley, Vol. 3, page181 and Vol. 6, pages 67–72

Owen Gingerich, "Newton, Halley, and the Comet," *Sky and Telescope,* Vol. 71, pages 230–232 (1986)

A. Rupert Hall, *From Galileo to Newton 1630–1720,* Chapters X and XI

J. M. Keynes, "Newton, the man," in *Essays in Biography,* pages 310–323, New York: Norton, 1963; extract in *Newton* (edited by Cohen and Westfall), pages 314–315

T. S. Kuhn, *The Copernican Revolution,* Chapter 7

John North, *Norton History of Astronomy,* Chapter 13

J. H. Randall, Jr., *The Making of the Modern Mind,* Boston: Houghton Mifflin, 1940; Chapters X–XV

S. Sambursky, *Physical Thought,* pages 296–300, 305–310

R. S. Westfall, "The career of Isaac Newton: A scientific life in the seventeenth century," *American Scholar,* Vol. 50, pages 341–353 (1981); "Newtonian cosmology" in *Cosmology* (edited by Hetherington), pages 263–274; "The scientific revolution of the 17th century: The construction of a new world view," in *The Concept of Nature* (edited by J. Torrance), pages 63–93, Oxford: Clarendon Press, 1992; *The Life of Isaac Newton*

For further discussion of the subject of this chapter, see the "Sources" at the website www.ipst.umd.edu/Faculty/brush/physicsbibliography.htm

PART D

Structure and Method in Physical Science

12 On the Nature of Concepts

13 On the Duality and Growth of Science

14 On the Discovery of Laws

Doing science.

"As in Mathematicks, so in Natural Philosophy, the Investigation of difficult Things by the Method of Analysis, ought ever to precede the Method of Composition. This Analysis consists in making Experiments and Observations, and in drawing general Conclusions from them by Induction, and admitting of no Objections against the Conclusions, but such as are taken from Experiments, or other certain Truths. For Hypotheses are not to be regarded in experimental Philosophy.

And although the arguing from Experiments and Observations by Induction be no Demonstration of general Conclusions; yet it is the best way of arguing which the Nature of Things admits of, and may be looked upon as so much the stronger, by how much the Induction is more general. And if no Exception occur from Phaenomena, the Conclusion may be pronounced generally.

But if at any time afterwards any Exception shall occur from Experiments, it may then begin to be pronounced with such Exceptions as occur. By this way of Analysis we may proceed from Compounds to Ingredients, and from Motions to the Forces producing them;

> and in general, from Effects to their Causes, and from particular Causes to more general ones, till the Argument end in the most general.
>
> This is the Method of Analysis: And the Synthesis consists in assuming the Causes discover'd, and establish'd as Principles, and by them explaining the Phaenomena proceeding from them, and proving the Explanations. . . ."
>
> Isaac Newton, Opticks

As long as there has been science, there has also been commentary on the tasks and procedures of science. From Aristotle to Galileo, from Roger Bacon to the present, almost all major scientists and philosophers have contributed their varying opinions. No room is left, it would seem, for remarks on a subject so lengthily and so competently examined. But even if this were true, we might permit ourselves to discuss, in an informal manner, representative opinions held by some writers who have prominently influenced scientific thought, and so attempt to reach a plausible point of view of science—not for the sake of a technical exposition of the problems involved, but for increased understanding and enjoyment of the scientific enterprise.

The three chapters in Part D, together with Chapter 33, carry on and complete a cycle begun with Chapter 3, "The Nature of Scientific Theory." It is suggested that you reread Chapters 3, 12, 13, and 14 together when you reach Chapter 33.

CHAPTER 12

On the Nature of Concepts

12.1 Introduction: The Search for Constancies in Change

When you ask "What is science?" you are in effect asking "What do scientists now do at their desks and in their laboratories, and what part of their past work is still useful to researchers in a given field?" Because all too many physical scientists have, over the last four centuries, found it more and more difficult to communicate what they do to nonscientists, let us first visualize the scientist's tasks in terms of an analogy involving a more immediately approachable colleague, the cultural anthropologist. In that case the laboratory can be thought of as an island populated by a little-studied tribe or community. The anthropologist immerses herself in the life of the community to observe it, to study its pattern—but, of course, the eye cannot see, the mind cannot grasp, meaningful patterns within the chaos of movement and sound until one has crystallized out of one's experience the necessary concepts with which to discover, think about, and describe relationships in the community life.

One set of concepts belongs to the prior professional background of our investigator, for example, *family, ruler, in-group,* etc.; this corresponds in the physical to the time-honored notions of *time, volume, speed,* and so on. Another and to us more interesting set of terms comes out of the phenomenon under study. The anthropologist learns the language and customs of this people and discovers and invents new concepts important to the organization of her observations and her understanding of the community, for example, the terms "taupo" and "uso," which denoted in Samoa, respectively, the ceremonial princess of the house and a sibling of the same sex. In the same manner the physical scientist, sometimes with great difficulty, must extract from his experience new guiding ideas like *electric field* or *valence.*

Significantly, our anthropologist may find that the group she is studying lacks the manifestations of many concepts without which our own society seemingly could not operate, for example, some elements of the taboos in Western civilization, the knowledge of one's own exact age, or even the simple recognition and naming of most colors. Again science, too, had to learn—often the hard way—that there are no useful equivalents in inanimate nature for such common-sense terms as human longing or absolute simultaneity.

The formulation or "discovery" of new concepts helps the anthropologist gradually to reinterpret the originally meaningless, aimless, or "primitive" life of the village, and a complex, rather rigid pattern may emerge. Her own early common-sense concepts themselves may undergo profound changes—the word *law,* for example, is surrounded by an entirely different atmosphere in different cultures, just as the idea of *force* in physics (and, in fact, every technical concept that grew out of the vocabulary of common sense) is now at variance with the original meaning of the same word as still used outside the sciences.

In the end, our anthropologist may succeed in decoding the original problem, bringing back, in terms that are meaningful to her professional colleagues, an account of that people's political and family organization, personal esthetic values, religious beliefs and practices, economic methods, and so on. Perhaps she will also be able to reconstruct the history of that people, or even to illuminate the behavior of another group, her own, in relation to that culture. At any rate, it is evident in this example that her job was not finished when she had collected the initial direct observables; on the contrary, that was the barest beginning, the stimulus for the really important part of her work. We may keep in mind, with all its limitations, the analogous picture of physical scientists as explorers in a universe of events and phenomena, attempting to find its pattern and meaning.

12.2 Science and Nonscience

Science, one may now say, is the ever unfinished quest to discover facts, establish the relationships between things, and decipher the laws by which the world runs. But let us go beyond this. While not trying to propose a one-sentence definition of the whole complex concept "science," we may perhaps agree at the outset that *the main business of science is to trace in the chaos and flux of phenomena a consistent structure with order and meaning,* and in this way to interpret and to transcend direct experience. "The object of all sciences," in Einstein's words, "is to coordinate our experiences and to bring them into a logical system." Another major twentieth-century physicist, Niels Bohr, agrees when he says, "The task of science is both to extend the range of our experience and to reduce it to order."

You may find these statements too all-inclusive; the same aim might well be claimed by art or by philosophy. Thus the poet T. S. Eliot said, "It is the function of all art to give us some perception of an order in life by imposing an order upon it," and the philosopher A. N. Whitehead defined speculative philosophy as "the endeavor to frame a coherent, logical, necessary system of general ideas in terms of which every element of our experience can be interpreted."

Indeed, in science, as in art and philosophy, our most persistent intellectual efforts are directed toward the discovery of pattern, order, system, structure, whether it be as primitive as the discernment of the recurring nature of seasons or as sweeping as a cosmological synthesis. In this sense, science is but one facet of the great intellectual adventure: the attempt to understand the world in each of its aspects. The search for constancies in the flux of experience is so fundamental and so universal a preoccupation of intelligent life itself that, in common with many of the Greek philosophers, we may regard mind as the principle that produces order. We note without astonishment among the great pioneers of science, an artist and a man of the church, Leonardo da Vinci and Copernicus; indeed, the very origin of the word *science* (Latin *scire,* to know, to learn) reveals the extent of its appeal and the depths of its roots.

Of course, one must not deny the fundamental and distinct differences separating the sciences from the nonsciences. There are obvious points that set one apart from the other, for example, the motivations of the investigators, to be discussed in more detail in Chapter 13. To predict nature and thereby not to be at its mercy, to understand nature and so to enjoy it, these are among the main motivations for the examination of nature through science, whereas the poetic understanding of nature is primarily the self-realization, the proclamation, and the ennoblement of one's own spirit.

But these are in a sense complementary, not contradictory motivations, illuminating the two sides of humans. Both coexist to some degree within each individual. How artificial would be a distinction between the triumph of the scientist and that of the artist within Johannes Kepler when he writes in the *Harmony of the World* (1619), on the occasion of publishing the third law of planetary motion:

> . . . What I prophesied 22 years ago, as soon as I found the heavenly orbits were of the same number as the five [regular] solids, what I fully believed long before I had seen Ptolemy's Harmonies, what I promised my friends in the name of this book, which I christened before I was 16 years old, I urged as an end to be sought, that for which I joined Tycho Brahe, for which I settled at Prague, or which I have spent most of my life at astronomical calculations—at last I have brought to light, and seen to be true beyond my fondest hopes. It is not 18 months since I saw the first ray of light, 3 months since the unclouded sun-glorious sight—burst upon me. Let nothing confine me: I will indulge my sacred ecstasy. I will triumph over mankind by the honest confession that I have stolen vases of the Egyptians to raise a tabernacle for my God far away from the lands of Egypt. If you forgive me, I rejoice; if you are angry, I cannot help it. This book is written; the die is cast. Let it be read now or by posterity, I care not which. . . .

A second point of difference between science and nonscience is much more clear-cut than the first; it is the kind of rules and concepts the scientist uses, the type of argument that will cause his or her colleagues to say, "Yes, I understand and I agree." This will occupy our attention to some degree, as will a third point of difference: the observation that in the course of time, despite great innovation and revolutions, there accumulates in science a set of internationally acceptable, basic, and fairly enduring conceptual schemes, whereas this can hardly be said for many other human endeavors.

At once the question arises as to how humans, so short-lived and so fallible, can penetrate such complex truths, discover lasting patterns, and obtain general agreement. For the scientist's work, like that of any other explorer, must involve the whole person, demanding as it does reflection, observation, experimentation, imagination, and a measure of intuition. Being human, we fail far more often than we succeed, and even our successes and failures may in the light of further progress reverse positions. Scientists often cannot fully explain their reasons for dedicating themselves to their work and the steps by which it progresses; and if scientists are vocal on such matters, they very likely are contradicted by the testimony of some of their colleagues. And yet, the *result* of this uncertain human activity, namely, the growing body of science itself, is undeniably a successful, vigorous enterprise, bountiful in discoveries, in distinct contrast to the frailties and confusions of its human creators. This paradox we shall explore, for it will be the key to an understanding of the activities of scientists and of the successful features of their work.

Immediately one answer suggests itself: There may be some trick, some *method* that scientists follow to solve their problems so well. After disposing of this opinion, we shall look quite closely at an alternative answer, namely, that there is perhaps some special merit in the idiosyncratic vocabulary, in the particular mental tools themselves, with which scientists analyze the world around them.

12.3 The Lack of a Single Method

If by scientific method we mean the sequence and rule by which scientists now do and in the past have actually done their work, then two truths soon become obvious. First, as for every task, there are not one but many methods and uncountable variants and, second, even those different methods are for the most part read into the story after it has been completed, and so exist only in a rather artificial and debatable way. The ever-present longing to discover some *one* master procedure underlying all scientific work is understandable, for such a discovery might enormously benefit all fields of scholarship; but like the search for the Philosopher's Stone, this hope had to be given up.

At about the beginning of the seventeenth century, the time of Francis Bacon and René Descartes, it was still quite reasonable to hope that both these all-powerful keys to wealth, health, and wisdom existed, but the verdict of the three centuries since then is plainly negative. Even the eighteenth-century chemist Joseph Priestley warned ". . . how little mystery there really is in the business of experimental philosophy, and with how little *sagacity,* or even *design,* discoveries (which some persons are pleased to consider as great and wonderful things) have been made. . . ."

Priestley's words, even though an extreme generalization that does not do justice to his own carefully prepared researches, are a proper antidote to the other extreme of opinion, which would present science to us as a special, infallible scheme relentlessly progressing from success to success with the precision of a smoothly moving machine. For as soon as one begins to look into the history of scientific discoveries, one is overwhelmed by evidence that there is no single, regular procedure.

By temperament and by characteristics of performance, scientists have always differed from one another as widely as composers. Some proceeded from one step to the next with the certainty and restraint of a Bach, others moved among ideas with the abandon of a Schumann. The Austrian physicist Ludwig Boltzmann (who had once studied piano with the composer Anton Bruckner), in his obituary for the German physicist G. R. Kirchhoff, wrote:

> Among Kirchhoff's papers just mentioned there are some of unusual beauty. Beauty, I hear you ask; don't the graces depart where integrals stretch their necks; can anything be beautiful when the author has no time for the slightest embellishment? Yet—just as the musician recognizes Mozart, Beethoven, Schubert with the first few bars, so would the mathematician distinguish his Cauchy, Gauss, Jacobi, Helmholtz after a few pages. The French are characterized by refined external elegance, although now and then the skeletal structure of their deductions is a bit weak; the English by the greatest dramatic force, above all Maxwell. Who doesn't know his dynamic theory of gases? At first the variations of the velocity are developed majestically; then the equations of state come in on one side and the equations of central motion on the other; ever higher the

> chaos of equations surges until suddenly four words resound: "Put $n = 5$." The evil demon V disappears,[1] just as in music a wild figure in the bass, which earlier undermined everything else, suddenly falls silent. As with a magic stroke everything that earlier seemed intractable falls into place. There is no time now to say why this or that substitution is introduced; whoever doesn't feel it, should put the book away; Maxwell is no composer of program music who has to add an explanation to the notes.
>
> (Broda, *Ludwig Boltzmann*)

Among the great scientists there have been adventurers and recluses, self-taught mechanics and aristocrats, saints and villains, mystics and businessmen, pious men and women, and rebels. In *The Study of the History of Science,* George Sarton said about them:

> . . . Their manners and customs, their temperamental reactions, differ exceedingly and introduce infinite caprice and fantasy into the development of science. The logician may frown but the humanist chuckles. Happily such differences are more favorable to the progress of science than unfavorable. Even as all kinds of men are needed to build up a pleasant or an unpleasant community, even so we need all kinds of scientists to develop science in every possible direction. Some are very sharp and narrow-minded, others broadminded and superficial. Many scientists, like Hannibal, know how to conquer, but not how to use their victories. Others are colonizers rather than explorers. Others are pedagogues. Others want to measure everything more accurately than it was measured before. This may lead them to the making of fundamental discoveries, or they may fail, and be looked upon as insufferable pedants. This list might be lengthened endlessly.

The process of discovery itself has been as varied. While it is perfectly true that individual research projects are almost always unspectacular, each investigation being fairly routine, logically quite consistent and sound within its prescribed scope, occasionally some most important theories have come from drawing wrong conclusions from erroneous hypotheses or from the misinterpretation of a bad experiment. Sometimes a single experiment has yielded unexpected riches, sometimes the most elaborately planned experiment has missed the essential effect by a small margin, and sometimes an apparently sound experiment was simply disbelieved by a scientist who, holding on to his "disproved" theory, emerged triumphant when the experiment was found to be faulty after all. Even the historic conceptual schemes in science themselves first captured the mind in seemingly the most casual or unpredictable way.

To cite two famous examples, Darwin said he got "a theory by which to work" while he happened to be reading Malthus' treatise on population "for amusement." Kekulé reported that one of his fundamental contributions to modern organic chemistry came to him suddenly during an idle daydream. Similar things have happened to scientists in every field.

On the other hand, some scientists had all the "significant facts" for an important finding in their hands and drew conclusions that were trivial or proved to be mistaken; sometimes a scientist proposed a correct theory although some facts before his very eyes were seemingly in violent contradiction. We shall have further examples of all of these. Even the work of the mighty Galileo, viewed in retrospect, occasionally seems to jump ahead from error to error until the right answer is reached, as if with the instinctive certainty of a somnambulist.

In Chapter 14 we shall read the opinions held by the scientists themselves concerning the methods of work in their field; but even this superficial recital of variety is enough to establish a feeling that the success of science lies not so much in some single method of work, but perhaps rather in some peculiar adjustment or mechanism mediating between two factors—the individual's achievement and the body of science in which all contributions are brought

[1]In Maxwell's theory a crucial role was played by the factor $V^{[(n-5)/(n-1)]}$, where V is the relative velocity of two colliding molecules and n is the exponent of the intermolecular distance in the force law: The repulsive force between two molecules at distance r is r^{-n}. He did not know how to calculate the velocity distribution except for a gas in thermal equilibrium (Section 22.6), nor did he know the actual value of n, so he "guessed" that $n = 5$. For this special case the annoying factor ("evil demon") becomes $V^0 = 1$, so the equation no longer depends on the relative velocity and can be solved directly. The system of point particles with inverse fifth-power repulsive forces turned out to be an extremely useful mathematical model, many (though not all) of its properties are similar to those of real gases (see the discussion of models in Section 14.2).

together after supplementing, modifying, co-operating, and competing with one another.

A very crude analogy may help here. If we compare the structure of science to that of an anthill, the successful construction of that astonishingly complex, well-arranged, and successful building must be explained not by the diverse, almost erratic behavior of the individual insects, but in larger measure by the marvelous coordination of innumerable solitary efforts. The mechanism of coordination among these social insects is still not fully understood. In the analogous case of science, something positive can be said. We approach it by turning to the alternative possibility posed at the end of the previous section, to the astonishing and many-sided power that resides in the concepts that scientists use.

12.4 Physical Concepts: Measurement and Definition

All intelligent human endeavor stands with one foot on observation and the other on contemplation. But scientists have gradually come to limit themselves to certain *types* of observations and thought processes. One distinctly striking "limitation" is the tacit desire among scientists *to assure that they are in fact discussing the same concepts in a given argument*. This desire for clarity is, of course, general, but here we find a particularly successful scheme of meeting it, even at the cost of some sacrifices.

To take a very simplified example: If the scientific problem were to find the dimensions or volume of a body, its electric charge, or its chemical composition, the specialists called in to solve this problem would all appear to understand clearly what is wanted because they all would go independently through quite similar manual and mathematical operations to arrive at their answers, even though these answers may not completely coincide. (If the word *similar* seems disturbing in the previous sentence, we may go so far as to claim that they will, on mutual consultation, very probably all agree on *exactly* the same operations to follow.)

As a consequence, their findings for, say, the length of one edge of a block may read 5.01 cm, 5 cm, 4.99 cm, and 5.1 cm. The impressive thing here is not that they seem to disagree on the exact length or on the accuracy, but that they do agree on the type of answers to give. They do not, for example, say "as long as a sparrow in summer," "five times the width of my finger," "as long as it is wide," and "pretty short." Evidently the words "length of one edge of a block" mean the same thing to all four experimenters, and we may be fairly sure that on consultation sooner or later all four would convince themselves of one single result, possibly 5.03 ± 0.04 cm. (Or perhaps they will agree that the measurement leading to the discordant value 5.1 cm should be repeated or rejected, and eventually conclude that the best value is 5.00 ± 0.01 cm.) It should not matter which meter stick they happen to pick up, for all manufacturers of measuring equipment are under legal obligation to check their products against carefully maintained universal standards.

On the whole, almost every measurement made in the laboratory is in a sense such a comparison between the observables and some well-known standard accepted by the whole community. Much of the success and rapid growth of science depends on these simple truths, for it is clear that in such circumstances the energies of investigators are not regularly drained off by fruitless arguments about definitions and rules of procedure; on the contrary, the labors of many scientists over centuries can combine in one advancing stream.

There remain, of course, large areas of possible disagreement among scientists, but such arguments can be settled, often by having recourse to some one series of measurements that all disputants (rightly or wrongly) acknowledge at the time to be decisive one way or the other. One of the impressive features of modern physical science is the rapidity with which most major differences of opinion in the field usually disappear.[2] The secret of this successful harmony and continuity in physical science (which we idealized only a little in the previous paragraph) lies to a large degree in the *nature of concepts and their definitions*. For example, the concept "length of an object" as used in science is ultimately defined by the very *operations* involved making the measurement. The problem, "What is the

[2]A comparison with the state of affairs in other disciplines may be misleading, but this is a good place to interject that the writings on the history and the philosophy of science themselves, in this book as in others, are traditionally and quite properly, ready subject matter for disagreement and debate, not only for a few years but perhaps forever.

length of a block?," is for all practical purposes identical with the question, "What is the difference between those two numbers printed on a specific meter stick that stand directly below two corresponding scratches, entered there to signify local coincidence with adjacent corners of the block?"

This last sentence contains an abbreviated example of what we shall call an *operational definition,* that of length, and although it seems ridiculously involved, and may indeed never be used explicitly, our four experts engaged to measure the length can regard it as the "true meaning" of the length of a block, available for examination if any dispute should arise. Ideally, each of the concepts used in the physical sciences can be made clear in terms of some such operational definition, and that perhaps is the most important of the mechanisms whereby mutual understanding among scientists is made possible. For it is clearly more difficult to misinterpret actions than words.

As has been implied throughout this discussion, not all concepts in science can be defined by *manual* operations. Instantaneous velocity, for example, was defined by the slope of a straight line, tangent to the distance-vs.-time graph; this is a mental or mathematical operation. But each such definition becomes a fairly unambiguous set of directives to scientists who have worked in the field, and this explains why their vocabulary is not regularly the subject of dispute.

You will find it revealing—more than that, almost indispensable to a true understanding of physical concepts—to make for yourself operational definitions of all important terms, such as we have made in previous chapters for acceleration, force, mass, and so on. As the influential American physicist and philosopher, P. W. Bridgman, said "The true meaning of a term is to be found by observing what a man does with it, not what he says about it" (*Logic of Modern Physics,* 1927).

At this point you are likely to have several disturbing questions. If you were tempted to object that the operational definition given here is far from the common-sense meaning of length, you would be quite right and would have hit on a very important truth: Everyday notions seem clear and scientific terms mysterious. However, a little thought shows the opposite to be the case. The words of daily life are usually so flexible and undefined, so open to emotional color and misunderstanding, that our first task here is to get used to the specific vocabulary of the sciences and to the apparently picayune insistence on its *rigorous* use—that enormously successful habit that scientists borrowed from scholastic logicians. (In literary writing, on the contrary, it is considered "bad style" to repeat the same noun or verb in a sentence or paragraph; instead one may use a synonym, which probably does not have *exactly* the same meaning.)

Then again, you are bound to have a vaguely uncomfortable feeling that an operational definition only shows us how to measure according to some man-made convention, and does not tell us what length "really" is; once more you are perfectly correct, and once more we must try to make our peace with the limitations of modern science. Here again we are touching on a problem that is at the very core of science, namely, what reality means to a person in a laboratory. Surely we shall have to explore this very thoroughly. For the moment let us take the answer of that great French mathematician and philosopher of science, Henri Poincaré, who at the beginning of the twentieth century illustrated the operational attitude toward physical concepts in this manner:

> When we say force is the cause of motion we talk metaphysics, and this definition, if we were content with it, would be absolutely sterile. For a definition to be of any use, it must teach us to *measure* force; moreover, that suffices; it is not at all necessary that it teach what force is *in itself,* nor whether it is the cause or the effect of motion.

Lastly, you may be puzzled that a simple measurable like length should not be given with perfect exactness, that we should have said "the edge of this block is 5.03 ± 0.04 cm long." The term ± 0.04 cm is called the *probable error* in the measurement; that is to say, the chances are fifty-fifty that the next measurement of the same edge will read between 4.99 cm and 5.07 cm. Evidently all measurements, apart from simple counting, contain some error or uncertainty, no matter how carefully the job is done.

But although this uncertainty may be slight, how can we dare build an exact science on concepts defined by necessarily uncertain measurements? The paradox is resolved by two recognitions. First, this word "error" does not have in science the connotations *wrong, mistaken,* or *sinful,* which exist in everyday speech. In fact, a measured value can be regarded as "exact" only if we know also just what range of values to expect on repeating the measurement. Second—and we must repeat this frequently—science is not after absolute certainties, but after relationships among observables. We realize that the observables can be defined and measured only with some uncertainty; and we shall not demand more than that of the relationships among them, thus leaving the search for absolute truths, if any exist, to other fields of thought, whether we like it or not.

12.5 Physically Meaningless Concepts and Statements

The consequence of even such a trivial-sounding operational definition as that of length may be quite startling. In the special theory of relativity, rigorous definitions of this type directed the course of thought to unexpected results, one being that the measured length of an object depends on how fast the object moves with respect to the observer, a finding that accounted for some most perplexing previous observations. As a result of the impact of Einstein's work, scientists have become aware of the operational nature of the concepts most useful in research. It was then discovered that some tacitly accepted ideas were leading to serious contradictions in parts of physical theory because, by their very formulation they could not be connected with any possible activities or operations in the laboratory. For example, the concepts of time and space were defined previously not by specific and exclusive reference to manipulations with meter sticks, light signals, pendulum clocks, and the like, but in some absolute, intuitive sense. The classic reference here is Newton's statement in the first pages of the *Principia:*

> Absolute, true, and mathematical time, in and of itself and of its own nature, without reference to anything external, flows uniformly and by another name is called duration.

Note that phrase "without reference to anything external," that is, without necessary relation to the rotation of a second hand on a real watch. By this very definition of "true" time, we could not hope to measure it. Or again, Newton wrote,

> Absolute space, of its own nature without reference to anything external, remains homogeneous and immovable.

From the point of view of modern practical scientists, such statements, inherently without operational meaning, are sometimes called "meaningless," perhaps a rather abrupt term, but accurate in this limited sense. You may well ask whether Newton's science was not invalidated by such meaningless basic postulates. The answer is no, simply because Newton did not, in fact, depend on the explicit use of these concepts in his own scientific work. (His reasons for stating them at all were complex and can be traced to his philosophical motivation.)

Galileo, although much nearer to the science of the ancients that he helped to displace, saw quite clearly that science should be based on concepts that have meaning in terms of possible observations. Recall again his preface to the definition of acceleration in *Two New Sciences:*

> We however have decided to consider the phenomena of freely falling bodies with an acceleration such as actually occurs in nature, and to make our definition of accelerated motion exhibit the essential features of this type of natural accelerated motion.

If we define as "meaningless" any concept not definable by operations, it follows that not only concepts but whole statements and even intelligent-sounding, deeply disturbing questions may turn out to be meaningless *from the standpoint of the physical scientist.* Here are a few examples:

- "What are the physical laws that would hold in a universe completely devoid of any material bodies?"
- "Which is *really* at rest, the sun, or the earth?"
- "Will this table cease to exist while it is not being observed?"
- "What is length in itself, apart from measurements?"
- "Are there natural laws that we can never hope to discover?"

Perhaps surprisingly we must also add to the list some statements such as

- "Is light made of corpuscles only or waves only?"

Many more such questions are listed by Bridgman, who speculatively ventures to extrapolate this operational view beyond science by saying,

> I believe that many of the questions asked about social and philosophical subjects will be

found meaningless when examined from the point of view of operations. It would doubtless conduce greatly to clarity of thought if the operational mode of thinking were adopted in all fields of inquiry . . .

(*Logic of Modern Physics*)

It is necessary to enter here a warning against carrying this mode of thinking too far. Rigorous application of operationism can be and in fact has been used to attack speculations that may well develop and turn out to be fruitful after all. For example, at the end of the nineteenth century the physicist Ernst Mach argued that the *atom* is a physically meaningless concept, because there was at that time no way to observe or measure its discrete, individual properties. Nevertheless, atomic theories being developed in those years (see Chapters 19–22) were to be of great value to science even though the atom at the time had to be defined in terms of observable physical and chemical interactions rather than in terms of measurable dimensions.

Another attempt to distinguish between scientific and nonscientific statements was made by the philosopher Karl Popper in his influential book *The Logic of Scientific Discovery* (1934). Basing himself on his principle that science proceeds by making bold conjectures and attempting to refute them (Section 3.1), he declared: a statement that cannot be "falsified" (that is, proven to be false) by any conceivable experiment is not a scientific statement. Proposed answers to many if not all of the "meaningless statements" quoted above would fail to satisfy Popper's *falsifiability criterion.* That does not mean they are wrong; rather, they belong to the realm of metaphysics or pseudoscience rather than science.

Popper's criterion may be too strict, if we interpret it to mean (as Popper himself originally did) that a theory is scientific only if leads to testable *predictions* about events or phenomena not yet observed. The domain of science cannot be—and in practice has not been—limited to what can be brought into a terrestrial laboratory for controlled experimentation. The falsifiability criterion has been used, quite absurdly, to argue that Darwinian evolution is not a scientific theory because we cannot use it to predict what species will evolve in the future. Even Popper eventually admitted that a theory can reasonably be tested in other ways.

The essential point that both Bridgman and Popper wanted to make, and which must not be obscured by any misapplications of their criteria, is that scientific concepts must *eventually* be connected with observations about the real world if they are to survive.

12.6 Primary and Secondary Qualities

If it be true that science as an established activity has its roots in operationally meaningful concepts, then there must be a large range of experience that is being neglected by science. In a sense one may compare a scientist with a person who looks at the night sky by means of a very powerful telescope and who is thus enabled to examine a small region with extreme penetration while, however, sacrificing the chance to scan a whole world at one glance. In addition, we are apt to think that with a telescope our scientist seems to observe only the number of stars, their relative brightness, and the like, instead of watching the skies for the sheer experience drawn from the beauty of the grand display that so impresses the casual observer.[3]

The comparison is apt. Galileo drew the distinction between those experiences and concepts that might safely serve as the foundation stones of science and those which, having a measure of more subjective meaning, should from the point of view of science be regarded as sources of illusion and debate. We now call them, respectively, *primary* and *secondary qualities.* The primary qualities are direct observables that can be mathematically symbolized and measured, such as position and motion, the very elements that could be quantified and at the time had simple operational significance, to use the modern phrase. The qualities regarded as secondary were those not then accessible to instrumental measurement, those that were largely qualitative in the modern sense.

By and large, the distinction between primary and secondary qualities, which impressed and was accepted by such followers of Galileo as Newton himself, still holds in the physical sciences. We might say that Galileo's distinction reduced

[3]The analogy is suggestive in another way: The prospect of inevitable, successive refinements of one's telescope reminds the researcher that one's findings are always tentative, whereas the unaided (nonscientific) observer can easily persuade himself that his observations are final and can yield final truths.

the extent of *eligible* experience to a small fraction of scientists' total experience, but to precisely that fraction that they could quantify and therefore share fairly unambiguously with their fellows. The ideas of physical science look so stylized and unreal just because we demand of them that they help us describe those features of experience that the common-sense view of reality by and large cares least about—measurement, mathematical manipulation, numerical prediction, and clear communicability—while failing to describe exactly those most prominent uses of everyday expressions, namely, our feelings, reactions, and other personal involvements.

Developments in branches of psychology and the social sciences are reminiscent of that early search for quantification in physics and chemistry. Throughout Europe, philosophers and scientists of the early seventeenth century were saying that the qualitative ideas inherited from the ancients would have to give way to a new, quantitative way of describing nature. A case in point is Aphorism XCVIII of Francis Bacon's *Novum Organum* of 1620, where we find this complaint concerning contemporary thinking:

> Now for grounds of experience—since to experience we must come—we have as yet had either none or very weak ones; no search has been made to collect a store of particular observations sufficient either in number, or in kind, or in certainty to inform the understanding, or in any way adequate. . . . Nothing duly investigated, nothing verified, nothing counted, weighed, or measured, is to be found in natural history; and what in observation is loose and vague, is in information deceptive and treacherous.

12.7 Mathematical Law and Abstraction

The insistence on quantitative concepts must, of course, appear incomprehensible until we recognize that the work of the physical scientist is based on a faith as ancient as it is astonishing, namely, that nature works according to mathematical laws and that the observations are *explained* when we find the mathematical law relating the observations. For example, little more is left to be said about simple freely falling bodies *in vacuo* once we have found that they move according to the law $s = v_0t + \frac{1}{2}gt^2$.

Galileo expressed it this way:

> Philosophy [we should call it science now] is written in that great book which ever lies before our eyes—I mean the universe—but we cannot understand it if we do not learn the language and grasp the symbols in which it is written. This book is written in the mathematical language, and the symbols are triangles, circles, and other geometrical figures [we should now add all other mathematical devices] without whose help it is impossible to comprehend a single word of it, without which one wanders in vain through a dark labyrinth.
>
> (*Discoveries and Opinions of Galileo,* translated by S. Drake)

To Galileo, as to his contemporaries (for example, Kepler and, to some extent, even the physician William Harvey) and to modern physical scientists, mathematical methods provide the technique *par excellence* for ordering and comprehending nature. (To logic, which for the medieval scholastic philosophers was the main tool of investigation, is now primarily delegated the task of establishing the consistency of the scheme of mathematical demonstration, hypotheses, and observations.) Although Galileo and Kepler still felt compelled to announce this faith loudly and often, by now through the unforeseeable fruitfulness of this view, it has become a fundamental, unexamined attitude among physical scientists.

By saying that laws in physics are generally mathematical relations between observables, we mean, of course, that the general descriptions of physics are usually stated in the form "variable X is related to variables Y, Z, etc. by such and such a mathematical function." (Consider again, for example, the equation for free fall or Newton's law of gravitation.) Laws may state that $(X/Y) + Z$ (or some other combination of the variables) always has a certain numerical value, or tends to attain a maximum or a minimum value under given conditions. Yet another example is the *superposition principle* (Section 8.1), which asserts that the two variables v_x and v_y in projectile motion are not interdependent.

Above all, there is that type of law most eagerly sought which says in effect "this function of these variables under given conditions is always *constant.*" The law of freely falling bodies as conceived by Galileo was of this kind, for it could be phrased: s/t^2 is constant if the falling body starts from rest. Kepler's third law is an obvious example. There are numerous others: the laws of conservation of mass, of momentum, of energy, of "caloric" in an earlier phase of science, etc. To us, laws of constancy are most highly prized; they combine the most successful features of science with its most persistent preoccupation—the mathematical formulation of concepts (s, t, etc.) that aid in the discovery of unchanging patterns in the chaos of raw experience.

The mathematical formulation of physical law has an incidental and initially disconcerting effect. Unless we are warned to think of a law as the relationship between variables, we are apt to regard it as a strict relationship between cause and effect. But in our law of the type $X = YZ$, we can just as well write $Y = X/Z$, or $Z = X/Y$, and there is no way of telling whether X or Y or Z is a cause or an effect. As a consequence, it is on the whole more fruitful to think of an interaction rather than a simple causation, and to ask, "to what factors is X related" instead of "what causes X." For example, Boyle's law (Section 19.3) states that for a given mass of gas, at a constant temperature, the absolute pressure P and volume V of the gas are related by PV = constant, i.e., P is proportional to $1/V$. Here the statement that pressure and volume are inversely proportional accomplishes more in a shorter space and with less ambiguity than the equivalent description of pressure changes *causing* contractions and expansions or of volume changes *causing* compressions or rarefactions.

A great aid to speedy understanding and manipulation of concepts is the simple fact that mathematically formulated ideas can be expressed symbolically in equations like the above. First consider the mere *convenience* of writing an equation like $s = v_0t + \frac{1}{2}gt^2$, instead of expressing this in a sentence. Note how extraneous word connotations disappear; how easy it becomes to communicate arguments and results clearly to others; and how such equations invite drawing further conclusions about the relationship between observables. There is a good parallel here in the field of chemistry, which was immeasurably advanced in the early nineteenth century by the simple expedient of expressing reactions in formulas.

However, still other consequences follow from the decision of scientists to restrict and direct their attention toward the measurables and the mathematical relations between them. If the important thing about a rolling ball on an inclined plane is no longer its composition, its history in the workshop, its color, or its usefulness for the sport of bowling, if all that really matters to the particular group of investigators is the relationship $s \propto t^2$, then this ball ceases to be an individual entity—it equally might have been a smooth loaf of bread or a round bottle of wine—and in the long run the real ball seems totally forgotten, its place being taken by "a point mass," an idealization abstracted from the experiment.

In this sense the laws of science and the controlled experiment do not directly deal with "real bodies" but with abstractions moving in a hypothetical pure space with properties of its own, in a world we can mentally shape at will, now thinking away all air resistance, now regarding the inclined plane as perfectly smooth and straight, now changing only one aspect, perhaps the angle of inclination, and leaving all other factors untouched. As seen from the outside, this world, rendered in an esoteric language of its own and filled with seemingly grotesque simplifications and exaggerations, is analogous in many ways to that of the modern painter, poet, or composer.

All the features of science conspire to transform real weights in beautiful Pisa into undistinguished particles moving in idealized space. It is true that much would be lost by this transposition if that were the end of the scientific process, although even this much, the first part, yields the unique and absorbing satisfaction of being able to reduce, order, and so to understand raw sense experience. But there is more: At no point is the contact with reality completely broken; the same rules by which the transition was made are also applicable on the return trip from the world of abstraction to the world of reality. This is guaranteed by the operational nature of the concepts employed in that abstract world.

For example, while you were deriving the equation $v^2 = (v_0)^2 + 2as$ from the other equations of motion, you may not have felt in touch with the real world of moving bodies, yet at the end the result, that equation, could again be referred to real motions. In fact—and this is most important—*our mathematical world in which the calculations could proceed is justified and taken seriously by physical science only insofar as it does yield new knowledge about the real world around us.* If it were to fail in this function, we should have to rearrange the rules within the abstract world until they finally gave a useful harvest. As it happens, the abstraction from a real ball to a point did give us laws of motion applicable not only to this ball but to all other balls, to all simple round and sliding objects, and the like.

This is indeed making a good profit in return for sacrificing temporarily the identity of the real spherical object. The philosopher Rudolph Carnap suggested a neat analogy at this point. The symbols and equations of the physicist bear the same relation to the actual world of phenomena as the written notes of a melody do to the audible tones of the song itself. The written notes, of course, do not make sounds by themselves, and yet the sounds are not altogether lost in them. At any time one can retranslate the marks on paper into the audible melody, provided one knows the convention that relates notes to tones.

Correspondingly, we might now visualize science as an arch resting on two pillars, observation and experience, with conceptualization and abstraction supported in between. The security of the arch depends on the firmness of the pillars.

A whole view of science can be developed along these lines. We find an early hint in Francis Bacon's words:

> But my course and method, as I have often clearly stated and would wish to state again, is this—not to extract works from works or experiments from experiments, as an empiric, but from works and experiments to extract causes and axioms, and again from those causes and axioms new works and experiments, as a legitimate interpreter of nature.

A more modern view along such lines is to be seen in James B. Conant's definition of science:

> Science is an interconnected series of concepts and conceptual schemes that have developed as the result of experimentation and observation and are fruitful of further experimentation and observations.
>
> (*Science and Common Sense*)

12.8 Explanation

We must now consider the validity of a fundamental point—the statement that a physical scientist finds *explanation* by discovering mathematical laws between observables. Someone might object: "I can see that this convention works in the sense that it permits one to form the laws of, say, projectile motion by means of which one can accurately direct missiles. However, I do not agree that anything is explained thereby. True, when Galileo pointed out that the paths of projectiles coincided with a type of geometric curve called a parabola, he did intuit an elegant and simple description of a large number of separate cases, and he may in fact thereby have discovered a simple device for predicting all kinds of trajectories. But I want to know the ultimate reason *why* projectiles do go in parabolic curves, and I shall remain dissatisfied with any of your explanations until you give me a picture or some intuitive necessity for the graceful arching of the paths of projectiles."

Now it is certainly simple enough to explain why trajectories are parabolic, but only in terms of the laws of gravitational force acting on massive objects, these laws being themselves mathematical relationships. Our interrogator, on the other hand, does not want more of that. He might well find satisfaction with the explanation current before the late sixteenth century, or perhaps with some other idea, such as a supposed existence of invisible curved tubes or tracks in ether, which guide projectiles.

If pushed far enough, our interrogator will have to make a startling concession: His only tools for understanding physical phenomena, his imagination's vocabulary, are pictures, allusions, and analogies involving the primitive mechanical events of everyday life. We might say he thinks predominantly with the concrete common sense of his muscles. For example, the gravitational or electrical attraction of one body for another across a vacuum is incomprehensible to him unless we allow him to imagine some invisible medium between those two bodies that somehow transmits a real mechanical pull or push from one to the other.

This feature of the human mind, this thirst for concreteness, characterizes not only the frequent preoccupation with mechanical models within science itself, but also the most primitive type of everyday explanation. We find it symbolized in the thought of the ancient people of India that the earth was supported in space on the backs of gigantic elephants, or in the belief of the Egyptians that Osiris weighed human souls in a hand-held balance. It would indeed be surprising if the imagery of religion, poetry, and even of early science had not found its original

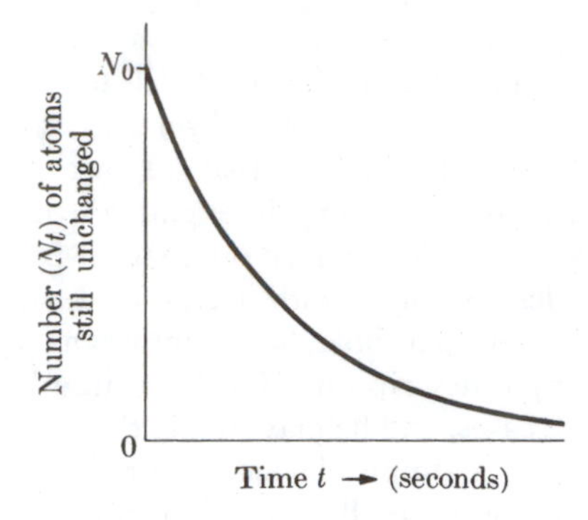

Fig. 12.1. Decay of radioactive element.

raw material predominantly in our ordinary, direct, everyday experiences. It is also true that some of the greatest scientific contributions were made by people whose physical intuitions were just so directed.

But as modern physical science has turned to problems more and more removed from the realm of common experience, it also has become necessary to enlarge the kit of tools with which to grasp and comprehend phenomena. Our discussion of the nuclear atom will serve later as a particularly striking example of the breakdown of those more naive types of understanding that insist on intuitive, visualizable explanations.

The best early statements of this theme were Galileo's insistence that motion was to be explained in terms of a particular mathematical law abstracted from experience and Newton's refusal to bolster the law of universal gravitation by a mechanical ether hypothesis, saying: "To us it is enough that gravity does really exist, and acts according to the laws which we have explained, and abundantly serves to account for all the motions of the celestial bodies and of our sea" (*Principia*).

To take another specific example, consider the phenomenon of radioactive decay of some one isotope of radium. The experimental fact is that a given pure sample containing initially the number N_0 atoms will gradually change into chemically different substances, each radium atom giving off alpha radiation and changing first into an atom of the element radon. After a time t seconds, only N_t atoms from the original quality N_0 are still unchanged (radium), as shown best in a graph (Fig. 12.1).

Again, it is an experimental fact, one that was originally difficult to get at and much prized, that this graph is, for all pure radioactive substances, an exponential curve, i.e., that the equation corresponding to this behavior is of the form

$$N_t = N_0 e^{-0.693t/T}. \qquad (12.1)$$

Here N_t, N_0, and t have been defined above, e = 2.718 . . . (base of natural logarithms), and T is a constant whose value depends only on the substance involved and tells of the rapidity of the decay (Section 27.2).

Now it is clear that as the equation given describes and summarizes the facts neatly we may refer to it as the *law of radioactive decay.* Although this empirical mathematical law is indispensable and enlightening, it does not in the least explain why atoms of radioactive elements emit radiation and change into other elements. That is a separate problem.

Perhaps it is not too frivolous to hold that "to explain" means to reduce to the familiar, to establish a relationship between what is to be explained and the (correctly or incorrectly) unquestioned preconceptions. The preconceptions of Aristotle and the scholastics were grandiose and sweeping (natural place, final causes, fundamental distinction between celestial and terrestrial phenomena, etc.), reflecting the impressive but perhaps overly ambitious universality of their questions and explanations. Modern scientists have trained themselves to hold somewhat more limited and specialized preconceptions, for example, that how nature works can be understood through simple models or mathematical schemes. To them the law that the mathematical sum of all energies in a system can be regarded as forever constant is the satisfying explanation for a host of phenomena, whereas the same law might be regarded as an unexplained mystery by nonscientists. It requires training and repeated personal success in solving physical problems to be really satisfied with a mathematical answer, just as it must surely involve a strenuous program of orientation to be fully content with the rewards and the rules of behavior in any profession or, for that matter, in our own civilization.

Having examined the nature of concepts in physical science, we are ready to consider how these concepts are integrated in the advance of science.

Problem 12.1. When Galileo and other supporters of the heliocentric system in the seventeenth century asserted that it is really the earth that moves rather than the sun and stars, did they violate the rule that physical concepts must have operational definitions?

Problem 12.2. Write down some commonly used statements that, from the viewpoint of physical science, are "meaningless" because not "falsifiable," not "operational," or not expressible in terms of numbers and equations. Does science suffer in any way because it limits itself by those exclusions?

Note: The recommended reading selections for this chapter will be found at the end of Chapter 14.

CHAPTER 13

On the Duality and Growth of Science

13.1 The Free License of Creativity

Our discussion of the nature of physical concepts has shown that a main reason for formulating concepts is to use them in connection with mathematically stated laws. It is tempting to go one step further and to demand that practicing scientists deal only with ideas corresponding to strict measurables, that they formulate only concepts reducible to the least ambiguous of all data: numbers and measurements. The history of science would indeed furnish examples to show the great advances that followed from the formation of strictly quantitative concepts—whether the measurements were concerned with the time of descent of rolling balls; or the tabulation of population statistics; or the weights of compounds in chemical reactions; or, in the case of the Russian psychologist Ivan Pavlov's experiments, with the rate of flow of saliva in experimental animals.

The nineteenth-century physicist Lord Kelvin commended this attitude in the famous statement:

> I often say that when you can measure what you are speaking about and express it in numbers you know something about it; but when you cannot measure it, when you cannot express it in numbers, your knowledge is of a meagre and unsatisfactory kind: it may be the beginning of knowledge, but you have scarcely, in your thoughts, advanced to the stage of *Science*, whatever the matter may be.
>
> ("Electrical Units of Measurement")

Useful though this trend is within its limits, there is an entirely different aspect to scientific concepts; indeed it is probable that science would stop if every scientist were to avoid anything other than strictly quantitative concepts. We shall find that a position like Lord Kelvin's (which is similar to that held at present by some thinkers in the social sciences) does justice neither to the complexity and fertility of the human mind nor to the needs of contemporary physical science itself—not to scientists nor to science. Quite apart from the practical impossibility of demanding of one's mind that at all times it identify such concepts as *electron* only with the measurable aspects of that construct, there are specifically two main objections: First, this position misunderstands how scientists as individuals do their work, and second, it misunderstands how science as a system grows out of the contributions of individuals.

Let us now examine the first of these two important points more closely. It would have been wrong to conclude from the previous pages that practicing scientists, while they are engaged in creative work, do—or even can—continually check the strict meaningfulness of their concepts and problems. As a rule an investigator is and should be unbothered by such methodological questions until some obstinate trouble develops. The operational view, in ordinary use, is almost instinctive; it usually becomes conscious and explicit only when something has gone wrong.

We are reminded of the case of a tightrope walker who undoubtedly is indifferent to an orderly inquiry concerning those physical laws by which he performs and survives; very likely he would fall off the rope if he worried much about them. *While a scientist struggles with a problem, there can be little conscious limitation on his free and at times audacious constructions.* Depending on his field, his problem, his training, and his temperament, he may allow himself to be guided by a logical sequence based on more or less provisional hypotheses, or equally likely by "feeling for things," by likely analogy, by some promising guess, or he may follow a judicious trial-and-error procedure.

The well-planned experiment is, of course, by far the most frequent one in modern science and generally has the best chance of success; but some men and women in science have often not even mapped out a tentative plan of attack on the problems, but have instead let their enthusiasms, their hunches, and their sheer joy of discovery suggest the line of work. Sometimes, therefore, the discovery of a new effect or tool or technique is followed by a period of trying out one or the other application in a manner that superficially seems almost playful.

Even the philosophical orientation of scientists is far less rigidly prescribed than might be supposed. A rough classification of choices is open to them. Perhaps most people would defend the position that the experiences that come to them through their senses are directly occasioned by and correspond exactly to an external world—a "real" world that exists regardless of our interpretations. This view is called *realism,* and it is in opposition to two major philosophical systems (among a host of others): First, *idealism,* which maintains that our ideas and impressions are the only realities, and that we use those to construct convenient concepts such as chairs, houses, and meter sticks. Second, *positivism,* which does not speak of such realities at all and regards nothing as ascertainable or meaningful beyond sense data (for example, pointer readings) or those concepts reducible to sense data.

Now it might be supposed that because the established part of physical science ideally should be directly verifiable in terms of measurements, all scientists must be positivists. The true state of affairs, however, is that the great majority of scientists, although in agreement with positivism's practical consequences for their own work, do not participate actively in the philosophical fight one way or the other.

Einstein explains very well that a practicing scientist, from the point of view of philosophy, may appear to an outsider,

> . . . as a type of unscrupulous opportunist: he appears as a *realist,* insofar as he seeks to describe the world independent of the act of perception; as *idealist* insofar as he looks upon the concepts and theories as the free inventions of the human spirit (not logically derivable from that which is empirically given); as *positivist* insofar as he considers his concepts and theories justified only to the extent to which they furnish a logical representation of relations among sense experiences. He may even appear as *Platonist* or *Pythagorean* insofar as he considers the viewpoint of logical simplicity as an indispensable and effective tool of his research.
>
> (*Albert Einstein Philosopher-Scientist,* edited by P. A. Schilpp)

13.2 "Private" Science and "Public" Science

The important problems of creative science do not present themselves with obvious routes to solutions, and so they differ in many fundamental ways from science problems that a student may encounter in homework or laboratory exercises. Above all, it is rarely known at the outset whether the problem can be solved at all with the initial conceptions or tools or where the new complications might lurk; all of science has a unity and, unlike the usual student's exercise, the work of even a specialist in modern science may involve a large range of different specific fields, sometimes in unexpected ways. An inquiry into the physical consistency of liquids, for example, may have to call upon such diverse topics as thermodynamics, physical chemistry, crystallography, acoustics, electronics, and even some practical metallurgy. Also, the methods and results of one field in physical science constantly suggest analogous procedures in another.

Consequently, the working scientist is ever alert for the slightest hints either of new difficulties or of their resolutions. She proceeds through her problem like an explorer through a jungle, sensitive to every clue with every faculty of her being. Indeed, some of the greatest theoretical scientists have stated that during those early stages of their work they do not even think in terms of conventional communicable symbols and words.

Only when this "private" stage is over and the individual contribution is formalized in order to be absorbed into "public" science (often called "textbook science"), only then does it begin to be really important that each step, each concept be made clear and meaningful. The difference between these two aspects of science, what we shall have occasion to call science-in-the-making or S_1, and science-as-an-institution or S_2, has thus been characterized by the American nuclear physicist, H. D. Smyth: "We have a paradox in the method of science. The research man may often think and work like an artist, but he has to talk like a bookkeeper, in terms of facts, figures, and logical sequence of thought."

For this reason we must not take very seriously the chronology or method outlined in the original publications of scientific results, including Galileo's own. It is part of the game to make the results in retrospect appear neatly derived from clear fundamentals until, in John Milton's phrase, "so easy it seemed/Once found, which yet unfound most would have thought/Impossible!" Much tortuous and often wasteful effort may be implied in a few elegant paragraphs, just as a sculptor removes his tools and clumsy scaffolding before unveiling his work.

Even in public science one will find widespread use of certain concepts that have not acquired a precise operational meaning. This situation sometimes gives rise to debates among philosophers and scientists concerned with methodology. From the seventeenth through the nineteenth centuries, there were three such controversial concepts in physical science: *atom, ether,* and *force.* As we mentioned at the end of Section 12.5, some of the strongest attacks on the atomic theory came toward the end of the nineteenth century, when the theory was on the verge of some of its most spectacular successes. Although Ernst Mach and his positivist colleagues may have had good reason to be skeptical about the real existence of atoms, it seems in retrospect that their criteria for even the tentative use of such theoretical concepts were unnecessarily strict. Insisting on excluding atoms from the formulation of physical theories, they deprived themselves of a good mental tool—and the rest of science of their possible additional contributions—by a doctrinal adherence to their position.

Even the ether concept, although judged to be superfluous by most physicists in the twentieth century, was not entirely useless in the history of science. It helped Maxwell, for example, to formulate a successful theory of electromagnetic fields and waves. Similarly the concept of force, although often criticized as vague or superfluous, has played an important role in the development of physics, in some cases just because of its ambiguity.

Most scientists tacitly agree that the creative activity must be unfettered. It has been said very well, "To set limits to speculation is treason to the future." Eventually, of course, all such speculation in the private phase of science must stand the trial of test and the skeptical assessment of colleagues before the community of science can accept the results into public science. Let us elaborate on this mechanism.

13.3 The Natural Selection of Physical Concepts

In the previous section it began to appear that there are no simple rules to lead us to the invention of new concepts or to the discovery of new ideas, none by which to foretell whether our invention or discovery, once found, is going to prove useful and durable. But science does exist, and is a vigorous and successful enterprise. The lesson to be drawn from history is that science as a structure grows by a struggle for survival among ideas; there is a marvelous mechanism at work in science, which in time purifies the meaning even of initially confused concepts, and eventually absorbs into what we may label S_2 (public science, science-as-an-institution) anything important that may be developed, no matter by what means or methods, in S_1 (private science, science-in-the-making).

We have seen that there are two general characteristics shared by those concepts that have contributed most to the growth of science.

First, each of the guiding concepts of physical science—acceleration, force, energy, valence, etc.—has a core that is clear and unambiguous or, at any rate, which through continual application to experimental situations has attained an operational meaning that is tacitly understood and communicable.

Second, by the same token, the great majority of physical concepts are quantitative, that is to say, they can be associated with numbers and measurements and therefore with manual or mathematical operations. Thus we say that a force is 16 newtons, a mass 10 kg, a temperature 100°C, a valence +2, the dimensions of space 3, an atomic weight 238.07.

This observation holds even for those concepts that superficially do not seem to take on such numerical meaning. For example, we cannot say that a body has 3, 4, or 5 units of *equilibrium.* On the other hand, equilibrium *can* be defined by saying that a body is in equilibrium when its acceleration is zero. The physical concept of "melting" can be defined as the change of state of a substance from a solid to a liquid, but this in turn has physical meaning only because the change from one state to the other can be recognized in terms of a large and abrupt change in the numerical value of measurable viscosity or hardness of the material.

Consider the example of *electron.* A physicist might say it is the smallest quantity of electric charge known to exist independently, but when he uses that concept he is really dealing with its numerical aspects, namely, that it has 1.6×10^{-19} units of charge, 9.1×10^{-31} kg mass, and so forth. The name "electron" is primarily a summary term for this whole complex of measurables.

These two characteristics of concepts are joined by a third, without which science would degenerate into a meaningless conglomerate of data. Evidently the chaos of experience allows us to formulate infinitely many ideas, all of them meaningful in the above sense. For example, we might arbitrarily wish to create some new concept X from observations of free fall, defining X, perhaps, by $X = (v - v_0)/s$. At first glance, X appears as worthy of a distinguishing name of its own and of wide acceptance, as did, for example, the quantity $(v - v_0)/t$; but pragmatically X does not happen to appear prominently in simple laws of nature. It is not, for instance, a constant quantity during free fall. Therefore X, though a possible, meaningful concept, actually falls by the wayside as useless and unwanted. (This X happens to be the very quantity that Galileo first and erroneously suspected of being the constant factor during free-fall motion.)

What makes certain concepts important, therefore, is their recurrence in a great many descriptions, theories, and laws, often in areas very far removed from the context of their initial formulation. The electron, first discovered in the current within discharge tubes rather like those now used for fluorescent fixtures, later reappeared prominently in the explanation for electric currents in wires and liquids, for photoelectricity, for thermionic phenomena (as in old radio tubes), for radioactivity, for the emission of light from hot bodies, and much else besides. That is the only reason and the full meaning behind the statement that scientists "believe in the reality" of electrons; the concept is needed so often and in so many ways. "The only justification for our concepts," said Einstein, "is that they serve to represent the complex of our experiences."

At the inception of an idea it is, of course, hardly possible to tell whether it will fulfill this criterion, whether it will stand the test of wide applicability and so survive in the body of science. In this sense we may say Galileo's work would have been meaningless without the successive growth of physical science. Galileo sensed, but he could not have known, that his work on mechanics would turn out to be as important as it actually did and that his special methods and special discoveries would (with Kepler's and Newton's work) find a general significance by transcending the immediate. There is a challenge in this for each individual scientist: the science of tomorrow may well need or even depend on your own work.

The most spectacular case of such generally useful ideas is, as we saw, that of the so-called "fundamental" concepts of physics (for example, length, time, mass). There are only a small handful of them, yet they are the building blocks from which all other concepts are constructed or derived (for example, velocity, being the ratio of a length and a time interval). One can never cease to marvel that nature should be so reducible—here must be either the basis or the result of the scientist's guiding preconception concerning the simplicity of nature about which we have spoken.

We shall meet an abundance of other important though "derived" concepts (for example, electric potential), which run like strong ropes through the maze of phenomena to give us support and direction under the most varied circumstances. Initially defined in some perhaps rather limited problem, they have become all-important over the years. In this evolution we find not infrequently that a concept changes its meaning as the context and the field of application widen; examples are such ideas as *force, energy,* and *element.* This possibility of change is again suggestive of the recurring discovery that the concepts of science are not absolute, that they have meaning only in terms of what scientists can do with them.

We can now attempt a preliminary summing-up of the main points so far. We see physical scientists as seekers of harmonies and constancies in the jungle of experience. They aim at knowledge and prediction, particularly through discovery of mathematical laws. Their individual contributions, however, are only half the story, for science has two aspects: One (which we have called S_1) is the speculative, creative element, the continual flow of contributions by many individuals, working on their own tasks by their own usually unexamined methods, motivated in their own way, and generally uninterested in attending to the long-range philosophical problems of science; and the other aspect, S_2, is science as the evolving consensus, science as a growing network synthesized from these individual contributions by the acceptance or adoption of those ideas—or even those parts of ideas—that do indeed prove meaningful and useful to generation after generation of scientists. The cold tables of physical and chemical constants and the bare equations in textbooks are the hard core of S_2, the residue distilled from individual triumphs of insight in S_1, checked and cross-checked by the multiple testimony of experiment and general experience.

This duality of science is one of its main sources of strength. In order to illuminate the duality in terms of concrete examples, we shall now briefly study two problems, that of motivation and that of factual interpretation. This will, incidentally, also prepare the ground for the bigger tasks that follow—examining the details of the mechanism whereby S_2 grows out of S_1 and explaining the vital role played by the existence of contradictory elements within science. Once an adequate distinction has been made between the several levels of meaning in the term "science," one of the central sources of confusion in methodological discussions has been removed.

13.4 Motivation

When the philosopher Alfred North Whitehead wrote, "Science can find no individual enjoyment in Nature; science can find no aim in Nature; science can find no creativity in Nature" (*Modes of Thought*), he touched on the ambivalent feeling of the present age toward science—the simultaneous fear and reverence for the reputedly emotionless yet deeply exciting enterprise. Perhaps we can see now that the opinion embodied in the quotation refers to only one of the two aspects of science, the stable, not the transient one—S_2 and not S_1. Modern science as a structure may say nothing about purposes and aims, but the scientist himself may be enchanted to some degree by nonrational preoccupations.

A few direct quotations will illustrate this point, for in other years, when scientists were freer with their human secrets, a nonrational, mystical, or religious conviction was often freely acknowledged. Galileo, a pious man, looked upon the laws of nature as proof of the Deity equal to that of the Scriptures (see Section 5.4), and ever since the Pythagoreans this faith, reflected in the recent phrase "the world is divine because it is a harmony," has been a motivating theme. Almost without exception, scientists as a group have at all times, including our own, manifested the same religious convictions or lack of convictions as did other contemporary groups of educated men and women; if anything, their position has more often moved them to be vocal on such questions than some other, equally well-informed groups. Until about 150 years ago, the typical scientist would quite openly assert that the physical world could not be understood without fundamental theistic assumptions.

In this, too, the scientists of the seventeenth century were the formulators of the fundamental concepts, and although the details of the argument have changed considerably since, its form has never been put more beautifully or stated more honestly than in Newton's description of his scientific activity as a prelude to religious knowledge (specifically, that of the "first cause," the Deity):

> . . . the main Business of natural Philosophy is to argue from Phaenomena without feigning Hypotheses, and to deduce Causes from Effects, till we come to the very first Cause, which certainly is not mechanical; and not only to unfold the Mechanism of the World, but chiefly to resolve these and such like Questions: What is there in places almost empty of Matter, and whence is it that the Sun and Planets gravitate towards one another, without dense Matter between them? Whence is it that Nature doth nothing in vain; and whence arises all that Order and Beauty which we see in the World?
>
> To what end are Comets, and whence is it that Planets move all one and the same way in Orbs concentrick, while Comets move all manner of ways in Orbs very excentrick; and what hinders the fix'd Stars from falling upon one another? How came the Bodies of Animals to be contrived with so much Art, and for what ends were their several Parts? Was the Eye contrived without Skill in Opticks, and the Ear without Knowledge of Sounds? How do the Motions of the Body follow from the Will, and whence is the Instinct in Animals? Is not the Sensory of Animals that place to which the sensitive Substance is present, and into which the sensible Species of Things are carried through the Nerves and Brain, that there they may be perceived by their immediate presence to that Substance?
>
> And these things being rightly dispatch'd, does it not appear from Phaenomena that there is a Being incorporeal, living, intelligent,

> omnipresent, who in infinite Space, as it were in his Sensory, sees the things themselves intimately, and thoroughly perceives them, and comprehends them wholly by their immediate presence to himself; Of which things the Images only carried through the Organs of Sense into our little Sensoriums, are there seen and beheld by that which in us perceives and thinks. And though every true Step made in this Philosophy brings us not immediately to the Knowledge of the first Cause, yet it brings us nearer to it, and on that account is to be highly valued.
>
> (*Opticks*)

In a letter to a friend, Richard Bentley, in 1692, Newton further explains his motivations for writing the *Principia:*

> When I wrote my treatise [*Principia*] about our [solar] system, I had an eye on such principles as might work with considering men for the belief of a Deity; and nothing can rejoice me more than to find it useful for that purpose.
>
> (*Newton's Philosophy of Nature,* edited by H. S. Thayer)

Consider another persistent "nonrational" trend in creative scientific work—the preoccupation with integral numbers, of which we have seen an example in Bode's law and shall see others. At the very outset of Galileo's historic work on the law of free fall, in the *Two New Sciences,* we find him drawing prominent attention to the fact that ". . . so far as I know, no one has yet pointed out that the distances traversed, during equal intervals of time, by a body falling from rest, stand to one another in the same ratio as the odd numbers beginning with unity." Throughout the development of physics and chemistry and to the present day we encounter this satisfaction with such simple numerical relations. Today's nuclear physicists have found that some of the characteristics of nuclei can be explained in a theory involving a few recurring numbers; these are generally called the "magic numbers," perhaps a little less jokingly than is often realized.

Examples can be multiplied indefinitely, but perhaps the following extract is sufficiently representative: The physicist Wolfgang Pauli, on accepting the Nobel prize for physics in 1945, was speaking of his great teacher Arnold Sommerfeld and his attempts to form an explanation for the particular colors found in the spectra of glowing gases, when he said,

> Sommerfeld, however, preferred . . . a direct interpretation, as independent of models as possible, of the laws of spectra in terms of integral numbers, following as Kepler once did in his investigation of the planetary system, an inner feeling for harmony. . . . The series of whole numbers 2, 8, 18, 32 . . . giving the length of the periods on the natural systems of chemical elements, was zealously discussed in Munich, including the remark of the Swedish physicist Rydberg that these numbers are of the simple form $2n^2$, if n takes on an integral value. Sommerfeld tried especially to connect the number 8 with the number of corners of a cube.
>
> (*Nobel Lectures in Physics*)

When thus presented, the motivation of many scientists, just like the process of discovery itself, appears perhaps surprisingly "unscientific." But because it has irrational elements, the drive toward discovery is powerful even under the most adverse conditions. How shall we explain that much great work has been done under the handicaps of extreme poverty and severe ill-health, sometimes even under the threat of imminent death? Or that to this day many scientists would probably reject the ever-present lure of increased standards of living in *uncreative* positions, and instead follow their chosen work without restrictions, although with relatively fewer material rewards?

We must recognize this symbiotic relationship: The progress of science has often depended on the almost unreasonable tenacity of its devotees and, on the other hand, the scientific activity yields a unique exhilaration and deep fulfillment. In support of this point the words of Poincaré are, as always, persuasive:

> The scientist does not study nature because it is useful; he studies it because he delights in it, and he delights in it because it is beautiful. If nature were not beautiful, it would not be worth knowing, and if nature were not worth knowing, life would not be worth living. Of course, I do not here speak of that beauty which strikes the senses, the beauty of qualities and of appearances; not that I undervalue such beauty, far from it, but it has nothing to do with science; I mean that profounder beauty which comes from the harmonious order of the parts and which a pure intelligence can grasp. This it is which gives body, a structure so to

> speak, to the iridescent appearances which flatter our senses, and without this support the beauty of these fugitive dreams would be only imperfect, because it would be vague and always fleeting. On the contrary, intellectual beauty is sufficient unto itself, and it is for its sake, more perhaps than for the future good of humanity, that the scientist devotes himself to long and difficult labors.
>
> (*Science and Method*)

These words, written at the beginning of the twentieth century, help to explain the powerful appeal of relativity—one of that century's most famous theories—to Einstein himself and to other physicists like Planck, Eddington, and Dirac: an appeal strong enough to sustain their belief in the essential truth of the theory in spite of experiments that initially seemed to refute it.

13.5 Objectivity

Exactly at this point we must answer a question that the last paragraphs must have raised in the minds of many readers: What of the much-vaunted *objectivity* required of scientists? We have been told of this demand so often that we have learned to call "scientific" any inquiry that claims to be systematic and unbiased. For example, Bertrand Russell said, "The kernel of the scientific outlook is the refusal to regard our own desires, tastes, and interests as affording a key to the understanding of the world." But how can this large degree of personal involvement, just demonstrated, fail to endanger the search for objective truth? Or, since it patently does not so endanger science, how can we account for the discrepancy between the nature of motivation and the nature of results?

The fact is that we have, by a second route, come back to the original dilemma, the apparent contradiction in function between S_1 and S_2. Now we can begin the resolution, first by pointing to an extreme example. The British astronomer and mathematical physicist, Sir Arthur Eddington, had such strong convictions that he once wrote a "Defense of Mysticism." Whether or not his persuasion was "meaningless" from the point of view of public science S_2, it may have been the mainspring, the deep cause, of his own devoted search for truth. What is now important is that in his voluminous and distinguished scientific writings you will find no overt expression of personal mysticism, nothing that might not equally well have been written by a gifted scientist with exactly the opposite metaphysical orientation, or even one belonging to the "hard-boiled" school and quite indifferent to such questions. In modern science, personal persuasions do not intrude explicitly into the published work—not because they do not exist, but *because they are expendable.*

We may go a step further. In a free society, the metaphysical tenets of individual scientists, though often quite strong, are generally so varied, so vague, and often technically so inept that in a sense they are canceled by one another, made ineffectual by the lack of a basis for general acceptance and agreement of such tenets. Perhaps our science would be weakened if this mechanism did not operate. It is only where there exists one explicit, widely accepted set of dogmas—as in some of the old medieval universities or in modern totalitarian states—that extraneous metaphysics in a scientific publication can survive the scrutiny of the scientists in that field and that location, with possibly detrimental consequences.

As contrasted with their largely unconscious motivations for which they have to account to no one, the intellectual disciplines imposed on scientists, like the form of publication of research papers itself, are now quite rigorously defined; they know that their inclinations may well not coincide with those of their colleagues and witnesses, and that the only points they might agree on are sound demonstrations and repeatable experiments. The personal satisfactions in S_1 must remain private and, therefore, from the point of view of S_2, incidental.

We can readily see how this severe convention, in such clashing contrast with the human side of the experimenter, is a kind of benign sociological device ensuring the rapid evolution of science. For it is exactly when enthusiasm for some result runs highest that the chances for mistakes arise. Therefore, all of the investigators will have to force themselves to the most searching reexamination before describing their work to that mercilessly impartial jury, their colleagues.

The French chemist and biologist Louis Pasteur gives us a good insight into this attitude:

> When you believe you have found an important scientific fact and are feverishly curious to publish it, constrain yourself for days, weeks, years sometimes; fight yourself, try and ruin your own experiments, and only proclaim

> your discovery after having exhausted all contrary hypotheses. But when after so many efforts you have at last arrived at certainty, your joy is one of the greatest that can be felt by the human soul.

This prescription is a fairly modern development, and even now is not universally followed. In many areas of science, several scientists fiercely compete to solve an important problem, knowing that the rewards (prestige, government grants, academic tenure, perhaps even the Nobel prize) will generally go to the one who is first to announce the correct solution. A few inevitably succumb to the pressure to publish a promising preliminary result without taking the time to do the kind of checking Pasteur advises.

Previously there were other proposed methods to enforce objectivity. For example, in attacking the qualitative and subjective science of his day, Francis Bacon held that in order to build on a sure foundation the scientist must suppress as long as possible the human element, his prejudices, premature hypotheses, his guesses and wishes. To this end Bacon outlined a procedure that prescribed the structure of scientific work so rigidly that the human element would indeed be kept at bay. In the historical context his solution was certainly not unreasonable; it might have worked, and it does seem to solve the problem of how to remain objective. But as it turned out, scientists found another, less rigorous, less safe, but far more fruitful procedure for overcoming human error, one that does so without stifling their enthusiasm and motivation, their main sources of both error and truth. This discipline is *superposed* on the free creativity of the individual; the inquiry is not controlled in its progress as if it were confined in a train moving on firmly laid out rails, but rather as though it were a rider on a horse whose obedience is assured by an occasional but firm tug on the reins.

13.6 Fact and Interpretation

Why, you may ask, should there be any need for an imposed discipline? Is there not discipline in the very *facts* the scientist observes? Are they not unambiguous in themselves? This raises another most interesting problem: the relationship between fact and interpretation.

Our very life, as well as our science, continually depends on the correct observation and classification of facts, and yet there is nothing more deceptive than facts. It is almost impossible to describe a fact without having some interpretation, some hypothesis, entering unseen—as careful cross-examination of witnesses in a courtroom often shows quite dramatically. If we follow the matter far enough, we might have to agree that the only real or basic facts are those that impress our crudest senses—the fact that we see a black or a white splotch, feel hot or cold, hear a painful or a soft sound.

However, those are exactly the "secondary qualities," which, as Galileo insisted, are only names for perceptions and feelings that exist in our own minds, not objective properties of the world; by themselves they could not conceivably have given rise to our science. Indeed, even simple animals must be able to transcend such raw observations, and thought surely does not deal with this type of isolated experience. It has been called the first law of thought that *perception* must pass into thought and knowledge via *conceptualization*. Starting from the vague impression "now-here-then-there," early humans had to invent—perhaps over a period of a thousand generations—such ideas as space, object, position, distance, time interval, motion, and velocity, in largely unanalyzed form. "Facts" cannot be the subject of conscious discourse by themselves, without intellectual tools for interpreting the sense impressions. Especially in the S_1 stage of science, scientists do have preconceptions or "themes" and do use unexamined associations, although these may be justified by a long history of successful results.

This admission seems to contradict the popular idea that the first step in science is to throw overboard all prejudgments. However, without some initial preconceptions one cannot conceive new thoughts; therefore, what is required in science, as in other phases of life, is to try to become aware of one's preconceptions in an emergency, to discard those that turn out to have been false or meaningless, to use the others with caution, and with willingness, if also found to be wrong by reference to actual phenomena, "to pronounce that wise, ingenious, and modest sentence 'I know it not'," as Galileo said.

Without our almost axiomatic key concepts we should be largely deprived of intelligence and communication. Imagine a geologist describing her observation of a mountain range in terms of "pure fact." She could not even begin to speak of a mountain range, for strictly speaking all she sees is a change in the level of terrain. Even that implies several derived concepts, so instead of saying, "The mountain is cut by glacial valleys," she would have to describe a pattern of light and darkness, and how her eyes had to focus differently on the various features of the field of view.

Thus our thoughts, when they deal with observation and fact, really must manipulate concepts, constructs, ideas, with all their hidden dangers—dangers of erroneous classification, of unwarranted extrapolation or analogies, of too-ambitious generalizations or too-timid specializations. No wonder physical science treasures those few concepts (for example, length, time, mass, charge) that, through long experience, have proved to be strong enough to carry the whole edifice.

The least ambiguous facts being those that involve only accounts of pointer readings and the like, physics should, as has been said, ideally start only with such descriptions. Even here there is room for quarrel, for we do have to decide in advance of the experiment which are the most important significant pointer readings to be collected and described. In this sense, and in this sense only, facts do not exist apart from the observer.

Allow a layman to observe the sky through a large telescope, and he will see little of interest and understand little of what he sees. But give a trained person only three good looks at a comet through a mediocre instrument, and he or she will call upon the theories in the field to tell you to a day how soon the comet will return, how fast it is traveling at any moment, what material it may be made of, and much else. In short, the pattern we perceive when we note "a fact" is organized and interpreted by a whole system of attitudes and thoughts, memories, beliefs, and learned theories and concepts. It is trained thought that gives us eyes.

Not only do brute facts alone not lead to science, but a program of enthusiastic compilation of facts *per se* has more than once delayed the progress of science. Politicians and administrators who demand "results" and timetables of findings to justify the funding of research may be more impressed by the sheer quantity of published data than by the hard-to-measure quality of theories and interpretations. The picture of the experimental scientist in a well-stocked laboratory subjecting matter to intensive and undirected observation is certainly absurd, though widespread. Particularly when instruments are to be used in making observations, even the first "preliminary experiments" are preceded by much nonexperimental activity, by study, thought, and calculation. Without this, the expensively obtained facts are meaningless.

We conclude once more that science cannot be made by individual scientists. To understand even the simplest observations each investigator must rely on the distilled wisdom of science as an institution. When Newton modestly pictured himself as standing on the shoulders of the giants who had done the earlier work of science, he pronounced a truth applicable to every individual scientist.

13.7 How Science Grows

We may now coordinate several observations on the growth of science to gain one overall view. The key recognition is the distinction between S_1 and S_2—if you will, the difference between the man Galileo Galilei and the statement $s \propto t^2$. This dual view of science has an analogy in considering a tribe either as a people (a classified abstraction with recognizable stable organization, customs, etc.) or as an interacting group of individually almost unpredictable persons. Better still, compare the difference between S_2 and S_1 with the double interpretation one may give to an animal species, either as an entity calmly catalogued and described in zoology texts at the present state of evolution or as exemplified by the diverse living, struggling individuals that it contains.

If we think about the last analogy more closely—and it is no more than a very helpful analogy—the mechanism of evolution for the species and for science appears to be similar in four ways: First of all, growth in both cases presupposes a mechanism of *continuity*—a species or a science can persist only if there is some stable means for handing on the structure from generation to generation in an unambiguous way. In biology the principle of continuity is found in the processes of heredity based on the highly specific nature of the genes—and in science it is identifiable chiefly with the specific operational and quantitative nature of the important concepts. Without this measure of unambiguous continuity, scientists could not communicate their work to one another and to their pupils.

Second, superposed on continuity is a mechanism of *mutation,* leaving open the constant opportunity for individual variations. In the case of biological species, of course, the process of mutation is made possible by various chemical and physical influences on the genes and on chromosome partition and recombination; in science, mutations are assured by the essential democracy of the institution and the boundless fertility of the individual, free human mind.

But perhaps we must reserve a special name for those rare mutations in science, those precious occasions when as in the work of Copernicus, Galileo, Newton, Einstein, and Bohr, a person transcends all human limitations and climbs to unsuspected heights in an "heroic effort," to use Sarton's phrase. Without this measure of heroism or supreme dedication, the normal evolutionary progress in science might be almost as slow as that of natural evolution.

A third mechanism is *multiplicity of effort.* To assure continuity and growth despite the low rate of occurrence of really good modifications, and in the absence of some obvious single master plan by which to proceed, science and the species alike must rely on a large number of individual attempts, from which may ultimately come those few types that are indeed useful. The uncountable bones of bygone members of the species, like the uncountable pages of the individual scientific researches of past years, are mute testimonies to the necessary wastefulness of the process of growth, both of the zoological type and of scientific knowledge.

This principle of multiplicity should not be confused with the thesis, going back to Francis Bacon, that anyone with moderate intelligence can make a contribution to physical science by working hard and following a prescribed method. That thesis, which appeals to our democratic sentiments, was formulated in a striking manner by the Spanish philosopher José Ortega y Gasset in his book *Revolt of the Masses* (1929):

> Experimental science has progressed thanks in great part to the work of men astoundingly mediocre, and even less than mediocre. That is to say, modern science, the root and symbol of our actual civilization, finds a place for the intellectually commonplace man and allows him to work therein with success. In this way the majority of scientists help the general advance of science while shut up in the narrow cell of their laboratory, like the bee in the cell of its hive.

But this thesis, now known as the *Ortega hypothesis,* has been refuted (at least for modern physics) by the sociologists Jonathan R. Cole and Stephen Cole. Their analysis of the influence of published papers (as indicated by how often they are cited) showed that "a relatively small number of physicists produce works that become the base for future discoveries in physics." Most papers are never cited at all by anyone except their authors. Contrary to Bacon and Ortega, the advance of science in a given specialty depends heavily on the "rare mutations" mentioned above, together with the substantial work of a relatively small number of researchers—not a multitude of busy bees.

Finally, there is a *selection* mechanism whereby certain of the seemingly innumerable contributions and unpredictable mutations are incorporated into the continuous stream of science, a struggle among ideas not greatly different from that fight for existence in nature that allows the species to adapt itself to the changing environment. The survival of a variant under the most diverse and adverse conditions is mirrored in science by the survival of those concepts and discoveries that are found useful in the greatest variety of further application, of those conceptual schemes that can withstand the constant check against experience. Here we see the value, stressed by Karl Popper, of conscientious attempts to *refute* theories rather than simply looking only for ways to *confirm* them. Perhaps an experiment cannot prove that a theory is correct, but a theory that survives several attempts to prove it wrong may emerge as superior to its untested competitors.

A skeptic might add: The secret of success in science is that, contrasted with art and philosophy, science has essentially an easy and clear-cut task. There are many ways to dig a well for water, for example, some no doubt very inexpert, but that problem is clear, and sooner or later water will be found almost everywhere. The job itself, once formulated, is inherently a job that can be done. If science has limited itself to discovering facts and relationships, then patient minds will search for them, and in time facts will indeed be found and related.

There may be much truth in this, but scientists are neither all-wise nor long-lived; hence what limited scientific knowledge they could gain by themselves would die with them if it were not for its communicability to others, and would forever be limited by the frailties of individual minds if it were not for some selective growth. *Continuity or communicability, individual variability, multiplicity of free effort, and selective growth—these are the four related principles on which is based the success of science.*

As an immediate consequence we see now another relationship between S_1 and S_2: Unlike the hive built from the tiny individual contributions of bees, science-as-an-institution does not exist by itself and outside the contributors, not in books nor in the abstract, but in the minds of those scholars who understand and work in the field. Each physicist, each chemist, each geologist, each astronomer fulfills two functions (and it is the easy confusion between these two that has produced the common erroneous picture of the scientist). On the one hand he is an unfettered creator in S_1, and on the other hand his mind, together with that of each of his colleagues, is the agency of transmission of S_2 from one generation to the next. Consequently the content of S_2, in passing through the ages from one group of rational beings to the following, continually undergoes a twofold change: It is ever more sharpened toward communicability and unambiguity; and it is incessantly expanded by tests for continuing meaning and usefulness against the widening experience and deeper experiments.

In retrospect we can now see the signs of this mechanism of evolution in the nature of physical concepts themselves. Recall those three characteristics of scientific concepts: The first is their operational meaning, the second the preferably quantitative nature, and the third the reappearance in diverse fields of application. Each of these can be regarded as a mechanism for assuring the continuation and increase of the scientific enterprise. They aid in the unambiguous communication of problems and results, make possible unambiguous agreement (or disagreement) among different workers on facts and their interpretations, and knit together the efforts of many independent scientists, even when widely separated in subject interest, time, and space.

13.8 Consequences of the Model

Several generally accepted features of scientific inquiry seem to be derivable from this view that science as an institution is continually filtered and modified by its flow through the minds of original creators, and that it exhibits the stated laws of evolution. Here is a summary of a few such points.

a) The convenience and freedom of communication is vital to the very existence of science *at each moment.* The jealous secrecy with which the alchemists hid their results doomed their efforts to stagnation and helped to delay the rise of chemistry. In our time, too, the imposition of secrecy in basic research would constrict the arteries of science. The right to pursue any promising lead, to publish and to exchange scientific information freely, must be vigorously defended.

From this point of view it is also apparent what an important role is played by free, disinterested research in academic institutions, where a scientist may follow the unpredictable path of knowledge without serious constraint and interference. For in science, as in society, truth can be found best in the free marketplace of ideas. The filter of years and of free debates, not the regulations of governments, can most effectively extract the real worth from each contribution. (As historians learn more about science under totalitarian regimes such as Nazi Germany and the Soviet Union, we come to recognize that it was not all worthless: A few scientists did manage to make significant discoveries despite the lack of freedom.)

b) The realization that science depends on communication between its practitioners partly explains the rise of organizations among scientists. From the early seventeenth century there has been a rapidly accelerating multiplication of the channels of communication—societies, institutes, journals, etc. National and, not infrequently, international meetings are held regularly each year by the practitioners of each specialty. Organizations have been formed for the single purpose of summarizing (abstracting), indexing, and distributing accounts of the growing avalanche of scientific developments. Scientists band together in committees and hold international congresses to decide on uniform nomenclature, definitions, and standards of measurement. They sponsor publication of monumental volumes of physical and chemical tables of constants. Above all, they are ever eager to share their latest findings and difficulties with one another.

c) Multiplicity of effort is not just a phrase applicable to the sum of all past work; it is a staggering reality in science today. Because there exists the very real danger that an account of this kind, in presenting necessarily only some of the more spectacular and historic advances of science, may distort the sense of proportion, look at least briefly in your library at a journal summarizing current work, for example, the monthly *Physics Today* or *Science News*. The point to note, above all, is the variety of topics, interests, contributors, and apparent importance of many of the papers. No one would venture to predict which of these contributions will be best remembered and used 30 years hence. The great revolutions, the spectacular breaks with the past, are rare indeed. But it may well be that one scientist or another has just planted the seeds from which a now entirely unsuspected harvest will be gathered.

d) Our view of science as an organism obeying laws of evolution helps to explain the truism that once "a situation is ready" for a discovery or for the formulation of a conceptual scheme, it frequently is made in several places by independent workers at the same time. When we speak about some general advance and say that the time was ripe for it, we usually mean that at least three conditions obtained: There were *reasons* for looking into the given problem, i.e., it now made sense to ask such a question and it may in fact have been generally discussed; second, there were *means* for looking, i.e., the needed experimental tools were obtainable; and third, there were mental schemes for *understanding*, i.e., the needed concepts and wider conceptual systems were readily assembled.

When all this happens at one time, several scientists may unknowingly compete; it is always possible that several researchers can expose themselves to the same heritage, can encounter and interpret the problem in the same way and, borne by the same wave, reach the same shore. Such "simultaneous discoveries" are striking when they occur. The sociologist Robert K. Merton has argued that they are actually common, but we never hear about most of them because the scientist who loses the competition to publish first is often forgotten or is discouraged from publishing at all.

e) We have had examples, and shall have others, of the very opposite phenomenon, of the rejection or long neglect that is experienced not infrequently by precisely the most startling innovators, even though their contribution later may come to be widely accepted. But if science is not simply the sum of individual contributions, if, as we saw, S_1 exists not outside but within the minds of practicing scientists, then it follows that a contribution, to become part of S_2, must find its way to acceptance by and into the minds of many scientists. That, however, must take time, particularly if the suggestion is very radically different from the climate of opinion in the field at the moment. In Section 3.2 we saw that this is indeed in the long run desirable for the evolution of science. And so it can happen (though rarely) that the contribution of a person regarded as something of a crackpot by contemporary colleagues reappears, even if much modified, as an important idea later on when the horizons of the field have changed in the course of evolution. Occasionally, science has needed the neglected recluse in the garret as much as the rare and widely acclaimed heroes and the army of competent foot soldiers.

f) In the same way that freedom in science increases the chances for both multiplicity of effort and individual variation, so also does a free science, ever productive of its own tools for further advancement, tend to become more and more a self-perpetuating or even an explosive chain reaction. As knowledge begets knowledge, it increases as though in a geometric series.

It is a source of constant wonder that once science (as we use the term) got its first real impetus in the early seventeenth century, it swept over Europe within a century like a long pent-up storm. The issue is, as we have seen, very complex, but one may speculate that two details contributed materially: European science had, from the fourteenth century on, expanded and progressed to the point where the principles of multiplicity of effort and individual variations were finally in effective operation. In addition to this, the work of Galileo and of his contemporaries now provided attitudes and concepts that, being quantitative and operational, allowed the principle of continuity to operate more fully than before, thereby modifying the science of the time toward the self-sustaining character it attained in the early eighteenth century.

g) Since the human mind is not divided into separate compartments and since science exists as a complex of conceptual schemes propagated through the minds of humans, we see that the one can interact with the other, that science can influence the whole culture pattern of a society and in turn can be shaped by it. It is only natural that, for example, Galileo's argument on the motion of heavenly bodies, like that of his adversaries, should be colored by and again expressed in contemporary theology or that the impact of Newton's work on the thinking of eighteenth-century Europe should have extended far beyond merely scientific questions.

When we study the long-range effects, we see ample evidence that alert persons of today, whether or not they have studied science, are intellectually children of Copernicus and Galileo, Newton and Faraday, Lavoisier and Dalton, Darwin and Mendel, Einstein and Bohr. Our imagination and intellectual tools have, to a large extent, been shaped by the advances in the knowledge of science they and their contemporaries made long ago. When the Copernican and Newtonian worldview triumphed in the West, the recognition that a uniform law holds sway over all matter everywhere helped to overcome hierarchical thinking and to prepare the mind for self-reliant democracy. The indirect influence of Newton on the creators of the United States' government and its Constitution has been shown by historians going back to F. S. C. Northrop (*The Meeting of East and West*) and more recently I. Bernard Cohen (*Science and the Founding Fathers*). We know also that in the nineteenth century the successes of statistics and of the concepts of energy prepared the ground for the modernization of the Newtonian worldview.

In addition to the kind of long-range influence that Newton's work had on the imagination of writers, poets, and philosophers from the eighteenth century on, there are also the more material long-range effects associated with the advances made by James Watt, Faraday, and Fermi. From an understanding of how the steam engine works flowed a century-long transformation of society that is now studied as the Industrial Revolution. From Faraday's "toys" came electric motors and generators and, in time, the electric-powered elevators, trains, and subways that facilitated the upward and sideways growth of cities. Similarly, the experiments of Fermi's group on neutron-induced artificial radioactivity prepared the way for the study of nuclear fission, and this in turn led to the design of new energy sources that may well, despite temporary setbacks, be the chief means for meeting our frantically growing energy needs. Of course the mention of nuclear fission immediately brings to mind the nuclear weapons that the United States had to build during World War II to preempt the work of Hitler's scientists, which threatened the very existence of western civilization; we return to that topic later in this chapter.

It is of course even more difficult to foresee the long-range effects of science on social change than it is to see the immediate practical influences. To avoid possible negative effects and to capitalize on positive ones, there is only one policy available to us: the exertion of uncompromising watchfulness as both citizens and scientists. We must call attention to existing abuses of scientific knowledge or skills and keep up to date on scientific advance so as to be ready to keep science from being derailed and abused in the future.

To return to the meaning of science not as merely a technical study but as a part of the general humanistic development of civilization, we can illustrate this sense of interconnectedness with two simple diagrams. The physics course, as traditionally given in many classrooms, is like a string of beads. One subject follows another, from Galileo's kinematics to the most recent advances in nuclear physics. This sequence is the usual one, which more or less parallels the historical development of the science, whether this parallel is made explicit or not. But few if any connections are shown with achievements of human beings who are not physicists, with sciences other than physics, or with studies and activities other than science. And all too often the materials studied in other courses (for example in chemistry, in biology, in literature) also hang like so many separate strings of beads (see Fig. 13.1).

There are some advantages in such a string-of-beads presentation. It is, for example, convenient for teaching and learning. But ignoring connections that do exist among all these fields does not do justice to the actual state of affairs.

A research project in experimental physics sooner or later may draw material not only from almost every part of a physics curriculum but also from mathematics, metallurgy, chemical thermodynamics, electronics, computer technology, and many other sciences; writing clearly and intelligibly about the work may call on skills in English composition and group psychology. Moreover, nobody who has engaged in actual scientific work can fail to see the influence that scientific advances can have in terms of social and practical consequences. "Pure" physics is an

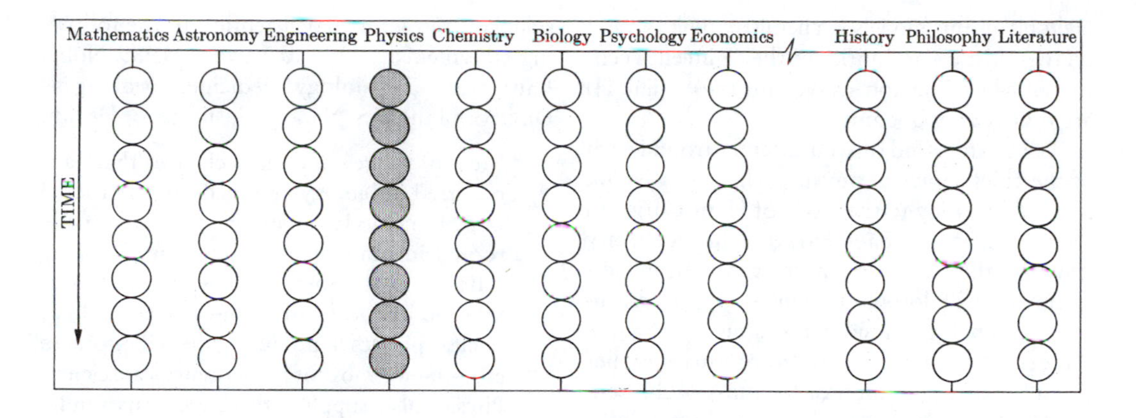

Fig. 13.1. Traditional view of physics as separate and independent of all other subjects, and as made up of separate sections, like beads on a string.

invention that exists only in the most old-fashioned classrooms. If you pick up a real research problem in physics (or in any other science), there extend from it connections to a number of expected and unexpected problems that at first glance seem to "belong" to other professions.

We believe this is one argument in favor of including occasionally in a physics course topics not usually referred to. Think back, for example, to the study of Newtonian mechanics as applied to planetary motion, a subject that is usually one of the "beads" on the physics chain. Newton had studied theology and philosophy, and those ideas emerge in the *Principia* in his sections about the nature of time and space (see Fig. 13.2, link A to philosophy). Within physics itself, Newton brought to a culmination the work of Kepler and Galileo (link B). Much of the established mathematics in Newton's work came from the Greeks (link C). New mathematics, particularly the basic ideas of calculus, were invented by Newton to aid his own progress, thereby advancing the progress of mathematics (link D).

Within physics, all who follow Newton would use his laws and approach (link E). His effects on the Deist theologians (link F), on John

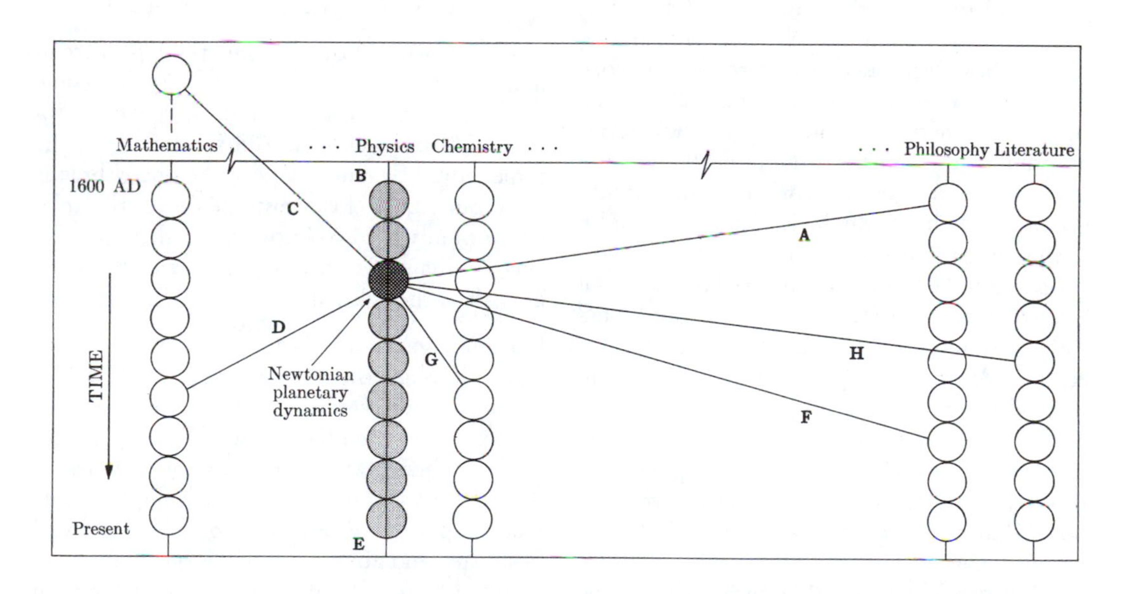

Fig. 13.2. An example of the interconnectedness of all fields. Newtonian dynamics is linked to past and present achievements, not only in physics but in other fields as well.

Dalton's atomic models in chemistry (link G), and on the artistic sensibilities of the eighteenth century in which Newton swayed the muses (link H) are quite easily documented.

The same kind of web extends around each of the chief topics of physics. Think of the link from philosophy to the work of Hans Christian Oersted and Michael Faraday in electricity (through their direct or indirect interest in German Nature Philosophy). Think of the link reaching from nuclear physics back along the chain to the classical physics of three centuries earlier (as in the determination of the mass of the neutron) and of the sideways links to biology, to engineering, and indeed to politics, owing to the various applications and by-products of nuclear reactors, for example.

Such links exist between main achievements in many fields. If we drew all the links among fields on the intellectual map, we would see instead of separate strings of beads a tapestry, a fabric of ideas. Science is in dynamic interaction with the total intellectual activity of an age. In a deep sense, science is part of the study of history and of philosophy, and it may underlie the work of the artist, just as it penetrates into the explanation a mother gives to her child to help him understand the way things move.

If we therefore tried to think away the achievements of physics, the course of modern history would be almost incomprehensible. We could not understand, and in fact would not have had much of the work of John Locke, Voltaire, and Alexander Pope, men who, among many others, were frankly inspired by the work of physicist contemporaries. Conversely philosophy, mathematics, and other fields would be far emptier studies without their fulfillment and extension through the work of philosopher-scientists such as Mach, Einstein, and Bohr. Eliminating physics would, of course, also make nonsense of the history of industrial development following Watt's steam engine, Alessandro Volta's electric battery, and Faraday's motors and generators. A neighboring science such as chemistry could not have developed without the models of gases and the theories of atomic structure that were largely the work of physicists. Biology and medicine would be primitive without the discoveries made with the aid of good microscopes.

It is sometimes said that our society should now channel its financial support primarily to biomedical research rather than to the physical sciences. But let us pay attention to the following statement by Harold Varmus, 1989 Nobel Laureate in Physiology/Medicine, speaking as Director of the U.S. National Institutes of Health:

> Most of the revolutionary changes that have occurred in biology and medicine are rooted in new methods. Those, in turn, are usually rooted in fundamental discoveries in many different fields. Some of these are so obvious that we lose sight of them—like the role of nuclear physics in producing the radioisotopes essential for most of modern medical science. Physics also supplied the ingredients fundamental to many common clinical practices—x-rays, CAT scans, fiber-optic viewing, laser surgery, ECHO cardiography, and fetal sonograms. Materials science is helping with new joints, heart valves, and other tissue mimetics. Likewise, an understanding of nuclear magnetic resonance and positron emissions was required for the imaging experiments that allow us to follow the location and timing of brain activities that accompany thought, motion, sensation, speed, or drug use. Similarly, x-ray crystallography, chemistry, and computer modeling are now being used to improve the design of drugs, based on three-dimensional protein structures. . . . These are but few of many examples of the dependence of biomedical sciences on a wide range of disciplines—physics, chemistry, engineering, and many allied fields.
>
> (AAAS Plenary Lecture, 13 February 1998)

In short, if you were to pull the thread marked "physics" from the tapestry—as some antiscientific persons wish to do—the whole fabric would unravel as an old sweater does; the same would be true if any of the other threads were pulled. In this view, therefore, the relevance of any field of knowledge, including science, is that it is an integral part of the total growth of thought, and of culture itself.

h) Let us consider one last point of universal concern, although others will suggest themselves readily. For the very same reason that one cannot foresee the eventual course of science, one can rarely evaluate at the time of discovery whether and where it will find uses, or how important its place in pure science will be, or whether it will be twisted and abused for evil purposes by those who guide the fate of society. As one common example, consider that radioactivity was discovered rather casually by Henri Becquerel when a photographic plate was exposed by the rays from a nearby piece of uranium salt (he had put them together in a drawer in preparation for an experiment that actually was based on a plausible but erroneous notion). Consider further that Becquerel did not properly understand the meaning of his observations; that through the lonely labors of Marie and Pierre Curie the new radioactive metals were isolated and identified—an unsuspected widening of the horizon of physics and chemistry; that the new knowledge of radioactivity then stimulated a generation of workers in all sciences, and so directly and indirectly revolutionized the conceptual models of the atom, of radiation, and of chemical reactions, with very practical consequences in such widely removed studies as the treatment for certain diseases. Then recall that today radioactivity is an integral part or indispensable tool of innumerable studies quite outside the field of discovery, in biology, geology, astrophysics, metallurgy, even archeology, and many others. Surely no one will dare prescribe limits to the process of discovery!

Finally consider the great contemporary dilemma: "atomic" (nuclear) energy may put the Philosopher's Stone into every citizen's pocket and, through the industrial *applications* of the eventual discovery of cheap and clean energy resources, may help lead mankind to a happier age for which it seems hardly prepared—or it may help, through the applications by those who make and wish to use nuclear weapons, end the short history of our species. All governments in the "civilized" world are simultaneously developing plans for engines of large-scale destruction ("dehumanization," as one official put it) of areas considered to be hostile.

But as our analysis has shown us, one must be careful not to confuse science with the industrial, technological, military, and other applications that are derived from it. Science and technology are related but separable. It is perfectly possible, and may often be wise, to limit or forbid some technological applications, as in arms-control treaties or in the decision not to build a supersonic transport plane that would merely drain the economy and threaten the environment. The progress of humanity and even of technology itself may well depend on making such self-denying decisions and not producing or doing everything that it is possible to produce or do.

The case is quite different with progress in the understanding of basic science itself. That progress, as even the example of Becquerel and Curie showed, often goes on in a quite unplanned and unexpected way. So it would be a grave threat to all science if fundamental research in any part of it were forbidden or discontinued. And though knowledge without deeper human wisdom may fail us, *less* knowledge hardly assures either wisdom or survival; moreover, one might at least doubt whether ignorance is now either possible or worth living for. For salvation and disaster are not reached by separate roads so that we may simply choose one and avoid the other; they both lie at the end of the same path, the choice being within ourselves. If we sought to solve the dilemma by fiat, by some official control of science, we should all have to submit to a censorship that, unable to foresee the result of curiosity and the future of discoveries, essentially would have to rule out all intelligent speculation. Surely, this system would guarantee the loss, along with science, of our basic freedom, while also forcing us to abandon all hope for an enlightened future. Both these results to our society would be quite comparable with the vengeance of savage conquerors.

If we look for an alternative solution, we might well try to emulate the lesson found in the growth of science itself: *our marvelous ability to arrive at truths by the free and vigorous exchange of intelligence.*

There is another reason why it is wrong to seek relevance for science only in the immediate benefits to technology. Not only have technological advances, all too often, brought with them major social problems as unforeseen byproducts, but also these problems cannot be cured or even properly understood through existing scientific, technological, or political means alone. Such cures, rather, depend to a large extent on new, basic advances, not least on advances in science itself. At the heart of social problems created by technological advance is the *absence* of some specific basic scientific knowledge. This realization gives a whole new mandate to basic scientific research.

Examples come readily to mind. It is quite customary, for instance, to say that the population explosion is in part caused by the advance of medical science (better sanitation, inoculation, antibiotics). But one can equally well claim that the population explosion is bound to overwhelm us precisely because we do not yet have at hand sufficient knowledge in pure science. That is to say, the complex problem of overpopulation is due in a large degree to our current (and long acknowledged) poor understanding of the basic process of conception—its biophysics, biochemistry, and physiology. No wonder that attempts at controlling the threat of overpopulation are so halting.

Similarly, it has sometimes been said that progress in physics was responsible for the threatening arms race during the "Cold War" between the western democracies and the communist countries, after World War II. But it is more accurate to say that nuclear arms-control treaties were difficult to achieve largely because insufficient knowledge of geophysics made inspection of suspected, illegal weapons tests with seismographs difficult. A better understanding of geophysics, it turned out, was needed before some nations considered it safe to enter arms-control treaties that outlaw underground weapons tests. When geophysicists achieved the ability to distinguish reliably between earthquakes and underground nuclear tests, it became possible to reach agreement and make the world a little safer.

The problem of having sufficient agricultural production to aid hungry people in arid lands near the sea, as in Peru, India, or Egypt, is to a large extent one of logistics and management. But it is also a problem of basic science: Before it was possible to design more economical desalination plants for irrigation, we needed a more fundamental understanding of liquid structure (one of the old but much neglected problems in physics and chemistry) and of flow through membranes.

These few examples serve to oppose two widely current but erroneous notions: First, that basic research in science is an unnecessary luxury and should be supported only if it is directed to immediate practical applicability, and second, that one way of stopping the abuses that are by-products of technical innovation is to stop science (whereas in fact curing existing abuses often depends on scientific advances yet to be made).

CHAPTER 14

On the Discovery of Laws

We have examined at some length the nature of concepts in physical science, and the relationship between scientists and their work. In this chapter we may now proceed to inquire more thoroughly whether anything positive may be said about the discovery of specific *laws* of science.

The familiar word "law" carries in its own history some important clues about the development of modern science. Suppose we ask: Why did the Scientific Revolution occur in Western Europe in the sixteenth and seventeenth centuries, rather than some other time and place? Historians who have studied the rise and fall of science in civilizations outside Europe identify two locations where such a revolution might have been expected to take place, based on the undoubted fact that they were technically and intellectually far superior to Europe and the rest of the world during the period from about the eighth to the thirteenth century: China and the Islamic countries in the Middle East. One of the remarkable features of both cultures is the *absence* of a strong belief in *universal laws governing both nature and society.*

By contrast, this belief was so strong in medieval Europe that one finds reports of animals being prosecuted and punished for violating the laws of nature: for example, in 1474 the magistrates of Basel, Switzerland, "sentenced a cock to be burned at the stake for the heinous and unnatural crime of laying an egg"! (Pyenson, "Without Feathers") The suggestion is that the concept of "laws of nature"—decreed by God, not invented by humans—played some role in the distinctive character of the European science established in the seventeenth century.

Following the example of Newton's *Principia,* physical scientists now usually reserve the term "law" for those statements of generality and trustworthiness whose descriptions involve mathematical relationships. Other words are frequently used for those statements that are mainly qualitative (for example, the statement that the horizontal and vertical components in projectile motion add up without disturbing each other) or when they are more in the nature of a *model* or a hidden mechanism to aid our comprehension of phenomena (for example, the idea that chemical reactions are explainable and predictable by considering matter as made of particles). In those cases various terms may be encountered: "principle" (for example, superposition principle), "hypothesis" in a larger sense, or "theory" in a limited sense (for example, Avogadro's hypothesis, or the atomic theory), "postulate," "rule," and so on. We shall not find it important to make much of possible differences between these words, and shall usually refer to all of them as laws for the time being.

It is revealing to consult first the opinions of active scientists. We then suggest a model to represent the process of discovery.

14.1 Opinions on Scientific Procedure

We remember the warning not to expect uniformity of procedure. While there has never crystallized a representative feeling among scientists in general, there are nevertheless several types of opinions about the main elements of scientific procedure that have been held prominently, and five facets have reappeared frequently enough to be of interest here. None of these by itself tells the whole story, none is meant to be in rivalry with the others. Rather we shall consider them as five complementary aspects of the same general opinion, the proportions being variables depending on the particular problem and the particular scientist.

a) The first of these views, symbolized in a quotation from Joseph Priestley, emphasizes that,

> More is owing to what we call chance, that is, philosophically speaking, to the observation of events arising from unknown [unsuspected]

> causes, than to any proper design or preconceived theory in the business. This does not appear in the works of those who write synthetically upon these subjects. . . .
>
> (*Experiments and Observations on Different Kinds of Air,* 1776)

Superficially one indeed cannot help but be impressed by the role of chance in scientific work. Every schoolboy is told how a falling apple directed Newton's mind toward the law of gravitation, or again how the law of the simple pendulum occurred to Galileo as he was watching a swinging lamp in the cathedral at Pisa. The initial observations leading to the epochal discoveries of the battery, many of the elements, x-rays, radioactivity, and contemporary chemotherapy were all made more or less by chance. Speaking of related sciences, W. I. B. Beveridge writes in his excellent book, *The Art of Scientific Investigation:*

> Probably the majority of discoveries in biology and medicine have been come upon unexpectedly, or at least had an element of chance in them, especially the most important and revolutionary ones.

This dovetails with what has been said before concerning the nonrational elements in the process of discovery. However, one must look at the whole picture. The huge majority of scientific work is unspectacular in achievement as well as in genesis; if chance enters at all, it is greatly outweighed by sheer hard thinking and hard work. Whatever else may be true of the methods of science, they are by no means shortcuts; they operate within the matrix of a life spent in the laboratory, at the desk, or in field expeditions.

As to the role of chance in those several famous cases, there is the unassailable dictum, "chance favors the prepared mind"; witness the obvious fact that great discoveries in science are in general not made by people previously ignorant of these fields. It is only the master of his subject who can turn the nonrational and the unsuspected to his advantage. Chance observations have usually become important if they triggered off trains of thought that from all the evidence might well have been arrived at more systematically by those minds. This trigger action of chance is by no means to be despised, but to interpret the stimulus correctly one must have had prior training and thought on the general subject of the discovery. In short, one must have acquired a sensitivity for recognizing a favorable though unexpected turn of events.

b) Although not trying to write off Priestley's statement as erroneous, we certainly do expect other elements to enter more prominently into the process of discovery. A second such main ingredient has been described by P. W. Bridgman, who stated straightforwardly,

> The scientific method, as far as it is a method, is nothing more than doing one's damnedest with one's mind, no holds barred. . . . This means in particular that no special privileges are accorded to authority or to tradition, that personal prejudices and predilections are carefully guarded against, that one makes continued checks to assure oneself that one is not making mistakes, and that any line of inquiry will be followed that appears at all promising. All of these rules are applicable to any situation in which one has to obtain the right answer, and all of them are only manifestations of intelligence.
>
> (*Reflections of a Physicist*)

Recalling what has been said before about the lack of a method, about the variability among scientists and their methods, one can hardly fail to agree with Bridgman's statement. It finds in the habit of using one's intelligence to the utmost a common ground for the diversity of individual approaches.

c) In much the same vein, others (among them Max Planck, the formulator of the quantum concept) regard their work simply as an extension and refinement of common sense. Initially this view must seem incomplete, if only for the reason that scientists in dealing with the common-sense problems of day-to-day living on the whole can hardly claim to show much more aptitude than do their nonscientific neighbors. We must admit that many relations exist between science and common sense, for example, that the primitive notions and techniques spring from prescientific experience, that science had its origin to some extent in the common sense of the early craftsmen, and that even today much research proceeds on a kind of technical common-sense or cut-and-try empiricism, particularly in the applied sciences and in the neighboring continent of technology.

But when we are dealing with men and women working in the modern physical sciences and with such conceptual schemes as elementary particle theory, their common sense is "common" only insofar as their special training and endowment allow them to find the way through their problems as naturally as carpenters and nurses doing their own work.

d) The foregoing is supplemented by a further opinion, which sees the essence of scientific procedure in some variation of Poincaré's terse characterization (duly qualified), "The scientific method consists in observing and experimenting."

This widely held view has a range of interpretations and misinterpretations. In the extreme it may refer to a technique close to that advocated by Bacon—the pursuit, as we saw, of an orderly experimental search for facts while carefully keeping one's mind from the influence of prior notions and from forming early hypotheses. The available facts were to be interpreted or explained only at the very end by a process of "true" induction, that is, induction unspoiled by premature guesses and hypotheses.

Now it is true that sometimes a research project in physical science reaches a stage where for months or even years theorizing on a large scale is held in check, pending the outcome of a single-minded search for specific experimental data; and it is further true that during the taking of data or the progress of a calculation scientists usually try to reduce the chances of mistakes by not becoming committed too early to a specific outcome, loving the game itself, so to speak, a great deal more than the final score. However, this is, after all, only the intelligent thing to do, and it also coincides with what we have called the discipline imposed on scientists; if your results have any real significance, sooner or later someone is sure to check and repeat your experiments, observations, and calculations, and will certainly discover faults and self-deceptions.

Whether in fact one can train one's mind to the Baconian level of control and asceticism, to completely uncommitted observation and experimentation, and whether the procedure is even fruitful as outlined, we have had reason to doubt before. Certainly the history of most scientific discoveries overwhelmingly denies these propositions. For example, Galileo's experiment with the inclined plane was surely not the result of an uncomplicated desire or plan to see what would happen to a rolling ball. On the contrary, as like as not Galileo knew rather well what he would observe. Some appreciation or foreknowledge of the type of solution to expect is almost inseparable even from the recognition of the *existence* of a problem. Galileo confessed his conviction that nature acted in the simplest possible way; hence he expected some relationship of the simplest possible kind between distances, times, and speeds.

Of interest here is that some criteria are indeed needed to focus attention on certain aspects of the total experience to the exclusion of the rest; the measurements of s, t, etc., in a controlled experiment become valuable for an eventual solution only if these factors and no others really do allow us to determine the features of free fall. Of course, not too much is expected of one's prescience at this stage; many elements may suggest themselves that will turn out to be unessential. For example, Galileo appears to have suspected initially that the material of the falling body determines its law of fall, but he finally decided against this factor. On the other hand, the major clue may appear only at the end of a long investigation.

On the whole, the problem here is similar to that of artists who must have at least some conception of the final product before deciding on the medium and tools for rendering their work. Experimentalists, particularly, must design their apparatus and procedure with a fair idea of the type and magnitude of the effect to be expected, including the allowable errors and the masking secondary effects. In short, scientists cannot just observe and experiment. They cannot just search; they have to search *for something,* and they must have an idea of what is to be expected—otherwise they may walk by the solution without recognizing it.

Now what has remained of the opinion, "The scientific method consists in observing and experimenting"? Much; above all, this: Whether we put our theorizing at the beginning or at the end, whether we induce or deduce the important laws, or use some interplay of both processes, sooner or later we must submit our work to the verdict of experience.

Even in that supreme example of "pure" theoretical work, the theory of general relativity, at the end of the theoretical development Einstein turned to the observational and experimental aspects of science by discussing several possible experimental tests of the theory. Moreover, he devoted considerable effort to recruiting a capable astronomer and providing him with the resources to perform these tests, as one can see from his correspondence (published in his *Collected Papers,* volume 8). Though he was usually rather skeptical of the claims and priority of experimentation, he says bluntly of one such proposed test, "If the displacement of spectral lines toward the red . . . does not exist, then the general theory of relativity will be untenable" (*Relativity*). Here we have again the persistent and striking difference between science and non-science—this appeal to quantitative observation and experiment as last authorities.

e) Finally, an element in scientific procedure held in highest esteem by those responsible for the spectacular achievements in contemporary theoretical physics is typified by Einstein's statement, "There is no logical way to the discovery of the elemental laws. There is only the way of intuition, which is helped by a feeling for the order lying behind the appearances."

We hear from these scientists that they immerse themselves completely in the problem at hand, and either speculatively try to postulate or induce a principle by which the problematical situation may be derived (explained) or, in fact, allow their thought processes to be manipulated without imposing a conscious direction. Whereas others seem to go from step to step with logical certainty, they make their large jumps accurately, as though borne by a guiding necessity. Sometimes they reveal that such solutions occur to them "in a sudden flash of insight" after long, even feverish study.

This method of solution by inspired postulation is rather close to the procedure advocated by Plato (for example, in the *Phaedo*) as well as by certain scholastics, although they did not share our convention of letting the entire validity of such induced postulates depend ultimately on the verdict of quantitative experience. One can, and many do, defend the view that this was the main approach of Galileo as well (although examination of his unpublished notes has shown that careful experiment did play a substantial role in his research). In that case we may imagine that he solved the problem of free fall by postulating the principle that all bodies fall such that $\Delta v/\Delta t$ is constant, and then deducing from this the experimentally confirmable behavior of falling and rolling bodies.

In the same way it used to be a fashion, particularly among French scientists of the last century, to organize their work on a model adapted from the later Greek mathematicians—beginning with clearly stated postulates or hypotheses, then demonstrating all deducible concepts rigorously and mathematically, and finally calling in experimental confirmations as though these were secondary and more in the nature of illustrations. In that type of account, and it has by no means disappeared, the important real experiment is sometimes replaced by the *thought* experiment, an argument that says in essence, "Let us imagine this or that plausible situation. Surely you must agree that things will happen thus and thus, exactly in accord with my deductions."

Similarly, Galileo would say, in his *Dialogue,* he does not actually need to perform an experiment to see whether a stone dropped from the mast of a smoothly moving ship will strike at the foot of the mast no matter what the speed of the ship; he knows this to follow from the principle of composition of motions, previously verified for the case of simple projectile motion.

But again, however true and important the role of postulation, the unexplained appearance of a key principle in the exceptional mind does not by itself tell the whole story, particularly in the experimental physical sciences. For the moment let us keep in the foreground of our minds all these distinguished opinions on the requirements for successful inquiry: *an intuitive feeling for nature, particularly for its quantitative aspects, a habit of using one's intelligence to the utmost, a deepened common sense, a reliance first and last on observation and experiment, and the sensitivity for recognizing a favorable though unexpected turn of events.*

Reviewing these opinions, we conclude that we may expect from the scientists themselves only the most general hints about their methods (what would a composer say if we asked by what method he created his concerti?), for in science, as in every field of endeavor, the most expert work is done by persons who have passed beyond the need and interest to rationalize each step. In the sense that each problem has its own difficulties and each scientist his own approach, we may say that the scientific inquiry has its art as much as its methods.

Perhaps the best single terms for the five qualifications cited above are "scientific orientation, or "scientific outlook," or "scientific attitude"; we may even venture to say that while scientific work in general is not characterized by any one single method, to a greater or less extent *scientists do share this complex of characteris-*

tic attitudes toward their life's occupation. This is shown by the ease with which persons trained in one specialized field can usually adapt themselves to another field, to new problems and a new set of collaborators; and it helps to explain why we find so frequently that a great scientist has made fundamental contributions to several widely separated branches of study.

14.2 A Sequence of Elements in Formulations of Laws

Having allowed the several evidences to dispel the last hope for an easy formula of discovery, we must take heed not to fall into the extreme skepticism of Priestley. From our now suitably modest point of view, we may still analyze how scientific laws grow out of observations, concepts, and hypotheses. To be quite silent on this subject is to disregard or dismiss the distinguished and varied testimony of influential writers on the philosophy of modern science.

What we shall do is analyze a *hypothetical* case of the formulation of a law, without claiming that formulations *in general* do actually follow this specific pattern; for in truth the development of an established law can rarely be traced completely, and with full accuracy, even in retrospect. Therefore the scheme to be discussed and its graphic representation in Fig. 14.1 are nothing more than a mnemonic device that may help us to think intelligently about the relationship between some recurring features in scientific procedure, particularly in experimental science. As an example we might keep in mind a concrete case, perhaps an interpretation of Galileo's discovery of the laws of motion of freely falling bodies and projectiles.

a) At the top of the list of comments we must surely put the importance of *the investigator's knowledge of contemporary science,* including the permissible tricks of the trade and the more or less tacit assumptions such as are embodied in every intellectual enterprise. Thus Galileo had available to him for his work on projectile motion the earlier and quite advanced inquiries of Tartaglia and Benedetti. Sometimes such prior knowledge proves erroneous and the assumptions misleading, but at the outset the existing conceptual schemes alone can give direction to the inquiry. They are the initial common sense from which the scientist must draw or perhaps eventually must break away.

At a later stage investigators will undoubtedly have to go back to their books or instruments to supplement their knowledge in unfamiliar fields that turn out to touch on the developing situation. Or they may even have to supplement existing disciplines by the addition of specifically invented tools of measurements and aids of mathematics. Galileo's inventiveness, of which we have heard much, extended also to the field of mathematics. We found that other great scientists had to face the same problems—Copernicus had to invent much of the solid geometry he needed, Newton his calculus.

b) A second point of departure, again principally in the experimental sciences, is a *first-hand acquaintance with nature through intelligent observation even before the problem has been formulated.* Not only do the great "chance" observations arise in this manner, but what is probably much more important, observations provide the

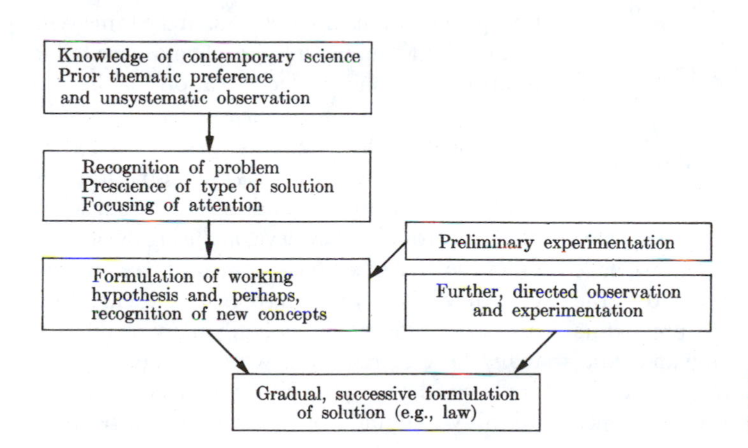

Fig. 14.1. Mnemonic diagram of relationships between recurring features in scientific procedure.

raw material with which to construct the great questions, and they provide the investigator with the necessary measure of nonrational insight and an almost intuitive feeling for the behavior of nature in the field of his specialty—what the significant variables are and how they interact.

To return to the work of Einstein, because in his mature years it was perhaps as far removed from personal participation in experimentation as is possible in physics, in his "Autobiographical Notes" we find this significant hint on the importance of early first-hand experience with "raw" natural phenomena: "[At the Polytechnic Institute in Zurich] I really could have gotten a sound mathematical education. However, I worked most of the time in the physical laboratory, fascinated by the direct contact with experience."

c) When the setting has been prepared, *the "problematic situation" may present itself to the mind in an "occasion of reflection"* (to use John Dewey's happy phrases), arising perhaps from some surprising chance finding, but more probably as the result either of some dissatisfaction with the degree of consistency, generality, or accuracy of the current explanation, or of some "hunch" where the rewarding new field of study lies.

There is no clear biographical account to tell us why and how Galileo first decided on his study of motion, but we may gather from his remarks and from what we know of his life and work that, being dissatisfied with contemporary qualitative modes of description, he wished to find quantitative relations in motion. Other scientists, notably the phenomenal nineteenth-century experimentalist Michael Faraday, seemed to have been motivated more by imaginative guesses and an intuitive feel for nature.

To add one example to the evidence on the variability of motivation and scientific temperament, we are told by Niels Bohr that his historic theory of atomic structure originated in his recognition of a similarity in the *form* of two equations from different fields of physics.

d) Intimately connected with this recognition of the existence of the problem is the *foreknowledge of the type of solution to be expected,* as we have seen. After the initial period of wonder, the trained but curious mind can hardly prevent its first rather vague flights of speculation and anticipation. It will suggest quite naturally some further observations and attention to specific details, and so point in a general way to a suspected answer to the problem, even though the real problem itself may on deeper thought turn out to lie quite far from the initial stimulus.

For example, Wilhelm Röntgen in 1895 was led to the discovery of x-rays by noting that a sample of mineral gave off a pale light while electric current was made to pass through a discharge tube 6 feet away. He quickly traced this effect to the action of previously unsuspected radiation coming from the tube. But he immediately abandoned the problem of why minerals fluoresce in this new radiation (for about 20 years this problem could not even begin to be understood), and turned instead to the far more interesting and researchable questions, for example, whether x-rays are a form of invisible light (Section 25.6).

e) This early stage of research may not only suggest new and interesting problems and some tentative hypotheses concerning the eventual solution of the problems, but it may also give rise to *new concepts with which to analyze the problematic situation.* To take our previous case, Röntgen found that objects are transparent to the passage of x-rays but that an increase in the thickness or the density of bodies decreased their transparency for x-rays significantly. It occurred to him to compare whether the transparency of various metals is equal when the product of thickness and density is constant. It so happens that this proved not to be the case (the complexity of the problem is great), but evidently if this simple constancy had been found, very quickly a new name for the concept "thickness times density of metals" would have been assigned, equipped with a symbol of its own, and no doubt would have figured prominently in the eventual solution of the problem of the degree of transparency of metals to x-rays. In the same vein, Galileo could analyze the problem of free fall in terms of the concept "acceleration" as he formulated it for that very purpose.

Evidently the great concepts, like the great problems on which physics is built, do not spring ready-made from the minds of individual scientists—far from it. They develop during the inquiry or, in the case of such concepts as force, energy, etc., they actually may have behind them a long history of evolution, involving many people and many false starts. Even when we give Galileo credit for defining the crucial concept "acceleration" we should not forget that in a sense he only completed a study that went back centuries before his time.

f) Proceeding from the initial preliminary experimentation and the first guiding ideas, there may begin to grow in the researcher's mind at first a rather modest *working hypothesis*—be it Röntgen's preliminary decision to regard x-rays as a form of ultraviolet light or Galileo's thoughts that the discrepancies in the free fall of bodies of different weight might be regarded as the minor effect of air friction.

The process of inducing an hypothesis or a postulate from the limited number of known facts has been compared with the unexplainable experience one may have while playing a game of chess, when one suddenly "sees through" the whole arrangement on the board and perceives in one stroke a pattern, a long chain of moves toward a successful ending. While the mechanism by which patterns are perceived and fruitful thoughts arise is still being debated vigorously among psychologists of various persuasions, we can at any rate point to the purpose of hypotheses in the scheme of scientific inquiry: namely, to stimulate some intelligent, directed activity, be it theoretical or experimental. To direct action, the hypothesis does not have to be perfect—we have long ceased to be surprised at the large number of correct theories first discovered while following a hypothesis that by itself proved to be wrong. But for that very reason the initial hypothesis should not be allowed to take hold of one's mind too firmly.

The very process of induction is a hazardous art, for no logical necessity exists that the predicted consequences should actually follow. To put it more bluntly, the reasoning is usually that if the product of induction, hypothesis—or for that matter a law—is true, then facts *a, b, c,* . . . should follow. When, on testing, these facts are actually observed, the hypothesis is perhaps true—but there may be numerous other and even better hypotheses that account for the same set of facts.

In his book *The Scientific Outlook,* Bertrand Russell gave a rather picturesque but stimulating example to show other dangers inherent in logical processes, particularly the limitations on experimental confirmations. He asks us to imagine that the statements to be tested are "Bread is a stone" and "Stones are nourishing." An elementary syllogism informs us that on the basis of such postulates it should follow that bread is nourishing; since the conclusion can be confirmed by the simple experiment of eating bread, we might be tempted to regard the premises as correct also! The result of induction is never unique, and deduction also may easily give faulty results if only, as here, by having assumed erroneous postulates.

A related form of fallacy is to mistake a temporal sequence for a causal relation, sometimes described by the Latin phrase *post hoc, ergo propter hoc* ("after this, therefore because of this"). You have probably had the experience of pushing the call button for an elevator and then, as soon as the door opens, rushing in only to encounter other people coming out. You unconsciously assumed that the elevator came to your floor *because* it arrived *after* you pushed the button; you may not have considered that it was already on its way because someone inside the elevator had previously pushed the button for your floor.

Science should never pretend to explain phenomena with absolute finality. Experiments and theories may seem to succeed or fail for an equally wide variety of wrong reasons. If you seemingly confirm a hypothesis by experiment, your measurements or the inferences from the experimental data may be faulty, or a series of canceling errors may deceive you, or the experiments may be right but the hypothesis wrong except for that singular prediction. Examples for all these combinations abound in the history of science.

On the other hand, when an experiment does not bear out an hypothesis, the effect you looked for in vain may indeed be there but too small for the instruments at your command, or you may be deceived by masking disturbances from an unrecognized source, or again there may be faults in the method or interpretation of the test.

The French scientist Pierre Duhem argued that it is logically impossible to refute an hypothesis by a single experiment, because one is always testing that hypothesis in combination with one or more assumptions needed to predict the result in a particular situation; if the prediction is refuted, all you know is that the hypothesis *or* at least one of the auxiliary assumptions is wrong. As an example Duhem cited the long-running battle between the wave and particle theories of light, apparently settled in 1850 when Fizeau and Foucault disproved the prediction from Newton's particle theory that the speed of light should be greater in glass than in air (Chapter 23). But Newton's prediction assumed not only that light is composed of particles but also that those particles have special properties that govern their refraction at an interface. Another particle theory might still be correct, and, indeed, a decade later Duhem was vindicated by the success of Einstein's quantum theory of light (Chapter 26).

The American philosopher Willard Van Orman Quine developed Duhem's view further, concluding that *any hypothesis can be considered correct no matter what results are observed in experiments, simply by making appropriate changes in the auxiliary assumptions.* This is now called the *Duhem-Quine thesis;* but unlike Duhem's original insight, Quine's version is only a theorem of abstract logic. In scientific practice one cannot rescue a refuted hypothesis indefinitely many times by inventing suitable auxiliary assumptions or *ad hoc hypotheses* (hypotheses for which there is no evidence other than their success in saving a refuted theory).

For all this there is no preventative. Nature's full complexity and subtlety may be far greater than our imagination can grasp or our hypotheses can accommodate. There is no substitute for the uncertain processes of thought or for the precarious business of confirming hypotheses by test. But there are several tricks that have shown themselves to be very helpful in the long history of thought: to try the simplest hypotheses first, to regard no answer or result as true beyond all possible doubt, to rely on several independent checks before accepting an experimental result of importance, and constantly to reevaluate and reformulate old knowledge in the light of new. We glimpse here again the fundamental truth that science depends on the continuing activities of generations.

Another trick, increasingly useful in modern physical science, is to formulate *theoretical models* of physical systems and processes. A model is *not* an hypothesis proposed as a true statement about the world, subject to experimental confirmation or refutation. Instead, it is a usually well-defined mathematical object, whose properties can be investigated by calculation or by computer simulation. It allows the scientist to investigate the logical consequences of certain assumptions. Such investigations may be useful even if the assumptions are obviously false, or if the consequences do not correspond to empirical observations. Making the model more "realistic" by including other factors that are obviously present in the real world (air resistance, friction) often reduces its utility by making exact calculations difficult or impossible. Theorists can develop several different models and then see which one works best; sometimes a model rejected at one time will be revived later when new observational data or changes in theoretical concepts make it look much better.

We have already discussed several theories that might be considered models:

1. Ptolemy's system of eccentrics, epicycles, and equants looks very artificial and complex but gave an accurate description of many of the observed planetary motions, though it predicted, obviously incorrectly, a large variation in the moon's apparent size.
2. The Copernican system was considered by some astronomers only a convenient model, useful for calculations, since an actual motion of the earth would contradict common sense as well as Aristotelian physics.
3. Galileo's mechanics was based on objects that can move indefinitely far at constant speed with no friction or air resistance—again, contrary to Aristotelian physics.
4. Newtonian celestial mechanics assumes that the sun, planets, satellites, and comets are point masses moving through empty space in order to calculate their orbits, yet acknowledges that they must be considered finite, nonspherical, heterogeneous objects when treating phenomena such as the tides and the variation of g on the earth's surface.

When we come to nineteenth-century physics we will encounter further examples:

5. The wave theory of light, in order to explain diffraction, interference, and polarization, employs the model of an elastic solid ether filling all space. It "works," but it can't possibly be true.
6. James Clerk Maxwell, in developing his electromagnetic theory, adopted another model for the ether: He imagined a collection of vortices to represent magnetic fields and a collection of "idle wheels" or ball bearings in between the vortices to represent electric current. No one, least of all Maxwell himself, pretended that such things really exist—but the model led him to his greatest discovery: Light propagates as electromagnetic waves.

Ernst Mach said, "Let us early get used to the fact that science is unfinished, variable." If we are reconciled to the fact that it is better to have a poor hypothesis than none, that an obviously absurd model may sometimes be more useful than a realistic one, that our thoughts may originally be grossly oversimplified, and sometimes they may need later and drastic modification, that sometimes several hypotheses may coexist—then we shall not be tempted to regard as foolish all work of the past, or as "certain" all work of the present.

Furthermore, if we regard an hypothesis as a stimulus for action and further thought, we shall not demand of it that it be immediately and directly confirmable. In the long run, of course, hypotheses or postulates do have to lead to some testable conclusions. Even though Galileo himself had no tools to check directly the relations involved in the motion of rapidly falling bodies, his assertion that their acceleration is constant was not inherently unoperational, and has, of course, since been tested directly. (Suggest such a check. Feel free to use high-speed cameras, etc. Just how direct is this test?)

Even the large-scale explanatory hypotheses may not lend themselves directly to experimentation, for example, the hypothesis that the agency by which the sun pulls on all planets is the same gravitational force that pulls a stone to the earth. But at any rate we shall insist on their verification by examining the degree of agreement between observed fact and the corresponding prediction deduced from our hypothesis—in our previous example, probably some astronomical observations.

g) As our work progresses, the initial working hypotheses will permit us to design and perform *more and more specific, directed, and controlled experiments*. Slowly our insight will progress, as if on two crutches—now the hypotheses helping to interpret one experiment and suggesting the next, now the experimental result modifying the hypotheses, in continual interaction. In this sense, the law of free fall (as expressed in the equation) is the end product, the one statement that best describes all the important features concerning free fall, whether executed by large bodies or small, whether applied to stone or lead, falling through long distances or short ones.

Indeed to be sure, it is often useful to guard against anticipated criticisms by performing tests that can quickly eliminate a seriously flawed hypothesis. But many hypotheses that we now recognize as valuable beginnings were "born refuted"—apparently contradicted by common sense or supposed experimental "facts"—and survived only because they found advocates convinced of their intrinsic value. Just as we suggested above that an initial hypothesis should not be too eagerly accepted, we now have to recognize that it may be a mistake to reject it too quickly.

14.3 The Limitations of Physical Law

In an important sense the word "law" is very misleading. Being built on concepts, hypotheses, and experiments, laws are no more accurate or trustworthy than the wording of the definitions and the accuracy and extent of the supporting experiments. Although the laws of nature are usually called inexorable and inescapable—partly because the word suggests the erroneous analogy with divine law—the human formulation of scientific laws certainly is neither eternally true nor unchangeable. The most certain truth about the laws of science is that sooner or later the advance of knowledge and experience will prove them to be too limited or too inaccurate. This lies in the tragic role of the law itself. It is induced from a necessarily limited number of experiments or situations, and yet, to be useful, it has to claim—cautiously—to apply to all situations of the same kind.

One example of this is the set of extensive tables that Galileo compiled for the range of projectiles. The direct experimental basis for this part of his work must have been quite small, but once he had discovered the parabolic trajectory, the essence of the law of projectile motion, he could derive and believe in consequences that had never been looked for experimentally. It follows that we must be alert for a new situation in which this ancient law will prove to break down—not only because it is an idealization never completely exemplified under ordinary conditions where air friction cannot be neglected, but also because there is, for example, no logical necessity for projectiles at extremely high speeds (as in the case of atomic particles) to obey the same laws as relatively slow cannon balls.

Here we remember the warning that all generalizations, all statements of physical law, carry a hidden "text"—including an explanation of the symbolism, a convention of how to manipulate the equations, definitions of the terms, statements of the accuracy and the range of experience supporting the law, and so forth. Almost invariably a limit is encountered sooner or later beyond which the law is no longer valid in its stated form; therefore the warning implied in the "text" should prevent continual misapplication of the law.

An analogy may be found in the awareness of a party of mountain climbers concerning the limits beyond which they can put no demands on their ropes, their gear, and their own bodies. As with scientific statements, the precise limits can usually not be clearly drawn until some failure brings them to attention. But whereas the climbers naturally dread such a failure, scientists are not helpless if occasionally a new phenomenon does not obey these established rules. For beyond the limits of one law may lie a hitherto undiscovered part of science—new and exciting knowledge. We shall see several great theories, such as Planck's and Bohr's, that had their beginning in the insufficiency of established laws, in some "unexplainable" observations.

Galileo's laws of projectile motion contain such a "text." We find in his work a thorough examination of the major limitations involved; for example, he points out that the trajectory will not be parabolic for extremely long ranges because then the acceleration of gravity, being directed toward the center of our round earth, would no longer (as assumed in our equations) be parallel for different parts of the paths.

Of course this failure applies, strictly speaking, even to small trajectories, but in such cases it is so minute as to make it permissible to neglect it for our purposes, just as we have already neglected the effect of air resistance, rotation of the earth, variation of the value of g over the whole path, etc. Of these limitations within which our laws operate we certainly need not be ashamed; for if we were forced to deal with actual motions in which these secondary factors were not negligible, we can be confident that we would surely be able to discover the complex laws guiding those smaller effects, and add them to our calculations. And in practice—for example, in computing the range of a projectile covering a large distance in actual use—such calculations can be now made with great precision.

But what if a law is found to break down at some point? Usually one need not become too concerned, for as a rule it is possible to rejuvenate and extend the law, perhaps by redefining some of the concepts used. Yet, one must pay the price for this flexibility in lawmaking in another way. The history of physics shows exciting cases where at one time from apparently the same set of experiments two different laws or theories were devised (for example, the inescapable classics, namely, the nineteenth-century wave theory vs. the Newtonian corpuscular theory of light). Either one of these is useful as long as no further experiments can be devised for which the two theories would predict sufficiently different answers to permit a clear choice between the rivals. However, once such a decisive experiment has been found—no easy matter—and carried out, one or the other (or possibly both) of the conceptual schemes must be conceded to be incapable of dealing with the new observation.

Yet no general scientific theory suddenly becomes regarded as completely "wrong," since it still is as useful as before in explaining and predicting the class of all the experiments prior to the decisive one that had incriminated the old theory or law. For instance, Newton's laws of mechanics as such are not "wrong" because they cannot deal alone with the motion of subatomic particles; Newton's laws are still as valid as they ever were in the huge range of experiences with larger bodies.

Here a famous analogy has been suggested: If we liken the facts to be explained to fish in a pond, then the law or theory is the net with which we make the catch. It may turn out that our particular net is not fine enough to haul in *all* the fish, large and small—but it may still be quite satisfactory for supplying our daily needs. We may go even further and maintain that to be useful at all, our conceptual schemes, like our nets, must contain some holes; if it were otherwise, we should not be able to distinguish between the significant and the trivial, the fish and the rest of the pond. To put it another way: Our most *reliable* knowledge of the world comes, paradoxically, from laws that have been refuted in a way that shows us their limits of validity (the size of the holes in the net) so that we can confidently use them *within* those limits; but if a law has never been refuted, we don't know its limits of validity and may someday be surprised by its failure.

14.4 The Content of Science: Summary

In this chapter we have tried to trace out the hypothetical development of a physical law. In previous chapters we dealt with the nature of theories and concepts. Let us in summary survey the three main elements that make up the established part of physical science. On this, at least, there is fairly good agreement among contemporary philosophers of science, although detail and phrasing may differ from one to the next.

First of all there are the *concepts* like velocity, mass, chemical element, etc.—the main ideas that the particular sciences use as vocabulary. They can be defined by operational definitions, the rules of correspondence between the concepts on the one hand and what can be observed (preferably measured) on the other. For example, we defined instantaneous velocity and acceleration by specific measurements and calculations, operations that involve distances, time intervals, and simple algebra.

Second, there are the *relationships between the concepts*. These relations may be simple factual observations (for example, ice under pressure melts below 0°C—a humble statement, but one that involves four concepts, of which at least one, temperature, is far more difficult than appears at first glance). Or they may be more general summaries of facts called laws or principles, and so forth (for example, Newton's second law of motion, relating the net force on a body, its mass, and its acceleration by $F_{net} = ma$). Or they may even be larger systems relating laws to one another (theories, which we examined in Chapter 3).

Last—although we tend to take it for granted—we must not forget at least to mention that part of science that contains the *grammar* for expressing, verbally or mathematically, definitions of concepts and relationships between concepts, i.e., the logic of language itself (rules on how to use words like *and, or*), and the logic of mathematics (how to use + or –, how to add vectors, etc.).

These three parts of science are obviously so completely interrelated that any one taken by itself is quite meaningless. A physical law means nothing without the definition of the concepts and the rules of mathematics or language. A concept is useless if it does not appear in relation to other definitions. This has helped us to an important insight about the way science grows: An advance in any one part of science is tentative until the rest of the system absorbs it. For example, a newly discovered law may disturb the hitherto accepted relationships between concepts or even the usefulness of some old concept. There will be then a period of rearrangement until the new discovery is fully incorporated into the ever-growing structure.

The development of a physical or chemical law is perhaps analogous to the progress of a body of water down a gentle slope where there are eddies and back currents from the advancing front and where here and there a sudden surge reaches out to embrace and temporarily hold some territory far ahead of the main advance. The process of gradual approximation leading toward a law may stretch over generations. How fortunate for the progress of science that individual scientists do not permit themselves to become discouraged by this prospect of lengthy struggles—that they find their main reward in the devotion to the day-to-day progress of their work.

Problems

Problem 14.1. What main points of difference are there between a *law of physics* (for example, the law of projectile motion as stated in the form of our several equations), on one hand, and, on the other, a *law of magic* (for example, the belief of the Hopi Indian that the ceremonial handling of rattlesnakes increases the chances of subsequent rainfall), a *law of phrenology*, a *civil law* (law against speeding on the highway), and a *moral* or *divine* law (the Ten Commandments)?

Problem 14.2. In a magazine article there appeared the following statement. Comment on the correctness or falsity of the *statement* and of its *implications* in a short essay.

> But the first thing to realize even about physics is its extraordinary indirectness. Physics appears to begin with very straightforward questions, but there are catches in it right from the start. For example: every high school student is told that according to Aristotle the heavier of two weights would fall faster, but that Galileo, by dropping two different weights from the Leaning Tower of Pisa, "proved" that Aristotle was wrong, for the weights were found to fall

in exactly the same time. And yet Aristotle was right. The heavier body does fall faster, because of air resistance, which slows up the lighter body more than the heavier. Only a very little faster, it is true, and the difference may not be detectable; but since scientists claim that they use words with such great precision, it is fair to hold them to it. If you press a physicist on this point, he will readily admit that what he means is that the two bodies would fall equally fast in a vacuum. If you press him further, he will very reluctantly admit that nobody has ever produced a complete vacuum. And so it turns out that not the physicist but Aristotle is talking about the actual world in which we live.

Problem 14.3. What exactly are the *raw observations* concerning the sun, stars, planets, satellites, etc., from which the theory of universal gravitation was derived? Try carefully to state observable facts and not to *assume* any theory in your description.

Problem 14.4. Examine a law outside physics with which you are *well* acquainted (for example, a law of economics, linguistics, biology, etc.) and decide whether or not one may trace its logical development by the same type of mnemonic diagram that we have used to analyze our hypothetical physical law.

Problem 14.5. Criticize these definitions: (a) The mass of an object is the measure of its quantity. (b) The range of a projectile is the horizontal component of its trajectory. (c) The "densage" of an object is its volume divided by its age. (d) The factor π is approximately equal to 22/7.

Problem 14.6. What is meant in science by the statements: (a) Atoms exist. (b) The ether does not exist. (c) Avogadro's (or any other) hypothesis is correct. (d) The findings do not represent absolute truths.

Problem 14.7. Why do you accept as a fact that the earth is round? That the moon is spherical? That there was once a scientist named Newton? Of which facts are you certain beyond all possible doubt? Why?

Problem 14.8. Try to account for the fact that astronomy was one of the earliest and best-developed sciences.

Problem 14.9. Assume that for his experiments Galileo had been using our most modern equipment for measuring the time and distance of freely falling bodies. Let us assume that his data had been accurate to 10^{-6} sec or 10^{-4} cm, respectively. Now write out in a few sentences: (a) what might have been the result of compiling a large set of experimental data, (b) what he actually would have had to do to discover the law of freely falling bodies by means of such data.

Problem 14.10. Write a short essay on one of these topics: (a) The scientist as a creative investigator. (b) Science as an evolving organism. (c) Important factors in the process of scientific discovery. (d) The discovery of truths by free and vigorous exchange of intelligence. (e) Complementary aspects of experience: scientific experience and poetic experience.

Problem 14.11. Write a short essay on P. W. Bridgman's proposal to extrapolate the operational view to fields other than physical science (Section 12.4).

Problem 14.12. Read again the two "deceptive" questions (about arguments from Euclidean geometry, and "What does mathematics tell us about physical events") following Problem 6.1, and the question "What does this reveal about the role of mathematics in physical science?" following Eq. (8.9). Then answer them in the light of Section 12.7.

Problem 14.13. Examine critically the analogy cited for the evolutionary processes (Section 13.7) and show where it can be extended (for example, dangers of overspecialization) and where it breaks down.

RECOMMENDED FOR FURTHER READING

P. W. Bridgman, *The Logic of Modern Physics*, Chapters I, II, and the first part of III

J. B. Conant, *Science and Common Sense*, particularly Chapters 1–3

Philipp Frank, *Modern Science and Its Philosophy*, New York: Arno Press, 1975

M. J. S. Hodge and G. N. Cantor, "The development of philosophy of science since 1900," in *Companion* (edited by Olby), pages 838–852

David L. Hull, "Studying the study of science scientifically," *Perspectives on Science*, Vol. 6, pages 209–231 (1998), on testing the principles of Popper and Planck and other hypotheses about how scientists respond to new ideas

Philip Kitcher, *The Advancement of Science: Science without Legend, Objectivity without Illusions,* New York: Oxford University Press, 1993

Ernst Mach, "On the economical nature of physical inquiry," in his *Popular Scientific Lectures,* pages 186–213

Henri Poincaré, *Science and Method,* Book I, Chapters 1 and 3

Karl Popper, *Conjectures and Refutations,* Chapters 1–3

Lewis Pyenson and Susan Sheets-Pyenson, *Servants of Nature*

George Sarton, *The Study of the History of Science,* to page 52

Stephen Toulmin, "The evolutionary development of natural science," *American Scientist,* Vol. 55, pages 456–471 (1967)

Eugene P. Wigner, "The unreasonable effectiveness of mathematics in the natural sciences," in *World of Physics* (edited by Weaver), Vol. III, pages 82–96

For further discussion of the subject of this chapter, see the "Sources" at the website www.ipst.umd.edu/Faculty/brush/physicsbibliography.htm

PART E

The Laws of Conservation

15 The Law of Conservation of Mass

16 The Law of Conservation of Momentum

17 The Law of Conservation of Energy

18 The Law of Dissipation of Energy

James Prescott Joule (1818–1889), British physicist whose experiments established the quantitative equivalence of heat and mechanical work.

If we now look back on the structure we have raised so far, we realize how greatly success in the pursuit of scientific knowledge depends on our ability to perform three closely interrelated tasks: to isolate the phenomenon from distracting or trivial other effects, to describe unambiguously what is happening, and to discern some specific permanence in the flux of events under observation. These are the beginnings of every scientific activity. And exactly these three functions are summarized most effectively in the various laws of conservation. These laws, perhaps the most powerful—certainly the most prized—tools of analysis in physical science, say in essence that no matter what happens in detail to a set of interacting bodies in some prescribed setting, the sum of this or that measurable quantity (the mass, or the momentum, or the energy, or the charge) is constant.

There is a spectacular beauty about these laws. They can break through the clamor and confusion of appearances and point to an underlying constancy so convincingly that the event changes at one stroke from chaos to ordered necessity. Moreover, so simple, general, and powerful are they that by the very extensiveness of their application the conservation laws unify the various physical sciences within themselves and with one another. Indeed, they have come to be called principles rather than laws, a terminology that betrays that they are no longer merely summaries of experimental facts, but instead have become the starting points of scientific understanding itself. For example, a famous theorem of the German mathematician Emmy Noether tells us that there is a direct connection between conservation principles and the symmetry of physical laws with respect to space and time variables.

In Part E we shall discuss specifically the laws of conservation of mass, of momentum, and of energy. (The law of conservation of charge is dealt with in Chapter 24.) We also consider a law asserting that an important physical quantity, "entropy," is *not* conserved, and that phenomena on the earth (and perhaps the entire universe) change irreversibly over time. This whole part also serves as a logical and historical bridge between the study of universal gravitation and subsequent chapters on atomic theory.

CHAPTER 15

The Law of Conservation of Mass

15.1 Prelude to the Conservation Law

In his great work *De Rerum Natura (On the Nature of The Universe)*, the Roman poet Lucretius, contemporary of Julius Caesar and Cicero, recorded and embellished the nature-philosophy of Greece's Leucippus, Democritus (both c. 450 B.C.), and Epicurus (c. 300 B.C.), and he reiterated what may be considered as one of the earliest hints of a profound general principle of science: "Things cannot be born from nothing, cannot when begotten be brought back to nothing." Everything now existing must have continual existence in past, present, and future, although form, appearance, and the like may indeed change.

Yet there is a very considerable distance from the panegyric of Lucretius to the modern law of conservation of mass—that almost axiomatic basis of much in our physical sciences—that teaches that despite changes of position, shape, phase, chemical composition, and so forth, the total mass in a given enclosed region remains constant. Indeed, it is quite futile to search Greek thought for direct parentage of the specific principles of modern physical science (except perhaps for certain astronomical ones). Lucretius' ultimate purpose, for example, was not to present a scientific text but to reduce the contemporary burden of superstition and fear through championing Epicurean philosophy and the rational explanation of natural phenomena. Although he documents his thesis throughout with penetrating observations from nature, he is not concerned primarily with a physical problem, but rather with a philosophical one, namely the basis for a particular type of atheism. This ideological intent is made clear in one of the first pages of *On the Nature of the Universe:*

> We start then from her [nature's] first great principle,
> that nothing ever by divine power comes from nothing.
> For sure, fear holds so much the minds of men
> because they see many things happen in earth and sky,
> of which they can by no means see the causes,
> and think them to be done by power divine.
> So when we have seen that nothing can be created
> from nothing, we shall at once discern more clearly
> the object of our search, both the source from which
> each thing can be created and the manner in which
> things come into being without the aid of gods.
>
> (translation by Ronald Melville)

15.2 Steps Toward a Formulation

Before the law of conservation of mass could reveal itself and appear in terms meaningful to modern science, three separate developments had to take place, each in itself the work of the greatest physical scientist of his generation. First had to appear the concept of the *ideally isolated system*. The dominant tradition of the Greeks and scholastics was the vision of a unified, completely integrated universe. In such a scheme, the behavior of a single body is determined by its relation to the rest of the cosmos, by the necessary role it must play in the whole drama. It was not meaningful, therefore, to think of events in isolation, for example, to interpret the behavior of a single object in terms of a physical law applicable to that particular region and uninfluenced by events occurring simultaneously in the universe surrounding this region.

We may regard the concept of the isolated system in part as one consequence of the work of Galileo on the motion of bodies. Specifically, Galileo's prescience of the law of inertia or, as we call it now, Newton's first law of motion, leads to thinking of the continuing, uniform, and unaccelerated motion of any object on a straight horizontal plane in the absence of external forces; and so it invites us to map out in thought a region containing only the body in equilibrium, a region at whose boundaries all causal connections with phenomena on the outside are broken.

Assuming, then, that there was now available the concept of an *isolated* or *closed* system by which to define the region of attention, there was next needed a criterion for measuring the quantity of matter before a conservation law could be formulated. And this was provided by Newton in the opening paragraph of the *Principia,* where, speaking of the term "quantity of matter," he states, "It is the quantity that I mean hereafter under the name of body or *mass.*" To know quantitatively the amount of matter in a system we need to know only its inertia (or else its weight, "for it is proportional to the weight, as I have found by experiments . . ."). With this definition in mind it was unavoidable to become aware that there is something enduring and constant about any given object—not necessarily its color or apparent size, its position or motion, or its volume or shape, or even its integrity, but rather its mass, anywhere, and therefore also its weight in a given locality.

A third contribution was clearly necessary, namely, the proof that the quantity of matter in a given system or, practically speaking, the weight of some material in a closed bottle, does not change during chemical transformations. The historian of science Charles Singer believed that Newton's explicit proof of the constancy of weight at the same place "gave a special impetus to the rationalization of chemistry" by revealing a simple and effective touchstone for checking on quantitative changes (*A Short History of Science*).

To us it would seem rather simple to proceed to a direct proof of the long-suspected law of conservation of mass: Enclose some air and a piece of wood tightly in a bottle, then focus the rays of the sun on the wood inside so that it may burn. After its transformation, the ashes and remaining gases together will be found to weigh neither more nor less than the initial components. Yet there was a period of just about 100 years between Newton's *Principia* and the work that provided this type of final and explicit proof, the memoirs on calcination and the text *Elements of Chemistry* (1789) by Antoine Laurent Lavoisier (1743–1794), often called the father of modern chemistry.

15.3 Lavoisier's Experimental Proof

The obstacles that had to be overcome in the interval between Newton and Lavoisier can now be stated so as to make them sound almost trivial. Lavoisier was one of the first to show conclusively that the most familiar of all chemical transformations, combustion of matter, is generally oxidation, that is, the combination of the material with that part of the ambient air to which he gave the name *oxygen;* that therefore the gas taken from the atmosphere has to be reckoned into the overall calculation. Before Lavoisier's time, neither the nature of gases nor that of the combustion process itself was clear enough. There was therefore little reason or facility for working with carefully measured quantities of gases or even for carrying out the reactions in closed vessels, isolated from the rest of the chemical universe. Instead of a general law of conservation of all participating matter during chemical processes, there impressed itself on the scientific mind first the clear, obstinate, and confusing facts that on combustion in the open some materials, like wood, would lose weight (because, as we would now say, more gases are given off than are taken on by oxidation), whereas others, like phosphorus, would conspicuously gain weight (because, in modern terms, more oxygen is fixed than vapors given up). During the reverse type of reaction, now called *reduction,* in which oxygen may be given up, a similar variety of changes would occur.

Since scientific theories quite rightly and naturally tend to deal initially with concepts derived from the most conspicuous aspects of the phenomena, there had grown up in the eighteenth century a conceptual scheme that attempted to deal with a large number of diverse observations on combustion, including the change in the physical structure and chemical nature of the burning object, the presence of heat and flames, the changes in the quality of the surrounding air, even the diverse changes of weight. For these purposes was invented the concept of *phlogiston,* a substance or "principle" whose migration into or out of the transforming bodies was to account for all such observations.

Fig. 15.1. Antoine Laurent Lavoisier (1743–1794), French chemist who replaced the phlogiston theory by the modern system of elements and compounds based on the principle of conservation of mass, and Marie-Anne Pierrette Paultze, Madame Lavoisier, who participated in the data-taking for his experiments and translated scientific papers from English for him. From the 1788 painting by Jacques Louis David.

The phlogiston theory died within a decade or so after Lavoisier's attack on it, not only because the concept of phlogiston had tried to explain so broad a range of phenomena that it failed to achieve a quantitative focus and became self-contradictory, but also because Lavoisier proved the concept to be unnecessary. By the incontrovertible evidence of a very accurate balance, he was able to show that shifting one's attention to the total quantity of matter undergoing a chemical reaction (including the gases and vapors) would be rewarded by attaining a rigorous law of conservation of matter.

A few translated excerpts from Lavoisier's text of 1789 will demonstrate something of his method in this respect, and will also hint at the relative novelty of this quantitative argument in chemistry:

> The elegant experiment of Mr. Ingenhouz, upon the combustion of iron, is well known. [Then follows a description of this experiment, the burning of iron wire in a closed vessel of oxygen-enriched air.] As Mr. Ingenhouz has neither examined the change [in weight] produced on iron, nor upon the air by this operation, I have repeated the experiment under different circumstances, in an apparatus adapted to answer my particular views, as follows. . . . [The modification is now given that permits accurate determination of weights.]
>
> If the experiment has succeeded well, from 100 grains [5.3 grams] of iron will be obtained 135 or 136 grains of ethiops [oxide of iron], which is an augmentation [of mass or of weight] by 35 percent. If all the attention has been paid to this experiment which it deserves, the air will be found diminished in weight exactly to what the iron has gained. Having therefore burnt 100 grains of iron, which has required an additional weight of 35 grains, the diminution of air will be found exactly 70 cubical inches; and it will be found, in the sequel, that the weight of vital air [oxygen] is pretty nearly half a grain for each cubical inch; so that, in effect, the augmentation of weight in the one exactly coincides with the loss of it in the other.
>
> (*Elements of Chemistry*)

Dozens of such experiments are described by Lavoisier, involving all types of reactions then known, and always repeating this lesson: In reactions within a closed system, the gain experienced by any one part of the system exactly counterbalances the loss by the rest of the system, or, in other words, *the total quantity of matter within the system remains constant.*

If we accept the principle of conservation of mass, as many scientists already did even before Lavoisier's definitive research, then we immediately see a difficulty with the phlogiston theory, exemplified by the above example. A combustion reaction that Lavoisier would write

$$\text{A} + \text{oxygen} \rightarrow \text{B},$$

where A might be iron and B would then be iron oxide, would be represented in the phlogiston theory by

$$\text{A} \rightarrow \text{B} + \text{phlogiston}$$

since burning was thought to involve the *expulsion* of phlogiston rather than the *absorption* of oxygen. But since Lavoisier's experiment showed that the iron oxide weighs significantly *more* than the iron when the burning takes place in a closed system, phlogiston would have to have *negative mass.*

Lavoisier's conceptual scheme of combustion and reduction was superior to the phlogiston theory in the form, the certainty, and the elegance of numerical predictions. It led to what historians now call the Chemical Revolution at the end of the eighteenth century. One of its significant characteristics was the emphasis placed on a *measurable quantitative* property of matter: *mass*. This turned out to be far more favorable to scientific progress—suggesting to chemists in the nineteenth century that elements should be characterized by their atomic weight rather than, for example, their color or smell.

Yet thereby a great many observations (for example, the presence of flames or the change in appearance of the material) were now left out of the scheme of immediate explanations. As we have pointed out in the preceding chapters, this is quite characteristic of modern science, and in fact a rather similar turn of events characterized several conceptual revolutions in science, as for example the rise and fall of the caloric theory of heat, which we shall discuss at some length in Chapter 17.

Needless to say, once the problems of the composition of air and of combustion were solved, the process of oxidation assumed its place as only one special, though sometimes unusually spectacular, case of chemical reactions, which all exhibit the conservation law as a general and basic proposition. As Lavoisier wrote in 1789:

> We must lay it down as an incontestable axiom, that in all the operations of art and nature, nothing is created; an equal quantity of matter exists both before and after the experiment, . . . and nothing takes place beyond changes and modifications in the combination of these elements. Upon this principle, the whole art of performing chemical experiments depends: We must [to cite one application of this principle] always suppose an exact equality between the elements of the body examined and those of the product of its analysis.
>
> (*Elements of Chemistry*)

15.4 Is Mass Really Conserved?

Despite Lavoisier's emphatic statement of the conservation law, there was still room for doubt. A modern experimental chemist, examining Lavoisier's reports of his experiments with some appreciation of the degree of accuracy that Lavoisier could have attained with his apparatus, might be somewhat skeptical about the claim that "the augmentation of weight in the one *exactly* coincides with the loss of it in the other." Nevertheless, the law was plausible, and therefore most nineteenth-century chemists were willing to go along with Lavoisier and accept it as an axiom until there was some definite reason to suspect that mass is not conserved. As long as the law of conservation of mass was thought to be consistent with other accepted views about the nature of matter, and as long as no obvious violation had been observed, there was no incentive to undertake further experimental tests.

In 1872, the German chemist Lothar Meyer (1830–1895) suggested that the rearrangement of atoms during chemical reactions might be accompanied by the absorption or emission of particles of "ether," the substance that was believed to transmit light waves (Section 23.3). Such particles might leave or enter the system even though it was sealed to prevent matter from coming in or going out. At that time it was still an open question whether such ether particles actually existed. If they did exist, then the mass of the system could change by a very small amount, depending on the mass of these ether particles and how many entered or left.

About the same time (in 1871) the Russian chemist Dmitri Mendeléeff, discussing the relation between his Periodic Law and Prout's hypothesis that all elements are compounds of hydrogen (Chapter 21), suggested that mass is a form of energy, and that the energy changes involved in forming these "compound" elements could slightly change the total mass of the compound.

Even though most chemists thought that the experiments of Lavoisier and others had established the validity of the mass-conservation law for chemical reactions, there was now interest in demonstrating it experimentally, at the highest level of accuracy. Mendeléeff's speculation, which did not seem susceptible to any empirical test in the nineteenth century, was ignored. But doubts such as those expressed by Lothar Meyer could not be simply dismissed without due consideration, as they challenged the accuracy of experiments already being done in the laboratory.

Hans Landolt (1831–1910), another chemist, decided that further experimental tests were needed. He conducted extensive research during the two decades after 1890, making very accurate measurements of the masses of systems in which chemical reactions were taking place. In 1909 he stated his conclusion:

> The final result of the investigation is that no change in total weight can be found in any chemical reaction. . . . The experimental test of the law of conservation of mass may be considered complete. If there exist any deviations from it, they would have to be less than a thousandth of a milligram.

Landolt's result is typical of the experimental data of the physical sciences: One can never claim to have proved that a difference or other measurement is exactly zero (or exactly any other number). One can only say experiment shows the number to be zero within a specified margin of error. (It may of course be convenient to assume for theoretical purposes that the exact value has been established.)

If there were no evidence available to us other than experiments on chemically reacting systems, we would have to conclude, even today, that the law of conservation of mass is correct. Modern chemists, repeating Landolt's experiment with the best available equipment, would still reach the same conclusion, even though their margin of error would be smaller.

However, we now know that in other kinds of experiments the mass of the matter in the system does change. The most important cases are those involving reactions of nuclei or elementary particles, such as radioactive decay, fission, or fusion (see Chapters 30 and 32). In some reactions, such as the annihilation of electrons and positrons, or protons and antiprotons, the mass of the *material particles* (the *rest mass*) disappears completely. On the other hand, the mass of an object may increase considerably when it is accelerated to speeds close to the speed of light.

In all these cases, the apparent change of mass is balanced by a corresponding change of energy, in accordance with Einstein's theory of relativity (Chapter 30). This could be interpreted by saying that energy itself corresponds to mass. Einstein's theory predicts that there will be a very small change of mass even in chemical reactions. This change is due to the fact that heat energy is absorbed or produced in the reaction, and such a change in energy reveals itself as a change in mass.

However, no chemical reactions involve a sufficiently large change of mass to permit this change to be measured. In the reactions studied by Landolt, the change of rest mass predicted by relativity theory is much less than a thousandth of a milligram. Landolt's conclusion, that the law of conservation of mass is valid for resting matter, is wrong in principle, even though his experimental data were as good as anyone has ever been able to obtain.

Here indeed is striking testimony to the power of a theory to help overthrow or amend a law just when it seemed to be firmly established by experiment. In Chapter 30 we shall tell in more detail how this happened and how eventually the law of conservation of mass was restored in a new form.

Problems

Problem 15.1. Read Chapter 7 in Conant's *Science and Common Sense;* then discuss three chemical reactions as they would each be explained by phlogiston chemistry and by modern chemistry (for example, the burning of coal, the calcination of a metal, the reduction of a metallic ore).

Problem 15.2. Read pages 1 to 10 in Freund's book *The Study of Chemical Composition;* then discuss Lavoisier's experimental procedure, specifically the extent to which he used the law of conservation of mass.

Problem 15.3. Read the Author's Preface and Chapters III, V, and VII of Lavoisier's *Elements of Chemistry;* then discuss three reactions in which the law of conservation of mass in chemical transformations is clearly demonstrated.

Problem 15.4. If the law of conservation of mass is valid, combustion reactions could be explained by the phlogiston theory only by assuming that phlogiston has negative mass or weight. Is there anything wrong with that? Discuss the motion of hypothetical negative-mass objects governed by Newton's second law, in a universe that contain objects of both positive and negative mass.

RECOMMENDED FOR FURTHER READING

Bernadette Bensaude-Vincent, "Lavoisier: A scientific revolution," in *History of Scientific Thought* (edited by M. Serres), pages 455–482
William H. Brock, *Norton History of Chemistry,* pages 78–127

Arthur Donovan, *Antoine Lavoisier: Science, Administration, and Revolution,* New York: Cambridge University Press, 1996

Richard P. Feynman, *The Character of Physical Law,* Chapter 3

Ida Freund, *The Study of Chemical Composition: An Account of Its Method and Historical Development,* New York: Dover, 1968, Chapters I and II

Henry Guerlac, "Lavoisier, Antoine-Laurent," in *Dictionary of Scientific Biography* (edited by Gillispie), vol. 8, pages 66–91

F. L. Holmes, *Antoine Lavoisier—The Next Crucial Year, or The Sources of His Quantitative Method in Chemistry,* Princeton, NJ: Princeton University Press, 1998

Carleton E. Perrin, "The Chemical Revolution," in *Companion* (edited by Olby), pages 264–277

Cecil J. Schneer, *Mind and Matter,* Chapter 6

Stephen Toulmin and June Goodfield, *Architecture of Matter,* Chapters 9 and 10

CHAPTER 16

The Law of Conservation of Momentum

16.1 Introduction

After everything has been done to describe the motion of bodies accurately, to analyze the path of projectiles, the orbits of planets, and the forces needed to account for such motions, there still remains, to the most basic common sense, a deep-lying and mystifying puzzle. While anyone can see intuitively that a body should move when it is somehow pushed or pulled by an agent in direct contact, it is far less clear to our primitive senses why a body should continue in motion in the absence of net forces or of any material agency in immediate contact with it.

The source of this organismic interpretation is probably not far to seek; after all, we each have learned from our first days that in order to make objects move toward us we have to take a hand in it; and to keep up the motion of anything on earth, we must be prepared to supply some effort.

To most of the Greeks and to most of the Aristotelian-Thomistic scholars of the Middle Ages, the proposition that the uniform straight-line motion of bodies requires neither force nor agent would have been patently absurd. And even the nonuniform progress of an arrow or a bullet was explainable only by the continual commotion in the surrounding air set off at the initial moment of release. The maxim was: Whatever moves must be moved by something else. Not until the days of Galileo and Newton did uniform motion without persistent and continual physical causation become generally conceivable. And significantly enough, even Newton could not free his conception of the physical universe from the need for constant association with a cause for a phenomenon. Of course, to him planets no longer moved by the incessant activity of God and the angels, but motion, in his cosmic philosophy, was thought of as proceeding in the "sensorium" (perceptible domain) of the Deity, by virtue of His continual though passive attention to and knowledge of the happenings.

Actually, the attempt to liberate the idea of motion and to discuss it as an independent phenomenon goes back to the fourteenth-century Franciscan friar William Ockham (d.1347). The University of Paris, where he taught, became a center of thinking on mechanics through the late Middle Ages, and although it presented a minority opinion at the time, it later influenced very strongly such researchers as Leonardo da Vinci and Galileo himself. Ockham's contribution has been succinctly described by Sir Edmund Whittaker in *From Euclid to Eddington:*

> In the example of the arrow, he proposed to explain the phenomena without depending on propulsion by the air, by assuming that throughout the flight the arrow carried a certain nonmaterial cargo, so to speak (corresponding to the modern notion of momentum), the possession of which ensured the continuance of its progress. This may be regarded as the first appearance in history of any sound dynamical principle . . . His idea was carried further by his disciple Jean Buridan, who became Rector of the University of Paris in 1327, and who asserted that the cargo in question was proportional to the product of the weight of the projectile and some function of its velocity (he did not distinguish between the concepts which we call momentum and kinetic energy). His doctrines were adopted, without improvement and indeed with some deterioration, by less able successors; however, eventually they reached Galileo, Descartes, and the other physicists of the seventeenth century, who succeeded in defining momentum precisely . . .

There is no doubt that Ockham deserves much credit for articulating a fruitful idea. Superficially it must seem that Ockham's own suggestion is not progress but a step back toward animism. He and his pupils after him suggest that during the act of setting-in-motion, something (whether we call it momentum or, as they did, impetus, matters little here) is imparted to the body; and if this quantity then grows or diminishes within, it makes the motion itself increase or decrease. But evidently this view, while it almost endows motion with the concreteness of matter, does help to concentrate attention on the moving object itself; and when it was eventually agreed upon how to measure this "cargo," namely, by the product of the mass m and the instantaneous velocity v of the object, then the pictorial element became unimportant. One could begin to think of motion in the abstract, and particularly of motion through a vacuum. One could forget about the question "What keeps a body in motion if not some continuous external push or pull," and instead ask the more fruitful question, "If we postulate that a moving body has at every instant a momentum mv, how does this momentum change under various interesting conditions?"

Thus the original questions on motion were to be answered finally in terms of the *law of inertia* (Section 5.3), by which an isolated body persists in a state of constant velocity and consequently of constant momentum, or in terms of Newton's second law, which allows one to calculate the effect of external forces on the motion and thus on the momentum. So the scientific mind in time solved an essentially barren problem by restating it from an entirely different point of view; and historically, by and large, this has always been one of the most powerful methods toward progress in science.

16.2 Definition of Momentum

The historical development leading from the early conception of momentum in the fourteenth century to the point that interests us now, namely, to the law of conservation of momentum, lasted almost 400 years and involved a galaxy of brilliant investigations. It will not be profitable for us to examine the complex details; instead, we shall mention only the final steps in the establishment of the law.

In Descartes' system of philosophy, widely acclaimed by scientists in the middle of the seventeenth century, the entire physical universe was likened to a clockwork mechanism. Once the machine had been constructed and set running by God, it was supposed to run indefinitely without any need for repairs or winding up (to think otherwise would be to imply that God was not a perfect clockmaker!). In order to ensure that the world machine would not run down, Descartes argued that there must be a principle of conservation of motion:

> It seems to me evident that is is God alone, who in His omnipotence created that matter with motion and rest, who now preserves in the universe in its ordinary concourse as much motion and rest as He put into it at creation. For although motion is nothing but a mode of moved matter, it exists in a certain quantity which never increases or diminishes, although there is sometimes more and sometimes less in certain of its parts.
>
> (translation by Marie Boas Hall in her book *Nature and Nature's Laws*)

Although one might tend to agree with this statement as a general philosophical principle, we cannot judge its scientific validity until we have a precise definition of the phrase "quantity of motion." Descartes recognized that this quantity must depend on something more than the *speed* of an object: A cannonball moving at 100 km/hr has more "motion" than a tennis ball moving at the same speed, if we judge by how much motion each can transfer to another object it strikes. Descartes proposed to define quantity of motion as the product: (mass) × (speed). The total quantity of motion of all the parts in the world, or in any isolated system, must then be constant from moment to moment, even though motion is continually being transferred from one body to another (from m_1 to m_2, etc.) by collisions within the system. In mathematical terms, the sum

$$m_1v_1 + m_2v_2 + m_3v_3 + \ldots \equiv \Sigma_i m_i v_i,$$

remains fixed even though $v_1, v_2 \ldots$ change.[1]

Descartes' law of conservation of motion seemed at first to provide a correct description of collisions, but closer examination revealed two serious flaws. *First,* his proposed law, even if it were correct, is not sufficient to determine the outcome of a collision. Descartes therefore had to add various auxiliary rules, some of which were obviously wrong. (For example: If a

[1]The symbol Σ (Greek, capital sigma) is used to indicate a summation of the quantities that follow it, in this case $m_i v_i$, where the index i runs from 1 up to N, the total number of terms in the sum. This abbreviation is useful in writing equations in which the number of summands is large or indefinite.

heavy object is at rest and a light object strikes it, the light one was said to rebound without transferring any motion to the heavy one.) *Second,* the speed v is a *scalar* quantity, whereas experiments indicated that the outcome of a collision depends also on the *directions* of motion of the colliding objects. An extreme example would be the head-on collision of two equally massive blobs of clay, moving toward each one at equal speeds, which stick together and remain at rest after the collision; clearly Σmv is not conserved in this case, since it is zero after collision but was a positive quantity before collision.

For those scientists who accepted Descartes' contention that the new science must be constructed mechanistically on the basis of the motions and collisions of small pieces of matter, it became imperative to find a correct version of his law of conservation of motion. When the Royal Society of London was organized in 1662, several of its members, especially Robert Hooke and Christopher Wren, were already interested in this problem, and had done some preliminary experiments. Christiaan Huygens, in Holland, was thought to have found a correct formulation of the laws of collision, but had not published it.

The Royal Society began to focus on the problem in 1666, with Hooke demonstrating collision experiments at the weekly meetings, Wren being urged to prepare an account of his earlier work, the mathematician John Wallis being asked to look into the theoretical aspects of the problem, and the secretary of the Society, Henry Oldenburg, being instructed to write to Huygens asking for a report of his discovery. By the beginning of the year 1669 this cooperative effort had borne fruit, and the Royal Society was able to publish in its *Philosophical Transactions* the result obtained independently by Huygens, Wallis, and Wren: The total *momentum* of a system is conserved; the momentum of any object is defined as the product *mass × vector velocity* for that object.

In modern notation, using $\boldsymbol{p}$ as the symbol for momentum, the definition is

$$\boldsymbol{p} \equiv m\boldsymbol{v}. \qquad (16.1)$$

Being the product of a scalar quantity (mass m) and a vector quantity (velocity $\boldsymbol{v}$), $\boldsymbol{p}$ is also a vector quantity, and representable by an arrow pointing in the same direction as the one that signifies the corresponding velocity vector.

The usefulness of the concept momentum, considered a vector compared to the scalar quantity "quantity of motion" (as in Descartes' definition), is most clearly revealed by considering our previous example: the head-on collision of two blobs of clay that stick together and remain at rest after the collision. Suppose the blobs have equal mass m and move with equal speeds v in opposite directions before the collision. Their vector velocities will then be $\boldsymbol{v}$ and $-\boldsymbol{v}$, respectively, and the total momentum of the system before the collision will be

$$m\boldsymbol{v} + (-m\boldsymbol{v}) = 0.$$

If both velocities are zero after the collision, then the total momentum at that time is also zero, and the law of conservation of momentum is obeyed. Descartes' supposed law of conservation of quantity of motion, on the other hand, is shown not to work, since $\Sigma_i m_i v_i$ is finite (equal to $2mv$) before the collision, but zero afterward.

This type of collision is called *completely inelastic;* the colliding objects do not bounce. At the other extreme would be the *completely elastic* collision, in which the two objects bounce back with the same speeds as before the collision, but in opposite directions. All the intermediate cases of partial elasticity may also occur in nature; but in each case the total momentum of the system is found to be conserved.

The mathematical statement of the law is: $\Sigma_i \boldsymbol{p}_i$ remains fixed from one moment to the next, even though the individual $\boldsymbol{p}_i$ may change as a result of collisions (elastic or inelastic) between all the particles in the system (labeled by the index $i = 1, 2, \ldots$).

Unhappily for the concept of the world machine, the new law of conservation of momentum does not by itself guarantee that the parts of the universe will keep on moving forever. We could imagine that the total momentum of all the parts was virtually zero even though initially most of the parts were moving in various directions. Unless all collisions of pieces of matter are completely elastic, the total quantity of "motion" (in Descartes' sense, defined as a scalar mv) will continually decrease, indicating the possibility that eventually the machine will grind to a halt, leaving all its parts at rest.

But as we shall soon see, an additional, quite different conservation law was found to hold as well, and it suggested that this is not the necessary fate. The first hint of the nature of that additional law was offered by Huygens in his rules for computing the effects of collisions. Huygens proposed that in the special case of perfectly elastic collisions, another quantity must be conserved in addition to momentum: the product (*mass*) × (*square of velocity*). In the seventeenth and eighteenth centuries this quantity was called *vis viva*, a Latin phrase meaning "living force":

$$\textit{vis viva} \equiv mv^2. \qquad (16.2)$$

The law of conservation of *vis viva*, it turned out, served to remedy—at least in the special case of elastic collisions—the first defect in Descartes' theory mentioned above, by providing a means for determining the outcome of a collision if the initial masses and velocities are known. Moreover, if one is willing to assume that *all* collisions are really elastic—if, for example, collisions between objects are ultimately collisions between elastic atoms—then the principle that the world machine must not "run down" might be rescued. But we must postpone this subject to the next chapter (and Chapter 22) and return to our discussion of the law of conservation of momentum.

16.3 Momentum and Newton's Laws of Motion

As we noted in Section 9.5, Newton originally stated his second law in the form: The change of momentum of a body is proportional to the motive (net) force impressed, and occurs in the direction of the straight line along which that force is impressed. In symbols,

$$\boldsymbol{F}_{\text{net}} \propto \Delta \boldsymbol{p} \quad (\text{where } \boldsymbol{p} = m\boldsymbol{v}).$$

(Remember that Δ, Greek delta, is not a quantity to be multiplied by $\boldsymbol{p}$ but means the *change* in $\boldsymbol{p}$. This change is itself a vector, which is not necessarily in the same direction as the original momentum $\boldsymbol{p}$.)

Although Newton did not explicitly include a time interval in stating the law—he was often dealing with impulsive forces that produce an almost instantaneous change of motion—we now find it preferable to write the law in the form

$$\boldsymbol{F}_{\text{net}} = \Delta \boldsymbol{p}/\Delta t, \qquad (16.3)$$

or, in words, "the net force on a body is equal to the change in momentum of the body divided by the time interval during which the net force is applied." If the force is not constant during Δt, this equation will apply only to the average force acting during that time interval. If we must deal with instantaneous force we must also take the value of $\Delta \boldsymbol{p}/\Delta t$ in the limit $\Delta t \to 0$; in that case the right-hand side becomes the rate of change of momentum with time.

It is not difficult to see that Eq. (16.3) is precisely equivalent to the more familiar form of Newton's second law given in Section 9.5. Replacing $\boldsymbol{p}$ by $m\boldsymbol{v}$, we have (since m is assumed constant)

$$\Delta \boldsymbol{p} = \Delta(m\boldsymbol{v}) = m\boldsymbol{v}_2 - m\boldsymbol{v}_1$$
$$= m(\boldsymbol{v}_2 - \boldsymbol{v}_1) = m\,\Delta \boldsymbol{v};$$

hence

$$\Delta \boldsymbol{p}/\Delta t = m(\Delta \boldsymbol{v}/\Delta t).$$

Recalling that acceleration was defined as

$$\boldsymbol{a} = \Delta \boldsymbol{v}/\Delta t$$

in Sections 6.5 and 7.4, we see that Eq. (16.3) reduces to

$$\boldsymbol{F}_{\text{net}} = m\boldsymbol{a}. \qquad (16.4)$$

Newton's second law, in the form of Eq. (16.3), together with Newton's third law, can now be used to *derive* the law of conservation of momentum. Suppose two bodies with masses m_A and m_B exert forces on each other, during a collision or in any other circumstances. We call $\boldsymbol{F}_{AB}$ the force exerted on body A by body B, and $\boldsymbol{F}_{BA}$ the force exerted on body B by body A. No other unbalanced force acts on either body, so the objects A and B can be considered an isolated system. By Newton's third law (Section 9.10), the forces $\boldsymbol{F}_{AB}$ and $\boldsymbol{F}_{BA}$ are equal in magnitude and opposite in direction:

$$\boldsymbol{F}_{AB} = -\boldsymbol{F}_{BA}. \qquad (16.5)$$

For any time interval Δt we can write the equivalent of Eq. (16.3) for each body (multiplying through on both sides by Δt):

$$\boldsymbol{F}_{AB}\Delta t = \Delta(m_A\boldsymbol{v}_A) \qquad (16.6a)$$

$$\boldsymbol{F}_{BA}\Delta t = \Delta(m_B\boldsymbol{v}_B). \qquad (16.6b)$$

But by Eq. (16.5), the left-hand side of Eq. (16.6a) and the left-hand side of Eq. (16.6b) are equal in magnitude and opposite in direction. Therefore the same must be true of the right-hand side, that is,

$$\Delta(m_A v_A) = -\Delta(m_B v_B). \qquad (16.7)$$

Let v_A and v_B stand for the velocities of the two bodies at the beginning of the time interval, and let v_A' and v_B' stand for their velocities at some later instant. Then (assuming that the masses m_A and m_B do not change), Eq. (16.7) becomes

$$m_A v_A' - m_A v_A = -(m_B v_B' - m_B v_B). \qquad (16.8)$$

Collecting the primed symbols on the right side, we find

$$m_A v_A + m_B v_B = m_A v_A' + m_B v_B'. \qquad (16.9)$$

This is just the law of conservation of momentum, applied to the closed system in which the two bodies interact, since the left-hand side of Eq. (16.9) is the total momentum of A and B at the beginning of the time interval, and the right-hand side is the total momentum at the end of the interval.

In general, with any number of bodies in a closed system, the law of conservation of momentum can be written

$$\Sigma \boldsymbol{p} \text{ before event} = \Sigma \boldsymbol{p} \text{ after event},$$

where Σ stands for vectorial summation of the individual momentum vectors.

Since the law of conservation of momentum can be derived from Newton's laws, it would appear that it is not really an independent physical law. On the other hand, it cannot be used by itself to derive Newton's second and third laws, since it does not by itself provide us with any concept of force. Would it not be preferable, then, to stick to Newton's three laws in an introductory general account such as this one, and leave the use of the law of conservation of momentum to more advanced courses dealing with special problems in mechanics?

No! We must not let the impressive structure of Newtonian physics delude us into thinking that all physical situations should or even can be analyzed in terms of forces and accelerations. Even in the most elementary collision problems, such as those to be discussed in the following section, it would be extremely difficult to measure the instantaneous forces accurately enough to permit a direct application of Newton's laws; yet the law of conservation of momentum gives us solutions simply. Later, when we discuss the interactions of subatomic particles, we shall see that instantaneous positions, velocities, and accelerations are meaningless, so that the moment-by-moment test of Newton's laws is not feasible; yet the law of conservation of momentum is still applicable. Moreover, the momentum concept will still be found to be useful even when other concepts such as mass lose their ordinary meaning, as in the case of light rays.

16.4 Examples Involving Collisions

Our first example illustrates how a law of conservation of quantity of motion was suspected even before Descartes' explicit statement of it in 1644.

EXAMPLE 1

The top part of Fig. 16.1, after Marcus Marci's *De Proportione Motus* (1639), may be taken to present the following (idealized) sequence of observable events in one diagram: A ball lies quietly on a table. A second ball, of equal mass, is shot horizontally toward the first, collides with it, and remains lying on the table while the other moves off with the same speed with which it was hit.

We may explain these observations (so well known to everyone acquainted with the games of croquet or billiards) directly from Newton's laws. For during the brief time interval Δt of contact between the balls, the mutual forces, by the third law, are equally large though oppositely directed; therefore the decelerating (stopping) force on sphere A is equal to the accelerating force on B; and because the two masses involved are equal, the actual deceleration of one is as large as the acceleration of the other. B gains as much speed as A loses; if A loses it all, B gains it all: $v_B' = v_A$.

But now we shall repeat this argument in terms of the law of conservation of momentum: Given are $m_A = m_B$; v_A; $v_B = 0$; $v_A' = 0$. Required: $v_B' = ?$

Solution: These two spheres form a closed system, for only the forces of interaction

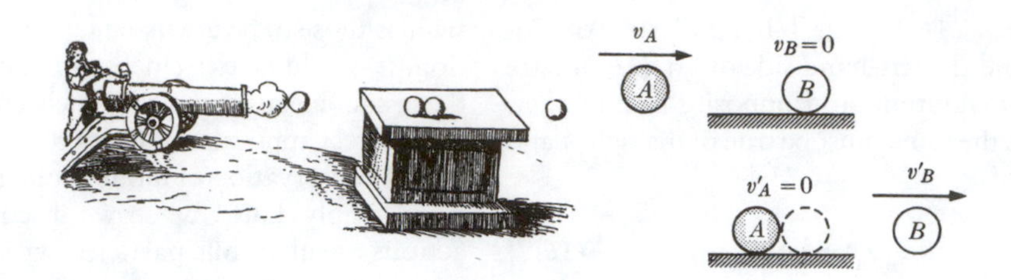

Fig. 16.1. A cannon ball collides with a stationary sphere of equal mass.

between them affect the phenomena. The law states that $\Sigma\mathbf{p}$ before collision = $\Sigma\mathbf{p}$ after collision, so

$$m_A\mathbf{v}_A + m_B\mathbf{v}_B = m_A\mathbf{v}_A' + m_B\mathbf{v}_B'. \quad (16.10)$$

(Here all vectors lie along the same line and in the same direction. Therefore vector addition becomes here just algebraic addition.). On substitution from the data, we get

$$m_A\mathbf{v}_A + m_B \times 0 = m_A \times 0 + m_B\mathbf{v}_B',$$

$$\mathbf{v}_B' = (m_A/m_B)\mathbf{v}_A.$$

But $m_A = m_B$; therefore

$$\mathbf{v}_B' = \mathbf{v}_A. \qquad \text{Q.E.D.}$$

Problem 16.1. In the more general case, the masses may not be equal, but then $\mathbf{v}_A'$, will not be zero. Nevertheless, the same type of calculation will apply. Show that then

$$\mathbf{v}_B' = (m_A/m_B)(\mathbf{v}_A - \mathbf{v}_A') + \mathbf{v}_B.$$

Note that this general case reduces to $\mathbf{v}_B' = \mathbf{v}_A$ if the conditions of the masses, etc. are as in Example 1.

EXAMPLE 2

The previous example represented a perfectly elastic collision. Now note what happens in a perfectly inelastic collision, that is, when the colliding bodies do not separate afterward but are wedged into each other and travel off together. For example, a car (mass m_A) goes at speed v_A, and tries to overtake a truck (m_B) going in the same direction at speed v_B. There is a misjudgment, and a collision. The car is embedded in the rear of the truck. What is their common speed just after collision?

Solution: The data are m_A; m_B; v_A; v_B; and $v_A' = v_B' = ?$ Here Eq. (16.10) becomes

$$m_Av_A + m_Bv_B = (m_A + m_B)v_B',$$

$$v_B' = (m_Av_A + m_Bv_B)/(m_A + m_B).$$

EXAMPLE 3

When interacting bodies, instead of moving in the same sense, approach each other with oppositely directed momenta or, in fact, at any angle whatever, then the vector nature of momenta must be more carefully considered than we have done in the last two examples. Let two spheres, moving oppositely hit; perhaps they are two pendulums released from different heights, meeting at the lowest point of their swing (Fig. 16.2). The collision is perfectly inelastic, that is, they interlock; now we ask: $\mathbf{v}_A' = \mathbf{v}_B' = ?$

Just as we are used to finding a net force by subtracting opposing tendencies, so it is clear also that here the sum of the momenta just prior to collision is obtained by subtracting the numerical value of m_Bv_B from that of m_Av_A. We still write the vector equation as before:

$$m_A\mathbf{v}_A + m_B\mathbf{v}_B = (m_A + m_B)\mathbf{v}_A',$$

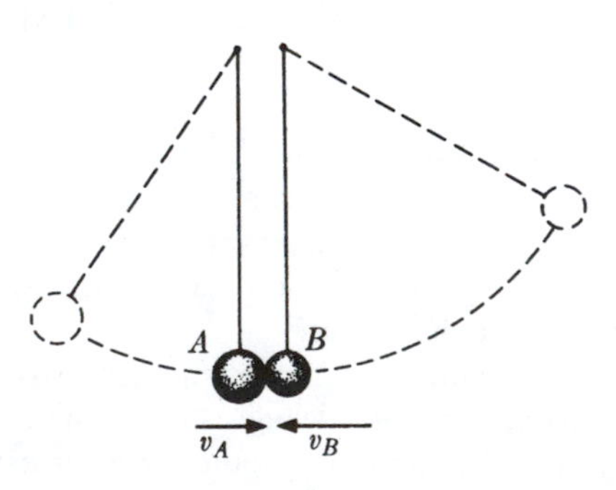

Fig. 16.2. Collision of pendulums of equal mass.

$$v_A' = (m_A v_A + m_B v_B)/(m_A + m_B). \quad (16.11)$$

But when we now substitute values for v_A and v_B, we must take care, as before, to assign positive va1ues for velocities or momenta toward, say, the right direction, negative ones to the opposite direction. If $m_A = 3$ g, $m_B = 2$ g, $v_A = 8$ cm/sec to the right, $v_B = 15$ cm/sec to the 1eft, then from Eq. (16.11),

$$v_A' = [(3 \text{ g})(8 \text{ cm/sec}) + (2 \text{ g})(-15 \text{ cm/sec})] /(3 + 2) \text{ g} = -1.2 \text{ cm/sec}$$

and the negative sign of the result informs us automatically that after collision both spheres move off together to the *left*. (This was, incidentally, just the type of experimentation by which scientists at the Royal Society established the validity of the law of conservation of momentum in 1666–1668.)

16.5 Examples Involving Explosions

In Chapter 9 we discussed the reaction-car experiment, useful for the determination of relative inertial masses. This, too, was a hidden example of the law of conservation of momentum: It was assumed that the only unbalanced forces acting on the two cars was the mechanical push on each by way of the spring. This symbolizes well what happens in an explosion (Fig. 16.3). Initially all parts of the system are at rest, the momenta are zero; then mutually opposite forces separate the members of the system. We write therefore: From

$$(p_A + p_B)_{\text{before event}} = (p_A + p_B)_{\text{after event}}$$

follows for this case

$$0 = m_A v_A' + m_B v_B',$$

$$m_A v_A' = - m_B v_B', \quad \text{or} \quad m_B/m_A = - v_A'/v_B'.$$

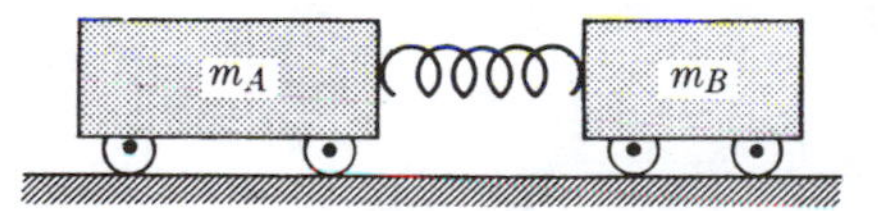

Fig. 16.3. The reaction-car experiment.

Thus we have derived the same relationship used to determine the mass m_B in terms of a standard mass m_A and observed velocities v_A' and v_B'.

Problem 16.2. (a) An old-fashioned, wheel-mounted 500-kg gun, initially at rest, fires a 5-kg projectile horizontally at 1000 m/sec. If the wheels are freely turning, what will be the speed of recoil of the gun? (b) If the gun is, however, firmly anchored to the earth, what will be the recoil speed of gun and earth together ($m_{\text{earth}} = 6 \times 10^{27}$ g)? (c) If the barrel of the gun is 2 m long and the projectile is accelerated uniformly inside from rest to muzzle speed, what was the average force on the projectile inside the barrel? (d) What was the average force exerted at the same time on the gun by the expanding gases of combustion in the barrel?

16.6 Further Examples

A great variety of other examples may be adduced to illustrate further the extent and power of the law of conservation of momentum.

EXAMPLE 4

Simply consider a man starting out on a walk. Before he begins, he and the earth are together at rest (as far as this problem is concerned). As soon as he moves forward, the earth below him must move backward, since $p_{\text{man}} + p_{\text{earth}} = 0$, and so $(mv)_{\text{man}} = (mv)_{\text{earth}}$. To be sure, the velocity change for the earth is, as shown by calculation, extremely small, because m_{earth} is so much larger than m_{man}. On the other hand, if the surface on which the man walks is only a plank on ice, the relative velocities will be more nearly equal. The point, however, is the same in both cases: The force that "couples" the earth or the plank into the system of moving bodies in such a problem is *friction*.

As a force of interaction, friction differs in no way from other forces. In those previous examples that involved pendulums and reaction cars, the forces of interaction were elastic forces; in those involving recoiling guns and projectiles, they were forces set up by gases under pressure. The presence of friction—which so far we have all too often had to consider negligible—does not reduce the applicability of the law of conservation of momentum; on the contrary, it extends it by widening any system of interacting bodies to include all those bodies that experience mutual forces of friction. If, for instance, friction under the wheels of the reaction cars had not been negligible, we should have written, instead of the first equation in Section 16.5,

$$0 \text{ at start} = [(m_A v_A' + m_B v_B') + m_{\text{earth}} v_{\text{earth}}] \text{ after separation.}$$

Although this may complicate the calculation, it is still a perfectly proper application of the law of conservation of momentum.

Problem 16.3. Assume that there exists a flat interstellar platform of mass 1000 kg, at rest as seen from some nearby star, and inhabited by a 70-kg man. Now with our supertelescope we see (from the star) that in 3 sec he walks a distance of 5 m, uniformly along a straight line. How far will the observed man think he walked (by his measurement along the platform)? Is this result dependent on how fast he walked? Explain. Describe the motion of the platform while he begins to walk, proceeds, and stops.

EXAMPLE 5

Lastly, we may mention the role of another force of interaction in the operation of our law: mutual attraction, whether gravitational, electric, or magnetic. If we apply the conservation law to a system consisting of an apple and the earth that attracts it, their $\Sigma \boldsymbol{p}$ before the apple's fall is zero. While the apple falls, by Newton's third law it attracts the earth upward with a force as large as that by which it is being pulled down, and the law of conservation of momentum demands that

$$\boldsymbol{p}_{\text{apple}} + \boldsymbol{p}_{\text{earth}} = 0.$$

Apple and earth will meet at some intermediate point that divides the total distance of initial separation in the ratio of the two masses. (Prove as a problem.)

In fact, it is this very type of problem in which we may find one of the earliest appreciations of the law of conservation of momentum. In 1609, Kepler wrote in his *New Astronomy:*

> If two stones were removed to any part of the world, near each other but outside the field of force of a third related body, then the two stones, like two magnetic bodies, would come together at some intermediate place, each approaching the other through a distance proportional to the mass of the other.

16.7 Does Light Have Momentum?

Another historic application of the law of conservation of momentum, though from twentieth-century physics, involves the interaction between a beam of negatively charged electrons and positively charged atomic nuclei. When a high-speed electron passes a nearby nucleus, the mutual forces of electrical attraction can strongly deflect the former from its path in exactly the same manner as the sun affects a passing comet (Fig. 16.4). It would be expected that the change of momentum of the electron, $\Delta \boldsymbol{p}_{\text{el}}$, is compensated exactly by a corresponding and opposite $\Delta \boldsymbol{p}_{\text{nucleus}}$. On checking experimentally, we find that the latter change is too small to account for $\Delta \boldsymbol{p}_{\text{el}}$. Is this not evidence that here at last the treasured law breaks down?

We should be loath to accept this conclusion without looking first for an explanation elsewhere. We remember well that the irregularities in the motion of the planet Uranus also offered such an alternative: either to consider a failure of the old law (of universal gravitation), or else to search for an explanation of the puzzling behavior by a bold hypothesis beyond the immediately observed facts; and it was the latter course that triumphantly proved to be correct and led to the discovery of Neptune.

Here also: Does nothing else happen that might carry away some momentum? In fact, during the deflection of the electron a flash of light is released in the general direction indicated,

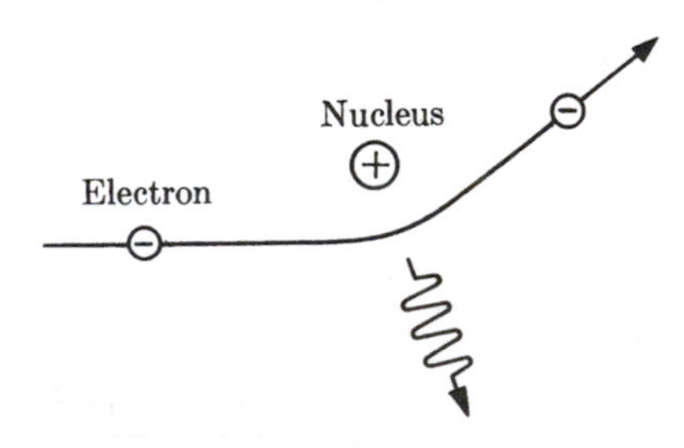

Fig. 16.4. Radiation of energy from an electron passing near a nucleus.

although of a frequency so far beyond the visible that instruments other than the eye are needed to detect it. If we could say that light has momentum, and assign it the missing quantity, all would be well and the law saved. But surely that is not so easily permissible! How can one conceive of momentum, (*mass*) × (*velocity*), where there is nothing but disembodied wave motion—radiant energy?

While some of the detailed arguments cannot be given here and others must be delayed to Part H of this text, the main facts are these:

i) One can no longer afford to reject any concept (such as the idea that light waves have momentum) just because it is not directly obvious to our mechanics-oriented thoughts. It is a dangerous limitation to insist on physical pictures. Nature has too often in the past forced us to accept what, for a time, could not possibly be imagined. After all, there was a time when it was nearly universally considered ridiculous and obviously absurd to propose that the earth might be considered to move or that the moon has mountains. Thus although one might, instead of the phrase "light has momentum," prefer to say "momentum is somehow associated with light waves," one must indeed say so if there is inescapable evidence that this is true.

ii) Apart from the experiment with high-speed electrons as just described, there was other evidence that light exerts pressure and can produce changes of momentum, most spectacularly the observation that the material in the tail of comets always points away from the sun at every point along the comet's path. The suggestion that it was the sun's light that pushed the comet's tail was first made by none other than—again—Kepler. Modern and most delicate measurements around 1900 showed clearly that light waves bring with them momentum, although the quantities are ordinarily small; a good beam can do no more than keep up the rotation of a little paddlewheel mounted in a vacuum with as little friction as possible. Moreover, and most importantly, the amount of momentum in a given light beam can be predicted from simple assumptions in the theory of light, and does indeed coincide with these measured values to better than 1%, despite the serious difficulties in experimentation. (More recently, it has been concluded that the greater part of the force acting to push a comet's tail away from the sun is due not to light pressure but to the solar wind, consisting of particles such as protons.)

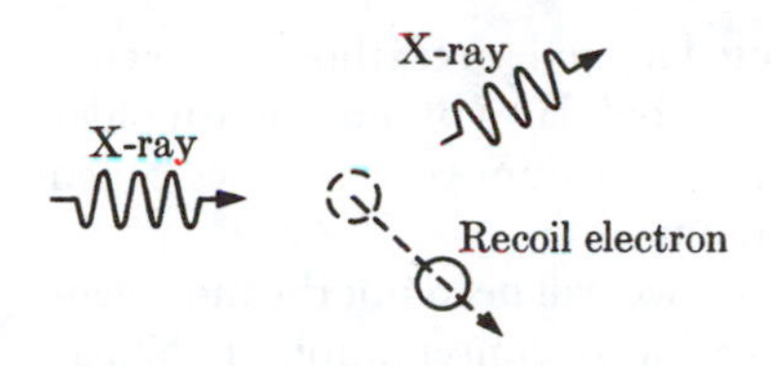

Fig. 16.5. Compton effect. The x-ray proceeds with reduced energy and longer wavelength after scattering from a recoiling electron.

iii) The concept that a flash of light waves has associated with it a momentum became important subsequently in several other independent contexts; for example, in 1923 the American physicist A. H. Compton discovered that a beam of x-rays (also a form of very high-frequency light) can collide with an electron, and bounce off with slightly changed characteristics in another direction while the electron "recoils" (Fig. 16.5) as though it were one billiard ball hit by another. Compton found that the observable momentum of the recoil electron was numerically just equal to the change of momentum of the x-ray, as calculated by the same relation needed to fulfill the conservation law in Fig. 16.4. Once more the conservation law was borne out, and at the same time the nature of light itself became better understood.[2]

16.8 Angular Momentum

The law of conservation of momentum is usually applied to the motion of objects in straight lines (before and after collisions). Another law, "conservation of angular momentum," governs the motion of objects, in curved or straight lines, with respect to a point. It was first used in a paper by the Swiss mathematician James Bernoulli (1654–1705) in a paper on pendulum motion in 1703, and was stated as a general principle by another Swiss mathematician, Leonhard Euler, in 1744. (Recall that it was Euler who first stated Newton's second law of motion in its modern form $F = ma$.) Bernoulli and Euler considered it to be an independent principle of mechanics, but it is

[2]According to Einstein's theory of relativity (Chapter 30), a photon has momentum even though it has no mass. Instead of the usual $\boldsymbol{p} = m\boldsymbol{v}$, he showed that the momentum is a vector in the direction of the photon's motion, with magnitude E/c, where E is the energy and c is the speed of light.

usually presented in modern textbooks as a consequence of Newton's laws of motion. (In older books angular momentum is sometimes called "moment of momentum.")

In this book we will need not the most general form of the law of conservation of angular momentum, only the simpler version that applies to circular motion. (We note, however, that Kepler's second law can be considered a consequence of the law of conservation of angular momentum applied to the motion of a point-mass in an elliptical orbit.)

We define the angular momentum l of an object with mass m moving with linear speed v in a circle of radius R to be

$$l \equiv mvR. \tag{16.12}$$

It can also be written in terms of the angular velocity ω (see Section 10.1):

$$l = mR^2\omega. \tag{16.13}$$

For an object of finite size, the different parts would in general be at different distances from the center. Consider an object that is rigid, so that each part has the same angular velocity; one can then calculate the total angular momentum from Eq. (16.13) by adding up the angular momenta for the different parts with their values of m and R.

If we now reduce the size of the object so that the values of R for all its parts decrease, without exerting any external force that could affect the rotation, Eq. (16.13) tells us that it must spin faster (ω increases) in order to keep the same total angular momentum; conversely if the average value of R increases, the object must spin more slowly.

The law of conservation of angular momentum in this simple version can easily be demonstrated without expensive equipment or precise measurements. In a standard classroom demonstration, a person stands on a small platform that can rotate freely with very little friction. If he is given an initial rotational motion, he will continue to spin at the same rate as long as he hold his arms in the same position; but by spreading them out (increasing the effective value of R) he can slow down the rate of spin, while by bringing them closer to his body he can spin faster.

For most people a more familiar example is the twirling ice skater. The skater can control her rate of spin by spreading her arms out to go more slowly or bringing them in to go faster.

Laplace's Nebular Hypothesis. In the development of physical science, the law of conservation of angular momentum found perhaps its most important application in the theory of the origin of the solar system proposed in 1796 by the French astronomer Pierre Simon de Laplace (1749–1827). Laplace postulated that the solar system began as a whirling spherical cloud or "nebula" of hot gas, as if the sun's atmosphere were extended beyond the present orbit of the then recently discovered planet Uranus. As the cloud cooled, it shrank to a smaller size, while at the same time becoming an oblate spheroid (see Section 11.6) and eventually a flattened disk like a pancake. Assuming that the nebula behaved like a rigid body so that all its parts had about the same angular velocity at any given instant of time, Laplace argued that it must spin faster as it shrinks.

Eventually, Laplace pointed out, the linear speed v of the outermost parts of the nebula would be so high that the centripetal acceleration $a = v^2/R$ needed to keep them moving in a circle would exceed the gravitational force of the inner parts of the nebula. At that point a ring of material at the edge of the disk would be "spun off"—it would continue to rotate at the same distance, while the remaining nebula would continue to cool and shrink. This process would be repeated several times, leaving a central mass (to become our sun) surrounded by several rings. The gas in each ring would, according to Laplace, later condense into a single sphere, which would continue to cool, condense, and eventually solidify into a planet. The larger planets would undergo a similar shrinking process, leading (by conservation of angular momentum) to increasingly fast rotation and spin-off of small rings that eventually would become satellites.

In Chapter 31 we will follow the fortunes of the nebular hypothesis during the following two centuries and its transformation into current ideas about the origin of the solar system.

RECOMMENDED FOR FURTHER READING

Herbert Butterfield, *Origins of Modern Science,* Chapter I.

A. H. Compton, "The scattering of X-rays as particles," in *Physics History* (edited by M. N. Phillips), pages 105–108

Richard P. Feynman, *The Character of Physical Law,* Chapter 4

S. Sambursky, *Physical Thought,* selections from Descartes and Laplace, pages 246, 349–355

The Law of Conservation of Energy

17.1 Christiaan Huygens and the Kinetic Energy (*Vis Viva*) Concept

The so-called fundamental quantities of mechanics are usually taken to be length, time, and mass. So far, we have used combinations of these three to set up derived concepts: velocity, acceleration, force, and momentum. And each of these has found a prominent role in some fundamental law of nature. Now we are about to introduce the last of the great mechanical concepts, energy, together with the law within which it reigns, the mighty principle of conservation of energy.

The full historical development of the concept and the law was lengthy and more involved than usual; it took more than 150 years from the first attempts at quantitative formulation to the point at which even the basic terminology itself was fully worked out. Yet the roots lie in the very same problem that gave rise to the concept of momentum and the law of conservation of momentum, namely, the problem of how the motion of bodies changes when they collide with one another. And significantly, in introducing a first major contribution, we encounter the same person who helped formulate the law of conservation of momentum in elastic impact: Christiaan Huygens (1629–1695), the Dutch physicist, who was in many respects the peer of Galileo and Newton.

Huygens was not only an outstanding mathematician (Newton referred to him as one of the three "greatest geometers of our time"); he also built an improved telescope, discovered a satellite of Saturn, and, building on Galileo's work, invented the first practical pendulum clock. Among his notable accomplishments in physics are the formula for centripetal acceleration (Section 10.2); the conservation laws in elastic impact; the theory of oscillating systems; and a treatise that laid the foundations for the wave theory of light (Section 23.1).

Fig. 17.1. Christiaan Huygens (1629–1695).

As we remarked at the end of Section 16.2, Huygens proposed as part of his solution of the collision problem, in 1669, the rule that a quantity equal to the product of the mass and the square of the speed of each moving body, summed over all the bodies in a system, remains the same after a perfectly elastic collision as before the collision. The quantity mv^2 was later given the name *vis viva* and was used as the basis for mechanical theories by the German philosopher and scientist Gottfried Wilhelm Leibniz and in other works by Huygens published around 1700 (especially in his posthumously published treatise on the motions and collisions of bodies, 1703). With a factor of ½ (to be explained in Section 17.4) it is our modern "kinetic energy."

We may discuss this postulate in terms of a concrete but general example, as in Fig. 17.2. Two unequal but perfectly rigid spheres approach each other with unequal speed, collide, and separate. To be sure, from our previous discussion we know that the law of conservation of momentum demands that

$$m_A v_A + m_B v_B = m_A v_A' + m_B v_B', \qquad (17.1)$$

but this is not the issue any longer. What Huygens is proposing here is that there is fulfilled at the same time *another* equation, namely,

$$m_A v_A^2 + m_B v_B^2 = m_A (v_A')^2 + m_B (v_B')^2. \qquad (17.2)$$

This is really a very astonishing proposal! Let there be no mistake about it—Eq. (17.1) does not by itself lead to Eq. (17.2). In the latter case the squares of the speeds are involved, not the velocity vectors themselves. Also, Eq. (17.2) is a scalar equation; that is, the directions of motion do not enter at all. Consequently, it holds true in this form even if body *A* were to overtake body *B*, or if it were overtaken by *B*, or if *A* and *B* were to meet at any angle whatsoever. A couple of examples will illustrate the usefulness of the new law.

EXAMPLE 1

Recall Fig. 1 in Chapter 16. A ball *A* of mass m_A was shot at velocity v_A toward a ball *B* of equal mass ($m_B = m_A$) resting on a table ($v_B = 0$). We inquired: $v_B' = ?$ With the law of conservation of momentum alone, this could not be solved unless we added to the four data a fifth, namely, $v_A' = 0$, the reason being that the momentum conservation law alone, applied to two bodies, gives only one equation [Eq. (16.10)] with *six* quantities:

$$m_A v_A + m_B v_B = m_A v_A' + m_B v_B'.$$

But if we also call on Eq. (17.2),

$$m_A v^2{}_A + m_B v^2{}_B = m_A (v_A')^2 + m_B (v_B')^2$$

we have two equations in six quantities; therefore we need only four data to solve for any two unknowns. For example, using what is given above concerning the two masses and the initial velocities, we may calculate directly the values of $(v_A')^2$ and $(v_B')^2$

$$\left.\begin{array}{l}\text{From Eq. (16.10): } v_A + 0 = v_A' + v_B'.\\ \text{From Eq. (17.2): } v_A^2 + 0 = (v_A')^2 + (v_B')^2.\end{array}\right\} (17.3)$$

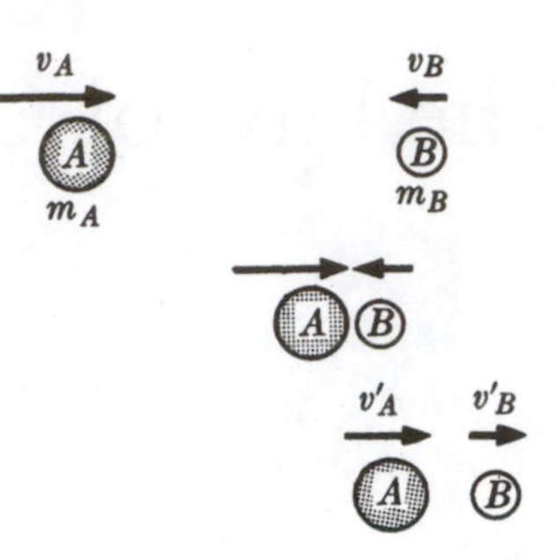

Fig. 17.2. Collision between two moving spheres.

Squaring the first and subtracting yields

$$0 = 2v_A' v_B'.$$

Therefore either v_A' or v_B' is zero, and the other one of the two final velocities, by Eq. (17.3), is equal to v_A. But if ball *B* is solid, v_B' cannot be zero while ball *A* passes forward with $v_A' = v_A$. Consequently, the other alternative is the correct one, namely, $v_A' = 0$, $v_B' = v_A$. Now this is a much more satisfactory solution, because we can calculate *both* final velocities exclusively from data that refer only to the initial conditions before impact.

This applies also in general for any problem involving direct collision of two perfectly elastic objects. With the aid of the law of conservation of momentum and Eq. (17.2), we can find the final velocities of both bodies from the initial conditions without further assumptions or data.

Problem 17.1. In the posthumous edition of Huygens' works (1703) we find this "Axiom": "If two equal bodies [bodies of equal mass] moving in opposite directions with equal velocities meet each other directly [in perfectly elastic collision] each of them rebounds with the same velocity with which it came." Show that this axiom is in accord (a) with the law of conservation of momentum, and (b) with the principle of conservation of *vis viva*.

EXAMPLE 2

In a later chapter we shall need to use the answer to the following question: A very light object strikes a very heavy object, the latter being initially at rest. What are the velocities of both objects after the collision, given the initial velocity of the light object? (The collision is still assumed to be perfectly elastic.)

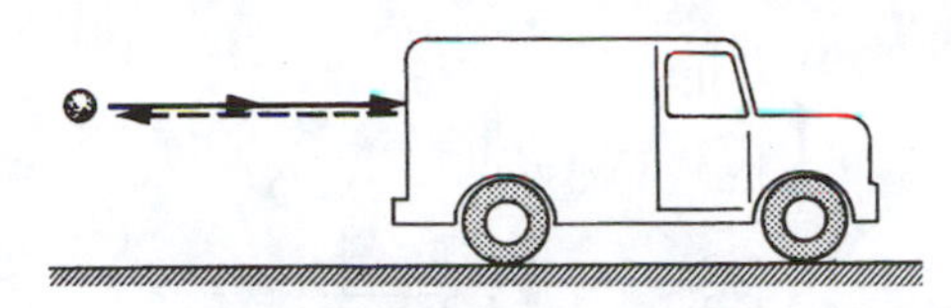

Figure 17.3. Elastic collision of a very light object with a very heavy object.

For the sake of concreteness, let us imagine that the light object is a rubber ball of mass m_A and initial velocity v_A, and that the heavy object is a large truck of mass m_B; the ball is thrown horizontally against the flat (vertical) back door of the truck, which is initially at rest ($v_B = 0$). If we ignore the force of gravity, there will be no reason for the ball to deviate in any direction from the line of its original path; it can only bounce back with some velocity v'_A, which we have to calculate. Hence this is just a one-dimensional problem.

You can probably guess the correct answer, but it is instructive to see how it emerges from the two conservation laws. To start with, we have to define what is meant by "very heavy" and "very light." Clearly we must have in mind a limiting case $m_B/m_A \to \infty$, but if we make that substitution immediately the equations contain indeterminate quantities and cannot be solved. So let us be patient and begin by putting the ratio $m_B/m_A = n$, solve for a finite n, and then take the limit $n \to \infty$ at the end.

The same procedure as that used above now gives, as a result of substituting into Eqs. (16.10) and (17.2), and dividing through by m_A:

$$v_A + 0 = v'_A + nv'_B \qquad (17.4a)$$

$$v_A{}^2 + 0 = (v'_A)^2 + n(v'_B)^2. \qquad (17.4b)$$

Squaring the first and subtracting yields

$$0 = 2v'_Av'_B + (n-1)(v'_B)^2.$$

As in the previous case, one solution of this equation is $v'_B = 0$, which must be rejected for the same reason as before, although, as we shall see shortly, the correct solution actually converges to that value in the limit $n \to 0$. But for finite n we must choose the other solution, obtained by dividing through by v'_B,

$$2v'_A + (n-1)v'_B = 0$$

or

$$v'_A = -\tfrac{1}{2}(n-1)v'_B$$

We can then eliminate v'_B by substituting its value given in terms of v_A and v'_A in Eq. (17.4a), namely,

$$v'_B = (v_A - v'_A)/n.$$

Solving for v'_A, we find:

$$v'_A = -\frac{(n-1)}{(n+1)}\, v_A. \qquad (17.5)$$

At last we can take the limit $n \to \infty$ and obtain the desired answer,

$$v'_A = -v_A$$

The ball simply bounces back from the enormously heavier truck with the same speed in the opposite direction! To complete the solution, we note that (by solving for v'_B from the above equations) the truck is still at rest after the collision.

Problem 17.2. Using the result of Example 2, show that: (a) If the truck were initially moving away from the ball, the ball would bounce back with a smaller speed. (b) If the truck were initially moving toward the ball, the ball would bounce back with a greater speed. [*Hint:* Redefine the velocities as relative to the truck's initial velocity v_B, for example, let $v_B{}^* = v_A - v_B$. You are now in a coordinate system in which the truck is at rest. After solving the problem in that coordinate system, translate the results into the original coordinate system and interpret them.)

The usefulness of the principle of conservation of *vis viva*, as demonstrated, is considerable, but *unlike* the law of the conservation of momentum *it is strictly limited to impacts between perfectly elastic bodies.* In truth, the matter is even more serious, for it is in general not possible to judge before the collision takes place whether it will be sufficiently close to a perfectly elastic one to allow use of the principle; and so we often decide whether the principle applied only after a tentative calculation has shown whether the theoretical results are reasonable. This *a posteriori* (retrospective) procedure is rather like diagnosing whether a sick child has malaria by seeing whether a malaria cure will put the child on her feet again.

Evidently what is needed is an extension of the principle of conservation of *vis viva* in some manner so that the extended principle will apply to all types of interactions, just as the law of conservation of momentum applies to all interactions in a closed system. This was brilliantly accomplished in the nineteenth century and, leaving the main details of the historical development for later, we turn first to an account of the results.

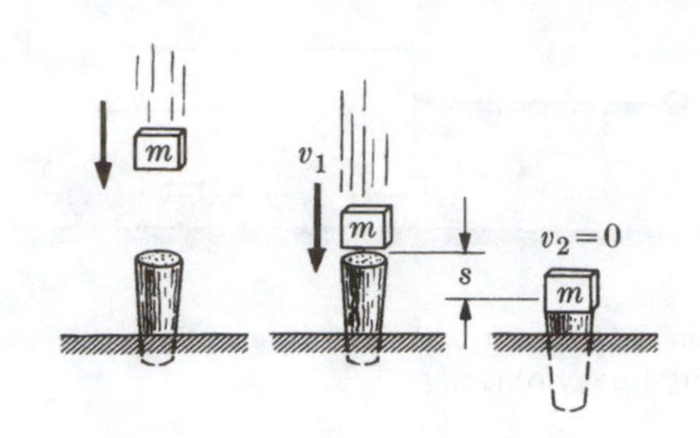

Fig. 17.4. Pile driver.

17.2 Preliminary Questions: The Pile Driver

Consider a body in motion, perhaps a block sliding on a horizontal plane. If you tried to stop it quickly by interposing your hand in its path, you would feel that the block exerts a force on you, perhaps a considerable one. Is this some new force we have not yet discussed? Do bodies exert forces on obstacles *by virtue of having been in motion,* as one might say springs exhibit forces by virtue of being compressed and magnets by virtue of having inherent magnetism?

First of all, note that this common-sensical and picturesque way of speech, which mirrors the phraseology of the seventeenth and early eighteenth century, explains little and can lead to serious semantic confusions. In the last analysis, as Newton was aware, it only serves to explain magnetism by magnetism, elasticity by elasticity, and so forth. Therefore, instead of speaking of an inherent force, we should rather seek the relationship between the force under examination and its observable effects. If we look in this light at the force that moving objects exert on an obstacle like our hand, we see at once that this is not some new "living force," as was sometimes thought, but rather that our hand experiences the force of reaction to the decelerating or braking force that, by Newton's second law, it must apply to the moving object to change its state of motion. This force of deceleration, and also the reaction to it, are of course both given numerically by the product of the mass m of the object and the deceleration a. And that is, strictly speaking, all there is to the original problem.

But it is very revealing to embellish this theme. For example, it has always been of interest to measure with some accuracy the speed of a moving object. Can this not be done by measuring the force needed to stop it? Galileo put the problem this way: Consider a falling body, such as a heavy sledge; after it has gathered some speed, let it fall upon a vertical pole, lightly impaled in some yielding sand (an arrangement that corresponds to the so-called pile driver). Then:

> It is obvious that if you now raise it one or two braccia and then it let fall on the same material, it will make a new pressure on impact, greater than it made by its weight alone. This effect will be caused by the [weight of the] falling body in conjunction with the speed gained in the fall, and will be greater and greater according as the height is greater from which the impact is made; that is, according as the speed of the striking body is greater. The amount of speed of a falling body, then, we can estimate without error from the quality and quantity of its impact.
>
> But tell me, gentlemen: if you let a sledge fall on a pole from a height of four braccia, and it drives it, say, four inches into the ground, and will drive it much less from a height of two braccia, and still less from a height of one, and less yet from a span only; if finally it is raised but a single inch, how much more will it accomplish than if it were placed on top [of the pole] without striking it at all? Certainly very little.
>
> (*Two New Sciences,* Third Day)

Indeed, are we not thus enabled accurately to estimate the speed of the falling body? Not so simply. By stopping the body, we change its momentum by mv (numerically). From Newton's laws,

$$mv = F_1 t, \quad \text{and} \quad v = F_1 t/m,$$

where F_1 is the (average) net force that decelerates the body (and numerically also gives the force of reaction that in turn pushes on the stake), t is the time during which the stake is in motion (starting from rest), and m is the mass of the falling body. Unless we know the factor t (hardly observable with ease), the force F_1 on the stake gives little information as to v. If the sand is tightly packed, t will be very small and F_1 very large; if the soil is loose, t is greater and F_1 is smaller for the same values of v and m.

But now, what can we learn about v from the distance s through which the stake moves while decelerating the pile driver from v_1 to $v_2 = 0$ (Fig. 17.4)? Let us assume that the retarding force F_1 on the falling mass was constant during the collision. Then, since

$$F_1 = ma,$$

$$F_1 \times s = mas.$$

But, from Eq. (6.7),

$$as = \tfrac{1}{2}(v_2^2 - v_1^2),$$

therefore

$$F_1 \times s = \tfrac{1}{2}mv_2^2 - \tfrac{1}{2}mv_1^2.$$

Since $v_2 = 0$,

$$F_1 \times s = \tfrac{1}{2}mv_1^2 \text{ (numerically).}$$

It appears now that for a given mass m, and for an obstacle that always offers the same retarding force (a condition nearly enough fulfilled by always using the same stake in the same soil), the distance through which the stake is moved is proportional to the *vis viva*, that is, to the square of the speed of the moving body!

Using Galileo's example, and assuming that 1 braccia ≈ 0.50 m and 1 finger-breadth ≈ 0.02 m, we should rephrase it thus: If a block is allowed to fall on a stake from a height of 2 m (with a final velocity v of $\sqrt{(2gh)}$ = 6.2 m/sec), and drives the stake into the earth a distance s of, say, 0.08 m, the same mass coming from a height of 1 m (with a final velocity v' of 4.4 m/sec) will drive the same stake a smaller distance s', namely (in the ideal case),

$$s'/s = (v'/v)^2, \quad \text{or}$$

$$s' = 0.08 \times (4.4/6.2)^2 \text{ m} = 0.04 \text{ m}.$$

It is only by such a calculation, and under such assumptions, that Galileo's pile driver permits a comparison of relative speeds.

17.3 The Concept of Work

The previous example, to which we shall return, has been chosen for several reasons, not the least being that it introduces, in the historical context, two concepts: work and energy. The definition of the first of these is as follows.

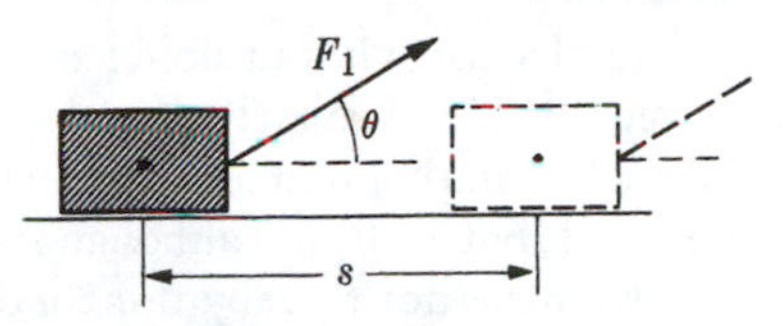

Fig. 17.5. The work done by force F_1 on the block is $F_1 s \cos\theta$.

When a force F_1 acts on a body during a displacement s, then we say that *the product* $F_1 \times s \times \cos\theta$ *is the work done by* F_1 *on the body,* where θ is the angle between the directions of the two vectors $\boldsymbol{F}_1$ and $\boldsymbol{s}$. Thus in the previous discussion, the work done by the force F_1, the action of the pile driver on the stake, was $F_1 \times s$, cos θ being unity in that case (why?). At the same time, the work done by the stake on the falling mass was $F_1 \times s \times (-1)$ or $-F_1 \times s$ (because the displacement s and the force F_1 were oppositely directed, and cos 180° = −1).

Or again, when a person pulls on a block in the manner indicated in Fig. 17.5, exerting a pull F_1 of about 45 N at an angle of 30° over a distance of about 0.3 m, the work done by the person on the block is 45 N × 0.3 m × 0.87 = 11.7 N·m. On the other hand, if a boy carries a load on his shoulders over a distance s on a horizontal plane and supplies to it a balancing force F_1 (upward), the work he does, as defined in physics, is zero; for here F and s may both be large, but cos 90° = 0 (Fig. 17.6). Again, the centripetal force F_c on an object in circular motion (Fig. 10.4) does no work either, for it is directed at 90° to the path at every instant.

Note carefully that, as in the example illustrated in Fig. 17.6, one may be "working hard"

Fig. 17.6. A boy carries a load but does no work.

in the colloquial sense without doing any work in the physical sense. Similarly, if one pushes hard on a brick wall that remains stationary, F_1 is large, $\cos\theta = 1$, but $s = 0$, and although it might tire us out, we have not thereby done *work* as defined in physical science. Evidently, we must not let our subjective notions of activity, achievement, or fatigue become confused with this very different use of the word "work."

Before we go into the usefulness of this concept, note that work is a *scalar* quantity, although it is a product of two vector quantities. The units of work are those of force times distance, that is, newton·meter, for which the name *joule* (symbol J) is used, so named in honor of the British scientist whose contributions will figure prominently in the following pages.

Note that since (as defined in Section 9.5) 1 newton = 1 kg·m/sec^2, the joule is equal to 1 kg·m^2/sec^2. The units newton, joule, kilogram, meter, and second are all part of the "SI" (for "Système Internationale") system of units, the modern version of the metric system generally used by the scientific community (see Appendix for details).

Problem 17.3. A girl pulls a sled with a force of 0.001 N using a horizontal string over a distance of 1 km on level ground. (a) What work does the girl do on the sled, in J? (b) What is the work done on the sled by the gravitational force that the earth exerts on it?

Problem 17.4. In the example involving the pile driver (Fig. 17.4) assume that the distance *s* was 0.08 m and the average force needed to stop the falling body is 10 N. (a) Find the work done by the pile driver on the stake. (b) Find the work done by the stake on the driver. (c) If the driver fell from a height of 4 m and had a mass of 20 kg, what work was done on it during the descent before collision by the gravitational force?

17.4 Various Forms of Energy

Work can be done on a body in many ways and with many results. The stake did work on the falling pile driver, thereby bringing it to rest; conversely, the driver did work on the stake and so drove it into the ground against the forces of friction. Work is done by the hand in lifting the driver to a high position before the descent, thereby setting up the condition needed for gravity to do work on the falling object and so to let it gather speed. When a spring is compressed or extended between two hands, the hands do work on it. When a flywheel is started or slowed down, forces act over certain distances, and again work is done. Evidently, we should investigate what happens here systematically and quantitatively, still by analyzing simple mechanical problems first, before we are ready to handle the full power of these concepts.

a) *Work done to overcome frictional resistance.* By far the largest expenditure of work on this earth would seem to be work done "against friction." We pull on a block horizontally with an applied force F_{ap} to move it through distance s on a level table. (This is just the same as the case in Fig. 17.5, except we call F_1 now F_{ap} and $\theta = 0°$.) The work done by F_{ap} on the block is $F_{ap} \times s$. If the force F_{ap} is just enough to counteract the force of friction f and no net force is left over for acceleration of the block, then the work done by F_{ap}, namely $F_{ap} \times s$, is numerically just equal to $f \times s$. If we now remove F_{ap}, the block will just sit on the table. The only reward for the work we have done on the block is the transfer of the block from one spot to another and a little heating up where the surfaces of contact rubbed against each other. If we call this "wasted" or "dissipated" work, it is so only by comparison with other circumstances where it will turn out that some or all of the work done can, on removal of F_{ap}, be recovered directly.

b) *Work done to overcome the inertia of bodies.* We concentrate next on a case where friction is absent, for example, on an ice-hockey puck on a (nearly enough) friction-free horizontal plane of ice, in equilibrium at some speed v_1 (which might be zero). Now we apply horizontally a force F_{ap} through distance s to accelerate the block to speed v_2. The work done thereby, $F_{ap} \times s$, is now

$$F_{ap} \times s = mas = \tfrac{1}{2}m(v_2^2 - v_1^2)$$

or

$$F_{ap} \times s = \tfrac{1}{2}mv_2^2 - \tfrac{1}{2}mv_1^2. \qquad (17.6)$$

It appears that the work done on it serves to change the *vis viva* of the object or, more properly, a quantity that is just half the *vis viva*, and which about a century ago received its current name, *kinetic energy* (KE). Putting Eq. (17.6) more generally, we can say that here

$$F_{ap} \times s = \text{Change in kinetic energy} \equiv \Delta\text{KE}.$$

For example, a cyclotron is to accelerate a nuclear particle, such as a deuteron (the nucleus of a heavy isotope of hydrogen, made of one neutron and a positive particle, a proton, of about the same mass; total mass of deuteron = 3.3×10^{-27} kg). The desired final speed is 1/10 of the velocity of light, namely about 3×10^7 m/sec. If it starts substantially from rest, what work must be done on the deuteron?

Solution: $\Delta\text{KE} = \frac{1}{2}mv_2^2 - \frac{1}{2}mv_1^2$; here $v_1 \approx 0$; $\Delta\text{KE} = \frac{1}{2} \times 3.3 \times 10^{-27}\ \text{kg} \times (3 \times 10^7\ \text{m/sec})^2 = 1.5 \times 10^{-12}\ \text{kg·m}^2/\text{sec}^2 = 1.5 \times 10^{-12}$ J. By Eq. (17.6), this corresponds also to the work needed (in the absence of other effects such as friction) and this amount of work is actually supplied by letting an electric force act upon the accelerating particle. Even though 1.5×10^{-12} J may seem little work, the difficulties of supplying it are as varied as the effects obtainable with such high-speed particles.

Thus work can be "converted" into energy, and work done exclusively to increase kinetic energy is in a sense directly "recoverable," for when a high-speed object is then allowed to fall upon some target, it can in turn do work on it while giving up its kinetic energy. One may say that *energy is the capacity for doing work.*

The pile driver is a case in point. While the driver of mass m falls through a distance h, the constant force of gravity, mg, does work on it ($mg \times h$), thereby increasing its kinetic energy [from Eq. (17.6), $mgh = \frac{1}{2}mv_2^2$ if there are no frictional losses on the way down and if v_1, the starting velocity, is zero]. The work done by gravity is, to use graphic terms, "stored up" in the moving object, and becomes available on impact with the stake to drive it into the sand.

Although such pictures are a little too concrete, these are important concepts: Work on an object can be converted into kinetic energy (or, later, other forms of energy); conversely the energy of a body may be given up in doing work on something else; and often work is done against friction and may be irrecoverably lost from the mechanical system, but does show up as heat.

c) *Work done to change the level of an object.* Consider this third method of doing work on and imparting energy to a body: Simply lift up a heavy object from the floor to a distance h above it. The constant applied force needed here is just mg, enough to balance the weight, provided we can neglect air friction and also proceed so slowly that there is essentially no increase of kinetic energy. If not to friction or to kinetic energy, where does the applied work go? It is not lost, for if we let go, the pull of gravity will accelerate the body while it falls through distance h and so will build up its kinetic energy at the end of the fall to exactly the same number of joules we had expended in lifting it. Here we may have recourse to an enormously useful imaginative metaphor and say that the work done in lifting the body was "stored" as *potential energy* (PE) in the general region around the body, in the gravitational *field,* and furthermore, that while the object is then allowed to drop, the potential energy is drawn or given back from that region into the body and converted into kinetic energy.

So on raising the level of a mass m by a distance h, the work we do goes into potential energy (mgh) and, on falling, the "work done by the gravitational force" represents only a withdrawal from a bank account of energy accumulated on the way up, until on returning to the original lower level all of the energy has become kinetic.

One should avoid trying to make too material a picture of this sequence or of the seat of potential energy, and regard it rather as a useful expository device at this stage, an aid to comprehension without intrinsic value. There is really one big difficulty in the new concept, namely the *level of reference.* If we see a book of mass m on a desk in the second-floor study and are asked point-blank, What is its potential energy?, it would be of no use to reply mgh, for there is no obvious way to decide whence to measure h: from the floor? or from the street below? or from the level of the shop counter where it was originally bought? The truth is that potential energy can always be reckoned with respect to an arbitrary zero or reference level, usually, for convenience, the lowest level to which the body descends in the course of a problem or example. And the reason this can be done is, as you will notice, that we shall always deal with *differences* or *changes* in the potential energy (ΔPE) between two levels. So no harm is done by calling the lowest level simply $h = 0$; on the contrary, calculations are much more simplified than if we were to reckon all potential energies from some fixed point, say from the center of the earth or from a point high above the earth.

Problem 17.5. Reconsider the case of the ball bouncing off the back of the truck (Example 2 and Problem 17.2 in Section 17.1) from the viewpoint of work. Show that in the three successive cases: (i) when the truck is stationary no work is done by either the ball or the truck; (ii) when the truck is initially moving away from the ball, the ball does work on the truck during the collision; (iii) when the truck is initially moving toward the ball, the truck does work on the ball during the collision.

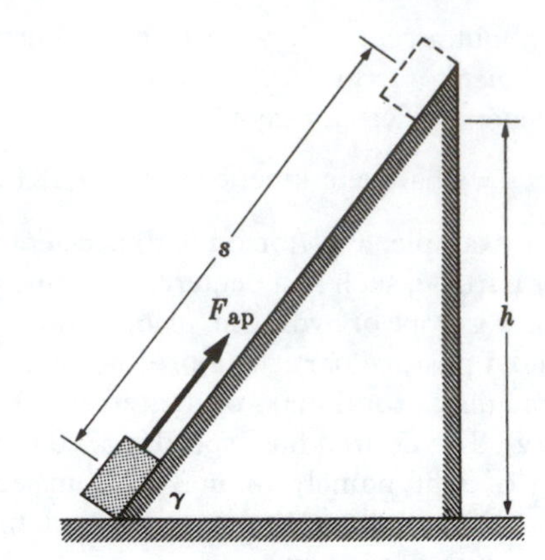

Fig. 17.7. Inclined plane.

17.5 The Conservation Law: First Form and Applications

a) *The inclined plane.* To summarize the three ways discussed here in which a body may dispose of work done on it by an applied force—against friction, against inertia, against gravitational forces—consider this very typical problem in mechanics. A block of mass m approaches the bottom of an inclined plane with a speed v_1 (moving from left to right) and ascends it for a running distance s while being pulled by an applied force F_{ap} as shown in Fig. 17.7. We are also given[1] the frictional force f, and the question is: What part of the work done by F_{ap} goes toward increasing the kinetic energy of the block? From this one might find the final speed v_2 at the top of the incline.

Solution: The work done by F_{ap} is $F_{ap} \times s \times \cos\theta$ (but here θ, the angle between the direction of s and of F_{ap}, is zero and $\cos\theta = 1$). This amount of work is split into three not necessarily equal parts: (i) work done against friction f, namely, $f \times s$; (ii) work done to change the kinetic energy ($\frac{1}{2}mv_2^2\ \frac{1}{2}mv_1^2$); and (iii) work done to change the potential energy, mgh. Note that it is the vertical level difference h that counts in calculating the potential energy, not the length of path s. In symbols,

$$F_{ap} \times s \times (\cos\theta) = \Delta KE + \Delta PE + fs \qquad (17.7)$$

or, here,

$$F_{ap} \times s = (\tfrac{1}{2}mv_2^2 - \tfrac{1}{2}mv_1^2) + (mgh) + fs.$$

Evidently, with F_{ap}, s, m, f at hand, ΔKE is quickly found. Note that F_{ap} refers to the pull by way of a rope or some other externally applied force, but leaves out of account entirely the force of friction and the force of gravity on the block, for those two are accommodated on the right side of the equation.

Now this was a quite plausible argument, and an experimental check would show that Eq. (17.7) is indeed correct. But is it self-evident or even proved? Neither! The fact that there are three obvious ways in which a body accepts and transforms work done on it does not mean that the sum of the three terms is necessarily exactly the amount of work done. There could be other drains on the available energy than we have thought of, or perhaps the moving body receives energy from some source other than the applied force. That this does not happen in simple mechanical phenomena, that Eq. (17.7) is correct, must initially be an experimental discovery, a discovery that amounts to a *law of conservation of energy;* for Eq. (17.7) says that every joule of work supplied to a body can be accounted for by corresponding changes in kinetic energy, potential energy, and work turned into heat by frictional losses.

---

[1]In a typical textbook problem one would be given the value of the coefficient of friction μ and the angle of incline γ; the frictional force is the push with which the block acts perpendicularly on the surface on which it sits (here equal to the component of the gravitational force in a direction perpendicular to the motion), multiplied by μ; that is, $f = \mu mg \cos\gamma$.

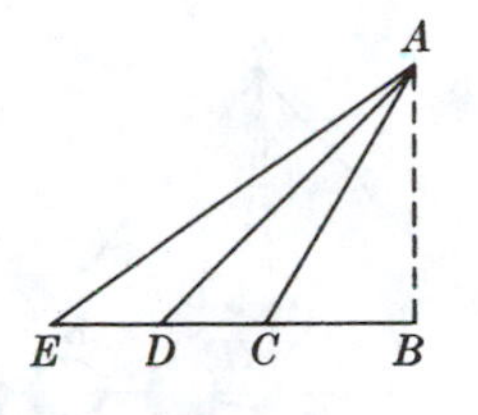

Fig. 17.8. A body moving from *A* down to a horizontal line *EB* acquires the same speed whether it follows the path *AB, AC, AD* or *AE.*

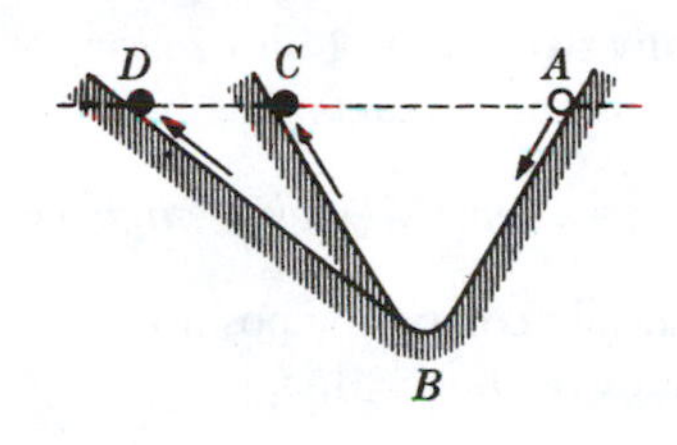

Fig. 17.9. A ball on friction-free planes ascends to its initial level.

More generally, imagine a closed system, say a box full of elastic balls, blocks, and inclined planes. Add some energy to the system, perhaps by dropping into the box a marble having a kinetic energy of x joules. After a while we look into the box again; it will be found that by mutual collision the bodies have so rearranged their speeds and positions that the energy we added is exactly equal to the sum of the potential energy, the kinetic energy, and frictional losses since the moment of intervention.

Dissipative system is the term applied to a system in which some of the mechanical energy is indeed dissipated irretrievably, as by friction. In contrast, *conservative systems* show no such losses; to them, the law of conservation of energy would apply in simpler form, for if the conservative system were totally isolated, unable to receive energy from or give it up to the rest of the world, then $\Delta KE + \Delta PE = 0$, a specialized case of the conservation law that was first expressed with most telling power and elegance in Lagrange's *Analytical Mechanics* of 1788, but was used much earlier in one guise or another.

Problem 17.6. Reexamine Galileo's argument in Section 17.2 again in the light of the law of conservation of energy.

Problem 17.7. Two crucial, almost axiomatic propositions on which many of Galileo's arguments in mechanics were based were the following: (a) "The speeds acquired by one and the same body moving down planes of different inclinations are equal when the heights of these planes are equal . . . provided, of course, that there are no chance or outside resistances, and that the planes are hard and smooth." (b) ". . . a body which descends along any inclined [frictionless] plane and continues its motion along a plane inclined upwards will, on account of the momentum acquired, ascend to an equal height above the horizontal, so that if the descent is along *AB* the body will be carried up the plane *BC* as far as the horizontal line *ACD;* and this is true whether the inclinations of the planes are the same, or are different, as in the case of the planes AB and BD." Use the law of conservation of energy law of conservation of energy to prove each of these two propositions separately (Figs. 17.8 and 17.9).

b) *Pendulums.* Yet another region of application of the law of conservation of energy relates to the varied and illuminating experiments with pendulums, favorites since the earliest days of modern mechanics. Consider an ideal pendulum, as in Fig. 17.10, released at its topmost position *a,* and swinging to the lowest position, *c,* before rising again on the other side. While it covers the left half of the arc, from *a* to *c,* the bob undergoes a vertical displacement of h_a cm, and changes its speed from $v_a = 0$ to v_c = maximum. Now we apply the law of conservation of energy: There is indeed a force F_{ap} on the bob (it is the tension in the string) but this acts at 90° to the path of the bob at every instant, and therefore can do no work. If we also neglect friction, the law of conservation of energy commands here

$$\Delta KE + \Delta PE = 0.$$

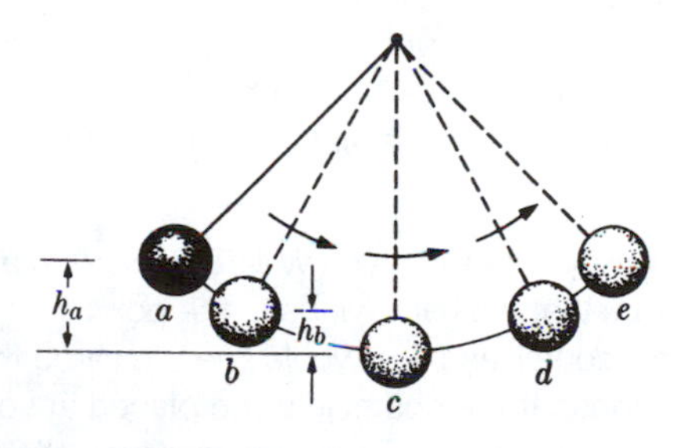

Fig. 17.10. The pendulum.

For any two points 1 and 2 along the path, this last equation predicts that

$$(\tfrac{1}{2}mv_2^2 - \tfrac{1}{2}mv_1^2) + (mgh_2 - mgh_1) = 0.$$

Specifically, comparing positions a and c, and noting that $v_a = 0$, $h_c = 0$,

$$(\tfrac{1}{2}mv_c{}^2 - 0) + (0 - mgh_a) = 0,$$

therefore

$$\tfrac{1}{2}mv_c{}^2 = mgh_a, \text{ or } v_c = \sqrt{(2gh_a)}.$$

Here it turns out that the speed of the pendulum bob at the lowest point is exactly the same as it would be if the bob were simply to *drop* through the same level difference h_a. Furthermore, we recognize that the pendulum has to lift itself through the same height on each side of the swing in order to convert all its kinetic energy at position c to potential energy at b or a. Once started, the pendulum must incessantly go through its symmetrical motions if there is no way for its energy to escape, continually exchanging its initial energy back and forth between potential and kinetic energy.

Problem 17.8. The bob of a simple pendulum (mass = 100 g) is let go from a vertical height h_a of 5 cm. (a) What will be its speed at the bottom of the swing? (b) What is its kinetic energy at that point? (c) What is its potential energy and kinetic energy at a point halfway up toward the top position on the other half of the arc?

Problem 17.9. One of Galileo's famous arguments involved an arrangement now called Galileo's pendulum. First let a simple pendulum swing freely back and forth between points *B* and *C* (Fig. 17.11). Now put a nail or a peg in position *D*, so that the pendulum, released again from *B*, cannot reach *C*; the string is bent at *D* and the ball or bob rises to *E* before returning. The significant thing is that *E* lies on the horizontal line *BC*, even though the pendulum on the right half of its trip now swings through the smaller arc *FE* rather than *FC*.

As Galileo has Salviati say in the *Two New Sciences:*

> Now, gentlemen, you will observe with pleasure that the ball swings to the point *E* on the horizontal, and you would see the same thing happen if the obstacle were placed at some lower point, say at *G*, about which the ball would describe arc *FH*, the rise of the ball always terminating exactly on the line *BC*.

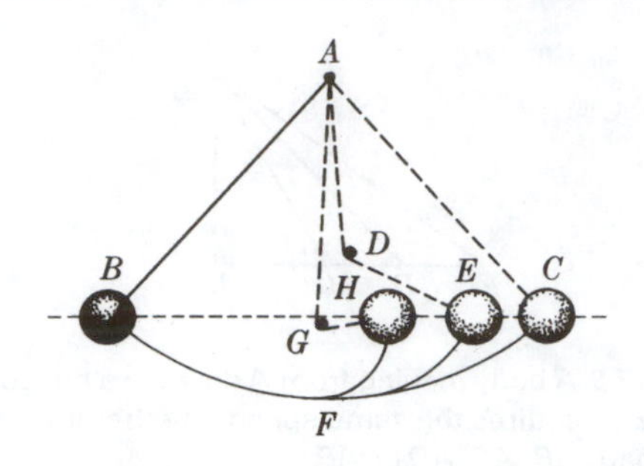

Fig. 17.11. **Galileo's pendulum.**

Explain these observations by means of the law of conservation of energy.

Problem 17.10. In Huygens' *Horologium Oscillatorium* (1673), we find this proposition: "If a simple pendulum swings with its greatest lateral oscillation, that is, if it descends through the whole quadrant of a circle [that is, starts, as in Fig. 17.12, from a horizontal position], when it comes to the lowest point in the circumference, it stretches the string with three times as great a force as it would if it were simply suspended by it." Prove this with the aid of the law of conservation of energy.

Problem 17.11. The measurement of the velocity of a fast projectile has always been difficult. One early solution was found in the so-called "ballistic pendulum" of Benjamin Robins (1742). A bullet of known mass m_b and unknown velocity v_b is fired horizontally toward, and embeds itself in, a bob (mass m_p) of a freely hanging simple pendulum. The maximum height h to which the pendulum rises after the inelastic collision with the bullet is noted. (a) With the aid of the law of conservation of momentum, derive the equation for the initial speed of the pendulum after impact. (b)

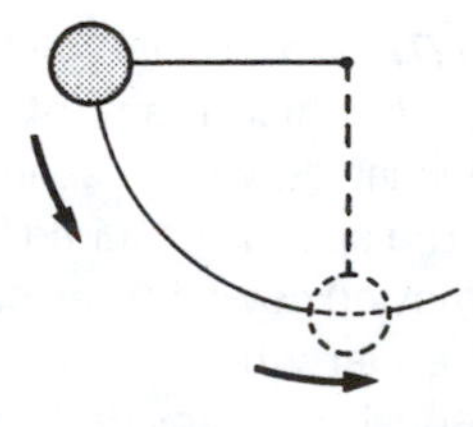

Fig. 17.12. **A pendulum at the bottom of its swing exerts three times as much force as it does at the initial horizontal position.**

Derive the expression for v_b before impact in terms of m_b, m_p, g, and h, using the law of conservation of energy for analyzing the motion of the pendulum.

c) *Collisions.* It is surely now evident that the law of conservation of *vis viva,* the original stimulus for the more general law of conservation of energy, is only a special case of the latter. For if a set of spheres has a perfectly elastic collision, then *by the very definition* of the phrase "perfectly elastic" we mean that none of the kinetic energy of the colliding bodies is lost to frictional processes, in other words, that the system of bodies is a conservative one. For such collisions, either in a horizontal plane or, more generally, of such brief duration that the change in level is insignificant during the time interval, the law of conservation of energy, which commands that

$$\Delta KE + \Delta PE + \text{Work lost to friction} = 0,$$

becomes simply

$$\Delta KE = 0.$$

For several colliding bodies, this means

$$[\tfrac{1}{2}m_A(v_A')^2 - \tfrac{1}{2}m_A v_A^2] + [\tfrac{1}{2}m_B(v_B')^2 - \tfrac{1}{2}m_B v_B^2] + \ldots = 0$$

Comparison with Eq. (17.2) shows that this is exactly the same as the law of conservation of *vis viva* (the factor $\frac{1}{2}$ may, of course, be canceled here throughout). Furthermore, we now have in principle a way to calculate what happens in all collisions, including those that are not elastic and in which *vis viva* (or kinetic energy alone) is not conserved. While it is, in general, not easy to judge how much energy is lost to friction, the case is not at all hopeless; one type of solution will suggest itself in connection with the following problem.

Problem 17.12. A certain ball is generally observed to bounce back from the floor to 1/3 of the height of release after free fall. If it has a mass of 5 g and is thrown vertically downward from a height of 100 m with an initial speed of 50 m/sec, where will the ball be after 2.5 sec? What will be its kinetic energy and potential energy at that moment?

Problem 17.13. A widely used demonstration involves a series of equally heavy white balls arranged in one line on a smooth table (Fig. 17.13). Now a black ball (like the others, a perfectly elastic one) is shot against them at speed v. If its mass is equal to that of each of the others, the last white ball flies off with the same speed v. But if its mass were twice as large, we would never see one white ball fly off at a greater speed, but all the white balls would move at different speeds. Show that this satisfies both the law of conservation of momentum and the law of conservation of energy. [If necessary, see J. V. Kline, "The Case of the Counting Balls."]

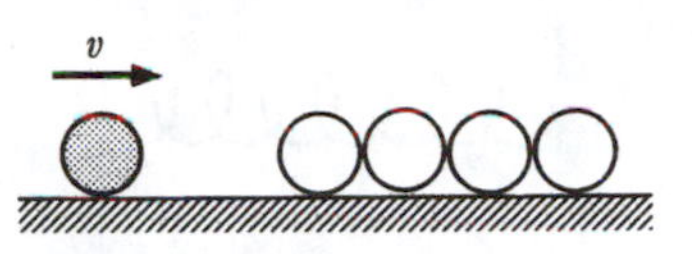

Fig. 17.13. **If the black ball strikes the first white ball, how will the white balls move?**

17.6 Extensions of the Conservation Law

For the sake of simplicity, we have so far avoided several sets of important problems to which the law of conservation of energy in its present form would not apply. This does not at all mean that it has only a limited validity! Rather, it means that we must extend the meaning of potential energy and kinetic energy a little to cover some other ways in which energy may appear or be stored.

Two cases frequently encountered in elementary mechanical problems are potential energy of elastic bodies and kinetic energy of rotation. We summarize here, because we shall need them in future chapters, the relevant formulas for these two cases, though detailed discussion will not be necessary.

a) *Potential energy of elastic bodies.* Work or kinetic energy is transformed into potential energy when it serves to lift a massive object through some distance. But besides gravitational potential energy there are other forms—for example, *elastic* potential energy. If a force acts on a spring (as in Fig. 17.14) to compress it for a distance x, the total work done on the spring is the average force F_{av} during the compression multiplied by the displacement x. This quantity

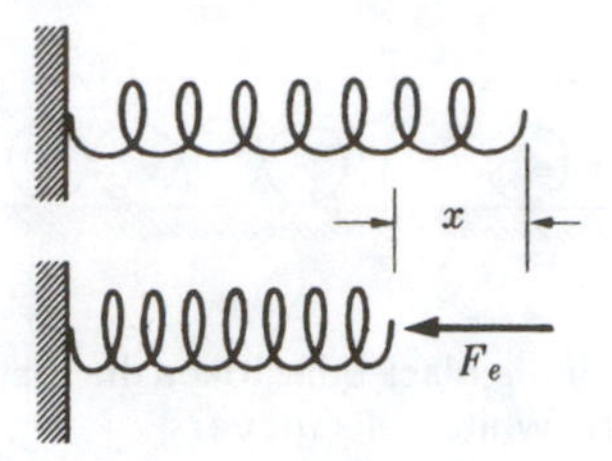

Fig. 17.14. Compression of a spring produces elastic potential energy.

$F_{av}x$ can be called potential energy stored in the spring, for as soon as the external applied force is removed, the spring will snap back and make available (to a machine, a clock, or in some other way) all the energy given to it in the initial compression.

The amount of energy stored in the spring can be calculated by using *Hooke*'s law (named for the seventeenth-century scientist Robert Hooke): Within reasonable limits, *the force needed to compress a spring is proportional to the amount of compression.*

Since the applied force F_e for a spring therefore changes linearly from $F_e = 0$ at the start (when $x = 0$) to F_e = (a constant k) $\times x$ at the end, the *average* force is $F_{av} = \frac{1}{2}kx$, and the work done is $(\frac{1}{2}kx) \times x = \frac{1}{2}kx^2$. (The mathematical reasoning used here is similar to that employed in proving the Merton mean-speed theorem in Section 6.6.)

Thus the potential energy of a spring compressed a distance x is

$$\text{PE} = \tfrac{1}{2}kx^2, \qquad (17.8)$$

where k is called the force constant of the spring.

The concept of elastic potential energy is also needed to understand how energy can be conserved in perfectly elastic collisions. In particular, for a head-on collision of identical bodies of mass m and velocities v_1 and $-v_1$, the total kinetic energy of both bodies before the collision is $2 \times (\frac{1}{2}mv_1^2)$, and if the collision is perfectly elastic, the total kinetic energy is the same after the collision. But at the instant of collision, when both bodies have momentarily stopped and are about to rebound, the kinetic energy is zero. If energy is to be conserved always, it must at that instant be temporarily stored as elastic potential energy. This would imply that the two bodies are compressed at the instant of collision, and will then expand, pushing each other off in opposite directions—as may indeed be observed by taking high-speed photographs of such events.

Problem 17.14. If an atom were to be defined as a particle that cannot be divided into smaller parts or change its dimensions, what would happen if two atoms approached head-on? What are the consequences of your answer for the cosmology of Descartes and the idea of the "world machine"?

Problem 17.15. Suppose that a cylindrical hole could be drilled straight from the North Pole to the South Pole of the earth, and assume that the earth is a perfectly uniform sphere. Describe the motion of a bullet dropped into the hole (see Problem 11.21).

b) *Kinetic energy of rotation.* A particle of mass m moving in a circle of radius r about a center O at angular speed ω (Fig. 17.15) has at each instant kinetic energy $\frac{1}{2}mv^2$, where v is its instantaneous linear speed. According to Eq. (10.6), the kinetic energy may also be written

$$\text{KE}_{\text{rot}} = \tfrac{1}{2}mr^2\omega^2. \qquad (17.9)$$

Just as the mass m represents the inertial resistance of an object to linear acceleration, we could say that the resistance of that object to being pushed around in a circle of radius r should be greater as r increases. It is convenient to define a new kind of inertia, the *rotational inertia I* of the particle, by the equation

$$I = mr^2, \qquad (17.10)$$

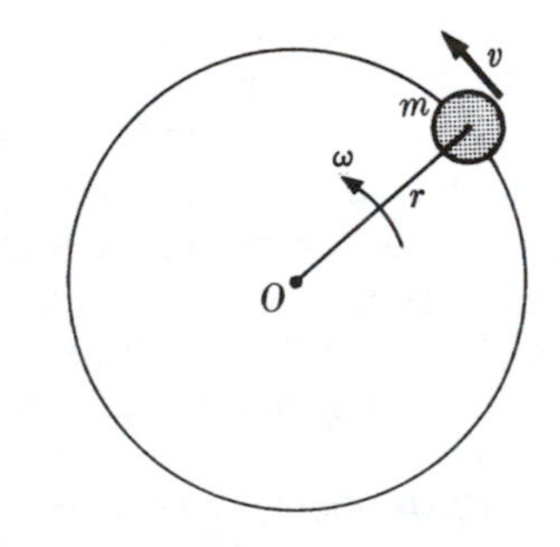

Fig. 17.15. Particle moving in a circle.

we can rewrite Eq. (17.9) in a form analogous to the formula for ordinary kinetic energy,

$$\mathrm{KE}_{\mathrm{rot}} = \tfrac{1}{2} I\omega^2. \qquad (17.11)$$

Here I plays the same role as the inertial mass, and ω the same role as the speed v in the formula $\mathrm{KE} = \frac{1}{2}mv^2$. Similarly, the quantity $\boldsymbol{l} = mR^2\omega$, defined as the *angular momentum* in Section 16.8, can now be written $\boldsymbol{l} = I\omega$, corresponding to the equation $p = mv$ for linear momentum.

Equation (17.11) can be made to apply not only to a rotating particle, but to any rotating object whatever, if we realize that the rotational inertia I will depend in each case on the shape and mass of the object (for example, for a *solid disk* of radius R and mass m it can be calculated to be $I = \frac{1}{2}mR^2$). So in general the total kinetic energy of any object can be written as the translational kinetic energy of the center of mass of the object, $\mathrm{KE}_{\mathrm{trans}}$, plus the kinetic energy of rotation around its center, $\mathrm{KE}_{\mathrm{rot}}$:

$$\mathrm{KE}_{\mathrm{total}} = \mathrm{KE}_{\mathrm{trans}} + \mathrm{KE}_{\mathrm{rot}}. \qquad (17.12)$$

The generalized concept of rotational inertia would also allow us to generalize the law of conservation of angular momentum, but we will not need that generalization here.

EXAMPLE 3

A disk of mass m and radius R begins to roll freely from the top of an incline to the bottom, that is, to a level h m below. What is its final speed, if rolling friction may be neglected?

Solution: We set up the law of conservation of energy in expanded form: $F_{\mathrm{ap}} = 0$, v_1 and ω_1 at top = 0, $f = 0$. Here,

$$\Delta\mathrm{PE} + \Delta\mathrm{KE}_{\mathrm{trans}} + \Delta\mathrm{KE}_{\mathrm{rot}} = 0$$

$$(0 - mgh) + (mv^2 - 0) + (\tfrac{1}{2}I\omega^2 - 0) = 0.$$

But ω, the angular speed at the bottom, is given by $\omega = v/R$, and $I_{\mathrm{disk}} = \frac{1}{2}mR^2$. Substituting and transposing,

$$mgh = \tfrac{1}{2}mv^2 + \tfrac{1}{2}(\tfrac{1}{2}mR^2)(v^2/R^2),$$

$$gh = \tfrac{1}{2}v^2 + \tfrac{1}{4}v^2, \qquad v = \sqrt{(4gh/3)}.$$

We note, perhaps with initial astonishment, that the final translational (forward) speed v of any disk is independent of its mass and radius. All disks attain the same final speed for the same descent, and v for any rolling disk is less [by $\sqrt{(0.67gh)}$] than that for an object sliding down without friction. [$\sqrt{2} - \sqrt{(4/3)} \approx \sqrt{(0.67)}$.]

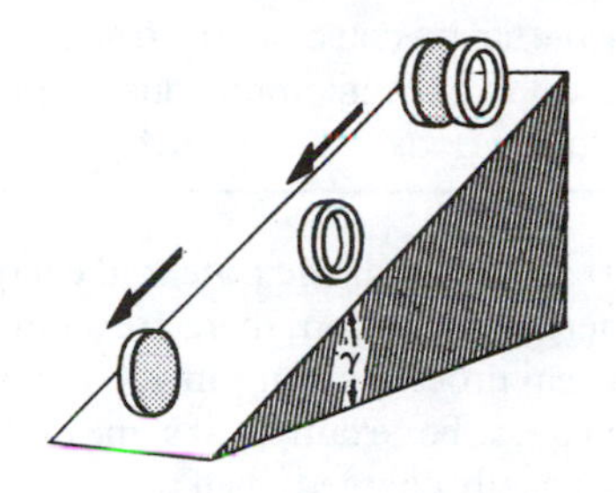

Fig. 17.16. Race between hoop and disk.

Problem 17.16. A hoop and a disk of equal radius r and equal mass m are released side by side along an incline. Soon the disk has outdistanced the hoop (Fig. 17.16). (a) Calculate and compare the respective final velocities. (b) Where is the hoop at the instant that the disk has reached the bottom?

Problem 17.17. Give a physical explanation as to why a sliding body reaches the bottom of an incline before a rolling body does.

Problem 17.18. Discuss why the following statement of Galileo is wrong and how it should be corrected: "All resistance and opposition having been removed, my reason tells me at once that a heavy and perfectly round ball descending along the lines *CA* [and] *CB* [a vertical free fall] would reach the terminal points *A* [and] *B* with equal momenta" (Fig. 17.17.).

Problem 17.19. By experimenting with *rolling* spheres on smooth planes, Galileo could indeed confirm that the acceleration is constant (see Section 7.4), but not having the concept of

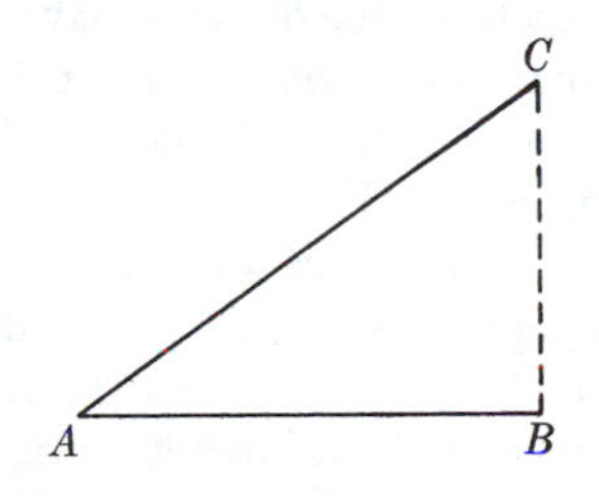

Fig. 17.17. Does a ball rolling down the inclined plane *CA* reach *AB* with the same momentum as if it dropped along *CB*?

rotational inertia, he could not from such data calculate the value of acceleration due to gravity (g) on freely falling bodies. Explain why.

Apart from the elastic potential energy and kinetic energy of rotation, there are a number of other physical processes that can add terms to the energy equation. For example, if some of the bodies are electrically charged, then there is a change in electrical potential energy as they approach or recede from one another, and this has to be included in ΔPE_{total}. Or if chemical energies are set free in a closed system, they must be accounted for. Heat, too, as will be seen at once, is to be counted as a form of energy, so that heat supplied to a system, or generated within by the transformation of other forms of energy to heat, must appear in our equations. So also must sound, and light, and other radiations that carry energy with them.

The complete story would seem to produce a very bulky equation for the law of conservation of energy, something like this:

External work done on, or other energy supplied to, a system of bodies = Σ [Δ(PE of all types) + Δ(KE of all types) + Frictional losses + Δ(Chemical energies) . . .].

Although the complete law of conservation of energy looks formidable indeed, in actual practice one almost always encounters only such cases as we have already seen, namely, cases in which only two or three of the many types of energy changes are of importance.

But every time we make use of it, we are struck by the magnificent generality of the law of conservation of energy. First of all, energy cannot be created or destroyed without a corresponding opposite change happening somewhere else in the universe (for example, the kinetic energy we impart to a cart or a baseball comes ultimately from a diminution of chemical energy within our muscles). One does not have to know (and often can't know) *how* the redistribution of energy happens in detail. The law applies anyway. Second, energy in any one form can be transformed into energy of some other form (for example, a magnet pulling an iron nail toward it shows how magnetic potential energy is changed into mechanical kinetic energy).

c) *Application to machines.* The law of conservation of energy applies to physics, astronomy, chemistry, biology, and other sciences. That is why we treasure it so, and why it helps to let the various sciences interact. But it can most simply be applied to machines, which are all essentially devices for transferring (more or less completely) the energy represented by work done upon one part of the system to some other part of the system, and making it available there is some convenient way.

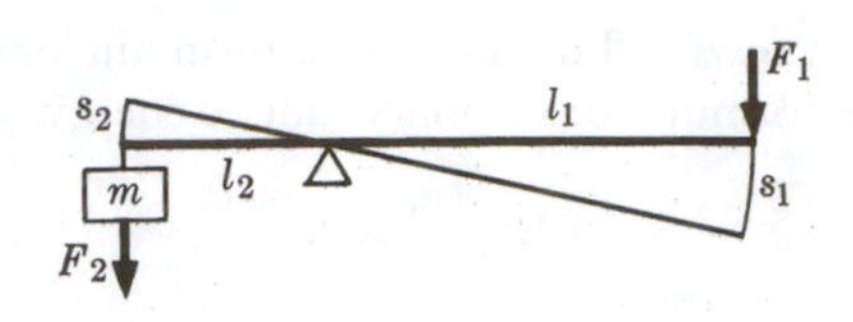

Fig. 17.18. The lever.

As one example, consider the lever in Fig. 17.18. The system here consists of a straight bar supported on a pivot or fulcrum that divides its length in the proportion l_1: l_2. A load F_2 (= mg) on the left is kept balanced by an applied force F_1 on the right. To lift the load, one need increase F_1 only infinitely little, provided that the friction at the fulcrum is negligible. While F_1 depresses one end through a small arc s_1, the other end rises through arc s_2. If the angle covered is very small, s_1 and s_2 are sufficiently close to being straight lines and perpendicular to the horizontal, so that the work done on the system by F_1 is F_1s_1. This work serves to change the kinetic and potential energies of the load and the lever.

But if, now, the speed of operation is kept very small, $\Delta KE = 0$ and the lever is relatively light, the only significant job that F_1s_1 can do is to increase the potential energy of the load ($\Delta PE = F_2 \times s_2 = mgs_2$). In the ideal case,

$$F_1s_1 = F_2s_2, \quad \text{or} \quad F_1 = F_2(s_2/s_1).$$

By similar triangles from Fig. 17.18, note that s_2: s_1 = l_2: l_1. On substituting,

$$F_1 = F_2(l_2/l_1) \quad \text{or} \quad F_1{:}F_2 = l_2{:}l_1. \qquad (17.13)$$

In words: The force needed to balance or slowly lift a load by means of a simple lever is to the weight of the load as the inverse ratio of the respective lever arms l_1 and l_2. This is the famous law of the lever, known empirically throughout antiquity and derivable also from other axioms, as is best illustrated by the famous work of Archimedes of Syracuse (287–212 B.C.). The problem was also discussed by Aristotle, Leonardo da Vinci, Galileo, and many others.

Problem 17.20. The Greek mathematician Archimedes is reported to have been so impressed with the power of the lever that he said: "Give me a place to stand, and I will move the world." Calculate first how much the earth might weigh; then calculate the dimensions and sketch the position of a lever that would have enabled Archimedes to carry out his promise of lifting the earth from its place in our solar system if he were given a long enough lever (and a fulcrum!).

An important principle from another branch of mechanics is hidden behind Eq. (17.13). If we write it as

$$F_1 l_1 = F_2 l_2,$$

then the meaning is: For equilibrium of the lever, the torque on the right-hand member in Fig. 17.18 (tending to produce a clockwise motion) must be balanced by the torque on the left side (which alone would give a counterclockwise motion). More generally, one can say that a body achieves rotational equilibrium, that is, has no angular acceleration, only if all clockwise torques balance all counterclockwise torques. This is the principle of equilibrium as regards rotational motion; it supplements the principle of equilibrium for translational motion, namely, Newton's first law, which commands that the forces themselves must balance if a body wishes to avoid linear acceleration. The latter is called the *first principle of statics;* the former is the *second principle of statics.*

Problem 17.21. In Fig. 17.19, a weightless bar is acted on by two equal forces. Show that although the forces cancel and therefore the first principle of statics (principle of translational equilibrium) is obeyed, the second principle of statics is not; that therefore the object will spin with increasing angular velocity. Then suggest where to apply additional forces, and of what magnitude, to fulfill *both* conditions of equilibrium.

Like the lever, the pulley is a simple machine well known in ancient times. A more useful and complex version, the simple block and tackle (Fig. 17.20), was considered by Aristotle and analyzed by Archimedes.

As your hand pulls a distance s_1 on the free end of the string, the lower pulley and the load

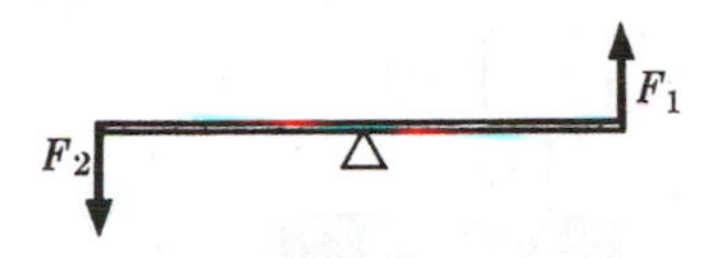

Fig. 17.19. Not only forces but also torques must be balanced to achieve equilibrium.

attached to it are raised through a distance s_2. Making the same assumptions as in the previous case, the work supplied is equal to ΔPE of the load, or

$$F_1 s_1 = F_2 s_2 \quad \text{or} \quad F_1 = F_2(s_2/s_1).$$

By measurement, we should find that in this particular pulley system, s_2 is always $\frac{1}{2}s_1$ (can you see why?), but in more elaborate arrangements the ratio s_2/s_1 may be much smaller. In every case, however, the last equation allows us to predict on the basis of an experimental determination of s_2/s_1 what force is ideally needed to lift a given load, even though we may be entirely mystified by the details of construction and operation of pulley systems! Here is another instance of the power of mathematization of physics.

We may carry this one step further and generalize on all simple mechanical arrangements in the following way. Such machines are equivalent to a black box (Fig. 17.21) to which we supply work through one opening and get work done on some load by a connection to a second opening. We may know or understand nothing about the gears inside. Yet the law of conservation of energy demands that the following be always true:

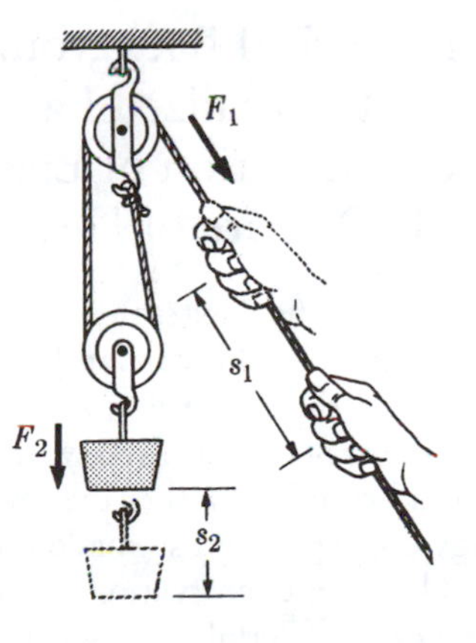

Fig. 17.20. Lifting a load with a block and tackle.

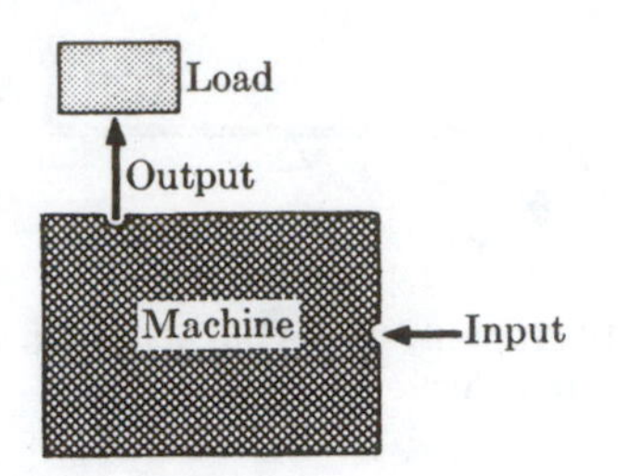

Fig. 17.21. Schematic representation of a machine.

Energy supplied to machine (INPUT) = energy supplied by machine to load (OUTPUT) + energy kept by machine (its own KE + ΔPE + friction losses).

Evidently, a good machine generally is one that does not divert much energy from the load, particularly by losses to friction. If we define the *efficiency* of a machine (symbol η, Greek letter eta) as the ratio of *output* to *input,* the ideal machine has η = 1.00 (or 100%); however, in actual practice we may find η = 0.1 (10%) or less.

Problem 17.22. An ideal, weightless and frictionless block and tackle as shown in Fig. 17.20 would require F_1 to be 10 N if the load is 20 N, but in an actual case F_1 may be 13 N. (a) How much work must be done to raise the same load by 1 m? (b) What is the efficiency of this machine in practice? (c) What is the reason for using the machine when it would evidently require so much less energy to lift the load directly?

17.7 Historical Background of the Generalized Law of Conservation of Energy: The Nature of Heat

There remains one final extension of the concept of energy: to problems involving *heat.* Historically it was just this extension, achieved in the middle of the nineteenth century, that in fact clarified and brought together the several separate energy concepts, and so made possible the first general law of conservation of energy. It is therefore worth undertaking a short digression on the development of ideas about the nature of heat before discussing the quantitative aspects of transformations between heat and other forms of energy.

We recall that Aristotle regarded Fire as one of the four elements, and that by this assumption much of the observed behavior of heated bodies could indeed be accounted for in a qualitative manner. The ancient Greek atomists explained differences in the temperature of bodies—their so-called intensity or degree of heat—by a closely related conceptual scheme, picturing heat as a special though not directly perceptible substance, atomic in structure like all the others, quick to diffuse through bodies, and presumably possessing some weight.

As a first step this picture is common-sensical, intuitively clear, even beautiful. Although we may quarrel about details, perhaps about the exact weight of the hypothetical heat substance, the concept on the whole is useful in explaining most of the casual observations, including the temperature equilibrium that two equally hot bodies will eventually attain while in contact with one another. By and by the picture became sharper: that of a tenuous fluid, free to pass in and out of the smallest "pores" of other materials, present in amounts depending on the temperature, yet otherwise imponderable and perhaps after all rather like Descartes' subtle, omnipresent fluid or the ethers then invoked by some to explain gravitation, the propagation of light and radiant heat, the transmission of magnetic and electric forces, and the like.

But those philosopher-scientists of the seventeenth century, including Bacon, Galileo, and Boyle, who—when they accepted a fluid theory at all—did so most hesitatingly, felt inclined to distrust the fluid theory of heat altogether and instead tended to the general though yet undefined view that heat was directly explainable as a vibration or similar small-scale motion of the atomic particles of heated bodies. The arguments were vague—for example, that heat evidently could be *generated* by motion, as by friction between two sticks or between an axle and the hub of a poorly lubricated wheel.

Throughout the seventeenth century and far into the eighteenth, neither the fluid view nor the motion view of heat became firmly established, and the reason was twofold. First, certain key concepts in the study of heat were still missing (for example, specific heat, latent heat, or even the consistent differentiation between heat and temperature) and, second, reliable methods of thermometry and calorimetry were still being developed. In the 1780s the great pioneer in the study of heat, Joseph Black of Glasgow and later Edinburgh (1728–1799), still referred uncertainly to heat as "this substance or modification of matter."

But one significant truth had begun to dawn in the first half of the eighteenth century. In experiments on mixtures, where bodies of varying temperature are brought together within a good enclosure, *heat is neither created nor destroyed;* that is, no matter how diverse the redistribution of heat among the different bodies in mixtures or in close contact, the total amount of heat remains constant. In modern terminology we should represent this *law of conservation of heat* in the following manner for the specific and simple case involving two differently hot bodies A and B, say a hot block of metal (A) plunged into a quantity of cool water (B): The loss of heat by A is equal to the gain of heat by B, assuming, of course, that the system containing bodies A and B is heat-insulated from all outside influences or losses. If A and B change only their temperatures, and not their chemical consistencies nor their physical states, the previous sentence can be symbolized by

$$\Delta Q_A + \Delta Q_B = 0$$

where ΔQ is the heat change for any sample. For simplicity we assume that a fixed amount of heat c must be added to raise the temperature of 1 gram of a substance by 1 degree on the Celsius scale (1° C); c is called the *specific heat* of the substance. Heat will be measured by the *calorie,* to be defined in the next section. The heat change may then calculated from the defining equation

$$\begin{aligned}\Delta Q \text{ (cal)} &= \text{mass of sample (g)}\\ &\times \text{specific heat [(cal/g)/°C]}\\ &\times \text{[final temp.} - \text{initial temp.(in °C)]}.\end{aligned}$$

Perhaps this, the basic rule of *calorimetry* ever since, helped most at the time to sway opinion toward the view that heat is a material substance, for such a law of conservation of heat very naturally brought to mind the tacitly accepted concept of the *conservation of matter.* In fact the fluid theory of heat now gave a simple and plausible picture of what happens when differently heated materials are brought together: The excess of "heat fluid" would flow from the hotter to the colder body until an equilibrium had been achieved. And truly, it was almost inconceivable that such a simple observation could be accounted for by the view of heat as a mode of motion of particles.

From the 1760s on, development proceeded at an accelerating pace, first mainly through the work of Black and his colleagues. The hypothetical *igneous* or *heat fluid,* named *caloric* by Lavoisier in 1787, began to take on greater plausibility when further properties were assigned to it. It was suggested that the particles of the caloric, unlike *ordinary* matter, repel one another but yet are attracted to the corpuscles of ordinary matter. (A somewhat similar hypothesis had recently been proposed by Benjamin Franklin and Franz Aepinus for electrostatics: Like particles repel; unlike particles attract [Section 24.2]. But in the caloric theory the relation between caloric and ordinary particles is not so symmetrical.)

If heat is applied to a material object, the caloric may be pictured as diffusing rapidly throughout the body and clinging in a shell or atmosphere around every one of the corpuscles. If these corpuscles (whose caloric shells repel one another) are free to move apart, as indeed they are in a gas, they will tend to disperse, and more strongly so the greater their crowding or the heat applied (just as one would expect from the gas laws to be discussed in Chapter 19). If, however, a heated object is in solid or liquid form, the mutual attraction among the corpuscles themselves (considered to be a gravitational one) so predominates that the caloric atmospheres can provide mutual repulsion sufficient only for the well-known slight expansion on heating. Or at any rate, the attraction predominates until enough caloric has been supplied for eventual melting or vaporization.

The same model made it plausible that on sudden changes of the physical or chemical states the particles of matter will exhibit a sharply changing capacity for accommodating caloric atmospheres, that is, they would release some caloric, thus changing it from a "latent" (hidden) state to a "sensible" (perceptible to the senses) one, where caloric would also communicate itself to neighboring bodies and heat them up. Obviously this would explain the heat generated in some chemical reactions or solutions.

Similarly it was not at all unexpected to find that "sensible heat" develops during mechanical deformations: For example, the sometimes very marked rise in temperature of a gas on rapid compression or of an iron bar by skillful and rapid hammering was regarded simply as evidence of a squeezing-out of caloric from among the atoms as they are forced closer to one another.

Other experimental facts, including those accessible through the newly developed concepts of specific and latent heat, were subjected to similar and quite successful analyses. The conceptual schemes gradually filled out on all sides, and the gathering weight of evidence persuaded almost all Europe from the 1780s on of the usefulness of the caloric theory.

Yet this theory hid two assumptions, one of which was to prove fatal. The first, which could not by itself discredit the whole structure when brought into the open, was the old problem of whether the caloric material also possesses weight. Naturally, this question was now repeatedly and often ingeniously examined by direct experimentation. But the findings were negative, where they were not indeed contradictory, owing to the inherent difficulties of weighing accurately at varying temperatures. The second assumption was that the caloric fluid is conserved in all processes involving heat.

The first serious challenges to these assumptions were offered by the experiments of Count Rumford (1753–1814), an amazing American who began life as Benjamin Thompson of North Woburn, Massachusetts. By 1798, a refugee (as Tory sympathizer) from his native country, he had been knighted by King George III and was in the employ of the Elector of Bavaria as Chamberlain, Minister of War, and Minister of Police, but still found time away from his many duties and schemes to perform some capable and historic experiments. After exhaustive and to us convincing experiments with some of the best balances in Europe, he stated in 1799: "I think we may safely conclude that *all attempts to discover any effect of heat upon the apparent weight of bodies will be fruitless.*"[2]

Granting this conclusion, we discover that we have been brought back to the same ground that for two centuries had been fought over: by the phlogistonists vs. Lavoisier, the vacuists vs. the plenists, Newton vs. the Cartesians, in a sense even by Galileo vs. the Scholastics—namely,

Fig. 17.22. Benjamin Thompson, Count Rumford (1753–1814). American-born physicist, who pursued his career of scientific research and invention in Europe. His experiments furnished evidence against the caloric theory of heat.

the question of whether science can accommodate an imponderable fluid with material properties, endowed perhaps even with atomic structure, but inaccessible to direct experimentation or even contemplation through its lack of that most general characteristic of materiality, weight.

Those seventeenth-century philosophers who had rejected the possible existence of "imponderables" (substances that have no weight)—and like them, Rumford—had done so in part because they simply could not imagine a weightless fluid, a mechanical entity lacking one of the most unforgettable properties of ordinary matter.

But calorists, on the other hand, could equally well say with Black: "I myself cannot form a conception of this eternal tremor [motion of corpuscles] which has any tendency to explain even the more simple effects of heat." Up to this point, Rumford may seem to have little cause for rejecting the imponderable caloric except a lack of imagination; and this is the more striking because the scientific world at the time generally did accept a whole set of imponderables and ethers.

Let us for a moment examine our own first impulses in this matter. We are probably tempted to reject the concept of imponderable fluids outright; and if urged to explain, we might say that a part of the very definition of the physical entity must be that we can discover it directly. But this would be a most naive attitude; we must discard

[2]See the fascinating biography by Sanborn Brown, *Count Rumford: Physicist Extraordinary.* Rumford's experiments, and the discussion of them, make most enjoyable and instructive reading in Roller's account, "The early development of the concepts of temperature and heat" in *Harvard Case Histories,* edited by Conant, pages 117–214.

it now before we go on to phenomena involving "imponderables" like photons and wave functions. We must be prepared to accept not just what sounds plausible or just what can be imagined by analogy, but must be prepared to welcome into science any concept that may be needed in our schemes if only it fulfills these requirements: It should be definable, without internal contradiction and ambiguities, in terms of observable and measurable effects, whether these be direct or not; and it should prove useful in a wide variety of descriptions and laws, long known as well as newly deduced. The caloric fluid fulfilled all these criteria at the time.

But this was not the reasoning that led the calorists to accept the imponderable, and Rumford was not just naive in rejecting it. On the one hand, they were not greatly bothered by Rumford's demonstrations, and, on the other, his objection to the caloric for lacking measurable weight was only part of his much more thorough experimental, 30-year attack on that concept. In time, Rumford succeeded in discovering the real Achilles heel—a failure of that most treasured function and main justification of the concept of the caloric, namely, *its usefulness in the general law of conservation of heat.*

The famous experiments leading up to this far-reaching discovery are described by him in 1798 as follows:

> [I] am persuaded that a habit of keeping the eyes opened to everything that is going on in the ordinary course of the business of life has oftener led—as if it were by accident or in the playful excursions of the imagination, put into action by contemplating the most common occurrences—to useful doubts and sensible schemes for investigation and improvement than all the more intensive meditations of philosophers in the hours especially set apart for study.
>
> It was by accident that I was led to make the experiment of which I am about to give account. . . . Being engaged lately in superintending the boring of cannons in the workshops of the military arsenal at Munich, I was struck by the considerable degree of heat [temperature] that a brass gun acquires in a short time in being bored, and with the still higher temperature (much higher than that of boiling water, as I found by experiment) of the metallic chips separated from it by the borer.
>
> (Reprinted in Brown, *Benjamin Thompson-Count Rumford*)

A commonplace type of observation, but heightened both by the uncommonly high temperature developed and by the exceptional gift of observation of Rumford, was therefore the effective beginning of the idea that a heat *fluid* can be generated by friction in enormous quantities. As we saw, that in itself is not altogether unexpected; on the basis of the caloric theory, the mechanical treatment of metals in boring might decrease the capacity for holding caloric in the metal, thereby yielding free, "sensible" heat. However, experiments with the metallic chips and comparison with the bulk metal of the cannons themselves showed Rumford that their capacity for heat (specific heat) had not changed.

What is much more significant, further experiments revealed that the source of heat generated "in these experiments appears to be inexhaustible. It is hardly necessary to add that anything which any insulated body, or system of bodies, can continue to furnish without limitation, cannot possibly be a material substance, and it appears to me extremely difficult, if not quite impossible, to form any distinct idea of anything capable of being excited or communicated in the manner in which heat was excited and communicated in the experiments, except it be MOTION." And throughout his voluminous writing Rumford continues to affirm those "very old doctrines which rest on the supposition that heat is nothing but a vibratory motion taking place among the particles of the body."

A similar conclusion was drawn the following year by the 21-year-old Humphry Davy, later a distinguished chemist, in original experiments in which he allowed two pieces of ice to rub against each other until they had both melted. The large amount of heat needed for melting the ice must have been produced by friction only, particularly in view of the large increase of specific heat on changing ice into water. Davy's conclusion that "it has just been experimentally demonstrated that caloric, or matter of heat, does not exist," although very influential on the later work of Julius Robert Mayer and James Prescott Joule, is rather an overstatement; he might have been satisfied with the claim that large quantities of caloric appeared to have been created by friction, contrary to the expectations of the calorists.[3]

---

[3]There is an ironic twist here. Reexamination of Davy's data and methods has shown that the ice must have melted owing far more to heat leaking into his apparatus than to friction; see Roller, "Early development," pages 86–87.

In brief, these and a great variety of other experiments, executed mostly by the indefatigable Rumford, had struck key blows against the caloric theory by 1807. But that was not enough to convert the calorists and topple their science, for the simple reason that there was no good constructive conceptual scheme to offer instead. Even though Rumford was able to show that his model of heat as a mode of motion in the substance would account qualitatively quite well for a variety of observations such as normal thermal expansion, heat conduction, change of state, and diffusion in liquids, his theory could not explain the undisputed facts summarized by the law of conservation of heat in mixtures and in other such simple cases in which, as we now would put it, by careful insulation the amount of heat in the system is kept constant, and is neither generated nor converted. The caloric theory was still most practical and plausible in those cases, and in a great region of other physical and chemical phenomena.

One of the most convincing arguments against Rumford's theory was provided by the phenomenon of *radiant* heat, on which Rumford himself had done pioneering work. The very fact that heat could travel across empty space (such as, for example, the region between the sun and the earth was thought to be) seemed to prove that heat could not simply be a mode of motion of matter, but rather a distinct substance. During the first three decades of the nineteenth century, numerous experiments by William Herschel in England, Macedonio Melloni in Italy, James Forbes in Scotland, and others indicated that radiant heat exhibits all the properties of light—reflection, refraction, interference, polarization, etc.—and is therefore probably qualitatively the same phenomenon as light. At that time the modern distinction between radiant heat and other forms of heat had not been established; it was therefore believed that any conclusion about the nature of radiant heat would be valid for the nature of heat in general.

Since many scientists in the first half of the nineteenth century accepted the principle of the qualitative identity of heat and light, we find the rather curious situation that opinions about the nature of heat depended to some extent on opinions about the nature of light. As we shall see in Chapter 23, before about 1820 the Newtonian particle theory of light was generally favored—that is, light was thought to be a *substance*. It was therefore reasonable to suppose that heat is also a substance, probably, like light, composed of particles. But as a result of the work of Thomas Young in England and especially Augustin Fresnel in France, scientific opinion after about 1825 swung around to the view that light is not a substance but a wave motion in the ether. Hence it was now reasonable to adopt, alongside a wave theory of light, a wave theory of heat—a theory in which heat was regarded as vibrational motion of the ether. This did not yet mean a return to Rumford's theory, but it did help to phase out the caloric theory of heat, though in some writings of the 1830s and 1840s one gets the impression that caloric has been reincarnated under the name "ether"! But more significant in the long run is the fact that, thanks in part to the wave theory of light, *heat was now being conceived as a form of energy.*

Several other contributions at that time also pointed to a relationship between heat and other forms of energy. The conversion of kinetic energy to heat, as by the quick hammering of a nail on an anvil, was, of course, common knowledge, as was the liberation of heat in chemical reactions. About 20 years before Rumford, Lavoisier and Laplace had introduced a note from physiology by their experiments to prove that a living guinea pig developed animal heat at about the same rate as does the chemical process of burning its food intake. In 1819, the French chemist and physicist, P. L. Dulong, showed that when gas is compressed quickly the heat developed in the gas is proportional to the mechanical work done on it. Research was begun on the heat produced by electric currents.

And all the while, two other influences develop. The energy concept becomes more clearly defined; the word *energy* itself (in its modern meaning) dates from 1807. And as particularly illustrated in the work of the British physicist Michael Faraday on the interaction of electric current, magnetism, and light (Section 25.2), there spreads increasingly a belief that the several types of natural activities are all connected and are ultimately all part of one unified story. In reviewing this period prior to the 1840s one is struck by a widespread groping toward the general law of conservation of energy from many different sides, often far beyond the context, the available evidence, or the clarity of the concepts themselves.

Heat not only changes temperature and physical state, but it can also move objects (for example, by making a gas expand) and so can be made to produce physical work. This was, of course, a commonplace after the spectacular development of steam engines in the eighteenth century, and their remarkable effect in promoting industrial development. But the relation between heat and work was by no means clear before the 1850s. Up to that time, it was believed that the flow of heat itself determined the motion of the parts of a heat engine in somewhat the same manner as the flow of water over a water wheel makes the mill turn, and that heat itself was no more consumed in the process than is water in the other case.

What was clearly called for (or so it seems to us now, with the advantage of hindsight) was to go beyond the qualitative idea that heat is a form of energy convertible into other forms, and to make specific determinations of the numerical equivalents or conversion factors of different forms of energy. Speculation about energy conversion and experimental data from which conversion factors could be computed were both available to scientists in the 1830s and 1840s. It is therefore not too surprising that the act of putting them together in a coherent way, which we now define as the "discovery" of the generalized law of conservation of energy, was accomplished almost simultaneously, and independently, by several people. We shall discuss two of the ways in which the discovery was made, thereby also illustrating the diversity of methods that can be successful in science.

17.8 Mayer's Discovery of Energy Conservation

Several basic ideas about the interchangeability of heat and work were pulled together in 1842 in an essay that for the first time suggested what we have assumed as basic in Sections 17.5 through 17.7, that is, a general equivalence and conservation of all forms of energy, the equivalence of heat and work being only a special case. That essay, full of imaginative, almost metaphysical, but largely unsupported generalizations to such a degree that its publication was at first refused by one of the leading journals of science, was the work of Julius Robert Mayer (1814–1878), then a young doctor in the German town of Heilbronn, his native city. He later said that his thoughts first turned to this subject when, as a ship's physician in the tropics, he discovered venous blood to be of a brighter red color than he had expected from experience in cooler climates; and that subsequently he began to speculate on the relation between "animal heat" and chemical reaction, and from there to the interchangeability of all energies in nature.[4]

In his 1842 paper "Remarks on the Energies[5] of Inorganic Nature," Mayer announces:

> Energies are causes: accordingly, we may in relation to them make full application of the principle—*causa aequat effectum* [the cause is equal to the effect]. If the cause *c* has the effect *e*, then $e = c$. . . . In a chain of causes and effects, a term or a part of a term can never, as plainly appears from the nature of an equation, become equal to nothing. This first property of all causes we call their indestructibility. . . . If after the production of [effect] *e*, [cause] *c* still remained in whole or in part, there must be still further effects [*f*, *g*, . . .] corresponding to the remaining cause. Accordingly, since *c* becomes *e*, and *e* becomes *f*, etc., we must regard these various magnitudes as different forms under which one and the same entity makes its appearance. This capability of assuming various forms is the second essential property of all causes. Taking both properties together, we may say, causes are quantitatively *indestructible* and qualitatively convertible entities. . . . Energies are therefore indestructible, convertible entities.

This, then, is the first voicing of conservation and equivalence of *all* forms of energy. But this way of arguing seems quite false to a modern scientist. "Causes" do not have the properties derived by this private logic, nor can one equate energy with "cause" unless there is prior *experimental* proof of the indestructibility and convertibility of energy. But Mayer believed that this proof was already available from previous experiments, provided they were interpreted from his viewpoint.

Further on, Mayer applies his doctrine to chemistry and mechanics, and he continues:

---

[4]As an illustration of the controversial nature of some of the simplest details in the history of science, we may note that it has been asserted, on the one hand, that this observation was "well known" before Mayer, and, on the other hand, that the observation must have been erroneous because venous blood is *not* bright red in the tropics!

[5]In the German original, the old word *Kraft* (force) is used throughout in this sense.

Fig. 17.23. Julius Robert Mayer (1814–1878), German physician who proposed the law of conservation of energy.

> In numberless cases we see motion cease without having caused another motion or the lifting of a weight; but an energy once in existence cannot be annihilated; it can only change its form; and the question therefore arises, What other forms is energy, which we have become acquainted with as potential energy and kinetic energy [modern terminology], capable of assuming? Experience alone can lead us to a conclusion.

There follows the argument along expected lines that if heat is generated by the motion of two surfaces against each other with friction, heat therefore must be equivalent to motion. Of course the idea that heat *is* a form of motion had frequently found expression throughout the history of science; Rumford and Davy had recently advocated it, and it was to be generally accepted within a decade. Curiously, Mayer held back from this conclusion; his philosophy did not lead him to make any one form of energy, such as motion, more fundamental than the others. Indeed he refused to accept the "materialistic" doctrine that refers all phenomena to a motion or property of *matter;* forces do not depend on the presence of matter but can exist independently. Indeed, if force has as much reality as matter, it must (like matter) be conserved; Mayer often used analogies with chemical transformations subject to the law of conservation of mass.

Mayer simply concluded: "If potential energy and kinetic energy are equivalent to heat, heat must also naturally be equivalent to kinetic energy." Consequently, "We will close our disquisition, the propositions of which have resulted as necessary consequences from the principle *causa aequat effectum* and which are in accordance with all the phenomena of nature, with a practical deduction. . . . How great is the quantity of heat which corresponds to a given quantity of kinetic or potential energy?"

This is indeed the crucial point. Mayer's presentation so far has been almost completely qualitative, and indeed rather reminiscent of scholastic science in many ways. This is partly understandable from the fact that Mayer himself had then no facilities for experimentation and found the physicists of his time quite uncooperative. But here, at the end of his paper, deceptively like an afterthought, is finally promised a derivation of the equivalence between heat and work, and by an approach opposite to that of Rumford. Here is the chance for empirical verification.

Mayer's solution was only sketched out, but one may reconstruct the steps. In the circumstances, he had to fashion his illustrative computation from some known data whose significance had by and large not been noted. To explain the nature of these data, we must first explain the units for measuring heat. We follow an old convention: The unit of heat, called the *calorie* (abbreviated "cal"), is defined as the amount of heat[6] that raises the temperature of a 1-gram mass of water through just 1 degree on the Celsius temperature scale.[7] After some

[6]The name calorie harks back to the old caloric theory of heat, but of course the unit for measuring heat is not dependent on a theory of what constitutes heat. The "calorie" used on food labels, sometimes called the Calorie (Cal), is actually a kilocalorie (10^3 cal). A more serious problem is that the phrase "amount of heat" suggests that a substance actually *contains* a certain amount of heat, whereas strictly speaking "heat" refers only to energy being transferred from one substance to another. In accordance with the law of conservation of energy, it is possible to change a substance from one physical state (characterized by specific values of temperature, pressure, and volume) to another state by different routes involving different amounts of heat transfer (or none at all).

[7]Named in honor of Anders Celsius (1701–1744), a Swedish astronomer who first proposed (in 1742) a temperature scale on which the boiling point of water was defined as 0°(and the freezing point as 100°. This was soon reversed, and the

experimentation one may then assign to every substance a specific heat (symbol *c*), a value that is numerically equal to the heat energy needed to raise the temperature of 1 g of that substance by 1°C.[8]

For water at 15°C and atmospheric pressure, *c* is 1 (cal/g)/°C, by definition; for copper, it is by experimental test about 0.09 (cal/g)/°C; for lead, 0.03 (cal/g)/°C; and for most other solids and liquids it has some intermediate value.

In the case of gases—and here we may include air, a mixture of gases—there is an interesting and important difficulty: The specific heat depends to a marked degree on the method of measurement. If, for example, we take a sample of air in a tight enclosure at a constant volume, we must supply 0.17 cal per g per °C temperature rise and, incidentally, we shall note that the pressure in the gas rises. Now we repeat this experiment with the same sample in a cylinder having one freely movable side or piston (Fig. 17.24). The pressure inside and outside the cylinder will always be the same, for the inflow of heat to the gas can now be attended by an expansion, and it will be found that one needs 0.24 (cal/g)/°C, that is, 40% more than before. We must specify two specific heats for gases, one for constant volume and another for constant pressure. These are usually denoted by the symbols c_v and c_p, respectively.

The difference between the two specific heats indicates that the expansion of a gas is associated with an exchange of heat with the surroundings, and this fact was generally known to scientists early in the nineteenth century. What

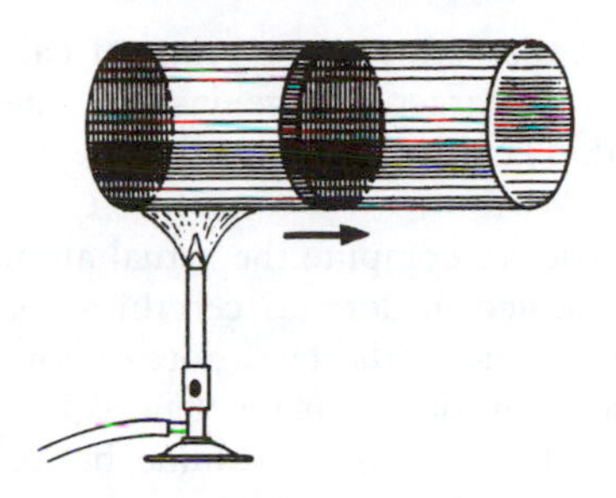

Fig. 17.24. Heating a sample of gas in a cylinder with freely sliding piston.

was not so generally known, however, is that if a gas is allowed to expand freely into a vacuum (instead of having to push away a piston as in Fig. 17.24), then there is no heat exchange with the surroundings, that is, no cooling. This was shown by the French chemist Gay-Lussac [known for his law of combining volumes (see Section 20.8)] in an experiment done in 1807. Gay-Lussac showed that a sample of gas will expand freely into the previously empty part of a set of connected vessels *without* measurably changing its overall temperature (this was a mystery from the viewpoint of the caloric theory)—whereas expansion against some surrounding medium or piston is possible without a temperature drop only if heat is supplied.

The free-expansion experiment was repeated by J. P. Joule in 1845, using more accurate thermometers, and a few years later by Joule and William Thomson [later known as Lord Kelvin], working together. In some cases a very slight cooling effect was observed, but the temperature change was always very much less than that experienced by a gas on expansion against a piston.

Mayer cleverly saw in these experimental facts a way to the law of conservation of energy. He used the free expansion experiment of Gay-Lussac to support his argument that heat actually disappears when it does mechanical work. According to the caloric theory, when a gas expands its temperature drops because the caloric is spread out over a larger volume, but the total amount of caloric was supposed to remain the same. However, Mayer pointed out that the temperature drops only when the gas does mechanical work by pushing a piston as it expands; if no work is done, the temperature remains the same. Therefore the drop in temperature that does occur when a gas does work cannot be explained

scale on which 0° is the freezing point of water and 100° the boiling point was known until recently as the *centigrade* (100-step) scale. However, since the Tenth General Conference on Weights and Measures in 1954 decided to redefine the temperature scale for scientific purposes by taking absolute zero and the *triple point* of water as the fixed points (see Section 19.3), the name centigrade is no longer appropriate and this scale should be called the Celsius scale instead. For scientific work the absolute Kelvin scale (Section 19.3) is generally used. The actual temperature on the Celsius scale is denoted by T_C when we want to distinguish it from the *difference* between one temperature and another on that scale given in Celsius degrees, °C.

[8]This definition is sufficiently precise for our purposes, but it may be noted that in accurate work the actual initial and final temperatures must be specified; for example, the calorie is the amount of heat needed to raise the temperature of 1 g of water from $T_C = 14.5$ to $T_C = 15.5$ (which is not quite the same as that needed to raise it from, say, $T_C = 95$ to $T_C = 96$).

by saying that the same amount of caloric is spread over a larger volume; instead, one must assume that heat actually disappears when work is done.

In order to compute the actual amount of heat consumed in doing a certain amount of work, Mayer used the fact, noted above, and known for some time, that 1 g of air requires 0.17 cal to gain 1°C at constant volume, but 0.24 cal if the piston is free to move and the chamber expands. Mayer assumed that in the second case also, 0.17 cal is needed to increase the temperature, but that the additional 0.07 cal is needed to move the piston out against the surrounding air. This air is normally at the constant pressure of 1 atmosphere (1 atm) or 1.013×10^5 N/m^2 of piston area, written in modern SI units as 1.013×10^5 pascals (Pa) (see Section 19.2).

In retrospect, one may see that this idea is inherent in Gay-Lussac's observation that no difference of specific heats exists if the gas can expand into a vacuum, against no resistance. Here, the work done by the gas on the piston (Fig. 17.25) is the perpendicular force (F) times the distance it moves (s). But since the pressure (P) in the gas is defined as force (F) ÷ area (A) on which F acts, the work done on expansion becomes $(PA) \times s$. Furthermore, $A \times s$ = change in volume during the expansion of this 1 g of air, and that was a known quantity, namely, 2.83 cm^3. Thus 0.07 cal of heat appears as 1.013×10^5 N/m$^2 \times 2.83$ cm$^3 \approx 0.286$ J of work, or *1 cal is equivalent to about 4 J*. (Values that are approximately equal to modern ones have been used here; the measurements known to Mayer gave him 1 cal ≈ 3.6 J.)

For a decade or more, little attention was paid to Mayer's work, despite the fact that similar ideas and experiments were being discussed by other scientists. Mayer's philosophical style of reasoning and writing was unconvincing to those who had become accustomed to defining science in terms of Newtonian mathematical calculation and quantitative experimentation. Moreover, this was not a case of a generally recognized conceptual crisis that Mayer's new ideas would have resolved; instead, he was presenting a new viewpoint, which could serve to unite previously separate branches of physical science, and the fruitfulness of this viewpoint had not yet become evident. Mayer did develop his principle of the general indestructibility and strict convertibility of energy further in several subsequent essays, presenting good illustrations of proof by calculation and extending the principle "with remarkable boldness, sagacity, and completeness" to chemistry, astronomy, and living processes (Tyndall, *Heat*).

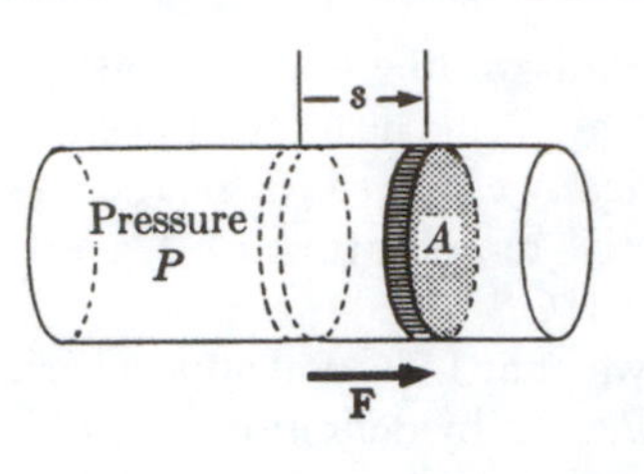

Fig. 17.25. When a gas is heated at constant pressure *P*, it can move a piston with area *A* through a distance *s*, exerting a force ***F***.

Although the work in retrospect was clearly one of genius and "on the whole, a production of extraordinary merit" (Tyndall, *Heat*), Mayer continued to be so utterly rejected that his mind, overtaxed by the strain of his work and his lonely fight, gave way for a time. He eventually recovered and witnessed growing recognition and honor in the closing years of his life.

This change of personal fortunes was by no means a sign that science had become sentimental about Mayer's condition; rather, it was the consequence of a general acceptance of the law of conservation of energy on the basis of an entirely different series of events—above all, the remarkable and persistent experimental work of a British amateur of science, James Prescott Joule (1818–1889).

17.9 Joule's Experiments on Energy Conservation

Joule was born near Manchester, the son of a well-to-do brewer to whose business he succeeded. But his dedication from the start was to science. At 17, he was a student of the great chemist John Dalton (see Chapter 20), and at 22, he had begun a series of investigations that was to occupy the greater part of his life: the proof that when mechanical work is turned to heat under any circumstance, the ratio of work done to heat evolved has a constant and measurable value. In part, his motivation seems to have been essentially the same as that of all other proponents of a conservation law; in his words, it was "manifestly absurd to suppose that the powers with which God has endowed matter can be destroyed."

The most powerful drive in Joule was undoubtedly the sheer love for experimental investigations. Independent of Mayer's work (with which he said he was "only imperfectly acquainted" as late as 1850), he announced in 1843 the first result of his labor, the comparison of mechanical work needed to operate an electric generator and the heat produced by the current so generated. He found that 1 J (in modern units) is converted to about 0.22 cal of heat, that is, 1 cal is equivalent to about 4.5 J. This was an average of his thirteen experiments with fairly large uncertainties, for in all such experiments it is extremely difficult to obtain precise and meaningful data, because of unavoidable leakage of heat and the like. Later the same year, Joule measured the relation between heat of friction for water flowing through thin pipes and the work needed to produce the flow (4.14 J/cal), and he concluded:

> We shall be obliged to admit that Count Rumford was right in attributing the heat evolved by boring cannon to friction. . . . I shall lose no time repeating and extending these experiments, being satisfied that the grand agents of nature are, by the Creator's fiat, *indestructible;* and that whenever mechanical force is expended, an exact equivalent of heat is *always* obtained.

But Joule certainly did not rest now. Despite the coolness with which his work was generally received, he continued with almost obstinate zeal. In 1844 he measured the ratio of work needed to compress a gas to the heat so created (4.27 J/cal), in principle rather close to the method Mayer had considered in 1842. A little later, he made the first of a long series of more and more accurate experiments to find the mechanical equivalent of heat by friction, for example, by stirring water with paddlewheels in a heat-insulated container. This is initially a delicate experiment because the temperature rise is fairly small, and his result (4.78 J/cal) is markedly different from the others. By 1847 the technical difficulties were in hand; stirring water and also sperm oil gave very closely the same result (≈4.20 J/cal).

When Joule wished to report on this particular experiment at a scientific meeting in 1847, a crisis arose. He was asked by the chairman to confine himself to a "short verbal description," partly because his previous communications, as Joule himself recalled later, "did not excite much general attention." "This I endeavored to do, and, discussion not being invited, the communication would have passed without comment if a young man had not risen in the Section and by his intelligent observations created a lively interest in the new theory." The young man was William Thomson (afterward raised to the peerage as Lord Kelvin), then 23 years old and destined to become one of the foremost British scientists, collaborator and staunch supporter of Joule's views.

The year 1849 brought further results on friction: in water, in mercury (where the temperature change is larger by virtue of its lower specific heat), and for two iron disks rubbing against each other. Joule compares his methods and results with those of Rumford, estimating a value of about 5.6 J/cal from the data given for one of Rumford's experiments. Considering the uncertainties in the available data, one must conclude that the two equivalence values are quite close.

Then Joule summarized all his previous work and assigned to the mechanical value of heat the value 4.15 J/cal. Today, after more than a century of increasingly refined precision methods, the value for the equivalent between work and heat is 4.184 *J*/cal. The italicized letter *J* was long used, with greatest justice, for Joule's equivalent. This usage has been abandoned and the letter J, *not* italicized, is now used as a abbreviation for the *unit* "joule," defined as 1 N·m, which is therefore numerically equal to 1/4.184 = 0.239 cal.

By 1850, after 10 years of devoted activity, Joule's work began generally to be taken more seriously, the gratifying result of a well-planned experimental assault. During the next 18 years, he carried through three further determinations of *J*. Without a doubt, it was the sum of his consistent experimental work that made acceptable the idea of an enlarged, general law of conservation of all forms of energy in place of the previous specific and limited application of the law solely to conservative mechanical systems.

But, as usually happens, Joule's triumph depended in large measure on developments beyond his own participation. It will be seen in the next chapters that at about the same time the general thinking on the nature of heat and of matter itself was undergoing profound changes toward a view more favorable to the conservation of heat as a form of energy, namely, the kinetic energy of motion of the molecules. This trend removed the cause of a great deal of the original resistance to Rumford, Mayer, and Joule. And second, the idea of a general law of conservation of energy found powerful and eloquent champions, particularly at first the young German physiologist and physicist, Hermann von Helmholtz (1821–1894), in an influential 1847 article. (It is probably no coincidence that most of the protagonists were quite young, under 30, when they became interested in the law of conservation of energy.) Helmholtz did what Mayer, then still in oblivion, had not quite done and Joule had never attempted, namely, to show by mathematical demonstration the precise extent of the conservation law in various fields of science—mechanics, heat, electricity, magnetism, physical chemistry, and astronomy. He derived by its aid, in terms that the professional scientist would at once recognize, explanations for old puzzles and new—confirmable—mathematical relations.

Indeed, as the law of conservation of energy gradually won more adherents and wider application, it was discovered that Mayer's vision had been prophetic—the law of conservation of energy provided striking general connections among the various sciences. It was perhaps the greatest step toward unity in science since Galileo and Newton, and a powerful guide in the exploration of new fields.

A footnote to this historical account may be in order to supply one rather typical detail to this sketch of the growth of a conceptual scheme in science. As Mayer's and Joule's work finally won supporters, there developed a sometimes rather bitter fight as to which of the two had been the first proponent of the general doctrine of conservation of energy in its various forms. A number of other claimants to the title were put forward, men who had expressed in some way the general idea of interconvertibility of forms of energy or had computed in particular the mechanical equivalent of heat: Ludvig Colding in Denmark; Karl Friedrich Mohr, Karl Holtzmann, and Justus Liebig in Germany; Sadi Carnot and Marc Seguin in France; Gustav Adolphe Hirn in Belgium; and W. R. Grove and Michael Faraday in England.[9]

When a discovery is obviously "in the air," as this multiplicity of discoverers suggests, it is meaningless to argue about priorities; yet such arguments are inevitable, for the hope of getting eternal credit for a discovery is one of the motivations that drives a scientist to work overtime on a job that (until quite recently) was paid poorly or not at all. Thus the Jouleites disparaged Mayer's contribution of 1842 that gave him technical priority over Joule, whose first but quite independent paper was dated 1843; and Joule himself wrote in 1864:

> Neither in Seguin's writing (of 1839) nor in Mayer's paper of 1842 were there such proofs of the hypothesis advanced as were sufficient to cause it to be admitted into science without further inquiry. . . . Mayer appears to have hastened to publish his views for the express purpose of securing priority. He did not wait until he had the opportunity of supporting them by facts. My course, on the contrary, was to publish only such theories as I had established by experiments calculated to commend them to the scientific public, being well convinced of the truth of Sir J. Herschel's remark that "*hasty generalization is the bane of science.*"

On the other side, and often not much more fairly, those partial to Mayer's claim to priority would say with Helmholtz:

> The progress of natural science depends on this: that new [theoretical] ideas are continually induced from the available facts; and that afterwards the consequence of these ideas, insofar as they refer to new facts, are experimentally checked against actuality. There can be no doubt about the necessity of this second endeavor; and often this second part will cost a great deal of labor and ingenuity, and will be thought of as a high accomplishment of him who carries it through well. But the fame of discovery still belongs to him who found the new idea; the experimental investigation is afterwards a much more mechanical type of achievement. Also, one cannot unconditionally demand that the inventor of the idea be under obligation to carry out the second part of the

[9]The merits of these various claims have been evaluated in Kuhn, "Energy conservation as an example of simultaneous discovery." The attitudes of scientists toward priority disputes have been discussed at length by R. K. Merton, in "Singletons and multiples in scientific discovery," *Proceedings of the American Philosophical Society,* Vol. 105, pages 470–486, and other articles reprinted in his book *The Sociology of Science,* Chicago: University of Chicago Press, 1973.

endeavor. With that demand one would dismiss the greatest part of the work of all mathematical physicists.

Here we see an old fight, or rather two fights: one for priority (and each great idea, after it is accepted, tends to give rise to such a dispute) and, second, the fight concerning the relative merits of the experimentalist and the theoretician. Both illustrate some of the human element behind the printed equation.

17.10 General Illustration of the Law of Conservation of Energy

As has been demonstrated, the law of conservation of energy may take on any one of many different forms, depending on the intended application. But broadly speaking, we have met three main groups of formulations:

1. *For an isolated system,* the sum of all forms of energy remains constant (the net changes add up to zero) although the internal energy may take on different forms. Examples are collisions between balls or molecules and disintegration in natural radioactivity.
2. *For a system to which energy is given,* including work done or heat supplied to it, *but which does not in turn give energy to the world outside the system,* the energy supplied to the system is equal to the total change in internal energy of all forms. Examples are a block being pulled by a string, gas in a cylinder being compressed by a piston, induced transmutation, and fission by neutron bombardment.
3. *For a system to which energy is supplied and that also does work on or loses energy to the outside world,* the energy supplied to a system is equal to the total change of internal energy in the system *plus* the energy given up by the system. Examples are machines, gasoline engines, and Joule's paddlewheel experiment with heat leakage.

But of course the law of conservation of energy is a *single law,* and the forms it takes depend only on how large we choose the "system" under consideration to be. Thus when a person hoists a load by means of a pulley, the form (3) applies if the pulley itself is the whole system, and the form (1) if the earth together with person, pulley, and load is considered one system. Furthermore, the first two (and one other possible form: which? examples?) are but special examples of the third form, with one or the other factor set equal to zero.

In scientific writings one encounters the law of conservation of energy phrased in many different ways. One equivalent name, *first law of thermodynamics,* is the law of conservation of energy in such form as applies in particular to cases in which exchanges between heat and mechanical energy enter. The first law of thermodynamics can be stated in the form:

$$\Delta E = H - W, \tag{17.17}$$

where E is the energy of a body, ΔE is the change of energy involved in going from one state (in other words, temperature and pressure) to another, H is the heat added to the body, and W is the work done by it on its surroundings.

In addition, it is assumed in thermodynamics that the energy of a body in a particular state does not depend on how it got to that state. Thus for example in Mayer's calculation of the mechanical equivalent of heat (Section 17.8), it was assumed that the net change in energy of 1 g of air is the same whether it is simply heated by 1°C with the pressure kept constant or first heated at constant volume (while the pressure rises) and then allowed to expand by pushing out a piston (thereby doing mechanical work and dropping the pressure) so that it ends up in the same final state. In the first case $\Delta E = H = 0.17$ cal; in the second case $\Delta E = 0.24 \text{ cal} - W$. Since in the second case the net energy change must also be $\Delta E = 0.17$, we can infer that W is equivalent to 0.07 cal.

The fact that a substance can be heated by compression or cooled by expansion—that is, its temperature changes without any actual exchange of heat with the surroundings—is a further illustration of the first law of thermodynamics. A given ΔE (signaled by a definite change of temperature or perhaps by a change of state such as melting or evaporation) can be accomplished by several different combinations of H and W. (Note that if the body loses heat instead of gaining it, H in Eq. (17.14) is negative. Similarly, if work is done on the body—for example, by compressing it—W is negative. This does not mean that heat and work are themselves negative quantities, but simply that they can either decrease or increase the energy of a body, depending on which process is being considered.)

Another, equivalent formulation of the law of conservation of energy is of particular interest: It is the statement that a *perpetuum mobile* (perpetual mover) is impossible; for by *perpetuum mobile* is generally meant a machine that has an efficiency of *more* than 1.00 (100%), one that yields more output than input. One may even identify the rise of the law of conservation of energy in mechanics with the continued unsuccessful attempts to build such a machine, for it should be remembered that this project was very popular for hundreds of years. The search, which engaged some of the greatest intellects, was in itself no more implausible than the earlier and equally fruitless search for the "Philosopher's Stone" (which was supposed to be able to turn all common metals into gold or silver); it may have reflected the preoccupation of an increasingly more machine-minded civilization. However, it probably points up equally that most pathetic of human traits, the wish to get something for nothing.

The inability to make a perpetual motion machine became itself a fact of physical reality, one example of a series of so-called *postulates of impotency,* which have profoundly influenced physical science. To this class belong the impossibility of transmuting matter chemically, which helped to shape the idea of chemical elements; the inability to discover the relative motion of the ether in optical experiments, which was embodied in a fundamental postulate of relativity theory; the impossibility of obtaining net mechanical energy from a medium by cooling it below the temperature of its surroundings,[10] which is the effective content of the second law of thermodynamics; and several others.

The following examples will provide additional familiarity with the law of conservation of energy in its most general form, and additional evidence of its wide-ranging power.

The Human Body as an Engine

Once it was thought that animals, including man, must eat only in order to grow and to rebuild used-up tissue. The realization that the food intake itself also provides heat and energy for the functioning of the body was attained only slowly. At the danger of simplifying a complex topic far too much, we may compare the body to a chemical engine: The oxygen we breathe burns the sugars, fats, and some of the proteins from our digested food intake, converting their complex molecules mainly into carbon dioxide, water, and other waste products in an exothermic reaction, that is, one that liberates heat and, in the contraction of muscles, provides mechanical work.

The balance of energy intake in the form of *combustible* food vs. work and heat output has been measured with precision, partly in an effort to check whether life processes also obey the conservation laws in the strict forms that prescribe the activities of inanimate nature. This question was relevant to the claims of the vitalists that biology requires the action of some "vital spirit" not subject to physical laws.

Since the 1890s this problem has been clearly settled in the affirmative: Life processes do conform to the law of conservation of energy. In a representative case, the daily food intake would yield about 3×10^6 cal if burned directly, and the heat given up together with the work done and the energy in the waste products come to the same figure within a small experimental uncertainty. But the body, considered as an engine, ordinarily has an efficiency of less than 20%; so if one has to do large amounts of physical work within a short period of time, the attendant heat output is correspondingly larger, and since the rate of dissipation of heat cannot easily be controlled, the body heats up drastically.

Photosynthesis

If the energy of the animal body were derived from a natural fuel like oil, the food supply on this earth would have been exhausted long ago. The present form of animal life depends on the fact that the body's fuels are continually synthesized by green plants and are readily available, one might say, in the form of either vegetables or vegetarians such as cattle. The green plant is uniquely able to draw carbon dioxide from the surrounding medium and water, both waste products of animal life, and to recombine the aggregates of carbon, hydrogen, and oxygen, plus those of some mineral salts, into more complex "organic" compounds, each having a large amount of built-in chemical "potential" energy

[10]An engine that functioned thus, if it were not impossible for other reasons, would not be in conflict with the law of conservation of energy. It is referred to as a *perpetuum mobile of the second kind* and was once as much sought after as the *perpetuum mobile of the* first kind described above. The second law of thermodynamics is discussed in Section 18.3.

that can be released later during decomposition inside the animal body. For good measure, the plant also releases oxygen into the atmosphere while it builds our food.

For each gram of carbon content, these "fuels" can deliver roughly 10^4 cal worth of work and heat, and on this depends the life of all animals except some lower organisms that live by fermentation. But is all this not a crass contradiction of the law of conservation of energy? Do the plant and animal worlds together not constitute a *perpetuum mobile?* The answer is no, because the energy in the manufactured organic molecules is originally taken by the plant from sunlight. The reactions depend on the availability of light energy (and on the presence of chlorophyll and other catalysts); technically this is termed photosynthesis. In this sense, our food is largely sunlight.

If not eaten, the energy-rich plant material in the open will soon decay into its original compounds; but if it is submerged or otherwise suitably kept from ordinary oxidation, it may turn into coal or other heating fuels by slow chemical decompositions that favor the retention of carbon-rich materials.

The Origin of the Solar System

In Section 16.8 we discussed Laplace's nebular hypothesis as an application of the law of angular momentum. Laplace assumed that the sun already existed as a large hot body at the beginning of the process of cooling, contraction, and spin-off of rings to form planets. But he did not attempt to explain where the sun got its enormous energy in the first place or how it could continually emit heat and light at such a staggering rate (about 3.8×10^{26} J or 9×10^{25} cal every second). If this were achieved by burning ordinary fuels or some such chemical reaction, it was difficult to see why the material for it had not been exhausted long ago. Mayer, Helmholtz, and others proposed promising alternatives—that the sun's radiated energy was derived from gravitational potential energy by a slow shrinking of the sun or by the capture of comets and meteors. But then it was calculated that the sun's diameter would have to shrink 1 part out of 10,000 every 2000 years to provide radiation in this manner at the present level. That is too fast to account for the long history of life on earth. Nor can enough meteoric matter be found to give the observed effect.

The now accepted source of the sun's energy, namely a nuclear fusion process (see Section 32.2), was of course not imaginable until much more recently. Yet, the original proposal is not altogether without merit, although it failed for the initial application. The same general mechanism of conversion of gravitational potential energy to heat is invoked today in theories concerning the creation of the solar system. (Recent theories are discussed in Chapter 31.)

17.11 Conservation Laws and Symmetry

In 1905 Albert Einstein proposed a remarkable generalization of the laws of conservation of mass and of energy, resulting from the special theory of relativity: Under certain circumstances, mass and energy are not conserved separately, but can be transformed into each other in accordance with the famous equation $E = mc^2$ (Chapter 30). A single quantity, which might be called "mass-energy," combines the total energy of a system and the energy-equivalent of the total mass of the system; this quantity is still conserved (Eq. 30.12).

The later development of the general theory of relativity involved some rather complex and abstract mathematical calculations, and it was not immediately clear whether mass-energy is necessarily conserved in the new theory. The German mathematician David Hilbert was also working out a general theory of relativity by a different method at the same time; he was so concerned about this uncertainty that he called upon a young mathematician, Emmy Noether (1882–1935) for assistance. This fact in itself is quite remarkable, since Hilbert had a broad knowledge of mathematics and was at that time (1915) one of the leading mathematicians in the world and a professor at Göttingen University, a major international center of mathematical research, while Noether did not even have a salaried academic position. She was, however, an expert on "invariant theory," a difficult and esoteric specialty that happened to be crucially important to the problem Hilbert wanted to solve.

In response to Hilbert's request, Noether proved a very general theorem, which states that if the fundamental equations of a theory have a certain kind of *symmetry,* then a corresponding quantity will be conserved. When applied to the

Fig. 17.26. Emmy Noether (1882–1935), German mathematician, proved a fundamental theorem on the connection between symmetry and conservation laws.

general theory of relativity, Noether's theorem states that since the equations are symmetrical or invariant with respect to a change in the time variable, for example changing from t to $t + \Delta t$, then mass-energy must be conserved. In other words, if we do a physics experiment today and then do exactly the same experiment in exactly the same way a year from now, we should get the same result.

The same conclusion is also valid in Newtonian mechanics as long as there are no frictional forces or inelastic collisions: Invariance under a shift along the time axis implies conservation of energy. It does not, however, imply that mass is conserved in this case. Moreover, Noether's theorem also says nothing about another kind of symmetry: time *reversal* (changing t to $-t$), which is the topic of our next chapter.

In addition to the connection between time-invariance symmetry and the law of conservation of energy, Noether's theorem also reveals an interesting fact about momentum. If a theory's equations are invariant under changes in the *position* variables, then *momentum is conserved.* If we do an experiment in Berkeley, California, and the same experiment in College Park, Maryland, we expect to get the same result. Similarly if the equations are unchanged when we rotate the system, then *angular momentum is conserved.*

For several decades Noether's theorem was known only to a few experts; Noether herself went back to her research in pure mathematics and made important contributions to algebra and topology as well as to invariant theory. Since the 1950s, however, her theorem has become an important tool for research in several areas of theoretical physics, especially elementary particle theory.

Noether's career, like that of many other women in science and the professions, suffered from the discrimination that was prevalent throughout the world until quite recently. After struggling to get permission to attend lectures and finally to get a doctoral degree at the University of Erlangen, where her father was a professor, she eventually moved to Göttingen at Hilbert's invitation. But the University of Göttingen had a rigid rule against allowing women to teach; even the powerful professor Hilbert could not persuade the University Senate to admit her to the faculty, despite his protest: "I do not see that the sex of the candidate is an argument against her admission . . . After all, the Senate is not a bathhouse!" Finally, after publishing her now-famous theorem and many other important papers, she was grudgingly allowed to give lecture courses and advise graduate students—but still with no salary.

In 1933, after Emmy Noether had attained international recognition for her mathematical discoveries, she was expelled from Germany by the Nazis and came to the United States, where she found a temporary position at Bryn Mawr College. She died unexpectedly after an operation in 1935.

Additional Problems

Problem 17.23. Before Joule obtained a more accurate value of the mechanical equivalent of heat (J) he wrote (1845): "Any of your readers who are so fortunate as to reside amid the romantic scenery of Wales or Scotland could, I doubt not, confirm my experiments by trying the temperature of the water at the top and bottom of a cascade. If my views be correct, a fall of 817 feet will of course generate one degree [1°F] . . . and the temperature of the river Niagara will be raised about one-fifth of a degree [F] by its fall of 160 feet." (This general prediction he later checked himself, during his honeymoon trip to a large Swiss waterfall.)

Now compute from this prediction what value of J he was using at the time. To convert temperature scales, recall that 0°C = 32°F, and 100°C = 212°F.

Problem 17.24. When a 1-kg ball of lead [specific heat = 0.03 (cal/g)/°C] falls freely through 10 m and is stopped suddenly by collision with the floor, what is the greatest temperature rise you would expect? Cite several causes that would contribute to making the observed temperature rise smaller.

Problem 17.25. (a) Calculate the total energy of the earth at one point of its orbit (PE with respect to the sun, plus KE_{rev} in its orbit and KE_{rot} about its axis). (b) Knowing the size and eccentricity of the earth's orbit ($e = 0.017$), find the difference in our actual distance from the sun at perihelion and aphelion, and compute from that plus part (a) the maximum difference in orbital speed.

Problem 17.26. In his sensationalistic book *Worlds in Collision* (Macmillan, 1950), Immanuel Velikovsky maintains that at about 1500 B.C. the planet Jupiter ejected material which, in the form of a comet, passed close to the earth; that the subsequent shower of meteorites falling on the earth stopped its rotation, at least for a short time; that the comet itself eventually turned into the planet Venus; and that later, in 747 B.C. and again in 687 B.C., "Mars caused a repetition of the earlier catastrophes on a smaller scale." (a) Given the mass and radius of the earth, as well as its angular speed of rotation ($m_e = 5.98 \times 10^{24}$ kg, $r \approx 6370$ km, $\omega = 2\pi$ rad per day), find the amount of angular momentum of the earth that must be carried off before it will stop rotating and that must be supplied again to start up the rotation afterward.(b) In the most favorable case the meteorites would have fallen on the equator at grazing incidence, traveling with a speed of perhaps 50,000 mi/hr relative to the surface of the earth. What must have been the approximate total mass of the meteorites? What is the total kinetic energy of rotation annihilated by the stopping of the earth and meteorites? What happened to that energy, and how could the earth's rotation start again without another collision? How could gravitational, therefore central, forces between passing spherical bodies radically and suddenly change their state of rotation? (c) What would happen to the oceans and the atmosphere if the earth were to stop rotating by Velikovsky's mechanism?

Problem 17.27. Explain by means of the law of conservation of energy the following observations: (a) A sample of naturally radioactive material maintains its temperature above that of its surroundings. (b) In a mass of air that is rising to greater height, the moisture content may freeze. (c) While a battery generates electric energy, it undergoes chemical decomposition at the plates or electrodes.

Problem 17.28. Reread the quotations from Joule and Helmholtz at the end of Section 17.9, and discuss, on the basis of the available evidence, the fairness of each.

Problem 17.29. Calculate the gravitational force exerted by the earth on the moon, assuming the moon moves in a circular orbit whose radius is the average earth-moon distance. What is the total amount of work done by the earth on the moon as it moves once around a complete orbit?

RECOMMENDED FOR FURTHER READING

S. C. Brown, "Benjamin Thompson, Count Rumford," in *Physics History* (edited by M. N. Phillips), pages 11–22

Allan Chapman, "Christiaan Huygens (1629–95): Astronomer and mechanician," *Endeavour,* Vol. 19, pages 140–145 (1995)

Auguste Dick, *Emmy Noether 1882–1935,* translated by H. I. Blocher, Boston: Birkäuser, 1981

Herman Erlichson, "The young Huygens solves the problem of elastic collisions," *American Journal of Physics,* Vol. 65, pages 149–154 (1997)

C. C. Gillispie (editor), *Dictionary of Scientific Biography,* articles by H. J. M. Bos on Huygens, Vol. 6, pages 597–613; by R. S. Turner on Mayer, Vol. 9, pages 235–240; by L. Rosenfeld on Joule, Vol. 7, pages 180–182; by Edna Kramer on Noether, Vol. 10, pages 137–139

P. M. Harman, *Energy, Force, and Matter,* Chapters II and III

Hermann von Helmholtz, "On the interaction of natural forces" and "On the conservation of force," translated from lectures given in 1854 and 1862, in his *Popular Scientific Lectures,* New York: Dover, 1962, pages 59–92 and 186–222

James Prescott Joule, "On matter, living force, and heat" (lecture given in 1847), reprinted in *Kinetic Theory* (edited by Brush), Vol. 1, pages 78–88

S. Sambursky (editor), *Physical Thought,* excerpts from Leibniz, Joule, and Helmholtz, pages 318–321, 394–405

Crosbie Smith, "Energy," in *Companion* (edited by Olby), pages 326–341

J. H. Weaver (editor), *World of Physics,* excerpts from Rumford, Mayer, Joule, and Ford, Vol. 1, pages 672–702, and Vol. 2, pages 762–790

Anthony Zee, *Fearful Symmetry: The Search for Beauty in Modern Physics,* New York: Macmillan, 1986, Chapters 8, 11 and 14 (on Noether's theorem and its applications)

CHAPTER 18

The Law of Dissipation of Energy

In the last three chapters we have seen how the conservation laws of mass, momentum, and energy came to be established. With certain modifications, such as the possibility of interconversions between mass and energy, to be discussed in Chapter 30, these laws are still considered valid today. Taken as a whole, they seem merely to confirm an opinion common in the seventeenth century: that the universe is like a perfect machine whose parts never wear out, and which never runs down. In such a "world machine," nothing really changes in the long run; the pieces of matter that constitute our universe can only perform simple or complex cycles of motion, repeated indefinitely. Moreover, Newton's laws of motion make no distinction between past and future: Any sequence of motions predicted by the laws can equally well be run backward.

This idea of the world as a machine (or perhaps as a mechanistic device analogous to a clock) does not in fact correspond to the general scientific viewpoint about the physical world today, nor to the facts, if one looks carefully. As the British physicist William Thomson (later known as Lord Kelvin) pointed out in 1874, Newton's laws alone do not explain why a possible state of affairs cannot in general be constructed by simply reversing the velocities of all the atoms in a system:

> If the motion of every particle of matter in the universe were precisely reversed at any instant, the course of nature would be simply reversed forever after. The bursting bubble of foam at the foot of a waterfall would reunite and descend into the water; the thermal motions would reconcentrate their energy, and throw the mass up the fall in drops reforming into a close column of ascending water. Heat which had been generated by the friction of solids and dissipated by conduction, and radiation with absorption, would come again to the place of contact, and throw the moving body back against the force to which it had previously yielded. Boulders would recover from the mud the materials required to rebuild them into their previous jagged forms, and would become reunited to the mountain peak from which they had formerly broken away. And if also the materialistic hypothesis of life were true, living creatures would grow backwards, with conscious knowledge of the future, but no memory of the past, and would become again unborn. But the real phenomena of life infinitely transcend human science; and speculation regarding consequences of their imagined reversal is utterly unprofitable.
>
> (Kelvin, "Kinetic theory")

The gradual decline of the world-machine idea in modern science has been a long and complicated process, which is not yet completed, and we shall not be able to tell the full story in this book (it involves, among other things, the rise of evolutionary biology). But we can in this chapter at least indicate one immensely important and fascinating principle that was added to the system of physical laws in the nineteenth century and which throws a different light on the idea of conservation: the *principle of dissipation of energy,* sometimes known as the generalized *second law of thermodynamics.*

18.1 Newton's Rejection of the "Newtonian World Machine"

It is one of the ironies of the history of science that the one man whose work by the nineteenth century had done most to establish the idea that the world is like a machine or clock was the most eager to reject that idea. Newton pointed out in his *Opticks* (Query 31) the consequences of adopting a principle of vector momentum in a world in which collisions are not always elastic: The particles can stop moving even though their total momentum is unchanged (see Section 16.2).

> By reason of the . . . weakness of elasticity in solids, motion is much more apt to be lost than got, and is always upon the decay. For bodies which are either absolutely hard, or so soft as to be void of elasticity, will not rebound from one another. Impenetrability makes them only stop. If two equal bodies meet directly in a vacuum, they will by the laws of motion stop where they meet, and lose all their motion, and remain in rest, unless they be elastic, and receive new motion from their spring. If they have so much elasticity as suffices to make them rebound with a quarter, or half, or three quarters of the force with which they come together, they will lose three quarters, or half, or a quarter of their motion. . . .
>
> Seeing therefore the variety of motion which we find in the world is always decreasing, there is a necessity of conserving and recruiting it by active principles, such as are the cause of gravity, by which planets and comets keep their motion in their orbs, and bodies acquire great motion in falling; and the cause of fermentation, by which the heart and blood of animals are kept in perpetual motion and heat; the inward parts of the earth are constantly warm'd, and in some places grow very hot; bodies burn and shine, mountains take fire, the caverns of the earth are blown up, and the sun continues violently hot and lucid, and warms all things by his light. For we meet with very little motion in the world, besides what is owing to these active principles. And if it were not for these principles, the bodies of the earth, planets, comets, sun, and all things in them, would grow cold and freeze, and become inactive masses; and all putrefaction, generation, vegetation and life would cease, and the planets and comets would not remain in their orbs.

Newton seems to be arguing here that the continual dissipation of motion in the world, which would occur in spite of the law of conservation of momentum, somehow is or must be counteracted by the effects of "active principles." It is tempting to interpret the mysterious phrase "active principles" as a suggestion of the role of energy in the universe; but the tenor and context of Newton's argument do not support the hypothesis that he was willing to let the continued operation of his system depend simply on the conservation of energy. Indeed, in an earlier edition of the *Opticks* there appeared the phrase "what is owing either to these active principles or to the dictates of a will," which was modified in the 1717 edition (perhaps as a result of the criticisms mentioned below) to read "what is owing to these active principles," as in the quotation above.

A further indication of Newton's reluctance to accept that his laws of motion would guarantee the existence of an eternal world machine is given by another passage from the same work:

> For while comets move in very eccentric orbs in all manner of positions, blind fate could never make the planets move one and the same way in orbs concentric, some inconsiderable irregularities excepted which may have arisen from the mutual actions of comets and planets upon one another, and which will be apt to increase, till this system wants a reformation.

What would a "reformation" be? A miraculous act of God, putting the comets and planets back into their proper orbits, from which they had strayed? Any such suggestion of a need for divine intervention would cast doubt on the adequacy of the laws of motion by themselves to maintain the world machine in perfect working order.

Newton was challenged directly on this point by Leibniz, the German philosopher-scientist who had previously developed the concept of *vis viva* (Section 17.1) and had invented the calculus independently of Newton. In a letter (soon made public) to Caroline, Princess of Wales, in 1715, Leibniz began by complaining that Newton and his friend, the philosopher John Locke, were undermining belief in God and the soul. He continued:

> Sir Isaac Newton and his followers have also a very odd opinion concerning the work of God. According to their doctrine, God almighty needs to wind up his watch from time to time; otherwise it would cease to move. He had not, it seems, sufficient foresight to make it a perpetual motion. Nay, the machine of God's making, is so imperfect, according to these gentlemen, that he is obliged to clean it now and then by an extraordinary concourse, and even to mend it, as a clockmaker mends his work; who must consequently be so much the more unskillful a workman, as he is often

> obliged to mend his work and set it right. According to my opinion, the same force and vigour [energy] remains always in the world, and only passes from one part to another, agreeably to the laws of nature, and the beautiful pre-established order. . . .
>
> (Alexander, *Leibniz-Clarke Correspondence*)

Rather than replying directly, Newton dictated his answers to a young theologian, Samuel Clarke, who had previously translated Newton's *Opticks* into Latin. Speaking through Clarke, he firmly rejected the idea of a world machine that can continue to run forever without divine intervention, along with Leibniz's suggestion that a law of conservation of energy is sufficient to account for all natural phenomena. He pointed out that the clockmaker, whose highest achievement would be to make a clock that runs forever without any need for adjustment, does not himself create the forces and laws of motion that enable the clock to run, but must take them as given. But since God, according to Newton, "is Himself the author and continual preserver of [the world's] original forces or moving powers," it is

> . . . not a diminution, but the true glory of His workmanship, that nothing is done without his continual government and inspection. The notion of the world's being a great machine, going on without the interposition of God, as a clock continues to go without the assistance of a clockmaker, is the notion of materialism and fate, and tends (under pretence of making God a supramundane intelligence) to exclude providence and God's government in reality out of the world.
>
> (*ibid.*)

Thus, he argued, the theory of the self-sufficient clockwork universe, far from being complimentary to God, as Robert Boyle, Leibniz, and others had claimed, was likely to lead instead to atheism. Skeptics, Newton predicted, could easily push the creation of the universe—the first and last act of God—further and further backward in time and might even conclude eventually that the universe has always existed as it does now.

As Newton's system of the world was further developed by European scientists in the eighteenth and early nineteenth centuries (especially Euler and the French mathematicians Lagrange, Laplace, and Poisson), it appeared that the instability that Newton had predicted would result from mutual gravitational interactions of comets and planets would not occur; these interactions would simply cause the average distances of the planets from the sun to oscillate periodically within finite limits (Section 11.6). Thus, as Newton had feared, the need for divine intervention or guidance to keep the system running smoothly seemed to disappear.

According to legend, when Napoleon met Laplace, after glancing through Laplace's masterpiece, *Celestial Mechanics,* Napoleon remarked that he found no mention of God at all in the book; to which the author replied, "Sir, I have no need of that hypothesis!"[1] The idea of the world machine had triumphed—at least for the time being; and it seemed to be established that Newton's laws by themselves guaranteed the stability of the solar system over indefinitely long periods of time, both past and future.

18.2 The Problem of the Cooling of the Earth

Although the successes of Newtonian planetary theory seemed to lead convincingly to the idea of the world as a machine, speculation about the earth's interior in the eighteenth and early nineteenth centuries gave rise to another viewpoint. We must now venture at least briefly into geophysics, an important though sometimes neglected part of physics.

Many writers in earlier centuries had suggested that the earth was originally formed in a hot, molten state, and has subsequently cooled down, solidifying at least on the outside. Leibniz endorsed this theory in 1693, arguing that many rocks found on the surface of the earth appear to have been formed by the action of fires, which may still be burning inside the earth. The prominent French naturalist Georges Louis LeClerc, Comte de Buffon (1707–1788)

---

[1]One source for this story is De Morgan, *Budget of Paradoxes,* Vol. II, pages 1–2. The usual interpretation of Laplace's remark is that he claims to have proved the solar system to be a perfectly cyclic clockwork mechanism, but it may also be read as a claim that the laws of physics can explain how the solar system reached its present state from a simpler initial state, as suggested by Laplace's nebular hypothesis.

suggested that the discovery of ivory tusks of elephants in polar regions, where these animals cannot now survive, supports the idea that the earth's surface was once warmer than it is now.

Buffon also conducted a series of experiments on the cooling of heated spheres of iron and other substances, in order to estimate the time required for the earth to cool down to its present temperature from the postulated molten state; his final published result was about 75,000 years (though privately he may have favored an even longer estimate). While this figure seems extremely low in the light of contemporary knowledge, it represented a shockingly long expanse of time to his contemporaries, who believed that the chronology of the Bible fixed the creation of the universe at a time only a few thousand years ago.

From the theory of the earth's central heat and gradual refrigeration, it was but a short step to the conjecture that all bodies in the universe are cooling off and will eventually become too cold to support life. This step seems to have been first taken explicitly by the French astronomer Jean-Sylvain Bailly (1736–1793) in his writings on the history of astronomy. According to Bailly, all the planets must have an internal heat and are now at some particular stage of cooling; Jupiter, for example, is still too hot for life to arise for several thousand more years; the moon, on the other hand, is already too cold. But Bailly predicts that all bodies must eventually reach a final state of equilibrium in which all motion ceases.

James Hutton (1726–1797), one of the founders of geological science in Britain, accepted the hypothesis that the inside of the earth is now much hotter than the surface as a convenient explanation for volcanoes and hot springs, but did not believe that there had been any cooling during past epochs. In his *Theory of the Earth* (1795), he wrote that "subterraneous fire had existed previous to, and ever since, the formation of this earth," but also "that it exists in all its vigour at this day." Hutton proposed a cyclic view of the earth's history, with periods of erosion and denudation leading to destruction of mountains and even entire continents, followed by consolidation of sediments and uplifting of new continents (powered by the subterranean fires).

Hutton's disciple John Playfair explained this cyclic viewpoint in a widely read book, *Illustrations of the Huttonian Theory of the Earth* (1802):

> How often these vicissitudes of decay and renovation have been repeated, is not for us to determine: they constitute a series, of which, as the author of this theory has remarked, we neither see the beginning nor the end; a circumstance that accords well with what is known concerning other parts of the economy of the world. In the continuation of the different species of animals and vegetables that inhabit the earth, we discern neither a beginning nor an end; and, in the planetary motions, where geometry has carried the eye so far both into the future and the past, we discover no mark, either of the commencement or the termination of the present order. It is unreasonable, indeed, to suppose, that such marks should anywhere exist. The Author of nature has not given laws to the universe, which, like the institutions of men, carry in themselves the elements of their own destruction. He has not permitted, in his works, any symptoms of infancy or of old age, or any sign by which we may estimate either their future or their past duration. He may put an end, as He no doubt gave a beginning, to the present system, at some determinate period; but we may safely conclude, that this great catastrophe will not be brought about by any of the laws now existing, and that it is not indicated by any thing which we perceive.

In a supplementary note, Playfair cited the mathematical investigations of Lagrange and Laplace, which have shown that:

> . . . all the variations in our [solar] system are periodical; that they are confined within certain limits; and consist of alternate diminution and increase. . . . The system is thus endowed with a stability, which can resist the lapse of unlimited duration; it can only perish by an external cause, and by the introduction of laws, of which at present no vestige can be traced.

The Huttonian system itself has the same stability; both systems, remarks Playfair, are just like machines designed by a superior intelligence.

The principle that geological phenomena are to be explained by using only those laws and physical processes that can now be observed to operate—as opposed to postulating catastrophic events in the past, such as the Biblical Flood—was known as *Uniformitarianism*. This principle was used by Sir Charles Lyell (1797–

Fig. 18.1. Jean Baptiste Joseph Fourier (1768–1830), French mathematician. His equation for heat conduction introduced a principle of time-irreversibility into the laws of physics (heat flows from hot to cold, not the reverse—later seen as an example of the second law of thermodynamics). His method for solving the equation, using infinite series of trigonometric functions, found many applications in mathematical physics.

1875) in his influential synthesis of geological knowledge, first published in 1830, and provided a basis for most Anglo-American geological speculation up to the twentieth century. Although uniformitarianism was an essential tool of the movement to make geology a science independent of theological dogma, some nineteenth-century scientists criticized it as being itself a dogma incompatible with the laws of physics. In particular, Lyell refused to accept the theory that the earth had previously been molten and has been cooling by losing heat; that would imply that some geological processes had been much more intense in the distant past than now.

By 1830 the doctrines of Hutton, Playfair, and Lyell were inconsistent with prevailing physical ideas about the earth's past. On the astronomical side, Laplace's nebular hypothesis (Section 16.8) was widely accepted; it implied that the earth had originally been formed as a hot gaseous ball, which later cooled and solidified. On the geophysical side, the theory of the cooling of the earth after its solidification had been put on a quantitative basis by the labors of the French mathematician Jean Baptiste Joseph Fourier (1768–1830). Motivated originally by the problem of terrestrial temperatures, Fourier had developed a theory of heat conduction in solids, together with a powerful method for solving the equations that arose in the theory.

Both Fourier's general theory of heat conduction and his method of solving its equations (by expressing an arbitrary mathematical function as the sum of an infinite series of trigonometric functions) have had enormous impact on modern theoretical physics. Although they cannot be discussed here, one significant fact must be remarked: Fourier's heat conduction equation, unlike Newton's laws of motion, is *irreversible* with respect to time. *It postulates that wherever temperature differences exist, they tend to be evened out by the flow of heat from high temperature to low.* (The contrary, the spontaneous flow of heat from cold to hot, is not part of nature's processes.) Conversely, the existence of such temperature differences now can be used to calculate the even greater differences of temperature that existed in the past.

Fourier used his theory to study the thermal history the earth may have had, taking account of the periodic variations in the heat received from the sun at various points on the surface, combined with a slowly diminishing internal heat. By using measurements of the rate of increase of temperature as one descends into the interior of the earth, in mines, together with estimates of the thermal conductivity of rocks, he was able to infer the earth's internal temperatures at earlier times. He concluded that the theory of a hot interior, gradually cooling down, is compatible with available data, although in recent geological epochs the temperature of the interior has not had any significant effect on the temperature of the earth's surface.

A more comprehensive theory was developed by William Hopkins (1793–1866), a British geophysicist who was also known as a brilliant teacher of mathematics at Cambridge University. (His students included Lord Kelvin and James Clerk Maxwell.) Hopkins pointed out in 1839 that when a hot liquid sphere cools down, it does not necessarily solidify on the outside first, even though the outer surface is most directly exposed to the lower temperature of the surrounding space. For a sphere the size of the earth, the weight of the outer layers produces enormous pressures on the inside, and these pressures may produce solidification even at temperatures far above the normal melting point of the substance.

In 1842, Hopkins published an elaborate mathematical study of the effect a partially fluid earth could have on the moon's motion, from which he concluded that a fairly large portion of the earth's interior must be solid. He proposed that the central core of the earth is solid and that a liquid region lies between this and a fairly thick crust.[2]

According to Hopkins' theory, changes in the surface climate have been produced by geographical changes rather than by cooling of the interior. This idea, previously proposed by Lyell, was reinforced by the evidence collected by the Swiss-American naturalist Louis Agassiz (1807–1873), showing that the earth's surface had been much colder during certain epochs in the past than at present and that ice had covered large parts of the continents.

Hopkins emphasized the point that if the earth's surface can warm up after an Ice Age, the older theories of the cooling of the earth cannot provide adequate explanations for most geological phenomena. But at the same time, while reviewing the situation in his presidential address to the Geological Society of London in 1852, Hopkins reiterated the principle that over very long periods of time the cooling of the earth *is* important, and imposes an overall irreversibility or progressive change on terrestrial phenomena:

> I assume the truth of the simple proposition, that if a mass of matter, such, for instance, as the earth with its waters and its atmosphere, be placed in a space of which the temperature is lower than its own, it will necessarily lose a portion of its heat by radiation, until its temperature ultimately approximates to that of the circumambient space, unless this reduction of temperature be prevented by the continued generation of heat.
>
> (Hopkins, "Anniversary Address")

Moreover, the earth cannot be saved from this fate by postulating an influx of radiation from the sun and stars to balance the heat it loses, for all heat in the universe must be subject to the general tendency to flow from hot to cold. For this reason, Hopkins insisted, "I am unable in any manner to recognize the seal and impress of eternity stamped on the physical universe"; thus he implied that the universe must be governed by this overall law of equalization of temperature, which makes it tend irreversibly toward an ultimate limit, the cold of outer space.

18.3 The Second Law of Thermodynamics and the Dissipation of Energy

Hopkins' enunciation of the principle of irreversible heat flow in geological processes did not attract much attention outside the audience of geologists to which he addressed it; but as it happened an even more general principle of irreversibility was proclaimed in the same year, 1852, by William Thomson. To understand the basis for his principle, we must go back to the origin of thermodynamics in the analysis of steam engines.

In 1824, a young French engineer, Sadi Carnot (1796–1832), published a short book entitled *Reflections on the Motive Power of Fire,* in which he raised the question: What is the maximum "duty" of an engine? (The duty of an engine was defined as the amount of mechanical work it could perform, using a given amount of fuel; it is related to the concept of *efficiency.*)

Carnot used the caloric theory of heat in his analysis of steam engines; he assumed that heat is not actually converted into work, but that the flow of heat from a hot body to a cold one can be used to *do* work, just as the fall of water from a high level to a lower one can be used to do work. Although the caloric theory upon which his results were based was later rejected by other scientists (and even by Carnot himself, in notes that remained unpublished until 1878), his conclusions are still valid. He found that there is a maximum value for the "duty." There is a fixed upper limit on the amount of work that can be obtained from a given amount of heat by using an engine, and this limit can never be exceeded regardless of what substance—steam, air, or anything else—is used in the engine.

---

[2]This central or "inner" core of the earth was discovered in 1936 by the Danish seismologist Inge Lehmann (1888–1993). Her analysis of records of a 1929 New Zealand earthquake indicated that the inner core has a radius of about 1400 km. Hopkins' model for the overall structure of the earth's interior has been qualitatively confirmed by twentieth-century research (see Brush, *History of Modern Planetary Physics,* Vol. 1, Part 2).

Fig. 18.2. Nicolas Léonard Sadi Carnot (1796–1832), French engineer whose analysis of the efficiency of steam engines provided a first step toward the second law of thermodynamics.

Fig. 18.3. William Thomson, Lord Kelvin (1824–1907), one of the most influential British physicists of the nineteenth century. He advocated mechanical models and quantitative measurements (see quotations in Sections 3.1 and 13.1). One of the founders of thermodynamics, he proposed the principle of dissipation of energy and used Fourier's heat conduction to estimate the age of the earth. Because of his expertise in electricity and in precise instrumentation, he was a frequent consultant for industry; thanks to his advice, the Atlantic Cable project was rescued from failure and telegraphic communication between Britain and America was established. He was knighted as a reward for this achievement, and later given the title Baron Kelvin of Largs (Lord Kelvin).

Even more significant than the existence of this maximum limit is Carnot's conclusion that all real engines fail to attain that upper limit in practice. The reason is that whenever a difference of temperature exists between two bodies, or two parts of the same body, there is a *possibility* of doing work by allowing heat to expand a gas as the heat flows from one body to the other. But if heat flows by itself from a hot body to a cold one, and we do not design our engine properly, we will lose the chance of doing work that might have been done.

Carnot's analysis of steam engines shows that the process of equalization of temperature by the flow of heat from hot bodies to cold ones is always taking place, not only in engines but in all of nature, and that when this equalization is achieved it represents a loss of the possibility of doing mechanical work, even though of course the total energy is the same before and after. This is what we mean when we say that energy is "dissipated"—*the total amount of energy in a closed system (or in the world) always stays the same, but energy tends to transform itself into less useful forms.*

After the discovery of the conservation law for energy (Chapter 17), Carnot's conclusions about steam engines were incorporated into the new theories of heat, and became known as the *second law of thermodynamics.* This law has been stated in various ways, all of which are roughly equivalent and express the idea that *the tendency of heat to flow from hot to cold makes it impossible to obtain the maximum amount of mechanical work from a given amount of heat.*

When Kelvin in 1852 asserted a "universal tendency in nature to the dissipation of mechanical energy," he relied primarily on Carnot's principle. But he also drew the conclusion that "within a finite time past the earth must have been, and within a finite period of time to come the earth must again be, unfit for the habitation of man. . . ." (Kelvin, "Universal tendency") Kelvin was especially interested in the theory that the earth has been very hot in the past and will be very cold in the future, for he was one of the first scientists in Britain to take up Fourier's theory of heat conduction and its geophysical implications.

In the 1860s Kelvin came to the conclusion that a period of 100 million to 200 million years might have been required for the earth, assumed initially to be at a uniform temperature of 7000° to 10,000°F, to have reached its present state. Before that, the temperature at the surface would have been above the melting points of all known rocks. At the same time, he estimated that the sun could not have been illuminating the earth for much more than a few hundred million years, assuming that its energy supply was limited to sources known in the nineteenth century (primarily gravitational contraction). But these conclusions seemed to be inconsistent with statements of Uniformitarian geologists that processes such as erosion had been continuing at the same rate for hundreds of millions of years, and that the history of the earth is essentially cyclic with no long-term changes (see Section 18.2). Kelvin therefore argued that the basic principles of Uniformitarian geology were contrary to established laws of physics, in particular the second law of thermodynamics and Fourier's theory of heat conduction.

Kelvin's attack on the geologists was also indirectly an attack on Charles Darwin's theory of evolution, for in his *Origin of Species* (1859) Darwin had conjectured, in support of his assumption that sufficiently long times have been available for natural selection to operate, that certain geological processes might have been going on for as long as 300 million years. Insofar as such remarks made it appear that the theory of evolution depended on the continuous existence of conditions favorable to life for hundreds of millions of years, Darwin had unwittingly given his opponents a new argument: Such long periods of time were not considered meaningful or acceptable to some physicists, such as Kelvin.

The debate between Kelvin (supported by other physicists) and the Uniformitarian geologists continued throughout the rest of the nineteenth century. The *problem of the age of the earth,* as it came to be called, seemed to call forth a bitter conflict between geology and physics, with each side asserting that its own evidence should not be ignored simply because it seemed to be incompatible with the other side's. Because of the prestige of physics, resulting from its past successes, the geologists were forced to revise their estimates of the rate of geological processes in order to fit the shorter time scale that seemed to be required by Kelvin's calculations. Supporters of Darwinian evolution tried to deflect Kelvin's attack by pointing out that the validity of evolutionary theory did not depend on any particular timescale; since there was as yet no definite measurement of the rate of evolutionary change, even 10 million years might be sufficient.

At the beginning of the twentieth century the situation changed radically, for a reason that neither geologists nor physicists could have anticipated—the discovery of radioactivity. After Marie and Pierre Curie isolated the new element, radium, from samples of rocks, it was realized that the heat generated by radioactivity inside the earth must be taken into account in any calculations of the rate of cooling, and might easily be sufficient to compensate for most or all of the loss of heat by conduction through the crust. Kelvin's conclusions would thus be invalid because he had incorrectly assumed that there are no sources of heat inside the earth. Moreover, as Ernest Rutherford and Frederick Soddy pointed out in 1903, the theoretical limit on the total heat available from the sun, which Kelvin had used to impose a further limit on the age of the earth, is also much too low if radioactive changes can contribute to solar energy.

An even more direct refutation of Kelvin's results came from a more detailed analysis of the process of radioactive decay (see Chapter 27). Measurements of the relative proportions of lead, helium, radium, and uranium in rocks could be used to estimate the "ages" (time elapsed since solidification) of those rocks. The first results of this radioactive dating technique suggested that the earth's solid crust was nearly 3 billion years old—a figure ten times as great as Darwin's 300 million years, which Kelvin had called much too high!

Evolution, cooling of the earth, and radioactivity are all processes that fall outside the neat framework of Newtonian physics. They are *progressive* or irreversible, but not cyclic. The failure of Kelvin's theory does not by any means prove that physical theory cannot deal with irreversible processes—it is easy enough to repair his analysis by inserting a heat-source term in Fourier's heat-conduction equation. But it does suggest that the physicist who is accustomed to working with laboratory experiments and mathematical analysis may have something to learn from scientists who are more experienced in observing and interpreting natural phenomena that take place over large spans of time and space. Astronomy and the earth sciences are not to be regarded merely as fields in which already-established physical laws may be applied, but rather as possible sources for new physical discoveries and concepts.

18.4 Entropy and the Heat Death

Although Kelvin's quantitative estimate of the age of the earth turned out to be much too small, his qualitative principle of dissipation of energy was not challenged. Hermann von Helmholtz, in a lecture delivered in 1854, pointed out that Kelvin's principle implied the cooling of the entire universe. All energy would eventually be transformed into heat at a uniform temperature, and all natural processes would cease; "the universe from that time forward would be condemned to a state of eternal rest." (Helmholtz, "Interaction of Natural Forces") Thus was made explicit the concept of the *heat death* of the universe, as foreseen nearly a century earlier by Bailly, but now apparently based much more firmly on the new science of thermodynamics.

Another way of stating the principle of dissipation of energy was suggested by the German physicist Rudolf Clausius (1822–1888) in 1865. Clausius had previously introduced a new concept to which he now gave the name *entropy*. It comes from the Greek words *energeia* (energy) and *tropy* (transformation). The concept had been used since 1854 under the cumbersome name, "equivalence-value of a transformation."

Clausius defined entropy in terms of the heat transferred from one body to another. More precisely he defined the *change* in entropy, ΔS_1, resulting from a transfer of heat H from a body 1, at temperature T_1 as

$$\Delta S_1 = -H/T_1, \quad (18.1)$$

where T_1 is the absolute temperature (see Section 19.3). When the same amount of heat is received by another body, 2, that is kept at temperature T_2, the change in entropy is

$$\Delta S_2 = +H/T_2. \quad (18.2)$$

From Eqs. (18.1) and (18.2) we can easily infer that if T_2 is less than T_1, the total change in entropy, $\Delta S = \Delta S_1 + \Delta S_2$, will be positive. The change in entropy would be zero only if $T_1 = T_2$. (The total amount of entropy in a body cannot be measured directly, but is usually defined in such a way that $S = 0$ at absolute zero temperature, $T = 0$.)

The above indicates that *the total entropy of the system always increases whenever heat flows from a hot body to a cold body.* It also increases whenever mechanical energy is changed into internal (thermal) energy, as in inelastic collisions and frictional processes, since here one body absorbs heat but no other body loses it.

A more general definition of entropy was proposed by the Austrian physicist Ludwig Boltzmann in 1878. According to Boltzmann's definition, entropy depends on the probabilities of molecular arrangements, and can change even when there is no heat flow, for example, if a system becomes more random or disordered (when you scramble an egg or mix two different pure substances). Thus entropy is a measure of the *disorder* of a system (see Section 22.9).

The principle of dissipation of energy (generalized second law of thermodynamics) can now be stated very simply: *The entropy of an isolated system always tends to increase.* In particular, this means that heat does not flow by itself from cold bodies to hot bodies; a ball dropped on the floor will not bounce back higher than its original position by converting heat into mechanical energy; and an egg will not unscramble itself. All these (and many other) events would not violate any of the principles of Newtonian mechanics, if they took place—but they would decrease the entropy of an isolated system and are thus forbidden by the second law of thermodynamics.

Clausius stated the two laws of thermodynamics in the form:

1. The energy of the universe is constant.
2. The entropy of the universe tends toward a maximum.

These laws have been verified directly only for terrestrial phenomena, and to extend them to the entire universe requires a rather bold

Fig. 18.4. The Heat Death, as portrayed in *La Fin du Monde* [("The end of the world") by Camille Flammarion (1894)]: (a) "The wretched human race will freeze to death." (b) "This will be the end."

application of Newton's rules of reasoning (Section 9.3). Indeed, within the last few decades, theories of the evolution of the universe have been proposed that would violate both laws. Nevertheless, these laws are so useful that they will probably continue to be accepted as working hypotheses until definitely disproved.

The second law of thermodynamics implies that *time* is not a neutral mathematical variable in the same sense as the space coordinates (x, y, z); it has a definite *direction*—what has been called "time's arrow"—pointing from past to future. This statement is rescued from the status of a ponderous triviality only by the recognition that no such directionality is associated with Newton's laws of motion. As Kelvin's statement quoted at the beginning of this chapter vividly illustrates, it is difficult to imagine a world in which the second law is not valid. In such a world nothing would forbid such reversals as the ones he describes.

The second law gained considerable notoriety in the last part of the nineteenth century; the concept of an eventual heat death for the universe had a certain morbid attraction for popular writers, caught up in the *fin de siècle* ("end of the century") mood of pessimism that swept some parts of European and American society. Since increase of entropy means more randomness and disorder, perhaps that was an explanation for social disintegration and environmental degradation!

The American historian Henry Adams tried to put all these ideas together in a series of essays on the application of thermodynamics to human history. He quoted Kelvin's statement of the principle of dissipation of energy, remarking that "to the vulgar and ignorant historian it meant only that the ash-heap was constantly increasing in size." If historians were to concern themselves, as Adams thought they should, with the future as well as the past, then they could hardly ignore the latest result of physics, which implied (according to him) that human society along with the physical universe must end in degradation and death, even though this prophecy was distasteful to the evolutionists, who preached nothing but eternal progress. Adams complained that evolution, together with Lyell's geological doctrine of Uniformitarianism, had already started to take the place of religious dogma, and "in a literary point of view the Victorian epoch rested largely—perhaps chiefly—on the faith that society had but to follow where science led . . . in order to attain perfection" (Adams, *Degradation of the Democratic Dogma*). But Adams could find numerous examples in anthropological studies and the popular press to confirm the notion that the human race was going from bad to worse, and would soon be extinct.

A more balanced interpretation of the second law will have to be postponed until Chapter 22; but it should be evident already why the writer C. P. Snow, in his famous book, *The Two Cultures* (1960), remarked that for a humanist to be completely ignorant of the second law of thermodynamics would be as shocking as for a scientist to be completely ignorant of the works of Shakespeare.

RECOMMENDED FOR FURTHER READING

M. Bailyn, "Carnot and the universal heat death," *American Journal of Physics,* Vol. 53, pages 1092–1099 (1985)

Joe D. Burchfield, "The age of the Earth and the invention of geological time," pages 137–143 in *Lyell: The Past is the Key to the Present* (edited by D. J. Blundell and A. C. Scott), London: Geological Society, 1998, and other articles in this book.

H. Butterfield, *The Origins of Modern Science,* Chap. 12

R. P. Feynman, *The Character of Physical Law,* Chapter 5

Camille Flammarion, "The last days of the earth," *Contemporary Review,* Vol. 59, pages 558–569 (1891)

C. C. Gillispie, *Dictionary of Scientific Biography,* articles by J. Z. Buchwald on Kelvin, Vol. 13, pages 374–388; by E. E. Daub on Clausius, Vol. 3, pages 303–311; by J. E. Challey on Carnot, Vol. 3, pages 79–84; by R. P. Beckinsale on Hopkins, Vol. 6, pages 502–504; by J. R. Ravetz and I. Grattan-Guinness on Fourier, Vol. 5, pages 93–99

H. von Helmholtz, "On the interaction of natural forces" (translated from a lecture given in 1854), pages 59–92 in his *Popular Scientific Lectures,* New York: Dover, 1962

Kelvin, William Thomson, "The kinetic theory of the dissipation of energy" (1874), reprinted in *Kinetic Theory,* Vol. 2, edited by S. G. Brush, pages 176–187

D. Kubrin, "Newton and the cyclical cosmos: Providence and the mechanical philosophy," *Journal of the History of Ideas,* Vol. 28, pages 325–346 (1967)

S. Sambursky, *Physical Thought,* excerpts from Carnot and Clausius, pages 389–394, 405–408

Stephen Toulmin and June Goodfield, *The Discovery of Time,* New York: Harper, 1965

S. S. Wilson, "Sadi Carnot," *Scientific American,* Vol. 254, no. 2, pages 134–145 (1981)

Refer to the website for this book, www.ipst.umd.edu/Faculty/brush/physicsbibliography.htm, for Sources, Interpretations, and Reference Works and for Answers to Selected Numerical Problems.

PART F

Origins of the Atomic Theory in Physics and Chemistry

19 The Physics of Gases

20 The Atomic Theory of Chemistry

21 The Periodic Table of Elements

22 The Kinetic-Molecular Theory of Gases

John Dalton (1766–1844).

The story of the development and the gradual acceptance into science of the atomic view of matter is astonishing in many ways—in its origins, in the length and vigor of the debate, in the diversity of scientists and fields involved, in the often quite unexpected concurrence of separate arguments, and in the ever-growing flood of consequences that finally swept over all branches of science.

The account in this Part will carry the development to about the end of the nineteenth century, but the main results are largely applicable to this day. In retrospect it is clear that the model of the atom at that point had grown out of three separate types of questions: What is the physical structure of matter, particularly of gases? What is the nature of heat? What is the basis of chemical phenomena?

Although at first glance these three questions may seem completely dissociated, the answers were obtained simultaneously, in a joint conceptualization of the nature of heat, of gases, and of chemical reactions, in terms of one quantitative atomic scheme. In fact, the atomic view and the conservation of energy were very closely linked concepts, and the fight for both involved the same set of protagonists. As in a fugue, the separate themes introduce themselves, approach one another in the development, superpose, grow apart, and in the end again coalesce.

The concepts and derivations we shall encounter in this story are not always simple, but hardly a better example exists on this level to show a conceptual scheme evolving from the labors of generations—a tapestry spun out of widely scattered experimental observations, shrewd assumptions, simplifying approximations, detected errors, penetrating intuitions, and mathematical manipulations. And there are three other important reasons why the closer study of this topic may be fruitful. First, it will give us an historically important and generally useful model of the elementary structure of matter. From this model, and from the outburst of successful speculations that it generated, there arose much of the physics and chemistry of the period immediately preceding the twentieth century—and its failures were the stimuli for much in the new physical science of our own time. Second, not only is the kinetic theory a connecting link between the Newtonian and the contemporary approaches to physics, but it leads from macrocosmic physics to the world of submicroscopic phenomena. And finally, this topic will introduce us, though of necessity rather sketchily, to the important type of physical law that, instead of certainty and the prediction of individual events, deals with probability and the discussion of a multitude of events.

CHAPTER 19

The Physics of Gases

19.1 The Nature of Gases—Early Concepts

From the time of the earliest recorded philosophical speculations to the present day, the human mind has been haunted by some paradoxical propositions: On the one side we can with our hands cut or subdivide gross matter—a rock or a quantity of water—into smaller and smaller parts, until the crudeness of our tools or the deficiency of our sight, but never the material itself, calls a halt to our experiments. Consequently, we might conclude that matter is inherently infinitely subdivisible. But on the other hand, our imagination encounters difficulty when we demand that it present to us matter as a collection of truly infinitely small parts, and this view makes for logical problems as well. So where our senses tend to one view of matter, our reason persuades us of another.

The great compromise was formulated about 2400 years ago by the Greek philosophers Leucippus and Democritus, and later extended by Epicurus. It is of course the atomistic idea (Section 3.2). Our senses and our reason would both rest satisfied if we assumed that matter is indeed divisible far beyond immediate experience, but that there does exist an ultimate substructure of infinitely hard, uncuttable, indivisible particles, which we might provisionally name *corpuscles* or *atoms*.

The Roman poet Lucretius, whose book *De Rerum Natura* (*The Nature of the Universe*) was cited before in connection with the origins of the law of conservation of mass, exemplified the teaching of those now-largely-lost Greek original works. He gave a summary of this atomistic science, which (according to George Sarton) marked the climax of Roman scientific thought. We have noted, of course, that Lucretius' work is not to be regarded as primarily a science text.

His leading theme is "All nature then, as it exists by itself, is founded on two things: There are bodies and there is void in which these bodies are placed and through which they move about. . . ." Ordinary matter is composed of these two realities: solid, everlasting particles and, on the other hand, the void, or what we might loosely call the vacuum.

Note that by their very definition these atoms are not directly perceptible, and until about 150 years ago there were very few fruitful and unambiguous consequences that could be drawn from the atomic hypothesis to serve as experimental tests of this challenging speculative idea. Even at the beginning of the twentieth century, when such confirmation was at hand, there were important scientists who still rejected the concept of atoms as too convenient a fable, unworthy of serious consideration by "hard-boiled" investigators.

It is therefore not surprising that the structure of matter was for a long time the subject of inconclusive discussions, whether the atomistic view was rejected (as by Plato, Aristotle, and their diverse followers) or in some form accepted (by Galileo, Gassendi, Bacon, Descartes, Newton, Leibniz, and many of their contemporaries). We might speculate that the idea of small particles became more convincing as a consequence of Newton's work, in which, as we have seen, the postulates leading to the law of universal gravitation provided that the *particles* of matter are the agencies of mutual attraction. Of course, these are not necessarily identical with ultimately *indivisible* particles or atoms. Yet the surpassing success of the treatment of gravitation in terms of the mechanics of particles could turn the thoughts again to the idea that the mechanics of small discrete bodies is the key to all phenomena. Newton himself wrote in the introduction to the *Principia:*

> Then the motions of the planets, the comets, the moon, and the sea are deduced from these [gravitational] forces, by propositions that are also mathematical. If only we could derive

> the other phenomena of nature from mechanical principles by the same kind of reasoning! For many things lead me to have a suspicion that all phenomena may depend on certain forces by which the particles of bodies . . . either are impelled toward one another and cohere in regular figures, or are repelled from one another and recede. Since these forces are unknown, philosophers have hitherto made trial of nature in vain. But I hope that the principles set down here will shed some light on either this mode of philosophizing or some truer one.

Later, in the *Opticks* (published 1704), we find this visionary, yet still unformalized opinion on atomism:

> All these things being considered, it seems probable to me that God in the Beginning formed Matter in solid, massy, hard, impenetrable, movable Particles, of such Sizes and Figures, and with such other Properties, and in such Proportion to space, as most conduced to the end for which he formed them; and as these primitive Particles being Solids, are incomparably harder than any porous Bodies compounded of them; even so very hard as never to wear or break in pieces; no ordinary Power being able to divide what God himself made one in the first Creation. . . . And therefore that Nature may be lasting, the Changes of corporeal Things are to be placed only in the various Separations and new Associations, and Motions of these permanent Particles; compound Bodies being apt to break, not in the midst of solid Particles, but where those Particles are laid together . . .
>
> God is able to create Particles of Matter of several Sizes and Figures and in general Proportion to the space they occupy, and perhaps of different Densities and Forces. . . . Now, by the help of these Principles, all material Things seem to have been composed of the hard and solid Particles above mentioned—variously associated in the first Creation by the Counsel of an intelligent Agent. For it became him who created them to set them in order. And if he did so, it's unphilosophical to seek for any other Origin of the World, or to pretend that it might arise out of a Chaos by the mere Laws of Nature; though being once form'd, it may continue by those Laws for many Ages.

Part of Newton's preoccupation here is theological, and for a most excellent historic reason that we must not pass over if we wish to understand fully the development of the atomistic view. Ever since its early inception, atomism has generally been regarded as atheistic; as the excerpt from Lucretius (Section 15.1) indicated, the atomist openly professed to give an explanation of matter and events not in terms of impenetrable designs of an ever-present Creator and Ruler, but by means of the interplay and structure of material bodies. In the seventeenth century, philosophers such as Pierre Gassendi had attempted to remove the taint of atheism by proposing that atoms are not self-animated, but are inert pieces of matter that required a divine agency to set them into motion at the creation of the world.

In the "mechanical philosophy," as it was advanced especially in England by Robert Boyle, God was assigned the role of designer and creator of the world machine, which, once set in motion, could run from then on by His pleasure without continual active intervention; all observable events should then be ultimately explainable by the configurations, sizes, and shapes of the eternal particles or by the mathematical laws of force and motion ruling among them.

As we have seen in the previous chapter, Newton himself did not accept the extreme version of the world-machine theory that would forbid God to take constantly an active part in the physical world. Yet the above quotation from the *Opticks* shows that in formulating the principles of his own system of the world, Newton was primarily concerned with specifying the results of God's initial creative acts, so that in practice the Newtonian system could easily be interpreted as a world machine.

Some of Newton's other remarks on the properties of atoms did give support to those who were dissatisfied with the mechanistic philosophy advocated by Descartes (Section 6.1). The Cartesians rejected the hypothesis of "action at a distance," preferring to believe that all apparent long-range forces, such as gravity, could be explained by contact actions propagated through an intervening material medium (the ether). Newton, while rejecting in principle the possibility of true action at a distance (see his letter to Bentley, quoted in Section 11.10), did explain many phenomena in terms of attractive and repulsive forces acting between atoms.

These remarks encouraged scientists like Boscovich and Priestley in the eighteenth century to develop an atomic theory of matter that relied more on the atoms' attractive and repulsive forces than on their hardness and impenetrability. The concept of an atom as a point center of attractive and repulsive forces survived into the twentieth century, along with the alternative concept of the atom as something like a billiard ball that could interact with other billiard balls only by contact.

19.2 Air Pressure

In the meantime the atomistic view of matter was extended from an entirely different direction—the research on gases.

Galileo, in his *Two New Sciences* (1638), noted that a suction pump cannot lift water more than about 10.5 meters. This fact was presumably well known by the time Galileo wrote his book; pumps were already being used to obtain drinking water from wells and to remove water from flooded mines, so their limitations must have become evident to many workmen.

One important consequence of the limited ability of pumps to lift water was that some other method was needed to pump water out of deep mines, and this need provided the initial stimulus for the development of steam engines, which allowed water to be conveyed out in buckets. Another consequence was that physicists in the seventeenth century became curious to discover the reason why the suction pump worked at all, as well as why there should be a limit on its ability to raise water.

If you remove air from a container and create a vacuum, there is a tendency for things to be sucked in. The Aristotelian philosophers explained this fact by saying that "Nature abhors a vacuum." It is unnatural, they said, for space not to be filled with matter, and therefore matter will move so as to fill up any empty space. This is an example of a *teleological* explanation, one based on an "ultimate cause" (another one would be: "Rain falls because crops need water"). One of the basic goals of the new mechanical philosophy of the seventeenth century was to eliminate such teleological explanations, and to explain phenomena instead in terms of immediate physical causes.

The Aristotelian theory seemed especially weak in this particular case: Even if one accepts the assertion that Nature abhors a vacuum, one finds it difficult to explain why this abhorrence

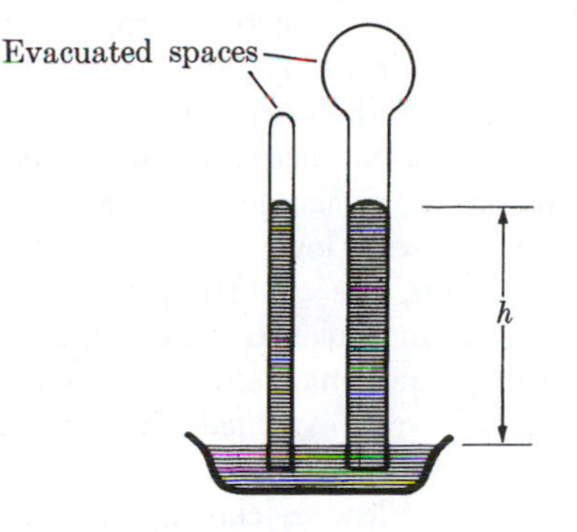

Fig. 19.1. Torricelli barometer (two forms). The mercury column in each vertical glass tube is balanced by the pressure of the atmosphere on the free mercury surface in the trough. Usually h is about 0.76 m at sea level.

is just sufficient to raise water by 10.5 m by means of a pump, no more, no less!

Evangelista Torricelli (1608–1647), once briefly the pupil of Galileo, realized that the atmosphere exerts a pressure at the surface of the earth, and suspected that this *air pressure* might be sufficient to account for the phenomena previously attributed to Nature's abhorrence of a vacuum. He made the fortunate guess that mercury, a liquid nearly fourteen times as dense as water, might be a more convenient medium for laboratory experiments; indeed, it turned out that the same vacuum pump that could raise a column of water 10.5 m could lift a column of mercury only 1/14 as high, 0.76 m.

The pump itself is not even necessary to hold up a column of liquid in the simple instrument that came to be known as the Torricelli *barometer.* One takes a straight glass tube somewhat more than 0.76 m long, open at one end and sealed at the other; a bowl; and enough mercury to fill the tube and the bowl. Fill the tube to the brim, close the open end with your finger, turn it upside down in the bowl, and remove your finger, and you will find that the mercury runs down into the bowl, but only until the difference between the level in the tube and the level in the bowl is about 0.76 m. The result is shown in Fig. 19.1 (the left tube). There is now found to be an evacuated space at the top of the tube above the mercury. Torricelli then repeated the same procedure with a tube that had a large bulb at the closed end, thereby making available an evacuated space of larger size in which small objects could be fastened.

The announcement of Torricelli's experiments in 1643 prompted other scientists (Otto von Guericke and Robert Boyle were among the first) to develop better vacuum pumps for experimenting on vacuum phenomena and the properties of gases at low pressures.[1] With a pump and a glass jar, one could prepare an experimental space in which to discern the effect of light, sound, and magnetic or electric forces within an atmosphere rarefied in varying degrees. One could attempt to discover, by the difference in weight of a hollow vessel before and after "exsuction" (sucking out), the weight of the gas within. Conversely, one could also pump gases into a vessel at higher pressures.

According to Torricelli and other followers of the mechanical philosophy, the force that holds the column of mercury up in the tube is simply the pressure of the atmosphere, transmitted through the mercury in the bowl. The air pushes down on the surface of the mercury, and since there is practically no gas in the space at the top of the tube, the mercury rises until its weight is sufficient to balance the force exerted by the air outside on the top surface of the pool of mercury.

In France, the scientist-philosopher Blaise Pascal (1623–1662) reasoned that if the pressure of the atmosphere is indeed the force that supports the column of mercury in the Torricelli barometer, then the height of that column should be less if the pressure of the atmosphere is reduced. He assumed that the atmospheric pressure at the earth's surface results from the weight of the rest of the atmosphere above it, just as the pressure one experiences under water increases at greater depths. Conversely, then, by taking the barometer to the top of a mountain one should find a lower atmospheric pressure, and consequently the height of the mercury supported by that pressure should be reduced. The experiment proposed by Pascal was successfully performed in 1648 by his brother Florin Perrier. It is often called the Puy de Dôme experiment, after the mountain on which it was conducted.

[1]These pumps were relatively crude and could not produce what we would now call a vacuum. For that matter, the complete removal of every trace of gas from a closed vessel is impossible even today. But considerable advances in the construction of vacuum pumps in the second half of the nineteenth century made possible evacuation good enough to allow, on the one hand, the discovery of subatomic particles (Section 25.5) and, on the other, the development of the incandescent electric light bulb.

While the height of the mercury in the tube depends on (and thus allows one to measure) the local atmospheric pressure, it does *not* depend on the diameter of the tube or the size of the bowl. This might at first seem rather strange, since the force exerted by the mercury depends on how much of it there is—that is, on its total volume, not just its height. Similarly, the force of the air outside will be greater on a large bowl than on a small one. Nevertheless, the mercury always rises to the same height whether the tube is thick or thin and whether the bowl is large or small. This fact is explained if we recognize that it is the balance of *pressure,* not *force,* that determines the height of the mercury column (and the equilibrium of fluids in general).

It is important to distinguish clearly between pressure and force. Pressure is defined as the amount of force acting perpendicular to a surface, divided by the area of the surface:

$$P = F/A. \tag{19.1}$$

A large force may produce only a small pressure if it is spread over a large enough area; for example, if you wear large snowshoes, you can walk on snow without sinking in. On the other hand, a small force may produce a very large pressure if it is concentrated in a small area; when spike heels were used on ladies' shoes some years ago, many wooden floors and carpets were damaged because the pressure at the place where the heel touches the floor—a very small area—can be greater than that exerted by an elephant's foot!

It is instructive to calculate the numerical value of the pressure exerted by the atmosphere, to see how large it is and why it does not depend on the diameter of the tube in the barometer. The column of mercury stands 0.76 m high (the average height of the barometer column at sea level at 0°C). If this column has a cross-sectional area of 1 m^2, the volume of the column is 0.76 m^3. A cubic meter of mercury has a mass of 13,600 kg. The mass of the mercury in the column is therefore

$$m = 0.76\ \text{m}^3 \times 13{,}600\ (\text{kg/m}^3) = 10336\ \text{kg}.$$

The force exerted by this amount of mercury is

$$F = mg = 10336\ \text{kg} \times 9.8\ \text{m/sec}^2$$
$$= 1.013 \times 10^5\ \text{N}.$$

(Remember that N is the abbreviation for the newton, the unit of force in the SI system.)

Since the pressure is defined as F/A, and $A = 1\ \mathrm{m}^2$, we find that

$$\text{Atmospheric pressure} = 1.013 \times 10^5\ \mathrm{N/m^2}.$$

Note that we had to multiply by the cross-sectional area to calculate the force and then divide by this same area to calculate the pressure, so that the final result is independent of the value assumed for the area.

The actual height of the mercury column in a barometer is not precisely 0.76 m; it varies from one day to the next and from one place to another, being itself an index of changing climatic conditions. It is now customary to define the "standard atmosphere" as $1.013 \times 10^5\ \mathrm{N/m^2}$. The unit of pressure in the SI system is the pascal, abbreviated Pa, which is defined as $1\ \mathrm{N/m^2}$; so atmospheric pressure is approximately 10^5 Pa. In common units this is about 15 lb/in.^{2}. Weather reports in the United States still use an older unit for atmospheric pressure, the *bar,* defined as 10^5 Pa. Weather maps usually give local atmospheric pressures in terms of millibars (1 bar = 1000 millibars).

The expression "gauge pressure" is sometimes used to indicate the excess of the actual pressure over atmospheric pressure. For example, if you inflate your tires and then check the pressure with a gauge and it shows 30 lb/in.^{2}, the actual pressure inside the tires is about 45 lb/in.^{2}.

It was hard for many people to believe that the atmosphere really exerts a pressure as great as 15 lb/in.^{2}. A famous experiment, conducted by Otto von Guericke in Magdeburg, Germany, in 1654, helped to dramatize the magnitude of atmospheric pressure. Two large hollow bronze hemispheres were fitted carefully edge to edge, and the air was removed from the inside of this space by a vacuum pump. A team of eight horses was harnessed to each hemisphere and the two teams were driven in opposite directions; they were just barely able to pull the hemispheres apart against the force with which the external atmospheric pressure pressed them together.

The properties of air and of atmospheric pressure were thoroughly investigated by the British scientist Robert Boyle (1627–1691), who heard about Guericke's experiments in 1657. Using an improved vacuum pump built for him by Robert Hooke, Boyle was able to obtain a large evacuated space in which several new experiments could be done. In one rather macabre test, he found that small animals became unconscious and eventually died when deprived of air.

Fig. 19.2. Robert Boyle (1627–1691), British physicist and chemist who advocated the mechanistic "clockwork universe" philosophy and established by his experiments that suction is due to air pressure rather than the Aristotelian "nature's abhorrence of a vacuum." Note in the background one of his pumps, set up to evacuate a glass globe.

In another experiment, Boyle placed a Torricelli barometer inside a closed container and then removed the air from the container. The level of the mercury in the tube fell as the air was removed from the container, until finally it reached the same level as the mercury in the bowl. Boyle realized that this result proved it was the pressure of the atmosphere outside that had been holding up the mercury in the tube, rather than the vacuum itself exerting some pull on the top of the tube. He had essentially done Pascal's experiment without leaving his own laboratory.

Soon after Boyle published his experiments in 1660, his conclusions were challenged by Franciscus Linus, a Jesuit scientist and professor of mathematics and Hebrew at the University of Liege, in Belgium. Linus asserted that the apparently empty space above the mercury really contains an invisible cord or membrane, which he called the *funiculus* (a Latin word meaning "small rope"). According to Linus, when air is stretched or rarefied, the funiculus exerts a violent attractive force on all surrounding objects, and it is this attraction that pulls the mercury up the tube. He would argue that if you put your finger over the end of the tube from a vacuum pump (or a modern vacuum cleaner), you can actually feel the funiculus pulling in the flesh of your finger. (Try it!)

The funiculus theory sounds like a fantastic idea, which no scientist would take seriously nowadays. But this kind of pseudomechanical explanation of physical phenomena was quite popular in the early days of modern science. Boyle did not simply ignore Linus, perhaps realizing that the new idea of air pressure needed further justification before it would be generally acceptable. Along with his refutation of the funiculus hypothesis, Boyle published in 1662 some quantitative measurements of the relation between pressure and volume of air, in support of what is now called Boyle's law, to be discussed in the next section.

While it is for his pressure-volume law that Boyle is chiefly famous, we should not forget that in his own time his work in establishing the qualitative importance of air pressure was more significant.

19.3 The General Gas Law

The compressibility of air, a fact realized generally for some time, was put on a quantitative basis through Boyle's celebrated experiments, from which it appeared that for a given mass of air trapped in a vessel at a constant temperature, any decrease of volume raises the gas pressure proportionally; conversely, any increase in pressure decreases the volume. In modern terminology, we write this result

$$P \times V = \text{constant} \quad \text{(at constant temperature)} \qquad (19.2)$$

where P refers to the actual pressure of the gas considered (for example, in N/m^2) and V to the volume of the gas (for example, in m^3).[2]

Following Boyle's work, continued effort was made to discover the effect of changes in temperature on the pressure or the volume of a gas. In the eighteenth century, Guillaume Amontons and Jacques Charles suggested that if a constant pressure is maintained in a sample of gas, its volume would increase proportionately to the rise in temperature ($\Delta V \propto \Delta T$ at constant P). This law was definitely established by extensive experiments on a number of gases conducted by the French chemist Joseph Louis Gay-Lussac, around 1800, and confirmed shortly afterward in England by John Dalton. For brevity we shall call it Gay-Lussac's law, as Gay-Lussac seems to have had the greatest share in establishing its validity, but recognizing that, just as in the case of Boyle's law, several scientists made important contributions by proposing and testing it.

To visualize the meaning of Gay-Lussac's law in an actual experiment, think of a quantity of gas contained in a very thin, easily stretched balloon. The pressures inside and outside tend to be the the same—if the pressure were higher inside, the balloon would expand, thereby reducing the inside pressure until it was the same as the outside pressure; if the pressure were lower inside, the balloon would be squeezed down to a smaller size, and its inside pressure would increase. So we can assume that the pressure inside is equal to atmospheric pressure (about 10^5 Pa).

As we now proceed to heat the surroundings (ΔT), the balloon expands (ΔV). What is so very startling is the fact that the fractional increase in volume per degree rise in temperature is exactly the same for all gases, even though their chemical constitutions are different.

To appreciate the unexpectedness of this finding, recall that on heating a solid or a liquid one ordinarily also obtains a length or volume expansion that is proportional to the temperature change; but there is a great difference in the relative amount of volume expansion per degree of temperature among different solids or liq-

[2] The complicated history of the discovery of Boyle's law has only recently been disentangled by historians. It appears that the law was first proposed by two other British scientists, Henry Power and Richard Towneley, on the basis of their experiments, begun in 1653. They did not publish their results immediately, but after Boyle's first experiments on air pressure had been published in 1660, Power sent a paper describing the results of their joint work to his friend William Croone, in London. The title of the paper was "Additional experiments made at Towneley Hall, in the years 1660 and 1661, by the advice and assistance of that Herorick and Worthy Gentleman Richard Towneley." But Power neglected to put his own name on the paper. Croone sent the paper to Boyle, forgetting to mention that Power was the author. Boyle was very careful to give proper credit for the information he had received, and in his monograph of 1662 replying to Linus, he stated that he had not realized that the simple relation PV = constant applied to his own data until Richard Towneley pointed it out. Later scientists, who read Boyle's works carelessly or not at all, assumed that Boyle had made the discovery all by himself.

uids (about 0.01% for ice, 0.0025% for quartz, 0.02% for mercury and for water, and 0.15% for acetone).

But under normal conditions, almost all gases expand roughly 1/273 (that is, 0.37%) of their volume when heated from 0° to 1°C. The same volume change is also observed at much lower temperatures, provided that the pressure of the gas is sufficiently low. (If a gas is continually cooled at a constant pressure, it will eventually condense to a liquid, and then, of course, cease to obey Gay-Lussac's law.) We can thus define an *ideal gas* as a gas that obeys both Boyle's law and Gay-Lussac's law with complete precision; the behavior of real gases approaches this as a limit at low pressures.

Assume that an ideal gas could be cooled to 273° below the ice point (that is, to $T_C = -273$); then since by Gay-Lussac's law—if it holds over such a large range—it should contract 1/273 of its volume at $T_C = 0$ for each degree of cooling, it would eventually have no volume at all. While this condition cannot be realized experimentally, it suggests that we should recognize the special temperature $T_C = -273$ (more precisely, -273.15) as the absolute zero of temperature, in the sense that no lower temperature can ever be reached.[3] Consequently, we can simplify the law $\Delta V \propto \Delta T$ (at constant P) by writing $V \propto T$ at constant P *if T is measured on a new temperature scale,* that is, the *absolute* or *Kelvin scale,* where $T = T_C + 273.15$. The unit of temperature is the *kelvin,* abbreviated K. Thus a temperature 5 degrees above absolute zero would be written T = 5 kelvins or T = 5 K.

Comparing the two scales, we see that for example, 0°C is equivalent to 273 K; room temperature (about 20°C) is 293 K; water boils at 373 K. Note that the size of the degree—the kelvin—is the same on both the Celsius and the Kelvin scale. However, in the SI system of units the degree symbol "°" is no longer used for K, but it is still used for Celsius degrees (°C).

The letter K was chosen in honor of William Thomson, Lord Kelvin, who first proposed this scale in 1848, and showed that it is consistent with the second law of thermodynamics. [Recall that the definition of entropy, given by Clausius in 1865, uses this temperature scale (see Section 18.4).]

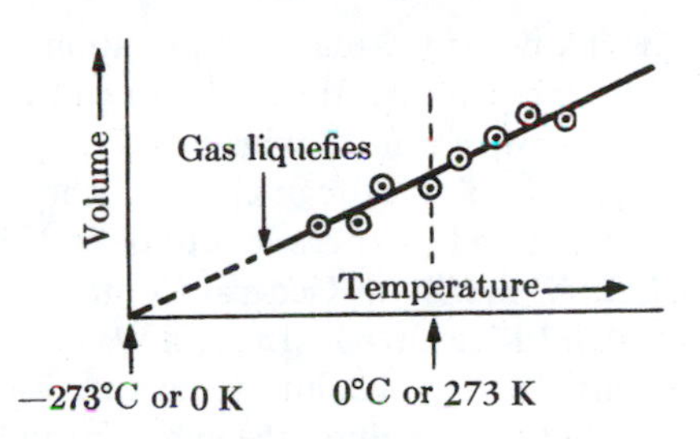

Fig. 19.3. Relation between volume and temperature for a gas at constant pressure. The dotted line represents an ideal gas.

Problem 19.1. Figure 19.3 is a hypothetical plot of experimental data for a sample of gas at constant pressure; here the volume V is plotted against the temperature T. The graph illustrates $\Delta V/\Delta T$ = constant (Gay-Lussac's law). Show by geometrical reasoning that if Gay-Lussac's law holds, then also $V \propto T$ (at constant P) provided T is measured on the absolute scale. Show that the proportionality $V \propto T$ is *not* true if T is measured on the Celsius scale.

What interests us most, however, is the obvious next step: combining the two laws into one. If for each gas, by Boyle's law, $V \propto 1/P$, and by Gay-Lussac's law, $V \propto T$, then $V \propto T/P$. This implies that the quotient PV/T is a constant (usually called r), whose value depends only on the sample. Thus we can write:

$$PV = rT. \qquad (19.3)$$

We may call the last equation, which is of great usefulness, the *ideal gas law;* it relates in a simple way the three variables of interest in any given sample of gas, and is approximately valid for all real gases under ordinary conditions.

Problem 19.2. Find r in Eq. (19.3) for a sample of hydrogen gas weighing 2 g and filling 22.4 L at 0°C and 10^5 Pa pressure (approximately atmospheric pressure). Then determine the volume of this sample at room temperature (20°C) and a *gauge pressure* of 10^7 Pa (= 10 MPa). (Remember: Gauge pressure is defined as the pressure *above* atmospheric.)

[3]According to the third law of thermodynamics, proposed by Walther Nernst in 1906, all thermal properties approach zero values as the absolute zero of temperature is approached, and as a consequence it is never possible to reach this temperature.

At this point we may call attention to an important temperature: the *triple point* of water, the unique temperature at which all three phases (gas, liquid, and solid) can coexist. The triple point was found by experiment to be at 273.16 ± 0.01 K. At the Tenth General Conference on Weights and Measures in 1954, it was decided to establish a new absolute temperature scale based on two fixed points: absolute zero and the triple point of water, which was defined to have the exact value T = 273.16 K. The new scale agrees with the old one for nearly all practical purposes, but has the advantage that the fixed points are no longer quite so arbitrary. (Since the melting and boiling points of a substance vary with pressure, one had to specify a particular pressure, namely the average pressure of the atmosphere at the earth's surface. The triple point, however, occurs at a definite pressure, 611.73 Pa.)

19.4 Two Gas Models

One is immediately tempted to interpret these observations and the experimental gas law that summarizes them in terms of some model or structure. Even qualitatively, the great compressibility of all gases and their "springiness" (tendency to expand again when the external pressure is reduced) seem to call for some explanation.

In 1660 Boyle himself, while favoring one model "without peremptorily asserting it," presented two opposing explanations by models, both atomistic, for he and his time had begun to accept atomism as part of the rising materialistic worldview. That view, expressed in his words, is: "whatsoever is performed in the material world, is really done by particular bodies, acting according to the laws of motion."

The two models that Boyle presents represent respectively the *static* and the *kinetic* views. If air consists of contiguous particles at rest, the corpuscles must themselves be compressible, rather like little springs or, as Torricelli had preferred, like pieces of wool. If, on the other hand, the corpuscles of the gas do not touch at all times, they need be neither variable in size themselves nor maintained by a static assembly, but then they must be in violent agitation, being whirled through all available space within a turbulent though "subtle" fluid. We recall that the subtle-fluid theory, then recently proposed by Descartes to account for the motion of planets, had captured the imagination throughout Europe; and indeed, it seemed to explain qualitatively the behavior of gases well enough. (Boyle attempted to find experimentally this subtle fluid, separated from the particles on which it supposedly acted. This was the start of a 250-year-long futile search for direct evidence of ethereal fluids.)

Having named the first of these the static model of a gas and the second the kinetic model, we must make a brief comment on each. As to the first, a springlike compressibility of static particles cannot easily account for one very conspicuous property of gases—their ability to expand infinitely in all directions, unless of course we allow the improbable, namely, that each atom itself can grow infinitely on expanding. For a pile of ordinary springs opposes compression, but does not disperse continuously when put into the open. Therefore some followers of the static view found it necessary to consider that gas corpuscles do not change in size but can repel one another at a distance. Evidently some special new agency of repulsion, some new force among gas atoms, would have to be invented, because the well-known mutual gravitational force between large bodies, also the logical candidate for intercorpuscular forces, was, unfortunately, always found to be a force of attraction, as is evidently also the nature of those forces that hold the particles together in *solids* and *liquids*.

The distinguished authority of Newton seems to lend some aid to the static picture, for in the *Principia* he had shown, by way of a mathematical demonstration, that *if* a sample of gas were made of mutually repulsive particles, and *if* a force of repulsion between any two particles were *inversely proportional* to the distance between their centers, then the gas pressure in this sample would increase with a small decrease in volume, just as found by Boyle's law.

Newton did not mean to prove that gases really do correspond to this model; on the contrary, he writes, "But whether elastic fluids do really consist of particles so repelling each other, is a physical question," in other words, one that has to be settled by appeal to experience. And indeed, this postulated type of force would cause gases to exhibit different behavior for different shapes of the container, which does not correspond to anything observed. Nevertheless, Newton's was the first *quantitative* deduction obtained for the behavior of gases on the basis of any atomistic hypothesis, and it appears to have been quite impressive.

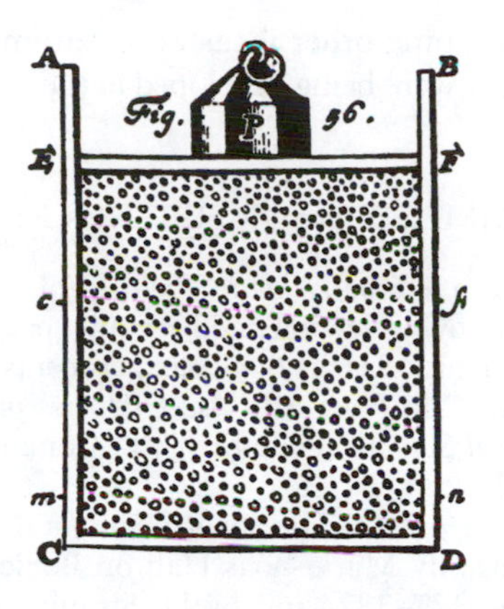

Fig. 19.4. The pressure on the piston exerted by the weight *P* is balanced by the impact of high-speed gas particles. From Daniel Bernoulli's *Hydrodynamica,* 1738.

The second or kinetic view of a gas might well have been formulated without recourse to an imponderable fluid (if it had not been for the inescapable, almost hypnotic hold of that idea), for it had been the opinion of Greek atomists, as reported by Aristotle and Lucretius, that particles, particularly those not joined in specific form, are "ever passing about in the great void," just as minute specks of dust can be seen dancing irregularly in a sunbeam that pierces a dark room.

In fact, the Swiss physicist Daniel Bernoulli in 1738 did publish a prophetic, quantitative development of a Lucretian type of kinetic model for gases. Bernoulli thought of the "corpuscles" of the gas as so minute as to be "practically infinite in number" under ordinary conditions, even in a small container. In their rapid motion hither and thither these corpuscles collide with one another and with the rigid walls of the closed vessel. But the collisions can be assumed to be perfectly elastic; therefore, as we would put it, the kinetic energy of the particles is conserved and the motion can continue undiminished. The pressure that the gas is expected to exert against all sides of the container is thus caused by the incessant impact of millions of high-speed particles—hence the name "impact theory of gas pressure."

Let us say that the gas-filled container used here is cylindrical, and that the top end can be made to slide in like a piston (see Fig. 19.4). If the volume is slowly decreased, the corpuscles are more crowded in the progressively smaller space and the number of collisions per second with the walls is larger, that is, the pressure should become greater, as is actually observed. Bernoulli even calculated the magnitude of this expected increase and found that it corresponds to Boyle's experimental law. Then he continued:

> The elasticity of air [the pressure in the container] is not only increased by the condensation [decrease of volume] but by heat supplied to it, and since it is admitted that heat may be considered as an increasing internal motion of the particles, it follows that if the elasticity of the air of which the volume does not change is increased [by heating the sample of gas], this indicates a most intense motion in the particles of air; which fits well with our hypothesis. . . .
>
> (*Hydrodynamica*)

This model, this line of reasoning, and this result, are all quite similar to the work that more than a century later finally clarified simultaneously the main problems of the nature of gases, heat, and chemistry. But at the time, Bernoulli's work was generally neglected. That lack of general attention and the long interval, are both remarkable, and cannot be entirely explained by pointing out that gases were not yet clearly understood by chemists, that density measurements were lacking that would have been needed to obtain the numerical values for the average speed of the particles, and so forth.

The other model, the Boyle-Newton theory based on repulsive forces between gas particles, proved to be more congenial to the eighteenth-century scientists who adopted the caloric theory of heat (Section 17.7). The repulsive force was now attributed to atmospheres of caloric fluid surrounding each atom. Heating a gas meant pouring in more caloric, expanding the "atmosphere" around each atom, and thereby intensifying the repulsive force. Although the caloric theory did not completely exclude the notion that heat might be associated with the motion of particles, it did seem to be incompatible with Bernoulli's hypothesis that heat is *nothing but* the motion of particles.

The discovery of latent heat by Joseph Black seemed to indicate that a simple proportionality between heat and temperature could not be generally valid, since a finite amount of heat must be added to a liquid in order to change it to a gas, without any change of temperature—as in the case of liquid water, boiling away into vapor at a steady 100°C. This was one of several phenomena that were thought to be more easily explicable by the caloric theory than by the kinetic theory. Again, Bernoulli's theory, by postulating free motion of the atoms through empty space, ignored the ether, which was thought to pervade all space; whereas caloric could easily be identified as a special kind of ether.

Perhaps we may compare Bernoulli's astonishing insight to some quick, isolated, premature stab from a general holding position, far through enemy lines, bringing no strategically valuable consequences until in time the whole main force has moved ahead to link up with this almost forgotten outpost. For Bernoulli, in effect, had made two enormous jumps in his thinking for which the great majority of scientists was not yet ready: first, the direct equivalence of heat and internal molecular motion, *ignoring* any interactions with the ether; and second, the idea that a well-defined numerical relationship, such as Boyle's simple law, could be deduced from the chaotic picture of randomly moving particles. We must postpone our account of the final victory of these ideas to a later chapter, turning now to examine some other ideas about atomic properties that were being developed in the meantime.

RECOMMENDED FOR FURTHER READING

S. G. Brush (editor), *Kinetic Theory,* Vol. I, excerpts from Boyle, Newton and Bernoulli, pages 43–65

J. B. Conant, "Robert Boyle's experiments in pneumatics," in *Harvard Case Histories in Experimental Science* (edited by J. B. Conant), pages 1–63

C. C. Gillispie, *Dictionary of Scientific Biography,* articles by Marie Boas Hall on Boyle, Vol. 2, pages 377–382, and by H. Straub on Daniel Bernoulli, Vol. 2, pages 36–46

Marie Boas Hall, "Robert Boyle," *Scientific American,* August 1967, pages 97–102

S. Sambursky *Physical Thought,* excerpts from Torricelli, Pascal, and Boyle, pages 256-263, 282-284

Martin Tamny, "Atomism and the mechanical philosophy," in *Companion* (edited by Olby), pages 597–609

Stephen Toulmin and June Goodfield, *Architecture of Matter,* Chapter 8

CHAPTER 20

The Atomic Theory of Chemistry

In the previous chapter we described how scientists studying the physical properties of gases were led to two different atomic models: kinetic and static. In the nineteenth century, further progress in understanding the nature of atoms came through two other areas of science: the physics of energy, especially the recognition that heat is a form of atomic motion (Chapter 17); and the atomic theory of chemistry, which grew out of Lavoisier's quantitative measurements of the weights of substances in reactions (Chapter 15). In this chapter and the next we follow the development of chemistry from Dalton to Mendeléeff, to show how a major part of the foundation for twentieth-century atomic physics was established; in Chapter 22 we return to the kinetic theory of gases, elaborated by Clausius, Maxwell, and Boltzmann with the help of thermodynamics and the chemical atomic theory.

20.1 Chemical Elements and Atoms

When in the preceding chapters we used the words *atom, corpuscle, particle,* and the like, it was not to be inferred that the writers of antiquity, or even of the seventeenth and eighteenth centuries, had in mind what we now mean by atom. The general idea of a "smallest particle" was a component of various conceptual schemes in physics and chemistry, usually mechanical models that postulated small indivisible bodies by which certain of the properties of matter could be explained. But, although theorists might assign particular properties such as shape or elasticity to their atoms, these properties were not susceptible to direct measurement or observation and could differ widely from one theory to the next.

As we have seen, the atomistic hypotheses up to the nineteenth century had not given rise to an unambiguous and precise theory either of matter or of heat. Yet, between the publication of Robert Boyle's *Sceptical Chymist* in 1661 and Lavoisier's death in 1794, there had developed a fundamental concept that was destined to give a new, vital, and precise meaning to the speculative atomistic hypothesis. This concept was the *chemical element.*

By the end of the eighteenth century, chemists had come to accept this operational definition of *element:* It is a substance that cannot be decomposed into other substances by chemical or physical means available at the present time. Whether these elements would in the future turn out to be really compounds of yet more basic substances was a question that could safely be left to metaphysicians, or to future scientists.

Lavoisier himself, in the preface to the *Elements of Chemistry* (1789), wrote this revealing analysis of the concept "element" and its relation to the atomistic hypothesis of his predecessors:

> All that can be said upon the number and nature of elements is, in my opinion, confined to discussions entirely of a metaphysical nature. The subject only furnishes us with indefinite problems, which may be solved in a thousand different ways, not one of which, in all probability, is consistent with nature. I shall therefore only add upon this subject, that if, by the term elements, we mean to express those simple and indivisible atoms of which matter is composed, it is extremely probable we know nothing at all about them; but, if we apply the term elements, or principles of bodies, to express our idea of the last point which [chemical] analysis is capable of reaching, we must admit, as elements, all the substances into which we are capable, by any means, to reduce bodies by decomposition. Not that we are entitled to affirm, that these substances we consider as simple may not be compounded of two, or even of a greater number of principles; but, since these principles cannot be separated, or rather since we have not hitherto discovered the means of separating them, they

> act with regard to us as simple substances, and we ought never to suppose them compounded until experiment and observation has proved them to be so.

In addition to the few elements known to the ancients (for example, metals, sulfur), a growing number of new ones had been identified (about one-third of our present list) although, of course, many recalcitrant compounds still posed as elements until electrolysis and other powerful means became available to separate their components. Not so very surprisingly, caloric (heat considered as a substance) also was generally listed among the elements.

The establishment of a relation between the two concepts *chemical element* and *atom* is now, in retrospect, perhaps all too obvious. Atoms of a chemical element would be the smallest physical particles that can participate in the identifying chemical reactions or physical tests. It seems natural to postulate, until evidence to the contrary appears, that each atom of an element is identical to every other atom of that same element ("like manufactured articles," as the scientists of industrialized nineteenth-century Britain used to say), although we may be prepared to allow that different elements may turn out to have physically quite different atoms (as indeed is the case).

It is also reasonable to expect that an element in the gaseous state, say oxygen or hydrogen, consists of individual atoms, and that a chemical compound, say water vapor, might be imagined as constituted of particles, each containing two or more atoms of different elements. In particular, it being agreed to make initially the simplest assumption, the smallest corpuscles of water vapor should be regarded as made of a close union of an atom each of hydrogen and oxygen, the two gases that yield water when mixed and exploded.

This, in brief, outlines the thoughts of the Englishman John Dalton (1766–1844), who, in his work from 1800 on, effectively performed the marriage of the ideas of chemical element and the ancient atomistic hypothesis. Dalton was largely self-taught. A retiring person, son of a handloom weaver, he supported himself poorly as a teacher (as early as his twelfth year) and general tutor in Manchester. Although he had a strong drive and a rich imagination, particularly along the line of mechanical models and clear mental pictures, his supreme gift was his astonishing physical intuition, which allowed him to arrive at correct conclusions despite being only "a coarse experimenter," as Humphry Davy (Section 17.7) called him.[1]

20.2 Dalton's Model of Gases

Like Kepler and Newton, whom he resembled in many ways, Dalton published on a great variety of subjects; but it appears that his main work originated in an interest in the physical structure of gases. He reports that through his interest in meteorology he had been led to wonder why the earth's atmosphere, being composed of a mixture of several gases of different density (for example, nitrogen, oxygen, and water vapor), should be so homogeneous—that is, why samples taken at rather widely differing altitudes should have the same proportion of chemical elements, even though nitrogen gas, being lighter per unit volume in the pure state than oxygen gas, might be expected to "float" on top of the latter, somewhat like oil on water. (Fortunately perhaps for the advancement of atomic theory, it was not yet possible to gather samples of air from regions higher than a few miles above the earth's surface; there *is* a noticeable difference in composition at higher altitudes.)

Count Rumford would have had a good answer to the puzzle; he might have said, as he did in an analogous case of mixed liquids, that by thermal agitation, by their "peculiar and continual motion," all the individual particles of the different kinds of gases must diffuse among one another and thus mix thoroughly. But Dalton could never accept this kinetic view of matter for a moment. He adopted the static model of a gas, being (wrongly) convinced that the great Newton had proved it in the brief passage from the *Principia* to which we referred in Section 19.4: "Newton had demonstrated," Dalton once wrote, "from the phenomena of condensation and rarefaction that elastic fluids are constituted of particles, which repel one another by forces which increase in proportion as the distance of their centers diminishes: in other words, the forces are reciprocally as the distances. This deduction will stand as long as the Laws of elastic fluids continue to be what they are."

Dalton was naturally forced to reject the kinetic view of heat, along with the kinetic view

[1]For an account of Dalton's work see Leonard K. Nash, "The atomic molecular theory," which was one of the main sources in the preparation of this chapter.

of gases, and to adopt the caloric theory, which was then, of course, still the general conceptual scheme for heat phenomena. The only satisfactory solution to the problem of homogeneity of gas mixtures based on a static model was, to Dalton's mind, the following set of assumptions and deductions, some of them taken from the work of predecessors:

a) *Each particle of a gas is surrounded by an atmosphere of caloric.* As Dalton put it in his main work, *A New System of Chemical Philosophy:*

> A vessel full of any pure elastic fluid presents to the imagination a picture like one full of small shot [bullets]. The globules are all of the same size; but the particles of the fluid differ from those of the shot, in that they are constituted of an exceedingly small central atom of solid matter, which is surrounded by an atmosphere of heat, of greater density next the atom, but gradually growing rarer according to some power of the distance.

b) *The particles are essentially at rest, their shells of caloric touching.* At the top of Fig. 20.1, taken from Dalton's notebook, is shown the case of two neighboring gas atoms. Throughout, Dalton accepts the notion that the only means for the action of forces among atoms is direct contact. He is as uncomfortable with the notion of action at a distance as Newton's contemporaries had been.

c) *The total diameter of each particle,* including the caloric atmosphere or shell, *differs from one substance to the next:.*

> ... the following may be adopted as a maxim, till some reason appears to the contrary: namely,—That every species of pure elastic fluid [gas] has its particles globular and all of a size; but that no two species agree in the size of their particles, the pressure and temperature being the same.

This point proved crucially important in the later development. Dalton thought he had deduced this maxim unambiguously from experimental evidences of the following type. One of the products formed when oxygen and nitrogen gas are combined is nitric oxide or, as Dalton called it, "nitrous gas." On other grounds, Dalton had decided that in this gas one nitrogen atom combines always with just one oxygen atom. However,

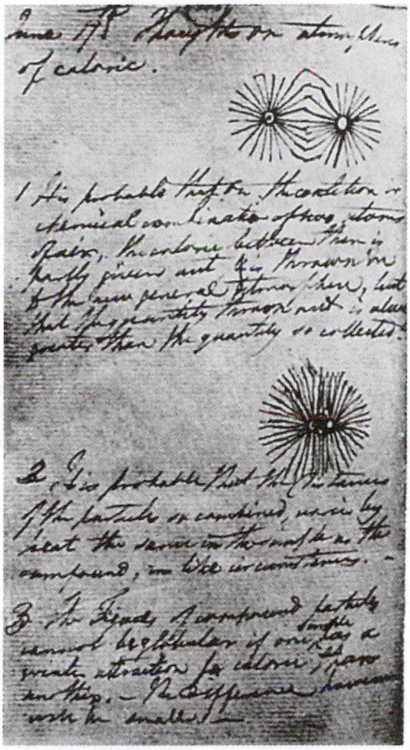

Fig. 20.1. From Dalton's notebook. At the top; two elementary atoms, as the would exist in a gaseous mixture. The straight lines indicate the region of caloric around each mass. Below: a molecule of a compound, consisting of two atoms in close contact, surrounded by a common, nearly spherical atmosphere of caloric.

in his own experiments, it seemed that the *ratio by volume* of oxygen gas to nitrogen gas needed in the reaction was not 1:1, which would have indicated equal volumes for both types of atoms, but more nearly about 0.8:1, pointing to the conclusion that nitrogen atoms were somewhat larger than oxygen atoms. (In actual fact, the experimental results here are easily misleading because several different compounds of nitrogen and oxygen may be formed concurrently.)

The same set of experiments seemed to furnish Dalton with another confirmation of his "maxim." The nitrous gas so formed took about double the volume of either of the component elements. In terms of a gas whose particles all touch, this seemed to indicate clearly that the particles or (compound) atoms of nitrous gas take about twice the volume of either oxygen or nitrogen particles.

Experiments on water vapor seemed to bear out the same type of conclusion:

> When two measures of hydrogen and one of oxygen gas are mixed, and fired by the electric spark, the whole is converted into steam, and if the pressure be great, this steam becomes water. It is most probable then that there is the same number of particles in two measure of hydrogen as in one of oxygen . . .

—that is, hydrogen atoms are about twice as large as those of oxygen.

d) Turning back to the initial problem, Dalton concluded on the basis of some qualitative speculations that there could be no stratification of the air into its elements *because the contiguous particles of several sizes would tend to push one another away* until stability is reached in a homogeneous mixture.

e) Therefore, he reasoned, the constituents of the atmosphere must have been mixed with one another in the disequilibrium during their original encounter, and since then have remained homogeneous This suggestive "pile of shot" picture of the atmosphere prompted Dalton to consider next the number and relative weights of atoms, particularly "of all chemical elementary principles which enter into any sort of combination one with another." But before we turn to look at the enormously fruitful results that followed this decision, let us in passing note the amazing fact that each and every one of the points (a) to (e) above is, from the present point of view and often even in terms of internal consistency, *wrong;* that furthermore the premise itself, Newton's supposed proof of the static model, is Dalton's misinterpretation.

This is a fine illustration of the position we argued in Part D, namely that science is not to be regarded as the inevitably successful result of following a clear, step-by-step method. On the other hand, neither can we conclude that science is a blind person's game, that we must grope from error to error, making a few valid discoveries only by chance, by sheer luck, or simply as by-products of a prodigious amount of largely fruitless labor being done everywhere and throughout the centuries.

Problem 20.1. Regarding as unquestioned the stimulating influence that Dalton's general atomic theory had on chemistry (the details will be presented in the following pages) write a brief analysis of Dalton's creative process in relation to the dual aspect of science, S_1 and S_2 (Sections 13.1 to 13.7)

20.3 Properties of Dalton's Chemical Atom

But to return to the growth of the atomic theory of chemistry: Dalton's great opus, *A New System of Chemical Philosophy,* from which most of the previous quotations were taken, was first published in two parts, in 1808 and 1810. From that work and from his other writings, we can abstract the several principles that reflect the basic ideas in his conceptual scheme.

a) *Matter consists of indivisible atoms.*

> Matter, though divisible in an extreme degree, is nevertheless not *infinitely* divisible. That is, there must be some point beyond which we cannot go in the division of matter. The existence of these ultimate particles of matter can scarcely be doubted, though they are probably much too small ever to be exhibited by microscopic improvements. I have chosen the word *atom* to signify these ultimate particles. . . .

b) *Atoms are unchangeable.* The atoms of different elements "never can be metamorphosed, one into another, by any power we can control," as the failure of centuries of alchemy had made clear. By continual *test,* transmutation of elements was found to be impossible—a *postulate of impotency* arrived at in a manner exactly analogous to the rise of the law of conservation of energy from the long failure to find a *perpetuum mobile.* (But of course this kind of "induction" is not a rigorous proof: one cannot be absolutely confident, on the basis of *past* experience, that it will never be possible to change what Dalton called atoms in the future.)

c) *Compounds are made of molecules.* Chemical compounds are formed by the combination of atoms of two or more elements into "compound atoms" or, as we now prefer to name the smallest particle of a compound, *molecules.* (Does that imply, conversely, that molecules are always made of *different* kinds of atoms?)

> I call an ultimate particle of carbonic acid a compound atom. Now, though this atom may be divided, yet it ceases to be carbonic acid, being resolved by such division into charcoal [carbon] and oxygen. Hence I conceive there is no inconsistency in speaking of compound atoms, and that my meaning cannot be misunderstood.

Although we may be tempted to think of the particles of compounds, the molecule, as made of two or more atoms, so to speak side by side, and although Dalton's symbols, as we shall see at once, at first glance give the same impression, he did not have this conception. Rather, he thought that the compound atom was essentially round or "globular," the centers of the combining atoms being "retained in physical contact by a strong affinity, and supposed to be surrounded by a common atmosphere of heat" (see Fig. 20.1).

In the specific example of the formation of water, Dalton asked that it should "be supposed that each particle of hydrogen attaches itself to a particle of oxygen, and the two particles so united form *one,* from which the repulsive energy emanates [by a redistribution of the combined caloric]: then the new elastic fluid may perfectly conform to Newton's Law. . . ."

d) *All atoms or molecules of a pure substance are identically alike.*

> The ultimate particles of all homogeneous bodies are perfectly alike in weight, figure, etc. In other words, every particle of hydrogen is like every other particle of hydrogen. . . .

We should here realize that Dalton was now simplifying even beyond the older atomists, who usually had felt it necessary to allow different sized atoms for the same element, just as pebbles made of the same material might have different shapes and sizes.

e) *In chemical reactions, atoms are only rearranged, not created or destroyed.*

> Chemical analysis and synthesis go no farther than to the separation of particles one from another, and to their reunion. No new creation or destruction of matter is within the reach of chemical agency. We might as well attempt to introduce a new planet into the solar system, or to annihilate one already in existence, as to create or destroy a particle of hydrogen. All the changes we can produce, consist in separating particles that are in a state of cohesion or combination, and joining those that were previously at a distance.

Here at last we have a strikingly simple physical conception to explain the law of conservation of mass, which Lavoisier had postulated and experimentally demonstrated decades before.

20.4 Dalton's Symbols for Representing Atoms

We have already noted a passage in which the combination of hydrogen and oxygen into water vapor is pictured. To picture to himself this and other chemical reactions, Dalton found it extremely helpful to make use of simple but ingenious symbols (Fig. 20.2). Whereas even the earliest alchemists had represented different substances by different symbols, Dalton's symbols

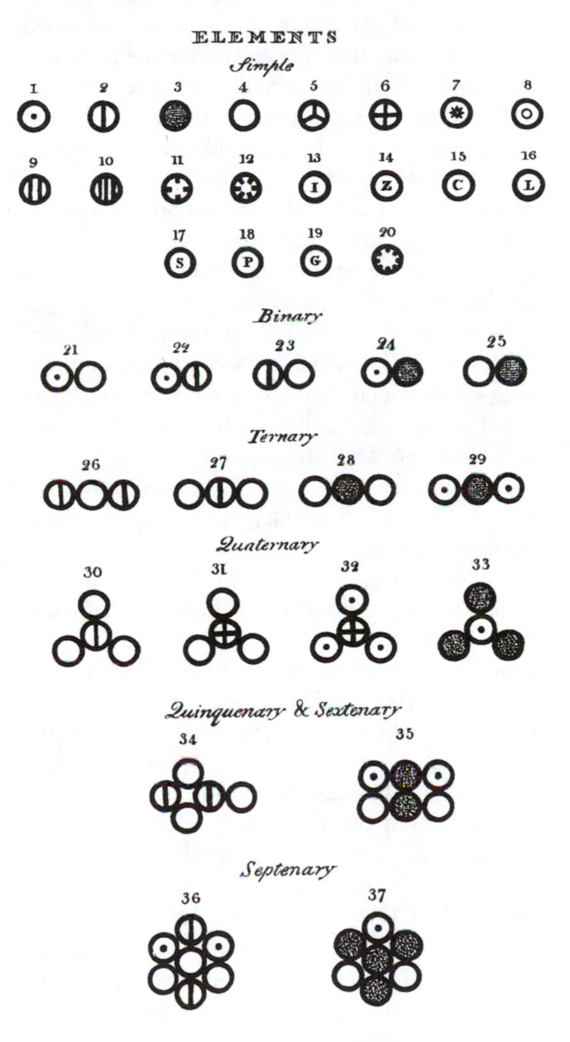

Fig. 20.2. Dalton's symbols for elementary and compound atoms, taken from his work, *A New System of Chemical Philosophy,* Part 1. Dalton writes, "This plate contains the arbitrary marks or signs chosen to represent the several chemical elements or ultimate particles." The modern names for the first few atoms are: 1, hydrogen; 2, nitrogen; 3, carbon; 4, oxygen; 5, phosphorus; 6, sulfur.

represented *individual atoms*. But note well that in Fig. 20.2 some of the first twenty symbols do not represent the atoms of elements but the molecules of compounds as Dalton understood them. For example, symbols 9 and 10, which stand for soda and potash, were known to Dalton from Davy's work of the same year (1808) to be "metallic oxides" (actually metallic hydroxides).

Because we shall make some use of these historic symbols, let us briefly note those for the compounds. For example, 21 depicts water, which Dalton considered to be a binary compound made of one atom of oxygen and one of hydrogen; 22, "an atom of ammonia, composed of one of azote (the then current name for nitrogen) and one of hydrogen"; 23 represents "an atom of nitrous gas (now called nitric oxide) composed of one of azote and one of oxygen"; 24 was Dalton's sign for what we now call methane; and 25 was for carbon monoxide, then called "carbonic oxide."

Among the ternary compounds 26 is nitrous oxide, still called thus, although the name laughing gas is more familiar; 27 is a compound atom of "nitric acid," which we would name a molecule of nitrogen dioxide; 28 represents "carbonic acid," now called carbon dioxide. In the same fashion, Dalton pictured to himself other compounds that he knew to contain more than three elementary atoms, namely quaternary, quinquenary, etc.

Nowadays, of course, we are used to other symbols: those introduced by the Swedish chemist J. J. Berzelius in 1819. Letters are used to represent the substances. Hydrogen, oxygen, nitrogen, carbon, and chlorine are, respectively, H, O, N, C, and Cl (the full list of all elements discovered by the end of the twentieth century appears in Appendix VI), and compounds are now rendered by the joining of the respective symbols of elements. For example, hydrochloric acid would be represented by HCl, and carbon dioxide by CO_2.

A chemical reaction, say the combustion of carbon (the formation of carbon dioxide), might then be given in one of these ways, the first being Dalton's and the last being the present convention:

● + ○ + ○ → ○●○

or

$$C + O + O \rightarrow CO_2$$

or

$$C + O_2 \rightarrow CO_2.$$

20.5 The Law of Definite Proportions

The accumulation of more and more accurate analytic data during the prior half century, and particularly the vigorous dispute between the French chemists J. L. Proust and C. L. Berthollet, had made increasingly acceptable what we now call the *law of definite proportions* of combining weights: No matter in which locality a pure substance—say a metallic oxide, a drop of some liquid, or a complex gas—is found, no matter whether it is artificially prepared or taken from a natural source, the proportions by weight of the constituent elements in every case are always the same. For water, this ratio by weight between hydrogen and oxygen gas is 1:8 (approximate modern value; Dalton's was first 1:6 and later 1:7).

Now, combining the simple pictorial view of atoms and the law of definite proportions by weights of combining elements, Dalton was forced to the hypothesis that in a chemical reaction there is always a fixed or definite proportion by *number* of combining atoms. To use a very crude analogy, if it were found on shelling peas that the weight of the total quantity of peas is a *constant fraction* of the weight of the empty pods, no matter how many pods are included or where the harvest is gathered, we should be quick to propose that the number of peas is the same in each pod, even if we do not know what that number actually is. To put Dalton's hypothesis simply, *in forming molecules during chemical reactions, the numbers of combining atoms of the different elements form simple, definite ratios.*

This part of Dalton's atomic theory denied a then widespread and distinguished opinion concerning chemical reactions, namely, that particularly in the case of two gases the compound is formed by a solution of one in the other, rather as salt and water form salt water. Dalton, a little querulously, gives several arguments "advanced to show the absurdity of this doctrine. . . ." Instead of regarding water as made of an indefinite or variable combination of hydrogen and oxygen particles, he conceived of each molecule of water vapor as being made the same way, namely, of one atom of each element.

But there's the rub; on the basis of what has developed so far, Dalton has, of course, no way of knowing *just how many* atoms of each element do in fact combine in one water molecule; as far as we can determine, the ratio might not be 1:1 but perhaps 2:1 or, for that matter, 23:61. The relative volumes of combining gases do not offer a clue, since Dalton has already assumed the atoms to have different and, of course, unknown sizes.

All that can be said is this: As the relative combining weights between hydrogen and oxygen are 1:8, if we *assume* that the combination proceeds by joining a single atom of each element *for every single molecule,* then we may conclude that the relative atomic weights of hydrogen and oxygen are as 1: 8 as well. On the other hand, if we assume that the molecular formula is not HO but H_2O (that is, $H + H + O \rightarrow H_2O$ or $H_2 + O \rightarrow H_2O$), then the relative atomic weights would be 1:16; and similarly for other formulas. The missing link to an understanding of chemical reactions, therefore, is the knowledge of some simple molecular formulas.

We must stop a moment to realize that like the law of conservation of mass, which was the fruit of a fierce struggle between Lavoisier and his followers in their fight against the phlogistonists, so also was the establishment of the law of definite proportions much speeded by the vigor of the dispute between able antagonists. Thereby were also established at the time several other most valuable results, for example, clearer distinctions between compounds and mixtures and a large quantity of data that provided the raw material for further theorizing. From our dynamic view of science we can well understand this new illustration of the evolutionary mechanism in the struggle of ideas.

Again, fortunately for the development of atomic theory, the weight of experimental evidence as of 1800 was in favor of Proust's theory of definite proportions; only much later was it established that there are *some* compounds (called "berthollides" after the chemist who believed in their existence) that do have variable composition.

20.6 Dalton's Rule of Simplicity

Dalton was finally driven to adopt a typical strategy of the early stages of science-in-the-making, an *assumption,* one that appears to typify the advantages and dangers of any hypothesis—his rule of *greatest simplicity* of the most common stable compounds:

> When only one combination of two bodies [elements] can be obtained, it must be assumed to be a binary one, unless some other cause appears to the contrary. . . . When two combinations are observed, they must be presumed to be a binary and a ternary. . . .

and so forth for more than two combinations. Specifically water, then the only known combination of hydrogen and oxygen, must be a binary compound (HO or ⊙ ○). (Hydrogen peroxide, H_2O_2, had not yet been identified.) In the same manner, the only two known combinations of carbon and oxygen were identified as follows: the lighter of the gases (*carbonic oxide*) as CO, the heavier (*carbonic acid*) as CO_2. The common compound of nitrogen and hydrogen, ammonia, was regarded as NH (we now know it to be NH_3). Ethylene and methane were identified as CH and CH_2 (whereas now they would be written C_2H_4 and CH_4).

One brief look at any modern table of chemical formulas will at once convince us of the fallaciousness of Dalton's maxim. For example, the only iron-iodine compound, iron iodide, is not binary but has the formula FeI_2; and the simplest iron chloride is not FeCl but $FeCl_2$. The iron phosphides are not FeP, FeP_2, and Fe_2P, as Dalton might have expected, but FeP, Fe_2P, and Fe_3P. The common oxides of iron are now known to be FeO, Fe_2O_3, and Fe_3O_4.

However, the analytic data in most cases (we might almost say luckily) were too ambiguous to point out a large-scale failure of Dalton's maxim, and in some cases it gave correct results (for example, for the carbon oxides and for the nitrogen oxides). Above all, Dalton's devotion to the time-honored idea of "the regularity and simplicity" in the laws of nature did make possible a start, and a fruitful one, toward the identification of molecular formulas. Where some early guess of a molecular formula was wrong, it could be corrected through internal consistency checks; we recall again the admonition by Francis Bacon that truth can come out of error more easily than out of confusion.

20.7 The Early Achievements of Dalton's Theory

We might discuss a variety of subtle consequences of Dalton's work: that his pictorial conceptualizations and his symbols must have produced a deeper feeling for the reality of atomic processes at the time; that his identification of chemical atoms helped in the final differentiation between the concepts of pure chemical compounds and mixtures; and other details of this sort. But above all, we must now summarize briefly the initial main achievements of Dalton's theory—its power to gather known facts and laws and to provide a simple, workable conceptual scheme; its fruitfulness in suggesting new relations and stimulating research; and its helpfulness in the solution of problems in contemporary science. Elements of each of these functions may be found in the following examples.

a) As we saw, Dalton's chemical atom is the concept that gives physical meaning to the law of conservation of mass, and that integrates it with the law of definite proportions. The latter, although essentially established at the time by empirical and exhaustive testing, was not yet explicitly and generally appreciated. Those experimental results were now given physical meaning and were supported by the inherent plausibility of Dalton's picture; in this sense we may say that Dalton did for the embattled law of definite proportions what Newton did for Kepler's established laws of planetary motion. Other established chemical results could be similarly accommodated (for example, the law of equivalent proportions).

b) Perhaps the best support of the atomic theory, by common consent at the time, was an extraordinarily powerful insight for the later progress of chemistry, one that followed directly and naturally from Dalton's conceptual scheme: *When one substance unites with another to form more than one compound and so forms more than one proportion* (for example, 2:1 and 3:2), *these different proportions bear a simple ratio to one another* (for example, 2:1/3:2 = 4:3). This is the *law of multiple proportion.* We see at once that by necessity it is fully implied in Dalton's assumptions.

For example, if we picture carbonic oxide and carbonic acid in Dalton's manner,

○● (carbonic oxide)

○●○ (carbonic acid),

it is clear that we should find associated with a given quantity of carbon twice as much oxygen by weight in the second case compared with the first; or, to put it into the language of the law, the respective weight proportion of oxygen to carbon in the second substance, *divided* by the proportion of oxygen to carbon in the first, should be as 2:1, a simple ratio indeed.

As to the experimental check of this prediction, the necessary data for combining weights had, significantly, been long available in about the following form:

- In carbonic acid: 73% oxygen and 27% carbon (by weight).
- In carbonic oxide: 57% oxygen and 43% carbon.

If we reduce these approximate percentages to proportions, using ratios of small integers, we obtain:

- In carbonic acid: ratio of oxygen to carbon is about 8:3.
- In carbonic oxide: ratio of oxygen to carbon is about 4:3.

Therefore the ratio between these proportions is equal to (8:3)/(4:3) = 2 (or 2:1).[2]

Similar confirmations of the law of multiple proportions could be obtained by an analogous treatment of other data from previously published scientific work. We might say that the facts had been in front of everyone's eyes, but the simplicity they implied remained hidden because there was previously no good reason to look for it. As has often been said in these pages, one cannot see much without the aid of a hypothesis to organize the perceptions; in a sense, the old saying "seeing is believing" should be supplemented by another, "believing is seeing."

As might have been expected, this discovery of the law of multiple proportions stimulated a great deal of new research. Given any one proportion of combining weights for two substances, one could now plan for and predict the analysis or synthesis of other compounds of the same elements. The results were impressive, although we now know that the postulate of "simple"

[2]The deduction does not proceed unambiguously in the opposite direction, however. From the fact that the ratio is 2:1 we could conclude either that the two molecular formulas are CO and CO_2 or that they are C_2O and CO. Dalton did not give any explicit rule for resolving this ambiguity in all such cases, but used various additional kinds of evidence to determine molecular formulas. See the discussion by A. J. Bernatowicz, "Dalton's rule of simplicity."

ratios is again too limited and that *large* whole numbers may be involved. For example, in the complex structure of the organic compounds, then still mostly inaccessible, each molecule may contain several dozens of atoms of a few substances, so that the ratio of carbon-to-hydrogen proportions is not such a simple one for two substances such as $C_{23}H_{22}O_6$ and $C_{22}H_{42}O_3$.

Problem 20.2. In 1810, Dalton gave the approximate results listed in the following table for the relative densities of five gases (his terminology), each made of molecules involving only nitrogen and oxygen. He also concluded that the lightest of these gases was made of molecules containing one atom each of N and O. (a) Decide on the molecular formula for each gas on the basis of this information. (b) Confirm the law of multiple proportions for these five gases.

		% by weight	
Substance	*Relative density*	*Nitrogen*	*Oxygen*
Nitrous gas	12.1	42.1	57.9
Nitrous oxide	17.2	59.3	40.7
Nitric acid	19.1	26.7	73.3
Oxynitric acid	26.1	19.5	80.5
Nitrous acid	31.2	32.7	67.3

c) A third achievement of the atomic theory of chemistry, one without which modern physics and chemistry would be unimaginable, was the start toward the *determination of atomic weights*. In Dalton's words, "Now it is one great object of this work, to show the importance and advantage of ascertaining the relative weights of the ultimate particles, both of simple and compound bodies . . ."—an undertaking previously neither possible nor, indeed, even meaningful. Note that Dalton only claimed to determine the *relative* weights of the atoms of different substances; no reliable method was known for determining absolute weights until 1865, when Loschmidt used the results of kinetic theory for this purpose. (We now know that the atoms of elements have masses of the order of 10^{-27} kg each.)

Historians have disputed for some time the exact way in which Dalton first came to compute his atomic weights. The most recent view is that he was attempting to develop a mechanical theory of the solubility of gases in water, and found that it was important to determine the sizes and weights of the particles of different gases. Entries in his laboratory notebook show that he had calculated a table of atomic and molecular weights by September 1803, although the table was not published until 1805, and a systematic explanation of the method of constructing it did not appear until 1808 in his *New System*. Our foregoing discussion has hinted strongly at how the relative weights of atoms could be obtained: from the *known combining weights* of the elements of a substance and from the *assumed molecular formulas*. For example, from the data given in Problem 20.2 it would appear, as you may check in five ways, that the relative weights of the atoms of nitrogen and oxygen could be taken to be nearly 6:8.2 (from better experiments we now have the values of 7:8, or more precisely, 14.008:16.000).

We may use other analytic data in the very same way: The combining weights of hydrogen and oxygen in Dalton's synthesis of water being as 1:6 (later 1:7, and by present measurements nearly 1:8), and the formula for water having been assumed to be HO (⊙ ◯), the relative atomic weights were also at first supposed to be as 1 to 6. By making a compromise between the various results of this type, Dalton could assemble a table of most probable values of relative atomic weights (although, for a Newtonian physicist, the words "relative combining *masses*" would have been more appropriate). By referring all values to the arbitrary value of unity for hydrogen, Dalton assembled tables for most of the elements and compounds known at the time.

One of the earlier tables, published in 1808, gave the following list of relative atomic weights: hydrogen 1, nitrogen 5, carbon 5, oxygen 7, phosphorus 9, sulfur 13, soda (a "compound atom," namely NaOH, sodium hydroxide) 28, and so forth. (Modern tables use carbon as the reference substance, so that hydrogen now has an atomic weight of about 1.008; see Appendix VII.)

It soon became obvious to Dalton that in using the rule of greatest simplicity, inconsistencies were bound to occur in the determination of atomic weights based on various compounds. In the specific case of water he said as early as 1810, "After all, it must be allowed to be possible that water may be a ternary compound" (H_2O or HO_2 instead of HO). In the former case the (modern) combining weights of 1:8 for the two gases would indicate that the relative atomic weight of oxygen is 16 if hydrogen is 1; in the latter case, if H = 1, O = 4. It is conceivable that by allowing such flexibility in the assumption of molecular formulas, and by eliminating contradictions, one might eventually have arrived at a set of consistent atomic weights (the process is illustrated in the next problem). But just then there appeared from France an experimental discovery and from Italy a theoretical insight, which at once aided this task, and which together were destined to replace the gradually less and less defendable Daltonian rule of greatest simplicity, putting atomic chemistry on a new basis.

Problem 20.3. In the table below is given a set of approximate *combining weights* for five substances, as determined by experiment. For each, Dalton's principle of greatest simplicity would demand a *binary* structure, although in fact most of them are more complex. Analyze these data and obtain, without recourse to any other aids and exclusively from the consistency of these data, (a) a set of the most consistent molecular formulas for all these substances, and (b) a set of the most consistent atomic weights for all the elements involved. *Then* consult a collection of constants for inorganic compounds (as for example in the *Handbook of Chemistry and Physics*) and find the correct formulas and the current names for these compounds.

	Combining Weights in grams			
Substance	*Iron*	*Sulfur*	*Hydrogen*	*Carbon*
A	14			9
B		16	1	
C	7	8		
D		16		3
E		48	1	6

[*Hint:* Begin by writing for each of the *first four* substances several possible molecular formulas, and the relative atomic weights for each of these cases as they would follow from the given combining weights. Next, select the one most consistent set of atomic weights, and you have thereby decided on the most probable molecular formulas as well. Then, as a test of your scheme, turn to the last substance and determine whether it can be accommodated.]

20.8 Gay-Lussac's Law of Combining Volumes of Reacting Gases

In 1808 Joseph Louis Gay-Lussac (see Section 19.3), a master of experimental chemistry, had noted that Cavendish's experiment on the explosion of hydrogen and oxygen gas to yield water vapor pointed to a beautiful simplicity. On repeating the experiment most carefully, he found that the *combining volumes* of hydrogen and oxygen are as 2:1, within 0.1%. For example, 2 liters of hydrogen would react with exactly 1 liter of oxygen, with no hydrogen or oxygen left over.

Now Gay-Lussac remembered that gases as a group exhibit simple and regular physical laws, that is, the ideal gas laws, which are approximately valid for all gases, and he speculated that therefore the combination of gases (by volume) in general might be guided by a set of simple rules. From his analysis of previously published work and his own careful research he concluded, "It appears to me that gases always combine in the simplest proportions when they act on one another; and we have seen in reality in all the preceding examples that the ratio of combinations is 1 to 1, 1 to 2, or 1 to 3." (To this we should now add other simple ratios by volume, such as 2:3.)

Furthermore, he noted that if the compound resulting from the chemical reaction remains a gas or a vapor at the temperature of the experiment, the volume of the compound is also simply related to the volumes of the participating elements. We might write results of this kind as follows (where "1 vol" might be "1 liter"):

1 vol nitrogen gas + 3 vol hydrogen gas
= 2 vol ammonia vapor,

2 vol hydrogen gas + 1 vol oxygen gas
= 2 vol water vapor,

1 vol nitrogen gas + 1 vol oxygen gas
= 2 vol nitric oxide gas,

1 vol hydrogen gas + 1 vol chlorine gas
= 2 vol hydrochloric acid gas.

Problem 20.4. Write out a concise statement of Gay-Lussac's law of combining volumes.

This marvelous simplicity would appeal to us at first sight; but for many reasons Dalton rejected these results utterly and to the very end. We recall that his model of gases, in order to explain the homogeneity of the atmosphere, required that the atoms of different elements be of different sizes. To accept the possibility of such general simplicity, even occasionally of equal combining volumes, seemed to him tantamount to accepting the notion that equal gas volumes contain equal numbers of atoms, or at any rate multiples thereof, that the atoms therefore might be of equal sizes, and that consequently he would have to give up both his static model and what he supposed to be Newton's "immutable law of elastic fluids."

Dalton found his arguments confirmed by the fact that his own (faulty) volumetric experiments contradicted Gay-Lussac's results, and that these, of course, were admittedly not *exactly* whole numbers (we recall that scientific measurements can never be made without some uncertainties; and in these delicate tests the likelihood of error was often considerable, though in fact Gay-Lussac is generally recognized to have been a much more accurate experimenter than Dalton). Therefore Gay-Lussac had had to round off his experimentally derived figures, and to Dalton this discrepancy was not a negligible experimental error, but the most significant truth.

This dispute summarizes for us brilliantly and compactly some typical features of the life of science: the search for simplicity, the fascination with integers, the inexactitude of measurement, the role of a theory to guide and interpret (or misinterpret) experimental results. Indeed, we are here reminded a little of the dispute between Galileo and the scholastics regarding the interpretation of free-falling motion.

Quite apart from Dalton's objections, Gay-Lussac's results did not at this point lend themselves directly to an interpretation by Dalton's atomic theory of gases. To make the two dovetail required an amendment of the theory, after which, however, the atomic hypothesis emerged stronger than ever—in fact, approximately in its present state.

Fig. 20.3. Amedeo Avogadro (1776–1856), Italian physicist and chemist. His hypothesis that every gas contains the same number of molecules under standard conditions introduced a clear distinction between atoms and molecules; a molecule of a single element may contain more than one atom. This became the basis for determining atomic weights later in the nineteenth century.

20.9 Avogadro's Model of Gases

The amendment was not long in coming; it was the work of the Italian physicist Lorenzo Romano Amedeo Carlo Avogadro di Quaregua e di Cerreto (1776–1856), usually known as Amedeo Avogadro (Fig. 20.3). Avogadro's obscurely phrased but penetrating publication of 1811 presents a model of gases that may be analyzed in five separate parts.

a) As had been often and persistently suggested by Gay-Lussac and others, Dalton's picture of a gas in which the particles are in contact is now replaced by another (it does not matter at this point whether a static or a kinetic model), in which the particles have only a very thin shell of caloric and are quite small compared with the distances between them. The volume actually taken up by the atoms is consequently very much smaller than in Dalton's model, being in fact a negligible fraction of the total volume of any sample. It is a return, if you wish, to Lucretius' "atom and void" picture of a gas.

With this assumption we must, of course, also give up the hope of obtaining any information on *relative atomic dimensions* on the basis of combining volumes. But this problem of dimensions, so important to Dalton, immediately becomes quite insignificant in this formulation.

b) Contrary to initial considerations of simplicity, or to the firm faith of Dalton, or, in fact, to every reasonable expectation at the time, the ultimate particles of gaseous elements are conceived to be not necessarily just single solitary atoms, but instead may be thought of as being made up of two or more chemically alike atoms "united by attraction to form a single whole." Calling the smallest such unit a *molecule,* we may then imagine that the molecules of element X in gaseous form all consist of two atoms each (a diatomic gas), or in the case of some element Y of three atoms (triatomic), or in gas Z of single atoms (monatomic), and so on. We still bear in mind that since the molecules of an element are made up of, at most, a few atoms they are negligibly small compared with the intermolecular distances.

c) In a *gaseous compound,* for example, water vapor, the molecules evidently each consist of definite numbers of chemically unlike atoms. But when two gases combine, as in the formation of water from hydrogen and oxygen gas, we must not just assume that one or more molecules of hydrogen and one or more of oxygen simply associate into a new or larger unit; we must examine the possibility that such a larger group forms only for a brief moment and then at once breaks up into two or more equal parts.

We may represent the various possibilities symbolically as follows:

$$H + O \rightarrow HO \qquad (20.1)$$

$$2H + O \rightarrow H_2O \qquad (20.2)$$

$$H_2 + O \rightarrow H_2O \qquad (20.3)$$

$$2H + O_2 \rightarrow [H_2O_2] \rightarrow 2HO \qquad (20.4)$$

$$H_2 + O_2 \rightarrow [H_2O_2] \rightarrow 2HO \qquad (20.5)$$

$$2H_2 + O_2 \rightarrow [H_4O_2] \rightarrow 2H_2O \qquad (20.6)$$

$$2H_3 + O_2 \rightarrow [H_6O_2] \rightarrow 2H_3O \qquad (20.7)$$

$$2H_4 + O_4 \rightarrow [H_8O_4] \rightarrow 2H_4O_2 \qquad (20.8)$$

etc.

d) But the most important part of Avogadro's proposal, his most daring insight, is that, at equal temperatures and pressures (conditions largely obeyed in Gay-Lussac's experiments), *equal volumes of all gases, whether elements or compounds, or even mixtures, contain equal numbers of molecules.* This statement is generally known as *Avogadro's hypothesis* or *Avogadro's law.* Avogadro was not the first scientist to propose it, but he did succeed in incorporating it into a coherent and convincing theory.

To turn to specific cases, let us first represent Gay-Lussac's experimental results for water in symbolic form:

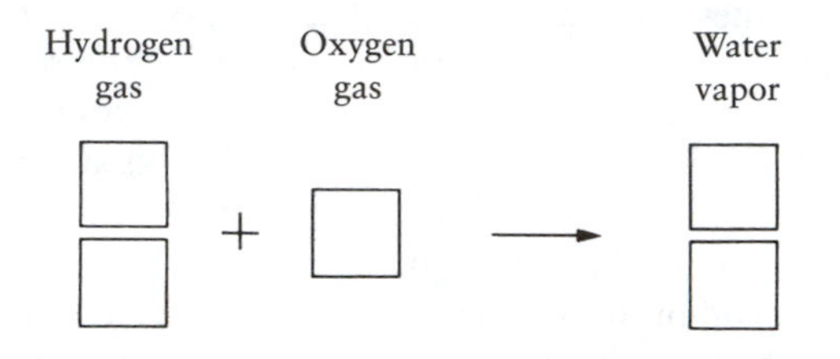

Each of these squares represents a unit volume (say 1 liter). In accord with Avogadro's postulates, we must draw into each square the same number of molecules; for simplicity, let this be just one molecule. If with Dalton we now assume that hydrogen and oxygen gases are monatomic, and if the reaction proceeds according to Eq. (20.1) above, we should at once find it impossible, without splitting the very atom of oxygen, to place into each of the two volumes of water vapor one whole molecule of the same material. We must also reject Eqs. (20.2) and (20.3) for a similar reason.

If we really believe in Avogadro's hypothesis deeply enough to entertain the possibility that one of the other equations represents the true reaction, we may next turn to the assumption that hydrogen is indeed monatomic but that oxygen is diatomic; then Eq. (20.4) applies, and filling in our squares,

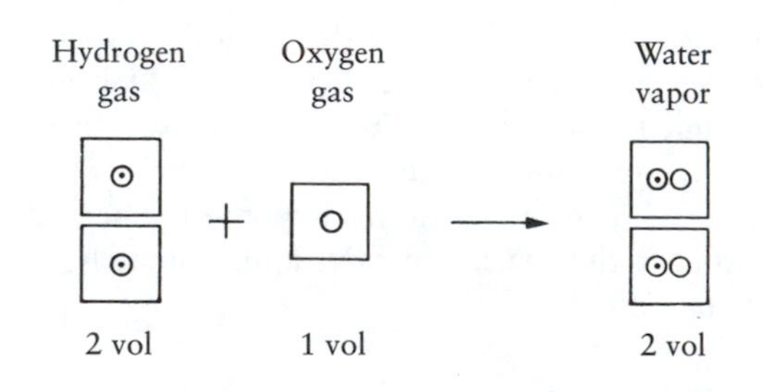

a reaction that is evidently in harmony both with the expectations of Avogadro's hypothesis and Gay-Lussac's experimental results.

But alas! Equations (20.6) and (20.7), and many others that may be imagined, would also be consistent with these two factors. Evidently there is an unlimited number of such possibilities; in abandoning Dalton's too simple reaction H + O = HO as wrong, we have discovered too many alternatives that may all be right.

However, we can at once exclude some of these by considering how these gases behave in other reactions. From Gay-Lussac's experiments we know of the following volume relation in the production of ammonia vapor:

3 vol hydrogen + 1 vol nitrogen
= 2 vol ammonia.

This result shows us at once that hydrogen cannot be monatomic if each unit volume of ammonia is to have the same number of molecules as each unit volume of hydrogen, since that would entail the impossible conclusion that a molecule of ammonia contains 1½ atoms of hydrogen. But the experimental result harmonizes with Avogadro's hypothesis if hydrogen is diatomic and nitrogen is also diatomic; for in this case we may write:

$$3H_2 + N_2 \rightarrow [N_2H_6] \rightarrow 2NH_3,$$

or

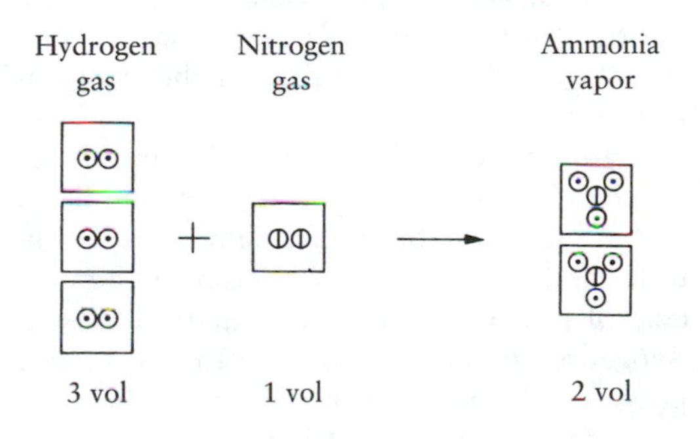

Having now ruled out that hydrogen gas is monatomic, we have automatically denied Eq. (20.4) for the production of water vapor. Equation (20.5) is impossible because it conflicts with Gay-Lussac's results on the combining volumes (demonstrate). So Eq. (20.6), involving the temporary formation of a single H_4O_2 molecule that splits into two H_2O molecules emerges as the simplest one that may still be correct:

$$2H_2 + O_2 \rightarrow [H_4O_2] \rightarrow 2H_2O,$$

or

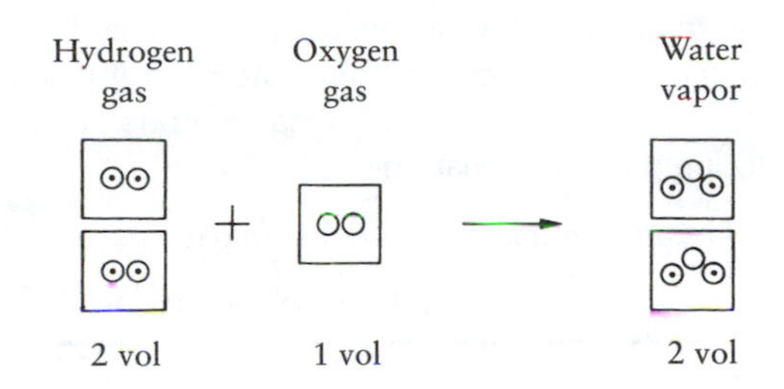

e) At this point Avogadro himself, quite properly, assumes also a "rule of simplicity"; that is, he regards as the correct process the one that is consistent with other chemical reactions, is the simplest, and is as well in harmony with Gay-Lussac's results and his own postulate. Thus he chooses Eq. (20.6) over (20.7), (20.8), and all the others that may be constructed along these lines.

But this is—and was—perhaps a bit disappointing; for all along we had hoped to do without any arbitrary rules of this kind. However, there is this difference: Dalton's rule, by assigning molecular formulas based on fairly incomplete data, had led at once to conflicts. For Avogadro, a rule of simplicity may be used as a final criterion of choice only *after* all the known conflicts have been considered.

Problem 20.5. Summarize the experimental and theoretical steps that lead to the result that the formula for the molecules of water vapor is H_2O, and not HO as Dalton thought originally. Is H_2O the only possibility on the basis of this evidence? If not, why might it be generally accepted as the correct formula?

Problem 20.6. A frequently used system for generating hydrogen on a large scale in a portable plant involves the following reaction: A unit volume of alcohol treated with 1 vol steam produces 1 vol carbon dioxide and 3 vol hydrogen gas. Prove that consequently the molecular formula of the alcohol is CH_3OH (wood alcohol).

Problem 20.7. Show that the reaction $N_2 + O_2 \rightarrow 2NO$ is consistent both with Gay-Lussac's volumetric experiment on nitric oxide and with Avogadro's hypothesis. Why are we sure that this reaction cannot proceed according to the equation $N + O \rightarrow NO$? If the production of nitrous oxide (N_2O) is given by the equation $2N_2 + O_2 \rightarrow 2N_2O$, what must be the volumetric relations in this reaction?

Problem 20.8. Accepting Avogadro's hypothesis, show that the following experimental result is harmonious with the proposition that hydrogen and chlorine gas are both diatomic but is in contradiction to the early idea that hydrogen and chlorine gas are monatomic:

1 vol hydrogen gas + 1 vol chlorine gas

= 2 vol hydrochloric acid vapor.

20.10 An Evaluation of Avogadro's Theory

In summary, let us note the following features of Avogadro's accomplishment.

a) Unlike Dalton, he accepted Gay-Lussac's results as correct, and showed that the work of both investigators was mutually consistent within experimental error.

b) He replaced the static model, in which atoms are in contact and fill up all the available space, with another model in which the molecules (possibly containing two or more atoms) are quite small compared to the spaces between them. This assumption, which was needed in order to explain why a gas of large molecules might occupy the same space as an equal number of small ones, turned out to be essential as well for the kinetic theory of gases developed later in the century.

c) In postulating molecules, Avogadro could offer no plausible explanation for the proposition that chemically identical atoms in the gas of some elements cling together in twos or threes, etc.; indeed, this proposition directly contradicted the prevalent idea that similar atoms in a gas exert repulsive forces on each other (Section 19.4).

If, however, forces of attraction could exist between like atoms, why, it was asked, do these forces not cause the whole bulk of the gas to condense? On such grounds Dalton had, in 1809, rejected as absurd the notion of gaseous elements other than monatomic.

Nor did others take more kindly to the idea. Chemists after 1811 began to become more and more convinced that the stability of molecules such as those of water could be explained by strong positive or negative charges attached to the atoms of different elements; therefore it was difficult to visualize that two hydrogen atoms, which on this model evidently would present each other with equal charges, should nevertheless be firmly bound together in a molecule.

As a matter of fact, it has been possible to answer such questions as why atoms of some elements do form diatomic, triatomic, etc., molecules only within the last few decades, with the help of the quantum theory (Chapter 29). But lack of such answers did not prevent scientists from accepting the existence of multiatom molecules for some elements, and developing useful theories on that basis.

d) At the outset there had been only four elements known in gaseous form, namely, nitrogen, oxygen, hydrogen, and chlorine, and for chemical reactions involving these, Avogadro's hypothesis worked well, provided diatomic molecules for each were assumed. Avogadro's theory might therefore have been accepted if it had simply required the postulate that all elements are diatomic in the gaseous state, even without a plausible explanation of why that should be so. But things were not as simple as that. Other elements were studied in gaseous form at higher temperatures, for example, mercury, phosphorus, arsenic, and sulfur; their chemical and physical behavior could be accounted for on the basis of Avogadro's hypothesis only by assuming that the molecules of these gases were not all diatomic, but that, instead, the molecular formulas were Hg, P_4, As_4, and, depending on the temperature, S, S_2, S_6, or S_8.

Why this diversity? It seemed to many that Avogadro, while trying to achieve simplicity, had produced confusion. Actually, an independent check on the number of atoms in a molecule of gaseous material eventually bore out Avogadro's hypothesis fully, but that was made possible only much later by the kinetic theory of gases, to be discussed in Chapter 22.

e) There seemed to be little plausibility attached to the idea—and less evidence for it—that during the reaction of gases, intermediate unstable molecules were formed that at once broke into two or more equal fragments. Although we know today that the intermediate stages in such reactions are far more complex, it is still true that Avogadro's remarkable intuition was closer to the facts than the conception of his contemporaries.

f) Avogadro's hypothesis that all gases at the same pressure and temperature have the same number of molecules in a given volume did not allow him to discover the exact value of that number, other than the vague suggestion that it would be generally very large. Experiments in many separate branches of physical science since that time have indirectly produced values, and they all agree fairly well on about 2.69×10^{19} molecules/cm^3 of any gas at 0°C and atmospheric pressure. As will be noted in Section 22.7, this is now called *Loschmidt's number* in honor of the scientist whose work led to the first accurate estimate of its value.

A more useful number is the number of molecules whose mass in grams is the same as the weight of one of those molecules relative to the hydrogen atom—for example, the number of molecules in a 32-gram sample of pure oxygen, if the molecular weight of oxygen is 32. This quantity is called the "mole" and will be defined more precisely at the end of this chapter. Avogadro's hypothesis states that the number of molecules in a mole is the same for any gas, and this number is therefore called *Avogadro's number,* even though he did not know its value.

g) At the time, two substantial obstacles to the immediate acceptance of Avogadro's work were the obscure manner of his presentation and the large admixture of other daring—but erroneous—conclusions.

20.11 Chemistry after Avogadro: The Concept of Valence

Some of these points may make plausible why Avogadro's work lay neglected for almost half a century, during which time chemistry, initially spurred on by Dalton's contribution, progressed vigorously. But this very activity had an inevitable result: It unearthed more and more evident data testifying to an insufficiency in Dalton's speculations. Consequently, while Avogadro's contribution remained in obscurity, other possible amendments to Dalton's conceptual schemes were strongly debated. They all, however, suffered from fundamental flaws, such as the persistent assumption that the atoms in a gas are contiguous. The discouragement of unavoidable contradictions seems to have produced in the 1840s a gradual loss of faith in the whole atomic theory of chemistry.

The solution to the dilemma came after 1858 when the Italian chemist Stanislao Cannizzaro (1826–1910) called for a return to Avogadro's work and showed, partly through his own experiments, that it indeed provided a rational basis for chemistry. At an international conference of chemists in Karlsruhe, Germany, in 1860, he presented arguments that eventually convinced other chemists to accept Avogadro's hypothesis as a basis for determining atomic and molecular weights. In fact, this was the final stage of the revolution of chemistry—thus were achieved the fundamental notions on which modern chemistry is built.

Cannizzaro was successful because of his persuasiveness but also because he had a powerful ally. By that time chemists had become more prepared to accept the ancient suggestion, the kinetic model of a gas, for in the meantime, a revolution had begun in physics—the abandonment of the caloric fluid and the growing acceptance of the kinetic model of gases in that field.

Dalton's atomic theory suggests a very simple interpretation of the empirical law of definite proportions. Carbon monoxide, for example, always consists of molecules made of one carbon atom and one oxygen atom each; each molecule of ammonia invariably consists of one nitrogen atom in close connection with three hydrogen atoms; and similarly for each molecule of any other compound. However, contemplation of this gratifying regularity soon brings up perplexing questions: Why do atoms have such fixed combining powers with one another? What determines the fact that hydrogen and chlorine always make HCl and not HCl_3 or H_5Cl_2? Why, given H and O, can we find either H_2O or H_2O_2, but not H_2O_3; or, given C and O, can we find CO or CO_2 but not C_2O?

Familiarity with the composition of many molecules might lead to a picturization of hydrogen atoms as having one "hand" or "hook" each by which to engage a similar "hook" on other atoms, as indicated by the formulas HCl and H_2O. Thus one hook should be assigned to the chlorine atom, two to oxygen atoms (see Fig. 20.4). The difference between the structure of H_2O and that of H_2O_2 is then directly visualizable. The nonexistence of H_2O_3 might be interpreted as the inability of these atoms to form stable molecules when the hooks or bonds are not all engaged. The formula CO_2 implies that the carbon atom has four such coupling mechanisms to engage the two of each oxygen atom; but in CO it would seem that two hooks of carbon are left unoccupied, after all, and ready to snatch more oxygen—a simple picture that coincides with the fact that CO is such a powerful "reducing" (antioxidizing) agent. (That property is what makes it so lethal in a closed space in which living organisms may be competing for the available oxygen.)

Ultimately this type of analogy is found to be incomplete and dangerous. The question of combining powers of atoms had to be answered in terms of the structure of the atoms themselves. Until that type of answer could be given, however, the general approach indicated in Fig. 20.4 formalized the subject in a powerful way, even though it remained at first mainly a descriptive device.

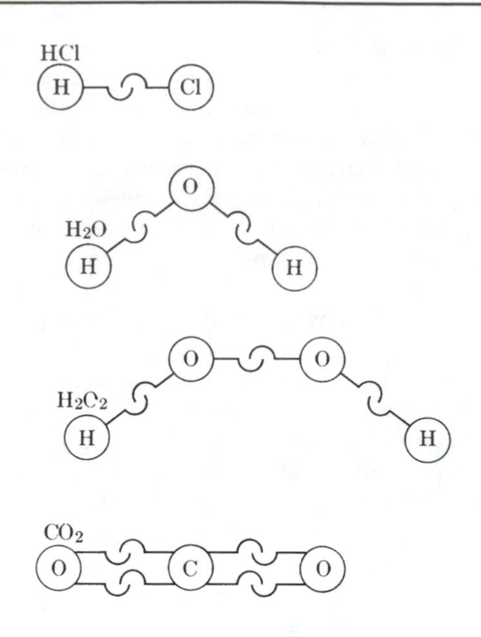

Fig. 20.4. Schematic representation of bonding in molecules. It is not to be imagined that all molecules are actually strung out in one line as represented here.

Thus we encounter, as of 1852, the concept of *valence* (or valency), which became indispensable for the description and understanding of chemical processes.[3] The valence of an element, speaking very roughly, is the number of "hooks" an atom can put out to enter into chemical partnership with other atoms or with stable groups of atoms.

The simplest way to determine the valence of an element is to look up the chemical formula of a substance involving only this element and hydrogen, which we define as having a valence of 1 (monovalent). For example, the formulas H_2O, NH_3, CH_4 indicate valence numbers of O = 2 (divalent), N = 3 (trivalent), C = 4 (quadrivalent). The formula H_2SO_4 shows that the sulfate group (SO_4) has the valence of 2 (divalent).

Next we may find further valences by noting the number of hydrogen atoms an element can *displace* in a molecule, for example, in CaO, Ca takes the place of two hydrogen atoms in H_2O, hence Ca has the valence 2. The occurrence of HF (hydrofluoric acid) and of OsF_8 (osmium octafluoride) shows that osmium can have a valence of 8, the highest valence number that exists. Similarly, from NaCl, $CaCl_2$, $AlCl_3$, $SnCl_4$, PCl_5, and WCl_6 we derive valences for Na = 1, Ca = 2, Al = 3, Sn = 4, P = 5, and W = 6.

This process of enumeration, however, leads to difficulties. Carbon is quadrivalent and oxygen is divalent by the above test. Yet we just saw that they do form CO. There the total number of available "hooks" is not engaged, and the valences of the parts of the molecule do not balance. We must now say either that two of the four valences of C are not used in CO or that the valence of carbon is still effectively 2 in CO but 4 in CO_2. In the long run the latter mode of expression is preferable, although it leads to the disappointing realization that carbon and many other elements can take on different valences depending on what molecule they help to form.

As shown in Fig. 20.4, we imagine that C uses two of its hooks, or as we will now call them, *valence bonds,* for each of the two O atoms. This arrangement is called a *double bond.* It is also possible to have a triple bond. Chemists call molecules with multiple bonds *unsaturated,* indi-

[3]Modern chemistry books have begun to abandon the term *valence* and use instead some phrase such as *combining capacity.* But the simple word survives in descriptions of the electronic structure of atoms, where one speaks of the *valence shell,* meaning the highest (partly or fully occupied) principal energy level: electrons in this shell are the ones available for chemical bonding (see Section 28.6).

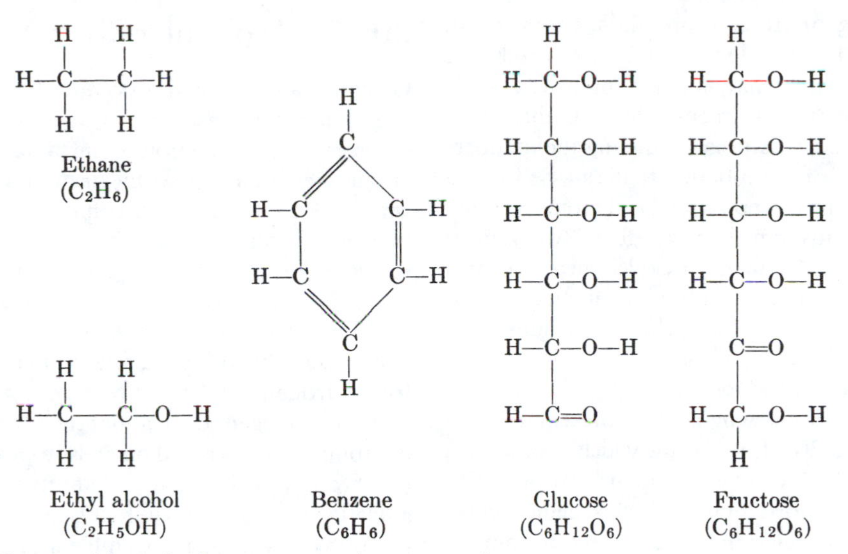

Fig. 20.5. Structural formulas for some organic compounds.

cating that the carbon atom has not exerted its full bonding capacity; unsaturated molecules tend to be more reactive, picking up more hydrogen or other atoms to saturate their valences. (This is the origin of the now-familiar term "unsaturated" or even "polyunsaturated" used to describe certain ingredients in foods.)

The oxides of nitrogen (N_2O, NO, N_2O_3, NO_2, N_2O_5) indicate that nitrogen can take on the valence numbers 1, 2, 3, 4, and 5, respectively (O = 2 throughout).

By the same argument, the valence of iron atoms is 2 in FeO, $FeCl_2$, and $Fe(NO_3)_2$, but is 3 in Fe_2O_3, $FeCl_3$, and $Fe(NO_3)_3$. Uranium can have five different valences, chlorine six, rhenium seven. As you can imagine, historically this element of variability brought on a great deal of debate throughout the last half of the nineteenth century.

Of the many valuable consequences of the valence concept, one is particularly noteworthy at this point. The device of valence bonds, in essence the same as the hooks used in Fig. 20.4, supplemented Dalton's pictorial representation of the atoms themselves and invited speculation on the structural arrangement in the molecule, particularly for the complex organic substances where the mind needed such a crutch badly.

This was all the more urgent because there had been discovered sets of *isomers* (Greek *isos* + *meros,* composed of equal parts). These are substances with precisely the same chemical formula but with very different physical and chemical properties. They occur frequently among the *organic* compounds, those containing carbon. The only hope for explaining the differences was the hypothesis of different placement of the atoms within the molecule. Hence the rise of the graphic structural formulas for representing chemical compounds such as those in Fig. 20.5 for C_2H_6 (ethane), C_2H_6O (ethyl alcohol, better written C_2H_5OH, to suggest its structure), and the isomers $C_6H_{12}O_6$, one being glucose (also called dextrose and grape sugar) and the other fructose (or levulose, fruit sugar).

Then there is the famous case of C_6H_6 (benzene; note the closed ring), which proved to be a severe test of the assumption that one can get away with using a double bond to meet the valence quota of the carbon atom. Contrary to what one might expect from the three double bonds shown in Fig. 20.5, benzene does not behave like a typical unsaturated compound: It is not especially reactive. Moreover, if one replaces the hydrogen atoms bonded to two adjacent carbon atoms by, for example, chlorine atoms, one would expect to have two distinct isomers, depending on whether those carbons were connected to each other by a single or double bond; whereas in fact there is only one such compound.

The German chemist August Kekulé (1829–1896), who established this general method of describing organic compounds, proposed in 1865 that the double and single bonds exchange places so rapidly that, in effect, all the CC bonds in benzene are equivalent. This hypothesis was brilliantly successful in predicting the numbers of isomers of compounds that can be formed from benzene by replacing the H atoms by other kinds of atoms or by joining together two or more benzene rings.Guided by Kekulé's theory, organic chemistry flourished in the late nineteenth century and helped German chemical engineers to develop major new industries for producing dyes and synthetic products.

In the 1860s, when such representations as those in Fig. 20.5 first became widely used among chemists, there was hesitation in regarding them as anything more than working schemes, without necessary relation to the actual three-dimensional and (presumably) unknowable structure of molecules. In truth, although the structural formula is indispensable as a rough sketch of nature's plan for the molecule, the correct disposition of atoms and stable groups of atoms in the molecule of a compound is generally difficult to determine experimentally, even with our present powerful tools of research. And even if the structure of all compounds were known with the greatest accuracy, this would still give us no clue to the initial puzzle: Why do atoms combine in such limited and specific compounds, and what is the physical mechanism that holds the components of a molecule together?

The fundamental *explanation* for the nature of chemical bonds that could also account for the stability of molecules like benzene had to wait for the twentieth century. These questions, like so many related ones, were answered only much later after the physical structure of the atom became clear. But in the meantime, the concept of valence was a keystone in the construction of another inspired scheme (to be taken up in the next chapter) that contributed to the advance of chemistry—almost as much as the atomic theory itself.

Problem 20.9. Evaluate Dalton's atomic theory of chemistry from the point of view of our criteria for theories in Chapter 3.

Problem 20.10. Discuss the growth of the atomistic hypothesis to this point in the light of Chapter 12.

20.12 Molecular Weights[4]

Given a more rational procedure for obtaining molecular formulas, we may at once expect an extension and correction of the table of relative atomic weights. For gases the method is extremely simple. If, for example, we compare (at the same temperature and pressure) the *relative weights of equal volumes* of several gases, remembering Avogadro's hypothesis that such samples contain equal numbers of molecules, we would find by experiment the following relative masses: 2.016 for hydrogen, 28.01 for nitrogen, and about 32.00 for oxygen. If we adopt the physical scale of atomic weights based on an isotope of carbon (see Section 27.1), we find that oxygen still has a relative mass of 32.00 to four significant figures. On this basis we may assemble a table of relative molecular weights of some gas molecules—for elements as well as compounds (Table 20.1).

It is sometimes useful to use the term "molecular weight" for gaseous mixtures such as air. If we take the composition of air to be N_2 (78%), O_2 (21%), CO_2 (0.03 to 0.04%), and less than 1% water vapor, rare gases, etc., we find its average molecular weight to be about 29.

The *relative atomic weights* are, of course, obtained at once by dividing the molecular weight of the element (cf. Table 20.1) by the number of atoms per molecule; the required molecular formulas are obtained from the nature of chemical reactions, particularly from the volumetric relations of gases, as has been demonstrated. A complete list of modern values for all elements will be found in the Periodic Table of Elements (Appendix VII).

Problem 20.11. What should be the relative molecular weight of wood alcohol (methyl alcohol, CH_3OH)? Ethyl alcohol (C_2H_5OH)? Diethyl ether [$(C_2H_5)_2O$]? Quinine ($C_{20}H_{24}N_2O_2$)? Glucose ($C_6H_{12}O_2$)? Clay [$Al_2O_3 \cdot 2(SiO_2) \cdot 2(H_2O)$]?

Problem 20.12. Did our chemical evidence exclude the possibility that hydrogen is triatomic? What type of experiment would you look for in order to prove or disprove the hypothesis that hydrogen is triatomic? If that hypothesis should be found to be

---

[4]This section completes the discussion of Dalton's atomic theory, but may be omitted without loss of continuity.

Table 20.1. **Relative Molecular Weights**

Gas or Vapor	Relative Molecular Weight (M)	Molecular Formula
Elements:		
Hydrogen	2.016	H_2
Helium	4.003	He
Nitrogen	28.01	N_2
Oxygen	32.00	O_2
Fluorine	38.00	F_2
Argon	39.95	Ar
Chlorine	70.90	Cl_2
Phosphorus	123.90	P_4
Mercury	200.59	Hg
Compounds:		
Water vapor	18.02	H_2O
Carbon monoxide	28.01	CO
Carbon dioxide	44.01	CO_2
Uranium hexafluoride	352.02	UF_6

correct, what do the experiments cited in this chapter tell us about the molecular structure of the other gases? How would the molecular weights given above be modified?

In Problem 20.3 you obtained the probable molecular formulas of five compounds exclusively from the internal consistency of data on combining weights. Note that of these five, two were solids, one liquid, and two gaseous. Later we saw that the work of Gay-Lussac and Avogadro supplemented and much simplified such calculations, at least for gaseous reactions. These approaches evidently made possible a complete determination of all atomic weights.

Now we must make more explicit the relationship between these three important factors in any reaction, namely, the combining weights, the formula for the reaction, and the atomic and molecular weights of the substances involved. *Given any two of these, the third can be calculated.* These two examples will help to illustrate this point.

EXAMPLE 1

In the reduction of the black iron ore, hematite, iron is obtained by burning carbon with the ore in the following reaction:

$$Fe_2O_3 + 3C \rightarrow 2Fe + 3CO.$$

Given this reaction equation, and the atomic weights (Fe = 55.85, O = 16.00, C = 12.01), find the combining weights.

Solution: The following must represent the proportions by weight of the substances involved:

$$\underbrace{[(55.85) \times 2 + (16.00) \times 3]}_{159.70}^{\text{Hematite}} + \underbrace{[3 \times 12.01]}_{36.03}^{\text{Carbon}}$$

$$\rightarrow \underbrace{[2 \times 55.85]}_{111.70}^{\text{Iron}} + \underbrace{[3 \times (12.01+16.00)]}_{84.03}^{\text{Carbon Monoxide}}$$

This schedule informs us that if we hope to obtain about 112 kg of iron, we must supply the furnace with about 160 kg ore, 36 kg pure coal (or equivalent), and must take precautions to dispose of 84 kg of carbon monoxide gas.

Problem 20.13. How many tons of carbon must we supply if we expect to obtain 80 tons of iron?

Problem 20.14. Depending on the temperature of the furnace, another possible reaction is

$$Fe_2O_3 + 3CO \rightarrow 2Fe + 3CO_2.$$

What are the relative weights of the participants in this reaction?

EXAMPLE 2

When 304 g of a certain oxide of chromium are ignited with 108 g aluminum (Al, at. wt. about 27), there is formed an aluminum oxide and 208 g of the pure metal chromium (Cr, at. wt. about 52). From these data, find the equation for this reaction.

Solution: First of all, there must be (304 + 108) − 208 g of aluminum oxide, that is, 204 g. Now we do not know the molecular formulas, but they can be written for the time being as follows:

$$Cr_xO_y + zAl \rightarrow Al_zO_y + xCr,$$

where x, y, and z may each stand for a whole integer (1, 2, 3, . . .), for evidently the equation, if it refers to combinations of atoms, must contain as many atoms of each element on one side as on the other; that is, here as always in chemical reactions, the "equation must balance."

We may now regard the relative combining weights as indicative of the relative molecular weights, and say

$$\underbrace{[(52x) + (16)y]}_{304}^{\text{Oxide of chromium}} + \underbrace{[z27]}_{108}^{\text{Aluminum}}$$

$$\rightarrow \underbrace{[(27)z + (16)y]}_{204}^{\text{Aluminum oxide}} + \underbrace{[x52]}_{208}^{\text{Chromium}}$$

$$z \cdot Al = 108 \quad \text{or} \quad z(27) = 108; \quad z = 108/27 = 4$$

$$x \cdot Cr = 208 \quad \text{or} \quad x(52) = 208; \quad x = 208/52 = 4.$$

Therefore

$$Cr_xO_y = Cr_4O_y = [(4 \times 52) + (y \times 16)] = 304;$$

$$y = (304 - 208)/16 = 6.$$

Our reaction equation is then

$$Cr_4O_6 + 4Al \rightarrow Al_4O_6 + 4Cr.$$

More simply,

$$2Cr_2O_3 + 4Al \rightarrow 2Al_2O_3 + 4Cr;$$

and dividing through by 2, finally

$$Cr_2O_3 + 2Al \rightarrow Al_2O_3 + 2Cr.$$

Problem 20.15. At one time, aluminum was produced by reducing aluminum chloride with metallic sodium (Na, at. wt. about 23), yielding sodium chloride and aluminum (Al, at. wt. about 27). If 34.5 kg sodium is needed to obtain 13.5 kg aluminum from 66.8 kg aluminum chloride, what is the simplest reaction formula? [*Note:* To obtain x, y, and z in integral form, one may have to multiply all these given quantities by some one number.]

Finally, we should mention a fruitful concept that makes these calculations very much more simple: the *gram-molecular-weight* or "mole." By definition, 1 gram-molecular-weight, or 1 mole, of any substance is an amount whose mass is numerically equal to its relative molecular weight. Thus 1 mole of oxygen (O_2) is nothing but 32.000 g of oxygen; 1 mole of hydrogen (H_2) = 2.016 g of hydrogen; 3 moles of mercury (Hg) = 601.77 g of mercury (that is, 3 × 200.59; note that for monatomic substances the term gram-molecular-weight is often replaced by "gram-*atomic*-weight"); n moles of water = $n \times 18.016$ g of water. Conversely, if a substance is known to have the molecular weight M, then y grams of that substance correspond to y/M moles.

It stands to reason that *all pure substances must have the same number of molecules in 1 mole,* no matter what the state, temperature, or pressure. The proof is simple: The total mass of a substance (m) is the mass per molecule (m_0) × number of molecules (N_m), or $m = m_0N_m$. If the mass is taken to be 1 mole, it is numerically equal to M, the molecular weight; then $M = m_0N_m$. If we compare two substances with molecular weights M' and M'', we may write

$$M' = m_0'N_m'$$

$$M'' = m_0''N_M''$$

Taking the ratio,

$$\frac{M'}{M''} = \frac{m_0'}{m_0''} \cdot \frac{N_m'}{N_m''}.$$

But by definition the ratio m_0'/m_0'' is the relative molecular weight of one substance with respect to the other, or: $M'/M'' = m_0'/m_0''$. Thus $N_m' = N_m''$ as was to be proved. Furthermore, the actual number of molecules per mole, by many indirect tests, is 6.02×10^{23} (called *Avogadro's number,* symbolized by N_A).

We recall that at 0°C and atmospheric pressure, all pure gases contain 2.69×10^{19} molecules/cm^3 (Loschmidt's number). It is now evident that under these standard conditions, 6.02×10^{23} molecules would occupy a volume of $(6.02 \times 10^{23})/(2.69 \times 10^{19}) = 22.4 \times 10^3$ cm^3 = 22.4 liters. This brings us to a common form of stating Avogadro's hypothesis: *At 0°C and atmospheric pressure, 1 mole of any pure gas,* *containing 6.02×10^{23} molecules, takes a volume of 22.4 liters.*

A useful and immediate application of the concept *mole* springs from this argument, which every experimental chemist lives by: If we know that in a certain reaction every molecule of A joins with x molecules of B, then we must provide for each *mole* of A exactly x *moles* of B (or corresponding fractions) to assure a complete reaction.

EXAMPLE 3

Of a substance X we are told that it is formed when equal numbers of moles of two elements combine. When a quantity of 40 g of X is analyzed, it is found to contain about 35.5 g chlorine and 4.5 g of an unidentified element Y; the stated amount of chlorine is $\frac{1}{2}$ mole of the chlorine gas Cl_2, because

$$n = \frac{m}{M} = \frac{35.5\text{ g}}{71\text{ g/mole}} = \tfrac{1}{2}\text{ mole;}$$

therefore the 4.5 g of Y represents $\frac{1}{2}$ mole of Y as well. This makes the molecular weight of $Y \approx 9$. If Y is monatomic, this is also its atomic weight; if diatomic, the atomic weight is 4.5; if triatomic, 3; etc. A glance at the table of atomic weights shows that, in fact, only one of these choices exists—namely, beryllium (Be) with an atomic weight 9 (or more accurately, 9.01). Thus we have identified the substance X as $BeCl_2$.

Problem 20.16. How many moles of ammonia is 1 g of that vapor? How many molecules are there in this quantity? How many *atoms?* What would be its volume at 0°C and 1 atm? at 100°C and 5 atm? (Use the general gas law.) How many moles of hydrogen does this vapor contain?

Problem 20.17. The reaction representing the rusting of iron may be written

$$3Fe + 4H_2O \rightarrow Fe_3O_4 + 4H_2.$$

How many moles of hydrogen are liberated for each gram of iron completely rusted?

Problem 20.18. Knowing Avogadro's number, the molecular weights of certain gases, and their molecular formulas, calculate the "absolute" mass (in grams) of a single atom of hydrogen, of oxygen, and of chlorine.

RECOMMENDED FOR FURTHER READING

O. Theodor Benfey, *From Vital Force to Structural Formulas,* Philadelphia: Beckman Center for the History of Chemistry, 1992

Bernadette Bensaude-Vincent and Isabelle Stengers, *A History of Chemistry,* Chapter 3

H. A. Boorse and L. Motz (editors), *The World of the Atom,* extracts and discussion of writings of Dalton, Gay-Lussac, Avogadro, and Prout, pages 139–191.

William H. Brock, *Norton History of Chemistry,* Chapters 4 and 5

Herbert Butterfield, *The Origins of Modern Science,* Chapter 11

L. K. Nash, "The atomic-molecular theory," in *Harvard Case Histories in Experimental Science* (edited by J. B. Conant), pages 215–231

C. C. Gillispie, *Dictionary of Scientific Biography,* articles by A. Thackray on Dalton, Vol. 3, pages 537–547; by M. P. Crosland on Gay-Lussac and Avogadro, Vol. 5, pages 317–327 and Vol. 1, pages 343–350; by J. Gillis on Kekulé, Vol. 7, pages 279–283

Cecil Schneer, *Mind and Matter,* pages 91–180

Mary Jo Nye, *Before Big Science,* Chapter 2

Cecil J. Schneer, *Mind and Matter,* pages 104–152

Robert E. Schofield, "Atomism from Newton to Dalton," *American Journal of Physics,* Vol. 49, pages 211–216 (1981)

J. H. Weaver, *World of Physics,* excerpts from Dalton and Avogadro, pages 613–629

CHAPTER 21

The Periodic Table of Elements

21.1 The Search for Regularity in the List of Elements

As the early list of chemical elements became longer, and as their physical and chemical properties became increasingly better established, a great variety of scientists rather naturally felt the urge to find some relation among the elements that would bring them into a logical sequence—a situation analogous to the Keplerian quest for principles that would relate the separate planets.

At first glance the task would seem hopeless. The most immediately striking thing is that each element seems to have unique properties that differentiate it from each of the other elements. Most are solids under ordinary circumstances, but many are gases and some are liquids; their relative abundances on earth have widely different values and so have their physical characteristics—boiling point, density, and the like.

Nevertheless, there are some clues for speculation. In 1815–1816 the English physician William Prout had suggested that the atoms of all elements were but the condensations of different numbers of hydrogen atoms, which themselves were the true fundamental particles. Consequently the atomic weights could be expected to be whole multiples of the atomic weight of hydrogen; and indeed the early determination of atomic weights by Dalton did allow this hypothesis.

But more accurate measurements were soon forthcoming, particularly those of the great chemists Berzelius, Dumas, and Stas, all demonstrating real deviations from Prout's rule. A glance at a modern and accurate table of elements shows that only a few of the atomic weights are even close to being exact multiples of the atomic weight of hydrogen. Thus Prout's hypothesis, despite its direct appeal, had to yield to the contrary evidence of quantitative analysis.

But much of this quantitative analysis was performed just for the purpose of testing Prout's hypothesis, and since accurate atomic weights turned out to be important to twentieth-century physics, his hypothesis must be judged as having been useful to science regardless of the fact that it was ultimately refuted: Score one for Popper's principle!

At the same time, these growing results favored another trend of thought toward bringing order into the apparently haphazard list of elements, namely, the attempts to connect the atomic weights of elements with their *chemical properties*—each typically the result of long research, some going back to the alchemists. It had long been known that dispersed throughout the list of elements are *families* of elements, that is, various elements of similar chemical behavior. An example of such a family is that of the *halogens*—a group that includes the elements fluorine (F), chlorine (Cl), bromine (Br), and iodine (I); their respective atomic weights are approximately 19, 35.5, 80, 127. Although they exhibit some striking dissimilarities (for example, the first two are gases, the third ordinarily a liquid, the last a volatile solid), these four elements have much in common. They all combine with many metals to form white, crystalline salts (halogen means "salt-former"). The boiling points change progressively as we shift attention from the first to the last one (F, –187°C; Cl, –35°C; Br, 59°C; I, 184°C), and similarly for their melting points. The solubility in water progressively decreases, and so does the chemical activity, as manifested by the violence of chemical reactions (speed of reaction, amount of energy liberated).

This list of properties relating the halogens by continuous changes is to be supplemented by a second list of properties that are the same for all these elements. Halogens form compounds of similar formulas in violent reaction with metals (for example, NaF, NaCl, NaBr, NaI; or AlF_3, $AlCl_3$, $AlBr_3$, AlI_3); that is, each has the same valence in analogous reactions. All four elements form simple compounds with hydrogen that dissolve in water and form acids. All four form a diatomic vapor under ordinary conditions.

Another family, referred to as the *alkali metals,* includes lithium (Li, approximate atomic weight 7), sodium (Na, 23), potassium (K, 39), rubidium (Rb, 85.5), and cesium (Cs, 133). Here, too, there is a long list of chemical and physical properties common to all five elements or changing in the same sense with increasing atomic weight.

In the restless search for some quantitative connection among the elements, there appeared as early as 1829 the discovery of the German chemist Johann Wolfgang Dobereiner, who found that within several chemical families the atomic weight of one element had a value equal, or nearly so, to the average of its two immediate neighbors; thus 23 for Na is the average of 7 for Li and 39 for K; and in the triad of Cl, Br, and I, 80 for Br is roughly the mean of 35.5 (Cl) and 127 (I). But no physical meaning could be attached to such numerical regularities (we are at once reminded of Bode's law). In fact, the underlying explanation of such relations eventually turns out to be extremely complex (involving the stability of nuclei) and has not yet been fully achieved.

However, the search for numerical regularities of this kind proved popular, as always, and was carried on with vigor. It is another example where research that proceeded for a time on the wrong track was nevertheless directly fruitful, for through this effort it became more evident that the list of elements did indeed have a meaningful general structure, although not according to such atomic weight groups.

From 1860 on, about the time when Cannizzaro's agitation on behalf of Avogadro's hypothesis began to succeed, it was possible in principle to fix with very good accuracy the atomic weights of almost all elements and the molecular weights of compounds. Now one could place the elements into a sequence based entirely on a property of the single atom itself, namely, its atomic weight. Such a sequence reveals a curious fact: One may discover a regular spacing between the elements belonging to one family, a recurrence of numerical position quite apart from Dobereiner's atomic weight triads. In the words of the English chemist Newlands, one of several to seek such a regularity of spacing, "the eighth element, starting from a given one, is a kind of repetition of the first, like the eighth note in an octave of music."

Newlands was probably the first to propose the idea of assigning each element a *number* (we now call it the *atomic number*) starting with H = 1, Li = 2, Be = 3, etc.) But his attempt to establish a general law was doomed to failure. Although it proved to be correct that chemically similar elements repeat periodically, and although it was possible to perceive several "octaves" in Newlands' arrangement, his analogies more often than not had to be rather farfetched, for he did not foresee that the periodicity in the end would not be regular throughout the list of elements, and he had not considered the necessity of leaving spaces for still undiscovered elements.

Here you might object that prior to the discovery of the new elements one could not possibly leave gaps for them in the proper places. However, it is exactly this achievement that is at the core of the work of Mendeléeff.

21.2 The Early Periodic Table of Elements

In the publications of the Russian chemist Dmitri Ivanovitch Mendeléeff (1834–1907) from 1869 on—and, starting about the same time, of the German chemist Julius Lothar Meyer—we find the culmination of the six decades of quantitative investigations and speculative inductions since Dalton's atomic theory first became generally known. A few of the most striking aspects of Mendeléeff's contribution can be given in his own words, even though a great deal of descriptive material between excerpts must here be omitted. Thus his initial paper on the subject (1869) gives us the following account of the background and motivation of his work:

> During the course of the development of our science, the systematic arrangement of elements has undergone manifold and repeated changes. The most frequent classification of elements into metals and non-metals is based upon physical differences as they are observed for many simple bodies, as well as upon differences in character of the oxides and of the compounds corresponding to them. However, what at first acquaintance with the subject-matter appeared to be completely clear, lost its importance entirely when more detailed knowledge was obtained. Ever since it became known that an element such as phosphorus could

Fig. 21.1. Dmitri Ivanovitch Mendeléeff (1834–1907), Russian chemist, shown here with his wife Feozva Nikitichna in the year of their marriage, 1862. Mendeléeff's form of the periodic table of elements led to several successful predictions of the properties of new elements, and was generally adopted.

appear in non-metallic as well as in metallic form, it became impossible to found a classification on physical differences. . . . In recent times the majority of chemists is inclined to achieve a correct ordering of the elements on the basis of their valency. There are many uncertainties involved in such an attempt . . .

Thus, there does not exist yet a single universal principle which can withstand criticism, that could serve as guide to judge the relative properties of elements and that would permit their arrangement in a more or less strict system. Only with respect to some groups of elements there are no doubts that they form a whole and represent a natural series of similar manifestations of matter.

. . . . When I undertook to write a handbook of chemistry entitled *Foundations of Chemistry,* I had to make a decision in favor of some system of elements in order not to be guided in their classification by accidental, or instinctive, reasons but by some exact, definite principle. . . .

The numerical data available regarding elements are limited at this time. Even if the physical properties of some of them have been determined accurately, this is true only of a very small number of elements.

However, everybody does understand that in all changes of properties of elements, something remains unchanged, and that when elements go into compounds this material *something* represents the [common] characteristics of compounds the given element can form. In this regard only a numerical value is known, and this is the atomic weight appropriate to the element. The magnitude of the atomic weight, according to the actual, essential nature of the concept, is a quantity which does not refer to the momentary state of an element but belongs to a material part of it, a part which it has in common with the free element and with all its compounds . . . . For this reason I have endeavored to found the system upon the quantity of the atomic weight.

The first attempt I undertook in this direction was the following: I selected the bodies with the smallest atomic weight and ordered them according to the magnitude of their atomic weights. Thereby it appeared that there exists a periodicity of properties and that even according to valence, one element follows the other in the order of an arithmetical sequence.

Li = 7, Be = 9.4, B = 11, C = 12, N = 14, O = 16, F = 19,
Na = 23, Mg = 24, Al = 27.4, Si = 28, P = 31, S = 32, Cl = 35.3,
K = 39, Ca = 40, . . . , Ti = 50, V = 51 (etc.)

Note that in this arrangement hydrogen, the first and lightest of the elements, is left out, for it has rather unique properties; that helium, together with the other elements of the family of rare gases, had not yet been discovered; and that Mendeléeff's values for atomic weights sometimes differ significantly from the more recent values. He writes the first seven elements from lithium to fluorine in sequence of increasing atomic weight, then writes the next seven, from sodium to chlorine, below the first line. The periodicity of chemical behavior is already obvious.

But even in the first row things are not quite so simple. In the 1870s there was a controversy about the atomic weight of beryllium, linked to a disagreement about its valence. The best method for determining atomic weights was based on Avogadro's hypothesis, which required measurements on a *gaseous* compound, but most common compounds of beryllium are solid. Some chemists favored atomic weight 9, corresponding to a valence of 2; others, especially Lars-Frederik Nilson and Otto Pettersson in Sweden, followed the earlier proposal of the Swedish chemist Berzelius that it should be trivalent with atomic weight 14.

Mendeléeff jumped into this controversy by asserting that it must be 9, because there is natural place for a bivalent element between lithium (valence 1, atomic number 7) and boron (valence 3, atomic number 11), whereas the space for an element with valence 3, atomic number 14, is already occupied by nitrogen. Mendeléeff already had enough confidence in his theory to suggest that the experiments done by the Swedish chemists were wrong because they disagreed with the theory!

And Mendeléeff was vindicated. In 1884 Nilson and Pettersson announced that they had succeeded in finding a gaseous compound of beryllium—its chloride—to which Avogadro's law could be applied and that beryllium's atomic weight is 9.1.

In the first vertical column of the table are the first two alkali metals, in the seventh column the first two halogens, and each of the columns between contains two chemically similar elements. Furthermore, the main valences for these columns are 1 for the first, 2 for the second, 3 for the next, then 4, 3, 2, and 1. Reading from left to right, any element (let us use a general symbol R) in each column may form higher oxides or hydrides according to the following scheme:

Column	Higher oxide or hydride
1st	R_2O
2nd	R_2O_2 (= RO)
3rd	R_2O_3
4th	R_2O_4 (= RO_2) H_4R
5th	H_3R
6th	H_2R
7th	HR

When Mendeléeff now adds to the two rows of elements the third, K comes to stand directly below Li and Na, and of course we know that K is indeed a member of the same family with the same valence of unity. Next comes Ca (calcium), divalent like Be and Mg above it. Next should come the next heaviest element then known, which was titanium (Ti), and previous workers trying to develop such tables had generally done this.

In Newlands' table, which coincides almost entirely with Mendeléeff's to this point, another element was placed in its stead whose atomic weight actually was much too high. This fact offers us accurate insight into the dominant thoughts of Mendeléeff. He recognizes that the chemical properties of Ti suggest that it appear under C and Si in the fourth vertical column, not in the third. If the classification is to be complete and meaningful, there ought to exist a hitherto unsuspected element with an atomic weight somewhere between that of Ca (40) and Ti (50, new value 47.9), and with a valence of 3. This amounts to a definite prediction, and later in the same work Mendeléeff finds other cases of this sort among the remaining elements.

In addition to beryllium, Mendeléeff found several other elements whose atomic weights had been determined incorrectly by chemists. The largest correction was for uranium, previously assigned an atomic weight of 120 (or sometimes 60). He gave it the value 240—near enough to the correct value—making it the heaviest of the elements. This value was accepted without much objection.

But Mendeléeff was not always correct in his assumption that the atomic weights of the elements must increase as one goes from left to right in a row, as well as from top to bottom in a column. He argued that the accepted value for tellurium, 128, must be wrong because it comes just before iodine, 127, in the seventh row, so he assigned it the value 125 (Table 21.1). Even his strongest supporters, like the Czechoslovakian chemist Bohuslav Brauner, could not confirm this value, and we now accept it as an exception to the rule (see Table 21.2).

The whole scheme to this point may be understood in terms of a very crude analogy. It is as if a librarian were to put all his books in one heap, weigh them individually, and then place them on a set of shelves according to increasing weight—and find that on each shelf the first book is on art, the second on philosophy, the third on science, the fourth on economics, and so on.[1] Our librarian may not understand in the least what the underlying explanation for this astonishing

[1]This idea is not as far-fetched as it may appear. The Library of Congress announced in 1999 that it would convert its bookstacks to a system (already followed by at least one other national library) in which all the books are arranged by *size,* in order to make the most efficient use of the available space. This arrangement would put most art books together because they are often published in a larger format.

Table 21.1. **Periodic Classification of the Elements (Mendeléeff, 1872)***

Group →	I	II	III	IV	V	VI	VII	VIII
Higher oxides and hydrides	R_2O —	RO —	R_2O_3 —	RO_2 H_4R	R_2O_5 H_3R	RO_3 H_2R	R_2O_7 HR	RO_4 —
Series 1	H(1)							
Series 2	Li(7)	Be(9.4)	B(11)	C(12)	N(14)	O(16)	F(19)	
Series 3	Na(23)	Mg(24)	Al(27.3)	Si(28)	P(31)	S(32)	Cl(35.5)	
Series 4	K(39)	Ca(40)	–(44)	Ti(48)	V(51)	Cr(52)	Mn(55)	Fe(56), Co(59), Ni(59), Cu(63)
Series 5	[Cu(63)]	Zn(65)	–(68)	–(72)	As(75)	Se(78)	Br(80)	
Series 6	Rb(85)	Sr(87)	?Yt(88)	Zr(90)	Nb(94)	Mo(96)	–(100)	Ru(104), Rh(104), Pd(106), Ag(108)
Series 7	[Ag(108)]	Cd(112)	In(113)	Sn(118)	Sb(122)	Te(125)	I(127)	
Series 8	Cs(133)	Ba(137)	?Di(138)	?Ce(140)	—	—	—	
Series 9	—	—	—	—	—	—	—	
Series 10	—	—	?Er(178)	?La(180)	Ta(182)	W(184)	— —	Os(195), Ir(197), Pt(198), Au(199)
Series 11	[Au(199)]	Hg(200)	Tl(204)	Pb(207)	Bi(208)	—		
Series 12	—	—	—	Th(231)	—	U(240)		

*Atomic weights are shown in parentheses. Horizontal lines in the boxes indicate "missing" elements, which Mendeléeff expected to be discovered in the future.

regularity is, but if he now discovers on one of these shelves a sequence art-science-economics, he will perhaps be tempted to leave a gap between the two books on art and science and look about for a missing philosophy book of the correct weight to fill the gap.

Mendeléeff had no illusions that he understood the reason for this symmetry of arrangement, but he saw clearly that his work would eventually lead to a physical explanation and that in the meantime "new interest will be awakened for the determination of atomic weights, for the discovery of new elements, and for finding new analogies among the elements." At a later time he added, "just as without knowing the cause of gravitation it is possible to make use of the law of gravitation, so for the aims of chemistry it is possible to take advantage of the laws discovered by chemistry without being able to explain their causes."

Table 21.1 gives Mendeléeff's arrangement in the version published in 1872. We note that Mendeléeff distributes the 63 elements then known (with 5 in doubt) in 12 horizontal lines or series, starting with the lightest element, hydrogen, at the top left, and ending with uranium at the bottom right. The others are written in order of increasing atomic weight, but are so placed that each element is in the nearest vertical column or group of chemically similar elements. Thus the halogens are in group VII; group VIII contains only ductile metals; groups I and II contain light and low-melting-point metals; and the specific family of alkali metals is in group I. The more detailed arguments for each particular placement, as Mendeléeff's extensive discussion shows, are often exceedingly delicate, guided by his profound and often almost intuitive knowledge of chemical nature.

The arrangement does, however, leave many gaps: "Vacant places occur for elements which, perhaps, shall be discovered in the course of time." And several of the elements toward the end of this table are in doubt: "The higher atomic weights belong to elements which are rarely encountered in nature, which do not form large deposits, and which therefore have been studied relatively little. . . . With respect to the position of some elements that are rarely encountered in

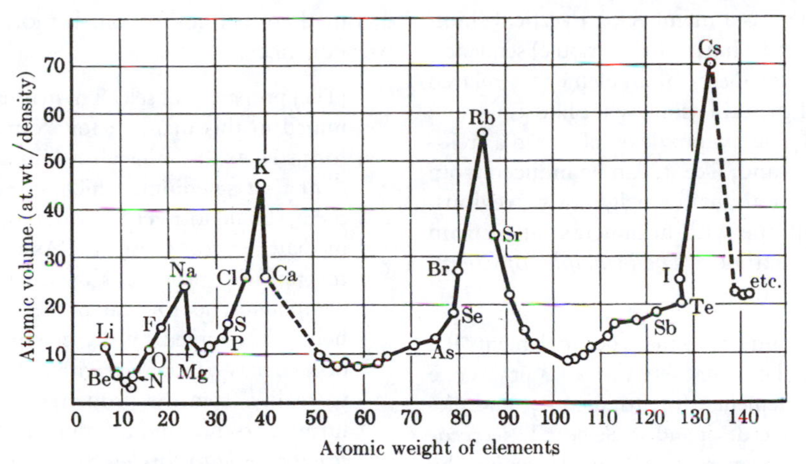

Fig. 21.2. Atomic volumes for the first portion of the table of elements. (After Lothar Meyer, but using modern values.)

nature, there exists, quite understandably, complete uncertainty."

But the main part of the table is fairly complete, and another feature of this scheme reveals itself: "For a true comprehension of the matter it is very important to see that all aspects of the distribution of the elements according to their atomic weights essentially express one and the same fundamental dependence—*periodic properties.*" By this is meant that in addition to the gradual change in physical and chemical properties within each *vertical* group or family, there is also a periodic change of such properties among the elements if they are scanned in a horizontal sequence. "The properties of the elements as well as the forms and properties of their compounds are in periodic dependence on, or (to express ourselves algebraically) form a periodic function of, the atomic weights of the elements."

This *periodic law* is the heart of the matter, a generalization of Newlands' suggestion (with whose work Mendeléeff was initially not acquainted). We can best illustrate it as Lothar Meyer did, by drawing a curve of the value taken on by some physical quantity as a function of atomic weight, for example, the *atomic volume* of each element [the ratio of atomic weight to density, the mass per unit volume, in the liquid or solid phase (Fig. 21.2)]. Note the irregular spacing of the "waves," yet the simultaneous periodicity of the position of analogous elements on the curve, entered here specifically for the halogens and for the alkali metals.

Problem 21.1. Use the data given in the *Handbook of Chemistry and Physics* or a similar source and plot the melting points of the elements against the atomic weights for the elements from hydrogen to barium (Ba) inclusive. Then discuss the periodicities that appear on the graph, for example, the positions of the two families of elements mentioned thus far.

21.3 Consequences of the Periodic Law

More detailed examination of the periodic scheme yields some additional information.

a) If an element R enters into chemical combinations with oxygen, the composition of one of its higher oxygen compounds is determined by the *group* to which R belongs, that is, the first group gives R_2O, the second R_2O_2 (or RO), the next R_2O_3, and so forth; similarly for combinations with hydrogen as indicated along the top of Table 21.1.

b) A distinction is to be made between the even and the odd series, indicated in Table 21.1 by a slight shift in the printing of the symbols to the right or left. Chemical and physical similarities are generally closer among all the elements in one group that belong to the even series only or the odd series only. Thus in group VI, S, Se, and Te are nonmetals, whereas Cr, Mo, and W are heavy metals.

c) Each element is a member of a vertical (family) sequence and a horizontal (periodic) sequence. Thus any one property of an element is related to the rest of the table somewhat like a piece in the middle of a jigsaw puzzle or a letter in a crossword puzzle—and, like it, can be induced from what is known about the neighboring regions. Here, to continue with arguments taken from Mendeléeff's textbook, *The Principles of Chemistry,* is one example:

> If in a certain group there occur elements R_1, R_2, R_3, and if in that series which contains one of these elements, for instance R_2, and element Q_2 precedes it and an element T_2 succeeds it, then the properties of R_2 are determined by the mean [average] of the properties of R_1, R_3, Q_2, and T_2. Thus, for instance, the atomic weight of $R_2 = \frac{1}{4}(R_1 + R_3 + Q_2 + T_2)$. For example, selenium occurs in the same group as sulfur, S = 32, and tellurium, Te = 127,[2] and, in the 5th series, As = 75 stands before it and Br = 80 after it. Hence the atomic weight of selenium should be $\frac{1}{4}(32 + 127 + 75 + 80) = 78.5$, which is near to that generally accepted, Se = 79, in which there is a possible error in the first decimal, so that 78.5 may be nearer the actual figure.

> We see at once what this leads to:

> . . . the periodic dependence of the properties on the atomic weights of the elements gives a new means for determining, by means of the equivalent weight, the atomic weight or atomicity of imperfectly investigated but known elements, for which no other means could as yet be applied for determining the true atomic weight. At the time [1869] when the periodic law was first proposed there were several such elements. It thus became possible to learn their true atomic weights, and these were verified by later researches. Among the elements thus concerned were indium, uranium, cerium, yttrium, and others. . . . Thus a true law of nature anticipates facts, foretells magnitudes, gives a hold on nature, and leads to improvements in the methods of research, etc.

[2]In this quotation Mendeléeff adopts the empirical atomic weight 127 for tellurium rather than the lower value 125 he estimated in order to place it before iodine (127), as explained in Section 21.2. If Te = 125, the rule would give Se = 78; these are the values recorded in Table 21.1.

d) But there is another, much more important consequence:

> [The] properties of selenium may also be determined in this manner; for example, arsenic forms H_3As, bromine gives HBr, and it is evident that selenium, which stands between them, should form H_2Se, with properties intermediate between those of H_3As and HBr. Even the physical properties of selenium and its compounds, not to speak of their composition, being determined by the group in which it occurs, may be foreseen with a close approach to reality from the properties of sulfur, tellurium, arsenic, and bromine. *In this manner it is possible to foretell the properties of elements still unknown,* especially when they are surrounded by well-known elements.

Mendeléeff then showed how one could predict the properties of the undiscovered element in group IV, series 5, to which he gave the name "ekasilicon," symbol Es, because it should have chemical properties similar to those of silicon; the preface *eka* means "one" in Sanskrit. By interpolating the properties of elements above and below, and to the right and left, in his table, he estimated that its atomic weight should be "nearly 72," and its density above 5.5. He also predicted the properties of some of its compounds, for example its oxide EsO_2 should have density about 4.7.

In 1887, Clemens Winckler of Freiberg, Germany, did indeed discover a metal, which he called germanium (Ge) in honor of his country, having an atomic weight of about 72½, a density of 5½, an oxide GeO_2 of density 4.7, and other compounds with properties similar to those predicted by Mendeléeff.

But that was not Mendeléeff's first successful prediction (nor his last). In 1875 the French chemist Paul Émile Lecoq de Boisbaudran discovered a new metal, which he named gallium (Ga) in honor of his native France.[3] Mendeléeff immediately suggested that gallium is the element

[3]It was later alleged by cynics that he had chosen this name because the Latin word *gallus* means "coq" in French ("cock," male bird, in English), thereby glorifying himself. This derivation is now found in dictionaries, and was cited as a precedent when it was proposed to name element 106 seaborgium, contrary to the supposed rule that an element is never named after a living person. The name was approved before the death in 1999 of Glenn Seaborg, codiscoverer of several elements.

"eka-aluminum," which he had predicted to fill the space in group III, series 5. It should have atomic weight 68 and density 6.0.

While this atomic weight was fairly close to the experimental value (about 69), Lecoq de Boisbaudran's value for the density was only 4.7. He initially insisted that gallium is *not* eka-aluminum. But he soon suspected that his sample might have contained cavities filled with water and determined the density again after heating the metal and solidifying it in a dry atmosphere. He was thus able to obtain a density of 5.935, confirming Mendeléeff's views. Here we have one of those remarkable cases in which the theory is (initially) more accurate than the experiment!

In 1879 Nilson (mentioned above in connection with the beryllium controversy) discovered a new element, which he named scandium in honor of Scandinavia. Another Swedish chemist, Per-Teodor Cleve, determined the atomic weight to be about 45 and pointed out that scandium's properties seemed to correspond to those predicted by Mendeléeff for eka-boron, to fill the place III, 4.

Even though not every aspect of Mendeléeff's work offered such marvelous agreement, these are indeed stunning results (again reminiscent of the discovery of the asteroids in the general orbits predicted for them by Bode's law, and of the planet Neptune predicted from Newtonian celestial mechanics by Leverrier and Adams). Numerical predictions of this sort are the very essence of science, and the chemical profession soon cast off its initial reluctance to embrace the main concepts of a periodic classification of elements.

In fact this was one of the few times when scientists accepted a new theory, not simply because it offered a better explanation of known facts, but because it successfully predicted previously unknown facts, or even facts *contrary* to experimental results. (The successful predictions of theories like Newton's gravitational theory often come *after* those theories have been accepted for other reasons.) The behavior of Nilson and Lecoq de Boisbaudran are striking illustrations of how we would like to think all scientists would act under such circumstances: They abandoned their earlier views when their own experiments turned out to support a different theory.

But success in predicting new facts, which according to Popper's principle is an essential justification for accepting a scientific theory, gives only a partial explanation for the rapid and widespread adoption of Mendeléeff's periodic table in the late nineteenth century. Almost every chemistry textbook published during that period, including Mendeléeff's own, gives much more attention to the *correlations* of properties of the *known* elements with their place in the periodic table than to the *prediction* of properties of *new* elements.

The reason seems to be the same as the reason why those new elements had not been discovered before 1869: their abundance at the earth's surface is very small. Of course this does not mean that rare elements are necessarily unimportant (think of radium). But in general they were of little practical value, and predictions of their properties were less useful than establishing the correct atomic weights of more common elements like beryllium and uranium. Nevertheless, it was the sensational confirmations of predicted new elements and their properties that first forced the chemical community to pay serious attention to Mendeléeff's periodic table.

The table was also a great benefit to *education* in chemistry: It was much easier for students to learn the properties of all the elements when they could be organized systematically, rather than memorized as so many unrelated facts. And, as Mendeléeff recalled in the passage quoted above, it was precisely the project of writing a *textbook,* not a laboratory research project, that led him to develop his table.

21.4 The Modern Periodic Table

It remains to tell briefly of some rather important rearrangements that had to be made subsequently in the periodic table. Table 21.2 is one of several modern forms, and comparison with Mendeléeff's table reveals these specific differences:

a) Hydrogen is placed uniquely so as not to be directly associated with any single family; it does in fact have a chemical relationship with both the halogens and the alkali metals.

b) The placement of the symbol in each rectangle is arranged so that elements along one vertical line belong to one chemical family. Thus in group I, the symbols for the alkali metals (Li, Na, K, Rb, . . .) are placed toward the left of the box

Table 21.2. Periodic Table of Elements

Period	Series	Groups→ I	II	III	IV	V	VI	VII	VIII			0
1	1	**1 H** 1.01										**2 He** 4.00
2	2	**3 Li** 6.94	**4 Be** 9.01	**5 B** 10.81	**6 C** 12.01	**7 N** 14.01	**8 O** 16.00	**9 F** 19.00				**10 Ne** 20.18
3	3	**11 Na** 22.99	**12 Mg** 24.31	**13 Al** 26.98	**14 Si** 28.09	**15 P** 30.97	**16 S** 32.07	**17 Cl** 35.45				**18 Ar** 39.95
4	4	**19 K** 39.10	**20 Ca** 40.08	**21 Sc** 44.96	**22 Ti** 47.87	**23 V** 50.94	**24 Cr** 52.00	**25 Mn** 54.94	**26 Fe** 55.8	**27 Co** 58.9	**28 Ni** 58.7	
	5	**29 Cu** 63.55	**30 Zn** 65.39	**31 Ga** 69.72	**32 Ge** 72.61	**33 As** 74.92	**34 Se** 78.96	**35 Br** 79.90				**36 Kr** 83.80
5	6	**37 Rb** 85.47	**38 Sr** 87.62	**39 Y** 88.91	**40 Zr** 91.22	**41 Nb** 92.91	**42 Mo** 95.94	**43 Tc** (98)	**44 Ru** 101.1	**45 Rh** 102.9	**46 Pd** 106.4	
	7	**47 Ag** 107.9	**48 Cd** 112.4	**49 In** 114.8	**50 Sn** 118.7	**51 Sb** 121.8	**52 Te** 127.6	**53 I** 126.9				**54 Xe** 131.3
6	8	**55 Cs** 132.9	**56 Ba** 137.3	**57–71***	**72 Hf** 178.5	**73 Ta** 180.9	**74 W** 183.8	**75 Re** 186.2	**76 Os** 190.2	**77 Ir** 192.2	**78 Pt** 195.1	
	9	**79 Au** 197.0	**80 Hg** 200.6	**81 Tl** 204.4	**82 Pb** 207.2	**83 Bi** 209.0	**84 Po** 209.0	**85 At** 210.0				**86 Rn** (222)
7	10	**87 Fr** (223.0)	**88 Ra** (226)	**89–103†**	**104 Rf** (261)	**105 Db** (262)	**106 Sg** (266)	**107 Bh** (264)	**108 Hs** (269)	**109 Mt** (268)	**110**	
	11	**111**	**112**	**113**	**114**	**115**	**116**	**117**				**118**

*Lanthanide series	**57 La** 138.9	**58 Ce** 140.1	**59 Pr** 140.9	**60 Nd** 144.2	**61 Pm** (145)	**62 Sm** 150.4	**63 Eu** 152.0	**64 Gd** 157.3	**65 Tb** 158.9	**66 Dy** 162.5	**67 Ho** 164.9	**68 Er** 167.3	**69 Tm** 168.9	**70 Yb** 173.0	**71 Lu** 175.0
†Actinide series	**89 Ac** (227)	**90 Th** 232.0	**91 Pa** 231.0	**92 U** 238.0	**93 Np** (237)	**94 Pu** (244)	**95 Am** (243)	**96 Cm** (247)	**97 Bk** (247)	**98 Cf** (251)	**99 Es** (252)	**100 Fm** (257)	**101 Md** (258)	**102 No** (259)	**103 Lr** (262)

Atomic weights are the most recent (1997) adopted by the International Union of Pure and Applied Chemistry. The numbers are all relative to the atomic weight of the principal isotope of carbon, C^{12}, defined as 12.00000. For some of the artificially produced elements, the approximate atomic weight of the most stable isotope is given in parentheses. The full names of the elements are given in Appendix V, and a system for constructing names of hypothetical heavy elements is given in Appendix II.

(these are identified as "group IA" in an expanded version of the periodic table), while the symbols for Cu, Ag, Au ("group IIA") are shifted to the right. In group VII the halogens (F, Cl, Br, . . .) are placed toward the right ("group VIIA) while Mn, Tc, Re, . . . are placed to the left ("group VIB). The elements have been given consecutive numbers, the so-called *atomic numbers,* to indicate their positions in the periodic table. Several elements have been assigned more correct atomic weights and new places.

c) Many new elements have been entered, and there are now no more additions expected within the table. In the 57th place belongs not one element but a whole group of 15 chemically almost indistinguishable elements, the *rare earths* or *lanthanide series,* written out for convenience in a separate row below the main part of the table, and given atomic numbers from 57 to 71 inclusive. These elements were largely unknown in Mendeléeff's time, and indeed they somewhat disturb the symmetry of the table. Similarly, the elements of the *actinide series* may best be placed into the one position marked 89–103, with the individual elements also written out at the bottom.

Because many *transuranic elements* (atomic numbers greater than 92) have been discovered—or perhaps one should say manufactured—since the time of Mendeléeff, it is necessary to keep adding more rows at the bottom of his table. Many of the heavier transuranic elements are unstable and decay so fast that it is difficult to determine their chemical properties experimentally. So one cannot be sure that element 111, for example, will behave like an alkali metal. But nuclear physicists expect, on theoretical grounds, to find an "island of stability" where nuclei may last for several minutes or hours before decaying; this region may be found in the 11th or 12th series.

In 1978 the International Union of Pure and Applied Chemistry proposed a system of temporary names for hypothetical elements, to be replaced by permanent names when the elements have been discovered. These names are created by using the syllables *nil, un* (pronounced as in *tune*), *bi, tri,* etc. for the digits 0, 1, 2, 3, etc. in the atomic number. Each element has a three-letter symbol derived from the initial letters of the syllables. Thus element no. 110 is *ununnilium,* symbol *Uun;* no. 111 is *unununium,* symbol *Uuu,* and so forth (see Appendix II for details, and Appendix VII for the placement of some of these elements in the periodic table). This system has not been generally adopted by physicists, who continue to refer to the elements by number ("element 114") until they have been discovered and permanent names assigned.

d) In 1894 the startling discovery was made that the atmosphere contained about 1% of a hitherto unnoticed gas, later called *argon* (Ar), which had escaped previous detection by chemists because of its almost complete unwillingness to enter into any chemical combination with other elements. Now other such gaseous elements have been discovered, forming a family of six inert gases (He, Ne, Ar, Kr, Xe, Rn). They all could be accommodated together in the periodic scheme by simply adding one vertical column, either interlaced with group VIII or, better, marked as a separate group with group number zero, in accord with the zero valence of each member. This harmonious fit of a whole group of unexpected additions was a considerable triumph for the scheme—too much of a triumph, indeed, for the neat conclusion that these elements have zero valence and hence form no chemical compounds seems to have prevented chemists until quite recently from looking for inert-gas compounds, some of which have in fact now been found to exist.

e) The lower end of the table has been pulled together for convenience in printing. Periods 4 and 5 are "long" periods, with 18 elements each instead of the 8 elements in the "short" periods 1, 2, and 3. Each of the long periods is divided into two "series" (thus period 5 consists of series 6 and 7); this is why the symbols for the alkali metals, for example, are placed to the left, to distinguish them from the other elements (Cu, Ag, Au, . . .) which are placed to the right because they do not have the same chemical properties. We could have avoided this (as is done in some textbooks) by making the table twice as wide so that each 18-element period is on a single line. But then we would have to increase the width again to accommodate the lanthanide and actinide series in periods 6 and 7, and perhaps yet again when the elements in period 8 are discovered.

f) Mendeléeff's basic scheme does break down in a few places. Note that Ar (argon) and K (potassium) are placed, respectively, in the 18th and 19th positions as demanded by their chemical properties, whereas on the basis of their respective atomic weights alone (A = 39.944 and K = 39.096) one should really transpose the two. Other inversions of this kind can be found in the table, for example, for elements 52 (Te = 127.6) and 53 (I = 126.9).

As mentioned in the previous section, Mendeléeff had confidently expected that the atomic weight for Te would on redetermination be found to be lower than that for I. The necessity of an inversion would have been to him not an inconvenience but a catastrophe. In connection with just such a point, he said:

> The laws of nature admit of no exception. . . . It was necessary to do one or the other—either to consider the periodic law as completely true, and as forming a new instrument in chemical research, or to refute it. Acknowledging the method of experiment to be the only true one, I myself verified what I could, and gave everyone the possibility of proving or confirming the law, and did not think, like L.

> Meyer, when writing about the periodic law that "it would be rash to change the accepted atomic weights on the basis of so uncertain a starting point." In my opinion, the basis offered by the periodic law had to be verified or refuted, and experiment in every case verified it.

Clearly, he overestimated the *necessity* of the periodic law in every detail, particularly as it had not yet received a physical explanation; and although the anomalous inversions in the sequence of elements are real, their existence did not invalidate the scheme. As is so often true, the complexity of nature is greater here than any one generalization can accommodate. Yet one must understand that it was just this utter, enthusiastic belief in the existence of a simple scheme that carried the undaunted Mendeléeff to his far-reaching conclusions.

The usefulness of the periodic law was not confined to the redetermination of some atomic weights and the prediction of properties of new elements. It had provided an inner order in the list of elements, revealing several previously unsuspected analogies among elements. The whole study of inorganic chemistry (dealing with elements and compounds not containing carbon) was thereby revitalized. And perhaps most important of all, just as the Keplerian discoveries of simple harmonies and mathematical order in planetary motions posed the tantalizing problem of how to account for them by a fundamental physical law, so was science now challenged to provide an explanation for the observed regularities in the periodic table in terms of a physical model of the atom. This task, and its solution in the twentieth century, so reminiscent of Newton's solution in the analogous case, will be recounted in Part G.

The atom of the nineteenth century by and large remained incapable of supplying a deeper insight into the fundamental processes of chemistry. However, it did achieve startling successes in the hands of the physicists. And we now turn to pick up that part of the story in the development of the atomic theory.

Additional Problems

Problem 21.2. In the 1897 English edition of his book *Principles of Chemistry,* we find Mendeléeff's own summary of his original work on the periodic table. Discuss each of the eight points, choosing further illustrations from the material in this chapter:

> The substance of this paper is embraced in the following conclusions:
>
> 1. The elements, if arranged according to their atomic weights, exhibit an evident *periodicity* of properties.
> 2. Elements which are similar as regards their chemical properties have atomic weights which are either of nearly the same value (platinum, iridium, osmium) or which increase regularly (for example, potassium, rubidium, cesium).
> 3. The arrangement of the elements or of groups of elements in the order of their atomic weights corresponds with their so-called *valencies.*
> 4. The elements which are the most widely distributed in nature have *small* atomic weights, and all the elements of small atomic weight are characterised by sharply defined properties. They are therefore typical elements.
> 5. The *magnitude* of the atomic weight determines the character of an element.
> 6. The discovery of many yet unknown elements may be expected. For instance, elements analogous to aluminum and silicon, whose atomic weights would be between 65 and 75.
> 7. The atomic weight of an element may sometimes be corrected by aid of a knowledge of those of the adjacent elements. Thus the combining weight of tellurium must lie between 123 and 126, and cannot be 128.
> 8. Certain characteristic properties of the elements can be foretold from their atomic weights. The entire periodic law is included in these lines.

Problem 21.3. In the *Handbook of Chemistry and Physics* there is printed, below one of the periodic tables, a plot of valence numbers for the elements vs. atomic number. Neglect the negative valence numbers and plot (to element 65) a graph of maximum valences observed vs. atomic weight. What periodicity is found? Is there any physical or chemical significance to this periodicity? Does there have to be any?

RECOMMENDED FOR FURTHER READING

Bernadette Bensaude-Vincent, "Mendeleyev: The story of a discovery," in *History of Scientific Thought* (edited by M. Serres), pages 556–582

H. A. Boorse and L. Motz (editors), *World of the Atom,* pages 298–312

W. H. Brock, *Norton History of Chemistry,* Chapter 9

R. E. Childs, "From hydrogen to meitnerium: Naming the chemical elements," in *Chemical Nomenclature* (edited by K. J. Thurlow), Chapter 2, Boston: Kluwer, 1998

C. C. Gillispie, *Dictionary of Scientific Biography,* articles by B. M. Kedrov on Mendeléeff, Vol. 9, pages 286–295; and by E. L. Scott on Newlands, Vol. 10, pages 37–39

David Knight, *Ideas in Chemistry,* Chapter 10

Vivi Ringnes, "Origin of the names of chemical elements," *Journal of Chemical Education,* Vol. 66, pages 731–738 (1989)

Eric R. Scerri, "The evolution of the periodic system," *Scientific American,* Vol. 279, pages 78–83 (September 1998)

CHAPTER 22

The Kinetic-Molecular Theory of Gases

22.1 Introduction

Having watched the establishment of the chemical atom through the labors of Dalton and Avogadro, we finally come, full circle, back to the topic treated by Democritus and Leucippus: the atomicity of physical matter and the nature of heat. In this chapter we shall see how, within a few years after the generalized law of conservation of energy and the equivalence of heat and mechanical work had been established (as described in Chapter 17), Bernoulli's gas model (Chapter 19) had been revived and transformed into a powerful theory of the thermal and mechanical properties of gases, using the atomic hypothesis.

We saw in Chapter 20 that in the years before 1847, Dalton's atomic theory of chemistry had raised the atomistic idea itself from the qualitative level of speculation; the densities and relative molecular weights of gases were becoming known with some accuracy; Rumford and Davy had called the caloric theory into question; the wave theory of heat had popularized the notion that heat is related to vibrations of a medium; Mayer had discussed the idea (in 1842 and 1845) that heat and mechanical energy both obeyed an overall law of conservation of energy; and Joule's laborious experiments had independently established Mayer's inspired postulate and had, in fact, yielded a value for the mechanical equivalent of heat. Yet during almost all of this time it was believed that atoms do not move freely through space, even in a gas, but, if they move at all, may only vibrate around fixed positions, caught in a medium composed of caloric or ether.

Although a few scientists before 1847 expressed views favorable to the theory that atoms can move quite freely in a gas, only one of them was able to develop such a kinetic theory into a serious challenge to the prevailing orthodoxy: John Herapath (1790–1868), a British scientist and editor of a railway journal. Without knowing of Bernoulli's work, Herapath proposed a similar theory, but developed it much further. He obtained from his model the fundamental relation between the pressure (P) of a gas and the speed (v) of the particles of which it is supposedly composed:

$$P = \tfrac{1}{3}\rho v^2 \qquad (22.1)$$

where ρ is the density (mass/volume) of the gas. (This equation will be derived from fundamentals in Section 22.4.) Since the pressure and density of a gas can be determined directly by experiment, this equation can be used to compute the speed the particles would have to have; Herapath did this, and found a result that is fairly close to the speed of sound in the gas (about 330 m/sec for air).

Herapath's calculation of the speed of an air molecule (first published in 1836) can be considered an important event in the history of science, but it was ignored by most of his contemporaries. His earlier papers on the kinetic theory, written in 1820, had been rejected for publication by the Royal Society of London, and despite a long and bitter battle (including letters to the editor of *The Times* of London), he had not succeeded in getting any recognition for his theory. Only Joule saw the value of Herapath's theory, but Joule's own attempt to establish the kinetic theory (based on Herapath's work and his new experiments) did not attract much attention either for another decade.

In the 1850s the kinetic theory of gases was again revived, and this time it was quickly established on firm foundations. For an account of the basic assumptions and some of the main results of the kinetic theory, we can do no better than quote passages (of which almost every word still corresponds to our current ideas) from the fundamental paper by Rudolf Clausius—the paper "On the nature of the motion which we call heat" (1857), which initiated the modern phase of development of the theory. Although

Fig. 22.1. Rudolph Clausius (1822–1888), German physicist (born in a part of Prussia that is now part of Poland), who was the principal founder of thermodynamics and, with Maxwell, of the kinetic theory of gases.

Clausius had adopted the kinetic theory some years before in connection with his studies on thermodynamics, he did not publish a systematic exposition until August Krönig, a German chemist, had announced the kinetic hypothesis, in a very rough form, in a short note published in 1856. Clausius had learned indirectly of Joule's work by 1857, but was not able to obtain a copy of the Englishman's paper until a year later. He did not yet know that both Krönig and Joule had been preceded by Bernoulli and Herapath.

Recall from Chapter 17 that the total kinetic energy of an object can be written as the sum of the "translational" kinetic energy of its center of mass plus the kinetic energy of rotation around its center (Eq. 17.12). That distinction played an important role in the development of the kinetic theory of gases.

Clausius wrote:

> Kronig assumes that the molecules of a gas do not oscillate about definite positions of equilibrium, but that they move with constant velocity in straight lines [translational motion] until they strike against other molecules, or against some surface which is to them impermeable. I share this view completely, and I also believe that the expansive force of the gas arises from this motion. On the other hand, I am of the opinion that this is not the only motion present.
>
> In the first place, the hypothesis of a rotatory as well as a progressive [translational] motion of the molecules at once suggests itself; for at every impact of two bodies, unless the same happens to be central and rectilineal, a rotatory as well as a translational motion ensues.
>
> I am also of opinion that vibrations take place within the several masses in a state of progressive motion. Such vibrations are conceivable in several ways. Even if we limit ourselves to the consideration of the atomic masses solely, and regard these as absolutely rigid, it is still possible that a molecule, which consists of several atoms, may not also constitute an absolutely rigid mass, but that within it the several atoms are to a certain extent moveable, and thus capable of oscillating with respect to each other.

Although Clausius could not provide a direct theoretical calculation of the amount of energy contained in these molecular vibrations and rotations, he argues that after a gas is put into an enclosure, an equilibrium must be established, as a result of collisions among the molecules, so that eventually a fixed proportion of the total energy will be found in rotation and vibration, and the remainder in the translatory motion of the molecules as a whole.

According to the kinetic theory, he held that, "the pressure of a gas against a fixed surface is caused by the molecules in great number continually striking against and rebounding from the same," and this pressure is, for constant volume, proportional to the kinetic energy of translational motion of the molecules (Eq. 22.1).

> On the other hand, from Gay-Lussac's law we know that, under constant volume, the pressure of a perfect gas increases in the same ratio as the temperature calculated from –273°C, which we call the absolute temperature. Hence, according to the above, it follows that the absolute temperature is proportional to the kinetic energy[1] of the translational motion of the molecules. But as,

[1]We have replaced the term *vis viva* [Eq. (16.2)], still in use in 1857, by its modern equivalent (except for a factor of $\frac{1}{2}$), kinetic energy.

> according to a former remark, the several motions in one and the same gas bear a constant relation to each other, it is evident that the kinetic energy of the translational motion forms a constant proportion of the total energy, so that the absolute temperature is also proportional to the total kinetic energy in the gas.

From this model it would follow that (as Clausius had already suggested in an earlier paper) the specific heat of a gas should be independent of temperature, and that the two specific heats (at constant volume and at constant pressure) differ by a constant term. Both of these consequences are in fact found experimentally to correspond to the properties of real gases over a large range of temperatures.

Although Clausius could perform detailed calculations only for a model corresponding to the gaseous state of matter in which molecules are presumed to move freely through a nearly empty space with only occasional collisions, he was willing to propose a plausible explanation of the difference between gaseous, liquid, and solid states, based on a qualitative kinetic-molecular viewpoint. In the solid state, he suggested that the molecules simply vibrate around fixed equilibrium positions and rotate around their centers of gravity. In the liquid state, the molecules no longer have definite positions of equilibrium, but can move from one place to another, though always acted on by the forces exerted by neighboring molecules. In the gaseous state, however, the molecules finally escape the influence of their neighbors and fly off in straight lines; if two molecules collide, they simply rebound in accordance with the laws of elastic collisions.

From this same qualitative picture, Clausius developed an explanation of changes of state. Thus, evaporation of a liquid could be explained by assuming that even though the average speed of a molecule may not be sufficient to carry it beyond the range of the attractive forces of its neighbors, a few exceptionally fast molecules may be able to escape from a liquid surface, even at temperatures below the boiling point. Inherent in this explanation is the notion that, even at a single temperature, molecules have a range of different speeds; this idea was soon to be developed with great effectiveness by Maxwell (see Section 22.6).

This summary has only hinted at the richness of conclusions and the generality of usefulness of the kinetic-molecular theory, which, through the further work of such scientists as James Clerk Maxwell in Britain, Ludwig Boltzmann in Austria, J. D. van der Waals in The Netherlands, and J. Willard Gibbs in the United States, eventually became a major triumph of nineteenth-century physical science. For our purposes it will be well to leave the strictly historical road, and to summarize in modern language some qualitative consequences before we turn to the quantitative side of the concept that heat is a mode of motion of submicroscopic particles.

22.2 Some Qualitative Successes of the Kinetic-Molecular Theory

a) *How work done on a gas raises its temperature.* Any work done on a substance (for example, by friction) may be converted into kinetic energy of motion or potential energy between the atoms or molecules. In gases, the increase in kinetic energy of the particles means greater random thermal motion; in solids, the energy may be stored in increased vibration of the crowded atoms; and in a liquid, both thermal motion and thermal vibration play their parts. *Heat conduction* from a hot body in contact with a cold body is thus to be thought of as the *transfer of energy of motion from one set of molecules to the other* at the interface, the surface of contact. *Expansion* of a solid or liquid *on heating* is then the consequence of *increased agitation of individual molecules and of their resulting greater separation* through more powerful mutual collisions.

Therefore, if a glass-in-mercury thermometer is inserted in a vessel containing a gas, and if now work is done on the gas (by compression or friction), the thermometer indicates a rise to a higher temperature reading on the scale because (i) the work done on the gas serves to increase the kinetic energy of the gas molecules, which in turn means (ii) stronger, more frequent collisions of these molecules with the mercury-containing glass bulb, (iii) a transfer of some of this energy to the mercury, and (iv) ultimately an expansion of the mercury column as the increased agitation of mercury molecules results in their greater mutual separation

b) *Brownian movement as visible but indirect evidence of molecular motion.* Our own sensations

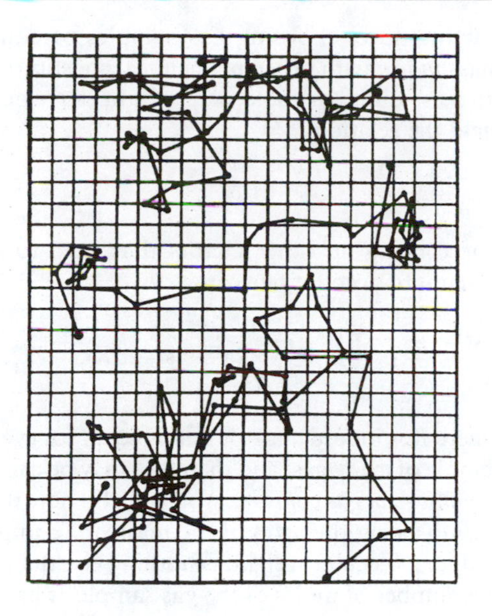

Fig. 22.2. Brownian movement. The positions of three grains on water, as seen through the microscope at 30-sec intervals.

of heat or cold by touching are not so simply explained by this or any other physical theory, but evidently we can neither feel nor see directly the individual agitation of the molecules, for their dimensions are a thousand times smaller than the wavelength of light or the sensitive receptors of our skin. Yet, if we consider a small dust particle or bacterium suspended in still air or on the surface of a liquid and bombarded from all sides by swarms of invisible but highly agitated molecules of the surrounding gas or liquid, we can understand why under high-power magnification the particle is seen to dance in perpetual, irregular motion, as though alive (see Fig. 22.2).

In fact, when first observed in suspensions of organic particles in the eighteenth century, this phenomenon was suspected of being evidence of "vitality"; but in 1827 the English botanist Robert Brown demonstrated that similar motions could be seen in suspensions of almost any kind of particle of a certain size, whether organic or inorganic. Thus what came to be known as "Brownian movement" was attributed to the effects of impacts of invisible molecules on the particles, though the development of a satisfactory quantitative theory had to wait until Einstein's work of 1905.

c) *Diffusion, expansion, and pressure of a gas.* The summary of Clausius' paper mentioned another qualitative success of the kinetic-molecular theory: an explanation of the difference between the three states of matter—solid, liquid, and gaseous—in terms of the relative agitation of the molecules. (Essentially the same explanation had been given earlier by both Davy and Joule.) The same fundamental ideas also help to make clear several other simple phenomena, e.g.: (i) the diffusion of one gas through another (one type of thermally agitated molecules working their way through a space filled with another type of molecules), or (ii) the immediate expansion of a gas to fill any vessel no matter how large (since the mutual attractive forces between gas molecules are supposedly negligible and the velocities of random motion very high), and (iii) the very existence of pressure exerted by a gas against its container (understandable as the cooperative effect of billions of collisions per second by all the agitated gas molecules with the walls of the container).

22.3 Model of a Gas and Assumptions in the Kinetic Theory

Turning now to quantitative achievements of the kinetic-molecular theory (and dropping "molecular" from the phrase, in conformity with custom), we shall concentrate mainly on the theory's applications to gases, this being most illustrative, mathematically easiest, and, in fact, physically most developed at the present time.

We shall follow a simplified derivation, in which our problem reduces to this: Let us construct a plausible model of a gas that will yield as a logical consequence (thus "explain") the known laws of the behavior of gases and that if possible will lead us to look for new verifiable effects. The model and the underlying hypotheses leading to the model together form the theory, which is useful to the extent that it explains previously known laws and generates new results.

In constructing the model of a gas, we now carefully examine all clues. Some may be well-known experimental facts about matter, others may be plausible but hitherto unchecked assumptions, and still others may be bold new hypotheses that are justifiable *a posteriori* ("after the fact") if they lead to verifiable results.

Let us summarize the ten major facts or assumptions from which we must build up our initial model, with the experimental evidence for each, where meaningful

a) *The general gas law.* The experimental facts about a given sample of a relatively rarefied and stable gas are, as we saw, well summarized[2] in Eq. (19.3):

$$\frac{PV}{T} = r.$$

b) *Molecular hypothesis.* One of the fundamental assumptions of our gas picture is, of course, the idea that gases are made of tiny submicroscopic particles—atoms or groups of atoms called molecules. The best experimental evidence originally was chemical—the success of Dalton's atomic theory, particularly in dealing with the laws of combining proportions. For our purposes we shall at first discuss only pure single gases made of identical building blocks (one type of molecule or one type of atom)

c) *The equation of state.* Another experimentally supported concept, which we may either build into our model at the start or else must demand as one of its derivable consequences, is Avogadro's hypothesis, which, simply stated, requires that all true gases at a given pressure, volume, and temperature have the same number of molecules. Having defined Avogadro's number N_A as the number of molecules per mole of substance (6.02×10^{23} molecules per mole of any substance), we can reformulate the general gas law, which reads

$$\frac{PV}{T} = r. \qquad (22.2)$$

and that has the inconvenient character that the value of r has to be determined experimentally for every particular sample of every gas under investigation (r changes both with the mass and with the chemical nature of the gas). Specifically, experiments with samples of the same gas but with different total masses m would reveal that

$$\frac{PV}{T} \propto m.$$

[2]At extreme conditions of pressure and temperature this simple equation is no longer sufficiently accurate, but can be replaced by a better approximation. Ideal, perfect, or true gases, by definition, are gases that behave accurately according to PV/T = constant—which fortunately applies to most actual gases far from such extreme conditions.

Furthermore, choosing gas samples of equal mass m but with different chemical constitution (that is, different molecular weight M) would yield the relation

$$\frac{PV}{T} \propto \frac{1}{M}.$$

There being no other factors that need to be taken into account, we can write

$$\frac{PV}{T} \propto \frac{m}{M} \text{ or } \frac{PV}{T} = \frac{m}{M} \times (\text{a constant } R), \qquad (22.3)$$

where R is indeed a factor that is equal for every sample of every gas, and therefore may be called the *universal gas constant.* You will also note that (m/M), being the ratio of the mass of a sample and its molecular weight, is nothing else than n, the number of moles of the gas sample at hand. Therefore we now write

$$PV = nRT. \qquad (22.4)$$

This most important equation, which is essentially a summary of experimentally observed facts, is the so-called *equation of state* for (perfect, ideal) gases. Since R is a *universal constant,* the equation $R = PV/nT$ should give us a numerical value for R if we know any one set of values of P, V, n, and T that any gas takes on simultaneously.

And indeed we do: At $T = 273$ K and $P = 1.013 \times 10^5$ Pa, all gases have, according to Avogadro's hypothesis, the same volume per mole ($V/n = 22.4$ liters/mol $= 2.24 \times 10^4$ cm^3/mol). Thus $R = 8.31$ J/mol/deg $= 8.31$ J/mol·K (check the value and the units).

This new form of the gas law has enormous practical advantages over Eq. (22.2). Now we can predict the physical behavior of a quantity of gas if we know only its chemical composition and its mass! This is evident in the following example, which could not have been solved directly before Eq. (22.4) was formulated.

EXAMPLE 1

What pressure will 10 g of helium exert if contained in a 500-cm^3 cylinder at room temperature (21°C)?

Solution: The relative molecular weight of He is approximately 4 (see Table 21.1); 1 mol would be 4 g, so x g corresponds to $x/4$ mol, therefore $n = 10/4 = 2.5$ mol.

$$P = nRT/V$$
$$= [2.5 \text{ mol}][8.31 \text{ J/mol·K}] \times [294 \text{ K}]/500 \text{ cm}^3$$
$$= 1.22 \times 10^8 \text{ N/m}^2,$$

which is more than 1000 atm (and therefore probably too much pressure for this container).

Problem 22.1. (a) A 50-cm^3 cylinder is safe only up to 10^7 Pa. Find the temperature at which 3 g of oxygen gas in the cylinder will burst the cylinder. (b) Determine the number of molecules of oxygen that must be present in a vessel of 10 cm^3 capacity at 20°C to produce a pressure of 50 atm.

To repeat: $PV = nRT$ is an experimentally verified equation for ideal gases, and therefore it must either appear axiomatically in our gas model or else must follow as a consequence of the model.

d) *Size of molecules.* The model of the gas we are piecing together must include the provision that the individual molecules are extremely small, perhaps of the order of 10^{-10} to 10^{-9} m. The experimental evidence for this point is extensive, varied, and conclusive. For example, vapors and gases occupy volumes that may be thousands of times larger than the liquids from which they can be obtained and, conversely, gases may be compressed to minute fractions of their original volume. They diffuse through invisibly small pores in unglazed porcelain. They combine in exact accord with the laws of chemical proportions even in extremely minute quantities. Even molecules as clumsy and large as those of oil (glycerin trioleate, $C_{57}H_{104}O_6$) cannot exceed 10^{-9} m in diameter because, by measurement, that is the approximate thickness of oil films spread very thinly on water.

We conclude therefore that ordinarily we may neglect the actual volume of all molecules in comparison with the total volume that the gas occupies by virtue of its unceasing motion, and indeed we might at the start consider them as point masses or, at any rate, as very tiny, smooth spheres.

e) *Number of molecules.* By the same evidence as in (d) we must suppose that there are enormous swarms of molecules in every cubic centimeter of ordinary gases; in fact, Avogadro's or Loschmidt's number expresses this point quantitatively. Even in the best vacuum ordinarily obtained in the laboratory (about 10^{-5} Pa) there are at ordinary temperatures still about 10^9 molecules/cm^3.

f) *Mean free path.* As the enormous expansion of a gas obtained from a liquid indicated, the average distances between molecules are correspondingly huge compared with their own size. On the average, a molecule may travel for distances hundreds or thousands of times its own dimensions before encountering another molecule. These distances, measured indirectly, are of the order of 10^{-7} m for ordinary densities, and are referred to as *mean free path.*

g) *Forces between molecules.* Although the mean separation of molecules may be large by atomic dimensions, it is so small an absolute quantity that we may wonder about the mutual forces between gas molecules (which we guess to be strong in the liquid and solid states because of the large latent heats).

First we may suspect conventional gravitational forces, but they are trivial, as a quick calculation will show. The masses of, say, oxygen molecules, are calculated from Avogadro's postulate. Since 32 kg of O_2 contain more than 6×10^{26} molecules, each molecule weighs about $32/(6 \times 10^{26})$ or about 5×10^{-26} kg. Two such molecules at about 10^{-7} m apart exert a mutual gravitational force of

$$F = G\,(m_1 m_2)/d^2$$

or

$$F = [(6.68 \times 10^{-11}) \times (5 \times 10^{-26})^2/(10^{-7})^2] \text{ N},$$

about 2×10^{-47} newtons, evidently a completely negligible quantity. (So also is the weight of each molecule by itself for all our purposes.)

But there may be among these particles forces other than gravitational—perhaps electrical ones. Although we may not readily imagine exactly what or how strong they are, we can make an experiment to see whether or not they exist. If they do, then a gas under pressure that is allowed to expand into a big *evacuated* vessel, on separating from its fellow molecules, will invoke such intermolecular forces to do work. If the forces are attractions, then work is done by the molecules on separating and since, according to the fundamental assumptions, the energy required for this process can come only from heat energy, the gas ought to cool itself on expanding into a vacuum. Conversely, if the intermolecular forces are repulsions, the total mass of the gas should heat up on expansion.

The experimental fact observed by Gay-Lussac (see Section 17.8), however, is that on expansion into a vacuum only a very small

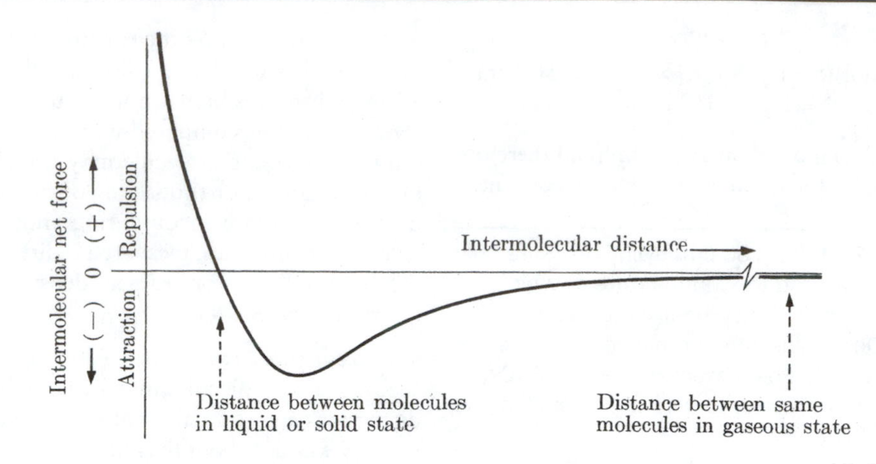

Fig. 22.3. Qualitative graph of the force between two molecules plotted against the intermolecular distance.

amount of net cooling or heating occurs in the whole mass of the actual gases; hence no significant intermolecular forces exist except during actual collisions, which presumably take very little time and on the average occupy a negligible span for each molecule compared with its whole history. Incidentally, you will recall that this is exactly the experiment that most directly refutes the old static model of mutually repelling gas molecules.

Here we may digress for a moment: Even if it be true that the molecules do not exert significant forces on one another while in gaseous form and widely separated, we must still expect that when they come closer together, as in liquids and solids, such intermolecular forces become important. We must postulate that slight increases in the intermolecular separation distances call into action a certain amount of mutual attraction in the case of liquids and solids, for otherwise we cannot explain cohesion, the resistance offered to efforts tending to separate the material in solid or liquid form into parts. There must also be the converse effect: mutual repulsion on decreasing the intermolecular distance; otherwise we cannot account for the large resistance of liquids and solids to the slightest compressions.

One may assemble all this information about intermolecular forces in one qualitative graph of distance vs. force (where positive forces represent repulsion among molecules and negative ones attraction; see Fig. 22.3).

Problem 22.2. Justify the indicated position in Fig. 22.3 for the intermolecular distances in solids and gases.

---

h) *Thermal agitation.* Strongly implied in the developing model, in the fundamental assumptions of the kinetic theory, and in the observations on Brownian movement and diffusion is the requirement of rapid motion of the gas molecules at usual and high temperatures. The picture of the gas then expands to the view of moving particles, passing through relatively "long" stretches of straight motion, and colliding briefly but many million times per second with other molecules or with the walls of the container (where each collision contributes to the maintenance of pressure). So brief are these mutual collisions between particles, so random the motions, that we shall in our analysis neglect the effect of such encounters entirely. We shall also see that thermal agitation, as implied by the observation of Brownian movement, must be omnipresent. All assemblies of atoms (if above 0 K) exhibit such agitation, not only in gases and liquids but even in solids, where it is expressed by a small-scale vibration among neighboring atoms.

i) *Nature of collisions—Newton's laws apply.* Now that we have allowed the molecules to undergo collisions, we must consider the mechanical laws governing such events. Naturally, we turn to Newton's mechanics, although we have no direct evidence that microscopic worlds obey these laws, which were developed from experiments with large-scale objects. But although we may have to keep our eyes open to retract this dangerous extrapolation if facts demand it, we are more or less forced to appeal to the principle of simplicity and start with the assumption that Newtonian laws hold among molecules, for we have no other mechanics available to us at this point.

In his pioneering work on this subject, James Clerk Maxwell discusses the relation between the atomistic model and the forces of interaction as follows.

> Instead of saying that the particles are hard, spherical, and elastic, we may if we please say that the particles are centers of force, of which the action is insensible [too small to be perceived] except at a certain small distance, when it suddenly appears as a repulsive force of very great intensity. It is evident that either assumption will lead to the same results.
>
> ("Illustrations of the Dynamical Theory of Gases," 1860)

j) *Conservation of kinetic energy.* We must also assume that the molecular collisions with one another and with the container are perfectly elastic, that is, that on the average the kinetic energy of a molecule before the collision is the same as immediately afterward; only *during* the brief collision will the kinetic energy pass into potential energy of elastic distortion, presumably just as in the case of an elastic ball bouncing off a wall. (But note that such molecules cannot be "atoms" in the original sense of rigid and undeformable particles.)

The reasons for this simplifying assumption are evident: If in the long run the collisions were not perfectly elastic, the kinetic energy of each molecule would gradually disappear even in a gas filling a heat-insulated container completely left to itself; with a diminishing total kinetic energy goes a reduced pressure, until eventually the speeds are so low that the feeble gravitational and other attractive forces induce liquefaction or solidification of the gas.

Nothing like that is observed at all. Nor can we at this point readily imagine what could happen to kinetic energy lost in a hypothetical inelastic collision of a molecule. Of course some energy could be—and sometimes is—somehow stored as potential energy, as in a change of position or increased vibration of the parts of a molecule with respect to one another or even in rotation of the molecule. But this is not an unlimited reservoir for energy, and in any case we should try out the simpler hypothesis of perfectly elastic behavior. The previous suggestion that the molecules be thought of as tiny, round, smooth spheres is simply an expression of this requirement of conservation of kinetic energy.

On the other hand, no large-scale body, not even a sphere of the best steel, shows perfectly elastic collisions; there are always some losses of mechanical energy through internal friction and a consequent rise in temperature through increased internal molecular agitation. But what could be the meaning of frictional heat losses in the collisions of the molecules themselves? Note that we have here a good example of the breakdown of a concept gained from large-scale experience when applied to the realm of the submicroscopic.

22.4 The Derivation of the Pressure Formula

We now have a detailed mental picture or *model* of a typical ideal gas. It is overpoweringly tempting to visualize it in terms of a three-dimensional assembly of agitated microscopic billiard balls. Therefore a warning is in order: We must be prepared for a breakdown of the analogy where it is carried beyond the evidence, be ready for new assumptions to be introduced where necessary, and be ready for surprising consequences of our model, for behaviors quite beyond anything an assembly of billiard balls would ordinarily dream of doing.

As an initial test of the validity of our model we now choose any well-established quantitative law of gases and try to deduce it from the model. Following the usual choice, we decide on the general gas law [in its modern form, Eq. (22.4)], $PV = nRT$. If we can derive this equation, we should be completely satisfied.

But actually we cannot do this directly, and instead we shall obtain from our analysis another equation, of the form $PV = XYZ$ (where XYZ stands for factors such as the mass and velocity of the molecules). The proof that our model is valid then depends on showing that $XYZ = nRT$, or that the equation $PV = XYZ$ actually yields experimentally verifiable consequences. This we shall do, with exciting secondary results.

To derive a relationship between the pressure, volume, and other variables of a gas, consider a strong and rigid cubical container with sides of length l, enclosing the molecules of a pure ideal gas (Fig. 22.4). Each molecule has the same mass, m_0, and moves rapidly back and forth in the box, frequently colliding with the six walls,

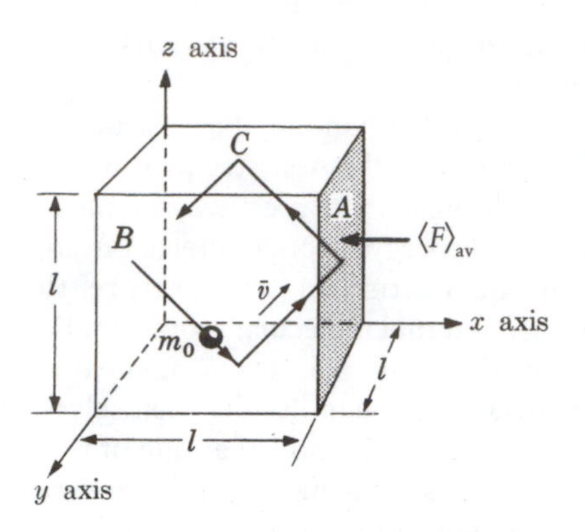

Fig. 22.4. Path of a molecule in a cubical container; collisions with the sides are perfectly elastic.

in accordance with the gas model which we have agreed on.

a) First imagine that there is only a single molecule rattling around in the box. For simplicity, we also assume that its speed v (and therefore its kinetic energy) does not change numerically except during actual collisions, which, however, take only a negligibly short time. Every time the molecule collides with one flat and rigidly fixed side of the box, say A (shaded in Fig. 22.4), it is reflected again with the same speed; evidently the component of the velocity perpendicular to side A is simply reversed, and the other two components of the velocity are not affected at all by that collision.

This situation reminds us of the result of Example 2, Section 17.1: If a very light object undergoes an elastic collision with a very massive object, it will simply bounce off with the same speed in the opposite direction. More generally, if the surface of the massive object is parallel to the zy plane and the lighter object strikes it (for example, at A) obliquely with a velocity vector $\boldsymbol{v}$ having components v_x, v_y, v_z, then the component v_x will be reversed by the collision, but v_y and v_z will remain unchanged. In this case the very light object is a molecule and the comparatively very massive object is one wall of the cubical container.

It is important to note that the kinetic energy of the molecule is conserved in these collisions, as long as the wall remains fixed and the speed of the molecules does not change. There is a continual transfer of momentum between the molecules and any wall with which it collides, and we shall calculate that at once for one wall. (But for a rigid box taken as a whole, the actions at the different walls will in time cancel out on the average as the molecule rattles around in the box.)

b) If v_x is the velocity component of the molecule in the x direction, and t is the time the molecule needs to travel the l cm from side B (opposite A) to side A, then

$$v_x = l/t \qquad \text{and} \qquad t = l/v_x$$

Then τ, the time interval needed for a complete round trip from one collision with side A to the next collision with that side, is $2t$. Thus we write:

Time elapsed per collision (with side A, for one molecule) $\equiv \tau = 2l/v_x$.

If τ is the number of seconds per collision, the reciprocal quantity $1/\tau$ is the number of collisions per second, given by $1/\tau = v_x/2l$.

c) During each of the $v_x/2l$ collisions occurring every second between the molecule of mass m_0 and the rigid side A, the velocity component v_x is changed from positive (to the right in Fig. 22.4) to negative (to the left), a total change of $2v_x$. The change in *momentum* of the molecule (being the mass times the change of velocity) is equal to $2m_0v_x$ for *each collision;* but during 1 sec, the sum of all such momentum changes is

$$(2m_0v_x) \times (v_x/2l),$$

that is, (momentum change/collision) × (collisions/sec). Of course,

$$(2m_0v_x) \times (v_x/2l) = (m_0v_x^2/l),$$

and this still applies for only a single molecule.

But let us now consider not just one but the actual number N_m of molecules in this box. Their total change of momentum per second at side A is

$$N_m(m_0v_x^2/l),$$

if they all ordinarily have the same speed. If they do not (and it would be too much of a limitation to put in such an improbable restriction) we need do no more at this time than write $\langle v_x^2 \rangle_{av}$ instead of v_x^2 in the last expression, where $\langle v_x^2 \rangle_{av}$ means the average value of the sum of the square of the x components of velocity for all N_m molecules present; that is,

$$\langle v_x{}^2\rangle_{av} = [(v_x)_1^2 + (v_x)_2^2 + (v_x)_3{}^2 + \ldots + (v_x)_{N_m}{}^2]/N_m \qquad (22.5)$$

This presupposes nothing about the way actual values of v_x are distributed among the N_m molecules, a question we may investigate later.

Thus we now write: The total change of momentum per second of all N_m molecules at side A is equal to

$$N_m m_0 \langle v_x^2\rangle_{av}/l.$$

(Watch how this expression is growing and changing in meaning in the development.)

d) By Newton's second law, the change of momentum is equal to the *average net force* $\langle F\rangle_{av}$ *exerted,* multiplied by the *time* during which the change of momentum occurs. Consequently, the change of momentum *per second* experienced by the molecules is given by $\langle F\rangle_{av}$ numerically; or, to put it differently, the average force exerted on the molecules by the side A as they rebound is numerically given by

$$\langle F\rangle_{av} = N_m m_0 \langle v_x^2\rangle_{av}/l. \qquad (22.6)$$

Since the hail of molecules against the wall is pretty thick, this force is what might be observed steadily if the wall or side A were a piston held in place (as indicated in Fig. 22.4) by means of a compression-type spring balance against the blows of the molecules. By the law of action and reaction it is, of course, also numerically equal to the average (perpendicular) force with which the molecules push on side A.

e) We are not after the force exerted on the wall, but after the *pressure.* We recall that pressure P is the perpendicular force/area. If we divide both sides of Eq. (22.5) by l^2, which is the area of side A (and is equally the area of every one of the other five sides), we now have

$$\langle F\rangle_{av}/l^2 \ (\equiv P \text{ on side } A) = N_m m_0 \langle v_x^2\rangle_{av}/l^3.$$

But l^3 is the volume V of the cubical container; hence

$$P = N_m m_0 \langle v_x^2\rangle_{av}/V, \text{ or } PV = N_m m_0 \langle v_x^2\rangle_{av}. \qquad (22.7)$$

This is a very promising development in our effort to derive the relationship between P, V, and other measurable quantities; all that remains to be done is to prove that the right-hand side of Eq. (22.7) is equivalent to nRT. But that will not be the easiest part of our work!

f) First of all, the expression $\langle v_x{}^2\rangle_{av}$ in Eq. (22.7) can be replaced by $\frac{1}{3}\langle v^2\rangle_{av}$; the argument is rather simple. From Eq. (22.7),

$$\langle v_x^2\rangle_{av} = PV/N_m m_0,$$

where P is the pressure on face A, and V, N_m, and m_0 are constants for this box and this gas sample. If we had chosen to pay attention to one of the other faces, say the top face C which stands perpendicular to the z axis, we should have obtained

$$\langle v_z^2\rangle_{av} = PV/N_m m_0.$$

The pressure P at C is no different from that on A or on any other face, and V, N_m, and m_0 are still the same for each case; therefore

$$\langle v_x^2\rangle_{av} = \langle v_y^2\rangle_{av} = \langle v_z^2\rangle_{av},$$

that is, we assume that there is no preference among the molecules for motion along any one of the three axes.

This is another example of the useful role of *symmetry* in physical science: The equations of a theory should not change if we simply rotate the coordinate axes x, y, z unless the special conditions of our problem have introduced a preferred direction.

By Pythagoras' theorem, extended to a vector v in three dimensions,

$$v^2 = v_x^2 + v_y^2 + v_z^2.$$

Therefore we may write

$$\langle v^2\rangle_{av} = \langle v_x^2 + v_y^2 + v_z^2\rangle_{av},$$

or in our case

$$\langle v^2\rangle_{av} = \langle 3v_x^2\rangle_{av} = \langle 3v_y^2\rangle_{av} = \langle 3v_z^2\rangle_{av}$$

Therefore

$$\langle v_x{}^2\rangle_{av} = \tfrac{1}{3}\langle v^2\rangle_{av}$$

and Eq. (22.7) can be rewritten

$$PV = \tfrac{1}{3} N_m m_0 \langle v^2\rangle_{av}. \qquad (22.8)$$

This is clearly reminiscent of Eq. (22.1), since the density $\rho = N_m m_0/V$.

Note that $\frac{1}{2}m_0\langle v^2\rangle_{av}$ would be the average kinetic energy of translation ($\langle KE_{trans}\rangle$) *per molecule,* and $N_m(\frac{1}{2}m_0\langle v^2\rangle_{av})$ is the *total* KE_{trans} for all N_m molecules in the sample of gas in our box. Therefore it becomes significant to recast Eq. (22.8):

$$PV = \tfrac{2}{3}N_m \times \tfrac{1}{2}m_0\langle v^2\rangle_{av}$$
$$= \tfrac{2}{3}(\text{total KE}_{trans} \text{ of gas}). \qquad (22.9)$$

In words, the product of the pressure and the volume of a gas is two-thirds of the total kinetic energy of translation of the molecules.

This is an astonishing consequence of our model. In a qualitative way we supposed all along that the pressure of the gas against the container was due to the incessant blows of highly agitated gas molecules, but we [that is, scientists before the time of Daniel Bernoulli, who derived the equivalent of Eq. (22.9) in 1738] had no idea that so simple and beautiful a relation between the pressure and the energy of the gas molecules could exist.

Equation (22.9) is the so-called pressure equation for an ideal gas. We may be chagrined that we did not obtain $PV = nRT$ by our calculations; but we have the promised relation $PV = XYZ$, and so the validity of our model can still be tested by examining whether the proposition $XYZ = nRT$ yields verifiable consequences. Moreover, if the factor $\langle v^2\rangle_{av}$ depends only on the temperature of the gas (to be proved), then Eq. (22.8) agrees with Boyle's law: PV = constant at given T. This would be, of course, most gratifying.

22.5 Consequences and Verification of the Kinetic Theory

a) First let us see whether Eq. (22.9) by itself is reasonable from the point of view of *direct experimental verification.*

It follows from Eq. (22.8) or (22.9) that $\langle v^2\rangle_{av} = (3PV/N_m m_0)$. But $N_m m_0$ is equal to the total mass of gas actually present, and $N_m m_0/V$ equals the mass density ρ of the gas (in kg/m³).

Therefore

$$\langle v^2\rangle_{av} = 3P/\rho \text{ and } \sqrt{\langle v^2\rangle_{av}} = \sqrt{(3P/\rho)}.$$

The expression $\sqrt{\langle v^2\rangle_{av}}$, for which we shall use the symbol c, is the square root of the mean (average) of the squares of the various velocities—in brief, the *root-mean-square* value of the molecular velocities (or "rms velocity"). It is mathematically not the same thing as a simple mean value $\langle v\rangle_{av}$, but, in fact, we find that the value of $\sqrt{\langle v^2\rangle_{av}}$ differs by only a few percent from the straight average[3] in actual gases, or $c \approx \langle v\rangle_{av}$.

Now the prediction following from our model is that the rms velocity c, or nearly enough the average velocity $\langle v\rangle_{av}$ of the molecules, should be equal to $\sqrt{(3P/\rho)}$. Substituting representative values for air at 1 atm pressure (1.01×10^5 Pa) and 0°C (where ρ = 1.29 kg/m³) we find that the rms velocity should be

$$\sqrt{(3 \times 1.01 \times 10^5/1.29)} = 4.85 \times 10^2 \text{ m/sec},$$

nearly 500 m/sec, or about $\tfrac{1}{3}$ mi/sec—comparable to the speed of a slow bullet!

Bernoulli might have carried his own work to this point if he had had good measurements of gas densities. Herapath, a century later, did so, and we may confirm his conclusions that the average velocity of the molecules of water vapor, say at 100°C, is in excess of 1 mi/sec. This is an enormous speed compared with the motion of everyday objects, but it is of the same order as the speed of propagation of *sound* at that pressure and temperature (331 m/sec). The rms velocities of molecules in other gaseous materials at that pressure and temperature have, by calculation, similarly high values, ranging from 1800 m/sec for hydrogen gas to 180 m/sec for mercury vapor, and in each case the velocity of sound is not much less.

But this is just what we should expect in terms of our model of a gas, for the obvious way to think of the propagation of sound waves is to visualize it as a directional motion of the molecules superposed on their random chaotic motion, so that the energy of the sound wave is carried as kinetic energy from one gas molecule to the next near neighbors with which it collides. The molecules themselves, despite their high speeds, do not move very far during any appreciable interval, being contained within a relatively small volume by multiple collisions, around a billion per second, with their neighbors—but the energy of the sound wave is communicated from one molecule to the next with that high speed.

Incidentally, this picture of molecules being restrained by mutual collisions despite their individual high speeds explains why when a bottle of perfume is opened, the scent (being nothing

[3]To appreciate the difference between $\langle v\rangle_{av}$ and c, note that if there were only three molecules with speeds $v_1 = 5$ units, $v_2 = 10$, and $v_3 = 15$, $\langle v\rangle_{av} = (5 + 10 + 15)/3 = 10$ units, and $c \equiv \sqrt{\langle v^2\rangle_{av}} = \sqrt{(5^2 + 10^2 + 15^2)/3} = 10.8$ units. This is a difference of 8%.

but a swarm of vapor molecules from that liquid) takes a relatively long time to travel from one corner of a room to the other. This familiar experience was one of the initial strong objections to the kinetic theory, raised by the Dutch meteorologist C. H. D. Buys-Ballot. Clausius responded by pointing out that his original calculation of the average speed was based on the simplifying assumption that the molecules are *mathematical points,* which could never collide with each other; as soon as one assumes, more realistically, that they are small spheres of finite diameter *d,* then they will frequently collide and change their direction of motion. It was for just this reason that Clausius introduced the *mean-free-path* concept. Buys-Ballot, like Franciscus Linus, who caused Robert Boyle to reinforce his theory of air pressure by introducing evidence for a quantitative relation between pressure and volume (Chapter 19), played the role of a "gadfly" who unintentionally advances science by fruitful criticism.

As to the numerical values involved here, we do not really expect that the speed of sound should be exactly as great as *c* (why not?); in fact, the numerical discrepancy is precisely predictable from the complete theory of sound propagation.

Problem 22.3. John Dalton, who first of all was a meteorologist, discovered experimentally a law concerning gas pressures, which we call Dalton's law of partial pressures, and which may be stated thus: If several (for instance, three) gases are mixed in one container at a given temperature, then the total pressure P exerted by all of them together is the sum of the individual pressure (for example, $P = P_1 + P_2 + P_3$) that each would exert if it were alone in that container at that temperature. Prove that our kinetic theory model is in harmony with this law. [*Hint:* Accept the general gas law as applicable even to gas mixtures. From Eq. (22.9), $P = (\frac{2}{3})\text{total KE}_{\text{trans}}/V$, and note that kinetic energies add by simple scalar addition.]

---

b) Returning to the main line of the argument, the fact that the theoretical result of our model $PV = \frac{1}{3} N_m m_0 \langle v^2 \rangle_{\text{av}}$ must coincide with the experimental facts about gases ($PV = nRT$), we proceed to look for derivable and *verifiable* consequences of the implication that the right-hand sides of both equations are equal, and that

$$nRT \stackrel{?}{=} \frac{1}{3} N_m m_0 \langle v^2 \rangle_{\text{av}}. \qquad (22.10)$$

This can be rewritten:

$$T \stackrel{?}{=} \frac{2}{3}(N_m/nR)(\tfrac{1}{2} m_0 \langle v^2 \rangle_{\text{av}}). \qquad (22.11)$$

That is, *the temperature of a gas (in K) is proportional to the mean kinetic energy of translation per molecule for the gas.* It is a most important result and a very fruitful interpretation of the concept of temperature.

Furthermore, note that the factor $\frac{2}{3}(N_m/nR)$ is a universal constant for *all* gases. The proof is as follows:

$$\frac{N_m}{n} = \frac{\text{number of molecules in given sample}}{\text{moles of gas in sample}}$$

But the number of molecules per mole is, by definition, Avogadro's number N_A, a universal constant. Therefore

$$\frac{2}{3}(N_m/nR) = \frac{2}{3} N_A/R.$$

But as we noted above, R is also a universal constant; and so it is convenient and customary to write the last expression as

$$\frac{2}{3}(1/k),$$

where

$$k \equiv R/N_A = 8.31 \text{ (J/mol·K)}/6.02 \times 10^{23} \text{ molecules/mol}$$

$$= 1.38 \times 10^{-23} \text{ (J/K)/molecule}$$

The universal constant k so defined is of greatest significance in many branches of physics, and is called *Boltzmann's constant,* named in honor of the great nineteenth-century Austrian physicist (Fig. 22.14) whose work paved the way for many advances in this field. We will discuss one aspect of his work at the end of this chapter.

We can now rewrite Eq. (22.11):

$$T = (2/3k)\, \tfrac{1}{2} m_0 \langle v^2 \rangle_{\text{av}}$$
$$= (\text{constant}) \times (\langle \text{KE}_{\text{trans}} \rangle_{\text{av}} \text{ per molecule}). \qquad (22.12)$$

In words (since k is truly a universal constant), *the temperature of a gas is directly proportional to the mean kinetic energy of translation per molecule for any gas.* This, then, is the consequence of the proposal that Eq. (22.10) is correct.

One further assumption lies behind the above results: The collisions among molecules bring about a state of *thermal equilibrium* in which the mean kinetic energy, and hence the temperature, are the same throughout all parts of a system that can freely exchange energy. (Equivalently we could assume that there is a tendency toward equalization of the temperature of the wall of the container and the gas molecules inside.) For example, if a container divided into two parts by a thermally insulated wall has hot hydrogen gas in one part and cold hydrogen gas in the other, and we then open a hole in the wall, allowing the gases to mix, we will find after a while that the temperature is the same in both parts. This result is consistent with Kelvin's principle of dissipation of energy (Section 18.3), but it is not easy to prove that it holds for our model system of colliding gas molecules—especially if the motions and collisions obey Newton's laws of mechanics. (Why not?)

c) Consider one of the many consequences of Eq. (22.12). At the same temperature T for two different types of gases (with values of m_1 and v_1 for the one, m_2 and v_2 for the other) our theory requires that

$$T = (2/3k)\tfrac{1}{2}m_1\langle v_1^2\rangle_{av} \text{ and}$$
$$T = (2/3k)\,\tfrac{1}{2}m_2\langle v_2^2\rangle_{av},$$

or

$$\tfrac{1}{2}m_1\langle v_1^2\rangle_{av} = \tfrac{1}{2}m_2\langle v_2^2\rangle_{av} \qquad (22.13)$$

Hence, at a given temperature, the molecules of any gas whatever have the same $\langle KE_{trans}\rangle_{av}$ as any other gas, or

$$\langle v_1^2\rangle_{av}/\langle v_2^2\rangle_{av} = m_2/m_1$$

and

$$\sqrt{\langle v_1^2\rangle_{av}}/\sqrt{\langle v_2^2\rangle_{av}} \equiv c_1/c_2 = \sqrt{(m_2/m_1)}. \qquad (22.14)$$

In words, *the ratio of rms speeds of molecules of two gases is equal to the square root of the inverse ratio of their masses.*

This conclusion is susceptible to immediate experimental test. If we take two different gases at equal temperatures, each in its own tight rigid container at equal pressures and then open in each container an equally tiny hole toward a vacuum for the molecules to escape, the gas with the lighter molecular mass, having a relatively higher rms speed, should escape faster than the heavier gas with slower rms speed. This experiment in *effusion* is in principle quite possible, and although the experiment itself obviously is difficult to perform, the result does, in fact, prove Eq. (22.14) to be correct.

Problem 22.4. The mass of each molecule of hydrogen gas is to the mass of a molecule of nitrogen gas as 2:28 (approximately). What are the relative speeds of effusion for two such gases under equal conditions of temperature, etc.? How might this result help to explain the fact that the earth's atmosphere contains a large amount of nitrogen but very little hydrogen, even though (according to some theories) the earth was originally formed from a nebula consisting mostly of hydrogen?

Another proof of Eq. (22.14) is to be found in the *diffusion* of two different gases trapped in one container with porous walls toward a vacuum; the lighter gas will again escape faster than the heavier one. This had been established empirically by Thomas Graham in 1830, and now was explainable in terms of a theory.

This process of diffusion is the basis of one type of large-scale separation of the light and heavy isotopes of uranium, the fateful metal. Specifically, uranium (chemical symbol U), as found in nature, consists primarily of an intimate mixture of two chemically equivalent but not equally heavy types of atoms, the isotope U^{238} (that is, relative atomic weight approximately 238, making up 99.3% of natural pure uranium), and the more potent but rarer isotope U^{235} (atomic weight approximately 235, abundance about 0.7%). Small traces of a third type of uranium atom, U^{234}, are inevitably also present, but are negligible for most practical purposes.

The historic problem of "enriching" natural uranium to make it useful for nuclear reactors or weapons, that is, of increasing the relative abundance of U^{235} with the help of the kinetic theory, was first discussed in the famous *Smyth Report* of 1945. For the complete understanding of the following extract, note that the ratio c_1/c_2 in Eq. (22.14) corresponds also to the ratio of the *number* of molecules with these respective speeds that will have found their way through a porous barrier of a vessel after a short time interval. This ratio c_1/c_2 is therefore also called the *ideal separation factor,* symbolized by α. The Smyth Report gives these details:

As long ago as 1896 Lord Rayleigh showed that a mixture of two gases of different atomic weight could be partly separated by allowing some of it to diffuse through a porous barrier into an evacuated space. Because of their higher average speed the molecules of the light gas diffuse through the barrier faster, so that the gas which has passed through the barrier (that is, the "diffusate") is enriched in the lighter constituent, and the residual gas (which has not passed through the barrier) is impoverished in the lighter constituent. The gas most highly enriched in the lighter constituent is the so-called "instantaneous diffusate"; it is the part that diffuses before the impoverishment of the residue has become appreciable. If the diffusion process is continued until nearly all the gas has passed through the barrier, the average enrichment of the diffusate naturally diminishes. . . . On the assumption that the diffusion rates are inversely proportional to the square roots of the molecular weights, the separation factor for the instantaneous diffusate, called the "ideal separation factor" α, is given by

$$\alpha = \sqrt{(m_2/m_1)}$$

where m_1 is the molecular weight of the lighter gas and m_2 that of the heavier. Applying this formula to the case of uranium will illustrate the magnitude of the separation problem. Since uranium itself is not a gas, some gaseous compound of uranium must be used. The only one obviously suitable is uranium hexafluoride, UF_6, ... Since fluorine has only one isotope [of atomic weight 19], the two important uranium hexafluorides are $U^{235}F_6$ and $U^{238}F_6$; their [approximate] molecular weights are 349 and 352 [respectively]. Thus if a small fraction of a quantity of uranium hexafluoride is allowed to diffuse through a porous barrier, the diffusate will be enriched in $U^{235}F_6$ by a factor

$$\alpha = \sqrt{(352/349)} = 1.0043.$$

Note that the enrichment is only 0.43% over the previous proportions.

Some of the difficulties of this process of separation are evident, not the least being the need for the final chemical extraction of uranium from uranium hexafluoride. Another serious point is that this separation factor α is far too low to permit use of the feebly enriched uranium

Fig. 22.5. Diffusion plant of the Clinton Works (now administered by the Department of Energy at Oak Ridge, Tenn.), formerly used for the separation of uranium isotopes.

hexafluoride after one such process. Therefore the first diffusate is made to go through another porous barrier, which again discriminates against U^{238} in favor of the prized U^{235}. Since α remains about the same, each single stage changes the relative abundance of U^{235} by only very little.

In Smyth's words:

> To separate the uranium isotopes, many successive diffusion stages (that is, a cascade) must be used. . . . [Studies] have shown that the best flow arrangement for the successive stages is that in which half the gas pumped into each stage diffuses through the barrier, the other (impoverished) half being returned to the feed of the next lower stage. . . . If one desires to produce 99 percent pure $U^{235}F_6$, and if one uses a cascade in which each stage has a reasonable over-all enrichment factor, then it turns out that roughly 4000 stages are required.

The construction by 1945 of many acres of such diffusion chambers in a Tennessee valley [Clinton Engineering Works, Oak Ridge, Tenn. (Fig. 22.5)] is described in the report. In a certain sense, the first atomic bomb, which used U^{235}, was then a gigantic confirmation of the kinetic theory that led to Eq. (22.14) and this diffusion process.

Yet a third proof of Eq. (22.14), on a rather more modest scale, is the closer observation of Brownian movement. If we watch the thermal agitation of a single tiny dust particle (mass m_1, rms velocity c_1), we note that its erratic dance movement never reaches the excessive speeds that we know the surrounding air molecules possess. In fact, Eq. (22.14) assures us that the rms speed c_1 of the dust particle is equal to $c_2\sqrt{(m_2/m_1)}$; and since m_2, the mass of the air molecules, is many million times smaller than m_1, the mass of our dust particle, we cannot be surprised that c_1 is so small. (Actually the direction of motion of the dust particle changes so rapidly that it is extremely difficult to observe c_1 directly. It was for this reason that attempts to base a quantitative theory of Brownian movement on Eq. (22.14) in the nineteenth century proved futile.)

Bodies larger than dust particles and the pollen of flowers are, of course, also randomly bombarded by the surrounding atmosphere, but because their masses m_1 are larger still, their c_1 is correspondingly slower and their Brownian movement becomes unnoticeable. Yet very delicately suspended light mirrors, as used in accurate instruments, do show the slight random jitter of Brownian movement.

22.6 The Distribution of Molecular Velocities

Consider another attack on Eq. (22.12),

$$T = (2/3k)(\tfrac{1}{2}m_0)\langle v^2\rangle_{av}.$$

For any given gas we can measure T, we know its mass per molecule m_0,[4] and we have calculated k before; thus

$$c = \sqrt{\langle v^2\rangle_{av}} = \sqrt{(3kT/m_0)}.$$

If we now were able to measure $\sqrt{\langle v^2\rangle_{av}}$ directly, we could check Eq. (22.12) at once. But how are we to make a measurement of the incredible speed of invisible molecules? To prepare for the answer, we must take a short excursion.

One of the most significant ways in which the development of the kinetic-molecular theory has influenced modern science is by showing that regular, predictable behavior on the macroscopic level can result from irregular, random behavior on the microscopic level. Just as an insurance company can predict how many 50-year-old people will die every year without attempting to predict *which* people will die, the physicist can accurately calculate the number of molecules in a sample of nitrogen gas that have speeds greater than 1000 m/sec without measuring the speed of any one of them. The use of statistical methods does not necessarily imply that molecules move randomly, or that a person's death does not have a specific cause, only that one can calculate the average properties of a large number of molecules or people by treating them *as if* they were governed by chance. But by their success in using a "statistical view of nature," Maxwell and Boltzmann helped to prepare the way for twentieth-century quantum theory, in which we do have to accept a degree of randomness at the atomic level.

At the beginning of the nineteenth century, the deterministic aspect of Newtonian mechanics had been epitomized in the following well-

[4]The mass m_0 in grams of a gas molecule is given by m_0 = (mass in grams/mol of gas)/N_A = (number of grams/mol)/(6.02×10^{23} molecules/mol) by Avogadro's law. For nitrogen, for example, this reduces to $m_0 \approx 28/(6.02 \times 10^{23}) = 4.65 \times 10^{-23}$ g/molecule.

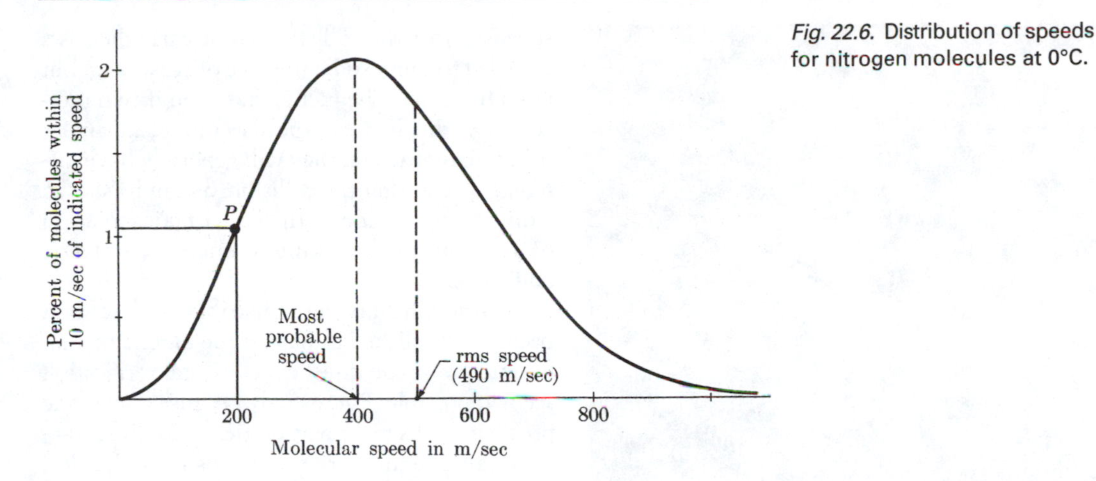

Fig. 22.6. Distribution of speeds for nitrogen molecules at 0°C.

known passage of the French mathematician, Pierre Simon, Marquis de Laplace (1749–1827):

> We ought then to regard the present state of the universe as the effect of its previous states and as the cause of the one which is to follow. Given for one instant a mind which could comprehend all the forces by which nature is animated and the respective situation of the beings who compose it—a mind sufficiently vast to submit these data to analysis—it would embrace in the same formula the movements of the greatest bodies of the universe and those of the lightest atom; for it, nothing would be uncertain and the future, as the past, would be present to its eyes.

Ironically this assertion was made in Laplace's *Philosophical Essay on Probabilities* (1814), serving as a preface to a major and influential contribution to statistical theory, in which determinism was not denied but simply set aside as useless in practical calculations. From the labors of Laplace and his contemporaries Poisson and Gauss emerged a powerful set of tools for dealing with events and processes that are either random or, though deterministic, so complicated that statistical assumptions are more convenient.

One of the most useful results of nineteenth-century statistical theory was the *law of errors* or *normal distribution*. According to the law of errors, if any quantity can vary randomly around a certain average value, then the curve showing the frequencies of deviations from the average value will be the familiar *bell-shaped curve*, with a peak in the region of the average value and a sharp decline on either side. Statisticians found that the law of errors applies in many cases involving large numbers of people, for example, in the distribution of heights and weights. Students are familiar with the law because it is supposed to apply usually to the distribution of scores on tests; instructors sometimes use the bell-shaped curve in deciding which scores should correspond to the various letter-grades; hence the expression "grading on the curve."

The suggestion that speeds of molecules in a gas might also be distributed according to a similar curve was made by Maxwell in his 1860 paper on kinetic theory. Even though the motions of the individual molecules were generally thought to be ruled by Newton's laws, it seemed to be impossible to determine them directly—or, even if that could be done, to use the information in calculating the properties of gases. But Maxwell realized that in physics, as in human affairs, we often have to draw conclusions even when we lack most of the relevant details. An insurance company must set its rates, not knowing of course when any particular person seeking insurance is going to die, but relying on mortality tables that give an estimate of the probable life span of people at various ages. Similarly, we can calculate the average pressure of a gas of many molecules using the kinetic theory, even though we do not know the motions of each individual molecule in the gas.

Maxwell proposed that molecular speeds are distributed in a way similar to the law of errors. The curve representing the magnitudes of the speeds (Fig. 22.6) is actually unsymmetrical, since there is a minimum value (zero) but no maximum. For the same reason, the rms speed (Eq.

Fig. 22.7. James Clerk Maxwell (1831–1879), British physicist, who put the kinetic theory of gases, proposed earlier by Bernoulli and Clausius, on a sound mathematical foundation and developed its applications to properties such as diffusion, viscosity, and heat conduction. His work on electromagnetic theory is discussed in Chapter 25.

22.14) is greater than the most probable speed (the higher speeds have relatively greater weight in the calculation of the root mean square).

We shall not need to use the precise mathematical formula for Maxwell's distribution[5] or provide a rigorous derivation. The crucial point to note here is simply that the collisions of molecules with each other do not tend to make all the speeds equal. You might think that if a slow molecule collides with a fast one their speeds after the collision would be closer to the average of the speeds before the collision. Some early theorists were led to think so by this line of reasoning, but it is fallacious. The fact is that even if two molecules with identical speeds collide, at angles other than head-on, they will generally have different speeds after the collision, as can be shown from an application of the laws of conservation of momentum and kinetic energy to elastic collisions.

Although Maxwell's distribution law soon became established as part of the kinetic theory, for many years no direct experimental verification seemed possible. But, exactly because a scientific problem looks most paradoxical and difficult—how can one measure the speeds of molecules, which are themselves invisible?—good minds will ponder it and frequently produce most ingenious solutions.

An especially interesting technique, developed by Otto Stern and others in the 1920s, has atoms of a metal evaporate in all directions from a hot wire of known temperature. The wire runs down the center of a close-fitting hollow cylinder that has a thin slit that can be opened and shut like a fast shutter. When the shutter is briefly uncovered, a short, thin, well-defined "squirt" of atoms escapes into the evacuated space beyond the cylinder; as they continue undisturbed in a straight line, the fast atoms outdistance the slower ones, thus spreading into a procession what started out to be a thickly packed crowd of atoms.

If now all atoms are allowed to impinge on a screen or glass slide, they will quickly condense on it and form a thin coating of metal at a well-defined spot. But in order to separate the fast atoms on the screen from the slow ones, the screen is pulled across the path of the atoms as they fall on it; thus the coating becomes a lengthy smear instead of one spot, with the fastest atoms falling at the head of the deposit, the slowest ones at the tail end.

This is rather like shooting off a load of buckshot against a moving train at the very moment when the first car passes by. The speediest pellets will indeed hit the first car, and the slowest might hit the last. From the distribution of hits along the side of the whole train, you could analyze the distribution of velocities of all the pellets.

One cannot simply and directly count how many atoms fall on any part of the moving plate, but from the geometry and known motion of the slide in this arrangement, we can calculate the

[5]According to his (now well-proved) distribution, the relative probability for a molecular speed v is proportional to $v^2e^{-3mv^2/2kT}$ where $e = 2.718$. When the exponent of e is a large negative number, corresponding to kinetic energies much greater than the average ($3kT/2$), the probability is extremely small but not zero.

Fig. 22.8. Schematic representation of apparatus for measuring the velocity distribution of gas molecules.

absolute speeds (in m/sec) needed by an atom to reach a given portion of the screen. By mounting the slide and the shutter on the inside of two concentric rotating drums with the oven at the center (see Fig. 22.8), one can accurately repeat the procedure every time that these drums make one revolution, and gather many such layers of spread-out "squirts" of atoms on the same plate. Then the relative opaqueness or the measurable *density* of the total smear tells us directly what *fraction* of atoms had what speeds.

The very first thing that a qualitative examination of such a smear reveals is the presence in the stream of atoms of all speeds, from practically zero to practically infinitely large. The percentage of atoms (that is, the relative density of the smear) at the extreme values is almost negligibly small but rises progressively as we shift our attention to the more central region of the deposit.

For silver atoms evaporated from a filament kept at a temperature several hundred degrees below the actual melting point of the metal, the maximum density of the silver coating in the original experiments fell on that portion of the slide that corresponds to a speed of about 5.5×10^3 m/sec. This speed we may call the *most probable speed,* but it is not necessarily the same as the rms speed; according to Maxwell's theoretical work, the most probable speed should be definitely lower than the rms speed (Fig. 22.6). To determine the latter, we divide the deposit on the slide into small regions and measure more exactly what fraction or percentage of the atoms had speeds of 0 to 1×10^3 m/sec, 1 to 2×10^3 m/sec, 2 to 3×10^3 m/sec, etc. We might find that for the given vapor and temperature, within the error of measurement, the following distribution of velocities exists.[6]

Velocity range in 10^3 m/sec	% of atoms in range
0–1	0.5
1–2	3
2–3	7
3–4	11
4–5	15
5–6	16.5
6–7	14
7–8	11.5
8–9	9
9–10	7
10–11	3
11–12	2
12–13	1
Above 13	0.5

This may be represented by a plot, as in Fig. 22.9. We see at once from the figure that: (a) the observed speeds are bunched strongly around

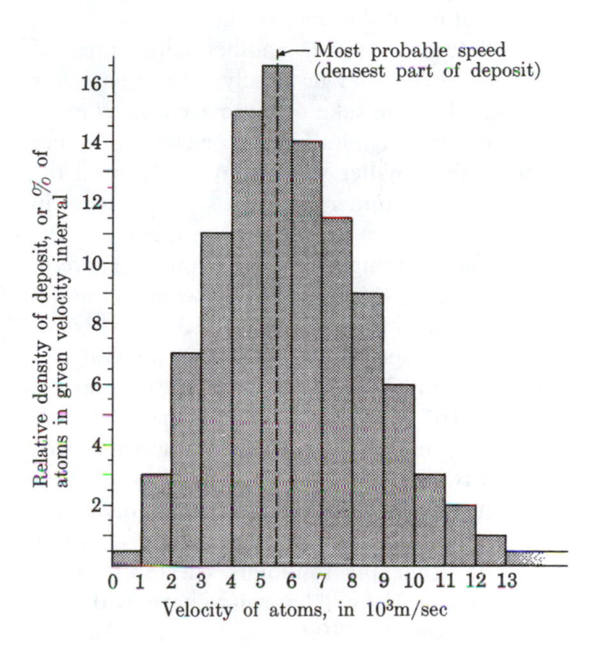

Fig. 22.9. Velocity distribution of atoms in silver vapor.

[6]These values actually apply only to truly random speeds such as exist in a gas or vapor in a closed box. The atoms in the beam described above have a slightly different distribution because the shutter arrangement automatically gives preference to one type of atoms—those with high velocity components toward the slit. The values in the table are thus really deduced indirectly from the actual density measurements, but this need not concern us here.

the maximum, the "most probable" value, dropping off to around 20% of the maximum at twice the most probable speed, and (b) that the distribution of speeds is quite lopsided toward the higher speeds. Even the straight average speed $\langle v \rangle_{av}$ is higher than the most probable speed—as you may ascertain by calculation.

In addition to confirming the shape of the velocity-distribution curve predicted by Maxwell, Stern's experiment also showed that the rms speed coincides with the theoretical value calculated from kinetic theory (see Eq. 22.12):

$$\sqrt{\langle v^2 \rangle_{av}} = \sqrt{(3kT/m_0)}.$$

Here is the final result. If we carry through the calculation of the rms speed based on these measurements (how is this to be done?), our experiment yields a value of about 6.5×10^3 m/sec; on the other hand, the calculated value of $\sqrt{\langle v^2 \rangle_{av}}$, using the derived equation for this gas and for this temperature, gives us, within the expected limits of error, the *same value!*

Before we turn to another achievement of kinetic theory, let us briefly reformulate our results. For the sake of more accurate computations, we may plot Fig. 22.9 again for thinner strips and smaller velocity intervals, until the stepwise graph merges into a smooth line as in Fig. 22.6, drawn here for nitrogen gas at 0°C.

The meaning of any point on the curve of Fig. 22.6, say *P*, is now the following: In this gas at this temperature only about 1.1% of all molecules have speeds in a 10^3 cm/sec interval centered around 2×10^4 cm/sec (that is, have speeds of 2×10^4 cm/sec $\pm$ 0.5×10^3 cm/sec). Figure 22.10 represents the speed distribution in a given gas at two different temperatures.

A useful concept arising from this discussion is *probability.* We may say that the probability is 0.011 that any randomly selected nitrogen molecule at 0°C will happen to have, within ±5 m/sec, a speed of 200 m/sec (see Fig. 22.6). As we use the words here, "probability of 0.011" means therefore that previous observations of the several speeds of many molecules make it safe to predict that of the next randomly selected 1000 molecules, about 11 will have the specified speed range. If we look at the speed of only 100 molecules, perhaps 1 will have the desired speed, but the smaller our sample becomes, the less predictable are the results. For a single molecule we cannot predict any definite speed, except perhaps to offer the platitude that the probability is 1.00 (100%) that its speed lies somewhere between zero and infinity; in fact, this whole concept of probability does not apply to such a singular situation.

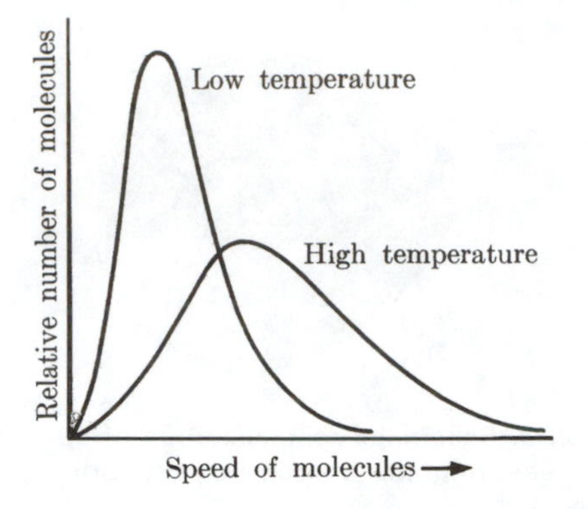

Fig. 22.10. Plot of speed distribution of the molecules of one gas at two different temperatures.

Problem 22.5. From Fig. 22.6, find the probabilities corresponding within the given range to the most probable speed, to the rms speeds, and to twice and three times the most probable speeds.

Note that although the probabilities are smaller and smaller at extreme (high and low) speeds, they are not totally zero; that is, a few precocious molecules could be found at all times (for temperatures above absolute zero) that will show enormous speeds and consequently enormous energies, whereas at any moment a few others will seem practically to stand still. Similar distribution curves apply to liquids and even to solids, and they explain why some molecules, even at low temperatures, will be able to escape ("evaporate") from the surface, as Clausius suggested in his qualitative theory (Section 22.2).

Problem 22.6. Equation (22.12) can be written as $\frac{3}{2}kT = m_0\langle v^2 \rangle_{av}$. Thus the $\langle KE_{trans} \rangle_{av}$ *per molecule* of any gas is $\frac{3}{2}kT$ and, for any gas,

$$\text{Total } KE_{trans} \textit{ per mole} = \tfrac{3}{2}kTN = \tfrac{3}{2}RT$$

(a) Calculate this energy (in joules) present in 1 mol of water vapor (superheated steam) at 200°C; also find the amount of energy given up (to the receding piston) by 9 g of expanding steam in a steam engine cylinder, if the steam cools itself during an adiabatic expansion to 100°C ("adiabatic" meaning that no heat is transferred to or from the surroundings). (b) Calculate also the amount of energy that must be removed and might be usefully available if 1 mol of ordinary air were cooled from room temperature to about 200°C, where these gases liquefy. Why don't we ordinarily utilize this energy to do useful work?

22.7 Additional Results and Verifications of the Kinetic Theory

a) An important implication of Eq. (22.12) is that the $\langle KE_{trans} \rangle_{av}$ of a perfect gas decreases linearly as the temperature (in K) is lowered toward zero; this gives physical significance to *absolute zero temperature* as the temperature at which—to put a subtle point very crudely—the random thermal motion of gas molecules has ceased. (This does not mean that *all* motion has ceased.)

b) Again on the qualitative side, one of the very important purposes of the kinetic theory was to explain why a gas in a thermally insulated container (adiabatic process) rises in (absolute) temperature in proportion to the work done on it by reducing its volume, for example, by means of a piston pushing in one side of the container. In fact, it was this type of observation that first served as a foundation for the comprehensive statement of the law of conservation of energy. The work done on the gas, or the energy given to it by the advancing piston, is equal to the increase of translational kinetic energy picked up by the gas molecules during collision with the moving piston. But by Eq. (22.12), any change in $\langle KE_{trans} \rangle_{av}$ is reflected in a proportional change in the temperature: $\Delta \langle KE_{trans} \rangle_{av} \propto \Delta T$; hence the proportional rise in temperature.

The same argument in reverse explains why an (idealized) gas cools itself when it is allowed to expand—not into a vacuum, in which case it does no work and, by experiment, suffers no total temperature drop, but when it expands against a receding piston, or the inside walls of a balloon, or against another gas at lower pressure, or some "obstacle" of that sort.

A familiar example is presented by a warm mass of air rising to a higher region, expanding against the more rarefied atmosphere, and so cooling itself to a temperature where the moisture brought along in the air is forced to condense into a rain or snow cloud.

Turning briefly again to the kinetic theory as applied to liquids, we now realize why evaporation from the surface cools the body of the liquid: since only the fastest molecules overcome the surface attraction and escape by evaporation, the average kinetic energy of the remaining ones drops correspondingly; however, a drop in $\langle KE_{trans} \rangle_{av}$ is, in fact, equivalent to a drop in T.

c) An important result of the kinetic theory, on which we can only touch briefly, was that it permitted a fair estimate of the size of individual molecules. Recall that Dalton tried to estimate the relative sizes of atoms on the basis of the mistaken model of a static gas in which the atoms touched. In the kinetic theory we have been assuming throughout that the volume of the molecules is negligible, seemingly destroying all hope of estimating sizes. However, there are several phenomena that can be treated by our model without making that assumption, specifically,

1. The pressure-volume behavior of denser gases and of vapors, where the molecular volume becomes of importance and where consequently Boyle's law breaks down.
2. The speed with which one gas diffuses through another, where evidently the size of the atoms will determine how often they collide and how easily and quickly they will pass by one another.
3. The speed with which heat can be conducted through a gas, another instance where the size of atoms will determine their progress despite collisions.
4. The measurable viscous drag that a gas exerts on a larger object moving through it, and which again brings the factor of molecular size into the calculations.

As it happened, the last phenomenon was the first to be used for a reliable estimate of molecular sizes. In his 1860 paper Maxwell, using Clausius' concept of the mean free path, derived a formula for the viscosity of a gas involving the *collision cross section*. He reasoned that a molecule of diameter d moving through a small region sparsely populated by other molecules of the same diameter will suffer collisions with a frequency or probability proportional to the number of molecules in that region, N, and to the cross-sectional area of one molecule, d^2. (We ignore numerical factors such as π for simplicity.)

So from a quantitative measurement of viscosity one can estimate the product Nd^2.

So far we don't know either N or d. But Loschmidt had the idea that one could estimate the product Nd^3 from a completely different property of the gas. If one condenses the gas to a liquid, and if one assumes that in the liquid state the molecules are nearly in contact, then the total volume of liquid occupied by N molecules is approximately Nd^3. (Imagine each molecule occupying a cube whose side is equal to the diameter of the molecule.)

Now Loschmidt had two equations for two unknowns: He had numerical values for both Nd^3 and Nd^2. He could thus divide one by the other to obtain d. His result, published in 1865, was that the diameter of an "air molecule" is about 10^{-10} m. While this is about four times as large as modern estimates, it was the first time anyone had made a *reliable* estimate of the correct *order of magnitude* of a molecular size.

Five years later Kelvin confirmed Loschmidt's estimate by using three other phenomena: the scattering of light, contact electricity in metals, and capillary action in thin films (such as soap bubbles). He argued that since we now can attach definite numbers to atomic properties such as speed and size, we must accept the atom as a legitimate scientific concept.

Evidently we must be on guard against conceiving of too literal a model for molecules as little spheres. The "diameter" may refer only to some average distance of closest approach among molecules, determined not by actual contact but perhaps by continuously changing forces (as shown in Fig. 22.3) or, more fundamentally, by quantum effects (Section 29.4). Nevertheless, Loschmidt's estimate of molecular diameters was certainly a major turning point in the long history of the atomic concept.

From Loschmidt's estimate we can also (by substituting back into an equation for Nd^2 or Nd^3) find N. The result, for a cubic centimeter of an ideal gas at atmospheric pressure and temperature 0°C, would be about 2×10^{18}. The modern value of this number, appropriately called *Loschmidt's number,* is $N_L = 2.69 \times 10^{19}/\text{cm}^3$. The corresponding value of *Avogadro's number* is found by dividing N_L by the volume of 1 mol of an ideal gas, 2.24×10^{-2} m^3, giving $N_A = 6.02 \times 10 \times 10^{23}$/mol.

Moreover, once we know Avogadro's number we can easily estimate the mass of an individual molecule (see footnote 4). For air (N_2 or O_2) it is about 5×10^{-26} kg.

d) Because it was Dalton's question of why the atmosphere's constituent gases are so thoroughly mixed that started him on his road to the atomic theory, we should surely take a brief look at the solution now given by kinetic theory to this problem. You recall that Dalton believed the mixing to be a result of a brief initial period of disequilibrium among unequally large, mutually repelling atoms and that he was wrong on several counts. The kinetic-theory picture of gases predicts that two or more gases, intermixed like the atmosphere, should diffuse one through the other by virtue of the thermal movement of the molecules. However, and quite ironically, the heavier gas (namely, oxygen) should indeed be expected to be found relatively somewhat more frequently per 100 molecules at lower levels in the atmosphere than at levels higher in the stratosphere (why?). While this would not correspond at all to the marked stratification between a heavy and a light *liquid*—say water and oil—it is nevertheless just the type of phenomenon that Dalton thought does not happen.

And there is a further twist: Whereas the expectations of the kinetic theory on that subject are fulfilled in a suitable quantity of gas under ideal conditions, our own atmosphere is so disturbed by wind currents that there is a fairly even mixing of all gases throughout the region accessible to direct and indirect measurements; so that it is after all a disequilibrium in the gas that explains the observed homogeneity, but a continuous disequilibrium of a type not looked for in the Daltonian context.

e) A startling set of observations to which the kinetic theory lends direct explanation concerns the *escape of the atmosphere* from planets and satellites. As everyone knows, the moon and the smaller planets, with their smaller gravitational pull, lost their atmospheres—if they ever had any—long ago and our earth is slowly doing the same, although we may be sweeping up enough interplanetary gas and vapor from meteorites to balance the loss. The process of escape is evident: The finite weight of the atmosphere "packs" it more densely near the earth, but the density decreases rapidly with height, in accordance with the result of Pascal's Puy de Dôme experiment (Section 19.2). The distance traveled by each molecule between collisions becomes correspondingly greater—at sea level it is only 10^{-7} m, but at a height of 1000 km it may be many miles. The paths of molecules will then be ordinary trajectories, but if their "initial speeds" are fast enough (more than 10^4 m/sec at that height above our earth) they may shoot too far out to be pulled back by the diminishing gravitational field.

One glance at Fig. 22.6 shows that for nitrogen at 0°C there are only very few molecules having such speeds. However, there are always some molecules at all speeds, and the temperatures at that height being much greater (perhaps 1000°C), the rate of escape for the common atmosphere is a little larger than would first appear, although still practically negligible.

Finally, the smaller the molecular weight, the larger the proportion of high-speed molecules at any given temperature, which means, in the extreme, that a hydrogen molecule is likely to escape from our earth only about 1000 years after it is liberated (in chemical decomposition). This may explain the apparent inconsistency between the modern nebular hypothesis (Chapter 31), which assumes that the sun and planets condensed from a cloud consisting mostly of hydrogen and helium (the present composition of the sun), and the fact that very little hydrogen is now present in the earth's atmosphere.

22.8 Specific Heats of Gases

We now come to one of the most striking and intriguing applications of the kinetic theory—the attempt to relate the internal motions of molecules to measurable thermal properties of gases. To find the specific heat of a gas (see Section 17.8), we may take 1 g of the gas, put it in a rigid, heat-insulated container, and feed it a measured amount of heat energy (in calories) until its temperature rises by 1 degree (on the centigrade or the absolute scale—in both, 1 degree corresponds to the same *interval,* the kelvin).

In the following discussion we consider only the specific heat at *constant volume,* thus none of the heat energy supplied is wasted in work done by an expansion of the gas. The difference between constant volume and constant pressure is crucial, especially for gases, and was in fact just what allowed Mayer to calculate the mechanical equivalent of heat (Section 17.8).

We could define the specific heat of the gas s_v as being equal to the calories needed per gram of gas, per kelvin of temperature rise. This would yield widely different values for different gases, 0.75 (cal/g·K) for helium, 2.41 for hydrogen, 0.15 for carbon dioxide, etc.

Instead of using 1 g, we might have used 1 mol of the gas, defining thereby another but related concept, the *molar* specific heat, C_v, that is, the number of (calories per *mole*) per degree (abbreviated cal/mol·K). C_v would correspondingly differ from the values of s_v; evidently $C_v = s_v \times$ number of grams per mole of gas $= s_v \times$ relative molecular weight of the gas.

In either case, our model of the gas tells us that for every calorie we supply, the total KE_{trans} of the molecules will rise by 4.18 joules (law of conservation of energy); for Q calories of heat energy supplied, KE_{trans} increases by $\Delta KE_{trans} = QJ$ joules (where J, printed in italics, is 4.18 joules/calorie, the mechanical equivalent of heat—not to be confused with J, the abbreviation for joule, the unit of energy).

Consider 1 mol of gas, about to receive Q calories. By Eq. (22.12), the total KE_{trans} per mole of this gas $= \frac{3}{2}RT$. Since the temperature and the total kinetic energy are linearly proportional, $KE_{trans} = \frac{3}{2}R\Delta T$; let us write this $\Delta KE_{trans}/\Delta T = \frac{3}{2}R$. From the law of conservation of energy, $\Delta KE_{trans} = Q \times J$; thus $QJ/\Delta T = \frac{3}{2}R$, and therefore we write[7]

$$\frac{Q}{\Delta T} = 3\frac{R}{2J} \qquad (22.15)$$

The left-hand side is the heat energy (cal) supplied to this 1 mol/deg rise in temperature but that is exactly the definition of C_v for a gas:

$$C_v = 3\frac{R}{2J} \qquad (22.16)$$

We note with astonishment that by our theory the molar specific heat of all gases should be the same, R and J being universal constants! Specifically, substituting values for R and J, we get

---

[7]Here some modern textbooks, carrying to an extreme the insistence on using only SI units, present the equation as $Q/\Delta T = 3R/2$, where the heat Q is no longer measured in calories but in joules; in other words the mechanical equivalent of heat J has been absorbed into the *definition* of heat! (The numerical values of molar specific heats of monatomic gases are then about 12.5 instead of 3.) This is one example of how science tends to obliterate its own past, forgetting the long struggle to understand the relation between heat and mechanical energy.

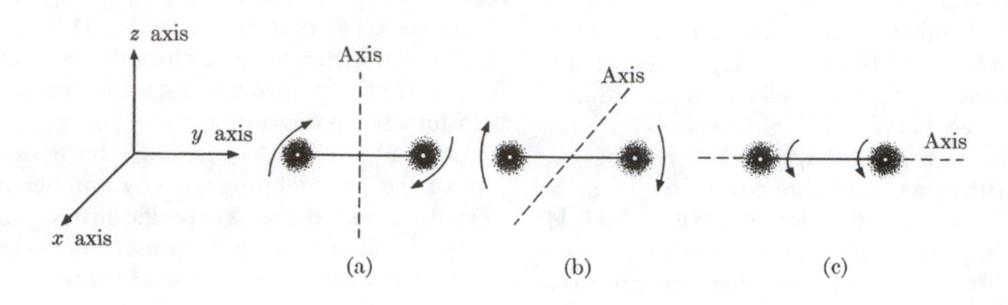

Fig. 22.11. Three independent ways in which a diatomic molecule can rotate.

$$C_v = (3 \times 8.31)/(2 \times 4.18) = 3 \text{ cal/mol·K}$$
(very nearly)

Can this really be true? Indeed we find by experiment that for helium, argon, and mercury vapor, C_v is 3.02, 2.95, and 2.94, respectively, and the values for the other monatomic gases also confirm our theory within tolerable limits.

But only the *monatomic* gases! Diatomic gases such as hydrogen, nitrogen, oxygen, carbon monoxide, and hydrogen chloride have experimental values of C_v near 5 (cal/mol·K) instead of the predicted value of 3. For these it is as though our theory must somehow be modified to yield

$$C_v \text{ (diatomic gases)} = \frac{5R}{2J}.$$

In short, diatomic molecules seem to have a way of absorbing (receiving and storing) more energy per degree of temperature rise than do monatomic molecules. Every calorie supplied to molecules made of only a single atom each did turn up as kinetic energy of translation, as expected; but every calorie supplied to a diatomic molecule seems to split five ways—only three parts go into KE_{trans} and occasion the observed temperature rise, and the remaining two parts somehow are stored away without affecting the KE_{trans} or the temperature. What is the physical explanation of this?

This is simply a breakdown of our model. Joule, who brought his work to this point, was stumped by the discrepancy in the specific heats of polyatomic gases, the more so as in his day no monatomic gases had yet been generally identified. In fact, his initial thought was that the experimental values of C_v then available might be in error; and when new determinations only widened the gap between prediction and experimental results, he seems to have been so discouraged as to leave the whole subject of kinetic theory. He had not made any provision for a molecule to receive any energy other than translational kinetic energy, and up until then there was no reason to revise this simple idea.

Now, however, we search for a plausible mechanism to account for the unsuspected heat capacity of diatomic molecules. And it might occur to us (this was done first by Clausius in 1857) that even if monatomic gases seem to fill the role of agitated point masses or small spheres without *rotational* energy, diatomic molecules might well be thought of as dumbbell-like objects, as two small spherical masses a short distance apart. This would be an arrangement with a much larger rotational inertia than either of the individual "point masses," and which therefore would be much more capable of being put into rotation and of containing a significant amount of rotational kinetic energy ($\frac{1}{2}I\omega^2$) at plausible speeds of rotation.

Figure 22.11 shows a very crude schematic presentation of a diatomic molecule and its three possible types of rotation, distinguished from one another by the positions of the axis of rotation—analogous to our method of differentiating between the three possible types of translational motions, that is, motion along the x, y, or z axis.

Problem 22.7. Compare the rotational inertia of a hydrogen molecule (H_2) conceived (a) as a small sphere, as in our first model, and (b) as a dumbbell molecule with two equal (spherical) hydrogen atoms about one atomic diameter apart. [*Note: I* depends on the axis of rotation (see Fig. 22.11). A dumbbell molecule has a different value in (a) than in (c).] Numerical values for *I* are obtainable if we make the assumption, based on other experiments, that the diameter of a single H atom is *about* 10^{-10} m, and that all but about $^1/_{2000}$ part of the total mass of the atom is concentrated at its center, the nucleus, whose radius is perhaps a 10^{-4} part of the atom's total radius.

It is very convenient to invent the term *degree of freedom of motion* to denote each one of the independent ways in which a molecule can receive energy. For example, a monatomic gas molecule that does *not* rotate, as we initially pictured it, can acquire kinetic energy only by virtue of its velocity along the x axis or y axis or z axis. These three motions are independent components, in the sense that we could imagine that one component is increased or decreased without the others being affected. In other words, the molecule has three degrees of freedom. If, however, we consider a particle made up of two or more atoms, capable of rotational as well as translational motion, it has three degrees of freedom for translation plus three more for the three possible independent ways of rotating (about the three different axes shown in Fig. 22.11), or six degrees of freedom altogether.

However, we must be prepared to face the possibility that a molecule for some reason may not make use of all the conceivable degrees of freedom to which it is entitled by simple geometry. Thus we have already implicitly denied a monatomic molecule the use of its own three geometrically quite conceivable degrees of freedom of rotation or spin by saying that its rotational inertia is too small to make them effective, or that it is a point mass. As it happens, this is a rather dubious trick, to be reexamined later, and is, in fact, a veiled way of saying that in order to explain the experimental facts of specific heat in terms of our model we need not and must not allow rotation for monatomic gas molecules in our model. Similarly, if rotation is neglected for a single atom, it should also be neglected for that one degree of freedom of rotation of a diatomic molecule that corresponds to motion about the lengthwise axis (Fig. 22.11c); therefore molecules of H_2, N_2, O_2, CO, HCl, etc., should have only five degrees of freedom (three for translation, two for rotation), or two degrees more than monatomic molecules.

Here, then, we have an explanation of why a mole of diatomic gas absorbs more energy per degree of temperature rise than does a mole of monatomic gas. Of any and every unit of heat energy supplied to the former, only three-fifths goes into translational motion [ΔKE_{trans}, which by Eq. (22.12) causes a proportional rise in temperature], and the remaining two-fifths of the supplied energy is stored in the two active modes of rotation ($\Delta KE_{rotation}$, which does not directly influence the rise in temperature). Therefore, to raise the temperature of 1 mol of diatomic gas by 1 degree, we must give it as much energy as that required for 1 mol of monatomic gas *plus* almost as much again (2/5} of the total) to satisfy the two active rotational degrees of freedom. So we understand that C_v (monatomic gas) = $3(R/2J)$, whereas C_v (diatomic gas) = $5(R/2J)$.

Reinspection of these two equations now reveals some interesting facts. If we were dealing with polyatomic gases whose molecules were so complex as to have significant rotational inertia about all three axes of symmetry, we should expect the third degree of freedom of rotation also to play its part. With a total of six active degrees of freedom, we expect C_v (polyatomic) = $2 \times C_v$ (monatomic) = $6(R/2J)$, or about 6(cal/mol·K).

Many triatomic gases *do* have values of C_v of about 6 (5.9 for CO_2 at low temperatures, 6.1 for water vapor at 100°C). Other polyatomic gases have even higher values, indicating that further hitherto neglected degrees of freedom begin to be important, namely, those representing *vibration* between the components of the molecule. For example, the complex molecules of ethyl ether ($C_4H_{10}O$) have a molar specific heat C_v of about 30.

Thus the idea of degrees of freedom of motion superposed on our initial model of the gas allows us to interpret the experimental facts of specific heat very satisfactorily. Within reasonable limits, we may write

$$C_v = \delta(R/2J), \qquad (22.17)$$

where δ is the number of active degrees of freedom of motion for the specified gas molecules.

This notion can be expanded to explain the observed specific heats of solids in terms of kinetic theory. But note now the astonishing fact that the specific heat of a gas has just given us a strong clue as to the molecular structure of gases, the very information Avogadro had lacked when trying to protect his theory from the worst doubts. For example, the fact that C_v for hydrogen gas is ordinarily 5(cal/mol·K) shows that it must be diatomic—not monatomic as Dalton had thought. Similarly, C_v for mercury vapor is about

3(cal/mol·K)—clear evidence for its monatomic structure, even though other (chemical) evidence may be ambiguous or difficult to obtain.

Implied in the last few paragraphs was an important idea, the *equipartition of energy* among all active degrees of freedom in a gas; we have assumed that any and every unit of heat energy supplied to a mole of gas is distributed in equally large shares to each active degree of freedom. In terms of individual molecules this means that at a given temperature each gas molecule has, on the average, energy in amount equal to $\frac{1}{2}kT$ for each of its active degrees of freedom (about 1.88 $\times 10^{-21}$ J at 0°C). Hence a typical average monatomic molecule (with $\delta = 3$) possesses a total energy of $\frac{3}{2}kT$ per molecule (translational energy only, and thus equal to $\frac{1}{2}m\langle v^2\rangle_{av}$), diatomic ones each have $\frac{5}{2}kT$, polyatomic ones have even more. If any external energy ΔE is supplied to the gas, each active degree of freedom absorbs an equal share of ΔE.

This principle of equipartition, which we have introduced to explain the discrepancy in specific heats, is derivable more rigorously from the experimental facts, and is of foremost importance in the study of gases.

Finally, we should examine the points where our kinetic theory frankly breaks down. We expected such a possibility from the beginning, because our gas model was admittedly very simplified, and the introduction of the idea of degrees of freedom and of equipartition of energy may not have had enough sophistication in the long run. And not only did we expect the inevitable deficiency that appears sooner or later in every theory, but we eagerly looked for it as a clue to the necessary modifications and perhaps as yet another key to the strange laws governing the world of atoms. One might say one-third of the excitement in physics research lies in constructing a theory, another third in confirming the theory, and yet another in pushing beyond the theory when some of its consequences are refuted by experiment.

Maxwell, who regarded all these attempts to explain the specific heat anomalies as unacceptable *ad hoc* fiddling with the laws of mechanics, insisted that in a consistent kinetic-theory model, *all* degrees of freedom must be equally active. His view was that in theoretical physics one must maintain a high standard of rigor, even if this "leaves us no escape from the terrible generality" of the results. That does not mean one should simply abandon the kinetic theory because it fails to provide an acceptable explanation of all the phenomena; one can accept a theory as the best now available while recognizing oneself to be in a "state of thoroughly conscious ignorance, which is the prelude to every real advance in knowledge" (Maxwell, "The Kinetic Theory of Gases," 1877).

The outstanding defect in our theory can be summarized in the following paradox: The principle of equipartition of energy requires that equal amounts of energy exist at a given temperature in each degree of freedom—but we want to make an exception for those (the inactive) degrees of freedom where *no* energy at all resides, for example, in the geometrically possible rotational degree for diatomic molecules about the lengthwise axis. (Strictly speaking, degrees of freedom are inactive if their energy is not *changed* by changes in temperature.) We are forced to ask, *exactly what determines whether a degree of freedom should be active or inactive?*

This question becomes even more intriguing when we note a discovery (made in 1912) concerning the amazing behavior of hydrogen gas,

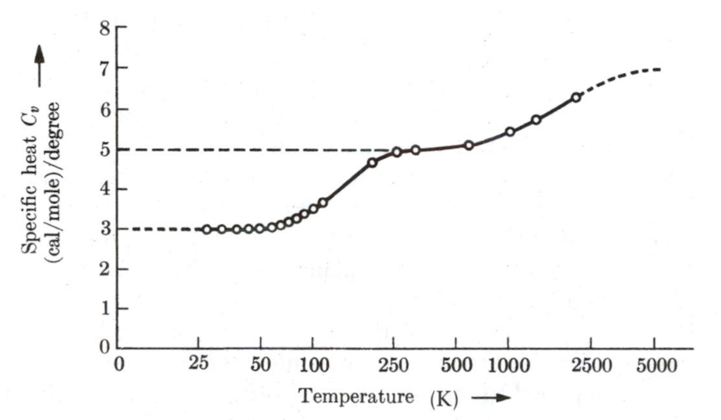

Fig. 22.12. Variation of specific heat C_v of molecular hydrogen gas. (To condense the graph, the temperature of the gas is plotted along the abscissa on a logarithmic scale.)

H_2. At ordinary temperatures its value of C_v is a little below 5(cal/mol·K), as becomes a diatomic gas. But at lower temperatures, C_v for this gas falls to about 3(cal/mol·K) (at about 100 K), the value appropriate for a monatomic gas (see Fig. 22.12)!

We must conclude that the two rotational degrees of freedom have "frozen in." Nothing in our theory has prepared us for such shocking behavior.[8] And what is worse, when H_2 gas or others such as O_2 are heated to extremely high temperatures, C_v increases to over 5(cal/mol·K) (at about 2000°C), as though other degrees of freedom begin to play a part not usual for diatomic molecules at ordinary temperatures.

The fact of the matter is that here we come into entirely new and unexpected territory; this will be our link between the "classical" concept of the atom as formed in kinetic theory and the "modern" concept that the next part of the text will develop. But we may indeed rejoice in how well our simple mechanical model now seems to serve us in so many contexts (with that exception of specific heats). So successful was the "billiard-ball" picture of the atom that it seemed, to most scientists except Maxwell, that any revision should prove only minor. That hope, however, was in vain—the revision was accompanied by a fundamental revolution in physics itself. This is the topic to which we turn in later chapters.

22.9 The Problem of Irreversibility in the Kinetic Theory: Maxwell's Demon

In Chapter 18 we discussed the tendency toward irreversible heat flow from hot to cold in connection with the second law of thermodynamics. In deriving the general gas law from the kinetic theory in this chapter we implicitly assumed a similar tendency toward equalization of temperature of two portions of gas allowed to mix or of the wall of the container and the gas molecules inside. Hence we accepted the observed fact that in these cases phenomena were not "symmetrical," in the sense that when a movie is taken of the temperature readings on the two thermometers, it corresponds to physically realizable phenomena only when run forward and not when run backward. Yet in the rest of our discussion of kinetic theory we have accepted the validity of Newtonian mechanics, which implies complete reversibility. That is, we assumed symmetry between positive and negative time directions in physical processes in the sense that a movie taken of the perfectly elastic collision of two billiard balls shows physically realizable phenomena whether it is run forward or backward.

This apparent inconsistency between the basic principles of mechanics, which we have assumed to apply to the molecular level, on the one hand, and the undoubted fact that observable natural processes are irreversible, on the other, is called the *reversibility paradox*. It was pointed out forcefully by Kelvin in 1874, in the paper from which we quoted at the beginning of Chapter 18, and again by Josef Loschmidt in 1876.

A fairly satisfactory solution of the reversibility paradox was proposed by Maxwell, Kelvin, and (much more elaborately) by Ludwig Boltzmann. It involves two recognitions: the surprising admission that the principle of dissipation of energy (second law of thermodynamics) is not an inviolable law of physics—it may occasionally, and for short time spans, be violated—and that in most ordinary circumstances the probability of such violations is extremely small.

Maxwell proposed a famous thought experiment to show how the second law of thermodynamics could be violated by an imaginary person who could observe individual molecules and, with virtually no expenditure of energy, could sort them out, thereby causing heat to flow from hot to cold. Suppose a container of gas is divided into two parts, *A* and *B*, by a diaphragm (solid partition). Initially the gas in *A* is hotter than the gas in *B* (see Fig. 22.13a, where we have indicated the speed by the length of the arrow). This means that the molecules in *A* have greater average speeds than those in *B*. However, since the speeds are distributed according to Maxwell's distribution law, a few molecules in *A* have speeds less than the average speed in *B*, and a few molecules in *B* have speeds greater than the average speed in *A*.

Maxwell pointed out that with such an arrangement, and without expending energy,

[8] If, on cooling, the two nuclei of H_2 came very close together, we might perhaps accept this as an explanation; but the molecular dimensions change only very little during cooling, as we find from other lines of investigation.

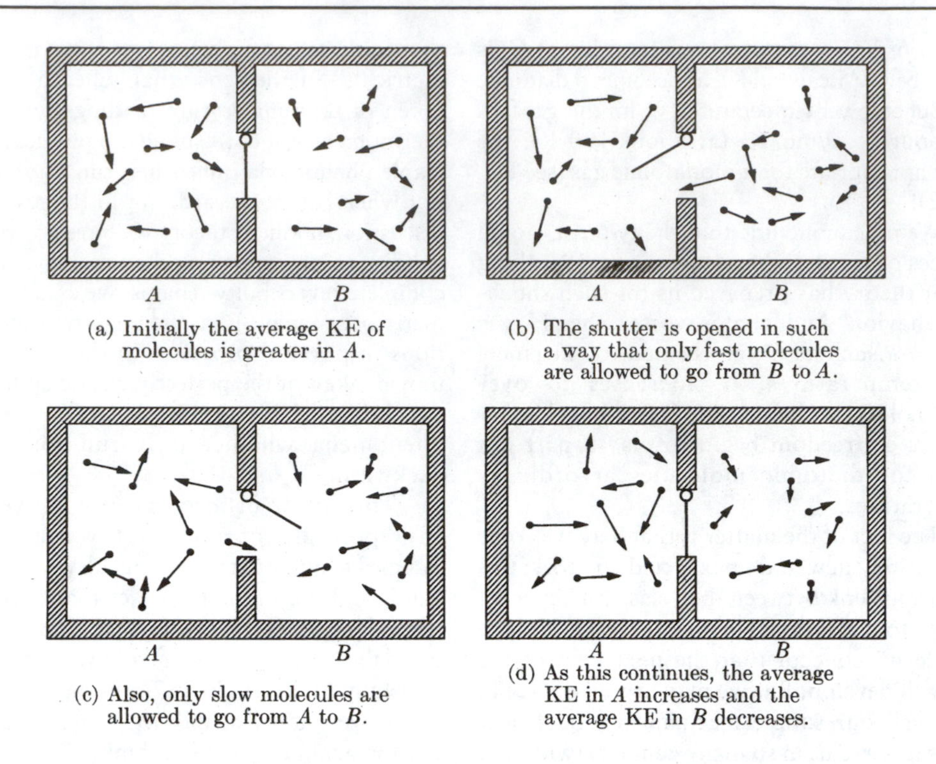

(a) Initially the average KE of molecules is greater in A.

(b) The shutter is opened in such a way that only fast molecules are allowed to go from B to A.

(c) Also, only slow molecules are allowed to go from A to B.

(d) As this continues, the average KE in A increases and the average KE in B decreases.

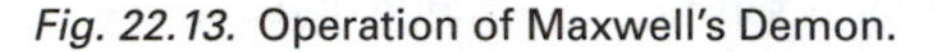

Fig. 22.13. Operation of Maxwell's Demon.

there would be a possibility of making heat flow "uphill" from a cold gas (from side *B* to side *A*), using this overlapping of the distributions for gases at different temperatures. "Now conceive a finite being," Maxwell suggested, "who knows the paths and velocities of all the molecules by simple inspection but who can do no work except to open and close a hole in the diaphragm by means of a slide without mass." (If the slide or shutter has no mass, no work will be needed to move it.) Let this "finite being"—who has come to be known as *Maxwell's Demon*—observe the molecules in *B*, and when he sees one coming whose speed is greater than the average speed of the molecules in *A*, let him open the hole and let it go into *A* (Fig. 22.13b). Now the average speed of the molecules in *A* will be even higher than it was before, while the average speed in *B* will be even lower. Similarly, let him watch for a molecule of *A* whose speed is less than the average speed in *B*, and when it comes to the hole let him draw the slide and let it go into *B* (Fig. 22.13c). This will have the same effect, that is, it will increase the average speed in *A* and reduce that in *B*. The net result is to collect more fast molecules in *A* and more slow molecules in *B* (Fig. 22.13d).

Maxwell concludes:

> Then the number of molecules in *A* and *B* are the same as at first, but the energy in *A* is increased and that in *B* diminished, that is, the hot system has got hotter and the cold colder and yet no work has been done, only the intelligence of a very observant and neat-fingered being has been employed.
>
> (Letter to P. G. Tait, 1867)

Thus Maxwell's thought experiment shows that if there were any way to sort out individual molecules by inspection, the principle of dissipation of energy could be violated. (Subsequent discussion in the light of modern physics has shown that Maxwell's Demon is more handicapped than Maxwell believed, and so it is much harder, if not impossible, for the Demon to violate the principle of energy dissipation.)

The fact that Maxwell's Demon operates by creating *order* on the molecular level—putting the fast molecules in one box and the slow ones in another—suggests conversely that the usual processes of dissipation of energy involve increasing the amount of *disorder* or *randomness* of a

Fig. 22.14. Ludwig Boltzmann (1844–1906), Austrian physicist who derived an equation governing the change in the velocity distribution function for a gas; his H theorem, based on that equation, provides a molecular version of the second law of thermodynamics. In attempting to explain the apparent inconsistency between irreversibility and Newtonian mechanics, he established a mathematical connection between entropy and disorder, implying that the second law corresponds (under certain conditions) to an increase in disorder.

system. Boltzmann made this idea explicit by developing a precise definition of entropy in terms of the probabilities of molecular arrangements. He found that if the entropy of a state is defined as

$$S = k \log W,$$

where W is the number of possible molecular arrangements corresponding to the state and "log" means natural logarithm, then the average entropy has the same properties as the entropy defined by Clausius (see Section 18.4). Disordered states of a system are those that can have a large number of molecular arrangements, so that W and log W are both large, as in an ordinary gas; ordered states can have only a small number of possible arrangements (for example, for a solid at absolute zero temperature, if there were only one possible molecular arrangement, $W = 1$, the entropy would be zero).

Boltzmann argued that if one were to list all the possible arrangements the molecules of a gas in a container can assume (for example, by taking many "snapshots"), nearly every arrangement would have to be considered "disordered." In only a few cases, for example, would all the molecules be in one corner of an otherwise empty container, or would all the fast molecules be on one side and all the slow ones on the other side.

It is to be expected that if we start from an ordered arrangement of molecules in a gas, the arrangement will in time become less ordered; random arrangements have the advantage of high probability. Similarly, if we put a hot body in contact with a cold one, it is almost certain that after a short time both will have nearly the same temperature, simply because there are many more possible arrangements in which fast and slow molecules are mixed together, than arrangements in which most of the fast molecules are in one place and most of the slow molecules are in another. But if we start from a disordered state, we will almost always end up in another disordered state after any specified time interval; it is extremely unlikely that we would just happen to evolve to an ordered state.

To illustrate Boltzmann's argument, consider what happens if one is shuffling and then dealing a pack of cards. If there were only two cards, A and B (analogous to two colliding particles), one of all possible arrangements (namely either AB or BA) is as likely to occur as the other. But if we use many cards, for example fifty-two cards to be dealt thirteen each to four players, the case looks different; most of the possible arrangements of the cards that occur are more or less disordered, and ordered arrangements come only very rarely out of a shuffled deck. If we *start* with an ordered arrangement—for example, the cards sorted by rank or suit—then shuffling would almost certainly lead to a more disordered arrangement. High probability is on the side of randomness. It does occasionally happen, and may even be likely if the game continues long enough, that a player is dealt thirteen hearts, even if one uses a well-shuffled deck each time. But it is *almost* certain that this will not happen in any particular deal in real life.

According to Boltzmann's view, it is almost certain that disorder (entropy) will increase in any physical process among molecules. The second law is therefore a statistical one that applies to collections of many molecules, and it has no meaning when applied to one individual molecule. Since it is a statistical law, however, there is a remote possibility that a noticeably large fluctuation may occur in which energy is concentrated rather than dissipated.

For example, the molecules in a glass of water are usually moving randomly in all directions; but it is not entirely impossible that at one moment they might all just happen to move in the same direction, upward. The water would then jump out of the glass. (If you held the glass in your hand at that moment you would expect that the glass would move downward at the same time, since momentum still must be conserved.) In such a case a disordered motion has spontaneously turned into an ordered motion; entropy has decreased instead of increased, and the second law of thermodynamics (if it is regarded as an absolute law of physics) has been violated. Such large fluctuations seem extremely unlikely; yet if they can occur at all we must recognize that the second law has limits that do not appear to be shared by the other fundamental laws of physics we have met with so far.

22.10 The Recurrence Paradox

In their attempts to prove the stability of the "Newtonian world machine," eighteenth-century mathematical physicists had arrived at the conclusion that the perturbations of planetary orbits owing to the gravitational attractions of other planets would not produce a continual change in the same direction on any one planet, but would have a cyclic effect. Thus the earth would not, on the one hand, gradually get closer to the sun over a long period of time, or, on the other hand, gradually get farther away; instead, its orbit would oscillate between definite inner and outer limits (Section 11.6). This result had been used by Playfair to justify Hutton's geological theories (Section 18.2), but in the 1860s Kelvin cast some doubts on it by pointing out that the calculations were only approximate.

In 1889 the French mathematician Henri Poincaré attacked the problem again using more accurate methods, and proved that *any* mechanical system subject to Newton's laws, whether the forces are gravitational or not, must be cyclic in the sense that it will eventually return to any initial configuration (set of positions and velocities of all the particles). However, the proof was subject to the condition that the system is restricted to a finite space as well as having fixed total energy.

Poincaré recognized that his *recurrence theorem,* if applied on the molecular level, would contradict the principle of dissipation of energy, since a gas enclosed in a container would eventually return to any initial ordered state if the recurrence theorem held on this level. Also, on the cosmic level, the theorem implied that while the universe might undergo a heat death as all temperature differences disappeared, it would ultimately come alive again. Poincaré wrote:

> . . . a bounded world, governed only by the laws of mechanics, will always pass through a state very close to its initial state. On the other hand, according to accepted experimental laws (if one attributes absolute validity to them, and if one is willing to press their consequences to the extreme), the universe tends toward a certain final state, from which it will never depart. In this final state, which will be a kind of death, all bodies will be at rest at the same temperature. . . .
>
> . . . The kinetic theories can extricate themselves from this contradiction. The world, according to them, tends at first toward a state where it remains for a long time without apparent change; and this is consistent with experience; but it does not remain that way forever; . . . it merely stays there for an enormously long time, a time which is longer the more numerous are the molecules. This state will not be the final death of the universe, but a sort of slumber, from which it will awake after millions of centuries.
>
> According to this theory, to see heat pass from a cold body to a warm one, it will not be necessary to have the acute vision, the intelligence, and the dexterity of Maxwell's Demon; it will suffice to have a little patience.
>
> ("Mechanism and Experience," 1893)

Though Poincaré was willing to accept the possibility of a violation of the second law after a very long time, others were less tolerant. In 1896, the German mathematician Ernst Zermelo (at that time a student of Max Planck) published a paper attacking not only the kinetic theory but the mechanistic conception of the world in general, on the grounds that it contradicted the second law of thermodynamics. Boltzmann replied, repeating his earlier explanations of the statistical nature of irreversibility and pointing out that the recurrences predicted by Poincaré's theorem would occur so far in the distant future that they could have no practical effect on the applicability of the second law to ordinary processes.

When these arguments failed to satisfy Zermelo, Boltzmann (half seriously) proposed the following hypothesis: The history of the universe is really cyclic, so that the energy of all its molecules must eventually be reconcentrated in order to start the next cycle. During this process of reconcentration, all natural processes will go backward, as described in the passage by Kelvin (Chapter 18). However, the human sense of time depends on natural processes going on in our own brains. If these processes are reversed, our sense of time will also be reversed. Therefore we could never actually observe "time going backward" since we would be going backward along with time!

In the dispute between Boltzmann and Zermelo it turned out that both sides were partly right and partly wrong. Zermelo and other critics of atomic theory such as Ernst Mach were correct in their belief that a complete microscopic description of matter cannot be based only on Newton's laws of mechanics. Gases are not collections of little billiard balls without internal structure, as Maxwell himself had concluded as a result of the specific heats discrepancy. But Boltzmann was right in his belief in the usefulness of the molecular model; the kinetic theory is correct except for those properties involving the detailed structure of atoms and molecules.

In 1905, Albert Einstein pointed out that the fluctuations predicted by kinetic theory need not be so rare as both Poincaré and Boltzmann had thought; they should be able to produce an effect that can be observed and measured quantitatively: in what had long been known as the *Brownian movement* of small particles suspended in fluids.

Subsequent studies inspired by Einstein's theory, in particular by the French physicist Jean Perrin, verified the quantitative predictions deduced from kinetic theory, and at the same time provided another method for determining molecular sizes. Perrin was successful in persuading the skeptics like Wilhelm Ostwald to accept the atomic theory as a useful working hypothesis—"molecular reality" as he called it—and thus in a certain sense it is Perrin who "discovered" the atom in modern physical science.

This new success reinforced the emerging statistical view of nature, but left open the question of whether the basic laws of nature are irreversible at the atomic level. This question is still a subject of lively interest among physicists today.

RECOMMENDED FOR FURTHER READING

H. A. Boorse and L. Motz (editors), *The World of the Atom*, extracts from the writings of Herapath, Brown, Waterston, Joule, and Maxwell, pages 195–274

E. Broda, *Ludwig Boltzmann: Man, Physicist, Philosopher,* Woodbridge, CT: Ox Bow Press, 1980

S. G. Brush, *Statistical Physics and the Atomic Theory of Matter,* Chapters I and II

S. G. Brush (editor), *Kinetic Theory,* Vol. 1, papers by Clausius and Maxwell, pages 111–171; Vol. 2, papers by William Thomson (Lord Kelvin), Boltzmann, Poincaré and Zermelo, pages 188–245

T. G. Cowling, *Molecules in Motion,* London: Hutchinson, 1950. A comprehensive but elementary exposition of kinetic theory.

William R. Everdell, *The First Moderns,* Chapter 4

C. W. F. Everitt, *James Clerk Maxwell*

C. C. Gillispie, *Dictionary of Scientific Biography,* articles by S. G. Brush on Boltzmann and Herapath, Vol. 2, pages 260–268 and Vol. 6, pages 292–293; by E. E. Daub on Clausius, Vol. 3, pages 303–311

Ian Hacking, "Probability and determinism 1650–1900," in *Companion* (edited by Olby), pages 690–701

P. M. Harman, *Energy, Force and Matter,* Chapter V

W. W. Porterfield and W. Kruse, "Loschmidt and the discovery of the small," *Journal of Chemical Education,* Vol. 72, pages 870–875 (1995)

C. J. Schneer, *Mind and Matter,* Chapter 12

Hans Christian von Baeyer, *Maxwell's Demon: Why Warmth Disperses and Time Passes,* New York: Random House, 1998

David B. Wilson, "Kinetic atom," in *History of Physics* (edited by S. G. Brush), pages 105–110

Refer to the website for this book, www.ipst.umd.edu/Faculty/brush/physicsbibliography.htm, for Sources, Interpretations, and Reference Works and for Answers to Selected Numerical Problems.

PART G

Light and Electromagnetism

23 The Wave Theory of Light

24 Electrostatics

25 Electromagnetism, X-Rays, and Electrons

26 The Quantum Theory of Light

Max Planck (1858–1947)

To most scientists of the late nineteenth century, the physical world, despite all its superficial complexity, appeared to be endowed with a truly superb unity. The mechanics of Galileo and Newton explained the motion of all bodies—whether everyday objects or planets. Furthermore, the kinetic-molecular theory had triumphantly extended the reign of this mechanics into submicroscopic worlds by interpreting most phenomena of heat as the motion of atoms. Sound, too, had become a branch of mechanics, having been analyzed as a vibrating disturbance traveling among the molecules, according to those same classical laws of motion.

Earlier there had been hope that the phenomena of light, electricity, and magnetism could also be explained mechanically, either by postulating special kinds of particles with appropriate force laws, or as influences propagated through space by the vibrations of an ethereal medium. But as theorists shifted from the former to the latter type of explanation, they found that while their equations became more and more accurate in describing and predicting the phenomena, the physical basis of those equations was becoming less and less comprehensible. The great triumph of Maxwell's electromagnetic theory of light, which finally unified an amazing range of phenomena previously considered unrelated (if they were known at all), was achieved only at the cost of erecting a mechanical model so complex and artificial that it was unanimously discarded as implausible.

What we have to describe in this Part, then, is the growth of a dinosaur that gobbles up so much food that it finally collapses under its own weight. But that is not quite an accurate metaphor; as we shall see in Part H, the skeleton of electromagnetic theory is still strong enough to provide the backbone for modern quantum mechanics and relativity.

CHAPTER 23

The Wave Theory of Light

In this chapter we shall not attempt a comprehensive survey of optics, but aim rather to introduce the chief ideas and experiments that shaped the development of the wave theory of light. The main theme will be a question that was long debated in the history of the development of optics: *Do the laws of mechanics, presented earlier in this book, apply also to the phenomena of light?*

23.1 Theories of Refraction and the Speed of Light

We begin with René Descartes' theory of refraction, not because it is now considered correct but because, as the first quantitative model to be published, it had a considerable influence on the ideas of Newton and other scientists. Descartes begins his *Optics* (1637) by asserting that the propagation of light is similar to the transmission of a mechanical impulse through a stick that a blind man uses to guide himself, poking it at various objects. (This is a traditional analogy, which can be found in scientific writings as early as the sixth century A.D.) Light is thus associated with motion in a medium, more precisely with a disturbance propagated quickly by mechanical means from one place to another (rather than, say, a bulletlike stream of

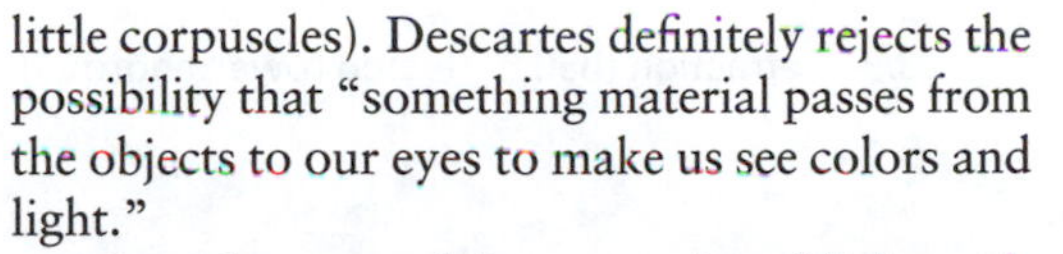

little corpuscles). Descartes definitely rejects the possibility that "something material passes from the objects to our eyes to make us see colors and light."

In order to explain properties of light such as reflection and refraction, Descartes proposes what now seems to be a rather artificial mechanical model. He compares the behavior of light when it strikes a surface (for example, the interface between two media such as water and glass) to that of a tennis ball. If a tennis ball strikes a perfectly flat hard surface it is reflected elastically in such a way that the angle of incidence equals the angle of reflection (Fig. 23.1). (You can convince yourself that this result follows from the law of conservation of momentum, assuming that the force acting on the object during the collision is perpendicular to the surface.)

To find an explanation for the refraction of light, Descartes now considers the case in which a tennis ball hits a cloth, "which is so weakly and loosely woven that this ball has the force to rupture it and to pass completely through it, while losing only a part of its speed." If the component of the ball's velocity perpendicular to the cloth is diminished but the component parallel to the cloth remains unchanged, the ball is deflected *away from the normal* (see Fig. 23.2).

At first it appears that Descartes is going to identify this behavior with that of a light ray

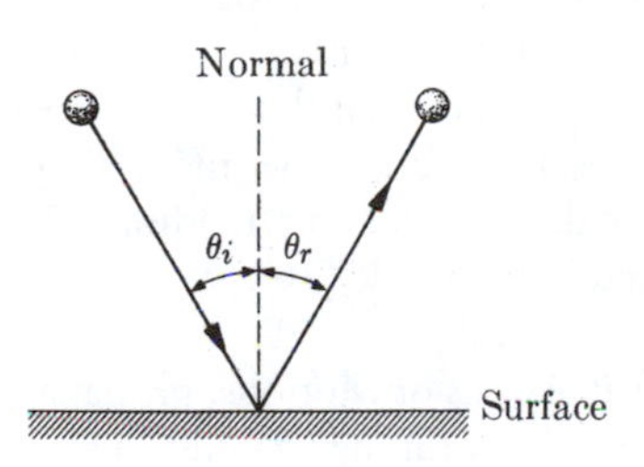

Fig. 23.1. "Reflection" of a ball showing equal angles of incidence and reflection.

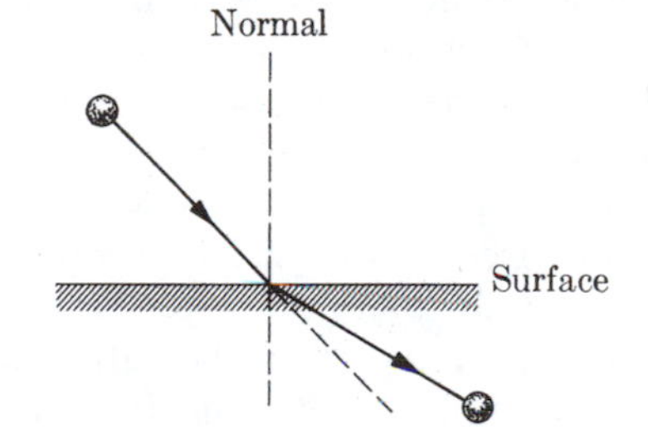

Fig. 23.2. Refraction (ball deflected away from normal).

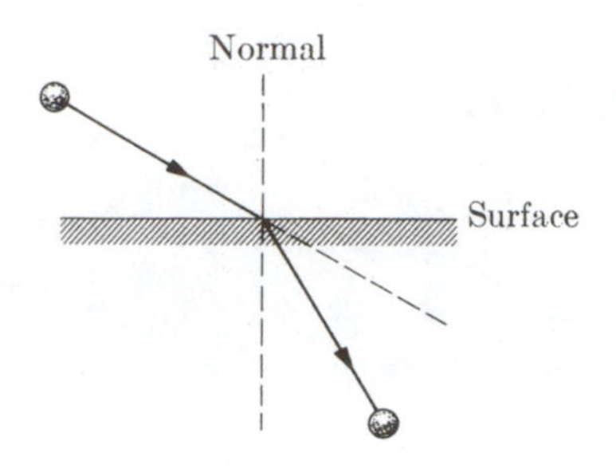

Fig. 23.3. Refraction (ball deflected toward normal).

going from a rare to a denser medium, for example from air into water, the natural assumption being that the ray is slowed down as it crosses such an interface. But, as Descartes is well aware, that would lead to error, for it is well known from common experience that when light rays pass from air into water they are refracted *toward the normal.*

The other possibility that Descartes considers is that the tennis ball, when it arrives at the interface, is hit by a racket in the direction perpendicular to the surface, increasing this component of its velocity but leaving the parallel component unchanged. The path is then bent *toward the normal.* This is in accord with the observed behavior of light rays going from air into water (see Fig. 23.3); of course it implies that light travels faster in water than in air, but that could not be tested until about 200 years later.

The major success of Descartes' model is that, once its premises are accepted, it accounts for the quantitative law of refraction discovered by the Dutch mathematician Willebrord Snell, around 1620. The law states that the sine of the angle of incidence is proportional to the sine of the angle of refraction (both angles being measured from the normal). Descartes was actually the first to publish the law; it is not known whether he learned it from Snell, to whom he makes no reference.

But Descartes must now accept the consequence of his model that light rays travel faster in dense media than in rare ones, whereas he had previously asserted that light is always propagated instantaneously (that is, at a speed $v = \infty$). This inconsistency is typical of the paradoxes that have afflicted theories of light since the seventeenth century. But it must also be remembered that Descartes did not take too seriously his picture of light rays as tennis balls moving at different speeds. It was more in the nature of a useful analogy. As he had written to a friend in 1634, he believed that light always does travel at infinitely high speed, that it "reaches our eyes from the luminous object in an instant . . . for me this was so certain, that if it could be proved false, I should be ready to confess that I know absolutely nothing in philosophy."

Newton took Descartes' model somewhat more literally as a theory of light. He concluded that if moving particles are accelerated or decelerated by forces acting at the interface between two media, then one explains refraction, at least provisionally, by assuming that light is actually composed of particles. Other properties of light seemed to favor a similar interpretation, for example, the fact that sharp shadows can be formed by solid obstacles placed in a beam of light. If light were composed of pulses in a medium, he reasoned, then it should be able to bend easily around corners, as does sound. (Though Newton was familiar with the diffraction of light, he considered it a less important phenomenon that could be explained by a particle theory of light if forces were assumed to act on the particles as they pass by the surface of objects.)

Newton also thought that an impulse theory could not account for *polarization,* a property of light unknown to Descartes. In 1669, the Danish scientist Erasmus Bartholinus discovered that crystals of Iceland spar had the curious property of splitting a ray of light into two rays. One of these rays can be split again if it strikes another spar crystal, but only for certain orientations of the second crystal; the ray seems to have a directional character. (This directional character of light is now familiar to anyone who has played with the lenses of polarizing sunglasses.) To Newton it was clear that this behavior could be explained only by assuming that the ray has "sides" so that its properties depend on its orientation with respect to the axis (direction of propagation) of the ray. This would be easy enough to understand if the ray is a stream of rectangular particles, but rather more difficult if light is a wave disturbance in a medium. As Newton remarked in his book *Opticks,*

> For Pressions or Motions, propagated from a shining body through a uniform medium, must be on all sides alike; whereas by those Experiments it appears, that the Rays of Light have different Properties in their different sides.

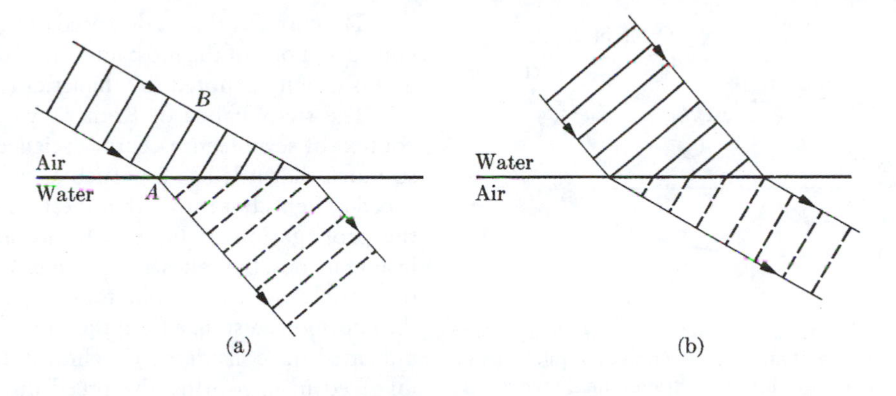

Fig. 23.4. (a) Refraction of an impulse on passing into a medium in which its speed is lower. (b) Refraction of an impulse on passing into a medium in which its speed is higher.

In spite of Newton's criticisms, other seventeenth-century scientists such as Robert Hooke and Christiaan Huygens continued to think of light in terms of impulses in a medium. This was not yet the "wave theory" in the modern sense, because the *periodic* nature of the pulses had not yet been recognized (see the next section); ironically it was Newton who suggested that light might have to be somehow assigned periodic properties as well in order to account for the phenomena of colors, even though such properties did not fit into his own particle theory very satisfactorily.

It was soon discovered that the pulse theory of light could after all account for refraction in a more plausible way than the particle theory—if one assumed that pulses are slowed down on entering a denser medium. This explanation was first published in 1644 by Thomas Hobbes, the political philosopher, and was presented in a more systematic and comprehensible manner by Huygens in his *Treatise on Light* (1690).

The chief idea is to consider what happens to the *wave front,* an imaginary surface representing the disturbance or pulse, which is usually perpendicular to the direction of propagation of light, that is, the direction of the ray. As shown in Fig. 23.4, when a pulse strikes an interface at an oblique angle, one part of the wave front enters the new medium and travels at its new speed, while the other part is still moving in the original medium. If the speed is less in the second medium, the ray is bent toward the normal to the interface, as one would expect when light passes from a rare to a dense medium; if the speed is greater in the second medium, the ray is bent away from the normal.

To decide between the Newtonian particle theory and the pulse theory of light, a "crucial experiment" is called for: Measure the speed of light in air and in water or glass, to see which is greater. But before this experiment could be done, it was necessary to settle the more fundamental question: Is the speed of light finite or infinite?

Galileo discussed this problem in his *Two New Sciences;* he pointed out that everyday experiences might lead us to conclude that the propagation of light is instantaneous. But these experiences, when analyzed more closely, really show only that light travels much faster than sound. For example, "When we see a piece of artillery fired, at a great distance, the flash reaches our eyes without lapse of time; but the sound reaches the ear only after a noticeable interval." But how do we really know that the light moved "without lapse of time" unless we have some accurate way of measuring whatever lapse of time there might be? Galileo could not find any satisfactory way to measure the extremely short time interval required for light to travel any terrestrial distance. He concluded that the speed of light is probably not infinite, but was not able to estimate a definite value for it.

The first definite evidence that light moves at a finite speed was found by a Danish astronomer, Ole Rømer. In September 1676, Rømer announced to the Academy of Sciences in Paris that the eclipse of a satellite of Jupiter, which was expected to occur at 45 sec after 5:25 A.M. on November 9, would be exactly 10

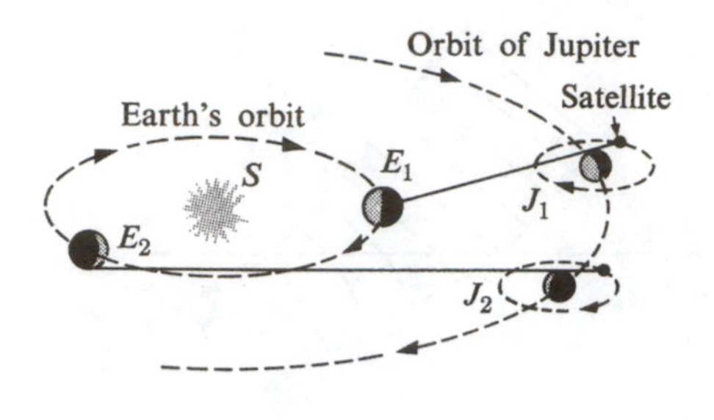

Fig. 23.5. Rømer's explanation of the delay in the observed time of an eclipse of one of Jupiter's satellites. It takes light longer to travel the distance J_2E_2 than the distance J_1E_1.

min late. On November 9, 1676, astronomers at the Royal Observatory in Paris, though skeptical of Rømer's mysterious prediction, made careful observations of the eclipse and reported that it occurred at 45 sec after 5:35 A.M., just as Rømer had predicted.

Two weeks later, Rømer revealed the theoretical basis of his prediction to the baffled astronomers at the Academy of Sciences. He explained that the delay in the eclipse was simply due to the fact that light from Jupiter takes a longer or shorter time to reach the earth, depending on the relative positions of Jupiter and the earth in their orbits (see Fig. 23.5). By comparing observed eclipse times at various points in the earth's orbit, he estimated that it would take 22 min for light to cross the earth's orbit.

Shortly thereafter, Huygens used Rømer's data to make the first calculation of the speed of light. Combining Rømer's value of 22 min for light to cross the earth's orbit with his own estimate of the diameter of the earth's orbit based on the latest astronomical observations (see Section 10.4), he obtained a value for the speed of light *in vacuo* that (in modern units) is about 2×10^8 m/sec. Although this is only about two-thirds of the presently accepted value of the speed of light, 2.998×10^8 m/sec, it was a considerable achievement of the time to find a value that good. (The discrepancy is mainly due to the fact that as we now know light actually takes only about 16 min to cross the earth's orbit; Huygens' value for the diameter of the orbit was fairly close to the modern value, about 3×10^{11} m.)

The correct value of the speed of light is now considered one of the most important constants, and is usually denoted[1] by the letter *c*.

The significance of Rømer's work in the context of seventeenth-century science was not so much that it led to a particular value of the speed of light, but rather that it established that the propagation of light in free space is *not* instantaneous, but takes a finite time. Yet the fact that it only takes a few minutes for light to cover the enormous distance from the sun to the earth indicated the considerable technical difficulties involved in measuring the speed directly in a terrestrial laboratory. This feat was accomplished in the middle of the nineteenth century; not until that time could it be shown that light does indeed travel faster in air than in water. Here was the long-sought confirmation of the wave theory; but when it came, most physicists had already abandoned the particle theory and accepted the wave theory for other reasons.

23.2 The Propagation of Periodic Waves

In order to understand the concept of periodicity, which is essential to the modern wave theory of light, let us consider the vibrations of a string that is clamped at one end. We take hold of the free end, as shown in Fig. 23.6, and move it rhythmically up and down in simple harmonic motion. As one particle of string is connected with the next, the motion is communicated from one to the next, but each lags behind its neighbor's motion, being coupled to it by elastic forces rather than by a rigid connection. We may visualize what happens in terms of the series of successive "snapshots" of Fig. 23.6.

Note how the motion of particle *A*, forced upon it by the hand, is communicated to *B*, *C*, *D* . . . and so on down the line. In particular, the maximum displacement, or *amplitude* of vibration, is reached by each particle in turn. So we

---

[1]In 1983, the Seventeenth General Conference on Weights and Measures *defined* the speed of light as $c = 299{,}792{,}458$ m/sec. The *meter* is therefore defined as the length of the path traveled in a vacuum in 1/299,792,458 of a second. "Thus the speed of light ceased to be a measurable constant. No further determinations of its value will be made" (Bailey, "Units, Standards and Constants"). The rationale for this decision was that further determinations of the speed of light would make no significant contribution to science.

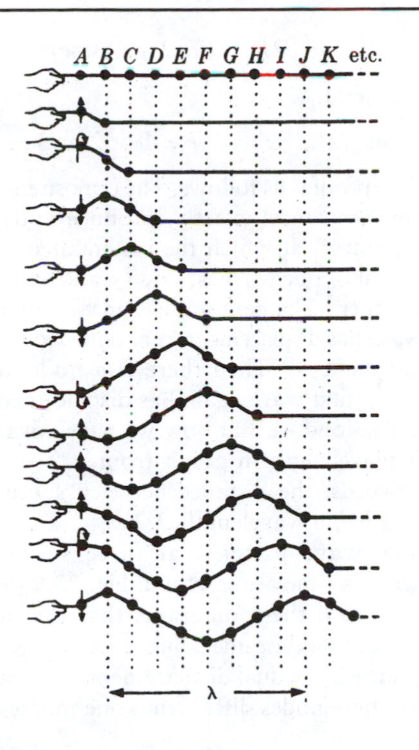

Fig. 23.6. Propagation of a transverse wave along a rope.

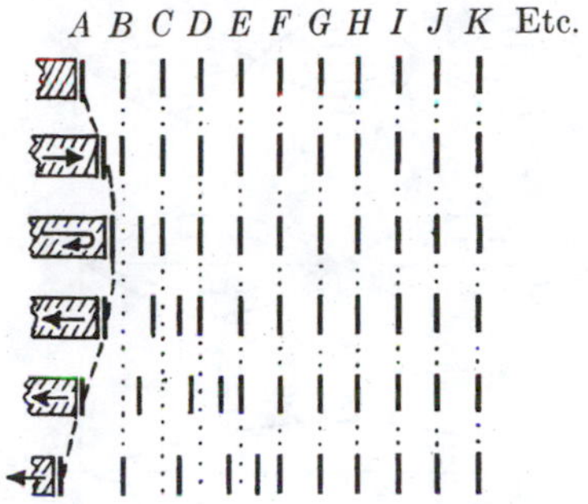

see a "crest" being generated, and after it a "valley"; and as both travel down the string to the right, they are followed by another crest, another valley, and so forth. In short, soon a complete *periodic wave* travels along the string, bringing motion from one end to the other, and with it, of course, energy. Note parenthetically that the wave travels here horizontally while the particles themselves move up and down. Technically, this mutually crosswise motion makes the wave a *transverse* one. If the motion of the particles is back and forth along the direction of motion of the wave itself, the wave would be called *longitudinal.*

The speed of propagation of the wave v_w, which we may here take to be the speed with which a crest travels, is of course the ratio of the distance s that a crest travels on the string to the time t needed to move that distance. We can evaluate v_w as follows: If we look at the last of the snapshots in Fig. 23.6, we note that particles B and J are both doing the same thing, that is, they are both at their maximum displacement upward and about to move down again. The two are "in phase" with each other; and the distance between successive points in phase is called

A B C D E F G H I J K Etc.

Fig. 23.7. Propagation of a sound wave in air. The displacement of the layers of gas molecules is greatly exaggerated.

the *wavelength* (λ, the Greek letter lambda). It is clear at a glance how long the crest at B will take to cover distance λ from B to J. Not until particle J has completed one full cycle of motion down and up will the crest arrive there. The time needed per cycle is, by definition, the observable period of vibration, T. Therefore

$$v_w = \frac{\lambda}{T} \qquad (23.1)$$

If we recall (from Section 10.1) that the frequency of vibration n is the reciprocal of T, we can write instead

$$v_w = \lambda n, \qquad (23.2)$$

a most useful relationship between velocity, wavelength, and frequency, applicable not only to waves in a rope, but to waves of any kind, including light waves.

Problem 23.1. Sound waves are longitudinal instead of transverse, that is, the particles in the path of the wave oscillate in the same direction as the motion of the wave itself. Consequently, the sound wave, instead of crests and valleys, produces successive condensations and rarefactions. Figure 23.7 shows layers of molecules, at first equally spaced, then acted on by a progressing wave, that is, by a piston or diaphragm that imparts its own motion to the gas in front of it. Copy this series and complete it in the same manner as Fig. 23.6 (13 lines altogether). Point out condensations (crowding of the planes) and rarefactions. Then mark off the distance corresponding to one wavelength.

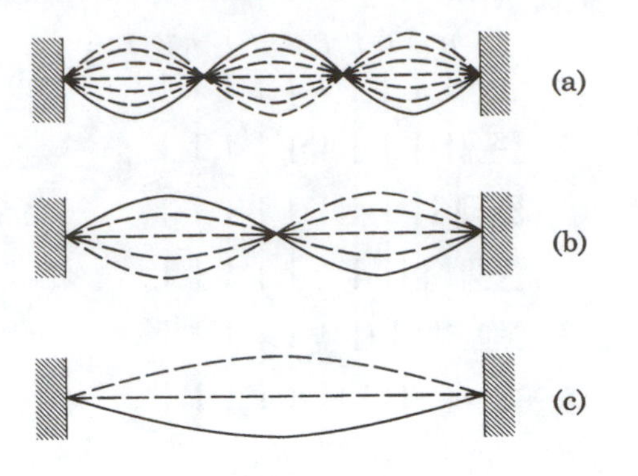

Fig. 23.8. Fundamental vibration and first two overtones of a string fastened at both ends.

Another property of vibrating strings should at least be mentioned here because it will turn out to help us later in the development of ideas in quantum mechanics (Chapter 29). Suppose a string of length L is fixed at both ends. If it is plucked in the middle, it is able to vibrate as shown in Fig. 23.8a. There is a *node* (point of zero displacement) at each end. If properly plucked or otherwise agitated, the same string can also take on other vibratory patterns, of which the first two are shown in Fig. 23.8b and c.

These vibrations are examples of *stationary waves* or *standing waves.* We can understand them if we think of them as the result of two overlapping *traveling* waves, of the same frequency and amplitude, going through the medium (the string) in opposite directions; for example, when the string is plucked in the middle, two waves go off along it to the right and the left, respectively, each with velocity u_w, and are reflected again and again at the two ends, crisscrossing on the same string. At some moments the displacements due to each of the two waves combine vigorously (interfere constructively); at other moments they cancel (interfere destructively), and for an instant the string is completely horizontal.

These standing wave patterns, however, can happen only when the waves' velocity (v_w), the frequency of each of the waves (f), and the length of the string (L)have the relation

$$f = v_w/2L \qquad \text{(for Fig. 23.8a)}$$

or

$$f = 2v_w/2L \qquad \text{(for Fig. 23.8b)}$$

or

$$f = 3v_w/2L \qquad \text{(for Fig. 23.8c)}$$

or, in general,

$$f = hv_w/2L, \quad \text{where} \quad h = 1 \text{ or } 2 \text{ or } 3 \ldots$$

The proof is as follows. The oppositely traveling waves cancel exactly (produce nodes) at some point x_1 at which the instantaneous displacement of one wave is, say, $+y$, and $-y$ is that of the other.[2] The next node is at x_2, where the first wave has displacement $-y$ and the second has displacement $+y$. Then there is a node at x_3, where the first wave again has displacement $+y$ and the second $-y$. But now we have advanced a full wavelength in going from x_1 to x_3. In other words, the distance between successive nodes is *half* a wavelength, $\lambda/2$.

The wave patterns at three successive moments in time are shown in Fig. 23.9. It can be seen from these diagrams that the nodes always occur at the same points $x_1, x_2, x_3 \ldots$ even though the individual displacements that cancel to give these nodes differ from one moment to the next.

Since the two fixed ends of the string must be nodes, the total length of the string L must be an integer multiple of the distance $\lambda/2$ between nodes,

$$L = \lambda/2 \text{ or } 2\lambda/2 \text{ or } 3\lambda/2 \ldots. \qquad (23.3)$$

Or, if we consider L fixed, the possible wavelengths for such standing waves are

$$\lambda = 2L, 2L/2, 2L/3, \ldots$$

as shown in Fig. 23.8. The corresponding frequencies are

$$f = v_w/2L, 2v_w/2L, 3v_w/2L, \ldots \qquad (23.4)$$

as stated above.

The frequencies following the fundamental frequency $v_w/2L$ are called *overtones.* The difference in tone quality of two musical string instruments playing the same note (same fundamental frequency) is due to the fact that overtones are always present to some degree along

[2]At a later time the first has displacement $y + \Delta y$, and the second displacement $-y - \Delta y$, so they still cancel; the changes in the displacements of the two waves must be equal and opposite, since they were stated to have the same frequency and amplitude.

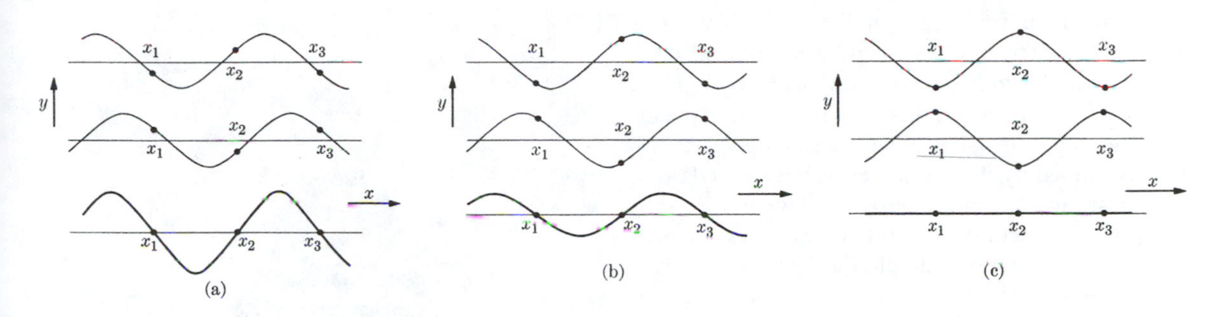

Fig. 23.9. Interference of two similar waves traveling in opposite directions, producing standing waves. Three parts, (a), (b), and (c), represent successive instants of time; in each part the first and second patterns (light lines) are the individual waves, and the third (bottom) pattern (heavy line) is their superposition.

with the fundamental, but are produced with different intensities in different instruments.

23.3 The Wave Theory of Young and Fresnel

From the time of Newton until the early years of the nineteenth century, the particle theory of light was favored by most physicists, largely because of the prestige of Newton—although the theory was not as dogmatically entertained by the master as by his later disciples. The English scientist Thomas Young (1773–1829) and his younger French colleague Augustin Fresnel (1788–1827) fought to put the wave theory of light on a sound theoretical and experimental basis during the years 1800–1820. They exploited a key feature that distinguishes waves from particles: When two or more waves are superimposed, in the same medium (as we have already seen in discussing standing waves) they may either reinforce or cancel out the displacement of particles at various places in the medium. This property is known as *interference;* it can be either *constructive interference* (when the waves reinforce) or *destructive interference* (when the waves cancel). Streams of particles, when they join, do not seem to have such behavior in all our normal observations.

The fact that interference effects can be demonstrated in optical experiments helped to establish the wave theory of light. The first of two famous experiments was Young's *double-slit*

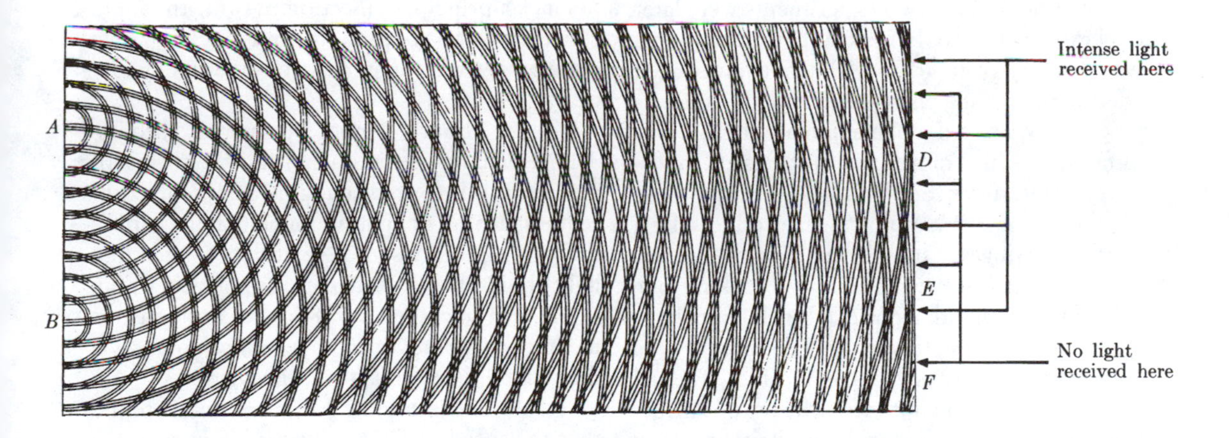

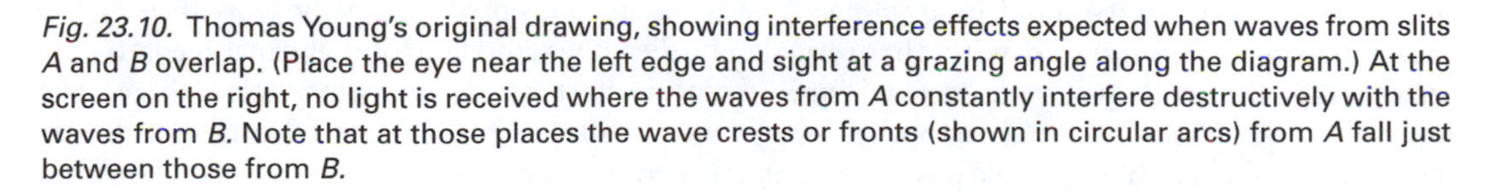

Fig. 23.10. Thomas Young's original drawing, showing interference effects expected when waves from slits *A* and *B* overlap. (Place the eye near the left edge and sight at a grazing angle along the diagram.) At the screen on the right, no light is received where the waves from *A* constantly interfere destructively with the waves from *B.* Note that at those places the wave crests or fronts (shown in circular arcs) from *A* fall just between those from *B.*

experiment, represented in Fig. 23.10. The interference pattern of alternating bright and dark regions—produced when a beam of light is passed through two slits or holes, and the spreading, separate parts then recombine—is easily explained by the wave theory, but seemed quite mysterious from the particle viewpoint.

The second experiment was actually proposed as a test of Fresnel's theory by a supporter of the particle theory, the French mathematical physicist Simon Poisson. Poisson was one of the judges appointed by the French Academy of Sciences to examine Fresnel's paper, submitted for a prize in 1818. Using Fresnel's mathematical equations for the diffraction of light waves around obstacles, Poisson showed that if these equations really did describe the behavior of light, a very peculiar thing ought to happen when a small round disk is placed in a beam of light. A screen placed behind the disk should have a bright spot in the center of the disk's shadow, because diffraction of the light waves spilling all around the edge of the disk should lead to constructive interference at the center of the shadow. According to Newton's particle theory, there could not be any such bright spot, and it had never before been observed. Poisson fully expected that this absurd prediction would lead to the refutation of the wave theory, and challenged Fresnel to do the experiment. Fresnel accepted the challenge and arranged for this prediction to be tested by experiment. The result: There *was* a bright spot in the center of the shadow!

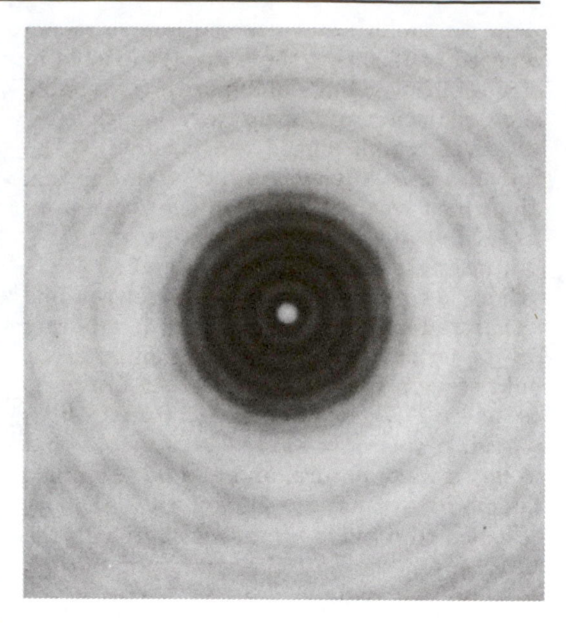

Fig. 23.11. The Poisson bright spot: a photograph of the diffraction pattern of a small circular obstacle, showing the bright spot predicted by Poisson from Fresnel's wave theory.

Although these two experiments were later celebrated in textbooks as definitive proofs of the wave theory of light, they did not have much effect in persuading physicists to abandon the particle theory. Fresnel, unable to read English, was unfamiliar with Young's publications and rediscovered interference for himself in 1814 while observing the diffraction fringes produced by a wire. The fringes inside the shadow of the wire could be made to disappear by blocking one side of the wire with a piece of black paper, thus showing that these fringes are produced by the combination of rays from both sides. Fresnel then developed a quantitative theory of diffraction, which he applied to several situations in which numerical results could be obtained by observation and compared with theoretical predictions. Rather than a single "crucial experiment," it was the wide range of optical phenomena explained by a few simple postulates that made Fresnel's theory so convincing to other physicists.

Another factor in the triumph of the wave theory in France was the ascent of a younger generation of pro-wave-theory physicists—Arago, Ampère, Fourier, Dulong and Petit—into positions of power and influence as professors and journal editors. To some extent we may credit "Planck's principle" (Section 3.4) for the rapid conversion of physicists, first in France and then elsewhere, from the particle to the wave theory of light.

What about "Popper's principle" (Section 3.1)? Could any experiment "falsify" the particle theory? In our discussion of seventeenth-century theories of light we found that the particle theory of Descartes and Newton predicted that light should travel faster in glass or water than in air or a vacuum. By 1850, scientific instrumentation had advanced far enough to allow a direct laboratory test of this prediction.

A. H. L. Fizeau and J. B. L. Foucault found that the speed of light in water is *less* than the speed in air (as predicted by the wave theory). But by then it was too late for that experiment to have any impact on the wave-particle debate. The validity of the Young-Fresnel theory was already generally accepted, and physicists had begun to

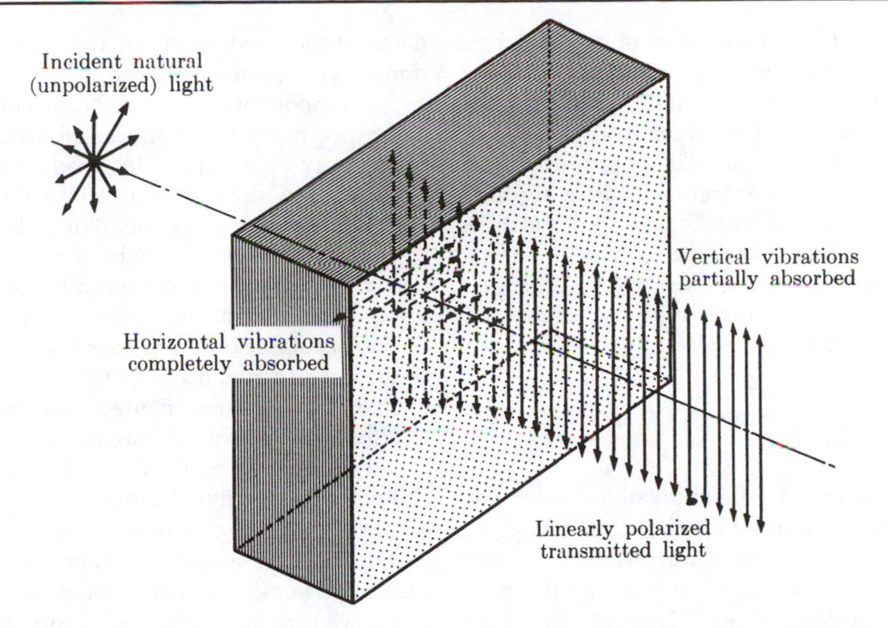

Fig. 23.12. An unpolarized beam of light (vibrations in all directions in a plane perpendicular to the direction of travel) passes through a crystal such as tourmaline, which removes the horizontal component of the vibrations, leaving only vertical vibrations.

work out the mathematical consequences of the theory and its application to all the different properties of light.

One property of light that had to be explained was *polarization.* Newton, as we noted earlier, had argued that a particle theory was needed to explain the fact that light rays seem to have directional properties because impulses propagated through a medium "must be on all sides alike." But that is only true of *longitudinal* waves, such as sound. Young and Fresnel pointed out that the phenomena of polarization could be explained by assuming that light waves are *transverse;* thus the vibrations could be vertical (as in Fig. 23.6) or horizontal, or in any other direction in a plane perpendicular to the direction of propagation. Ordinary light from the sun or light bulbs is a mixture of waves vibrating in all these directions, and hence is said to be *unpolarized,* but on reflection or refraction by certain substances the waves polarized in some directions are selectively removed, leaving a beam of light polarized, as shown, for example, in Fig. 23.12.

But the assumption that light consists of transverse waves brought new difficulties, for the only transverse waves known to Young and Fresnel were waves that travel in a *solid* medium. In order to transmit transverse waves, the medium must offer some resistance to "twisting" motions; it must have a tendency to return to its original *shape* when deformed. This is true of a solid (and of a taut string, as we saw in Section 23.2) but not, in general, of liquids or gases. It therefore seemed necessary to assume that the "luminiferous" (light-bearing) ether—the medium that was thought to occupy all otherwise "empty" space, that penetrated all matter, and whose vibrations were identified at the time with light waves—is a solid!

It was also known that ordinarily the speed of the waves is equal to $\sqrt{(\varepsilon/\rho)}$, where ε is a measure of the elasticity of the medium and ρ its density; this formula was confirmed by experiments on the speed of sound in solids, liquids, and gases, and therefore was assumed to apply to light waves as well. But for a speed as great as that of light, the medium of propagation, the ether, would have to show simultaneously a very high elasticity ε and an extremely low density ρ. In order to explain the properties of light, the nineteenth-century physicists were therefore compelled to assume that all space is filled with a highly elastic but rarefied solid. How could one believe that such an ether really exists, when it was well known that the planets move freely through space just as if they were going through a vacuum that offered no resistance at all?

Scientists of today, who more than their ancestors have grown up without feeling the need to account for phenomena by mechanical models, would find it harder to conceive of a medium with such contradictory properties than to think of the propagation of light through space without the benefit of an intermediate material medium. But in the nineteenth century, instead of discouraging the proponents of ether models, such paradoxes only increased their efforts to find better models.

23.4 Color

Up to now we have said little about *color,* one of the most important properties of light. Here the best place to begin is Isaac Newton's paper on a "New Theory of Light and Colors," published in the Royal Society's *Philosophical Transactions* in 1672, based mostly on his research done in his early 20s, during the period 1665–1666. It was well known at that time that a beam of white light entering a prism (triangular piece of glass) comes out as a "spectrum" of colors, but it was generally assumed that this effect was the result of some definite *change* that the prism produced in the originally pure white light. Newton claimed on the contrary that white light is itself a mixture of colored rays; all the prism does is separate them out by refraction. (Each color is refracted at a slightly different angle.) The least-refracted rays are red, the most-refracted are violet, and there is a continuous gradation in between.

In support of this theory Newton reported two experiments. In the first, he allowed a single colored ray of the spectrum produced by one prism to pass through a hole in a board and strike a second prism. If one color had been due to some change produced by the prism, then one would expect that the second prism would produce *another* change, that is, the color of the ray would be different after being refracted by the second prism. Instead, there was no change in color, only a change in direction (Fig. 23.13). In other words, once the color had been *selected* out of the mixture by the first prism, it was not subject to further modification.

In the second experiment, Newton used a lens to focus the spectrum of rays coming from the first prism, so that they converge on a second prism; when properly arranged, the second prism would combine the colors to produce a beam of white light similar to the original one. This showed that white light could be *created by mixing colors* and supported his claim that white light *is* a mixture of colors.

Critics argued that Newton's theory was "only an hypothesis" that happened to explain the phenomena, and that some other hypothesis might do so as well. According to Popper's principle, an hypothesis can never be proved correct by any experiment; at best it can become more credible by surviving attempts to refute it. Newton indignantly replied that he had not adopted

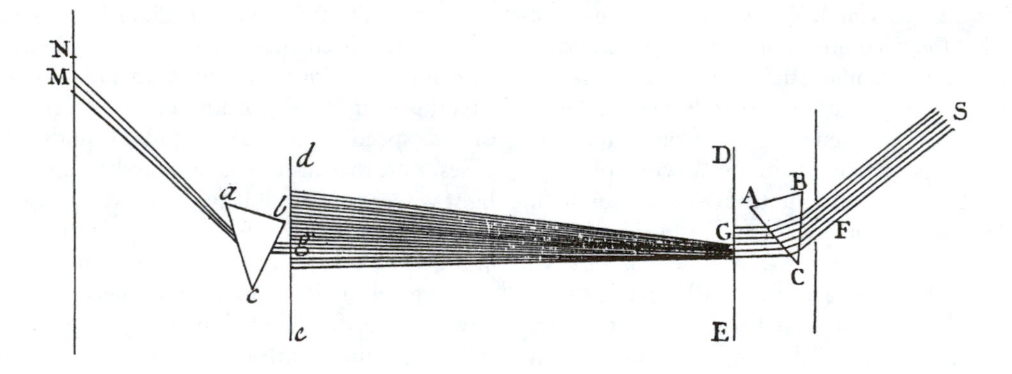

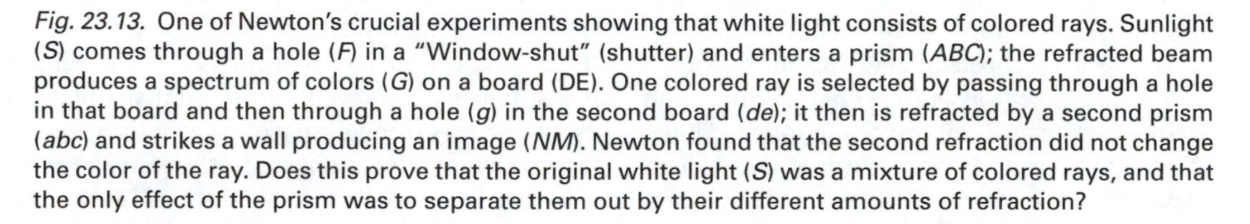
Fig. 23.13. One of Newton's crucial experiments showing that white light consists of colored rays. Sunlight (*S*) comes through a hole (*F*) in a "Window-shut" (shutter) and enters a prism (*ABC*); the refracted beam produces a spectrum of colors (*G*) on a board (DE). One colored ray is selected by passing through a hole in that board and then through a hole (*g*) in the second board (*de*); it then is refracted by a second prism (*abc*) and strikes a wall producing an image (*NM*). Newton found that the second refraction did not change the color of the ray. Does this prove that the original white light (*S*) was a mixture of colored rays, and that the only effect of the prism was to separate them out by their different amounts of refraction?

any hypothesis about the nature of light (for example, particle or wave) and his conclusion that white light is a mixture of colors had been proved without regard to the validity of any such hypothesis.

One hypothesis, hinted at by Robert Hooke in his comments on Newton's paper and made more explicit by the French physicist Léon Gouy in 1886, is that white light may be a series of pulses that can be decomposed into vibrations of different wavelengths. These vibrations are thus *potentially* present in the beam of white light but don't really exist until the pulse is decomposed by the prism.

These criticisms seem reasonable, but history rejects them and indeed casts doubt on the universal validity of the philosophical principles employed against Newton's claim. Scientists found in the nineteenth century that the spectrum of light coming from a source such as a hot vapor in a terrestrial laboratory or the atmosphere of the sun contains characteristic bright and dark lines that can be identified with those emitted or absorbed by particular chemical elements. They concluded that those lines *were present in the light before it reached the prism,* and that they give us reliable information about the source of the light, such as its chemical composition. Spectrum analysis provides the experimental basis for modern atomic physics and astrophysics (Chapter 28).

The wave theory also held out the hope of explaining phenomena other than light by considering vibrations of different wavelengths. Newton had suggested in 1672 that the different colors of the spectrum, formed by refracting a beam of white light through a prism, might be imagined to correspond to ether vibrations of different sizes. (At that time, Newton had not committed himself to the particle theory of light.) Thomas Young was able to measure the wavelengths corresponding to the colors of the spectrum, and found that they ranged from about 7×10^{-7} m for red to about 4×10^{-7} m for violet (see Table 26.1). But William Herschel had discovered in 1800 that there always existed in the spectrum of hot solids—along with the visible light—some radiation of longer wavelengths (beyond the red, a region named the *infrared).* The suggestion that this "invisible light" was identical with radiant heat was supported when heat rays were shown to exhibit all the characteristics of light—refraction, interference, polarization, and so forth.

Already there is a useful simplification here: Newton had thought that radiant heat might need a separate ether for its propagation; now the luminiferous ether could serve a double purpose. And, as was noted in Section 17.7, once the wave nature of *light* was accepted, it was only natural to suppose that *heat* also has a wave nature, and can be explained in terms of motions of matter. Thus the wave theory of light seemed to provide an essential link in the chain binding together all the forces of nature, including later (as will be seen in Chapter 25) electricity and magnetism.

RECOMMENDED FOR FURTHER READING

G. N. Cantor, "Physical optics," in *Companion* (edited by Olby), pages 627–638

I. B. Cohen, "The first explanation of interference," in *Physics History II* (edited by French and Greenslade), pages 85–93

Eugene Frankel, "Corpuscular optics and the wave theory of light: The science and politics of a revolution in physics," *Social Studies of Science,* Vol. 6, pages 141–184 (1976).

C. C. Gillispie, *Dictionary of Scientific Biography,* articles by H. J. M. Bos on Huygens, Vol. 6, pages 597–613; by E. W. Morse on Young, Vol. 14, pages 562–572; and by R. H. Silliman on Fresnel, Vol. 5, pages 165–171

Frank A. J. L. James, "The physical interpretation of the wave theory of light," in *History of Physics* (edited by S. G. Brush), pages 64–77

David Park, *The Fire within the Eye,* Princeton, NJ: Princeton University Press, 1997

S. Sambursky, *Physical Thought,* excerpts from Galileo, Rømer, Huygens, Newton, Euler, and Fresnel, pages 225–226, 284–294, 310–317, 342–346, 374–379

Dennis L. Sepper, *Newton's Optical Writings: A Guided Study,* New Brunswick, NJ: Rutgers University Press, 1994, Chapters 1, 2, and 3

William Tobin, "Léon Foucault," *Scientific American,* Vol. 279, no. 1, pages 70–77 (July 1998)

Arthur Zajonc, *Catching the Light,* New York: Oxford University Press, 1993, Chapters 4 and 5

CHAPTER 24

Electrostatics

24.1 Introduction

The physical atom of the nineteenth century was discussed largely in terms of mechanical force, momentum, kinetic energy, elastic collision, and the like. But as the century drew to a close, there was discovered the electron, that carrier of a fixed negative charge, endowed with a definite mass, a particle much smaller than the atom—and undeniably a part of its structure (Section 25.7). Then radioactivity brought out more charged particles that had to be somehow accommodated in the atom, reenforcing long-standing demands from other branches of physics that electric phenomena must be associated with atoms directly. Finally, the twentieth century gave us a model that stresses concepts like charge, electric potential, and electric fields at least as prominently as its predecessor had stressed mechanical concepts. Therefore, as we now turn toward the modern atomic theory, it is not only fitting but necessary to undertake first of all a rather detailed study of the basic ideas of electricity.

In reviewing the development of electrostatics we note for the first time a major contribution by an American scientist in the early formulation of a physical theory. Benjamin Franklin (1706–1790) was and deservedly is as well known as a statesman and writer as a scientist. He was greatly interested in the phenomenon of electricity; his famous kite experiment and invention of the lightning rod gained him wide recognition. His other inventions include the Franklin stove, bifocal glasses, and the glass harmonica. While in Philadelphia, Franklin organized a debating club that developed into the American Philosophical Society (one of the oldest scientific and humanistic societies still in existence), and he helped to establish an academy that later became the University of Pennsylvania. One of his most honored titles, however, is "the Newton of electricity."

24.2 Electrification by Friction

The word *electricity* is derived from the Greek word for amber, *electron,* for it was in ancient Greece that the observation seems to have been first made that after amber has been rubbed with cloth it *attracts* small pieces of lint, as we now know, by electrostatic forces. Not until the seventeenth century were forces of repulsion observed in such simple experiments as with bodies electrically charged by rubbing. We still use Franklin's terminology to describe the main observations:

1. When two different bodies such as a glass rod and a silk cloth are rubbed against each other, they will thereafter attract each other with a new, nongravitational force, because they have developed electric charges of opposite sign. By an arbitrary (and unfortunate) convention we call the charges on the glass positive (+), those on the silk negative (–). Note well that the observed attraction defines two things: that charges are present and that they are of different kinds.
2. Two glass rods, rubbed in turn with a silk cloth, both become charged, but repel each other. We note that positive charges repel one another.
3. When a rod of resin (or amber, rubber, etc.) is rubbed against a new silk cloth, it also becomes charged, as seen by the appearance of a force of attraction between the resin and silk. But this charged resin rod also attracts the previously charged glass rod; wherefore we decide to assign a minus sign to charges on the resin.
4. Finally, two resin rods charged in the same way repel, and we conclude that negative charges repel one another. (The postulate that negative charges repel each other was not part of Franklin's original theory but was added by the German physicist Franz Aepi-

nus in his systematic treatise on electricity published in 1759.)

Further experiments with similar combinations reveal no case in which the cloth is repelled by the material that it has freshly charged, nor any cases in which a charged body is attracted to (or repelled from) both + *and* – charges. All these observations can therefore be summarized by the following hypotheses (and the deduction from the above experiments is a useful exercise in clear thinking):

1. Two neutral bodies can charge each other by friction, but always with opposite charges.
2. There exist only two kinds of charges that interact, *positive* and *negative*. Positive charges appear on glass when it is rubbed with silk (definition of + charge), negative ones appear on resin when it is rubbed with silk or, more effectively, with fur (definition of – charge).

If suitable care is taken, any solid object can be charged by friction with a suitable cloth, and the charges thereby produced turn out to be + or –, but never of a third kind.

Problem 24.1. What is the action of a silk cloth that has charged a glass rod on another silk cloth that charged a rubber rod? Can you predict what will happen if you rub silk against silk? Silk against fur?

Problem 24.2. Although they are not found in nature, some kinds of charges other than + and – are logically possible. Describe what properties they would have, that is, how you would *know* when charges different from both + and – are present.

24.3 Law of Conservation of Charge

Even before we define charges quantitatively, we may suspect a law, to be confirmed later more rigorously, of *conservation of charge,* that is, a law stating that electric charges can be produced only in pairs of equally strong positive and negative charges. The preliminary experimental proof, also due to Franklin, is a simple observation: Wrap an uncharged cloth around an uncharged glass rod. Rub the rod and separate the cloth and the rod; individually each exerts, as expected, strong electric forces of repulsion or attraction on some third object, perhaps a charged leaf of tinfoil. But now wrap the charged cloth again around the charged rod; the electric forces exerted on some third object disappear.

Our conclusion is that the effect of the opposite charges cancels when they are brought near each other; therefore cloth and rod charged each other equally strongly though with opposite signs. *No net* charge (excess of + or of) was created by mutual friction. As usual, the existence of such a law of conservation is the deepest justification for the power of the concept "electric charge" itself.

Recalling Noether's theorem (Section 17.11) we might ask if the law of conservation of charge corresponds to any symmetry property of the equations of physical theory. The answer will be given in the first footnote of Chapter 29.

24.4 A Modern Model for Electrification

Simple experiments like these led to some reasonable and fruitful early theories concerning the nature of electric charges. Our modern model cannot be presented as a logical development without reference to many more subtle and quantitative recent experiments, but let us simplify and speed up our discussion of electrostatics by stating the modern concepts now and justifying them later.

All materials—solid, liquid, or gaseous—are atomic in structure, as we convinced ourselves some time ago in discussing the successes of the kinetic-molecular theory. Each atom can ordinarily be considered as a neutral, uncharged object, but actually it is made of a strongly positively charged nucleus (the small central kernel of the atom that represents most of its mass within a radius of about 10^{-14} m), surrounded at a relatively large distance (about 10^{-10} m) by negatively charged electrons (from 1 to more than 100 around each nucleus, depending on the particular element). Electrons are all alike, and very light compared with the nuclei.

The neutrality of the ordinary atom is the consequence of an exact balance between the positive charges of its nucleus and the negative charges of its extranuclear electrons. When an electron is somehow torn from its parent atom, the remaining atom lacks one negative electron-charge, that is, it is a *positive ion* (see Section 24.13). The separated electron meanwhile represents a free negative charge; it may in time attach itself to a positive ion and so restore the neutrality of that atom, or it may attach itself to a neutral atom and so transform it into a *negative ion*.

In solids, the atoms at the surface are at rest except for thermal vibrations. As a simple model we can imagine their electrons to be held near the nuclei by electrostatic attraction, but in some substances it is possible to interfere with this attraction (even by simple surface contact with another material) and to snatch some surface electrons, leaving a positively charged surface on one material and carrying away an equally strong negative charge on the other. This is our picture of electrification by intimate contact, as by friction, although we are not prepared to say just how or why some materials, like glass, joyfully allow their electrons to be carried off, whereas others, like amber and resin, tend to grab extra electrons very copiously.

Here then is a simple model to explain the experimental facts of electrification by friction. Furthermore, it accounts for the observation that there are only two kinds of charge, and it will offer an adequate explanation of all other observed phenomena in electrostatics, for example, the fact that in principle a practically limitless number of electric charges can be separated by friction, understandable now in terms of the huge quantity of atoms available on every small portion of surface.

Note that the charges of ions and electrons are not something *apart from* those bodies. For example, it is not possible to discharge an electron and leave a small neutral mass; the whole electron *itself* is a carrier of negative charge. It is best to realize right here that there is no easy picture for visualizing the concept "charge." An electron has a definite mass (this statement undoubtedly and unfortunately invokes a picture of a tiny, hard, round sphere), and the electron also carries a constant negative charge (this statement adds nothing to our picture except that we must endow that little sphere with the ability to repel other electrons and to be attracted by positive ions). In fact, the concept "electric charge" has no meaning apart from some of these nongravitational forces that bodies sometimes exert on one another. The following quotation illustrates the conceptual difficulties in electricity very well.

> Some readers may expect me at this stage to tell them what electricity "really is." The fact is that I have already said what it is. It is not a thing like St. Paul's Cathedral; *it is a way in which things behave.* When we have told how things behave when they are electrified, and under what circumstances they are electrified, we have told all there is to tell. When I say that an electron has a certain amount of negative electricity, I mean merely that it behaves in a certain way. Electricity is not like red paint, a substance which can be put on to the electron and taken off again; it is merely a convenient name for certain physical laws.
>
> (Bertrand Russell, *ABC of Atoms*)

24.5 Insulators and Conductors

Experimentally we can distinguish between two types of materials: (a) those that retain an electric charge wherever on the surface the charge was placed initially, and (b) those that allow equal charges placed on their surfaces to repel one another, to "flow" and distribute those charges all over the body no matter where they were placed initially. The first behavior characterizes *insulators* (or nonconductors, dielectrics), such as glass, fur, silk, paper, resin, etc. The second type of materials, of course, are *conductors,* such as metals and graphite.

We can also differentiate between conductors and insulators by the relative resistance that they offer to an electric current. In good conductors like metals, some of the electrons more loosely connected to their atoms quite easily drift away from their parent atoms when some external electric force acts on them; hence they are called *conduction electrons,* and while they move they constitute an *electron current* through the conductor (or along its surface). In a good dielectric, however, the electrons are held more tightly by the parent atoms and are not free to flow away when acted on by external electric forces of attraction or repulsion.

Many of the experiments leading to these conclusions can be presented in concise diagrammatic form. Let us agree to indicate by a + or – the sign of the excess of + or – charges at any point on a solid. Then Fig. 24.1 tells the following simple story.

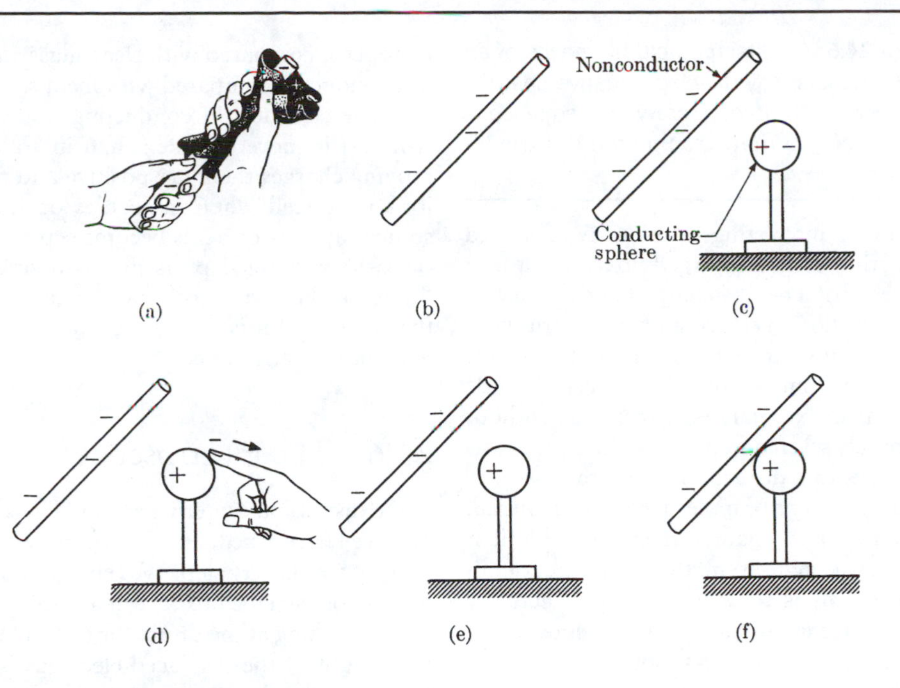

Fig. 24.1. Charging by induction.

(a) When it is rubbed vigorously, a piece of fur loses electrons to a resin or hard rubber rod; the fur is charged positively and the rod negatively.

(b) The fur is withdrawn; the rod remains charged.

(c) When a conducting metal sphere on an insulating stand is brought near, the conduction electrons are repelled to one side, leaving the sphere differently charged at the two sides; this separation of charges on a body by external electric forces is called *induction.* The rod and sphere now attract each other, because the rod is nearer to the unlike charges on the left part of the sphere than to the like charges at the right.

(d) We touch the right side of the metal sphere with our finger or with a wire connected to a large conducting body such as the (moist) earth (*grounding*). At once almost all the "free" negative charges at the right flow off, still repelled by the charged rod and by one another, and eager to distribute themselves as far away from the rod and from one another as possible. The positive charges at the left are *bound* charges, that is, they cannot be neutralized by grounding at that side as long as the presence of the negatively charged rod holds them in place. (Explain why.)

(e) Just withdrawing the finger or ground connections at the right leaves that side still neutral and the left still positively charged. If the rubber rod were now also withdrawn, the + charges, now free, would freely distribute themselves all over the whole metal sphere, which is now *charged by induction.*

(f) Touching the rod against the sphere at one point also changes practically nothing: None of the excess electrons on the nonconducting rod except those by chance on the point of contact can flow to annihilate the excess + charges on the sphere. Only if we twisted and turned the rod, so that in time most of the rod has made contact with the sphere, would each + charge on the sphere be able to pick up its neutralizing electron from the rod.

Problem 24.3. What would have happened if step (c) had been omitted? (Draw the sequence.) Step (f) would then be called *charging by contact.* How does the sign of this charge on the sphere differ in the two cases? What would happen if in step (f) the negative charge on the rod were very large compared with the positive charge on the sphere?

Problem 24.4. Repeat this series of pictures to illustrate the same sequence, starting, however, with a glass rod and a silk cloth.

Problem 24.5. A metal rod held by means of an insulating handle may develop negative charges when rubbed with cloth. Redraw the sequence in Fig. 24.1 using a metal rod instead of a nonconductor.

Another interesting problem is presented by the action of an uncharged *non*conductor in the presence of a nearby charged body. There is, of course, no flow of electrons on or in a true nonconductor, but we may imagine that the negative parts of each atom or molecule are pushed a little away from an external negative charge without being entirely severed from the positive parts; this corresponds to a distortion rather than a separation in the atoms or molecules of the material. Although such a small shift is negligible if it affects only a single atom, the total effect of billions of atoms is to *induce* charges, actually bound surface charges, on the two ends of the dielectric (see Fig. 24.2). As soon as the external charge is withdrawn, the induced surface charges usually also disappear; but until it is withdrawn the two bodies in Fig. 24.2 attract each other. We see therefore that a charged body attracts a previously neutral one, whether they are both conductors or not; and this explains the observation of the Greeks that a rubbed piece of amber attracts small pieces of lint.

Problem 24.6. Redraw the sequence in Fig. 24.1 using a nonconducting sphere.

The exact degree of mobility of charges in different bodies depends a great deal on the experimental circumstances, and no really sharp division is possible. Cool, dry gas acts like a nonconductor, but very hot gases are good conductors. Perfectly pure water does not carry electric currents, but tap water containing certain impurities is a fairly good conductor and becomes an excellent one if a pinch of salt or any other substance (electrolyte) is added that separates in the solvent to form ions. The human skin is a fair conductor compared with clean glass and a fair nonconductor compared with metals.

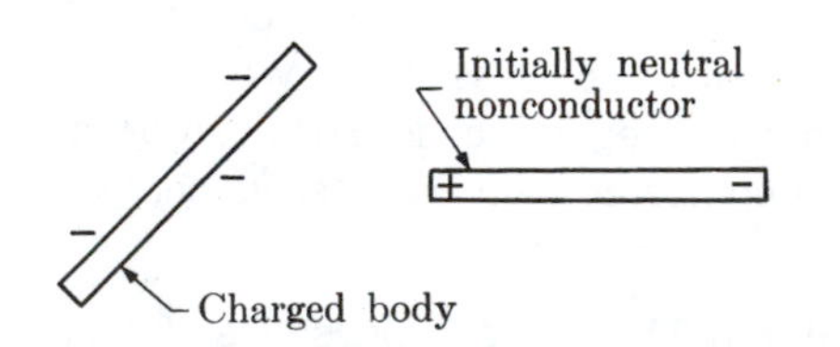

Fig. 24.2. Effect of induction on a nonconductor.

The situation in conducting liquids (electrolytes) is more complex than in solids. The moving charges there are positive and negative ions produced when molecules of such substances as salts or acids become separated into oppositely charged parts on dissolving in the liquid. In heated or otherwise agitated gases, the currents consist of electrons as well as positive and negative ions.

24.6 The Electroscope

In the discussion centering around Fig. 24.1 you may well have asked, How do we know what the charges are at every step? We certainly cannot tell the sign or even the presence of an electric charge by just looking at the object (or by weighing, for the weight of the displaced electrons is inconceivably small compared with the bodies involved).

One possible device for detecting the presence of a charge might be a tiny glass sphere at the end of a thin, long nonconductor. If we rub the sphere gently with silk we place on it a small positive charge. This sphere is then useful for exploring the neighborhood of another charged body: A strong repulsion indicates the presence of a like (positive) charge and strong attraction the opposite.

However, we should soon find that we need a much better instrument. An early and still useful device of this sort is the leaf electroscope (Fig. 24.3). The metal rod *B*, which is separated from the protecting case D by means of a good insulator *C*, is terminated at the top by a metal knob and at the bottom by very light leaves of gold or aluminum, *A*. If a charge (+ or) is "placed" on the knob and allowed to distribute itself over the stem and leaves, then the charged leaves will fly apart by mutual repulsion. The same effect is seen if the neutral knob, instead of being charged by contact, is simply approached by another charged body; then, by induction, there is a separation of charges in the knob and one type of charge is "driven" to the leaves, while the other remains on the knob. Again the leaves deflect. In general, the deflection or divergence of the leaves indicates a presence of charge in the leaves, and we assume that the divergence of the leaves, which is a measure of the repulsion between them, also is a qualitative measure of the

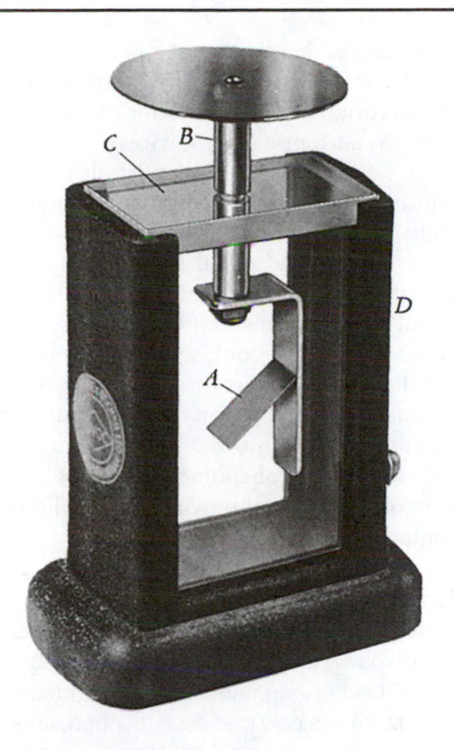

Fig. 24.3. The leaf electroscope.

quantity of charge, to be defined more rigorously at once.

Problem 24.7. Figure 24.4 shows without text how an electroscope is charged by *induction.* Supply text for each picture; then repeat this series of pictures for the case in which an initially positively charged rod is used. [*Note:* Only the sign, not the quantity, of the *excess* charges is indicated.]

A good electroscope, once charged, holds its charge extremely well, as seen by the steady deflection of the leaves. If, however, the air surrounding the knob is heated or is made conducting in some other way, the charge leaks off and the leaves collapse. In fact, the discharge of an electroscope at high altitudes in a balloon expedition in 1911 provided the first solid evidence for the existence of cosmic rays.

From the rate of collapse of the leaves one may judge the strength of a source of x-rays or cosmic rays or of radioactive emission, for these types of high-energy radiations *ionize* the gases through which they pass. Hence electroscopes are important instruments in modern research, although the details of construction may differ widely from the instrument pictured in Fig. 24.3. For example, one type of rugged electroscope, as small and slim as a fountain pen, is nowadays carried in the pockets of many research workers whose tasks require exposure to nuclear radiation; the electroscope is charged at the beginning of the day, and periodic checks of its deflection reveals whether the carrier's exposure has exceeded the experimentally known safe dosage of high-energy radiation.

Problem 24.8. Explain the following simple phenomena or devices in terms of the discussion of electrostatics: Walking across a carpet is followed by a shock when one's hand touches a metal object. Violent motion of moist air or raindrops is followed by lightning. Gasoline trucks often drag chains, and cars are brushed by metal strips as they cross a toll bridge.

24.7 Coulomb's Law of Electrostatics

Place some electric charges on two very small metal spheres; the relative amounts of charge, by measurement with the can arrangement, are q_1 and q_2 in magnitude (for example, q_1 = 2 units, q_2 = +2 units). Both spheres—so small and so far apart that we may call them *point* charges—are

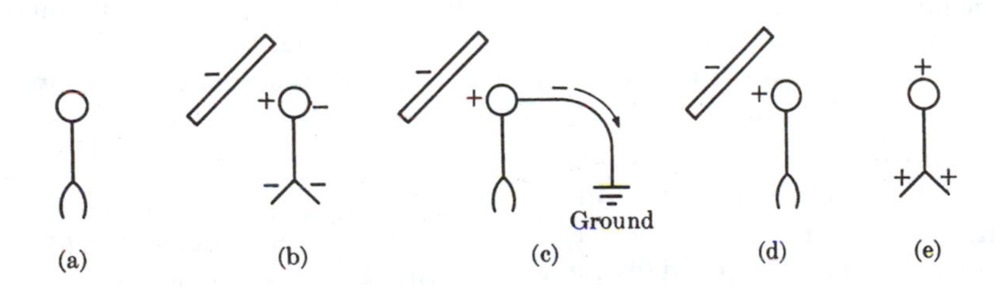

Fig. 24.4. Charging an electroscope by induction.

suspended in a vacuum, and are so far away from other materials that the observed forces on the charged spheres are in effect only the electric force F of q_1 on q_2 and the force of the reaction of q_2 on q_1.

By experiment, $F \propto 1/r^2$, where r is the distance between the two point charges. Also, $F \propto q_1$ (q_1 may be increased to 3, 4, 5, . . . units by recharging and referring to the electroscope calibration); and $F \propto q_2$.

In short, $F \propto q_1q_2/r^2$, quite analogous to the law of gravitation. Introducing a constant of proportionality, k_e, we can write:

$$F = k_e \frac{q_1 q_2}{r^2} \qquad (24.1)$$

There had been numerous attempts to find the quantitative form of the electric force law, going back to Robert Hooke and Isaac Newton in the seventeenth century. By about 1770 most scientists believed that both the force between electric charges and the force between magnetic poles should be inversely proportional to the square of the distance, like Newton's law of gravity except that the force could be either repulsive or attractive. The final experimental proof was provided by the French engineer-physicist Charles Augustin de Coulomb (1736–1846).

In his engineering work Coulomb had developed the theory of the stretching (elasticity) and twisting (torsion) of solids. He first used his knowledge of solid mechanics to develop a torsion balance to measure the earth's magnetic field in 1777; the success of this work led him to develop a more accurate balance, capable of measuring the forces between electric charges and magnetic poles at various distances. In 1785–1787 he published his data confirming Eq. (24.1) for electric charges and a similar equation for magnetic poles.

Equation (24.1) for the electric force between two point charges is called *Coulomb's law;* modern measurements have indirectly confirmed its accuracy to within one part in a billion. The unit of electric charge is called the *coulomb,* abbreviated "coul" (The single-letter abbreviation "C," used in some modern texts, seems likely to cause confusion with other things denoted by the same letter—see Appendix I.)

The coulomb is officially defined (by international agreement) in terms of the unit of electric current, the *ampere:* The coulomb is the amount of charge that flows past a point in a wire when the current is equal to 1 ampere. The ampere, in turn, is defined in terms of the force between two current-carrying wires.

Since we cannot go into the details of electrical measurements here, we must simply state the result: The constant k_e in Coulomb's law [Eq. (24.1)] is very nearly[1] equal to 9 billion newton·meters squared per coulomb squared (9×10^9 N·m²/coul²). This means that two objects, each with a *net* charge of 1 coul, separated by a distance of 1 m, would exert forces on each other of 9 billion N. As can be seen from this number, a coulomb is an enormous amount of charge; most of the familiar phenomena of static electricity involve charges of only about a millionth of a coulomb.

Problem 24.9. (a) How do we know when q_1 and q_2 are each 1 coul? (b) From accurate experiments it is known that each electron carries about 1.6×10^{-19} coul of negative charge. The neutral hydrogen atom has only one such electron, about 0.5×10^{-10} m from its positively charged central nucleus. What is the electric force holding the electron to the nucleus? (c) If the mass of an electron is approximately 10^{-30} kg and of a hydrogen nucleus is about 2×10^{-27} kg, what is the magnitude of the gravitational force compared with the electrostatic force? (d) How many electrons make up 1 coul of charge? (e) How many electrons must be placed on each of two spheres 0.1 m apart to cause each sphere to exert a force of 0.001 N on the other?

By experiment, it has also been found that if there is air instead of a vacuum between q_1 and q_2, the force these charges experience is essentially unchanged. However, if the space between and around the two charges is completely filled with another homogeneous isotropic dielectric such as paraffin or pure water, the force experienced by each charge is reduced by a factor of about ½ and 1/81, respectively. This is not because Coulomb's law is invalid there, but because in the presence of q_1 and q_2 the charges that are always present in the dielectric separate and in turn exert forces

[1]More precisely, $k_e = 10^{-7} c^2 = 8.9874 \times 10^9$ N·m²/coul². The reason k_e is related to the speed of light (c) will be explained in Section 25.3. The constant k_e is sometimes replaced by $1/4\pi\varepsilon_0$, where ε_0 is called the permittivity constant and is equal to 8.85×10^{-12} coul²/N·m².

of their own on q_1 and q_2 that tend to counteract the direct effects of q_1 on q_2 and q_2 on q_1. However, this is a complex situation, and we shall usually have little occasion to talk about any cases other than charges *in vacuo*, for which Eq. (24.1) describes accurately the total force actually observed.

Problem 24.10. Plot a graph of relative force on a point charge q' vs. distance from a second charge q if there are no other charges present.

To find the electrostatic net force on a point charge q in the neighborhood of a large charged body, we must add vectorially the several forces exerted individually on q by each of the point charges distributed on the large body. Evidently this is tedious, if not impossible, in a general case with the mathematical methods at our command, but there are two special cases that come up so often in our later studies that we might now at least look at the obtainable results.

1. The net force F on a single small point charge q in air or vacuum *outside* a uniformly charged *spherical* body is given by $F = k_e q \times Q/r^2$, where Q is the net charge on that spherical body, and r is the distance between q and the center of the sphere. Note the similarity between this conclusion and Newton's results for the gravitational attraction exerted by a spherical mass on a small point mass. There, too, we heard that the effect of a (homogeneous) spherical body is the same as if the bulk of the sphere were shrunk to a point at its center.
2. The net force on a single small point charge is constant in direction and magnitude anywhere inside the air gap between two oppositely charged large parallel *plates*, an arrangement called a condenser or capacitor. (Note that the force on a charge is *not* constant along the line between two opposite *point* charges; cf. Problem 24.10.)

Problem 24.11. Compare Newton's law of universal gravitation as applied to two small point masses and Coulomb's law of electrostatics applied to two point charges; consider the similarities and outstanding differences. What conclusions do you draw from the repeated appearance of the inverse-square law in the various branches of physics? (Do not assume that a dogmatic answer to this question exists.)

24.8 The Electrostatic Field

We must now present some information that may seem rather technical and uninteresting when phrased in terms of abstract charges and forces. But we need it to prepare for a good understanding of how the atom works—just as, in Chapters 6–10, we needed to develop the mechanics of accelerated masses in order to understand in Chapter 11 how the solar system works.

If we attach a small body to the end of a thin nonconducting stick and give it a positive charge, say +1 coul, the body (a "test point charge") will experience a force F when placed anywhere in the vicinity of other electric charges. We may use this probe to explore every point in space, and to find the electric force (in magnitude and direction) at every point. We could populate this whole space mentally with arrows to a given scale to show the situation at every point at a glance.

In Fig. 24.5 a few such electric force vectors have been represented in the neighborhood of simple charge distributions, namely (a) around a positively charged sphere, and (b) around two oppositely charged spheres. Obviously the force vectors become negligibly small far away from these charges, but even in their immediate neighborhood we realize that there must be an infinity of such vectors, one for each point of space.

It is inconvenient that these force vectors depend on two factors, that is, on both the charge Q that sets up the action and the charge q that is acted upon. The English physicist Michael Faraday developed a very useful concept, which, by a standardization of the test charge, used to explore the action of Q, is independent of q. This is called the *electric intensity* at P, or *electric field strength* at P, and is directly defined as the electrostatic force per positive unit charge placed at that point P. The letter E is usually assigned to this concept. By definition,

$$\mathbf{E} \text{ at } P = \frac{(\text{net force } \mathbf{F} \text{ on positive charge } q \text{ at } P)}{q} \tag{24.2}$$

usually expressed in units of N/coul. The arrows in Fig. 24.5, if they referred to forces *per unit charge*, now represent directly the vector $\mathbf{E}$ at each selected point. *The totality of such* $\mathbf{E}$*-vectors in a given region we call "the electrostatic field" in that region.*

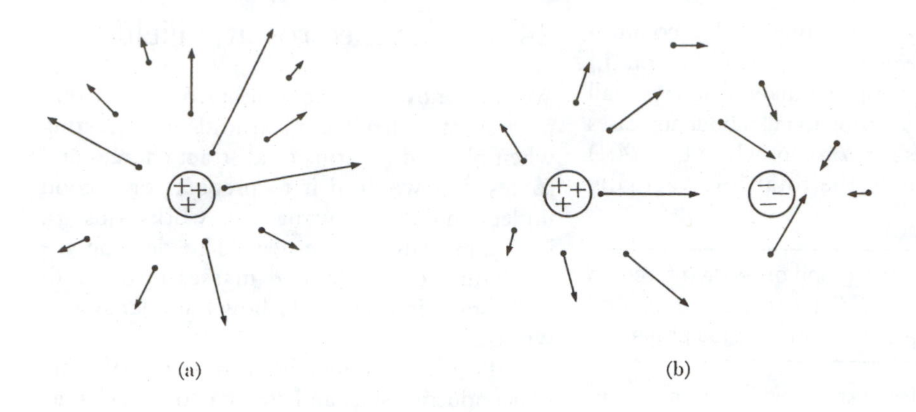

Fig. 24.5. Force vectors on a small positive test charge (q coul) at selected points near fixed charges.

This is a loose definition of one of the most fruitful concepts of physics (gravitational field, magnetic field, etc., are similarly conceived).

Note three important characteristics of the concept "electric intensity."

1. The *magnitude* of $\boldsymbol{E}$ (written as E) at a point P may be determined by measuring the force F on a test charge of, say, +15 coul placed at P, and dividing this force F by 15 coul, since E is the *ratio* of force by charge. If 30 coul had been chosen for the test charge, F would have been twice as large, but the ratio E at that point comes to the same value no matter how large the test charge. (But the test charge still must not be so large that it induces perceptible charge rearrangements in its neighborhood.) The magnitude of $\boldsymbol{E}$ is determined by the charges that *set up* the field, and not by the small test charges used to explore its strength. In fact, that is the main usefulness of the concept.
2. The *direction* of $\boldsymbol{E}$ at P, by convention, is the same as the direction of the force on a *positive* test charge at P. We might equally well have chosen the opposite convention, but once we agree on the present one we must stick to it consistently.
3. The *units* of $\boldsymbol{E}$ are the units of force/charge, usually N/coul.

EXAMPLE 1

(a) The force F_P on a +4-coul test charge at a point P, 10 cm from a small charged sphere, is 12 N toward the sphere, by test. Find the electric intensity E_P at P.

Solution. E_P = 12 N/(+4 coul) = 3 N/coul (toward sphere).

(b) Now find the total charge Q on the sphere.

Solution. $F_P = 12\text{ N} = k_e Q \times (+4\text{ coul})/100\text{ cm}^2$; therefore

$$Q = k_e\,[(12/4) \times 100]\text{ coul},$$

and Q is *negative* since it attracted a + test charge.

The defining equation

$$E = \frac{\text{(force on positive test charge)}}{\text{(magnitude of test charge)}}$$

becomes particularly simple if the field-producing charge Q is a point charge at a distance r from the location of test charge q, or if Q is uniformly distributed on any sphere whose center is r cm from q. In either case (see Fig. 24.6), using Eq. (24.1) for F_P, we get

$$E_P = F_P/q = (k_e[Q \times q/r^2])/q = k_e\,(Q/r^2). \quad (24.3)$$

But for all other cases, the value of $\boldsymbol{E}$ at point P must be found by experiment from the defining

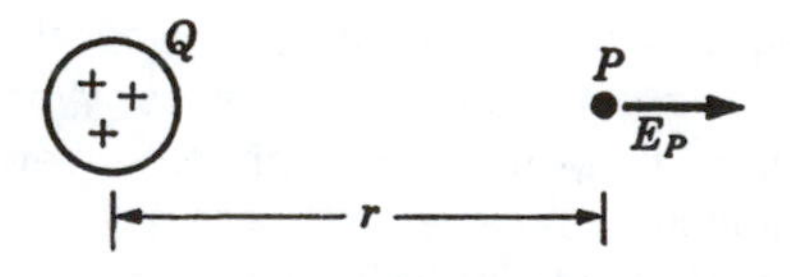

Fig. 24.6. Field at P owing to charge Q.

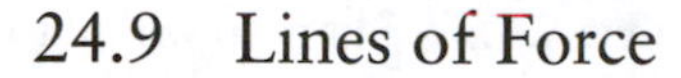

equation (force/charge), or else by adding vectorially the contributions to $\boldsymbol{E}$ at P of each point charge Q at its particular distance r, each contribution calculated by Eq. (24.3). We have here neither the tools nor the need for carrying out this mathematically advanced exercise.

Problem 24.12. At a distance of 1 m from the center of a uniformly charged large sphere the electric intensity is 4 N/coul. What is the force on a point charge of 10^{-9} coul at that distance from the sphere, and what is the total charge in coulombs on the sphere?

As an aside: The field concept can be used to describe other kinds of forces as well. The simplest case is the gravitational field, for which [by analogy with Eq. (24.2)] we may define a gravitational field strength as follows:

$$\text{Gravitational field strength } f_G \text{ at a point } P = \frac{(\text{Gravitational force } \boldsymbol{F}_G \text{ on mass } m \text{ at } P)}{m}$$

Thus the magnitude of the gravitational field strength at a distance R from a point mass m_2 would be, according to Eq. (11.6),

$$f_G = Gm_2/R^2.$$

Conversely, the gravitational force $\boldsymbol{F}_G$ acting on a mass m_1 at point P can be expressed in terms of m_1 and the value of the field strength at P, without reference to the mass or masses that may have created that field in the first place:

$$\boldsymbol{F}_G = m_1 f_G.$$

Note that f_G is in every sense equal to the concept we previously identified as the gravitational acceleration $\boldsymbol{g}$.

24.9 Lines of Force

Knowing how to find $\boldsymbol{E}$ at any point is going to be essential. But to visualize the electric field around a charged body we must try mentally to supply each and every point in space with an arrow indicating the force per unit charge at that point; and this is obviously a theoretically feasible but practically impossible task. As an aid to our limited powers of imagination we may picture the field in another way.

We know that if we bring a small positive test charge q into an electric field created by some other charges, q will experience repulsive forces from + charges, attractive forces to − charges. Figure 24.7 shows the paths of a small positive test charge q *released* at various points in the neighborhood of some other charges. All forces except electric ones are neglected, and the inertia of the body carrying q is also negligible. Therefore the direction of travel (indicated by the arrow on each line) follows the direction of the *force* acting on q at every point along the path, and, following Faraday, we may call these lines the *lines of force*. At every point along each line, the force acting on q is tangential to the line, but the line only tells us the direction of the force vector on the positive charge, not its magnitude.

Problem 24.13. Invent a modification of this idea to present at one glance both direction and magnitude at every point along the path.

Clearly it is possible in every case to draw infinitely many different lines of force around a charged body. Every point in space lies on a line of its own; for example, you can readily imagine (and draw now) a curve from Q_1 to Q_2

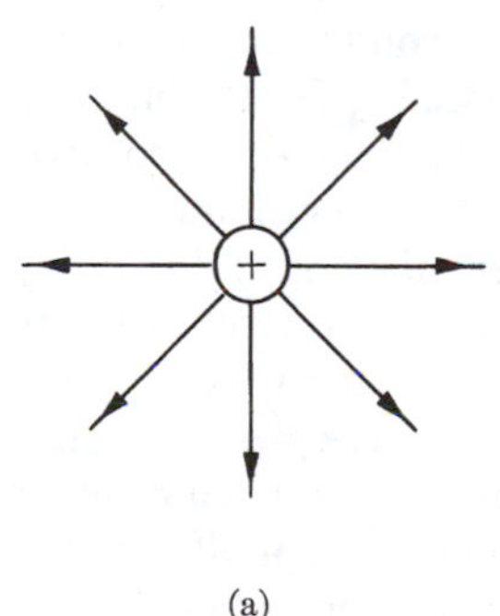

(a)

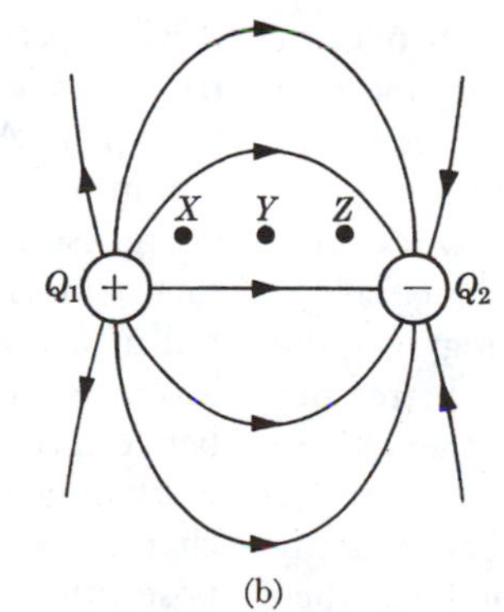

(b)

Fig. 24.7. Lines of force (a) in the neighborhood of a single positive charge, (b) around equally large point charges of opposite sign.

through point X in Fig. 24.7b. But even without filling in all possible lines, we get a good feeling for the state of affairs by looking at only a few such lines. As Fig. 24.7a informs us, a crowding of the lines of force means numerically a relatively stronger field (near the charged body), while a larger spread means numerically a relatively smaller field strength E. Thus point X is in a stronger field in Fig. 24.7b than point Y; and in this direct relation, so pictorial and helpful to our imagination, lies much of the usefulness of this concept of lines of force, as we shall see.

Problem 24.14. (a) Note that a line of force, starting at a positive charge, either goes to infinity or terminates on a nearby negative charge. Why? (b) Can lines of force intersect each other? (c) Draw some lines of force (i) around two positive charges placed like Q_1 and Q_2 in Fig. 24.7b, (ii) between a parallel-plate condenser charged oppositely on the two plates. (d) Is the electric field around a charged body *real?* Are the lines of force real?

Problem 24.15. It is a fact that if a conductor carries electric charges that have found their *equilibrium* position (steady state), the lines of force join the conductor perpendicularly at every point on the surface. Explain why. [*Hint:* Consider what would happen to the charges on the surface if the lines of force were oblique instead of normal.]

24.10 Electric Potential Difference—Qualitative Discussion

We add now to our tool kit another concept of fundamental importance in electricity—electric potential difference, for which the symbol V will be used.

The word *potential* reminds us at once of potential energy, defined in mechanics (Section 17.4). In fact, the new concept also refers to work done by virtue of a change of position (of charged bodies), although it is now work done by or against electric forces rather than gravitational or elastic ones. But the analogy is quite deceptive in many respects and is better not invoked at all. For a definition we rather choose the following.

The potential difference V_{AB} between any two points A and B is the ratio obtained when the work done in moving a positive electric charge q from B to A is divided by the magnitude of the charge q; V_{AB} is the work done per unit charge moved, in joules/coul.

By definition, 1 joule/coul = 1 volt.

For example, refer again to Fig. 24.7b. When we attempt to carry a positive charge q from Y to X our muscles must do work against the repulsion from Q_1 on the left and the attraction to Q_2 on the right. The net work done per unit charge can be found experimentally (suggest a way), and turns out to be independent of the exact path chosen between Y and X just as the net work done in raising a weight depends on the total displacement rather than on the exact path taken.

By convention, we say that X is at a more positive potential than Y or Z because our muscles or some other nonelectric forces must do work on a positive charge to carry it from Y or Z to X. On the other hand, Z is at a more negative potential with respect to Y or X because on carrying a positive charge to Z from Y or X all the work is done by the electric field instead of our muscles. In brief, the relative positive and negative potential at a point is recognized by the behavior of a positive charge that, "on its own," tends to go *away* from a point at relatively *positive* potential and toward one at relatively *negative* potential.

While in this manner the words "potential difference between A and B" are easily given operational significance, the frequently encountered phrase "the potential at point A" is not so clear, because no mention is made in it of the starting or reference point. By *convention,* the *implied* reference point is some infinitely distant uncharged body or, in practice, for example, the surface of the earth (*electrical ground*) at some distance below point A. Therefore if we hear "the potential is positive at point A," we know that a positive test charge brought to point A will experience a net force of repulsion and will try to escape to ground if a conductor is stretched from point A to ground (see Fig. 24.8). Similarly, a point A' is at a negative potential if a positive charge is readily attracted from ground toward that point A'. The numerical value of the potential at a point A (no second reference point given) is then simply the potential difference between A and any point on the ground, or the work done per unit charge on taking a positive charge from ground to point A. For all our purposes, then, we may always replace the words

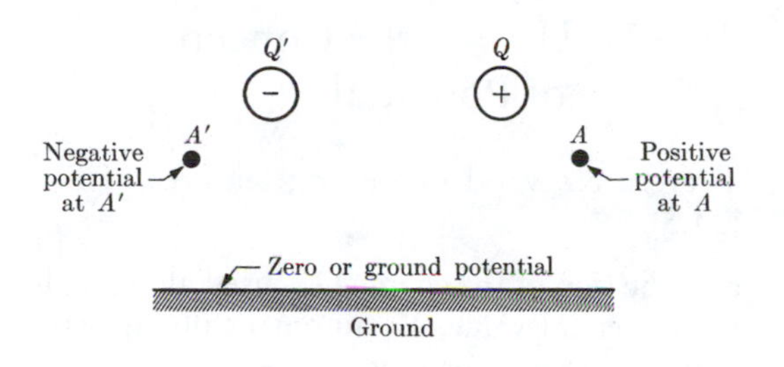

Fig. 24.8. "Potential at a point" implicitly defined relative to ground.

"potential at a point" by "potential difference between that point and ground."

Problem 24.16. What is the operational meaning of the phrase "a point *P* has a negative potential of 1 joule/coul"?

Problem 24.17. Decide whether the potential is +, –, or 0 at points *X*, *Y*, and *Z* in Fig. 24.7b, providing $Q_1 = Q_2$ numerically but is of opposite sign.

Problem 24.18. At electrostatic equilibrium (no current), the potential difference between any two points on the surface of a conductor is zero. Explain why.

Problem 24.19. In Fig. 24.8 there will also be points between *Q* and *Q'* where the potential is zero. Indicate a few. [*Hint:* The potential at any point *P* with respect to ground is zero if no work is done moving a charge from ground to *P*. Furthermore, no work is done moving always perpendicular to lines of force; why?]

24.11 Potential Difference—Quantitative Discussion

So far we have developed only the qualitative and the experimental aspects of the concept of potential difference; now we ask the inevitable question, How can one *calculate* and predict the potential difference between two points?

Consider the simplest possible case, the potential difference V_{AB} between points *A* and *B*, in the vicinity of a charged sphere *Q* and away from all other charges (Fig. 24.9). Let us assume the charge *Q* is positive. Point *A* is therefore at a higher positive potential than *B*, since one has to supply work to transport a small positive

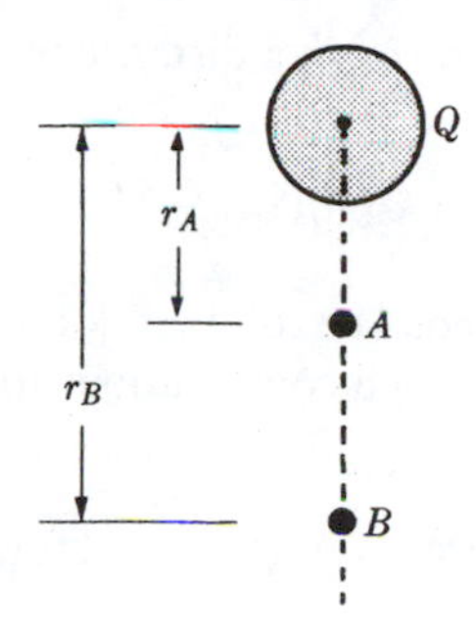

Fig. 24.9. The difference in potential between points *A* and *B* near uniformly charged sphere *Q* equal to $V_{AB} = k_eQ/r_A - k_eQ/r_B$.

charge *q* from *B* to *A* against the repulsion of *Q*. V_{AB} is this work per unit charge; and if the force on *q* between *B* and *A* were constant (say *F*) the work in this simple case would be $F(r_B - r_A)$, V_{AB} being consequently $F(r_B - r_A)/q$.

But the force on *q* is not at all constant during its transit from *B* to *A*. In vacuum or air the force changes from k_eQq/r_B^2 at *B* to k_eQq/r_A^2 at *A*, and not linearly either (where k_e is the electrostatic force constant). Frankly, we do not know how to calculate the exact work done under such circumstances without calculus, but we can make an instructive approximate calculation of V_{AB}, which will point to the simple result

$$V_{AB} = k_e(Q/r_A) - k_e(Q/r_B). \qquad (24.4)$$

Derivation. To find an approximate value for the work done on *q* from *B* to *A*, we mentally subdivide the path *BA* into very small parts, each of length Δs. The force does not change much during a small displacement Δs. For example, on going from *B* at distance r_B to a near point at distance r_1 from *Q* (whence $\Delta s = r_B - r_1$), the average force is nearly k_eQq/r_B^2 or also k_eQq/r_1^2 or k_eQq/r_Br_1; the smaller Δs, the more nearly true this is. But the work ΔW_1 done during this small displacement Δs is simply the average force × displacement (forces and displacements being here parallel):

$$\Delta W_1 = (k_eQq/r_Br_1)\Delta s = (k_eQq/r_Br_1)(r_B - r_1)$$

$$= k_eQq(1/r_1 - 1/r_B).$$

On proceeding toward *A* over the next small interval to a near point at distance r_2

from Q, that is, through a distance $(r_1 - r_2)$, the work done is

$$\Delta W_2 = k_e Qq(1/r_2 - 1/r_1),$$

similarly obtained. The total work on q from B to A, the sum of such contributions for n small intervals, is

$$W_{\text{total}} = \Delta W_1 + \Delta W_2 + \ldots + \Delta W_n$$
$$= k_e Qq(1/r_1 - 1/r_B) + k_e Qq(1/r_2 - 1/r_1)$$
$$+ \ldots + k_e Qq(1/r_A - 1/r_n)$$
$$= k_e Qq(1/r_1 - 1/r_B + 1/r_2 - \ldots + 1/r_A - 1/r_n)$$
$$= k_e Qq(1/r_A - 1/r_B),$$

all other terms canceling. *By definition,*

$$V_{AB} \equiv W_{\text{total}}/q = (k_e Q/r_A) - (k_e Q/r_B);$$

Q.E.D.

This last result applies only to the region (in air or vacuum) around a single point charge Q, or to a single sphere on which Q charges are uniformly distributed.

An important consequence of Eq. (24.4) becomes apparent if we let r_B become very large. As $r_B \to \infty$, V_{AB} now means the potential difference of point A with respect to an infinitely distant uncharged body (for example, to ground); V_{AB} is now "the potential at a point A that is r_A meters from a single point charge Q or a uniformly charged sphere." In this case, since $1/r_B = 1/\infty = 0$, let us write simply V_A instead of V_{AB}, and we have

$$V_A = k_e Q/r_A. \qquad (24.5)$$

Problem 24.20. In Problem 24.9 we considered the force at 0.5×10^{-10} m from a nucleus of a hydrogen atom. Now find the potential at that distance. Also find the potential difference between two points 0.5×10^{-10} m and 2×10^{-10} m from the nucleus, respectively.

Problem 24.21. Find the potential at a point halfway between Q_1 and Q_2 in Fig. 24.7b if Q_1 = +3 coul, Q_2 = 5 coul, and the distance between them is 10 cm.

24.12 Uses of the Concept of Potential

There are several interesting features of Eq. (24.5).

a) If the sign of the charge Q is used along with its numerical value, V_A automatically appears with its correct + or – sign.

b) Since the units of k_e are N·m²/coul², the units of V_A are N·m/coul; this is consistent with our previous statement that potential difference has units of joules/coul, since 1 joule = 1 N·m.

c) Like the concept of work itself, potential is a scalar; therefore simple algebraic addition gives the total potential at any point in the vicinity of more than a single point charge. Hence a point equally far from equally strong opposite charges, like point Y in Fig. 24.7b, is at zero potential with respect to ground (but the *force* is *not* zero there! Is this a contradiction?).

d) In our derivation there was no restriction on how close to the charged sphere the point A may be for the equation $V_A = k_e Q/r_A$ to hold. Actually, point A may even be right on the surface of the sphere, in which case the potential of the sphere, V_s, is the ratio of its charge Q to its radius r_s, that is, $V_s = k_e Q/r_s$. Thus 1 coul of charge deposited on a 2-cm sphere (r_s = 1 cm) raises the sphere to $(9 \times 10^9$ N·m²/coul²)(1 coul)/(0.01 m) = 9×10^{11} J/coul = 9×10^{11} volts potential. It is clear that relatively small bodies can easily achieve large potentials if supplied with enough charge; a rubber or glass rod can quickly be given enough charge by friction to develop several thousand volts on a dry day.

The usefulness of the concept *electric potential* may be further illustrated by two other instances:

e) We can now quickly calculate the potential energy of a point charge q at distance r_A from a point charge Q. The potential at that distance is V_A; by definition this is the potential energy of a charged body *per unit charge*. Therefore the total potential energy of q at A is the product $V_A \times q$. If, for example, Q refers to the nucleus of a hydrogen atom and q refers to its electron at $r_A = 0.5 \times 10^{-10}$ m, the potential energy of the electron is

$$\text{PE} = (k_e Q/r_A) \times q$$

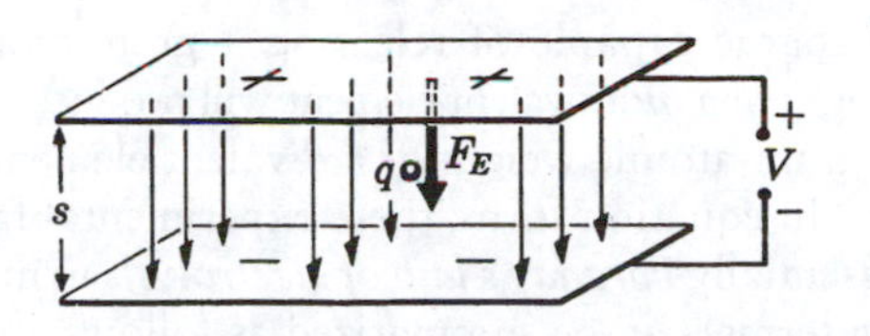

Fig. 24.10. **The electric force F_E on a small charged body q between oppositely charged plates.**

$$= (9 \times 10^9 \text{ N·m}^2/\text{coul}^2)(1.6 \times 10^{-19} \text{ coul}) \times (-1.6 \times 10^{-19} \text{ coul})/0.5 \times 10^{-10} \text{ m}$$

$$= -4.6 \times 10^{-18} \text{ joules.}$$

(What is the meaning of the negative sign?)

f) We found before that the electric force F_E on a charge between two equally and oppositely charged plates is constant. As a consequence, the electric intensity E in such a gap is also constant. The work that must be done *per unit charge* to transport a small positive test charge q from the negative plate a distance s straight across the gap to the positive plate (Fig. 24.10) is, by definition, V (its value is read off most conveniently from the label on the battery that might be used to provide the potential difference). In this situation, the work that must be done on q against the electric forces is work per unit charge × total charge, or

$$F_E \times s = V \times q.$$

Or transposing, and noting that $F_E/q = E$,

$$E = V/s. \qquad (24.6)$$

If V is in J/coul (= volts) and s is in m, E is in N/coul = volts/m.

The importance of Eq. (24.6) is that since s is easily measured and V is known from the battery used, E can be readily calculated at all points of the space between the plates.

Problem 24.22. Two parallel horizontal plates are 2 cm apart in vacuum, and are connected to a supply providing a 10,000-volt potential difference. An electrically charged oil drop, which from other measurements appears to have a mass of 10^{-4} g, is seen to be suspended motionlessly between the plates. What is the electric force on that oil drop? What is its charge q? What would be the charge on the drop if it were in vertical motion in the electric field with an acceleration of 2 cm/sec² up? 10 cm/sec² down?

Problem 24.23. A Van de Graaff generator is essentially a metallic sphere on an insulating stand, charged to a high electric potential, which is used to repel and accelerate charged atomic particles along an evacuated tube to reach extremely high speeds; when they crash into a "target" they will have an energy suitable for "atom smashing." If we wish to use a sphere 10 ft in diameter in this accelerator at 2,000,000 volts, how much charge must be supplied? (Watch the units.) What might be the practical limits to reaching much higher potentials with such a device?

Problem 24.24. If an electron is repelled from a point near the surface of this same Van de Graaff generator and is allowed to accelerate freely, what will be its final kinetic energy as it hits a target far from the generator? What will be its final speed? (Mass of electron = 9.11×10^{-28} g, charge = 1.6×10^{-19} coul.) What was the potential energy of the electron just as it began to leave the surface of the sphere?

Problem 24.25. How much kinetic energy (in joules) does an electron gain when it freely passes through a potential difference of 1 volt? [This quantity of energy has been given the unfortunate name *1 electron volt* (1 eV), an important unit of energy in atomic physics.]

24.13 Electrochemistry

The history of electricity shows that the fundamental notions of charge, conductor, insulator, induction, and the like were acquired only slowly and sometimes after lengthy struggles. It was not until about 1800, the year of Alessandro Volta's discovery of the voltaic cell, that the subject of electricity really opened itself to widespread and fruitful investigations touching on the structure of matter. (A "battery" is a sequence of such cells.)

One of the first consequences was the discovery of the spectacular decomposing effect of current on certain liquids. Electric current proved to be a tool capable of breaking up into their elements dissolved or molten materials whose molecules previously had resisted separation by chemical or physical means. It was by this process, called electrolysis, that Davy in 1807 broke down potash and soda, until then considered to be elements, and obtained from them the metals potassium and sodium.

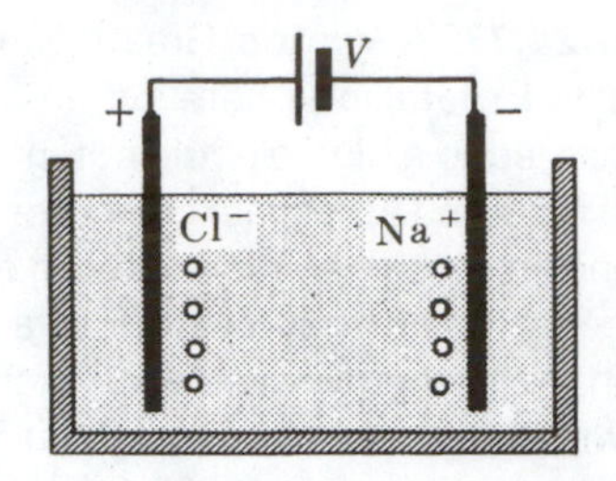

Fig. 24.11. Migration of ions to electrodes during electrolysis.

While the study of electric currents led physicists to the subject of electromagnetism, which is part of our next consideration, further research on electrolysis itself led chemists to the atomic hypothesis, by a door other than Dalton's. In a systematization and extension of Davy's methods, Faraday proved that the passage of current through a liquid (for example, a diluted acid or a salt solution) would dissociate the liquid at the two electrodes in a regular and accurately repeatable manner. The mass of an element liberated from the liquid, either in the form of gas bubbles (as for hydrogen or chlorine) or as metal deposited at or near the electrode (for example, copper), is proportional both to the molecular weight of the material freed and to the quantity of charge transferred, but inversely proportional to the valence of the material liberated.

For example, charge passing through a molten sample of common salt (NaCl) caused the sodium metal to collect at the negative electrode [the *cathode* (Fig. 24.11)], and liberated chlorine gas bubbles at the positive electrode (*anode*). The mass of metallic sodium is found to be about 24×10^{-5} g per coulomb of charge so transferred, and the mass of chlorine gas about 12×10^{-14} g. It follows that in order to obtain 1 gram-atomic weight of Na (namely, 23 g), one needs 23 g/(24×10^{-5} g/coul) or 96,500 coul; at the same time, the transfer of this quantity of charge (occasioned, say, by a 26.8-ampere current lasting 1 full hour) also releases 1 gram-atomic weight of chlorine (35.5 g) at the other electrode. The transfer of twice that charge yields twice the respective amounts of these monovalent substances.

If we now electrolyze water, 96,500 coul will release 1 g of hydrogen but only 8 g of oxygen, for oxygen has the valence 2, and the experimental law of electrolysis teaches that a quantity of charge capable of releasing 1 gram-atomic weight of a *mono*valent element will produce only ½ gram-atomic weight of a *di*valent element.

In equation form, these experimental facts (essentially *Faraday's law of electrolysis* in modern terms) can be summarized as follows:

$$\text{Mass of material liberated at one electrode (g)} = \frac{\text{charge passed (coul)}}{\text{96,500 (coul per g-at.wt)}} \times \frac{\text{at. wt of material}}{\text{valence of material}} \tag{24.6}$$

Incidentally, the recurring quantity of 96,500 coul is given the convenient name *1 faraday* of electric charge, so that we may phrase Faraday's law of electrolysis simply: In electrolysis, 1 faraday releases 1 gram-atomic weight of any monovalent material, ½ gram-atomic weight of divalent materials, etc., or, in terms of concrete examples, 1.008 g hydrogen, 31.78 g copper, 40.59 g antimony. As you can imagine, as electrolysis is of the highest importance in industrial production, chemical and biomedical research, etc., Faraday's law is an enormous aid.

Problem 24.26. The element barium (Ba, at. wt 137.36, valence 2) was also discovered by Davy by electrolysis. If you repeat an experiment of this kind and for 1 hour pass a current of 1 ampere through molten barium chloride ($BaCl_2$), how much Ba and how much Cl do you expect to collect? [*Note:* 1 ampere = 1 coul/sec.]

24.14 Atomicity of Charge

Now that the hard-won concepts of atom and electron are so familiar to physical scientists, it is easy to visualize a mechanism that will explain the main phenomena of electrolysis. In such simple cases as the breakdown of molten sodium chloride one can think of the liquid as made of separate atoms of sodium and chlorine. But since Na migrates to the cathode and Cl to the anode, we must account for the influence of electric forces on the atoms; each Na atom after the breakup of the molecule must somehow be positively charged, and each Cl atom must be negatively charged. The passage of electricity through the liquid is then accomplished by the movement of *charged* atoms, or *ions* (from the Greek word for *to wander* or *to go*).

At this point enters a theoretical picture of the molecule: We are led to the assumption that both Na and Cl, in the pure state of the element sodium and the element chlorine, are ordinary electrically neutral atoms by virtue of a perfect balance between the positive and negative components that make up the atom. Specifically, the negatively charged parts of the atom are the electrons, each having the same definite mass m_e and charge e.

Now when one neutral atom of Na and one of Cl together form one molecule of NaCl in the solid salt, one of the electrons from the Na atom attaches itself to the Cl atom (later we must explain why this should happen), so that on separation during electrolysis the Na ion lacks one electron and the Cl ion has one electron more than it does when in its neutral state. These ions can be symbolized by Na^+ and Cl^-. Similarly, a molecule of copper sulfate, $CuSO_4$, splits during electrolytic dissociation into Cu^{++} and SO_4^{--} ions by the transfer of two electrons.

An explanation of Faraday's law of electrolysis as summarized by Eq.(24.6) suggests itself. Every time one Cl^- ion reaches the anode, it gives up its excess electron to that electrode, becomes neutral, and rises eventually in a bubble with other such chlorine atoms. At about the same time, an Na^+ ion picks up one of the electrons at the cathode and settles out as neutral sodium metal; we may imagine that this neutralizing electron was made available when the Cl^- ion gave up its excess electron to the circuit on the other electrode. Thus for each molecule of NaCl electrolyzed into its elements, the charge corresponding to *one* electron moves through the external circuit (wires, battery, and current meters).

This hypothesis leads at once to an estimate of the value of the charge e on one electron: We know that the passage of 96,500 coul of charge results in the electrolysis of 1 mol or 6.02×10^{23} (= N) atoms of a monovalent element. At the rate of transfer of one electron per atom, we have

96,500 coul/mol = (N, the number of atoms/mol) × (e, charge in coul/electron)

or

$$e = 96{,}500/N, \tag{24.7}$$

from which we can calculate that the charge per electron is 1.6×10^{-19} coul.

The confirmation of this value for e in several independent and entirely different experiments (for example, Millikan's oil-drop experiment, spectroscopic data) gives us confidence that the whole physical picture leading to this point is plausible. Or conversely, Eq. (24.7) together with separate determination of e enables us to compute a value for Avogadro's number N, a value that is indeed in agreement with the results of other experiments.

This concordance of numerical results obtained by widely different methods shows us again the interconnectedness of physical science. From isolated observable facts and limited hypotheses we fashion, as it were, a net or web, and the constants of nature—the constant of universal gravitation, the values of N and e, the atomic weights, the faraday, and the like—are at the points of intersection of different strands, the knots that hold the whole structure together.

The two ideas of importance introduced here are, first, the atomicity of matter and, second, the "atomicity" of electric charges (that is, charges in liquids are transferred by ions bearing 1, 2, 3, . . . electrons more or less than when the atoms are neutral). A third idea is also implied, namely, that the electron is ordinarily a part of the atom.

But all three postulates were far from evident when Faraday phrased his generalizations on electrolysis in 1833. Atomicity of matter soon became acceptable on other grounds (through chemistry and kinetic theory), but the existence of discrete charges called electrons was first proposed in definite form only in 1874 by G. Johnstone Stoney, and was not completely acceptable for another 20 years, when precise measurements of the mass and charge on the electron became available through the study of beams of electrons in cathode-ray tubes (Section 25.5).

The part that electrons play in atoms was then the next major puzzle; its solution, which is the substance of Chapters 28 and 29, formed one of the most exciting segments in the recent advances of physical science.

RECOMMENDED FOR FURTHER READING

I. B. Cohen, *Benjamin Franklin's Science,* Cambridge, MA: Harvard University Press, 1996

C. C. Gillispie, *Dictionary of Scientific Biography,* articles by I. B. Cohen on Franklin, Vol. 5, pages

129–139, and C. S. Gillmor on Coulomb, Vol. 3, pages 439–447

J. L. Heilbron, *Elements of Early Modern Physics,* Berkeley: University of California Press, 1982

Albert E. Moyer, "Benjamin Franklin: "Let the experiment be made" in *Physics History* (edited by M. N. Phillips), pages 1–10

D. Roller and D. H. D. Roller, "The development of the concept of electric charge: Electricity from the Greeks to Coulomb," in *Harvard Case Histories in Experimental Science* (edited by J. B. Conant), pages 641–639

CHAPTER 25

Electromagnetism, X-Rays, and Electrons

25.1 Introduction

The two previous chapters have been devoted to subjects that so far may appear to be entirely separate: light and electricity. We must now introduce yet a third apparently distinct topic: magnetism. By exploring the relationships between electricity and magnetism, we shall arrive at a general theory (Maxwell's theory of electromagnetic fields), which turns out, perhaps unexpectedly, to provide an explanation for the propagation of light, as well as being a modern model of a grand synthesis of previously separate parts of physics.

Just as the concept of *energy* provided a unifying link for mechanical and thermal phenomena (Chapter 17), the concept of *field* brought electricity, magnetism, gravity, and light into a common framework of physical theory. And the two concepts themselves are closely related, for, as we shall see, a field can be regarded as a form of energy in space. If it could be established that "space" really contained (consisted of) an ether having mechanical properties (so that all energy is really mechanical energy, traveling in a medium, the ether, as Maxwell and other nineteenth-century physicists believed), then this finding would have to be declared the ultimate triumph of Newtonian mechanics, and the evolution of concepts and theories in physical science would have reached the goal of a complete mechanical explanation of nature.

That indeed was the judgment of a few physicists in the 1870s and 1880s, echoed in A. A. Michelson's famous 1894 statement:

> It seems probable that most of the grand underlying principles have been firmly established and that further advances are to be sought chiefly in the rigorous application of these principles to all the phenomena which come under our notice. . . . An eminent physicist [not identified] has remarked that the future truths of Physical Science are to be looked for in the sixth place of decimals. (Michelson, "Some of the Objects and Methods")

But not everyone was so confident that all the fundamental laws of physics were known; some physicists wanted to replace mechanics by electromagnetism as the source of basic concepts. Other leading scientists, like Lord Kelvin in a 1901 paper, perceived "clouds over the dynamical theory of heat and light," foreshadowing the coming revolution in physical theory that will be the subject of the last chapters of this book.

25.2 Currents and Magnets

For most phenomena of electrostatics described in the preceding chapter there exist similar counterpart phenomena in *magnetostatics* (the interactions of magnets at rest relative to each other). The principal difference is that magnetic poles, unlike electric charges, are always found in *pairs,* called north and south poles. (Single poles or *monopoles* might exist on the subatomic level, but at the present time the search for them remains fruitless.) The obvious similarity between electricity and magnetism suggested long ago that there should be some interaction between electric charges and magnetic poles, but no such interaction could be demonstrated experimentally until the early part of the nineteenth century.

The first concrete evidence of an interaction between electricity and magnetism came in 1819–1820, when the Danish physicist Hans Christian Oersted (1777–1851) performed a series of experiments of momentous consequences. The original inspiration for these experiments, according to Oersted's own account, was the metaphysical conviction of the unity of all forces in nature, which he derived from the German "nature philosophers," in particular Friedrich Schelling. Although nature philosophy was generally characterized by mystical,

Fig. 25.1. Hans Christian Oersted (1777–1851).

antimathematical speculation of a type that has generally been of little benefit to science, it did in this case encourage the trend toward unification, which had some outstanding successes in the nineteenth century. As we insisted in Section 12.3, there is no single method that all scientists can be expected to follow; the route to some of the greatest discoveries in science has been amazingly unorthodox.

In his famous experiment, Oersted placed a magnetic compass needle directly beneath and parallel to a long horizontal, electrically conducting wire. He had placed the wire along the earth's magnetic north-south line, so that the magnetic needle was aligned parallel to the wire. When the wire was connected to the terminals of a battery, the compass needle swung to an east-west orientation—perpendicular to the wire! Although it had been shown much earlier that electric charge at rest does not affect a magnet, now it was clear that charge in motion (a current) does exert a strange kind of "sideways" (orthogonal) force on a magnet needle placed near the conducting wire.

Oersted's results were the first instance in which a force was observed that did not act along a line connecting the sources of the force. That is to say, the force that the current-carrying wire exerts on a magnetic pole placed below the wire is not along the straight line from the wire to the pole: the force on the pole is *perpendicular* to any such line. The magnetic needle is not attracted to, or repelled by, the current; it is *twisted* sidewise by forces acting on its poles.

The unusual way in which a compass needle is affected by an electric current helps to explain why it took so long before anyone found this interaction between electricity and magnetism. To begin with, there is no effect between static electric charges and magnetic poles. Also, steady electric currents had not been readily available for laboratory experiments until Volta developed the battery in 1800.

But even if one has currents and compass needles available, one does not observe the effect unless the compass is placed in the right position, so that the needle can respond to a force that seems to act in a direction *around* the current rather than toward it. The force in this case is neither attractive nor repulsive but acts at right angles to the line between the current and the magnet.

We are dealing here with what had been a psychological obstacle to discovery: the presupposition that all forces are like gravitational and electrostatic forces, acting directly along the line between the centers of pieces of matter or charge. These are now called *central* forces. Although no such presupposition is explicitly stated in Newton's writings, later scientists imbued with Newtonian methods and ideas seem to have unconsciously accepted it.

The announcement of Oersted's discovery in July 1820 produced a sensation in scientific circles in Europe and America. One of the first to jump into the field of electromagnetic research was the French physicist André-Marie Ampère (1775–1836). Ampère reasoned that since magnets exert forces on each other, and magnets and currents also exert forces on each other, one might expect that currents exert forces on other currents—a case of reasoning by analogy or extrapolation or symmetry, often used, but only sometimes successful.

He quickly tested the hypothesis, and on September 30, 1820, within a week after word of Oersted's work reached France, Ampère reported his result to the Paris Academy of Sciences: two current-carrying wires do indeed exert forces on each other. The nature of the force depends not only on the distance between the wires but also on their relative orientations and the amount of current they carry. For the special case of parallel wires, the force is *attractive* when the currents flow in the same direction, contrary to what one would expect from the analogy of electrostatics.

Ampère insisted on preserving the "central" character of Newtonian forces in his formula for the force between two infinitesimal current elements; his formula is no longer used, although it gives the correct result for the observed total force when integrated over the entire current.

Ampère's force between two currents has now been adopted as the basis for electrical units in the MKSA system (meter, kilogram, second, ampere). One ampere (abbreviated amp) is defined as the amount of current in each of two long straight parallel wires, one meter apart, that causes a force of 2×10^{-7} newtons on each meter of wire. (The numerical factor of 2×10^{-7} was chosen, somewhat arbitrarily, in order to get a unit of convenient size for practical use. A force of 1 N corresponds to a weight of about 0.225 lb.)

As we noted in Section 24.7, the *coulomb* is then defined as the amount of charge that flows in one second when the current is one ampere. The *volt* is defined as the potential difference between two points such that 1 *joule* of work (see Section 17.3) is done in moving 1 coulomb of charge between those points.

Beginnings of Electrical Technology

The excitement about Oersted's discovery was caused not simply by the fundamental nature of his work, which showed a relationship between two hitherto separate physical phenomena, but perhaps even more by an interest in technological applications of electric currents. It was already suspected that electricity might offer an effective way of transmitting and using the large amounts of energy available from steam engines, thereby making energy available at a distance through wires, instead of having to transport the engine itself.

If a current can exert a force on a magnet, as in Oersted's experiment, one could expect (by Newton's third law, if for no other reason) that a magnet should also exert a force on a current, and it does not take too much imagination to speculate that a magnet might somehow *produce* a current. If this could be done significantly more cheaply than by Volta's battery (which consumed large quantities of expensive metals to yield only small amounts of current), one would find oneself at the threshold of the "Electrical Age," the fulfillment of the Industrial Revolution. And indeed it turned out that only one more bold spirit was needed to open that door.

Fig. 25.2. Michael Faraday (1791–1867).

Faraday's Discovery

In 1821 the editor of the British journal *Annals of Philosophy* asked a young man, Michael Faraday, to undertake a historical survey of the experiments and theories of electromagnetism, to summarize the burst of activity in this field inspired by Oersted's discovery in the previous year. Faraday, who was at that time merely a laboratory assistant to the well-known chemist Humphry Davy, did not yet have a reputation in science, but he was eager to learn all he could. Faraday agreed to accept the assignment, but soon found that he could not limit himself to simply reporting what others had done; he felt that he had to repeat the experiments in his own laboratory. Not being satisfied with the theoretical explanations proposed by other physicists,

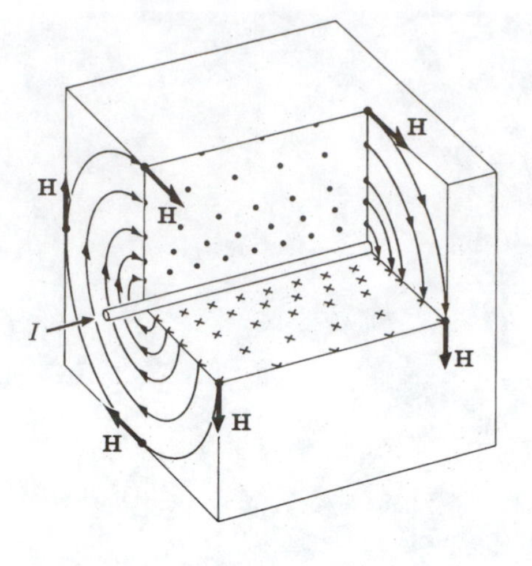

Fig. 25.3. Magnetic field around a long straight conductor.

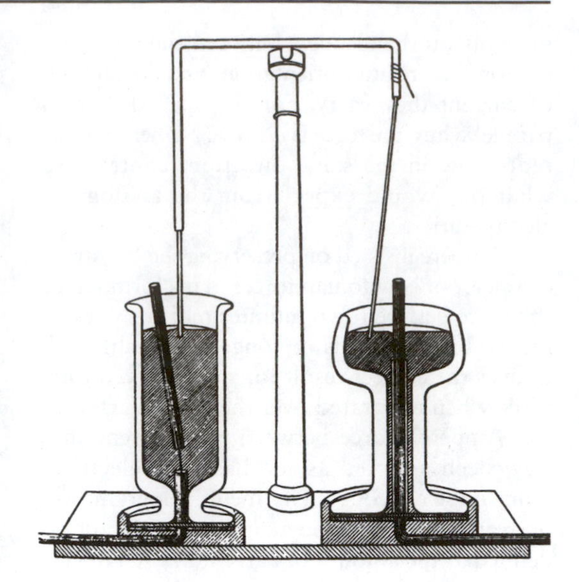

Fig. 25.4. Two versions of Faraday's electromagnetic rotator. In each, the cup is filled with mercury so that a large electric current can be passed between the top wire and the base. At the left, the south end of a bar magnet is fixed, and the north end is free to revolve along one of the circular magnetic lines of force surrounding the current. At the right, the rod carrying the current revolves around the firmly fastened bar magnet.

Faraday started to work out his own theories and plans for further experiments. Before long he had launched a series of researches in electricity that was to make him (despite his lack of formal training in science) one of the most famous physicists of his time.

Faraday's first discovery in electromagnetism was made (according to the careful records he kept) on September 3, 1821. Repeating Oersted's experiment by holding a compass needle at various places around a current-carrying wire, Faraday realized that the force exerted by the current on the magnet is circular in nature. As he expressed it a few years later, using his great skill of visual imagination, the wire is surrounded by an infinite set of concentric circular *lines of force,* so that a magnetic pole that is free to move experiences a push in a circular path around a fixed conducting wire. The collection of these lines of force is called the *magnetic field* of the current, a term introduced by Faraday. (See Fig. 25.3; the symbol ***H*** is used for the magnetic field vector at every point around the current I. ***H*** is everywhere tangential to the line of force.)

Faraday immediately constructed an "electromagnetic rotator" based on this idea, whereby a bar magnet pivoted at one end could rotate around the wire along the line of force acting on the freely moving pole. He used mercury to complete the electric circuit (see Fig. 25.4, left side).

Faraday also designed an arrangement in which the magnet was fixed and the current-carrying wire rotated around it, thereby showing that, in accordance with Newton's third law, the magnet does exert a force on the wire (Fig. 25.4, right side). As in many other cases, Faraday was guided by the idea that for every effect that electricity has on magnetism, there must be a converse effect of magnetism on electricity—although it was not always so obvious what form the converse effect would take!

Armed with his lines-of-force picture for understanding electric and magnetic fields, Faraday joined in the search for a way of producing currents by magnetism. Scattered through his diary in the years after 1824 are many descriptions of such experiments. Each report ended with a note: "exhibited no action" or "no effect."

Finally, in 1831, came the breakthrough. Joseph Henry at the Albany (New York) Academy in America was apparently the first to produce electricity from magnetism, but he was not able to confirm and publish his results immediately because of his heavy teaching duties. He was not able to publish his work until a year later, and in the meantime Faraday had made a similar discovery and published his results. Faraday is known as the discoverer of *electromagnetic induction* (production of a current by magnetism) not simply because he established official priority by first publication, but primarily because he conducted exhaustive investigations into all aspects of the subject.

Faraday's earlier experiments and his thoughts about lines of force had suggested to him the possibility that a current in one wire ought to be able to induce a current in a nearby wire, perhaps through the action of the magnetic lines of force in the space around the first current. But again, presuppositions based on earlier ideas delayed the discovery by restricting the range of possible combinations. Oersted had shown that a *steady* electric current produced a *steady* (constant) magnetic effect around the circuit carrying the current. By arguing from "symmetry," one could easily expect that a steady electric current could somehow be generated if a wire were placed near or around a magnet, although a very strong magnet might be needed to show an effect. Alternatively, a steady current might be produced in one circuit if a very large current was flowing in another circuit nearby. Faraday tried all these possibilities, without success.

At last, on August 29, 1831, Faraday found an effect. He had put two wires, *A* and *B*, near each other and sent current through one of them (*A*) from a battery. He noticed that a current did appear in the other wire (*B*), but only while the current from the battery in the first wire *started* or *stopped*. The current "induced" in wire *B* lasted only for a moment, just while contact was being made with the battery. But as long as there was a steady current in wire *A*, there was no current in wire *B*. When the current in wire *A* was stopped, again there was a momentary current induced in wire *B*. To summarize Faraday's result: A current can induce another current only while it is *changing*. A steady current in one wire cannot induce a current in another wire.

Faraday was not satisfied with merely observing and reporting this result, which he had reached by accident (that is, without having looked for this particular form of the long-suspected effect). Guided by his concept of lines of force, he tried to find out what were the *essential factors* involved in electromagnetic induction.

In Faraday's first successful experiment, wire *A* was not a straight segment, but was wrapped around one side of an iron ring (the *primary coil,* shown as *A* in Fig. 25.5) and could be connected to a battery. The other wire (the *secondary coil B* in Fig. 25.5) was wrapped around the other side of the ring and connected to a current-indicating device (as shown in Fig. 25.5, but a compass needle placed near wire *B* would serve the same purpose). Faraday reasoned that the change in the current in the primary coil would cause the lines of magnetic force throughout the iron ring to change, and that this change in some way served to induce a current in the secondary coil. But if this was the correct explanation of induction, Faraday asked himself, should it not be possible to produce the same effect in another way? In particular:

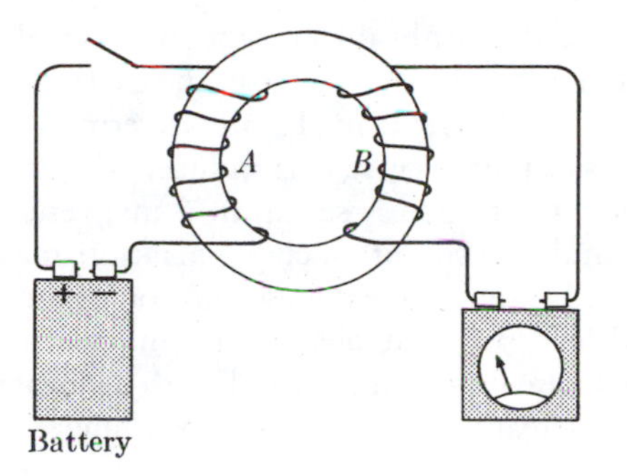

Fig. 25.5. Faraday's electromagnetic induction experiment.

1. Is the iron ring really necessary to produce the induction effect, or does it merely intensify an effect that would occur anyway?
2. Is a fixed primary coil and a switched current in *A* really necessary, or could current be induced in *B* merely by changing the magnetic field in which it is bathed in some other way, such as by doing entirely without wire *A* and moving a bar magnet relative to wire *B*?

Faraday answered these questions almost immediately by further experiments. First, he showed that the iron ring was not necessary; starting a current in one coil of wire would induce a momentary current in a nearby second coil, separated from the first only by air (or even empty space). Second, he found that when a bar magnet was inserted into the end of a coil of wire, a pulse of current was induced in the coil at the instant of insertion.

Having done these and many other such experiments, Faraday stated his general principle of electromagnetic induction: Changing lines of magnetic force can cause a current in a wire. The "change" in the lines of force can be produced either by (a) a magnet moving relative to the wire, or (b) a change in the current going in a second wire. Using Faraday's term *field,* we can say that a current is induced in a circuit when a variation is set up in a magnetic field around the circuit; such a variation may be caused either by the relative motion of the wire and field (for example, when a magnet near a fixed wire is pushed back and forth by means of, say, a steam engine), or by any change in the intensity of the field.

Consequences

Let us note at least briefly the enormous consequences for all of society that can spring from such "disinterested" research in basic (or "pure" or "fundamental") science. Faraday's rotators were the first motors. From these toylike objects came the development of larger, more effective motors over the next half century and to this day—all of which work on essentially the same original principle.

From Faraday's experiment on induced current followed the whole set of "dynamos," generators of electric current, each again working essentially on the same principle as his original simple and clumsy current generator. Together, the generator and motor made possible a transformation of society through the "electrification" of most tasks that require transmission and use of energy or of information. To give just one example: Once engineers used the advancement of electrical knowledge by scientists to make efficient electric motors, those motors were used effectively to create streetcars, subways, and elevators, which, in turn, in the hands of architects, city planners and political decision makers, enabled the expansion of the size of cities (in all three dimensions)—for good or ill.

The effective generation of electricity and the invention of the electric light bulb also helped to change the quality of life, for example, by bringing cheap and safe light into the farmer's house—one of the reasons President Franklin Delano Roosevelt gave when he urged the construction of huge electric generating facilities such as the Tennessee Valley Authority.

Leaving such practical applications of his discovery to others, Faraday continued his basic research. Among his many other results was one that hinted at, and perhaps helped to inspire, Maxwell's great breakthrough, described in the next section: Faraday found that a *magnetic field can rotate the plane of polarization of a beam of polarized light.* Does this imply that light itself has magnetic properties?

25.3 Electromagnetic Waves and Ether

We now turn to the work done in the 1860s by the great British scientist James Clerk Maxwell. We have already met Maxwell through his contributions to the kinetic theory of gases (Chapter 22); Fig. 25.6 is a portrait of him at a later period in his relatively short life.

It has often been said that the importance of Maxwell's electromagnetic theory for physical science approaches that of Newton's work, but we must be content here with the briefest summary. His theory of the propagation of light waves is now referred to as the *classical theory of light.*

Maxwell first considered what happens when an electric current oscillates along a straight piece of wire or circulates in a wire loop. To visualize the interactions between electric currents and magnetic fields, he worked out an elaborate theoretical model, in which magnetic fields were represented by rotating vortices in a fluid and charges by tiny spheres—like ball bearings or idle wheels in a machine, whose function is to transmit the

Fig. 25.6. James Clerk Maxwell (1831–1879).

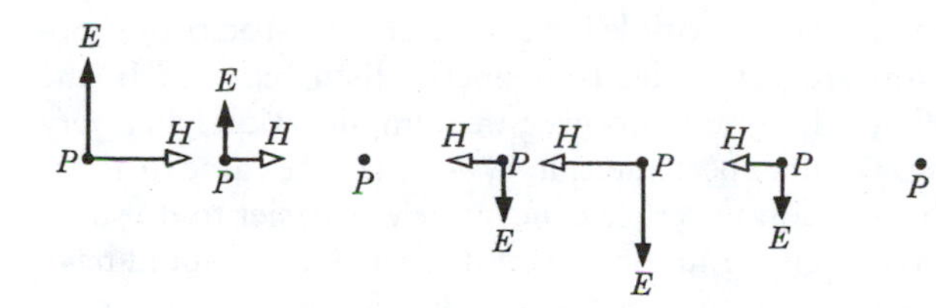

Fig. 25.7. Electric (***E***) and magnetic (***H***) field intensity vectors at the same point *P* and at eight consecutive moments.

rotation from one vortex to its neighbors. The model allowed him to generalize the ideas of Oersted, Ampère, and Faraday so that they applied to electromagnetic interactions in a region of space where no current-carrying wire is present. He postulated that such regions contain charges (the ball bearings) that can be moved, or *displaced,* by changes in the magnetic fields (the vortices). Even though the charges snap back to their previous positions if the magnetic field variation ceases, their motion amounts to a current, which may be called a displacement current.

The vortex-ball bearing model suggested that, just as a varying magnetic field could generate an electric current in a wire (Faraday's "electromagnetic induction"), it could also produce a motion of charge in space. This displacement current then acts like a "real" current in a wire: it produces a magnetic field (Oersted's effect). That field can then displace other charges, producing more displacement currents.

Maxwell's theoretical conclusion, based on his model, was that an electric current in a wire (or any motion of charge) must send energy out through space, in the form of electric and magnetic fields; this energy is radiated away from the electric current and spreads out, wavelike, in all directions. By this is meant that any electric charges in a distant object in the path of the radiation, say, in a second wire, are set into oscillation at the same frequency as the original source, the first wire; any magnetic poles present in the obstacle also undergo similar vibrations, and the obstacle itself will experience a slight mechanical pressure.

Once he had used his vortex-ball bearing model to arrive at the displacement current hypothesis, Maxwell could drop the model and formulate his theory simply in terms of electric and magnetic fields. (Nevertheless he did not abandon the fundamental premise that space is filled with an ether, and he insisted that electrical and magnetic phenomena are essentially mechanical.) A vibrating or circulating electric charge (called a transmitter or oscillator) sets up a fluctuating electric and magnetic field in the region all around it; when other electric charges or magnetic poles are introduced into this field (as in an antenna or receiver) they are acted on by electric and magnetic forces that change with the same periodicity as the transmitter's original oscillations.

Figure 25.7 shows the electric and magnetic field intensity vectors at one point *P* in the electromagnetic field, at eight successive moments. If you concentrate your attention on the behavior of the electric field vector exclusively—it is the only one that we shall pay much attention to—a useful analogy suggests itself: If some receiver, perhaps a metal wire, which contains electrons that are free to move, is placed at point *P,* the fluctuating electric field acts on these charges in the same manner as a wave passing along a rope agitates the particles of one section of the rope. In fact, it is perfectly correct to say that an electromagnetic wave is passing by this point *P* in space, and to refer to the energy responsible for the motion of the charges as electromagnetic energy or electromagnetic radiation.

Maxwell's theory also predicted that metallic objects would reflect a beam of incident electromagnetic energy like a mirror; that on entering other obstacles, such as a sheet of glass, the path of the beams would be bent; and, most startlingly, that this radiation must travel through vacuum or air with the speed of about 3×10^8 m/sec, a value equal to the already known speed of propagation of light.[1]

---

[1]In Maxwell's theory, the ratio of the electrostatic force constant k_e to the electromagnetic force constant k_m is shown to be equal to the square of the velocity of propagation of electromagnetic waves. The constant k_m arises in the law of force for two parallel currents I_1, and I_2 at distance R, established by Ampère: $F = 2k_m I_1 I_2 L/R_{12}$. (This is the force on length L of the wire.) The Eleventh General Conference on Weights and Measures, in 1960, decided to adopt the arbitrary value $k_m = 10^{-7}$. This choice serves to define the

These and similar predictions of Maxwell's theory all agree on one point: The radiation expected on theoretical grounds from vibrating or rotating currents should behave in every way *like light.* The analogy between the known behavior of light and the expected (but not yet proved) behavior of Maxwell's theoretically predicted electromagnetic waves was indeed tantalizing. At the time Maxwell was developing his electromagnetic theory, the theory of light as a mechanical wave propagating in some ether was generally accepted (Chapter 23). Furthermore, in order to account for the polarization of light, people had assumed that light waves are transverse rather than longitudinal, and it was thought, as we noted, that transverse waves could be propagated only in a solid ether. As can be seen from Figs. 25.5 and 25.6, the electromagnetic waves whose existence Maxwell predicted are also transverse, since the "disturbance" being propagated—the pattern of changing electric and magnetic field vectors—would be perpendicular to the direction of propagation of the wave. Different directions of polarization of light could correspond simply to different orientations of the ***E*** and ***H*** vectors in the plane perpendicular to the direction of propagation.

The question naturally arose as to whether the propagation of disturbances in electric and magnetic fields in space would not also depend on the action of a mechanical medium, an ether. It seemed that the mind could not contemplate waves without visualizing a specific medium. In fact, Maxwell's own thinking and his derivations depended on his assumption of certain rather complex mechanical models for the ether. To this day there are echoes of this conceptual struggle, which we recognize as having its equivalent in physics since the time of Aristotle.

Earlier, the merger of the two ethers postulated for radiant heat and for light respectively into one (Section 17.7) had been the first step toward unification, making one ether do the job of several. Maxwell took a second, big step: He proposed that the propagation medium for magnetic and electric effects also takes on the function of transmitting light and heat waves. (We shall see in the next section that experiments later proved that Maxwell's assumption was right—visible light is merely one species of propagating electromagnetic disturbances.) In the long run, however, this simplification had a very unexpected result: When the time came to abandon the ether concept, it was easier to do so by giving up one general ether than it would have been to disprove and dismiss several ether concepts separately.

That time came in 1905, and it invites a brief but important digression. The one feature all hopeful ether models had in common was that they provided a medium for the propagation of light that existed apart from the source and the observer, just as air is a medium for sound transmission. To pursue this analogy further, consider how one measures the speed of sound: One might station a loudspeaker and a microphone many meters apart and time the progress of a short pulse of sound along that distance. At a temperature of 0°C, the speed of sound in air is ordinarily 330 m/sec. But if we did this experiment on an open railway flatcar, or in some other "laboratory" that moves with respect to the medium (air), then the speed measurement for sound would be affected by our own laboratory's motion. For example, if the sound is directed to a microphone at the rear end of the flatcar, it will get there sooner if the car moves forward than if the car stands still.

Similarly, the measured velocity of light should vary depending on the motion of the "laboratory"—or, more technically, of the coordinate system—through the ether and the placement of the equipment relative to the motion. Our earth itself is a fast-moving "platform," revolving around the sun and rotating on its own axis; light moving along a meter stick pointing east is caught up in the relative eastward motion of the ether through which the earth was imagined to be moving, and arrives at the other end sooner than if it had to travel west or along a north-south line.

This expected difference in the speed of light, which should allow one to measure the velocity of the earth through the ether or, equally well, the relative velocity of the ether past the earth, was essentially what the American experimenters A. A. Michelson and E. W. Morley attempted to find in 1887 (Section 30.2). On the basis of their technique one could have expected a clear and convincing positive result—and, yet, no difference in speeds was observed! Here was a new "postulate of impotency": One cannot measure the relative velocity of the ether. (That

ampere as the unit of current (Section 25.2). Since, according to Maxwell's (now-established) theory, $k_e/k_m = c^2$, we obtain the value $k_e = 10^7c^2$ in the appropriate units of N·m^2/coul^2, as given in Section 24.7.

the earth was not simply "dragging" along the ether in its vicinity was made clear in other experiments.)

After a period of adjustment and speculation, with varying degrees of success, this realization was incorporated into a more far-reaching postulate, one of the two basic pillars of relativity theory (Chapter 30), as announced by Einstein as a principle of physics in 1905: The measured velocity of light in space is constant, no matter what the motion of the observer or of the source may be (and no matter whether one can *intuitively* understand how such a principle can be part of nature).

On the one hand, this principle extends the previous discussion on the relativity principle (Section 8.5) to include optical in addition to mechanical experiments in the statement, "All laws of physics observed in one coordinate system are equally valid in any other coordinate system moving with a constant velocity relative to the first." Moreover, the acceptance of the postulate of the constancy of the speed of light as a fundamental principle relieves us of the need to account for the propagation of light by an ether model. For if the principle is right, as was eventually found by checking experimentally all consequences derivable from it, then we are better off accepting it as a basic point of departure than we would be if we continued the long and dubious quest to discover an ether free from all internal contradictions and in accord with all experiments. The chief point is that one must allow that a useful principle, even one unfruitful of convenient mental images, is to be preferred over a clear, visualizable model that is not logically consistent and not in accord with empirical facts.

This point reminds us of the fate of the caloric, the medium whose function was to provide a visualizable mechanical model for the processes of heat transfer. It was challenged by Rumford and Davy on the grounds that it failed to exhibit the behavior expected of material media—specifically, that it had no measurable weight, and could, in friction, be created without limit and without drawing from the materials out of which it arose. The comparison between the fates of the caloric and the ether is faulty in one detail, however. The caloric was eventually replaced by a scheme built on another visualizable model, the kinetic theory picture of the atom, whereas the ether was replaced by a principle and by the habit of speaking of "the field" as the arena for physical processes that occur in a vacuum, in the absence of ponderable matter.

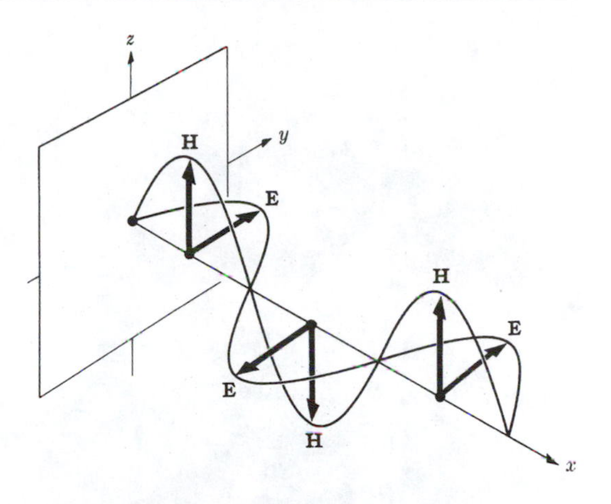

Fig. 25.8. Instantaneous pattern of electric and magnetic field vectors in an electromagnetic wave. The entire pattern moves through space in the positive *x* direction.

To return to Maxwell's achievement: if we wish, we may conceive of light for the present as progressing through space by the propagation of electric (and magnetic) field vectors (see Fig. 25.8 for a "snapshot" of a region through which light travels, with some field vectors drawn in). The distance between successive instantaneous maxima is one wavelength λ, and the frequency of the wave f is the number of crests passing a given point per second. The velocity of light c is given by the usual relation (equivalent to Eq. 23.2), $c = \lambda f$. Our elimination of the ether from this discussion does not invalidate the rest of the Maxwellian theory of electromagnetic waves.

If Maxwell's theory could be verified, then he would have achieved an immense step forward in the unification program of nineteenth-century physics, by joining to the unity of electricity and magnetism (established by Oersted and Faraday) the behavior of light and other electromagnetic radiation. And that is just what happened.

25.4 Hertz's Experiments

If Maxwell's theory of light as an electromagnetic radiation emitted by the periodic fluctuation of currents is correct, it stands to reason that a heated object, e.g., the metal filament of an

Fig. 25.9. Heinrich Hertz (1857–1894).

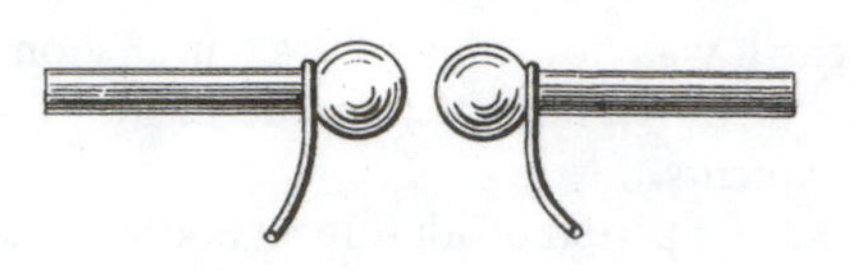

Fig. 25.10. Spark gaps in Hertz's oscillator.

incandescent lamp, sends out light by virtue of the agitation on the filament of submicroscopic electric charges such as one expects to be distributed throughout neutral bodies in balancing amounts.

We cannot decide at this point whether these charges are parts of the atoms in the emitter, or separate entities such as electrons, or positive particles; so permit us to call them for the moment, in a general way, *oscillators*. Somehow these oscillators are able to emit electromagnetic energy that our eye interprets as light, the frequency ν of the light waves being equal to the periodic frequency f of the oscillators.

This "classical" interpretation of light emission might be more acceptable to us if we could refer to direct experimental confirmation. One very obvious test might be to produce an oscillating current in a wire at a frequency equal to that of, say, green light, and watch whether green light is actually released into the surrounding region. Even before the days of radio and microwaves it was easy enough to produce electric oscillations in wires, but the frequencies achieved were relatively low in Maxwell's day, and even today the higher frequencies of electric oscillation in the most modern equipment can hardly be pushed above 10^{12} sec^{-1}, far below the frequency of visible light (~10^{15} sec^{-1}). Consequently, a direct test seemed out of the question, and partly for this reason Maxwell's work did not attract a great deal of attention for two decades.

There was, however, another point of attack. The gifted German physicist Heinrich Hertz (1857–1894) (Fig. 25.9) undertook in 1887–1888 to show that at least the main premise of Maxwell's view was correct, namely, that an oscillating electric current (he attained a frequency of about 10^9 sec^{-1}) does indeed send out electromagnetic waves that have all the characteristics of light except, of course, visibility. Hertz's proof succeeded so brilliantly that hardly a single important physicist remained unconvinced of Maxwell's equivalence of light and electromagnetic radiation.

In essence, Hertz's experimental method was simple. Two rods, each about 5 in. long and terminated at one end by a small polished metal sphere, were fixed along a line so that a small air gap separated the spheres (Fig. 25.10). A pair of wires led from the spheres to a device supplying short pulses of very high potential difference, and during the intervals between pulses brought large quantities of electric charges of opposite signs to the two spheres until a spark jumped across. The air in the gap remained conducting for a short time and so provided a path for these charges as they oscillated from one rod to the other until electrical equilibrium was achieved. Then, the air being restored to nonconductivity, the stage was set for the next "voltage" pulse and another spell of oscillations by the same processes.

(Let us take note of a short and somewhat mystifying remark of Hertz's, which will assume strange importance later, namely, that he found that "it is essential that the pole surfaces of the spark gap should be frequently repolished" to assure proper operation of the spark.)

If Maxwell was right, this arrangement should, during each spark, give off electromagnetic waves of the same frequency as the oscillations, up to about 5×10^8 sec^{-1} by Hertz's estimate, corresponding to a wavelength of about 60 cm. Furthermore, as has been pointed out, this radiation should spread out from the rods and be detectable by its production of a fluctuating current in a wire some distance away.

Hertz's first triumph was to observe this very effect, even when he placed the receiving wire (antenna) many yards from the sending oscillator. This new, invisible electromagnetic radiation, still sometimes called Hertzian waves, is of course identical with our now familiar radio waves.

Hertz showed next that the radiation between sender and receiver could be made to go through all the characteristic behavior of light: reflection, focusing into a beam, refraction, interference, polarization, and so forth. One interesting test, the direct determination of the velocity of this radiation, was beyond the instrumentation available to Hertz. Experimentation conducted after 1895 confirmed the assumption that the speed was the same as that of light.

One conclusion was soon almost universally accepted even before detailed additional evidence came in: Since Hertzian waves and light waves behave so much alike, light too must be caused by the rapid oscillatory motion of charged particles, perhaps those in the atoms of the light emitter. Here then was the model for light emission. And in terms of this picture, a great many details became clear and could be integrated into one scheme—the mechanism for transmission of light through crystals and other (transparent) objects, the reflection from metal surfaces, and so on.

A particularly impressive, even sensational, confirmation of the theory was provided by P. Zeeman, who discovered (1896) that a glowing gas sends out light at changed frequencies if the emitting material is placed in a strong magnetic field. This effect could be explained for many cases by a straightforward adoption of the fundamental Maxwellian idea that light energy is released by the vibration of charged particles, since the observed change in light frequency brought about by the magnetic field is equal to that calculated from Maxwell's theory.

The demonstration of Hertzian waves proved to be not only the triumph of Maxwell's theory but also the unintended beginning of the technology of radio and television. It is not amiss to contemplate briefly that strange plant, which emerged from a seed falling from the tree of science. Consider an ordinary television set from all the following points of view—symbolically, as it were: how the first hint of radio waves originated in the mathematical speculations of Maxwell, and through the ingenious experiments of Hertz came to the notice of the young Italian inventor, Guglielmo Marconi, the first to send radio waves across large distances. Consider how even these three men, in the cooperatively cumulative nature of their work and in the differences in their nationalities, religion, and training, exemplify the interdependence of humanity, the basis of the humanistic ideal. Then consider the meaning of television today—using of course radio waves—as a huge industry in the economics of the nation and how society uses this tool, sometimes for mass education or genuine entertainment, more often for mass propaganda, mass tranquilization, commercialization, and portrayals of violence.

25.5 Cathode Rays

What we call the "discovery of the electron" is not a single historical event but rather a combination of experiments, debates, and theoretical speculations in the last part of the nineteenth century. The work of the English physicist Joseph John Thomson (1856–1940) in 1897 is usually identified as the definitive discovery, although it is sometimes claimed that others deserve to share the credit. (We must call him J. J. Thomson to avoid confusion with William Thomson, Lord Kelvin—no relation.)

Thomson, building on the work of such predecessors as William Crookes in England and Jean Perrin in France, established the modern view that the so-called "cathode rays" are streams of discrete particles with negative electric charge and small but finite mass, moving at speeds that are high but definitely less than that of light. The opposing theory was that cathode rays are some form of electromagnetic waves or ether-disturbance.

The discovery of the electron was in several ways a direct outgrowth of the work of Michael Faraday. It was Faraday who in 1821 constructed an electric motor showing that a magnetic field could deflect an electric current into a circular path (Fig. 25.4); the deflection of cathode rays by magnetic fields was one of the principal methods for studying the properties of the electron. As we noted, it was Faraday who in 1833 discovered the quantitative law of electrochemical decomposition, and in publishing this research in 1834 introduced the modern terminology:

electrode, cathode, anode, ion, anion, cation, electrolyte. Here Faraday established the idea that a definite quantity of electricity is associated with each atom of matter (Eq. 24.7). Finally it was Faraday who in 1838 studied electric discharges in a vacuum (or rather, as good a vacuum as was then available).

Maxwell, in his *Treatise on Electricity and Magnetism* (1873), recognized that one could use Faraday's law of electrolysis to identify an elementary "atom" of electric charge, although he warned that "this tempting hypothesis leads us into very difficult ground" (i.e., the then-difficult-to-test hypothesis that "the molecules of the ions within the electrolyte are actually charged with certain definite quantities of electricity, positive and negative, so that the electrolytic current is simply a current of convection"). One could thereby calculate a definite quantity of electricity, the "molecular charge," which would be, if known, "the most natural unit of electricity." One might even call it, Maxwell wrote, a "molecule of electricity" though he immediately labeled this a "gross" phrase, "out of harmony with the rest of this treatise." In fact Maxwell formulated the main part of his electromagnetic theory without assuming that electricity is composed of atomic particles. As a result, some of his followers, such as Hertz, were reluctant to accept the electron as a basis for electrical theory.

Although Maxwell could have calculated a value for the unit of electric charge in 1873, he did not; this was left to the Irish scientist George Johnstone Stoney (1826–1911), who, in 1874, used the value of Avogadro's number, which he himself had estimated from kinetic theory in 1868, to compute an electric unit equivalent to 10^{-20} coulombs. (This is about 1/16 the modern value of 1.6×10^{-19} coulombs.) It was his 1894 article on this subject that established the use of the name *electron.*

The other main line of research leading to the discovery of the electron was the observation of vacuum discharges. Here we see the technology of instruments and apparatus playing a crucial role in the development of physics, as it often does: only after the middle of the nineteenth century was it possible to make observations at extremely low pressures, where the most interesting electronic phenomena occur. The invention of a new vacuum pump by Geissler in 1855, and its improvement by Hermann Sprengel in 1865, led directly to the flourishing of atomic physics as well as to the developments in electrical technology initiated by Thomas A. Edison (electric light) and electronic tubes for communications, etc.

Johann Heinrich Wilhelm Geissler (1815–1879) was a glassmaker and mechanical technician employed at the University of Bonn in Germany. He provided instruments for the mathematician-physicist Julius Plücker (1801– 1868) and other scientists at Bonn and participated in their experimental work. In 1852, at the invitation of a Bonn industrialist, he constructed an instrument called a *vaporimeter* to measure the alcoholic strength of wine by determining the pressure of vapors of alcohol and air. This led him to further study of the properties of vapor at low pressure and the invention of his mercury vapor pump in 1855. Geissler's pump was based on the idea of repeatedly creating a Torricelli vacuum (Section 19.2). Geissler was able to make small glass tubes with electrodes melted into the ends, filled with gases at very low pressure, and these became known as *Geissler tubes.*

What followed was, as we now see in retrospect, a series of small steps, made by many dif-

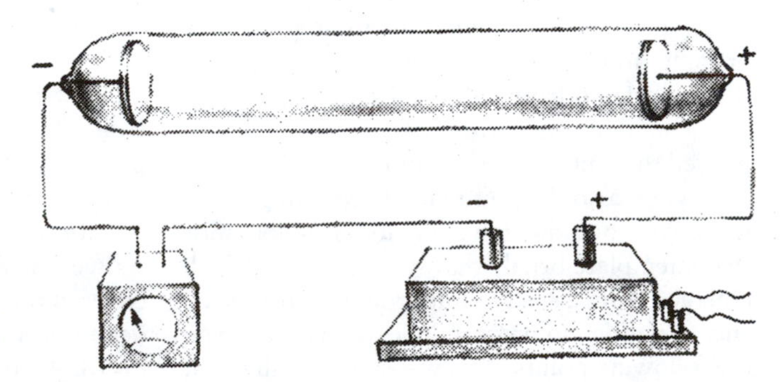

Fig. 25.11. Cathode ray apparatus, using Geissler tube. Fluorescent lights are essentially Geissler tubes with an inner coating of fluorescent powder.

ferent scientists, none of whom really knew what was causing their fascinating observations—yet all the while moving toward a discovery one day in 1895 of such momentous importance that the rise of all "modern physics" is usually identified with that moment.

Plücker used Geissler tubes to pursue Faraday's study of electrical discharges in a vacuum, and found that at the lowest pressures he could reach, the inside of the tube became darker and darker and that there was an extended glow on the glass walls around the cathode. The glow was found to be affected by an external magnetic field. This research was done in 1858, and it seems that rather a long time elapsed before the study of vacuum discharges was followed up by other scientists.

The first of these was Plücker's student Johann Wilhelm Hittorf (1824–1914), professor of chemistry and physics in Münster, Germany. In 1869 Hittorf found that any solid body placed in front of the cathode cut off the glow on the surrounding glass walls, as it if were blocking some kind of rays emanating from the cathode. By making an L-shaped tube Hittorf established that these rays seemed to travel in straight lines from the cathode to the glass wall; the rays themselves are invisible but produce a phosphorescent glow on the glass wall when they strike it. Furthermore, the rays are bent by a magnetic field.

Eugene Goldstein (1850–1930), a Polish physicist, did extensive experiments on cathode rays; he established that they are only emitted perpendicular to the surface of the cathode, can cast sharp shadows, produce chemical reactions, and are independent of the nature of the cathode. The name cathode rays is due to him.

What are these cathode rays? One might suppose that Plücker and Hittorf considered the possibility that they are some kind of atomic particles, but the first published suggestion along these lines is usually attributed to an English scientist, Cromwell Fleetwood Varley. In 1871 Varley proposed that the rays are "composed of attenuated particles of matter projected from the negative pole [cathode] by electricity in all directions."

William Crookes (1832–1919), an English chemist and physicist, took up the study of cathode rays a few years later, as an outgrowth of his research on the precise determination of atomic weights by measurements in an evacuated chamber. He used a bent Geissler tube (Fig. 25.12) and found that the most intense green glow appeared on the part of the tube directly opposite the cathode, as if the rays were being repelled from it rather than attracted to the anode. The rays did not seem to be affected by an electric field, but could be deflected by a magnetic field. In 1879 he proposed that the rays are molecules that have picked up a negative electric charge from the cathode and are then repelled by it. Perhaps his best known experiment is the one that shows the shadow produced on the glass wall by a barrier in the form of a Maltese cross (Fig. 25.13).

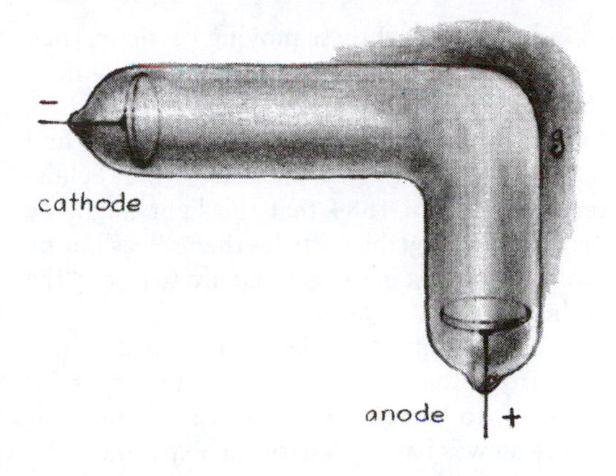

Fig. 25.12. Bent Geissler tube used by Crookes. The most intense green glow appeared at g.

While the fact that the cathode rays can be bent by a magnetic field suggests that they are charged particles, that hypothesis also leads to some other consequences. First, as pointed out by the Scottish physicist Peter Guthrie Tait, if one assumes that the illumination of the walls is due

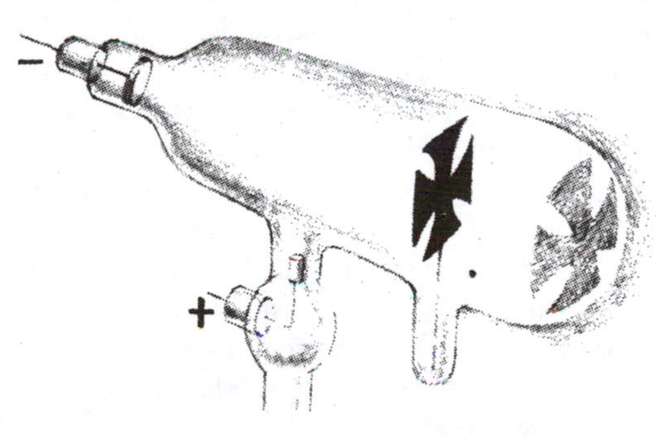

Fig. 25.13. Crookes tube with Maltese cross.

to light emitted by these moving particles, then one should be able to observe a Doppler shift in the spectrum of this light. Goldstein and others looked for such a shift and were unable to find it. However, the English physicist Arthur Schuster suggested in 1884 that the light might be produced not by the particles themselves but by molecules at rest in the gas that are struck by the cathode rays.

A more damaging piece of evidence against the particle theory was the fact that the rays did not seem to be deflected by electric fields. This objection was based primarily on experiments by Hertz in 1883, a few years before his discovery of electromagnetic radiation. After Hertz succeeded in producing electromagnetic waves in 1888, a debate developed between German and English physicists on the nature of cathode rays. The Germans, following Hertz, believed them to be some kind of electromagnetic waves in the ether, while the English scientists, following Crookes, were convinced that they are particles (though not necessarily molecules as Crookes had thought).

Hertz's followers, who tried to account for cathode rays as ether waves, fell into the same trap as many other scientists who invoked ether to explain phenomena that seemed otherwise mysterious. Since there was no unique accepted theory of the ether, one could attribute to it practically any property. This can be seen in the case of cathode rays, where it was apparently established that they were deflected by magnetic but not by electric fields. For the particle theory that was a paradox since if the electron has a charge it should be deflected by both kinds of fields. But for the ether theory, one could counter the criticism that light and other known types of electromagnetic waves are not deflected by magnetic fields with the extra hypothesis that cathode rays may be some other kind of ether wave that is deflected by magnetic fields.

This is the kind of situation Karl Popper had in mind when he proposed his "falsifiability" criterion. He argued that if a theory is so flexible that it can explain any observations that appear to refute it, then it is not really an acceptable scientific theory. Nevertheless the "wave theory of cathode rays" is not necessarily wrong just because it cannot be definitely tested at a particular time; we must not be too hasty in using Popper's criterion to throw it out.

25.6 X-Rays and the Turn of the Century

On November 8, 1895, in the physics laboratory of the University of Würzburg, Germany, Wilhelm Conrad Röntgen noticed that a barium platinocyanide screen, which happened to be lying about a meter away from a Hittorf-Crookes tube covered with black cardboard, became mysteriously fluorescent when a high-voltage discharge passed through the tube (Fig. 26.11). One may say that because of what followed, for science as a developing human enterprise involving a widespread community of scientists, (S_2; see section 13.2), modern physics was born on December 28, 1895. On that day, after he had systematically investigated the properties of the mysterious rays coming from the tube and even used them to produce a photograph of the bones in his wife's hand, Röntgen presented a paper "On a New Kind of Rays" to the secretary of the Würzburg Physical-Medical Society, for publication in its proceedings. This paper, announcing the discovery of what Röntgen called "x-rays," was quickly printed, and on New Year's Day 1896 the author mailed out reprints, along with prints of some of his x-ray pictures, to several important physicists in Germany, Austria, France, and England.

The story of the discovery itself, of the immediate popular sensation it caused in Europe and America, and of its revolutionary impact on medicine has been told many times. (At this point you might reread the part of Chapter 14 in which we used Röntgen's research as an example of scientific procedure.) Here we will focus on the role of x-rays in the transition from classical to modern physics, and comment briefly on the transition to a modern technological society that was going on at the same time.

On the same day that Röntgen announced his discovery of x-rays, a few hundred miles away at the Grand Café on the Boulevard des Capucines in Paris, 33 people paid 1 franc each to see the first public demonstration of the *cinematograph*. This device, constructed by Auguste and Louis Lumière, projected a series of photographs on a screen fast enough to create the illusion of continuous motion in human eyes. Thus was born the modern "cinema" or "motion picture." While the viewers of Röntgen's x-ray photograph of a hand experienced a thrill of

Fig. 25.14a. Wilhelm Conrad Röntgen (1845–1923).

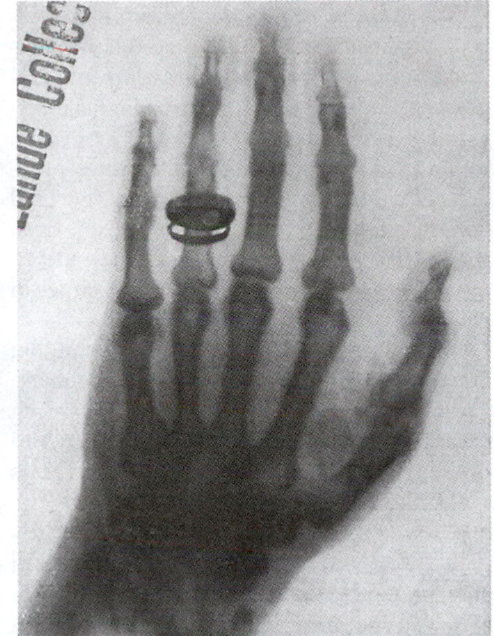

Fig. 25.14b. X-ray photograph of the hand of Röntgen's colleague Albert von Kölliker, published in Röntgen's second article on x-rays.

horror on seeing the skeleton of a living person—the Prince of Wales was reported as saying "how disgusting"—the audience at the Lumières' cinema "fled in panic as a locomotive pulling into a station appeared on the screen, thinking that it was really bearing down on them and would crush everybody in its way" (Lesak, in Teich and Porter, *Fin de Siècle,* 1990). Yet they returned to see the moving pictures again; and the opportunity to have one's own body x-rayed proved to be an added attraction.

Perhaps already in 1896 some people suspected that twentieth-century physics and technology would tell them more than they wanted to know about the world, while others could not resist exploiting the invasion of privacy offered by the "naughty, naughty x-rays." Yet x-rays, used with suitable precautions, would prove to be the greatest gift from physics to medicine in the twentieth century.

Just as the construction of the cinematograph relied on previous technical advances (Edison's kinetoscope, Reynaud's *théâtre optique*) but leaped beyond them into something qualitatively new and startling, so the discovery of x-rays was made possible by earlier inventions, such as Gramme's generator, the first successful commercial development of Faraday's dynamo (electric generator), which allowed mechanical energy to be transformed into electrical energy, and Geissler's mercury pump which allowed the production of a high vacuum. The same two inventions also led to the development of a practical electrical (incandescent) light. Edison's lamp, invented in 1879, was widely available by 1900; its popularity created a demand for electrical generating systems, which in turn enabled the development and marketing of many other applications of electricity.

The 1880s and 1890s saw the introduction and proliferation of the telephone (based on electricity) and the automobile (incorporating the internal combustion engine and rubber pneumatic tires). The Atlantic Cable allowed rapid telegraphic communication between Europe and America. Increasing literacy created a market for popular newspapers, a few of which reached a daily circulation of about a million by 1900. Thus the news of Röntgen's discovery and its amazing applications could be disseminated to a wide public within a few weeks.

During the same period, rapid communication and transportation facilitated the domination of countries by powerful governments and large corporations; people became increasingly dependent on centralized systems for power, food, and information. Colonial empires prospered. The politics of mass society fostered nationalism, anti-Semitism, and socialism. Even sports became bureaucratized and internationalized with the revival of the Olympic Games in 1896.

For many observers, society was changing too fast; modern life was becoming too exhausting and dangerous. Materialism was crowding out tradition and spiritual values. Science and technology seemed to be pursuing novelty for its own sake without regard for consequences. The phrase *fin de siècle* ("end of the century") was largely associated with feelings of pessimism about the course of civilization. Some artists were self-consciously "decadent" and aimed to "shock the bourgeoisie" with obscenity and perversions. "Hereditary degeneration" was considered a respectable biological explanation for all the ills of society.

Physicists no longer regarded Newtonian mechanics as the firm foundation of physics. Ernst Mach rejected the fundamental concepts of space, time, and mass as defined by Newton. Hertz, just before his premature death, expressed his dissatisfaction with mechanics as usually presented; he wanted to eliminate the concept of force, alluding to "the statements which one hears with wearisome frequency, that the nature of force is still a mystery" (*Principles of Mechanics,* 1894). Others such as H. A. Lorentz, Wilhelm Wien, and Max Abraham, wanted to dethrone mass as a fundamental entity and instead derive it from electromagnetic interactions; they speculated that the mechanical worldview might be replaced by an "electromagnetic worldview." Mach, Hertz, Wilhelm Ostwald, Pierre Duhem, Gustav Robert Kirchhoff, and other influential scientists rejected the premise that science should explain the nature of the real world in terms of unobservable entities such as atoms and ether. Instead, pointing to the inconsistencies in popular models of the ether and atoms, they argued that the purpose of science is to describe observable phenomena as accurately as possible, so that useful predictions can be made.

By a sequence of events that could not have been imagined, the next giant step to modern ideas was the accidental discovery of radioactivity. The physicist Henri Becquerel was present at the meeting of the Paris Academy of Sciences on January 20, 1896, when two French physicians displayed an x-ray photograph of the bones in the human hand. He asked Henri Poincaré, who had received a copy of Röntgen's paper, how the x-rays were produced; Poincaré replied that they were produced where a beam of cathode rays played on the wall of a discharge tube, a spot marked by a lively fluorescence of the glass. It occurred to Becquerel that the same mechanism that produced the visible light might also produce the invisible x-rays.

Although the assumed connections among x-rays, fluorescence, and phosphorescence were later found to be accidental, they led Becquerel to investigate more closely the possibility of the emanation of rays from a phosphorescent compound of uranium. Eventually he established that various salts of uranium emit penetrating radiation even after being subjected to chemical and physical processes that destroy their phosphorescence and even after being kept isolated for several years from light or any other external source of energy (Section 27.1). In this investigation he used gelatin-bromide photographic plates sold by the Lumière brothers, thus continuing the subterranean connection between x-rays and the cinema.

Physicists now had to understand the nature and properties of several kinds of rays. The work of Maxwell and Hertz had suggested that light, ultraviolet, infrared, and the radio waves discovered by Hertz were all transverse-wave motions of electric and magnetic fields. Cathode rays were thought by some to be electromagnetic waves and by others to be electrically charged particles, called electrons; the experiments of J. J. Thomson in 1897 would soon establish the latter view. The emanations from uranium and other "radioactive" substances such as radium and polonium, investigated by Marie and Pierre Curie, were shown by Ernest Rutherford to consist of two kinds, α (with positive electric charge) and β (negatively charged), and these were subsequently found to be the same as doubly ionized helium atoms and electrons, respectively. Another kind of rays produced in cathode ray tubes, *canal rays,* was shown by Wilhelm Wien to consist of positively charged particles with the same mass as hydrogen atoms; they were later called *protons.* Then a third kind of radioactive emanation, with no electric charge, was discovered by Paul Villard and called γ-rays; he initially thought that they were x-rays but they were later found to have distinctive properties (owing to their being electromagnetic waves of much higher frequency, as we would now put it).

As it happened, the general question of the nature of electromagnetic radiation eventually depended on the specific question of the nature of x-rays. We can see intimations of this in the earliest speculations. On one hand x-rays travel in straight lines like light—one could take photographs with them, and the visual impact of these photographs was compelling. On the other hand, their wave properties, such as diffraction and interference, could not be demonstrated at the time.

X-rays could not be deflected by either magnetic or electric fields, so they could not be charged particles. Röntgen himself suggested that his rays are *longitudinal* ether waves, and this view was initially supported by Boltzmann and Lord Kelvin. The fact that they apparently could not be polarized seemed to indicate that they are not transverse waves like ordinary light. As a compromise, G. G. Stokes and other physicists suggested that they are not periodic waves but localized "pulses" of electromagnetic radiation—an idea reminiscent of Huygens' theory of light, which preceded the periodic wave theory (Section 23.1). If they could later be shown to have some properties of electromagnetic waves and also some properties of particles, x-rays would form a bridge between waves and particles and thus a clue to the nature of each.

Physicists were finally convinced that x-rays are electromagnetic waves by the 1912 discovery by Max von Laue, Walter Friedrich, and Paul Knipping that they can be diffracted by a crystal lattice. This discovery soon proved to be even more important because x-ray diffraction techniques, developed by W. H. Bragg and his son W. L. Bragg, could be used to determine the structure of crystals and large molecules, including those of biological importance. X-ray crystallography, like radioactivity and nuclear physics, was a field in which women were especially successful: Dorothy Crowfoot Hodgkin, Kathleen Lonsdale, and Rosalind Franklin are a few of the more famous names. Franklin's x-ray photographs of DNA, used by James Watson and Francis Crick to determine the structure of that molecule in 1953, show the historical link between the beginning of modern atomic physics and the beginning of modern molecular biology.

Although the *wave nature* of x-rays seemed securely established by the 1920s, that was by no means the end of the story. In 1923 the American physicist Arthur Holly Compton showed that x-rays can collide with electrons in a way that clearly shows them to have a definite momentum (Section 16.7)—thereby suggesting that x-rays, and by inference all kinds of electromagnetic radiation, also have a *particle* nature. We will begin to explore this puzzling result in Chapter 26.

25.7 The "Discovery of the Electron"

Joseph John Thomson (1856–1940) settled the debate about the nature of cathode rays and was largely responsible for making the Cambridge University Cavendish Laboratory, founded by Maxwell in 1871, the leading center of experimental physics in the world in the first part of the twentieth century. Thomson was born in Manchester, the son of a bookseller and publisher; like Maxwell, Kelvin, and other leading British physicists, he studied mathematics at Cambridge. His early work was a combination of mathematical physics and electrical speculation. He developed the idea that all forms of energy can be reduced to kinetic energy, but later favored the view that mass is not a fundamental entity but an effect of electromagnetic interactions.

Thomson was appointed director of the Cavendish Laboratory in 1884. Although not exceptionally skilled as an experimenter himself, he had a good sense of how to do significant experiments, and proved to be very effective as a leader of a research group. When Cambridge University changed its regulations in 1895 to allow graduates of other universities to earn a B. A. after two years of residence with a thesis instead of passing an examination, several first-rate students came to do research with Thomson at the Cavendish, the best known being Ernest Rutherford.

In 1894 Thomson attempted to measure the speed of cathode rays. His value of 200,000 m/sec was later found to be much too low (he later obtained 3,000,000 m/sec), but at the time seemed to be evidence against the theory that the rays are ether waves, which would presumably have to move at the speed of light, according to Maxwell's theory.

Fig. 25.15. J. J. Thomson (1856–1940).

Evidence that Cathode Rays are Particles

Thomson's famous 1897 paper, "Cathode Rays," was published in the *Philosophical Magazine* like many other major discoveries of nineteenth-century physics, rather than in the proceedings of a scientific society. (The results were first announced at a meeting of the Cambridge Philosophical Society on February 9, 1897.) This work is generally called the "discovery of the electron" because Thomson accomplished three major tasks. First, by placing a charge collector out of the direct path of the cathode rays, he showed that an electric charge was collected only when a magnetic field was used to bend the cathode rays into a path leading to the collector. The charged particles must have followed the same curved path as the cathode rays, thus making it difficult to maintain, as some did, the position that the particles were not the same thing as the rays.

Thomson's second achievement in the 1897 paper was to show that, contrary to Hertz's conclusion, the rays are after all deflected by an electrostatic field. He was able to do this by using a better vacuum pump (improved thanks to the demands of the electric light bulb industry) to remove nearly all the gas ions from the neighborhood of the plates, so they could not neutralize the potential difference by collecting on the plates. (Thomson suggested that Hertz did not observe a deflection of cathode rays by electric fields because ions, produced by the discharge in the gas, shielded the rays from the field.)

Third, Thomson was able to obtain a reasonably good value for the charge to mass ratio, by balancing the magnetic and electric deflections of the beam. The result was a charge/mass ratio of about 10^{11} coul/kg (modern value 1.76×10^{11}).

Around 1910 the American physicist Robert A. Millikan developed a new direct method to determine the charge of the electron. He found that a small object such as an oil drop can pick up electric charges in multiples of a certain minimum value and that the charge on the drop can be determined by suspending it in such a way that the gravitational force pulling it down is just balanced by an electric force pushing it up (see Problem 24.22). Millikan found that this minimum charge is about 1.6×10^{-19} coul, in agreement with estimates from electrolysis (Section 24.14). Thus the charge of the electron was added to the list of fundamental constants of nature.

Problem 25.1. Using the results of Thomson and Millikan, calculate the mass of an electron and compare it with the modern value.

The Proton

Positively charged rays, called *canal rays,* were studied by J. J. Thomson and other physicists in connection with cathode-ray experiments. In 1919 Rutherford found these rays to have the same properties as the hydrogen nuclei knocked out of the nitrogen nucleus by α-particles in his famous experiment on artificial nuclear transmutation (Section 32.1). Soon afterward he began to call them protons, and proposed that they are a basic constituent of the nucleus of all atoms. He occasionally suggested *prouton* in honor of the British chemist who had proposed that all elements are compounds of hydrogen (Section 21.1), but that name did not stick.

RECOMMENDED FOR FURTHER READING

Olivier Darrigol, "Baconian bees in the electromagnetic fields: Experimenter-theorists in nineteenth-century electrodynamics," *Studies in History and Philosophy of Modern Physics,* Vol. 30, pages 307–345 (1999)

C. W. F. Everitt, *James Clerk Maxwell*

Alan Franklin, "Are there really electrons?" *Physics Today,* Vol. 50, no. 10, pages 26–33 (1997), reconstructs "the argument that a physicist in the early twentieth century might have used."

A. P. French and T. B. Greenslade, Jr. (editors), *Physics History II,* articles by J. L. Spradley on Hertz and W. Thumm on Röntgen, with a translation of Röntgen's paper, pages 42–47, 137–144, 145–148

C. C. Gillispie, *Dictionary of Scientific Biography,* articles by L. P. Williams on Ampère, Faraday, and Oersted, Vol. 1, pages 139–147, Vol. 4, pages 527–540, and Vol. 10, pages 182–186; by R. McCormmach on Hertz, Vol. 6, pages 340–350; by G. L'E. Turner on Röntgen, Vol. 11, pages 529–531; by A. Romer on Becquerel, Vol. 1, pages 558–561; and by J. L. Heilbron on J. J. Thomson, Vol. 13, pages 362–372

P. M. Harman, *Energy, Force and Matter,* Chapter IV

James R. Hofmann, *André-Marie Ampère,* New York: Cambridge University Press, 1996

Helge Kragh, "J. J. Thomson, the electron, and atomic architecture," *Physics Teacher,* Vol. 35, pages 328–332 (1997)

Mary Jo Nye, *Before Big Science,* Cambridge, MA: Harvard University Press, 1999, Chapter 3

Melba Newell Phillips (editor), *Physics History,* articles by J. R. Nielsen on Oersted, S. Devons on electromagnetic induction, and by E. C. Watson introducing a translation of Hertz's lecture on light and electricity, pages 23–53

Robert D. Purrington, *Physics in the Nineteenth Century,* Chapter 3

S. Sambursky, *Physical Thought,* excerpts from papers by Oersted, Ampère, Faraday, Maxwell and Hertz, pages 379–388, 408–444, 461–464

Emilio Segrè, *From X-Rays to Quarks,* San Francisco: Freeman, 1980, pages 1–60

Howard H. Seliger, "Wilhelm Conrad Röntgen and the glimmer of light," *Physics Today,* Vol. 48, no. 11, pages 25–31 (November 1995). Other articles in this special issue on the centennial of x-rays describe applications to condensed matter physics, molecular biophysics, medicine, and astronomy.

Harold I. Sharlin, *The Making of the Electrical Age,* New York: Abelard-Schumann, 1966

Robert C. Stauffer, "Speculation and experiment in the background of Oersted's discovery of electromagnetism," in *History of Physics* (edited by Brush), pages 78–95

Charles Susskind, *Heinrich Hertz: A Short Life,* San Francisco: San Francisco Press, 1995.

J. W. Weaver (editor), *World of Physics,* excerpts from papers by Faraday, Maxwell, and J. J. Thomson, Vol. I, pages 839–858 and Vol. II, pages 715–724

M. Norton Wise, "Electromagnetic theory in the nineteenth century," in *Companion* (edited by R. C. Olby *et al.*), pages 342–356

Arthur Zajonc, *Catching the Light,* Chapter 6

CHAPTER 26

The Quantum Theory of Light

26.1 Continuous Emission Spectra

The study of light led into modern physics by three separate paths. The problem of the earth's motion through the hypothetical ether was mentioned in Section 25.3; we shall return to that topic when we discuss Einstein's relativity theory in Chapter 30. The second path goes from x-rays through radioactivity (Chapter 25) to nuclear and elementary particle physics (Chapters 27 and 32). The third path now brings us, by the somewhat roundabout route of real history, to the quantum theory.

Our newly gained—though so far rather vague—model of the emitter of light, the submicroscopic electric oscillator, raises our hopes for the solution of some long-standing puzzles about radiation: Can we not now begin to *explain the colors (frequencies) and distribution of light energies* sent out by different sources of light? (The identity of the character of x-rays, ultraviolet rays, visible light, infrared (heat) rays, and longer wavelength (Hertzian) waves having been established, we shall from now on refer to all these as "light.")

Before we can grapple with this problem we must become aware, at least by a brief account, of the large amount of experimental knowledge available by the beginning of the twentieth century. Newton had shown, in a famous experiment, that light from the sun is composed of colored rays, which can be resolved by a prism into a spectrum (Newton's word) of these colors arranged in order of their refrangibility (Section 23.4). Just such a "rainbow" spectrum had subsequently been obtained by examining the light from *all* glowing solids (filament, carbon in arc) and glowing liquids (molten metals). With only a little idealization, these emission spectra all have three features in common:

a) The emission spectra are *continuous,* that is, there are no gaps, no "color" bands missing from the spectrum, even if measurements of the radiation are carried into the infrared or ultraviolet region.

b) All glowing solids and liquids, *no matter what their chemical nature,* send out light with about the same color balance if they are at the same temperature. A red-glowing coal has that color not because there are no other colors save red being radiated from it, but because of all the visible radiation that in the "red" region is most intense, and an iron poker put into the coals, after reaching the same temperature, glows with the same light.

This must be put more precisely. The wavelength of each color (and, by means of the relation $c = \lambda\nu$, the light's frequency ν also) can be measured with great accuracy by means of a spectroscope (although not a prism but a *diffraction grating* is used for the determination of absolute wavelengths). Approximate ranges in wavelength for different colors are given in Table 26.1.

Note that the *angstrom* (Å) (named in the honor of the Swedish spectroscopist A. J. Ångström), formerly used as a unit of length, equal to 10^{-10} m, for electromagnetic waves and atomic sizes, has now been replaced by the nanometer (nm, 10^{-9} m) in the SI system of units.

The exact amount of energy contributed by the components of different wavelengths in a beam of light can now also be measured with precision, although originally the details of experimentation were extremely delicate and difficult. For the infrared ranges, the energy can be detected conveniently by sensitive heat-measuring devices (thermopiles, bolometers), and in the visible and ultraviolet ranges by photographic exposures (so used since about 1840), photocells, and the like.

A typical distribution curve of radiated energy vs. wavelength, specifically for a temperature corresponding to that on the surface of the sun (6000 K) is seen in Fig. 26.1. In this case there is relatively little energy at very short wave-

Table 26.1. **Wavelengths of Colors in the Spectrum**

Color	*Range of λ in nm (=10^{-9} m)*
Red	610 to about 750
Orange	590 to 610
Yellow	570 to 590
Green	500 to 570
Blue	450 to 500
Violet	about 400 to 450
(Infrared)	Longer than 750
(Ultraviolet)	Shorter than 400

lengths. The peak of the curve falls into the green part of the visible region, and the largest share of the total energy emitted is in the infrared; our eye perceives only the particular mixture of "visible" radiation, and compounds it to give the impression of white light.

The chief points of interest here are that such a curve can be drawn for every temperature of the emitter and that, remarkably, all emitters of the same temperature, regardless of chemical composition, yield about the same curve. To be more accurate, the radiation has the distribution shown for emitters that, when cool, are perfect absorbers of light. These are called *blackbodies,* but we can here disregard deviations from blackbody radiation.

c) The third feature that all hot solid or liquid emitters have in common is that their energy distribution curve shifts with changing temperature T, as shown in Fig. 26.2 for four temperatures. With increasing T, more energy is radiated in each wavelength region, and the peak in the curve moves toward shorter wavelengths.

The exact data were summarized by the German physicist Wilhelm Wien in 1893 in the so-called "displacement law," which in part states that for a perfect emitter of continuous spectra, the product of λ (in cm) at the peak and its temperature (in K) is a constant with the empirical value of 0.2897 cm·K:

$$\lambda_{peak}\ (\text{cm}) \times T(\text{K}) \approx 0.0029\ \text{m·K}. \qquad (26.1)$$

For example, λ_{peak} for the sun is, by experiment, about 5.5×10^{-7} m. It follows that $T = 5300$ K. But this figure is in fact a little too low, partly because some of the energy radiated by the sun, particularly at short wavelengths, is absorbed in the atmosphere before reaching our instruments, so that the true λ_{peak} is lower; and also because the sun is not the ideal radiator (so-called *blackbody radiator)* for which Eq. (26.1) is designed. Allowing for these difficulties, one is led to a value of about 6000 K for the sun's surface.

Provided Wien's displacement law holds even in extreme temperature ranges, we may at once make some surprising calculations: The radiation from distant stars can be analyzed to find where λ_{peak} falls. From this we may

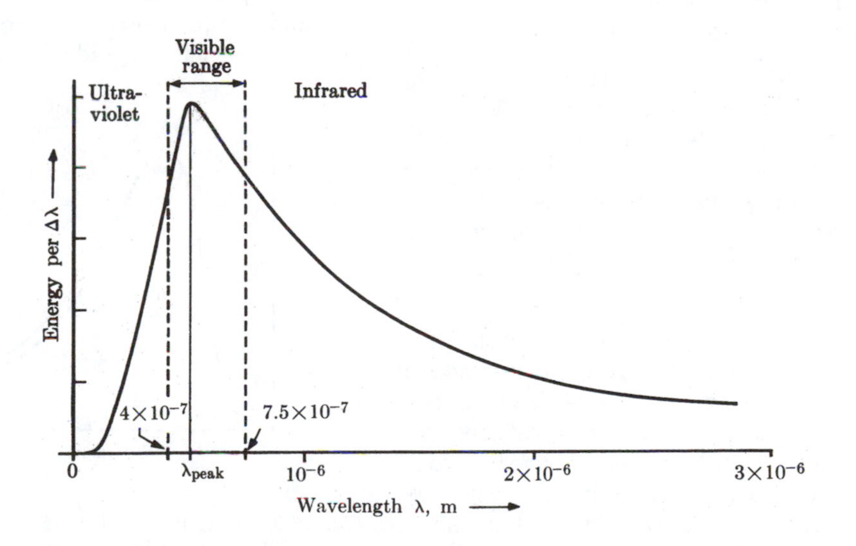

Fig. 26.1. Distribution of energy in light emitted from a glowing body at 6000 K.

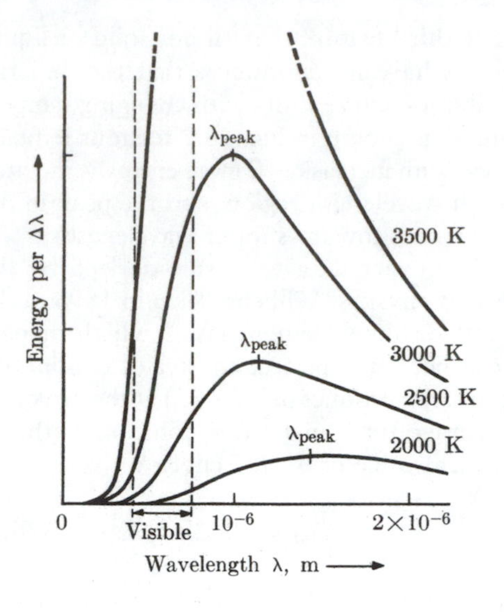

Fig. 26.2. **Distribution of energy in light radiated from a glowing solid at different temperatures.**

calculate approximate stellar surface temperatures. For the "hot" stars, whose λ_{peak} is small and lies at the blue end of the spectrum or even beyond, the temperatures are correspondingly higher than that of the sun (for example, 14,000 K for Vega). The cooler, reddish stars have larger values for λ_{peak}, and the calculated values of T are consequently lower (for example, 2500 K for Antares).

Problem 26.1. What is the wavelength of the most intense radiation for Vega, for Antares, and for a body at room temperature? Are there qualifications to your answer?

Problem 26.2. The total area under the curve for intensity of radiation vs. wavelength is proportional to the total energy radiated throughout the spectrum. Trace the curves in Fig. 26.2 and find an approximate law connecting the total energy radiated and the temperature of the emitter. (In the ideal case the relation is total energy $\propto T^4$.)

Problem 26.3. On the basis of our model of light emitters, would you expect to find a distribution curve such as shown in Fig. 26.1, or instead that all energy is sent out at *one* frequency? Should chemical and physical consistency influence the light emitted? What might temperature have to do with the distribution of intensity vs. wavelength? (Defend your conclusions, but be frank about vague guesses.)

The displacement law could actually be derived theoretically from the classical (Maxwellian) theory of light emission, and that was indeed another impressive victory! But at that very same point we encounter serious trouble. Surely we should also be able to explain the exact shape of the characteristic emission curve as a whole by derivation from the basic model of emission. This would involve first a calculation of the fraction of individual oscillators for each range of emitted frequencies, then the addition of their individual contributions to find the overall pattern.

But all such attempts proved fruitless. Specifically, Fig. 26.3 compares the actual emission curve with the prediction made by two modifications of the classical theory that differed from each other by some of the assumptions introduced in the classical model. Before 1900, the most successful theory, proposed by Wien, gave fairly good agreement with experiment except at the longest wavelengths, where data were sparse (Theory I in Fig. 26.3). Later it was pointed by Lord Rayleigh that a formula based on the *equipartition theorem* from the kinetic theory of gases (Section 22.8) would give better agreement for very long wavelengths (Theory II in Fig. 26.3). But that formula would imply that the emission energy goes to infinity as the wavelength goes to zero, so Rayleigh did not recommend it. (This difficulty, subsequently known as the "ultraviolet catastrophe," apparently had no influence on Planck's blackbody theory discussed in the next section.)

As experimentalists obtained more accurate and comprehensive data for emission at longer wavelengths, it gradually became clear to

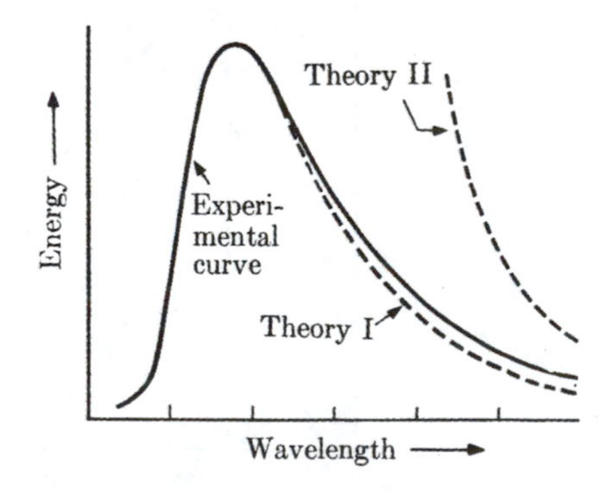

Fig. 26.3. **Comparison of experimental emission curve with predictions from two classically based theories.**

a few theorists like the German physicist Max Planck (1858–1947) that the discrepancy between theory and experiment was real and could not be removed by any plausible adjustment. We recall instances of similar dilemmas in the history of science. The most memorable case is perhaps Kepler's attempt to derive from the Copernican theory a path for the planet Mars that would fit the observations. He failed, just as we now find ourselves unable in this case to fit theory and fact.

Planck's solution was not to forget about the fairly small discrepancy—such neglect being permissible only if one can later account for it exactly by computation of the secondary effects at work, or at least tentatively by a consistent plausibility argument, as Galileo accounted for small deviations from the law of free fall by citing the effects of air friction. In Kepler's case, the small discrepancy provided a real challenge, the point of departure for a profound change of outlook in the whole field. You will recall that Kepler turned from the unfulfilled theoretical prediction to find a mathematical formulation of the actual (elliptical) planetary path, and to adopt that as the starting point of a modification of the Copernican heliocentric theory. This led him next to the great discovery of the three mathematical laws of planetary motion. Finally, a good physical interpretation of Kepler's laws was given by Newton in terms of the gravitational forces, thereby "explaining" Kepler's empirical findings.

A very similar turn of events will very soon be revealed in our coming discussion of the emission problem: We shall see that the obstinate mismatch of the curves in Fig. 26.3 prompted first producing a mathematical formulation of the actually observed curve; then came a physical interpretation of the mathematical formula that modified the old theory of light emission as drastically as Newton's work had improved on the Copernican concept of the universe. For the moment, however, we must look at more of the experimental results that sparked the revolution in physical science—the quantum theory.

26.2 Planck's Empirical Emission Formula

On October 19, 1900, Planck disclosed that he had found for continuous emission spectra from ideal (blackbody) radiators an equation that accurately represented the experimental energy-distribution curves at every temperature (Fig. 26.2). It is easy for most curves to find a mathematical expression that fits that particular curve. But he found one equation for an infinite number of different curves.

Fig. 26.4. Max Planck (1858–1947), German physicist. Winner of the Nobel Prize in Physics in 1918, for discovery of energy quanta.

One way of formulating his result is this: If we designate by the symbol E_λ the radiation energy emitted per second, per unit area of emitter surface, per unit wave length interval, then

$$E_\lambda = C\lambda^{-5}/(e^{D/\lambda T} - 1), \qquad (26.2)$$

where C and D are constants, and e is a number (the base of natural logarithms, about 2.718).

But Planck, one of the greatest physicists of his time, was not at all satisfied with this accomplishment. Finding Eq. (26.2) was no more of an explanation of the physical mechanism responsible for the phenomenon of emission than was Bode's law an explanation of the construction of the solar system (Section 11.8). Planck was, in fact, already working on the problem of deriving the energy distribution curve by assuming that light was emitted from a Maxwell-Hertz type of oscillator, but he now attacked the problem with new urgency.

In his autobiographical notes he relates that the problem appealed to him deeply because of the feature we have had cause to marvel at—that the continuous spectrum from (ideal) emitters is entirely independent of their chemical properties. We are reminded of the law of universal gravitation, which also was found to hold regardless of the chemical nature of the participating masses. Here again there seems to be a universal law at work. In Planck's own words, this fact concerning emission "represents something absolute, and since I had always regarded the search for the absolute as the loftiest goal of all scientific activity, I eagerly set to work." We note parenthetically the philosophic kinship between Planck and Newton; both, contrary to some nonscientists' views about science fashionable at the end of the twentieth century, believed that they could discover universal laws of nature, independent of human opinions and social interests.

As had all others who struggled with the emission problem, Planck failed at first, but at any rate his empirical formula of intensity vs. wavelength was a startling success. The very night of Planck's public announcement in a scientific meeting, a colleague extended the measurements beyond the range and accuracy previously obtained, and confirmed a complete agreement between the precise facts and Planck's formula. Spurred on by this success, Planck once more set to work to provide a conceptual scheme from which his formula would follow as a necessary consequence—"until after a few weeks of the most strenuous labor of my life the darkness lifted and a new, unimagined prospect began to dawn." December 14, 1900, when Planck presented his solution, marked the day when physicists began to follow the new path into what came to be called the Quantum Revolution.

26.3 The Quantum Hypothesis

Hertz's work had convinced Planck that radiation must be explained by the action of submicroscopic electric oscillators. (Planck referred to them as *resonators.*) The wall of a glowing solid emitter may thus be imagined as populated by such oscillators (perhaps the vibrating electrons, but this need not be specified), all at different frequencies. Since Maxwell one knew how such an oscillator may send off energy. But how does such an oscillator gain energy? One mechanism might be the incessant collisions with neighbors as heat is supplied to the wall to keep it emitting. And how is light radiated? By sending out electromagnetic waves during the vibration, in varying amounts of energy and at the frequency of the particular oscillator.

But just here Planck found it necessary to introduce a radical change in the classical picture. To calculate the distribution of energy among the submicroscopic oscillators, he has to make use of the statistical methods developed by Maxwell and Boltzmann for the kinetic theory of gases (Section 22.6). Boltzmann had shown how to calculate statistical distributions by assuming first that each particle has a variable number of small energy-units, and then letting the size of the energy-units go to zero at the end of the calculation. Planck found that to fit the observed facts in the case of blackbody radiation, the last step must be omitted: One must *not* assume, as seemed a matter of course until then, that an oscillator of some frequency f can send out any part of its total amount of energy, no matter how small.

What had seemed initially to be just a matter of mathematical convenience (the distributions must be described in terms of discrete entities in order to count them) gradually turned into a physical postulate: The energy content of the oscillator, its kinetic and potential energy at any moment, can only be an integral multiple of the quantity hf, where f is the frequency of its vibration and h is a universal constant, now known as Planck's constant, whose value remained to be found by experiment.

For example, the total energy at any one instant may be zero, hf, $2hf$, . . . nhf (n = any integer), but never, say, 1.85 hf. That is, the energy of the oscillator is *quantized,* limited to integral multiples of one given amount, or *quantum,* of energy, hf. For if the energy changes are in steps of hf, the total energy E can be considered given by $E = nhf$ (n = 0, 1, 2, . . . any integer). If an ordinary pendulum, on losing its energy by friction during vibration, behaved according to such a rule, its amplitudes of vibration would not fall off gradually, but would visibly decrease in a stepwise manner, each step representing an energy interval proportional in magnitude to the frequency of vibration or a multiple of that frequency.

A more concise rendition of the quantum hypothesis can be given in terms of the concept

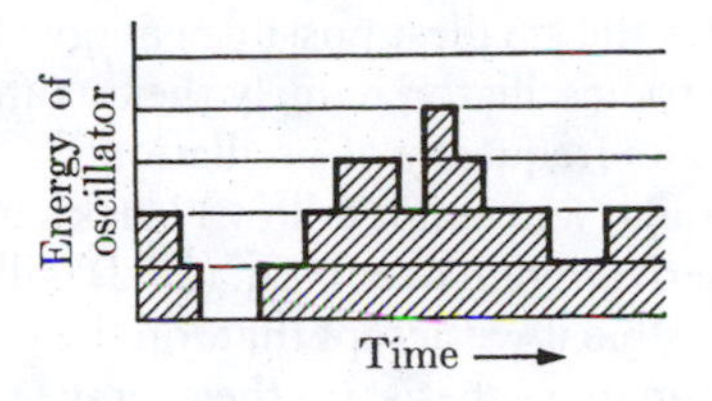

Fig. 26.5. The energy content that a single Planckian oscillator may have changes discontinuously from one level to the next.

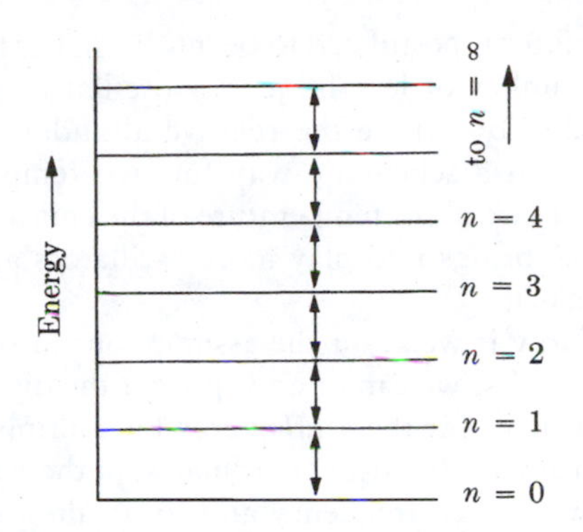

Fig. 26.6. Energy level scheme for one oscillator of frequency *f*. For any level, the energy is *nhf;* transitions are permitted between neighboring levels, as indicated.

energy levels: If the frequency *f* of an oscillator is constant despite changes in amplitude, then we might represent its energy content over a short time interval by a graph like the one in Fig. 26.5. Each of these steps or changes of levels is occasioned by absorption or emission of energy corresponding to a single quantum of magnitude *hf*. If we now draw all the different energy levels that the oscillator may reach at one time or another (Fig. 26.6), we obtain a ladder with equal steps. At any moment, the oscillator, as far as its energy content is concerned, is "located" on one or another of these energy levels, if it is not just then changing levels.

Planck knew nothing about the physical nature of the oscillator, or about its actual behavior in time; therefore the energy level diagram is the best representation of oscillators, for it avoids premature or erroneous pictorializations of the physical situation within the emitter.

To this point we have been looking only at a single oscillator of frequency *f*. Now we extend our view to encompass all the uncountable oscillators of different frequencies on the surface of the glowing body. This body as a whole, in Planck's view, emitted the continuous spectrum of electromagnetic (light) waves by virtue of containing simple oscillators of all frequencies, each emitting light of a frequency ν identical with its own frequency of vibration *f*. As the energy *E* of each oscillator is quantized according to $E = nhf$, energy can be emitted only during a sudden change in the amplitude of oscillation, for this amplitude is directly related to the energy content.

To Planck's fundamental postulate that the electric oscillator can change its energy only between neighboring energy levels E'' and E' according to $\Delta E = E'' - E' = nhf$, we must add the requirement that at any moment the likelihood is much greater that the emitting body contains more elementary oscillators at low energies (low *f*) than at high energies. From assumptions such as these, fully and quantitatively developed in a brilliant series of publications, Planck was able to deduce the emission formula [Eq. (26.2)] that he had previously found empirically, and he derived as well the other well-established characteristics of continuous emission spectra, as for example Wien's displacement law.

Although the details of the argument are on a more advanced level of mathematical physics, we may, in a qualitative way, make Planck's work plausible by means of the following curves. Figure 26.7a shows the energy emitted by individual oscillators during a drop in energy level as a function of the frequency of vibration or of emitted light; it is a straight-line graph, according to the assumption $\Delta E = hf$, and so shows that the high-frequency radiation carries away the largest amount energy during emission, whereas low-frequency emission does not deplete the emitter of much energy. On the other hand, the *number* of high-frequency oscillators present in

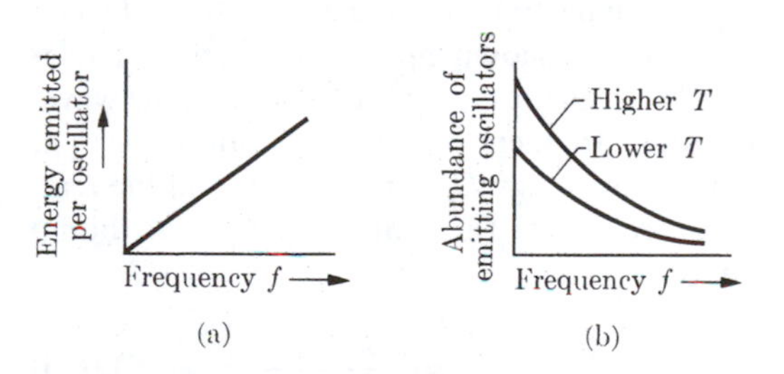

Fig. 26.7. (a) energy emitted by an oscillator, and (b) number of oscillators emitting energy; both are plotted as functions of frequency.

an emitter is postulated to be much smaller than the number of low-frequency oscillators (see Fig. 26.7b), where the relative abundance is shown in a schematic way for two temperatures; a rise in the temperature of the emitter, of course, brings into play more oscillators at all frequencies.

Now if we grant the assumptions in these two graphs, we can at once predict the distribution curve for the *total* energy by multiplying the ordinates (vertical coordinates) of these two graphs at each frequency and so obtain a new graph. The new ordinate is (energy emitted per oscillator) × (number of oscillators at given frequency), that is, total energy radiated at (or within a small frequency interval around) a given frequency; the abscissa (horizontal coordinate) is still frequency.

When we make such a graph, two things become apparent. The maximum intensity of radiation must be in the range between the small and high frequencies, and for higher temperatures the peak of intensity shifts to the right, toward higher frequencies, that is, shorter wavelength, exactly as expected by Wien's displacement law. In short, the graph derived from Planck's quantum hypothesis exhibits the main features of the corresponding graph (Fig. 26.2) obtained by experiment.

Problem 26.4. (a) Copy Fig. 26.7a and b, superimpose on each a grid of vertical and horizontal lines for reference purposes, and construct the total energy distribution graph (for both temperatures). (b) But note that in Fig. 26.2, with which our new graph is to be compared, the abscissa is not f but λ. Therefore draw a second graph from the data for the first with the same quantity as ordinate but with λ as abscissa (recall $\lambda = c/f$). Compare its shape with Fig. 26.2.

From his fundamental assumptions, Planck deduced the following precise mathematical relation between the energy radiated (E_ν), the wavelength or frequency of the radiation (λ or ν) at which the energy E_λ is measured, and the temperature (T) of the whole (ideal, blackbody) emitter:

$$E_\lambda = \frac{8\pi\lambda^{-4}h\nu}{e^{h\nu/kT} - 1} \qquad (26.3)$$

where k = Boltzmann's constant (see Section 22.4), and h is the constant of proportionality that relates the smallest possible energy change of an electric oscillator, namely the quantum of energy, to its frequency of oscillation f.

Although we cannot follow Planck's mathematical derivation in detail, Eq. (26.3) will seem more plausible if we accept that, on the basis of the quantum hypothesis: (a) the average energy per degree of freedom of the emitter was not kT, as in the classical theory based on the principle of equipartition,[1] but is given rather by the expression

$$\frac{h\nu}{e^{h\nu/kT} - 1},$$

and that (b) the number of degrees of freedom of the emitter at wavelength λ is $8\pi\lambda^{-4}$. Multiplying the last two expressions, we obtain Eq. (26.3).

By inspection, we can reach at once two conclusions:

1. The theoretically derived Eq. (26.3) is indeed perfectly equivalent to the empirical Eq. (26.2), for if we replace ν by c/λ, where c is the velocity of light, we can identify C with $8\pi ch$ and D with ch/k, the quantities c, h, and k being universal constants.
2. In the last statement it is implied that we may use the relations there to calculate h. Given the measurements of E_λ, λ, and T, and knowing the factors c and k, there remains only h as unknown. On solving, we find

$$h = 6.626 \times 10^{-34}\ \text{J·sec.}$$

Planck's own calculations of h at the time gave him a value correct to within 1% of the present well-confirmed figure. It is one of the most fundamental constants in atomic physics, of the same importance as G, of the law of universal gravitation, is in astrophysics, or as N is in chemistry. The quantity h, initially evaluated as discussed above, made its appearance thereafter in many other contexts and connections, so that there are now several independent methods of calculating and confirming its value.

Before discussing the role of Planck's emission formula in the development of quantum theory, we may note that Eq. (26.3) has found

---

[1]At very low frequencies Planck's expression does reduce to kT, as one can see by using the expansion of the exponential function, $e^x = 1 + x + \ldots$ for small values of x.

important applications in astrophysics, the most recent being the discovery that the whole universe is filled with residual radiation produced by the "primeval fireball," whose explosion—the Big Bang—is believed to have begun the present epoch of cosmic history (Section 32.5). The measured frequency distribution of this background radiation has been found to agree with Planck's formula if the temperature is set equal to about 2.7 K. That does not mean the universe originally had that temperature, but that it has cooled down over billions of years while maintaining thermal equilibrium. (This itself is a puzzle: What keeps distant parts of the universe at the same temperature?)

Problem 26.5. An atomic oscillator sends out light at 700 nm (red). At what frequency does the oscillator vibrate? What is the quantum of energy for this oscillator? What is it for another oscillator that emits violet light (at 400 nm)? *Note well that the quantum of energy depends on the frequency of the oscillator;* unlike electrical charge, energy is not quantized in uniformly large bundles under all conditions.

Problem 26.6. An atomic oscillator sends out light at 500 nm. Draw some portion of its energy level diagram (to a stated scale) according to the quantum hypothesis. What is the magnitude between successive steps on the diagram?

Problem 26.7. A small object of mass $m = 3$ kg at the end of a spring oscillates about an equilibrium position with simple harmonic motion, an initial amplitude x_1, and zero initial speed; the force constant is $k = 100$ N/m. Using the formula for the potential energy given in Section 17.6, and assuming that the period of vibration is given by $T = 2\pi\sqrt{m/k}$, what is the frequency f of vibration? If the energy can change only by quanta of hf, calculate the actual change of amplitude when the energy changes by one quantum. Why does the decay or increase of the oscillatory motion of springs, pendulums, and the like generally seem to us continuous when it is really quantized?

We might think that these successes must have gratified Planck deeply. However, he later confessed that his own postulate of energy quantization, to which he had been driven by the facts of emission, always remained to him deeply disturbing. He had indeed attained his goal of finding an "absolute" as nearly as one can—that is, he had found a universal and basic process—but at the cost of introducing a new hypothesis that did not seem to be in harmony either with Newtonian mechanics or with the wave theory of light. Furthermore, like most physicists of the time, he was still fundamentally convinced that natural processes are continuous; as Newton had expressed it, *natura non saltus facit* (nature does not make jumps). Indeed, both the ancient faith in a sequence of cause and effect and even the usefulness of the mathematical calculus itself seemed to depend on the proposition that natural phenomena do not proceed by jumps.

Not unexpectedly, the quantum hypothesis was regarded as rather too radical and unconvincing by almost all physicists at the time. Planck relates that he himself spent years trying to save science from discontinuous energy levels while the quantum idea "obstinately withstood all attempts at fitting, in a suitable form, into the framework of the classical theory." From a broader point of view, the general initial hesitation to plunge into quantum theory is only symptomatic of a useful law of inertia in the growth of science. One had to scan carefully the full implications of a step into such strange territory.

Moreover, Planck's initial treatment was not free from ambiguities that had to be resolved. There was, for example, a subtle paradox in Planck's reasoning. On the one hand, he jealously held on to the Maxwellian wave theory of light, considering it unquestionable that the light energy given off by a quantized oscillator spreads continuously over an ever-expanding set of wave fronts; and on the other hand, he apparently repudiated the Maxwellian theory by implying that the subatomic oscillators would radiate energy *not* while oscillating with a certain amplitude, but only while suddenly decreasing that amplitude.

Thus one can interpret his 1900 paper to mean only that the quantum hypothesis is used as a *mathematical* convenience introduced in order to calculate a statistical distribution, not as a new *physical* assumption. (He even writes that if the number of quanta contained in a certain amount of energy is not an integer, one should take it to be the nearest integer.) Lastly, Planck's work did not contain a physical picture or model for explaining the nature of the "oscillator," or for solving such puzzles as the difference between continuous and line emission spectra (the latter will be discussed in Chapter 28).

If one chose, one could regard Planck's scheme as a formalism, which at best left the "real" questions of physical mechanisms unanswered. The general acceptance of Planck's work depended therefore not just on experimental verifications, but on further demonstrations of the usefulness, even the necessity, of the quantum hypothesis in scientific thought. For some years the quantum hypothesis was uneasily suspended between doubt and neglect. Then Einstein extended and applied Planck's conception in rather the same manner that Avogadro developed Dalton's, and thereafter the doubts slowly began to fall away. Because Einstein's specific contribution to this topic depended on the clarification of a long-standing puzzle, the *photoelectric effect*, we shall begin by tracing the relevant facts.

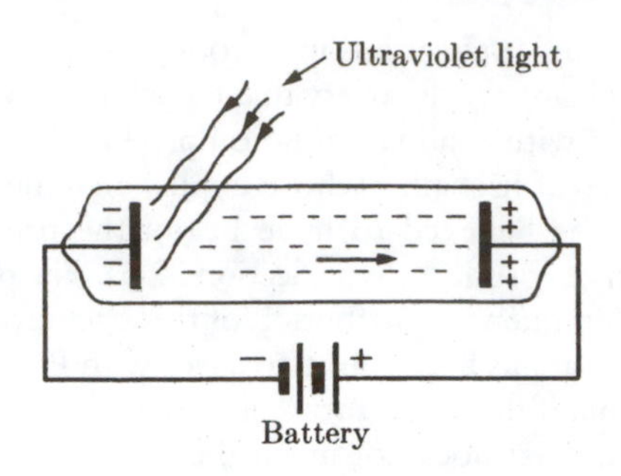

Fig. 26.8. A photoelectric current is set up through the circuit when ultraviolet light falls on the electrode at the left.

26.4 The Photoelectric Effect

Ironically, the first facts leading to the photoelectric effect, and through that eventually to the recognition that the classical theory of light had to be fundamentally revised, were by-products of the very same research that seemed at the time to furnish the most convincing proof for the classical theory—namely, Hertz's experiments on the propagation of electric waves.

As we noted, Hertz mentioned in passing that an electric spark would jump more readily between the metal spheres if they were freshly polished. It soon became clear, through the work of various contributors, that the ultraviolet component of light—as from a spark itself—had the effect of drawing or expelling negative charges from the clean surface of metals, and that it was these charges that aided in maintaining a suitable transient current between the spheres in Hertz's equipment. Indeed, some metals, notably the alkali metals, exhibited the same effect even with visible light. The air around the metal was not involved; on the contrary, when two clean metal plates (Fig. 26.8) are sealed in a transparent, well-evacuated quartz tube and are connected to a battery that charges one of the plates negatively, the latter, if illuminated, copiously emits negative charges that are then attracted to the positive plate and recirculated by way of the battery. The current so obtained clearly deserves the name *photoelectric current.*

The exact nature of those negative charges was not clarified until 1897. At that time, principally through the work of J. J. Thomson, electricity was found to be carried in fixed quantities of charge by discrete particles much smaller than the lightest atoms. This identification of electrons was first made from a study of the current through an almost completely evacuated tube when a sufficiently high potential difference was applied between two electrodes (Section 25.7). Although the electron concept was eventually found to be essential in explaining every phenomenon involving electric charges, the existence of free electrons was initially a radical hypothesis, in serious contradiction to the classical view of the structure of matter, for it had previously not been necessary to entertain the notion that any material particle smaller than the atom should exist by itself. J. J. Thomson himself wrote:

> At first there were very few who believed in the existence of these bodies smaller than atoms [that is, the electrons]. I was even told long afterwards by a distinguished physicist who had been present at my lecture at the Royal Institute that he thought I had been "pulling their legs." I was not surprised at this, as I had myself come to this explanation of my experiments with great reluctance, and it was only after I was convinced that the experiments left no escape from it that I published my belief in the existence of bodies smaller than atoms.

However, the electron quickly explained and coordinated a wide variety of phenomena. The flow of electrons accounted not only for currents in cathode-ray tubes and photoelectric cells, but also in solid metallic conductors; in a completely evacuated tube, a heated metal wire (filament) "boils off" electrons that are attracted

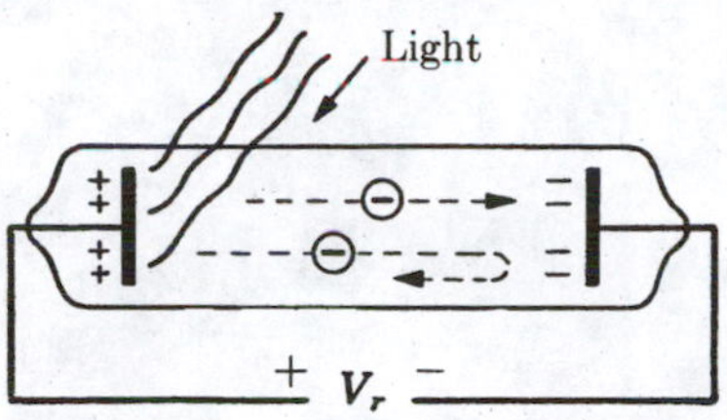

Fig. 26.9. To overcome the retarding potential V_r, the photoelectron must have a sufficiently large initial kinetic energy.

by a positively charged plate nearby and can be recirculated by a battery connected between plate and filament (*thermionic* tube, or diode as once used in radio circuits). The overabundance or deficiency of one or more electrons would turn an atom into a negatively or positively charged ion, and so help to account for the facts of electrolysis. The electrons associated with a neutral atom, one in which the positive and negative charges balance, could be identified more definitely with the oscillators responsible for the emission of light in the classical (and later on the quantum) model of radiation.

More detailed studies of the photoelectric effect itself revealed the following experimental facts.

a) When, in the apparatus shown in Fig. 26.8, the light incident on the left-hand plate is of one given frequency ("monochromatic" light), the number of electrons ("photoelectrons") emitted per second per unit area of plate, as measured by the current through the tube and the wires, increases with the intensity of the incident light. This is not unexpected, for the energy of the incident light must somehow be absorbed by the electrons as they leave the cathode; in this conversion from light energy to the energy in an electric current, an increase in the former should be attended by a corresponding increase in the latter.

b) When photoelectrons of mass m_e are "knocked out" of the metal by the incident light, they must leave with some initial kinetic energy $\frac{1}{2}m_e v^2$. To find this energy experimentally, we simply remove the battery connections in Fig. 26.8 that put an accelerating potential between the electrodes and apply instead a potential difference in the opposite sense, a *retarding voltage* V_r (see Fig. 26.9). Under these conditions only those electrons whose initial kinetic energy $\frac{1}{2}m_e v^2$ is as large as or larger than $V_r e$ will reach the right side. In such an experiment we can increase V_r, until no photoelectric current flows at all. At that point the "retarding voltage" has become a "stopping" potential, V_s; that is, even those electrons emitted by the action of the light with the highest speed v_{max} just fail to reach the plate; in equation form,

$$V_s e = \tfrac{1}{2} m_e (v_{max})^2 \qquad (26.4)$$

We should expect to find that the maximum kinetic energy of photoelectrons increases in some way with the intensity of the incident light, just as high ocean waves will not only move more of the pebbles on the shore, but also give them greater speeds. However, the experiments showed that *the maximum kinetic energy of photoelectrons is not a function of the intensity of incident light of a given frequency.* This was unexpected. It implied that an electron in the cathode can draw from the incident light energy up to a fixed maximum only, rejecting the remaining energy if the light waves are very intense, and, if the light is very feeble, presumably accumulating that maximum energy before leaving the surface.

Problem 26.8. In 1902 Philipp Lenard determined that photoelectrons released from a metal on which the ultraviolet portion of sunlight was falling needed a stopping potential of 4.3 volts. What was the maximum speed of these electrons? (Be careful to use consistent units.)

c) It is natural to inquire next whether the KE_{max} of photoelectrons depends on the frequency ν of the incident light; and experiments showed, though at first only qualitatively, that the kinetic energy does in fact increase with frequency. But it was not a case of simple proportionality: For each metal there was some frequency of light ν_0 below which *no photoelectric emission could be obtained, no matter how strong the incident radiation.* This *threshold frequency* ν_0 is for example, for zinc about 8.5×10^{14} sec^{-1} (λ_0 is about 350 nm), and for metallic sodium is around 5×10^{14} sec^{-1} (λ_0 below 600 nm). This observation raised the same type of conceptual difficulties as noted in (b) above concerning the mechanism by which the electron absorbs energy from the light.

d) Finally, it was also observed that whereas no amount of illumination below the threshold frequency would yield any photoelectrons from a given metal, even the feeblest intensity of light of a higher frequency than ν_0 sufficed to obtain photoelectric currents *without any delay*. Modern experiments have confirmed that the time interval between the arrival of light and the departure of electrons from the surface is less than 3×10^{-9} sec. That, however, is in flat contradiction to our expectations: We have already measured the energy photoelectrons may have, and we can calculate how long we should have to let light fall on a metal surface in a typical experiment before that much energy is available to the electron. One result of such a calculation is that an electron would have to wait for some seconds before it could accumulate the necessary energy from a weak beam of light—always assuming, as we do in the Maxwellian picture of light propagation, that the energy of a light wave is uniformly distributed over the whole wave front.

Fig. 26.10. Albert Einstein (1879–1955), German physicist, in 1912. He received the Nobel Prize in Physics in 1922 for his discovery of the law of the photoelectric effect. His theory of relativity is discussed in Chapter 30.

26.5 Einstein's Photon Theory

The resolution of these problems came in a 1905 paper of the German-born physicist Albert Einstein (1879–1955), entitled simply, "On a Heuristic Point of View Concerning the Generation and Transformation of Light."[2] (This is one of three famous papers he published that year, the others being on Brownian movement, mentioned in Section 22.10, and on relativity, to be discussed in Chapter 30).

Einstein begins by paying tribute to the wide usefulness of Maxwell's theory of light; it "has proved itself excellently suited for the description of purely optical phenomena [reflection, refraction, interference, polarization, etc.], and will probably never be replaced by another theory." But this classical theory of light was designed and tested above all with reference to the *propagation* of light, its progress through space measured over relatively long time intervals. Therefore, he continues, one should not be surprised if contradictions with experience arise when the investigation turns to problems in which everything is contingent on a *momentary* interaction between light and matter. Two prominent examples of the latter case in which troubles have developed are the origin or emission of light from an oscillator and the transformation of light energy into the kinetic energy of photoelectrons.

Now Einstein proposes the following daring conceptual scheme: *The energy of light is not distributed evenly over the whole wave front, as the classical picture assumed, but rather is concentrated or localized in discrete small regions* or, if we may say so, in "lumps" or "bundles." A mental image is difficult to come by, but this analogy might help: If the same statement could be applied to the energy of water waves, we would expect that an incoming wave could move a floating cork up and down (that is, give the cork energy) only at certain spots along the wave front; between

[2]*Heuristic* means useful as an aid in the advancement of a conceptual scheme, but not necessarily in the final correct form. An example of a heuristic device is the use of analogies. It is curious to note that Newton, in communicating to the Royal Society in 1675 a corpuscular view of light, entitled his paper "An Hypothesis Explaining the Properties of Light," and wrote that this hypothesis was also a heuristic one: ". . . though I shall not assume either this or any other hypothesis, not thinking it necessary to concern myself whether the properties of light discovered by men be explained by this or any other hypothesis capable of explaining them; yet while I am describing this, I shall sometimes, to avoid circumlocution . . . speak of it as if I assumed it."

those spots, the passing wave would leave the cork undisturbed.

One can see the roots of Einstein's idea in Planck's work. We remember that the Planckian oscillator can lose energy only in quanta of hf, and that the light thereby emitted has the frequency $\nu = f$. But in introducing a new concept of light emission Planck did not suggest in the least that a change is also necessary in the classical view concerning the nature and propagation of light waves after the moment of emission; the Maxwellian picture of expanding electromagnetic waves with evenly and continuously distributed energy remained effective.

However, this raised conceptual problems of the following kind: If a Planckian oscillator can change its energy only in steps of hf, how can it absorb energy from light in which the energy is thinly distributed? If not by a slow continuous increase in oscillation, which is prohibited by the quantum theory, how can the oscillator store up the stream of energy given to it by incident light until there is enough for a jump to a higher energy level? Even if a storage mechanism could be imagined, the fact of instantaneous emission of photoelectrons [item (d) in Section 26.4] shows that no such storage actually occurs.

Roughly speaking, Einstein therefore extended the quantum concept to the energy in the emitted light. The quantum of energy (hf) lost by the emitter does not spread out over the expanding wave front, but *remains intact as a quantum of light with energy* $h\nu$, *where* $\nu_{light} = f_{oscillator}$. (The name *photon* was later given to the quantum of light.) Each wave front of light emitted by a glowing solid, or the sun, etc., must therefore be imagined to be studded with photons, like little dots on a balloon, traveling with the wave at the speed of light; and when the wave expands as it progresses, the light becomes weaker because the distance between neighboring photons increases, thereby decreasing the density of photons along the wave front.

Ordinarily the light beams we observe in experiments on reflection, refraction, polarization, and the like make use of beams with substantial cross section; and then we are dealing with enormous numbers of photons on each wave front. Therefore the individuality of those photons is masked, just as an ordinary stone appears to our eyes to be made of continuous matter rather than of individual atoms. But in the absorption and transformation of light energy by an individual submicroscopic oscillator, as in the photoelectric effect, the finer "structure" of the light wave becomes important. That is why the classical theory of light remains completely valid and most appropriate for dealing with the ordinary problems of optics, whereas the quantum theory of light must be used to understand the interaction between light and individual atoms.

An application of these fundamental ideas by Einstein yielded at once a quantitative explanation of the photoelectric effect. A quantum of light, a photon, falling on an atom may or may not give it enough energy—all in one "bundle"—to loosen an electron; if the photon's energy is enough, the electron can free itself by doing the work W that must be done against the force of electric attraction binding it to the atoms of the cathode; if the photon's energy is less than W, the electron will not be emitted. This at once explains why emission is instantaneous if it takes place at all [item (d)]. (But note that no mental picture is supplied for imagining how the photon energy is converted to the energy of the electron.)

But now the meaning of the threshold frequency ν_0 is clear also: To break the electron off from an atom at the surface, the photon energy $h\nu$ must be at least equal to W, or

$$h\nu_0 = W. \qquad (26.5)$$

Furthermore, we expect chemically different atoms to have different internal structures, so that W, and consequently also ν_0, should vary from one element to another, as indeed it does.

For light of higher frequency than ν_0 each photon has an energy $h\nu$ larger than W, so that energy is left over to let the photoelectron leave with some initial speed. Because light can penetrate the topmost layers of atoms, the photoelectric effect may take place below the surface. If the photoelectron then has to pass through layers of atoms before reaching the surface of the metal plate, it may lose some of its speed; but some electrons suffer no such losses and come off with the full speed v_{max}, directly toward the anode. For these we write

$$h\nu = \tfrac{1}{2}m_e(v_{max})^2 + W. \qquad (26.6)$$

This is *Einstein's equation for the photoelectric effect*, an application, on the level of individual atoms, of the law of conservation of energy to the quantum theory of photoelectricity. It contains the algebraic expression of the experimental fact noted in Section 26.4(b) and (c) that the KE_{max} of photoelectrons increases with the frequency ν, but does not increase with the intensity of the incident light. On the other hand, a greater intensity means more photons and larger *numbers* of photoelectrons per second, in accordance with observation (a). Thus Eq. (26.6) summarizes all the observations in *one* mathematical law.

But again, it did more. We must realize that at the time of Einstein's proposal, the data on hand were incomplete and often only qualitative. Precise and reproducible measurements of the photoelectric effect are still difficult; for example, roughness or small impurities on the metal surfaces greatly affect the results for ν_0 and v_{max}. Therefore Einstein's equation was essentially still a prediction rather than a summary of detailed quantitative facts. It was the fulfillment of these predictions in later experiments by others that gave Einstein's work such importance and changed the photon from a "heuristic point of view" to a fundamental part of contemporary scientific thought. (Nominally, it was mainly for this work on the quantum theory of light rather than for his relativity theory that Einstein was awarded the Nobel Prize in Physics.)

In Eq. (26.6), the constant *W* is often called the work function of the metal, and *h* is, of course, Planck's constant. The equation indicates how both quantities can be found by experiment, for if we were able—by careful experimentation with two different frequencies of light ν_1 and ν_2—to ascertain in terms of the necessary stopping potentials V_{s1} and V_{s2} what the term $\frac{1}{2}m_e(v_{max})^2$ is for the photoelectrons in each case [see Eq. (26.4)], then we could write

$$h\nu_1 = V_{s1}e + W, \qquad h\nu_2 = V_{s2}e + W.$$

Therefore

$$h = (V_{s1} - V_{s2})e/(\nu_1 - \nu_2). \qquad (26.7)$$

After *h* is obtained from Eq. (26.7), the constant *W* can be quickly calculated from Eq. (26.3) or (26.6). And of course here is a chance to check the value for *h* from Eq. (26.3), that is, from Planck's theory of energy quantization in the radiating oscillator, a theory supplemented but by no means invalidated by Einstein's work.

The credit for making the necessary precise measurement under extremely difficult experimental conditions belongs to the same man who had previously measured the electronic charge *e*, namely the American physicist Robert A. Millikan. The result, obtained in 1916, was a coincidence of the values of *h* by these two very different methods, within the experimental error.

Problem 26.9. In a typical experiment, 10^{-6} J of light energy from a mercury arc are allowed to fall on 1 cm^2 of surface of iron during each second. Clean iron reflects perhaps 98% of the light, but even of the remaining energy only about 3% falls into the spectral region above the threshold frequency of the metal. (a) How much energy is available for the photoelectric effect? (b) If all the energy in the effective spectral region were of wavelength 250 nm, how many photoelectrons would be emitted per second? (c) What is the current in the photoelectric tube in amperes? (d) If the threshold frequency is 1.1×10^{15} sec^{-1}, what is the work function *W* for this metal? (e) What is the stopping potential (in volts) for photoelectrons emitted by light at 2500 Å?

Problem 26.10. Sketch the graph you expect to obtain when plotting ν, the frequency of light incident on a metal surface (abscissa), vs. $\frac{1}{2}m_e(v_{max})^2$ or V_s for the photoelectrons (ordinate). Should the line be straight? What is the physical meaning of the intercept on the horizontal axis? What is the equation for the line? What is the meaning of the slope of the line?

Problem 26.11. The photoelectric effect eventually became the basis for a large number of technological devices. Describe a couple of them.

26.6 The Photon-Wave Dilemma

Einstein's photon model of light was soon found fruitful in explaining other phenomena, for example, photochemistry and anomalous specific heats, two topics to which Einstein himself greatly contributed. As has been aptly said:

> It is striking how thoroughly the theory affects our judgment of the importance of one field or another. Strictly from the experimental point of view, the photoelectric effect seems to be a rather remote corner of physics with [initially] technical applications of no great significance. However, Einstein's theory makes it clear that the photoelectric effect gives evidence of the

> nature of light, one of the greatest problems of physics.
>
> (Oldenberg, *Introduction to Atomic Physics*)

By showing that the energy of light was quantized, Einstein contributed directly toward the acceptance of Planck's original quantum theory, then still without general support. But, a little like Dalton, who repudiated Gay-Lussac and Avogadro, Planck himself was far from pleased with Einstein's photon. By its acceptance, he wrote in 1910, "the theory of light would be thrown back by centuries" to the time when the followers of Newton and Huygens fought one another on the issue of corpuscular vs. wave theory. He thought all the fruits of Maxwell's great work would be lost by accepting a quantization of energy in the wave front—and all that "for the sake of a few still rather dubious speculations."[3]

Indeed, the inconvenience of having to deal with photons is very real. They represent bundles of energy without having ordinary *rest mass* (a term to be explained in Chapter 30); in this they differ from the Newtonian *corpuscles* of light, leaving only a faint analogy between them, even though it is customary to refer to Einstein's photon theory as the *corpuscular* or *particulate* theory of light. But our minds tend to insist on a good picture, and it requires great self-control to visualize a quantum of energy without bringing in some matter to which it can be attached. It was a little easier to think of Maxwell's light energy spread evenly through the "field" along a wave front.

Then, too, there are other questions: How large in volume and cross-sectional area is the "spot" on the wave front where the photon is located? What could be the meaning of "wavelength" and "frequency" of light, which determine the energy content of the photon by $E = h\nu$, if the photon is, so to speak, only a dot on the wave front and not part of a whole wave train? By what mechanism does the wave determine the path of the photon in such wave phenomena as interference and polarization? How does an electron absorb a photon?

In the past, one partial answer to such questions has been to hold concurrently two separate views on light—the wave and the photon model—and to apply one or the other as required by the problem. The more recent answer, which allows, in principle, a solution of every problem with physical meaning in this field, involves combining both views and assuming that the photons are distributed over the wave front in a statistical way, that is, not individually localized at a particular point.

But at our level the more practical solution is, first of all, to realize that some of these questions, while possibly very disturbing, are asked on the basis of a mechanical view of atomic phenomena that may stem from an erroneous transfer of experience with large bodies obeying simple Newtonian laws. Thus the "size" of the photon is not a concept that we should expect to have the same meaning as the size of marbles and projectiles. Furthermore, photons (and atomic particles as well) differ from water waves, pebbles, and other large-scale entities in that one cannot make various experiments on the *same* subatomic entity. One can localize and measure and weigh a stone, find its velocity, etc., and all the while it is the same unchanged stone. But a photon, after it has been sent into a Geiger counter or a photographic emulsion, is no more; two photons on which we impress different experimental conditions—for example, in searching for wave and for corpuscular properties—are not, strictly speaking, the same entities.

As the physicist Max Born said:

> The ultimate origin of the difficulty lies in the fact (or philosophical principle) that we are compelled to use the words of common language when we wish to describe a phenomenon, not by logical or mathematical analysis, but by a picture appealing to the imagination. Common language has grown by everyday experience and can never surpass these limits. Classical physics has restricted itself to the use of concepts of this kind; by analyzing visible motions it has developed two ways of representing them by elementary processes: moving particles and waves. There is no other way of giving a pictorial description of motions—we have to apply it even in the

[3]It is amusing to note that the work of Thomas Young in 1801, which introduced Huygens' wave theory and became the basis for Maxwell's view, was generally attacked at the time by the proponents of the current corpuscular theory on the basis that the wave theory, as one critic put it, "can have no other effect than to check the progress of science and renew all those wild phantoms of the imagination which. . . . Newton put to flight from her temple." (Henry Brougham in the *Edinburgh Review,* 1803.)

region of atomic processes, where classical physics breaks down.

Every process can be interpreted either in terms of corpuscles or in terms of waves, but on the other hand it is beyond our power to produce proof that it is actually corpuscles or waves with which we are dealing, for we cannot simultaneously determine all the other properties which are distinctive of a corpuscle or of a wave, as the case may be. We can therefore say that the wave and corpuscular descriptions are only to be regarded as complementary ways of viewing one and the same objective process, a process which only in definite limiting cases admits of complete pictorial interpretation. . . .

(*Atomic Physics*)

However, no matter how uncomfortable the wave-photon duality, or how great the disparity between concept and intuition at this stage, the photon theory stands on its proven power to explain, to predict, to stimulate further discoveries. We may illustrate this power by brief references to a few examples.

26.7 Applications of the Photon Concept

a) *Fluorescence.* A material is fluorescent if it absorbs light and immediately reradiates it. It had long been noticed that in general the frequency of such reradiated light is equal to or less than the original incident light. This is at once plausible when we imagine the following processes. The incident photon of energy $h\nu$ raises the atomic oscillator to a new level of higher energy than before. There is a general tendency of all oscillators to remain only briefly at high energy levels and to spend their time mostly at the lowest levels; thus the oscillator quickly reradiates the energy, though it may do so in two or more stages. If the absorbed photon is again given up in one process, we have resonance radiation (see Section 28.2), but if the original energy $h\nu$ is to be divided and emitted in, say, two photons, the frequency of the reradiated light must be lower than the original frequency.

Problem 26.12. What does the above account reveal concerning the relationship between the frequency f of the oscillator and ν of incident, absorbed light?

Problem 26.13. In some cases light is reradiated at a *higher* frequency. How can the model account for this?

---

b) *X-ray photons.* We noted earlier that, in 1895, W. C. Röntgen discovered that a beam of cathode rays (soon after identified as high-speed electrons) could create a new type of radiation when it was allowed to fall upon an obstacle such as the glass of the tube itself. In time, these so-called x-rays were shown to be high-frequency electromagnetic radiation of the same fundamental nature as radio waves, gamma-rays, and all other forms of "light" (Section 25.6; see Table 26.2).

Nowadays x-rays are generated by the bombardment of a metal target forming the anode in a hot-filament cathode-ray tube. In a sense this is the reverse of the photoelectric effect; now photons are "knocked out" of the substance by incident electrons. We may provisionally imagine that the energy of the incident electron is first absorbed by the atomic oscillators, and then is reemitted as a photon when the oscillator returns to its original energy level. Therefore we expect that the energy of the x-ray photon is equal to that of the electrons in the beam, or

$$h\nu = \tfrac{1}{2}m_e v^2. \quad (26.8)$$

In practice, we find that Eq. (26.8) holds, giving the highest frequency of x-rays obtainable for a given electron beam. Furthermore, when such an x-ray is allowed in turn to fall on a clean metal

Table 26.2. Spectrum of Electromagnetic Radiation

Name of Radiation	*Approximate Range of Wavelengths*
Radio waves	A few m and up
Microwaves	A few mm to a few m
Infrared waves	750 nm to 0.01 cm
Visible light	400 to 750 nm
Ultraviolet light	10 to 400 nm
X-rays	0.01 nm to 50 nm
Gamma-rays	Less than 0.05 nm

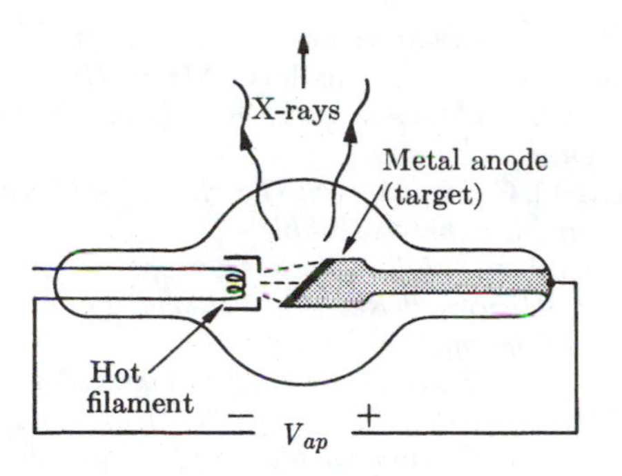

Fig. 26.11. X-rays are generated when a stream of high-speed electrons falls on a target.

surface, it causes the emission of photoelectrons of about the same velocity as that of the original beam (the work function W being very small compared with the energies involved in x-rays).

Problem 26.14. Figure 26.11 shows schematically an apparatus for producing x-rays. If V_{ap} = 50,000 volts, (a) what is the speed of the electrons incident on the anode, and (b) what is the energy of the x-ray photons emitted? [*Note:* this figure gives only the maximum energy, for the incident electron may not give all its energy to an oscillator.]

Problem 26.15. Look again at Fig. 26.1, which has been constructed for the temperature of the sun's surface, about 6000 K. Is it just a coincidence that the peak occurs in the *visible* range? Explain.

26.8 Quantization in Science

The enormous change in the natural sciences during the hundred years following 1808 can be correlated with a strangely similar discovery in each of such widely different fields as biology, chemistry, and physics—the discovery that there exists a structure, an elementary discrete quantization as it were, in the material on which each field builds. The case just discussed, that of energy quantization in systems that exhibit vibratory motion or periodic fluctuation (as in a Planck oscillator or in the electromagnetic field), is but the last of a series of such developments.

We recall that Dalton showed in the 1800s that chemical matter is not to be regarded as an infinitely divisible continuum, but that it consists of atoms that maintain their integrity in all chemical transactions. In biology, the cell theory of Schleiden (1838) and Schwann revealed that all living matter consists of individual cells and their products. Gregor Mendel's work (1865) led to the immensely fruitful idea that the material governing heredity contains a structure of definite particles, or genes, which may be handed on from one generation to the next.

Meanwhile the physics of heat, electricity, magnetism, and light—that is, that part of physics that the eighteenth century had visualized largely in terms of the action of imponderable, continuous fluids—was being rephrased in a similar way. Perceptible heat was identified with the motion of discrete atoms. Electrical charge was found to be carried by the electron (whose motion would serve to set up magnetic fields, thereby reducing two topics to one). Finally, the energy of light and of atomic oscillators was found to be quantized as well.

In short, it was as if these new views in the various branches of science stemmed from a similarly directed change of the mental model used to comprehend fundamental quantities and processes, a change in the guiding idea from a *continuum* to a well-defined, discrete particle or *quantum.* In part, one can explain such changes by noting that the development of better instrumentation and techniques opened up a new level of observation. But this is largely arguing *a posteriori,* for such technical developments frequently were stimulated by the proposal of the idea which they later helped to fortify. One must allow the possibility of a change of tone and mood in the whole field, of a general conceptual movement not related to any single cause—yet another expression of the organic and dynamic nature of science.

Henri Poincaré, writing in 1912 after an international meeting on quantum theory, wrote in some despair:

> "The old theories, which seemed until recently able to account for all known phenomena, have recently met with an unexpected check. . . . Is *discontinuity* destined to reign over the physical universe, and will its triumph be final?

As it turned out, the answer, which had been building up throughout several sciences, was to be Yes. The fundamental theme of continuity, which had served so well from Galileo on, was now joined and challenged by the anti-theme, discontinuity. Writing in 1905 on his great autobiography, *The Education of Henry Adams,* Adams has foreseen, as he put it, that "after the nineteenth century the course of history itself would be away from unities and toward multiplicity and fragmentation."

RECOMMENDED FOR FURTHER READING

H. A. Boorse and L. Motz (editors), *World of the Atom,* pages 462–501, 533–567. Extracts from the writings of Planck and Einstein, and biographical sketches.

William R. Everdell, *The First Moderns,* chapters on Planck and Einstein

Phillipp Frank, *Einstein: His Life and Times*

J. L. Heilbron, *The Dilemmas of an Upright Man: Max Planck as Spokesman for German Science,* Berkeley: University of California Press, 1986.

Gerald Holton, "On the hesitant rise of quantum physics research in the United States," in *Thematic Origins,* pages 147–187

Paul Kirkpatrick, "Confirming the Planck-Einstein equation $h\nu = \frac{1}{2}mv^2$," in *Physics History II* (edited by French and Greenslade), pages 149–152

M. J. Klein, "Thermodynamics and quanta in Planck's work," in *History of Physics* (edited by S. Weart and M. Phillips), pages 294–302); "The Beginnings of the Quantum Theory" in *History of Twentieth Century Physics* (edited by C. Weiner), pages 1–39, New York: Academic Press, 1977.

Thomas S. Kuhn, "Revisiting Planck," in *History of Physics* (edited by S. Brush). pages 132–153

Max Planck, "A scientific autobiography," translated from German by F. Gaynor, in his book *Scientific Autobiography*

J. H. Weaver (editor), *World of Physics,* excerpts from papers by Planck and Einstein, Vol. II, pages 284–292, 295–312

GENERAL BOOKS ON THE HISTORY OF MODERN PHYSICAL SCIENCE (PRIMARILY TWENTIETH CENTURY)

(Complete references are given in the Bibliography at the end of this book)

Benjamin Bederson (editor), *More Things in Heaven and Earth: A Celebration of Physics at the Millennium;* a collection of review articles, some historical.

Laurie M. Brown *et al.* (editors), *Twentieth Century Physics*

S. G. Brush and L. Belloni, *The History of Modern Physics: An International Bibliography*

Jed Z. Buchwald (editor), *Scientific Practice: Theories and Stories of Doing Physics*

Robert P. Crease and Charles C. Mann, *The Second Creation: Makers of the Revolution in 20th-Century Physics*

Michael J. Crowe, *Modern Theories of the Universe, from Herschel to Hubble*

J. T. Cushing, *Philosophical Concepts in Physics: The Historical Relation between Philosophy and Scientific Theories*

E. A. Davis (editor), *Science in the Making: Scientific Development as Chronicled by Historic Papers in the Philosophical Magazine;* four volumes covering 1798–1998.

David DeVorkin, *The History of Modern Astronomy and Astrophysics: A Selected, Annotated Bibliography*

Albert Einstein, *Collected Papers* (edited by John Stachel *et al.*)

Albert Einstein and Leopold Infeld, *The Evolution of Physics*

Peter Galison, *Image and Logic: A Historical Culture of Microphysics*

Peter Galison, *How Experiments End*

Elizabeth Garber, *The Language of Physics*

J. L. Heilbron and B. R. Wheaton, *Literature on the History of Physics in the 20th Century*

Norriss S. Hetherington, *Encyclopedia of Cosmology*

Dieter Hoffmann *et al.* (editors) *The Emergence of Modern Physics*

F. L. Holmes (editor), *Dictionary of Scientific Biography,* Supplement II (Vols. 17 & 18), includes biographies of scientists who died since 1970

Aaron J. Ihde, *The Development of Modern Chemistry*

Christa Jungnickel and Russell McCormmach, *Intellectual Mastery of Nature,* Vol. 2, *The Now Mighty Theoretical Physics 1870–1925*

Daniel J. Kevles, *The Physicists: The History of a Scientific Community in America*

Helge Kragh, *Quantum Generations: A History of Physics in the Twentieth Century*

John Krige and Dominique Pestre (editors), *Science in the 20th Century*

Kenneth R. Lang and Owen Gingerich (editors), *Source Book in Astronomy and Astrophysics, 1900–1975*

David Leverington, *A History of Astronomy, from 1890 to the Present*

Seymour H. Mauskopf (editor), *Chemical Sciences in the Modern World*

Albert E. Moyer, *American Physics in Transition;* "History of Physics" [in 20th Century America], *Osiris,* new series, Vol. 2, pages 163–182 (1985)

Mary Jo Nye, *Before Big Science*

Mary Jo Nye, *From Chemical Philosophy to Theoretical Chemistry*

R. C. Olby *et al.* (editors), *Companion to the History of Modern Science*

John S. Rigden (editor), *Macmillan Encyclopedia of Physics*
Wolfgang Schirmacher (editor), *German Essays on Science in the 20th Century*
Richard H. Schlagel, *From Myth to Modern Mind*, Vol. 2
Emilio Segrè, *From X-Rays to Quarks*
Joan Solomon, *The Structure of Matter*
H. Henry Stroke (editor), *The Physical Review: The First Hundred Years*
Curt Suplee, *Physics in the 20th Century*, a popular history, lavishly illustrated.
Spencer R. Weart and Melba Phillips (editors), *History of Physics*
Bruce R. Wheaton, *The Tiger and the Shark: Empirical Roots of Wave-Particle Dualism*
E. T. Whittaker, *A History of the Theories of Aether and Electricity*, Volume II, *The Modern Theories*

Refer to the website for this book, www.ipst.umd.edu/Faculty/brush/physicsbibliography.htm for Sources Interpretations and Reference Works and for Answers to Selected Numerical Problems.

The Atom and the Universe in Modern Physics

27 Radioactivity and the Nuclear Atom

28 Bohr's Model of the Atom

29 Quantum Mechanics

30 Einstein's Theory of Relativity

31 The Origin of the Solar System and the Expanding Universe

32 Construction of the Elements and the Universe

33 Thematic Elements and Styles in Science

Niels Bohr (1885–1962), Danish physicist, with his fianceé Margrethe Norlund shortly after the announcement of their engagement. Bohr's model of the atom showed how quantum concepts might be fruitful, and inspired the development of quantum mechanics, for which he then provided the influential "Copenhagen Interpretation."

The first years of the twentieth century saw the publication of a handful of short papers by Max Planck and Albert Einstein, which, in retrospect, appears to have been among the chief events to usher in a transformation in the fundamental concepts of physics. As we saw in Part G, the groundwork for what might be called the *Second Scientific Revolution* (the first being the one initiated by Copernicus and essentially completed by Newton) had already been laid in the nineteenth century through studies of light and electromagnetic radiation, although scientists had not yet fully realized how the classical concepts were being undermined. Planck's quantum hypothesis seemed at first not a new physical postulate but only a somewhat arbitrary modification of Maxwell's electromagnetic theory, and Planck himself devoted considerable effort to trying to fit it into the framework of classical theory. Similarly, when the Michelson-Morley experiment appeared to indicate that the earth's motion through the ether could never be observed, G. F. FitzGerald and H. A. Lorentz suggested as an *ad hoc* hypothesis that bodies moving through the ether might simply contract in length, in such a way that a measuring device would fail to indicate any motion.

It was Einstein who first turned from all *ad hoc* attempts to patch up the old, failing theories and insisted on the need for new physical laws, for example, to explain the properties of radiation. His work was soon paralleled in related fields by that of Bohr, Rutherford, and many others in developing detailed theories of atomic structure. Out of the mathematical formulations of Schrödinger and Heisenberg came a new model of the atom, far less easily visualizable in mechanical terms than the classical models, but far more effective. The revised model gave us the new, larger unity. That is, the contemporary theory of the atom is the meeting ground of modern science, where such fields as physics, chemistry, biology, astronomy, and engineering often join or overlap. The growth of science became virtually explosive, the harvest in pure and applied knowledge unparalleled, and the promise for future developments ever more challenging.

In the meantime, as most other physicists concentrated on understanding the microstructure of matter, Einstein and a few other physicists, astronomers, and mathematicians were generalizing his relativity theory to deal with models of the structure and evolution of the entire universe. In returning to the subject of astronomy and cosmology, with which we started this book, we find that this is an adventure without end, for today's scientific journals and newspapers continue to announce discoveries that show that the universe is even more remarkable and manifold than we had so recently thought.

CHAPTER 27

Radioactivity and the Nuclear Atom

27.1 Early Research on Radioactivity and Isotopes

The account of the progress of physics to this point has given us an atom that has some well-defined and some still mysterious features. Around 1910, just about one century after Dalton's announcement of the atomic theory of chemistry, one could be fairly certain of a few facts. For example, the size of atoms was known, though only very roughly, from kinetic theory; and although the internal structure of the atom was not known, it was certain that some of the components had to be charged, for electric currents of all kinds needed the electron, and all theories of light agreed on electric oscillators in the atom. Also, it had been shown by Lenard in 1895 that a stream of cathode rays (soon to be identified as electrons) would easily penetrate foils or sheets of material and, although very thin, such foils still represented many thousand layers of atoms, so that the fact of penetration led one to think of the atom no longer as a ball-like solid, but as a porous, open structure.

A set of astonishing new facts came to light after the discovery of radioactivity in 1896 by the French physicist Henri Becquerel (1852–1908). Becquerel was trying to find substances that emit the mysterious x-rays just discovered by Röntgen (Section 25.6). He found a substance—potassium uranyl sulfate—spontaneously emitting radiations without any stimulus and without apparently undergoing any change itself. While he considered them to be x-rays, we know them to be due to radioactivity, which was heralded and identified in the popular press as an amazing new source of energy—one that might even overthrow the principle of conservation of energy.

Becquerel's initial discovery itself would have been worth little interest except that it drew the attention of the Polish-French physicist Marie Sklowowska Curie (1867–1934). Born in Warsaw, Poland, she went to Paris in 1891 to study science, earned a degree in physics in 1893, and met the physicist Pierre Curie (eight years older, and already known for his work on piezoelectricity) in 1894. They were married in 1895.

She realized that Becquerel's rays were not x-rays but a completely new phenomenon, and searched for its cause by the tedious work of isolating, from uranium ore, any new element that might be responsible for the intense radiation Becquerel had found.

In collaboration with her husband, she isolated *radium,* a highly radioactive element found in minute quantities in some uranium-bearing minerals. In 1902 they had a sample large enough and pure enough to determine the atomic weight of radium, and the next year they were jointly awarded, with Becquerel, the Nobel Prize in Physics.

Marie Curie made radium available to the scientific community as a source of intense radioactivity. Other scientists with better facilities than the scant ones made available by the authorities in Paris to the Curies (even after their Nobel prize) quickly made a number of major discoveries and proposed bold new theories about the nature of radioactivity.

Pierre Curie was killed in a horse-cart accident in 1906. Marie continued her strenuous experimental work, although she was beginning to suffer the effects of radium exposure. Though a heroine of science to the public, she was still denied full recognition by the establishment in France; as late as 1911, the Academy of Science refused to elect her to membership. But the international scientific community showed its growing understanding of the continuing importance of her work through the award of the 1911 Nobel Prize in Chemistry.

The leadership in research on radioactivity was soon taken over by Ernest Rutherford (1871–1937), who came from his native New Zealand to work with J. J. Thomson at the Cavendish Laboratory at Cambridge University in England. During the early decades of the

Fig. 27.1. Marie Sklodowska Curie (1867–1934), Polish-born physicist who did her research in Paris, pioneer of research on radioactivity, and the first scientist to win two Nobel prizes (Physics in 1903, Chemistry in 1911).

Fig. 27.2. Ernest Rutherford (1871–1937), physicist, born in New Zealand, who did his research in England and Canada. He discovered radioactive transmutation and the atomic nucleus. Nobel Prize in Chemistry in 1908.

twentieth century, the Cavendish became the world center for research in experimental atomic physics, although Rutherford himself made some of his most important discoveries earlier while on the faculty at McGill University in Montreal, Canada, and then later at the University of Manchester.

Two kinds of radioactivity were first identified, called *alpha* (α) and *beta* (β) rays. Rutherford and Frederick Soddy argued that atoms of the heavier elements such as uranium, thorium, and radium must be assumed to disintegrate spontaneously by the emission of either a high-speed α-ray (a particle with a positive charge numerically equal to that of two electrons and a structure that was later found to be that of the helium nucleus), or a high-speed β-ray (a particle with the same rest mass and negative charge as any ordinary electron). In many elements, the emission of either particle was attended by that of a *γ-ray,* a very penetrating type of short-wavelength electromagnetic radiation.

The most remarkable conclusion of the 1902–1903 research of Rutherford and Soddy was that radioactive decay involves the actual *transmutation* of one element to another. After uranium has emitted an α-particle, it is no longer uranium but thorium! This conclusion, which seemed to overturn the foundations of chemistry as it had been understood since the time of Lavoisier, and harked back to the days of alchemy, was quickly accepted with little resistance. But to understand what it really means we must learn more about the nature of radioactivity and isotopes.

The most central facts concerning natural radioactivity, which will serve as background for the following discussion, are not difficult to summarize in modern terms. Above all, an element (for example, uranium) as prepared by a chemist is not the simplest starting point of investigation for radioactivity, because such a sample generally consists of a mixture of isotopes (atoms of the same chemical behavior but differing from one another in atomic weight, in natural radioactivity, and in the reaction to nuclear bombardment).

All our references to isotopes so far have emphasized their differences in weight, and this is the difference on which the first separation of isotopes by J. J. Thomson in 1912 was based. If

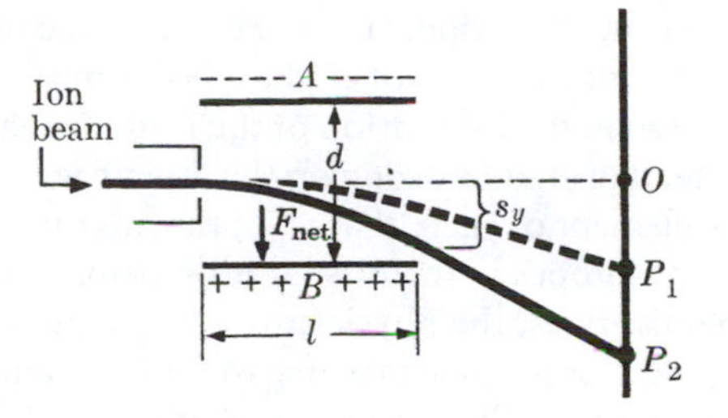

Fig. 27.3. The path of a beam of negative ions bends in a strong electric field. For given velocity and charge, light ions fall at P_2, heavy ones at P_1, nearer O.

the gas or vapor from one element is ionized so that the individual atoms carry a charge, a beam of such ions can be deflected by an electric field, a magnetic field, or both fields applied simultaneously. But the amount of deflection depends on the mass of the ions. A brief hint of the operation of an isotope separator, called a *mass spectrograph*, working on such a principle will suffice.

Figure 27.3 represents an ion beam emerging with initial speed v from the box at the left into the gap between plates A and B, held at a potential difference V_{AB}. If each ion has the same (negative) charge q, each experiences a constant force $F_{net} = (V_{AB}q)/d$ and an acceleration $a = F_{net}/m$ while traversing the distance l. Just as in the case of a projectile with horizontal initial velocity in a constant gravitational field, the displacement s_y of the ion during the time of transit $t = l/v$ under the action of force F_{net} is

$$s_y = \tfrac{1}{2}at^2 = \tfrac{1}{2}(F_{net}/m)t^2 = \tfrac{1}{2}(V_{AB}q/dm)(l^2/v^2). \qquad (27.1)$$

Note that there is no initial velocity in the y (vertical) direction; a and v in the above equation refer only to the horizontal (x) direction.

It follows from Eq. (27.1) that for a beam of ions with equal charge q and equal initial speed v, the displacement in transit between the plates is inversely proportional to the mass m of the particular ion; all other factors (V_{AB},*q*,*l*,*d*, *and* v) can reasonably be considered to be constant for all ions by hypothesis. Thus, in continuing their flight, the "heaviest" ions hit a detector (for example, a photographic plate) at the spot P_1 near spot O in Fig. 27.3, whereas the lighter ions may come to P_2. Conversely, from the points of incidence of the beam on the plate and from the construction of the apparatus, the mass m of the ion may be calculated.

In practice, it is hard to make this method work precisely; for example, the equality of v and q for all ions is by no means easily obtained. The actual mass spectrograph is therefore designed on much more refined principles; but the final results are not different in kind from those given above.

Now we are in a position to summarize some of the findings concerning isotopes. The most striking fact is probably the great variety. Some elements have only a single isotope (and, strangely, their atomic numbers, that is, their placement numbers in the periodic table of elements, counting from hydrogen = 1, tend to be *odd*); examples are fluorine, sodium, aluminum, and phosphorus. Other elements have a great number of natural isotopes; six or seven are not uncommon. (Many more can usually be made artificially by nuclear bombardment or as fission products, but these are never stable.)

Second, with the exception of some elements that have radioactive isotopes, the ratios of abundances for the isotopes are generally quite constant in a given element, no matter where it was obtained. For example, neon gas, whose isotopes were in fact the first to be separated, has three isotopes of approximate relative mass 20, 21, and 22, in respective amounts of 90.48, 0.27, and 9.25% of a chemically pure sample from any natural source. Oxygen, the gas that chemists have selected to give them a standard of relative atomic weight (16.0000 exactly), also has three isotopes, with atomic weights very nearly equal to 16, 17, and 18.

We identify an isotope by printing its atomic weight, to the nearest integer, as a superscript. Thus the three isotopes of oxygen, with their abundances in parentheses, are: O^{16} (99.76%), O^{17} (0.04%), and O^{18} (0.20%). We will occasionally designate the atomic number as a subscript written to the left of the element symbol (although this information might be considered superfluous since the atomic number is determined by the element symbol through the periodic table): thus we can write ${}_{8}O^{16}$, ${}_{9}F^{19}$, ${}_{11}Na^{23}$, etc.

Deviations from the standard abundances, such as those given above for oxygen, are known as "isotopic anomalies," and they may provide useful clues to the origin of these isotopes during the earlier history of the earth or before its formation (Chapter 32).

Third, the mass of an isotope can be obtained with very great precision (six or even seven significant figures are now usually given),

Table 27.1. Isotopes of a Few Elements

Element	Natural isotopes	Isotopic mass, amu	Abundance, atom %
Hydrogen	$_1H^1$	1.007825032	99.985
	$_1H^2$	2.014101778	0.015
	$_1H^3$	3.010604927	(trace)
Helium	$_2He^3$	3.01602931	1.37×10^{-4}
	$_2He^4$	4.00260325	~100
Carbon	$_6C^{12}$	12.0 (defined)	98.89
	$_6C^{13}$	13.00335484	1.11
Oxygen	$_8O^{16}$	15.99491462	99.762
	$_8O^{17}$	16.9991315	0.038
	$_8O^{18}$	17.999160	0.200
Neon	$_{10}Ne^{20}$	19.99244018	90.48
	$_{10}Ne^{21}$	20.9938467	0.27
	$_{10}Ne^{22}$	21.9913855	9.25
Uranium	$_{92}U^{234}$	234.040945	0.0055
	$_{92}U^{235}$	235.043922	0.720
	$_{92}U^{238}$	238.050784	99.2745

(Source: N. E. Holden, "Table of Isotopes")

and invariably is very close to a whole number. This is, of course, the justification for writing the atomic mass number, that is, the superscript in H^1 or U^{238}, as a whole number, usually denoted by *A*. The mass number, which may best be defined as the true isotopic mass (in amu) rounded off to the nearest whole integer, is in fact always within less than 1% of the true isotopic mass, as Table 27.1 shows.

The mass number of an isotope gives the actual mass as a multiple of the *atomic mass unit* (symbol "amu"), 1 amu = 1.66×10^{-27} kg. This is approximately the mass of a hydrogen atom.[1]

It is evident from Table 27.1 that the isotopes of an element are generally not too far removed from one another in mass. The table also uses a convenient symbolism, namely, the subscript at the left of each element symbol, the *atomic number,* which refers to the place of the element in the chemist's periodic table. However, the physicist has not strictly adhered to the chemist's convention in the calibration of the isotopic masses. Instead of standardizing on the atomic weight of the element oxygen, that is, of the mixture of its three isotopes in their usual proportions, as the chemist does, the physicist needs a standard of his own for isotopic mass measurements, and has chosen to call the isotopic mass of $C^{12} \equiv 12.0000$ amu exactly. By this convention, the weighted average of the oxygen isotopes as found in nature is 15.9994 amu on the physical scale as compared with the *chemical atomic mass* of 16.0000, by definition, on the chemical scale. The ratio between these figures is 0.99996, which is the conversion factor to use in going from one scale to the other. But actually we shall make little use of the conversion factor, and instead shall use the physical scale for discussions concerning isotopes and nuclei and the chemical scale for problems in the chemical behavior of elements and compounds, as indeed we have tacitly done so far.

Problem 27.1. On the principles given, design a mass spectrograph to separate the three isotopes of uranium, whose masses are approximately 238, 236, and 234 amu. Assume a singly ionized beam (q = 1 electronic charge) of initial speed v = 10^6 m/sec. You are free to choose V_{AB}, l, etc. What is the separation between the points on the detector (in cm) where these three isotopes arrive? If there appear traces of the arrival of some ions just halfway between the points of incidence for U^{238} and U^{235}, what is the mass of these unknown ions? What assumptions underlie the last answer? How can we check them?

Problem 27.2. Using the data in Table 27.1, (a) calculate the average atomic weight of the element uranium on the physical scale, and (b) convert the result to the chemical scale.

Problem 27.3. The isotopes of ordinary argon are $_{18}A^{36}$ (35.9675 amu, 0.34%), $_{18}A^{38}$ (37.9627 amu, 0.06%), $_{18}A^{41}$ (39.9624 amu, 99.60%). It is by no means clear at present why the abundances are distributed as they are, but we expect that a complete theory of nucleosynthesis in stars—from which all the materials making up the earth and other planets originated in the early history of the universe—will explain them quantitatively. Calculate (approximately) what the molecular weight of argon would be if the abundances of $_{18}A^{38}$ and $_{18}A^{41}$ had been exchanged. Consult the

---

[1]In some recent textbooks the abbreviation "u" is used instead of amu. We consider that the saving of two letters is less important than the advantage of a more easily recognizable abbreviation; "u" could mean any "unit."

periodic table, and note how this would have straightened out one of the perplexing inversions.

Problem 27.4. Ordinary hydrogen gas consists largely of $_1H^1$ atoms (in molecules of H_2), but with some traces of the $_1H^2$ isotope (called *heavy hydrogen* or *deuterium,* from the Greek word *deuteros,* second) as indicated in Table 27.1. (a) Explain how the abundance of the latter could be increased by a physical process. (b) In the formation of water molecules, various isotopic combinations are possible; list them and give the molecular weight of each on the physical scale. (c) What is the molecular weight, on the chemical scale, of "heavy water," that is, of water formed by the combination of $_1H^2$ atoms with the usual mixture of oxygen isotopes?

Problem 27.5. Note the doubly astonishing facts that the isotopic masses are all nearly *whole* numbers, but that they all do deviate a little from exact whole numbers. What might be the physical reasons for these two observations?

27.2 Radioactive Half-Life

After the recognition of the existence of isotopes, the next important step toward an understanding of radioactivity is the concept of *half-life.* Ernest Rutherford, who originally noted the distinction between α- and β-rays and who named them, suggested in 1902–1903 together with Frederick Soddy that each of the radioactive elements has its own characteristic rate of emission of α- or β-rays. If we make an experiment with any number (N_0) of atoms of a specific radioactive isotope (radioisotope), the number of atoms N left after a time interval t without having yet experienced radioactive decay follows a curve similar to Fig. 27.4. There eventually elapses a time interval T when just half the original atoms ($N_0/2$) survive without having yet emitted a ray. Rutherford called T the half-life of the particular isotope, since an atom that has emitted has already been transmuted and so no longer belongs to the type of isotopes under observation.

The principal advantage of the concept lies in the *experimental* fact that the half- life can be defined at any point in the total existence of the radioisotope; no matter how old the sample is now, in another T seconds half the present sample will still survive. Note that this is very different for a population of, say, humans rather than of

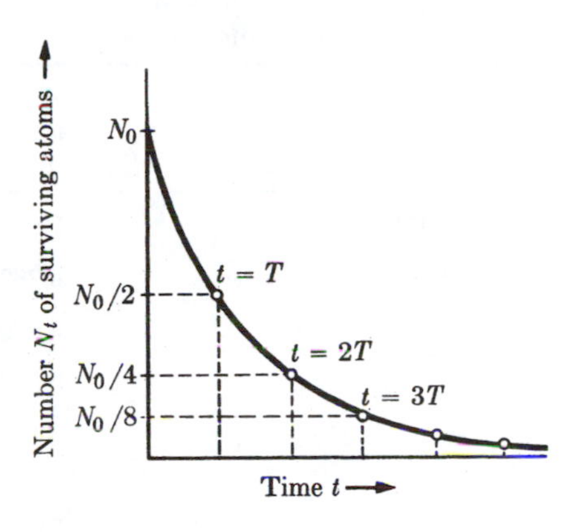

Fig. 27.4. Radioactive decay curve. After a time interval *T,* one-half the initial number of atoms has survived without radioactive decay.

radioactive atoms. If we select N_0 babies, half that number may survive the 70th birthday, but of the $N_0/2$ oldsters none is likely to celebrate a 140th birthday. However, a radioisotope sample with a half-life of 70 years will still have $N_0/4$ atoms intact after 140 years, $N_0/8$ after 210 years, and so forth.

Note that half-life is a *statistical* concept. It is determined by the *probability* that an atom will decay radioactively during a given time interval. This probability is a fixed number independent of the age of that atom and almost entirely independent of its surroundings (except at very high pressures or temperatures). The success of the concept in describing observations, and the absence of any known cause that would *determine* the exact moment when a particular atom decays, gave rise to the notion that radioactive decay is a *random process.* This was one of the first indications that the deterministic laws of Newtonian mechanics may not be completely valid on the atomic level.

There are several ways of writing an expression relating N_0, N_t, T, and the elapsed time t of observation in a specific experiment (compare Section 12.8). We shall adopt the following form:

$$N_t = N_0 e^{-0.693t/T} \qquad (27.2)$$

This is also the equation for the curve in Fig. 27.4. For a very active radioisotope, the half-life T as determined by experiment is short, the exponent is large, and the curve of N_t vs. t drops rapidly to smaller values; conversely for slowly decaying substances.

Table 27.2. Half-Lives of Some Isotopes

Radio-isotope	*Half life T*	*Type of particle emitted*	*Total decay energy, MeV*
$_{88}Ra^{223}$	11.4 d	α	5.979
$_{88}Ra^{224}$	3.66 d	α	5.789
$_{88}Ra^{226}$	1.599×10^3 y	α	4.870
$_{88}Ra^{228}$	5.76 y	β	0.046
$_{92}U^{234}$	2.45×10^5 y	α	4.856
$_{92}U^{235}$	7.04×10^8 y	α	4.6793
$_{92}U^{238}$	4.46×10^9 y	α	4.19694
$_{94}Pu^{239}$	2.411×10^4 y	α	5.244
$_{107}Bh^{261}$	12 ms	α	10.4
$_{109}Mt^{266}$	3.4 ms	α	11.1
$_{110}Uun^{269}$	0.17 ms	α	
$_{111}Uuu^{272}$	1.5 ms	α	

(Source: N. E. Holden, "Table of Isotopes")

Some representative experimental values for T of naturally occurring isotopes of uranium and of radium are assembled in Table 27.2. Note the variety of half-life values. There exist, for other elements, some larger and many far smaller values than those shown here; it is possible to speculate that some radioactive elements, present in great quantities at the time of their synthesis in stars, have since decayed so rapidly that no measurable traces are now left. On the other hand, other elements decay so slowly that only the most refined experiments could reveal that they do not have a constant, unchanging activity.

Following the proposal by Rutherford and Soddy that radioactive decay involves the transmutation of one element into another, Rutherford and others realized that measurements of the relative proportions of these elements, along with their half-lives, could be used to estimate the ages of rocks and even the age of the earth's crust. At the International Congress of Arts and Sciences, held in conjunction with the World's Fair at St. Louis, Missouri, in 1904, Rutherford announced an estimated age of 40 million years for a sample of fergusonite (a mineral containing uranium). This number was soon revised to 140 million years, and the next year the British physicist R. J. Strutt published an estimated age of 2400 million years for a sample of thorianite (a mineral containing uranium and thorium).

These early estimates were based on measuring the amount of helium remaining in the rock from α-decay, and were only a *minimum* age since some of the helium had probably escaped from the sample. But they were good enough to suggest that Kelvin's estimates of the age of the earth were much too low (Section 18.3). Research by the Curie group in Paris suggested the reason why: Radium, though relatively rare, is abundant enough on the earth that its radioactivity generates a substantial amount of heat, and this can replace some (or perhaps all) of the heat lost by the earth by the cooling of its interior.

Thus radioactivity removed one of the major scientific objections to Darwin's theory of evolution (the objection that there had not been enough time for the process of natural selection to work), and at the same time forced geologists to reconsider their long-held belief that the earth had cooled down from a high temperature over a long period of time.

We also introduce in the right-hand column of Table 27.2 a new unit of energy, widely used in nuclear and particle physics: the MeV, an abbreviation for "million electron volts." As noted at the end of Section 24.12 (Problem 24.25), the *electron volt* (abbreviated "eV") is defined as the energy gained by an electron in crossing a potential difference of 1 volt. It is equal to 1.60×10^{-19} joule, and this value is also used to specify the kinetic energy of other particles. Multiples of the electron volt are frequently encountered: 1 keV = 1000 eV (k for the prefix "kilo" in the metric system); 1 MeV = 1,000,000 eV (million electron volts).

The next level up is 1 GeV = 1,000,000,000 eV (= 10^9 eV), formerly called, in the United States, 1 BeV = one billion electron volts (as in the "Bevatron" at Berkeley). The phrase "billion electron volts" was unsatisfactory for use in international circles because until recently the British used the word "billion" to mean a million million (10^{12}) while the Americans defined it as a thousand million (10^9), so the unambiguous term GeV (pronounced "gee-e-vee") was adopted. (The British now follow the American usage.) The G stands for the metric prefix "giga" (see Appendix II for other metric system prefixes). Similarly, the ages of very old rocks (or of the earth itself) are specified by using the unit "giga annum," abbreviated Ga, equal to 1000 million years.

As we will see in Chapter 30, a particle such as an electron can be completely transformed into energy (for example, by annihilation with a positron). If its mass when "at rest" is m_0, its *rest mass energy* is m_0c^2 according to Einstein's famous equation $E = mc^2$. Thus instead of giving the rest mass in kilograms or amu or (as in some older textbooks) as multiples of the electron mass, it is more convenient to give the rest mass energy in MeV, using a shorthand expression like "the mass of the electron is 0.51098 MeV."

Problem 27.6. You are given 1 g of radium $_{88}Ra^{226}$, the long-lived and useful one among the radium isotopes. (a) On the basis of Avogadro's hypothesis, how many atoms of radium are there in the initial sample? (b) Make a plot of N_t vs. time for a period of about 8000 years. Roughly how many atoms of $_{88}Ra^{226}$ are left at that time? How much total mass has been lost because of this particular decay? (c) What is the speed of the α- particles emitted by Ra^{226}? [*Hint:* Use Table 27.2.]

Problem 27.7. The factor $0.693/T$ in Eq. (27.2) is the probability p that one atom will decay in time $t = 1$ (using the units of time in which T is given, for example, in seconds or years). Thus, starting with N_0 atoms, the number that decay after a time t that is very short compared to T is

$$(N_t - N_0)/t = -pN_0.$$

Verify this statement by showing from Eq. (27.2) that in a rock containing 10^{10} atoms of $_{92}U^{235}$, approximately 10 atoms will decay in 1 year. [*Hint:* use the first two terms of the expansion $e^x = 1 + x + \ldots$]

Problem 27.8. Justify the statement that one can compare N_0 with N_t after t sec for a sample by comparing the number of particles emitted at the start and after t sec.

27.3 Radioactive Series

The previous two sections have seriously upset our original idea, derived from chemistry and kinetic theory, of the atom as a stable unit, uniquely given for each element. It is no wonder that the concepts of radioactivity, named by the Curies; of radioactive decay, established by Rutherford; and of isotopes, now formalized by Soddy, were initially difficult to assimilate. But by early 1913, all these concepts had been fitted together by several workers, principally by Soddy, in a truly beautiful theory that illuminated all the different aspects of the complex picture.

Fig. 27.5. **Frederick Soddy (1877–1956), British physicist, who received the Nobel Prize in Chemistry in 1921 for his research on radioactive substances and isotopes.**

One must accept, on *physical* evidence, that generally (but not always) the product of radioactive decay is itself radioactive. Thus Ra^{226} decays, with the emission of an α-particle, into a substance that is itself an α-particle emitter, although of much shorter half-life and much greater energy of emission.

This in turn means a new decay product, which is again radioactive in a very individual way. But on *chemical* evidence we must also accept that each of these decay products behaves chemically in a different manner than its immediate parent and its daughter product. For instance, the metal radium (Ra^{226}) is a decay product of thorium (Th^{230}) and in turn gives rise to a rare gas, radon (Rn^{222}).

The great wealth of chemical and physical facts was summarized by recognizing that each radioisotope falls on one of three long chains, or *radioactive series,* stretching through the last rows of the periodic table. These natural series are named after an element at or near the head of each, the *Uranium,* the *Thorium,* and the *Actinium Series.*

Table 27.3. The Uranium Series*

	$_{92}U^{238}$	$_{90}Th^{234}$	$_{91}Pa^{234}$	$_{92}U^{234}$	$_{90}Th^{230}$	$_{88}Ra^{226}$	$_{86}Rn^{222}$	$_{84}Po^{218}$
Energy (MeV)	4.20	0.19	2.32	4.77	4.68	4.78	5.49	6.0
Emission	α	β, γ	β, γ	α	α, γ	α, γ	α	α
Alternative symbol	UI	UX_1	UX_2	UII	Io	—	—	RaA
Half life	4.5×10^9 yr	24.1 days	1.18 min	2.5×10^5 yr	8×10^4 yr	1620 yr	3.82 days	3.05 min

	$\rightarrow {}_{82}Pb^{214}$	$_{83}Bi^{214}$	$_{84}Po^{214}$	$_{82}Pb^{210}$	$_{83}Bi^{210}$	$_{84}Po^{210}$	$_{82}Pb^{206}$
Energy (MeV)	0.7	3.17, 1.8	7.68	0.02	1.16	5.3	
Emission	β, γ	β, γ	α	β, γ	β	α, γ	
Alternative symbol	RaB	RaC	RaC′	RaD	RaE	RaF	RaG
Half life	26.8 min	19.7 min	1.69×10^{-4} sec	19.4 yr	5 days	1.38 days	(stable lead)

*There are in this and in all other series points at which a small fraction of the total radioactive decay proceeds in a branch line with different characteristics. These have been omitted for simplicity.

Table 27.3 gives the first of the chains by showing in some detail the inescapable life history of each atom of $_{92}U^{238}$. After a time T of 4.5 $\times 10^9$ years the chances are equal that it will have decayed. Having lost an α-particle, the atom is now about 4 amu lighter, hence its mass number is 234. Although we may name it uranium X_1, the chemist will prove to us that the new atom is thorium, and therefore belongs in the ninetieth place in the periodic table (atomic number 90). This atom of $_{90}Th^{234}$ in turn emits a β-ray, which does not change the mass number materially, but that, again on chemical evidence, leaves us with an atom of protactinium, $_{91}Pa^{234}$.

The rest of the chain shows that eventually the atom is for a time radium, then radon, and eventually stable lead ($_{82}Pb^{206}$), although on the way it occasionally reverses its progress through the periodic table to become temporarily another isotope of an element it had represented previously. Each decay is accomplished by the emission of an α- or β-particle of stated energy, usually associated with a γ-ray.

Examination of this and the other two series discloses a regularity of behavior that has the name *displacement law of radioactivity: When an atom undergoes α-decay, its mass number A decreases by 4 and its atomic number decreases by 2; however, when an atom undergoes β-decay, its mass number does not change and its atomic number increases by 1.*

Note that the full law was a generalization from an immense number of experimentally obtained facts and, like the existence of isotopes or of radioactive decay as such, was not comprehensible in terms of any model of the atom as derived from the physical sciences of the preceding era.

Problem 27.9. The Actinium Series begins with $_{92}U^{235}$; each atom sends out in succession the following particles: *α, β, α, β, α, α, α, α, β, β, α.* From this information and by consulting the periodic table, write out an account of the Actinium Series on the model of Table 27.3, complete with the necessary symbols and the figures for atomic mass and atomic number.

Problem 27.10. From plutonium ($_{94}Pu^{241}$), artificially prepared by bombarding uranium in a nuclear reactor, stems one of the many artificial radioactive series ("Neptunium Series"). The first members of the chain are $_{94}Pu^{241}$, $_{93}Am^{241}$, $_{93}Np^{237}$, $_{91}Pa^{233}$, $_{92}U^{233}$, and $_{90}Th^{229}$. Unlike all three natural series, this one does not end in a stable isotope of lead, but in $_{83}Bi^{209}$. For the first six members, draw the series as in the previous problem or Table 27.3.

Problem 27.11. The third natural series originates from thorium, $_{90}Th^{232}$. On the basis of the following information, replace the symbols used here and write the complete series. Indicate alternative possibilities where they can exist:

$${}_{90}\mathrm{Th}^{232} \xrightarrow{\alpha} {}_{?}\mathrm{MsTh}_1{}^{?} \xrightarrow{?} {}_{89}\mathrm{MsTh}_2{}^{?} \xrightarrow{\beta} {}_{?}\mathrm{RaTh}^{?} \xrightarrow{\alpha}$$

$$\rightarrow {}_{?}\mathrm{ThX}^{?} \xrightarrow{?} {}_{?}\mathrm{Tn}^{?} \xrightarrow{?} {}_{84}\mathrm{ThA}^{?} \xrightarrow{?} {}_{?}\mathrm{ThB}^{?} \xrightarrow{\beta}$$

$$\rightarrow {}_{?}\mathrm{ThC}^{?} \xrightarrow{?} {}_{84}\mathrm{ThC}^{1\,?} \xrightarrow{?} {}_{?}\mathrm{ThD}^{208}\ \text{(stable)}$$

Describe briefly what experiments might have been done to determine the series as given above. (Note that this example indicates the type of interpolation that was initially often required.)

Problem 27.12. Make a plot on graph paper or computer screen of atomic number vs. mass number for the elements in the uranium series. Enter each element as a labeled dot, and draw a straight arrow from each element to its immediate decay product. This gives one of the useful and widely current representations of radioactive series.

27.4 Rutherford's Nuclear Model

We return now to the state of affairs at about 1910. The chemical and physical facts of radioactivity presented a stupendous challenge. The existence of radioelements, of α- , β- , and γ-radiation, and of separate and measurable half-lives was well established. The idea of isotopes, arising both from the deflection of ion beams and from the first glimpses of radioactive series, was just beginning to emerge. Although the need to accommodate such findings in any one model of the atom seemed to complicate the picture beyond the power of imagination, the solution was hidden in the problem itself after all. The swift α- and β-rays, the by-products, as it were, of radioactivity, provided exactly the tools needed to take the next important step, the means for probing the perplexing interior of the atom.

Soon after Lenard's experiment on the transmission of cathode rays through thin foils, several attempts were made to construct an atomic model that would incorporate the porosity of the atom as indicated by such experiments, as well as the presence of electrons revealed by the facts of electricity and light emission. It was variously suggested that each atom was made of a myriad of massive fragments, each having a positive and a negative charge of equal amounts (Lenard's *dynamids,* 1903), or that it was a homogeneous sphere with thinly distributed mass and positive electricity, inside which there were negatively charged electrons either placed like seeds in a pumpkin or moving in circular orbits (J. J. Thomson, 1898, 1904). The attempt was made to arrange the electron positions or orbits in a stable fashion so that the mutual electric forces would not lead one to expect a dispersal or collapse of the whole structure (Thomson, Nagaoka, 1904).

But the persistent and fatal shortcoming of such models was that there was no way of using them to explain the old puzzle of line emission spectra. Even if one wished to identify the source of emission of light with a small vibration of the electrons to either side of their equilibrium position or orbit, the known, complex spectra did not follow.

An entirely new basis for all speculation along such lines was provided in 1909–1911 by Rutherford, who had already shown an unmatched virtuosity in experimental physics by his work on radioactive emission in Canada. He had moved in 1907 to Manchester University in England, where he headed a most productive research laboratory, and was awarded the Nobel Prize in Chemistry the next year. Rutherford had noticed that a stream of α-particles, when it passed through a thin film of mica or metal, was somewhat broadened (*scattered*) by the material through which it had passed. This is interesting, for the amount of scattering might give a hint concerning the disposition of the atom's mass, then generally thought to be fairly evenly distributed, as in Thomson's ill-defined sphere of positive electricity. But just how Rutherford was led to the discovery that almost all the mass of the atom exists in a small central *nucleus* has been best summarized by Rutherford himself:

> . . . I would like to use this example to show how you often stumble upon facts by accident. In the early days I had observed the scattering of α-particles, and Dr. Geiger in my laboratory had examined it in detail. He found, in thin pieces of heavy metal, that the scattering was usually small, of the order of one degree. One day Geiger came to me and said, "Don't you think that young Marsden, whom I am training in radioactive methods, ought to

begin a small research?" Now I had thought that, too, so I said, "Why not let him see if any α-particles can be scattered through a large angle?" I may tell you in confidence that I did not believe that they would be, since we knew that the α-particle was a very fast, massive particle, with a great deal of energy, and you could show that if the scattering was due to the accumulated effect of a number of small scatterings the chance of an α-particle's being scattered backwards was very small.

Then I remember two or three days later Geiger coming to me in great excitement and saying, "We have been able to get some of the α-particles coming backwards. . . ." It was quite the most incredible event that has ever happened to me in my life. It was almost as incredible as if you fired a 15-inch shell at a piece of tissue paper and it came back and hit you. On consideration, I realized that this scattering backwards must be the result of a single collision, and when I made calculations I saw that it was impossible to get anything of that order of magnitude unless you took a system in which the greater part of the mass of the atom was concentrated in a minute nucleus. It was then that I had the idea of an atom with a minute massive center carrying a charge. I worked out mathematically what laws the scattering should obey, and I found that the number of particles scattered through a given angle should be proportional to the thickness of the scattering foil, the square of the nuclear charge, and inversely proportional to the fourth power of the velocity. These deductions were later verified by Geiger and Marsden in a series of beautiful experiments.

("Development of the Theory of Atomic Structure")

This, then, was the birth of the nuclear atom—a conception almost immediately accepted into science despite the initial uncertainties of detail. Because it gave the picture of a central concentration of a (positively charged) mass about which the light electrons can swarm at relatively great distances, the whole problem of atomic theory was immediately simplified. For now one could think in separate terms about the very different phenomena for which the atom had to be made responsible; the extranuclear electrons would be concerned with chemical behavior, electricity, the emission of light, etc., whereas the nucleus would serve for the phenomenon of radioactivity, the explanation of isotopes, and the like.

The implications of Rutherford's nuclear atom were perhaps best caught in the artist Wassily Kandinsky's outburst that, now that the old atom had been destroyed, the whole existing world order was annihilated and so a new beginning was possible. Kandinsky's autobiographical sketch (in his book *Rückblick*) indicates how he overcame a block in his artistic work at that time:

> A scientific event removed the most important obstacle: the further division of the atom. The collapse of the atom model was equivalent, in my soul, to the collapse of the whole world. Suddenly the thickest walls fell. I would not have been amazed if a stone appeared before my eye in the air, melted, and became invisible.

Now we turn to the quantitative details of the work of Rutherford and his co-workers, Hans Geiger and Ernest Marsden. The projectiles used were α-rays with speeds of about 2×10^7 m/sec, fired in a narrow beam against metallic films, for example, a gold foil only 4×10^{-7} m thick. On the basis of the approximate atomic dimensions deduced from kinetic theory, one may expect such a film to represent about 10^3 layers of atoms.

Most of the α-particles do go right through the foil without noticeable deviation, and therefore we have additional evidence for the proposition that the atoms are *not* "solid." They must contain electrons, but these can hardly be held responsible for changing the path of those α-particles that are scattered so drastically as to come almost straight back. At each encounter with an electron, the α-particle can experience only a very minute change in momentum because of the great disparity between the respective masses.

The α-particle has a mass of $m_\alpha \approx 6.85 \times 10^{-27}$ kg, whereas the mass of the electron, m_e is only 9.11×10^{-31} kg, that is, about 1/7500 as large; even if we assume a head-on collision, the forward component of the velocity of the α-particle changes only very little each time—to be precise, about 0.01%. The probability is vanishingly small that an α-particle, in thousands of successive encounters with electrons, will experience small deviations from its original direction that are so predominantly in one sense that the net result is an actual turning back.

The alternative explanation of the observed scattering is, therefore, that an α-particle may

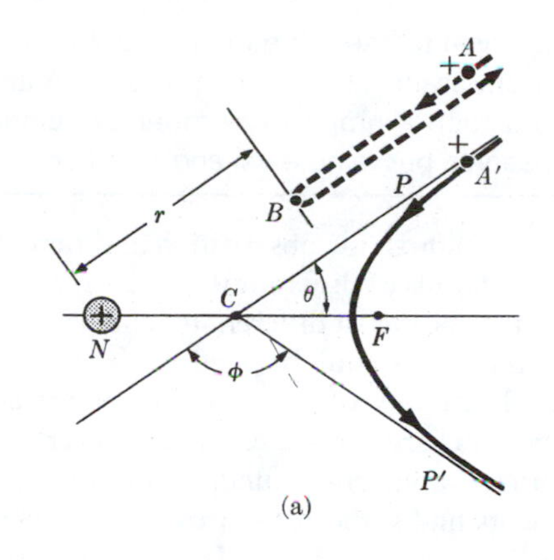

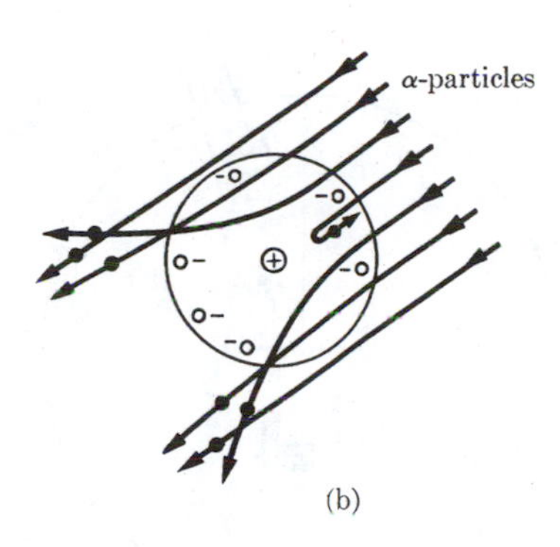

Fig. 27.6.

meet a single object in the foil that is heavy enough to turn it back in one encounter, just as a bullet ricochets back from a boulder but not from a loose pile of sand particles.

Problem. 27.13. An α-particle of mass m_α with initial speed $v_{\alpha 1}$ approaches an electron with mass m_e, the latter initially at rest. After the head-on (perfectly elastic) encounter, the α-particle continues with speed $v_{\alpha 2}$ and the electron moves with speed v_e. (a) Write the two equations for conservation of momentum and of energy for this event, and prove that

$$v_e = [2m_\alpha/(m_\alpha + m_e)] \times v_{\alpha 1},$$

$$v_{\alpha 2} = [(m_\alpha - m_e)/(m_\alpha + m_e)] \times v_{\alpha 1}$$

(b) Calculate the actual change in speed and in momentum for an α-particle in one encounter (in percentage).

In the paragraphs above we have assumed that an α-particle interacts with an electron as though this were a mechanical collision between marbles, but of course we should consider now that both are oppositely charged particles. The mechanism of interaction is then primarily not an elastic collision but rather a mutual attraction, the same effect by which the sun causes a comet that is passing close by to turn around after half-circling the sun. To be precise, the comet's path is hyperbolic if the comet does not remain in the solar system but rather escapes from the gravitational attraction between it and the sun. But as for the momentum and energy changes, those are the same whether the mechanism of interaction is elastic or electric—the laws of conservation of momentum and energy apply universally.

The fact of occasional large-angle scattering of α-particles is not explainable by multiple encounters with electrons even if we consider that the force of interaction is electric attraction.

On the other hand, visualize the hypothetical behavior of an α-particle if it encounters a small heavy nucleus of positively charged mass. Figure 27.6a is based on a figure in Rutherford's publication "The Scattering of α- and β-Particles by Matter and the Structure of the Atom," in the British scientific journal, *Philosophical Magazine* in 1911, the article that may be said to have laid the foundation for the modern theory of atomic structure. The α-particle at A' approaches a nucleus at N, is repelled by the (Coulomb's law) force between the positively charged bodies, and follows a path PP' that must be a hyperbola.[2]

Now one may inquire how large the angle of deviation or scattering from the original direction should be for a parallel beam of α-particles approaching a layer of distributed nuclei. Particle A, for example, is heading directly toward N, approaches only up to point B, and then returns

[2]On the basis of Newton's *Principia* it had long been known that if a body A moves toward another body X against a force of repulsion that is inversely proportional to the square of the distance between them, as is the case with these positively charged particles, then the path taken by A is the same as if, instead of X, there were a similar but oppositely charged body at F, the focus of the curve. This defines a hyperbola with its center at C. This is another illustration of the continuity of science, despite the major and minor upheavals.

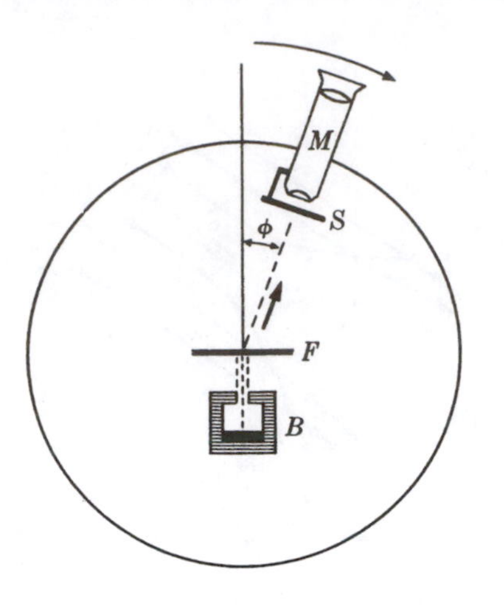

Fig. 27.7. Diagram of apparatus for verifying Rutherford's theory of α-particle scattering.

directly back; most other particles do not pass so close to any nuclei and so do not experience such serious deflections (Fig. 27.6b).

Rutherford considered the effect on α-particle scattering of all important factors such as charge on the nuclei, speed of incident particle v_α, and thickness of film *t*, and derived the following predictions. If scattering occurs by single encounters with small positive nuclei, the fraction of α-particles scattered by angle ϕ should be directly proportional to (a) $[\sin(\phi/2)]^{-4}$, (b) the foil thickness *t*, (c) the magnitude of the nuclear charge, and (d) $(v_\alpha)^{-4}$.

Predictions (a), (b), and (d) could be checked quickly; in fact, the previous year Geiger had published scattering measurements that were most favorable, and a more complete and fully convincing check was provided by Geiger and Marsden in 1913.

Figure 27.7 represents schematically the apparatus used by Geiger and Marsden. *B* is a lead box from which an α-emitter (radon) sends out a beam through a small diaphragm; the foil *F* scatters the rays in varying degrees through different angles ϕ. The number of scattered particles at each angle is found by letting them fall on a zinc-sulfide screen *S*, where each incidence causes a momentary flash of light (scintillation) observable and countable through the microscope *M; S* is fixed to *M* and can be rotated in this plane up to ϕ = 150°.

Problem 27.14. A beam of α-particles of equal speed is directed at a metal foil; about 100 are observed to penetrate the foil each minute without any scattering. Plot on the basis of Rutherford's prediction a graph of number of scintillations observed per minute vs. angle ϕ of scattering.

Although the observation at different angles ϕ, with foils of different thickness *t*, and with α-particle beams of different v_α, confirmed three of Rutherford's four predictions, there appeared no direct way of checking on the effect of the nuclear charge on scattering. However, one could turn the argument around, assume that the experiments had sufficiently proved the general supposition, and then calculate the nuclear charge from the data. This was done for the carbon atom (in a sheet of paraffin wax) and for metal foils from aluminum to gold, and it turned out that for these materials the nuclear charge, expressed in multiples of the numerical electronic charge *e* (1.6×10^{-19} coul) was roughly $\frac{1}{2}$(at. wt) × *e*. This would be 6*e* for carbon, 13*e* or 14*e* for aluminum, 78*e* or 79*e* for gold, and so on.

These experiments therefore provided simultaneously a vindication of Rutherford's hypothesis, and an important piece of new information; for if the nucleus carries a charge equal to $\frac{1}{2}$ at. wt units of electronic charges, there must be equally many electrons around it to provide for the electric neutrality of the whole atom. In fact, previous experiments on the scattering of x-rays and β-particles had given the number of electrons as about equal to $\frac{1}{2}$ at. wt as well.

As so often, data from entirely different laboratories and taken in different research pursuits came together to weave a net of scientific coherence. It was pointed out soon afterward by Antonius van den Broek in 1913 that the important comparison was not between nuclear charge and half the atomic weight, but between the former and *atomic number* (Z). (As noted in Section 27.1, the atomic number identifies an element's place in the periodic table.) It is an astonishing fact that for most substances, $Z = \frac{1}{2}$ at. wt within narrow limits; for example, *Z* for carbon is 6, the atomic weight is 12.010 amu. Thus the experimentally founded suggestion that the number of positive charges on the nucleus or, equally well, the number of electrons around the nucleus be simply equal to the atomic number *Z* at once made the picture of the atom more precise and complete.

Thus a hydrogen atom (*Z* = 1) in its neutral state has one electron moving around its positively charged nucleus, a helium atom (*Z* = 2) has two electrons, and so on up to uranium ($Z = 92$) with ninety-two extranuclear electrons.

This simple scheme was made more plausible when singly ionized hydrogen (H^+) and even doubly ionized helium (He^{++}) were found in ion beams in the mass spectrograph but never H^{++} or He^{+++}, evidently because H atoms had only a single electron to lose, and He atoms only two.

A second important consequence from scattering experiments concerned the size of the nucleus, which Rutherford estimated by an ingenious argument. Since the rest of his deductions depended on the validity of Coulomb's law of electric forces between the nucleus and α-particle, and since these deductions (as that of the variation of scattering with angle) had been borne out, we may assume Coulomb's law to hold during the scattering process. Now surely it would not hold if the projectile were to come so close to the nucleus as to deform it elastically or touch it in any way. Therefore the closest distance of mutual approach between the centers of the target nucleus and the α-particle must be less than the sum of their radii.

The actual distance of nearest approach can, however, be calculated, for in such a case the α-particle must be heading straight toward the nucleus (see path *AB* in Fig. 27.6a) and penetrates to point *B* at a distance $NB \equiv r$ from the nucleus, where the kinetic energy, by the work against mutual repulsion, is all converted to potential energy; after this the projectile turns back like a stone at the top of its trajectory. At *B*, the point of momentary rest, $KE_{init} = PE$ or, numerically,

$$\tfrac{1}{2} m_\alpha v_\alpha^{\,2} = k_e(Ze) \times (2e)/r,$$

where v_α is the initial speed of the α-rays, ($2e$) its charge, and (Ze) the charge on the target nucleus *N*. [Here we have assumed that the potential energy of the α-particle at distance *r* is given by Eq. (24.5) for point charges.]

For a gold foil bombarded with α-particles of about 2×10^7 m/sec speed, the distance *r* can be calculated to be about 3×10^{-14} m (see Problem 27.15). Compared to the overall size of the atom ($\sim 10^{-10}$ m), the nuclear size must therefore be of the order of 10^{-14} or 10^{-15} m, that is, less than 1/10,000 as large!

The mass of electrons being almost negligible compared with that of even the lightest atom as a whole, most of the mass of the atom is thus concentrated within a distance that bears to the total atomic radius a relation similar to that between the radius of the sun and the radius of the total solar system measured to Pluto's orbit! One may assume that it is the presence of moving electrons around the nucleus at the relatively large distance of 10^{-10} m that determines the "size" of the atom. Evidently this scheme presents us with an atom that is mostly empty space, as indeed it must be to account for the relative ease of penetration of α- and β-radiation through thousands of atoms, as in a metallic foil or in the gas of a cloud chamber.

Problem 27.15. Determine the distance *r* of nearest approach between an α-particle with $v_\alpha = 2 \times 10^7$ m/sec and a nucleus of a gold atom. (This calculation was made by Rutherford in his 1911 paper, but he considered $Z \approx \frac{1}{2}$ at. wt.)

Immediately, the developing picture of the atom raises three issues: (a) What is the structure of the nucleus? (b) What keeps the nucleus in general from exploding by the mutual repulsion of its components if it is made of densely packed charges of equal sign? (c) How are the electrons arranged around the nucleus?

To the first of these questions, Rutherford initially could do no more than to suggest that at least in some elements the charge and mass of the nucleus can be explained by assuming a grouping of a number of helium nuclei. "This may be only a coincidence, but it is certainly suggestive in view of the expulsion of helium atoms [α-particles] carrying two unit charges from radioactive matter."

Problem 27.16. According to this suggestion, how many helium nuclei would make up one nucleus of carbon, of nitrogen, of oxygen? Does the scheme work for all elements in the periodic table?

While the first question was actually unsolvable at the time, the second one, concerning nuclear stability, was even more puzzling, and had to be dismissed by Rutherford as follows: "The question of the stability of the atoms proposed need not be considered at this stage, for this will obviously depend on the minute structure of the atom, and on the motion of the constituent charged parts." Rutherford's later work on artificial transmutations opened the way to a better answer to both problems, as we shall see at once.

Finally, the third question raised above was destined to be solved soon by Bohr, another coworker of Rutherford's. At the time, Rutherford simply noted that the Japanese physicist Hantaro Nagaoka had speculated in 1904 on a model of the atom in which a central nucleus was pictured as surrounded by electrical charges revolving in rings similar to those seen around Saturn.

Problem 27.17. If in a gold foil the atoms are so close as to touch one another, with a distance of about 3×10^{-10} m between their nuclei, and if the radii of the nuclei of the atoms and of the α-particles are for this rough calculation each assumed to be about 3×10^{-14} m, calculate the probability of an α-particle approaching a nucleus virtually head-on in any one layer of metal atoms. [*Hint:* Divide the diameter of a gold atom into small sections, each equal to the diameter of the α-particle; then assume that the chances for the α-particle to pass through any section are evenly distributed.]

Fig. 27.8. **Henry Gwynn Jeffrey Moseley (1887–1915), English physicist, who established the correspondence between chemical atomic number and nuclear charge by his X-ray research.**

27.5 Moseley's X-Ray Spectra

In the year 1913, when Geiger and Marsden's scattering experiments completely verified the main features of the theory of the nuclear atom, there came yet another profound contribution from Rutherford's laboratory. There the young physicist H. G. J. Moseley (1887–1915) had been investigating the wavelengths of x-rays emitted when a stream of high-speed electrons (cathode rays) impinges on metallic targets (see Fig. 26.11). A plot of the intensity of x-rays vs. wavelength gives a graph rather like Fig. 27.9. It reminds one somewhat of a continuous emission spectrum of light from a hot solid, but with a line spectrum superimposed on it.

Indeed, we might analyze the graph in two parts. The continuous spectrum of so-called *general radiation,* which serves as "background" for the sharply defined spikes, is generated by the conversion of part of the kinetic energy of cathode rays into photons as they pass through the atoms of the target. Those electrons that give up all their energy in one interaction with the target atoms give rise to the highest-frequency x-ray photons by the relation $h\nu = \frac{1}{2}m_e v^2$ [Eq. (26.8)], but as $\lambda = c/\nu$, the shortest wavelength in Fig. 27.8 is $\lambda_{min} = hc/\frac{1}{2}m_e v^2$, and is well defined for a given electron beam.

Problem 27.18. (a) What voltage must be applied to the x-ray tube to obtain a continuous spectrum with $\lambda_{min} = 0.05$ nm? (Recall that, in consistent units, $\frac{1}{2}mv^2/e = Ve$, where V is the potential difference between electrodes, and e is the charge on an electron). (b) How would you use these relations to measure the ratio h/e with precision? (c) Show that $\lambda_{min} = 12{,}400/V$, where V is in volts.

Moseley, however, was not concerned with the continuous spectrum (which was not fully understood at the time). His work concentrated on the other part of the total x-ray spectrum, the "spikes" or lines of *characteristic radiation.* For each of his targets, the result was similar to Fig. 27.9; there were several such lines, usually in two groups named K and L, although later, better equipment brought out more groups (M, N, . .) and, in each group, usually many more lines. Nevertheless, even the initial data revealed an astonishing simplicity. The lines in the characteristic x-ray spectrum progressively shifted to shorter wavelengths as the *atomic number* of the target material increased. Table 27.4 compares the prominent K_α and K_β lines of five neighboring

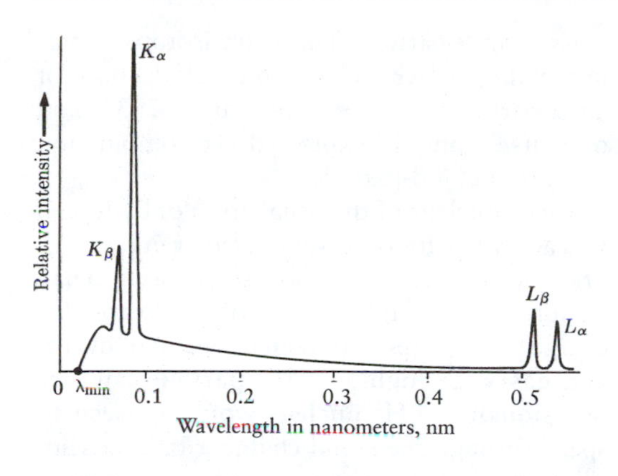

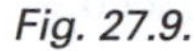

Fig. 27.9.

metallic elements in the periodic table; in each row, the wavelength of K_α or K_β, is shorter than in the preceding row.

At once, several conclusions could be drawn from this discovery of regularity.

a) From his initial data on the characteristic radiation for nine elements from calcium to copper and from his further work in 1914, which embraced elements between aluminum ($Z = 13$) and gold ($Z = 79$), Moseley found that the frequency ν of the K_α line could generally be given by an empirical equation that may be written

$$\nu\ (\text{sec}^{-1}) = 2.48 \times 10^{15}\ (Z - 1)^2 \qquad (27.3)$$

An analogous equation was given for L_α lines.

Such an equation has immediate uses. For example, although the metal scandium ($Z = 21$) was not then available as a target material, the frequency of K_α could be calculated with considerable confidence. When that line was later measured, the agreement was, in fact, found to be good.

Table 27.4. Wavelengths of Prominent Characteristic K Lines (in nm)

Element	*Atomic Number*	*At. Wt.*	*λ of K_α*	*λ of K_β*
Manganese (Mn)	25	54.94	21.0	19.1
Iron (Fe)	26	55.84	19.3	17.5
Cobalt (Co)	27	58.93	17.9	16.2
Nickel (Ni)	28	58.69	16.6	15.0
Copper (Cu)	29	63.55	15.4	13.9

Problem 27.19. The prominent K_α line of scandium has a wavelength of 30.3 nm. Check to what extent Moseley's empirical formula holds for this line.

b) The predictive value of Moseley's formula is, of course, most gratifying. So much confidence could later be placed in it that the identification of a newly discovered element depended on its having characteristic x-ray lines that fitted properly between those of its immediate neighbors in the periodic table; and conversely, wherever a great gap occurred between the wavelengths of K_α for neighboring elements, there, and only there, could one suspect the presence of a yet unknown element.

Yet Moseley had a rather different point in mind in originally undertaking his research, a problem that he illuminated brilliantly. It must be remembered that in Moseley's time the concept of *atomic number* was not yet of importance in physics; the idea was in the air that it corresponded to the number of positive charges on the nucleus, as Rutherford said later, but that had not been announced. In fact, it was Moseley who seems to have first used the term atomic number explicitly. It requires only a glance at Table 27.4 to see that the progressive regularity in the K lines of elements depends on the atomic number and not on the atomic weight. Cobalt and nickel represent an inversion in the periodic table of the sort that disturbed Mendeléeff's contemporaries, yet the K lines follow regularly. (The same point was established for the other inversions.)

Here is comfort for the chemist who was forced to make inversions on the basis of chemical behavior and macroscopic physical properties, and here is also proof that whatever the cause of the characteristic x-ray spectrum, it is connected with the atomic number, that is, with the number of positive charges on the nucleus.

This is quite reasonable, for we cannot help but guess, pending a full explanation, that x-rays are generated when the penetrating cathode ray sets the electrons near the nucleus into such motion as to cause the emission of high-energy photons. But the behavior of an electron near the nucleus surely must be governed, above all, by the field strength in that region owing to the charges on the nucleus.

Moreover, as Moseley pointed out, "The very close similarity between the x-ray spectra of the different elements shows that these radiations originate inside the atom, and have no direct connection with the complicated light spectra and chemical properties governed by the structure of its surface."

c) That x-rays are purely the property of atoms was also made clear by noting that if the target in the x-ray tube were made of brass, an alloy of copper and zinc, the undisturbed characteristic radiations of both Cu and Zn atoms were observed simultaneously.

To summarize the central points in Moseley's own words,

> We have here a proof that there is in the atom a fundamental quantity, which increases by regular steps as we pass from one element to the next. This quantity can only be the charge on the central positive nucleus, of the existence of which we already have definite proof.

A great deal of research soon made it quite certain that this charge, measured in units of electronic charge but with a positive sign, was in every case exactly equal to the atomic number. Moreover, here we have at last been given one property that changes in a regular way from one element in the periodic table to the next. The chemical properties, such as valence, and the physical behavior, such as melting points, do not vary in continuous gradation but in definite periods, with sudden breaks between one and the next. Even the atomic weight progresses unevenly, as the inversions prove. Some other features, such as line-emission spectra, establish no simple relation whatever between neighboring elements. Only the progressively increasing charge on the nucleus gives us a truly secure foundation for the arrangement of elements in a unique order.

27.6 Further Concepts of Nuclear Structure

Because this chapter is to be in the main an account of the theory of nuclear structure, we may briefly deviate from the historical development to summarize a few of the more recent concepts of nuclear physics.

The emission from natural radioactive elements of α- and β-particles suggested naturally that a model of the nucleus start with these building blocks. But the very first element, hydrogen, offered difficulty, having but a single positive nuclear charge and about one-fourth the mass of an α-particle. The heavy isotope of natural hydrogen, which has about half the mass of an α-particle, was not known until 1931, and of course cannot be explained as a combination of an α- and a β-particle.

The nucleus of the usual atom of hydrogen was accessible for observation, either in the form of H^+, the ionized atom whose single electron had been removed in the violence of an electric discharge or as the result of bombardment of hydrogen gas with high-speed α-particles, where occasionally a $_1H^1$ nucleus would be seen to speed through the cloud chamber or to a scintillation screen after having experienced a collision. The name *proton* was generally adopted for the particle in 1920 (see Section 25.8). The previous year, Rutherford had reported that such particles appeared to be emitted when nitrogen is bombarded by α-particles: ". . . we must conclude that the nitrogen atom is disintegrated under the intense force developed in a close collision with a swift α-particle, and that the hydrogen atom which is liberated formed a constituent part of the nitrogen nucleus." He added prophetically, "The results on the whole suggest that, if α-particles—or similar projectiles—of still greater energy were available for experiment, we might expect to break down the nuclear structure of many of the lighter atoms."

Now the problem of nuclear structure seemed much nearer to solution. Each nucleus carries a number of protons equal to its atomic number Z. Each proton has 1 unit charge and a mass of about 1 amu, so that this scheme so far accounts for the proper positive charge but, in general, for only half the mass; one could therefore imagine the nucleus to contain sufficient neutral pairs of protons and (negative) electrons to make up the difference between the observed atomic weight and the atomic number.

Thus a helium nucleus would be composed of 4 protons and 2 nuclear electrons, to give a charge of +2 and a mass of about 4 (the electrons having only 1/1836 of the mass of a proton, therefore not adding significantly to the total nuclear mass). Similarly, a nucleus of $_{92}U^{238}$ would contain 92 unneutralized protons and (238 – 92) neutral proton-electron pairs.

This hypothesis has two appealing features. It places in the nucleus electrons that are known to come from there in radioactive decay as β-radiation, and it allows one to understand at last why an element, after α- or β-decay, is found in a different position in the periodic table. Since the atomic number is given by the number of positive nuclear charges, the loss of two positive units by α-emission should move the element down two places, and the gain of one positive unit through the loss of a (negative) nuclear electron in emission should move the element up by one place—all just as observed.

In 1920 Rutherford predicted the existence of a proton-electron combination, later called a *neutron,* and a wide search was made to discover it experimentally in a free state, for example, as a product of inelastic collision of fast α-particles with other nuclei.

None of the experiments planned specifically for this purpose succeeded, but in 1930 it was reported that energetic α-particles, when stopped by light elements (lithium, beryllium, boron) could create a radiation so penetrating that they were considered to be high-energy γ-rays. Indeed, the energies obtained were greater than of any then known γ-radiation. In an historic experiment, the French physicists Frédéric Joliot and his wife Iréne Curie (daughter of the discoverers of radium) noted in 1932 that this new radiation could in turn eject protons of very high speed from hydrogen-containing materials such as paraffin wax.

Immediately after the publication of this result, the British physicist James Chadwick (1891–1974), who had been an assistant of Rutherford's, showed that the protons could not have been ejected by a γ-radiation. In his paper "The Existence of a Neutron" he proved that the observations of Joliot and Curie could be best explained by radiation that "consisted of particles of mass nearly equal to that of a proton and with no net charge, or neutrons" (see Problem 32.3 for a quantitative discussion of this reaction).

Having no charge, these neutrons are nonionizing; therefore they do not lose their energy easily in passing through matter, can penetrate thick sheets of lead, and can approach a charged nucleus directly without experiencing a force of repulsion. In a head-on collision with a hydrogen nucleus, of approximately equal mass, the neutron's energy would be virtually completely transmitted to the proton.

In 1932 the German physicist Werner Heisenberg (1901–1976) proposed that the neutron, together with the proton, would account completely for the nuclear structure of all elements. But one no longer thinks of the neutron as a particle composed of a proton and an electron; instead, the neutron is regarded as a fundamental particle that may, under certain conditions, decay into a proton, an electron (β-ray), and a neutrino. The word *nucleon* is used more generally for a particle that is either a proton or a neutron.

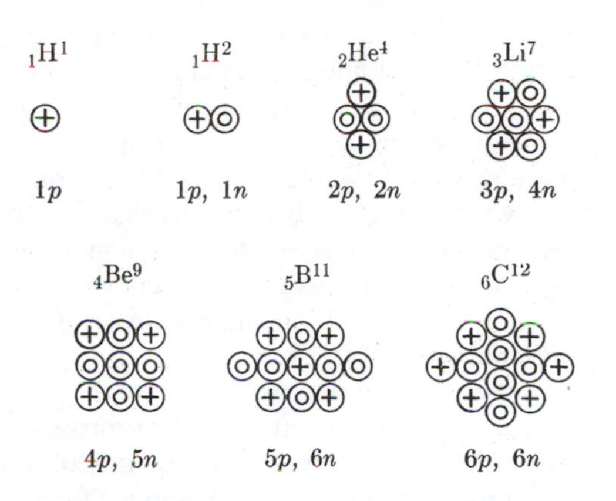

Fig. 27.10. Purely pictorial representation of some nuclei; ◎ = neutrons, ⊕ = protons.

Figure 27.10 represents schematically the structure of several nuclei (n = neutron, p = proton). For $_1H^2$, the "heavy" isotope of hydrogen, or *deuterium,* one neutron and one proton are required, a combination referred to as a deuteron. The helium nucleus or α-particle comprises two neutrons and two protons. Similarly for the other nuclei.

Problem 27.20. (a) Draw similar representations for $_8O^{16}$, $_8O^{17}$, $_8O^{18}$, $_{15}P^{31}$. (b) How many protons and neutrons are there in each of the natural isotopes of uranium?

We may conclude the account of the nuclear atom with a brief summary of some typical concepts that are useful in contemporary research. Most important is probably the question of the stability of the nucleus, the problem that Rutherford originally wisely bypassed. Mechanisms have been invented that account adequately for the main facts of stability and for the general facts of natural and induced radioactive decay, although by no means in the framework of the classical physics that we have largely relied on so far in describing the nuclear model. For us, one most intriguing point is that the neutron and proton, having about equal mass, very close to 1 amu, necessarily create nuclei that have almost integral masses on the amu scale.

Thus we are led back to Prout's hypothesis (Section 21.1), although in a manner he did not envisage at all, and we now understand the previously most mysterious fact that all isotopic masses are nearly whole integers. On the other hand, we are quite ready to expect that the isotopic masses are not all *exactly* integral multiples of the mass of ${}_1H^1$, because the neutron has a slightly larger mass than the proton (1.00867 compared with 1.00728 amu).

But even when we take account of the slight difference between the mass of the proton and that of the neutron, we cannot quantitatively explain the masses of isotopes without a further, very significant addition. For example, we find that the nucleus of a helium atom has a smaller mass (4.0014 amu) than the sum of the individual particles in their free states (two protons = 2 × 1.00728 and two neutrons = 2 × 1.00867, together 4.0319 amu), a difference Δ of about 0.03 amu. What has happened to this small but measurable amount of mass in the process of forming the helium nucleus from its constituents?

The answer, which will emerge from Einstein's theory of relativity (Chapter 30), is that this missing mass corresponds to energy (4.5×10^{-12} J or 28 MeV) lost or given up during the formation of the atom out of its parts. Only if this much energy is supplied to the atom can the helium nucleus break up again into its constituents. The absence of this much energy binds the protons and neutrons together in the nucleus, just as the absence of extra kinetic energy keeps our moon from breaking out of its orbit around the earth.

But a detailed discussion of nuclear structure is better left suspended temporarily, until we introduce the radical changes in physical theory that led to the generalization and combination of the laws of conservation of mass and energy. Now that we have reached a model of the nucleus that is actually as close to pictorialization as is permissible in a subject largely given over to mathematical rather than physical models, we must turn to another set of questions: What is the structure of the swarm of electrons that we have tentatively postulated to exist around the nucleus? What are the laws governing their behavior, and to what extent can they account for our remaining puzzles of chemistry and physics, for example, valences and emission spectra? When these questions have been answered, we shall indeed be in possession of the key to a unified understanding of many of the major questions of physical science.

RECOMMENDED FOR FURTHER READING

Lawrence Badash, "'Chance favors the prepared mind': Henri Becquerel and the discovery of radioactivity," *Archives Internationales d'Histoire des Sciences,* Vol. 18, pages 55–66 (1965); "How the 'newer alchemy' was received," *Scientific American,* Vol. 215, no. 2, pages 89–94 (August 1966); "Radium, radioactivity, and the popularity of scientific discovery," *Proceedings of the American Philosophical Society,* Vol. 122, pages 145–154 (1978)

H. A. Boorse and L. Motz, *The World of the Atom,* pages 385–407, 427–61, 641–46, 693–733, 779–824. Extracts from the writings of Roentgen, Becquerel, the Curies, Rutherford, etc., and biographies.

John Campbell, "Ernest Rutherford: Scientist supreme," *Physics World,* Vol. 11, no. 9, pages 35–40 (Sept. 1998)

Marie Curie, *Pierre Curie,* translated by Charlotte and Vernon Kellogg, New York: Dover, 1963, including autobiographical notes by Marie Curie.

C. C.Gillispie, *Dictionary of Scientific Biography,* articles by A. Weill on Marie Curie, Vol. 3, pages 497–503; by Jean Wyart on Pierre Curie, Vol. 3, pages 503–508; by L. Badash on Rutherford, Vol. 12, page 25–36; by T. J. Trenn on Soddy, Vol. 12, pages 504–509 and on Geiger, Vol. 5, pages 530–533; by C. A. Fleming on Marsden, Vol. 18, pages 595–597; by J. L. Heilbron on Moseley, Vol. 9, pages 542–545; by Hendry on Chadwick, Vol. 17, pages 143–148

C. W. Haigh, "Moseley's work on x-rays and atomic number," *Journal of Chemical Education,* Vol. 72, pages 1012–1014 (1995)

Marjorie Malley, "The discovery of atomic transmutation: Scientific styles and philosophies in France and Britain," in *History of Physics* (edited by S. G. Brush), pages 184–194

Naomi Pasachoff, *Marie Curie and the Science of Radioactivity,* New York: Oxford University Press, 1996

Melba Newell Phillips (editor), *Physics History,* articles by L. Badash on radioactivity before the Curies, and by M. Malley on the discovery of β-radiation in radioactivity, pages 75–90

Marelene Rayner-Canham and Geoffrey W. Rayner-Canham (editors), *A Devotion to Their Science: Pioneer Women of Radioactivity,* Philadelphia: Chemical Heritage Foundation, 1997

Ernest Rutherford, "The development of the theory of atomic structure" (1932), in *Background to Modern Science,* New York: Macmillan, 1940.

J. H. Weaver, *World of Physics,* excerpts from papers by Marie Curie, Rutherford, Soddy, in Vol. II, pages 28–59

CHAPTER 28

Bohr's Model of the Atom

28.1 Line Emission Spectra

As we saw in Chapter 26, it was the study of continuous emission spectra that led Max Planck to formulate his blackbody radiation law in 1900. We were concerned only with the spectra from glowing solids or liquids and, incidentally, also from gases at those extreme conditions of density and temperature found in our sun and other stars, but not in our laboratories. Such spectra were found to depend only on the *temperature* of the source, not on its chemical composition. And of course it was precisely this *universal* character of emission spectra that attracted Planck's attention and made him suspect that a fundamental law of physics must be involved.

Having identified, at least provisionally, the general features of this law (Section 26.3) and having seen how it was successful in providing an interpretation of the photoelectric effect (Sections 26.4 and 26.5), we must now examine how it might apply to the more specific problems of atomic structure. For this purpose we turn to another type of spectrum.

It had long been known that light is emitted by gases and vapors when "excited" by the passage of an electric spark or arc or, as in the case of neon lights, by a continuous electric current established through a thin tube filled with some of the gaseous material (Fig. 28.1) or when volatile matter is put into a nonluminous flame. Furthermore, the light so emitted, when resolved into its components by the prism spectroscope, was found to have a spectrum startlingly and fundamentally different from the continuous emission spectra of glowing solids and liquids: Gases and vapors have *line emission spectra,* that is, instead of a solid band of changing "rainbow" colors they show light only at some well-defined wavelengths, with dark gaps between neighbors. As seen through the spectroscope or photographed on a plate in a spectrograph, the spectrum thus appears as a set of lines, usually irregularly spaced, some very bright, others less so.

The second point of difference with respect to continuous emission spectra is that the line emission spectra are markedly different for different radiating elements. Each substance has its own characteristic pattern of wavelengths throughout the whole observable region (Fig. 28.2). The unaided eye synthesizes these separate lines and recognizes the mixture of colors as reddish for neon, pale blue for nitrogen, etc. Some materials reveal a most complex emission spectrum, others are far simpler; iron vapor shows about six thousand bright lines in the visible range, whereas sodium has only two strong, close yellow lines there. The great variety

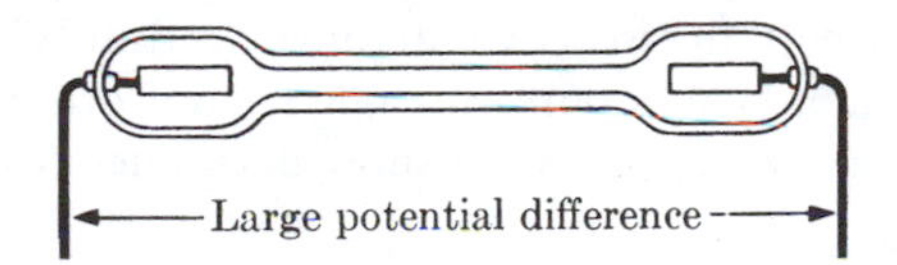

Fig. 28.1. Electric discharge tube. Current from one sealed-in electrode to the other causes the gas to emit light.

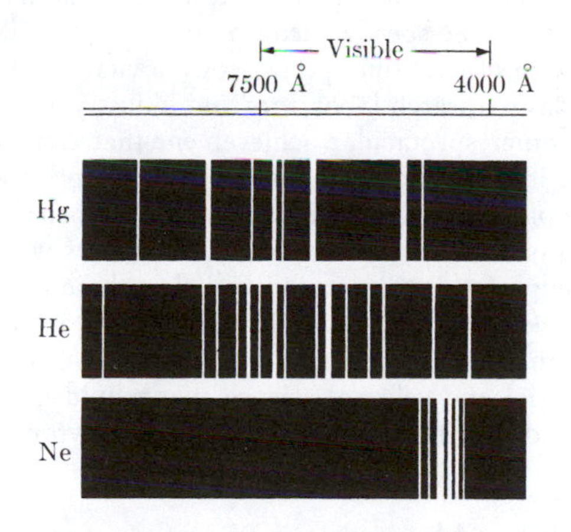

Fig. 28.2. Sections of the line emission spectra of mercury, helium, and neon. (redrawn from photographic record).

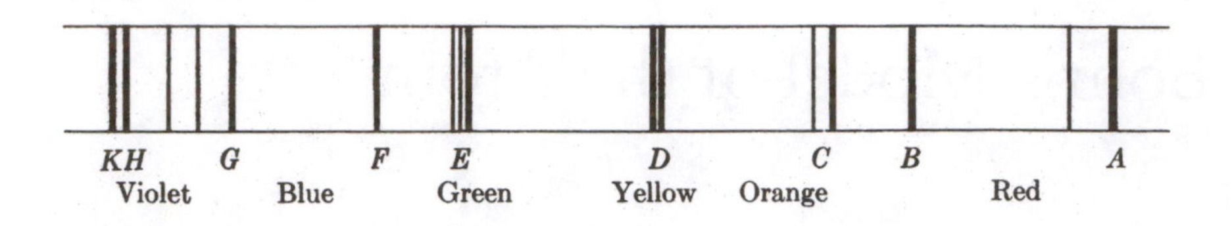

Fig. 28.3. Fraunhofer lines in the sun's continuous spectrum; only some of the most prominent lines are shown.

of patterns and the individual spacing of lines seemed quite unaccountable; why should gases emit line spectra at all, and why should closely related elements present such differing line patterns? All this was destined to remain most puzzling for nearly three generations. But at least each material could be identified from its unique line emission spectrum as surely as criminals can be traced by means of their characteristic fingerprints.

Rapidly, a whole art of spectroscopy arose; more and more powerful prism and grating spectroscopes became available to chart the exact wavelengths of these patterns. As early as the 1830s, the suggestion was made that the presence, identity, and abundance of materials in ores, etc., might be determined from the type and intensity of the emission spectra. And in fact, the physicist G. R. Kirchhoff (1824–1887) and the chemist R. W. Bunsen (1811–1899) at the University of Heidelberg, Germany, who were the foremost pioneers in the development of spectroscopy, discovered two elements, rubidium and cesium, in the vapor of a mineral water by noting hitherto uncharted emission lines.

It was the first of a brilliant series of such discoveries, and the origin of a technique that has given us the speedy chemical analysis of small samples by routine spectroscopy, a vital tool in today's research laboratory and industry alike. Another spectacular achievement that caught the imagination of all Europe was the spectroscopic analysis of meteorite material. When the emission spectra of vaporized meteorites were compared, it was established that these contained only well-known elements present on our earth, a direct confirmation of the unity of terrestrial and celestial science that would have gladdened the hearts of Galileo and Newton.

28.2 Absorption Line Spectra

In 1814, the German optician Joseph Fraunhofer studied a strange and unexpected peculiarity in the appearance of the continuous spectrum of sunlight. When light was passed through a narrow slit and its visible spectrum very carefully examined with good prism systems, the continuity of colors appeared to be disrupted by a series of fine, steady, irregularly spaced *dark lines* (Fig. 28.3). Fraunhofer counted more than 700 such lines (now, with better instruments, we know of more than 15,000) and assigned the letters *A, B, C,* . . . to the most prominent ones. He also observed many such lines in the spectra of some bright stars.

It is significant that Fraunhofer's interest in these dark lines (absorption spectrum lines) was initially a practical one; they identified and labeled portions of the solar spectrum, and so provided reference points useful for comparing the degree of dispersion of colors attainable with different types of glass. Here, as so often in the history of science, the skills of the craftsman and the needs of technology eventually bring to the notice of scientists a new natural phenomenon and techniques for dealing with it; and from such beginnings, in the hands of a gifted scientist, profound and unexpected contributions may follow.

The key observation toward a better understanding of both the absorption and the emission spectra came in 1859. By that time it was generally known that the emission line from the heated vapor of sodium metal had the same wavelength as the very prominent dark line[1] in the solar spectrum to which Fraunhofer had assigned the letter *D*.

What Kirchhoff now found looks simple in retrospect. In essence the argument is this: When the light from a pure, glowing solid is ordinarily analyzed, it does not show dark lines cross-

[1]There are two lines, with wavelengths of 588.9953 and 589.5923 nm, but only modern instruments of great precision can resolve them that well, and so we shall speak of *a* sodium line at about 589.0 nm.

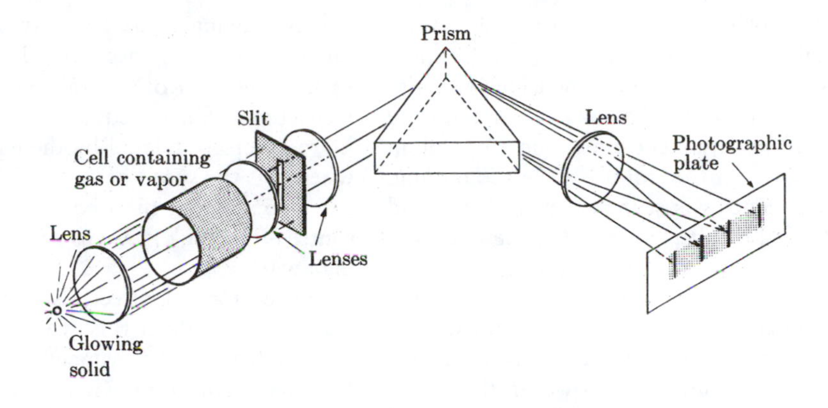

Fig. 28.4. Diagram of spectrometer set up for recording an absorption spectrum of a vapor.

ing the continuous emission spectrum; but if (as shown in Fig. 28.4) that same light on its way to the prism is now allowed to pass through a suitably large volume of gas or a cloud of vapor, for example, vaporized sodium metal, then the spectrum does exhibit one or more dark lines of exactly the same wavelength and appearance as do those in the Fraunhofer observation. Each vapor was found to have a characteristic line absorption pattern (Fig. 28.5); evidently each vapor strongly absorbs certain characteristic frequencies from the passing "white" light. And, most interesting of all, the wavelengths of the *absorption* lines in each case correspond exactly to some lines in the catalog of *emission* patterns for the particular gaseous substance.

We may therefore attempt a preliminary interpretation of the dark lines in solar and other stellar spectra: The missing lines correspond to light that was withdrawn or somehow scattered from the total beam on passage through the (relatively cooler) atmosphere surrounding the star and, to some extent, through the atmosphere of our earth. This view leads us to regard the Fraunhofer lines as evidence of the presence of specific elements in the atmosphere of a star. We see at once the enormous potentialities of this theory for astrophysical research.

Problem 28.1. How would you decide, on an experimental basis, which absorption lines in the solar spectrum are due to absorption near the sun's atmosphere rather than to absorption by gases and vapors in our own atmosphere?

Problem 28.2. How might one decide from spectroscopic observations whether the planets shine by their own light or by reflected light from the sun?

Among the consequences following from this explanation of Fraunhofer lines, three are of particular interest to us now:

a) Each single Fraunhofer line ought to be assignable to the action of some specific gas. Therefore, when with better techniques a new line was discerned in the green part of the solar spectrum that corresponded to none of the lines known to be emitted or absorbed by terrestrial material, it was plausible to ascribe it to the action of a gas in the atmosphere of the sun not yet discovered on earth. The appropriate name *helium* was chosen and, indeed, many years later a gas with such an absorption line was found, first as a constituent of many minerals and later also in natural gases and in our own atmosphere.

Fig. 28.5. Absorption spectrum of sodium compared with the line emission spectrum of the same substance.

b) After absorbing its characteristic colors from white light, a gas may reradiate it, and generally at the same frequency. This conclusion was original with Kirchhoff. We may understand the essence of his argument quickly in terms of an analogy: A beam of sound waves containing many frequencies is sent past a mounted and silent tuning fork tuned to 440 cycles/sec. Any sound of that particular frequency does not simply pass by the tuning fork as do all the other frequencies; instead, being "in resonance" with the instrument, its energy is absorbed as it sets the prongs vibrating. However, the tuning fork now vibrates, and sends out the sound energy it received—but in all directions, not just in the direction in which the absorbed sound was traveling. The proportion of sound energy that happens to be reradiated *toward* the observer is very small compared with the total energy absorbed. A distant listener along a line between sound source and tuning fork might therefore notice that this one component reaches him with less than the full energy.

The conclusions from this analogy are, first, that a dark line in an absorption spectrum appears dark only because of the relatively much greater intensity of the "background" of unaffected radiation at either side; *some* light is always also reradiated in the original direction of propagation. Second, we expect to see some light reradiated toward us from a volume of gas after it has been absorbing light, even if we view it at 90° to the direction of propagation of the original radiation; for example, we expect from the edges of the absorbing layer of the sun's atmosphere a bright line spectrum corresponding exactly to a *reversed* Fraunhofer spectrum. This should happen during those few moments that the main disk of the sun itself is just hidden in a total eclipse; and this deduction was first and spectacularly fulfilled during an observation of this kind in 1870.

c) If the source of the light is moving relative to the medium in which it propagates, the wavelengths of the observed spectral lines will be changed. This is the well-known *Doppler effect,* widely used by astronomers to measure the speeds of galaxies and by police officers to measure the speeds of automobiles.

Johann Christian Doppler (1803–1853), an Austrian scientist, predicted in 1842 that if a star is moving toward the observer the wavelengths of the light it emits will appear to be shorter, while if it is moving away the wavelengths will be longer. These two cases would correspond to a shift in the color of the star toward blue or red, respectively. Since frequency ν and wavelength λ are inversely related by the equation $\nu\lambda = c$, where c is the speed of light [cf. Eq. (23.2)], we can also say that the frequency is higher if the source is moving toward us and lower if it is moving away.

The Doppler effect also applies to sound waves, and indeed this was how it was first observed, in an experiment conducted by the Dutch meteorologist C. H. D. Buys-Ballot in 1845. He arranged for a railway locomotive to draw an open car with several people playing trumpets and found that the pitch (frequency) of a given note does change as its source approaches or recedes from the listener.

The predicted change in wavelength is proportional to the ratio of the speed of the source, v, to the speed of the signal, c: The wavelength is changed from λ_0 to $\lambda = \lambda_0(1 - v/c)$. The frequency is changed from ν_0 to $\nu = \nu_0/(1 - v/c)$. The effect is extremely small for light emitted from macroscopic objects that can be studied in the laboratory, and thus the optical Doppler effect was not directly confirmed by terrestrial experiments until around 1900.

In 1848, the French physicist Armand-Hippolyte-Louis Fizeau (1819–1896) pointed out that while the Doppler effect would shift all wavelengths of radiation emitted by a star in relative motion and thus would have little visible effect on its overall color, one should be able to observe directly the shift in wavelength of a Fraunhofer dark line known to occur at a precise wavelength for the sun. The English astronomer William Huggins (1824–1910) used the Doppler effect to estimate in 1868 that the star Sirius is moving away from the sun at a speed of 29 mi/sec.

Doppler later returned to Austria and became a professor of physics at the University of Vienna in 1850; one of his students there was Gregor Mendel, the famous founder of modern genetics.

d) It is not difficult to visualize the basic mechanism for this *resonance* interpretation of absorption in terms of the classical theory of light as it developed soon after Kirchhoff's work. We recall that an "oscillator" corresponds to an electric charge vibrating or revolving at the same frequency as the light radiation thereby emitted.

When such an oscillator is inactive because no energy is being supplied to it (by heating or by an electric discharge), it nevertheless remains "tuned" to its particular frequency, ready to receive energy from a passing light ray of that same frequency, and once in full vibration, it must give up (reradiate) energy as fast as it is supplied.

These points, most of which followed from the tuning fork analogy, are in general accord with the Maxwellian scheme. But if the classical theory could furnish no explanation for the distribution of intensities in the relatively simple and uniform case of continuous emission spectra, the explanation of bright line emission spectra seemed infinitely more hopeless. Here are some of the conceptual difficulties:

One might argue that the atoms of different elements may well have characteristic structures of their own, so that each elementary atom by itself would emit at the frequency of its particular oscillatory vibration. But the spectra from solids or liquids or very hot and dense gases do *not* show any differences from one element to another; this can only mean that in those physical states the nearness and incessant mutual collision among atoms blurs the individual vibration characteristics, just as pendulums or springs can no longer vibrate with their free frequencies when connected together in large groups.

This picture qualitatively accounts for the difference between the continuous and the line emission spectra, but there is this serious flaw: Even in the rarefied gas in a glowing discharge tube, atoms are continually colliding billions of times per second, according to our kinetic theory model of gases, and usually with enormous speeds; yet they do give us definite line spectra. What could account for the complete blurring of spectra in one case and the lack of interaction among colliding "oscillators" in the other?

Quite apart from this last difficulty, how are we to account for the exact line pattern itself? In some of the more involved spectra, one might have to visualize an internal structure of "oscillators" in the atom that would make a grand piano exceedingly simple by comparison (to use an analogy then current). Equally disappointing was the fact that the line spectra, although definite for each element, still were apparently without order. They showed no progressive change from one element to the next; as noted, even chemically related elements often exhibited strikingly different arrangements of lines.

A further complexity was introduced by the observation that the number of absorption lines in every case depended on the violence of the electric discharge; that further, although every absorption line indeed corresponded to some emission line for that same gas, yet there were always many lines in the emission spectrum without equivalents in absorption spectra. This point conflicted with the idea of resonance radiation, for surely a musical instrument, to return to our analogy, should freely absorb every sound to which it is "tuned" and which, on being played, it can emit.

Another most serious difficulty was that each emission line corresponds to a definite state of vibration for the atomic oscillator. According to the kinetic theory of gases, these internal motions of the atom ought to share the energy of the entire molecule, thereby increasing the theoretical value of the specific heat of the gas. But the theoretical value even without these internal motions was already higher than the experimental value! Maxwell and other physicists were well aware that any attempt to give an atomic interpretation of spectral lines would intensify the difficulty of the kinetic theory in explaining specific heats (Section 22.3), but they could not find a way to resolve the problem within the framework of Newtonian mechanics.

Because the field of experimental spectroscopy after the work of Kirchhoff and Bunsen grew so vigorously, the avalanche of data tended to point up the chaotic features with more and more discouraging certainty. True, there were a few hints of order, by themselves also unexplainable: The general appearance of the spectra of alkali metals showed *some* features of similarity when the regions beyond the visible were included; the wavelength difference between neighboring lines for some parts of the sodium spectrum and the zinc spectrum was constant, and groupings of two or three lines tended to recur in some other spectra; but not much else.

This was a period of almost obsessive searching for some hint of a numerical relation among the tantalizing lines, for some mathematical key to decode the message that surely had to be behind the appearances. It is as if we watched again the students of Plato trying to reduce the offensive disorderliness of planetary motion to a series of circular motions or Kepler poring endlessly over his data to find a relationship between the radii of planetary orbits and the times of revolution. And, most prominently, we recall the chemists from Döbereiner to Mendeléeff trying to find order in the list of chemical elements. We are here very close to a vital process in the evolution of the sciences.

28.3 Balmer's Formula

As sometimes happens, the first important break in this problem came from a fairly obscure corner. In 1885 a Swiss schoolteacher, Johann Jakob Balmer (1825–1898) published a paper entitled simply "Notice Concerning the Spectral Lines of Hydrogen." His starting point was Anders Ångström's fine measurements of the wavelengths of the four quite well-known visible hydrogen emission lines, commonly named H_α (red), H_β (green), H_γ (blue), and H_δ (violet). Probably by a straight cut-and-try method, Balmer hit on a formula that would relate these four wavelengths, and which we may write as

$$\lambda \text{ (in meters)} = C\frac{n^2}{(n^2 - 2^2)}, \quad (28.1)$$

where C is a constant quantity, equal to 364.56 nm, and n is a whole number. Specifically, $n = 3$ if λ for H_α is to be found, $n = 4$ for H_β, $n = 5$ for H_γ, and $n = 6$ for H_δ. Table 28.1 shows the excellent agreement (to within 0.02%) between Balmer's empirical formula and Angström's measurements on which it was based.

Table 28.1. Data from Balmer's Paper (in Modern Units)

Name of line	*Wavelength λ, in nm* — *Experimental*	*From Balmer's formula*
H_α	656.210	656.208 ($n = 3$)
H_β	486.074	486.08 ($n = 4$)
H_γ	434.01	434.00 ($n = 5$)
H_δ	410.12	410.13 ($n = 6$)

With the sublime self-confidence of a numerologist Balmer comments at this point: "A striking evidence for the great scientific skill and care with which Ångström must have gone to work."

Balmer speculated next what the wavelength of a fifth line might be if it existed. Using $n = 7$, he found λ of $H_\varepsilon = 396.965$ nm, from Eq. (28.1), and he wrote, "I knew nothing of such a fifth line, which must lie within the visible part of the spectrum . . . and I was compelled to assume that the temperature relations [in the glowing gas] were not favorable to the development of this line or that the formula was not generally applicable." But then, he continues, a colleague mentioned that more lines had been found in the violet and ultraviolet, in the analysis both of emission from discharge tubes and of stellar (absorption) spectra. The fifth line did indeed exist, and had the predicted wavelength! In a postscript Balmer relates that just then he received the measured wavelengths for nine more lines in the ultraviolet region, and they all fitted his predictions within better than 0.1%.

His formula seemed completely vindicated; he was equally correct in his prediction that all higher lines must crowd more and more together as $n \to \infty$, for according to his formula, the shortest wavelength (λ at $n = \infty$) is 364.56 nm, the so-called *convergence limit.*

Increasingly improved techniques in spectroscopy soon made it possible to record more and more of the ultraviolet region, as for example by photographing the emission spectrum of the hot gases in the sun's chromosphere during an eclipse; the 35 consecutive "Balmer" lines so recorded were all in good agreement with his formula. The convergence limit came to 364.581 nm. Figure 28.6 represents a Balmer spectrum on a plate sensitive into the ultraviolet region.

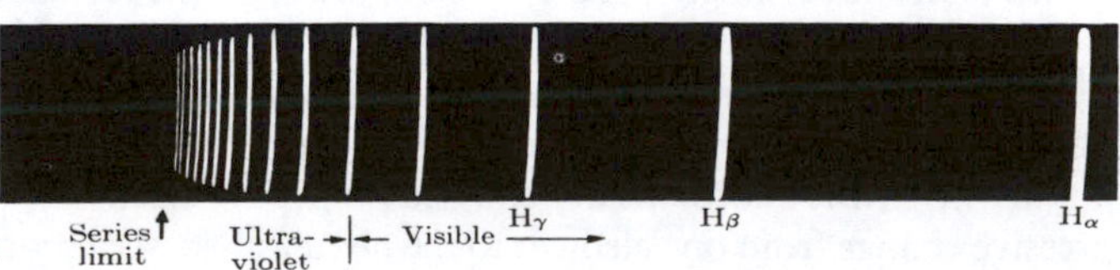

Fig. 28.6. Balmer series of hydrogen emission spectrum.

Problem 28.3. (a) Check Balmer's calculation for the fifth line H_ε and for the convergence limit. (b) How many hydrogen lines should there be in the range between 365.00 and 366.00 nm?

But there was more. Balmer thought that on purely speculative grounds one might expect several complete new series of hydrogen lines, for which

$$\lambda = C\frac{n^2}{(n^2 - 3^2)},\ \lambda = C\frac{n^2}{(n^2 - 4^2)},\ \text{etc.}$$

and so on; and he also expressed the hope that this formula might guide the way to discoveries of series relationships among other spectra apart from that of hydrogen. It soon became evident that Balmer's suggestions were quite correct—although there was as yet no hint of a *physical* basis on which to arrive at this formula.

It is now customary to write Balmer's formula in a perfectly equivalent but more suggestive way, which we shall adopt from now on:

$$1/\lambda = R(1/2^2 - 1/n^2) \qquad (28.2)$$

where the value for the constant R is, according to modern measurements, 1.09678×10^7/m. The letter R has been chosen to honor the Swedish spectroscopist J. R. Rydberg (*Rydberg constant, R*).

Problem 28.4. Derive Eq. (28.2) from (28.1), and show that $R = 4/C$.

Now we can restate Balmer's daring prediction of the existence of other series of hydrogen lines, by writing speculatively

$$1/\lambda = R(1/3^2 - 1/n^2),\ 1/\lambda = R(1/4^2 - 1/n^2),\ \text{etc.,}$$

perhaps also

$$1/\lambda = R(1/1^2 - 1/n^2).$$

In general, all these possible series can be summarized in one formula, namely,

$$1/\lambda = R[1/(n')^2 - 1/(n'')^2], \qquad (28.3)$$

where n' is a fixed integer (1, 2, 3, . . .) that distinguishes one series from the other (for the Balmer series, $n' = 2$), and where n'' also represents integers (namely, $n'' = n' + 1, n' + 2, n' + 3$ . .), those integers distinguishing one line from the next in a given series, so that for H_α and H_β, n'' is 3 and 4, respectively.

Table 28.2.

Name of Series in hydrogen line spectrum	*Date of discovery*	*Values in Eq. (28.3)*
Lyman series	1906–1914	$n' = 1$, $n'' = 2, 3, 4, \ldots$
Balmer series	1885	$n' = 2$, $n'' = 3, 4, 5, \ldots$
Paschen series	1908	$n' = 3$, $n'' = 4, 5, 6, \ldots$
Brackett series	1922	$n' = 4$, $n'' = 5, 6, 7, \ldots$
Pfund series	1924	$n' = 5$, $n'' = 6, 7, 8, \ldots$

And indeed, in 1908, Friedrich Paschen in Germany found two new hydrogen lines in the infrared whose wavelengths were given by the first of the speculative formulas or, in terms of Eq. (28.3), by $n' = 3$ and $n'' = 4$ and 5. Many other lines in this series have been identified since.

To the Balmer series and the Paschen series there have gradually been added others, as by the improvement of experimental apparatus and techniques new spectral regions could be explored. In the above list, the name of the series corresponds to its first observer, except, of course, that the *existence* of Balmer's lines was common knowledge at the time he formulated the series expression.

Problem 28.5. Find the wavelength of the third line (always counting up from the longest wavelength) in the Lyman series and the convergence limit for that series.

Problem 28.6. Draw a long line and divide it into 20 equal intervals, about 1 cm apart. Let this represent a wavelength scale from 0 to 8000 nm. On this scale enter the first, second, third, and last line of each of the five series for hydrogen. Do they overlap? How many series could one *conceive* of between $\lambda = 0$ and $\lambda = \infty$? How many lines in all?

The rest of Balmer's initial suggestions began to be fulfilled even sooner than those concerning other series for hydrogen emission; his formula did indeed prove suggestive in the search for relationships among the spectra of other gases. In 1896 E. C. Pickering observed a series of hitherto unknown lines in the absorption spectra of a star that virtually coincided with that of the Balmer series for hydrogen, except that there was an additional line between every couple of Balmer-type lines. The whole set of wavelengths fitted the relation

$$1/\lambda = R[1/(n'/2)^2 - 1/(n''/2)^2] \qquad (28.4)$$

where $n' = 4$ and $n'' = 5, 6, 7, \ldots$ (later, other series of this type were found, for example, one with $n' = 2$ and one with $n'' = 3$).

The thought was voiced at first that this was evidence for some celestial modification of common hydrogen. The true solution to this intriguing puzzle was one of the earliest triumphs of the new model of the atom, to which the historical development here is gradually leading us; it is such a good story that we shall save it for the last section of this chapter.

While Balmer's formula itself was not found to serve directly in the description of other spectra, it inspired formulas of similar mathematical form that were very useful in discerning some order in portions of a good many complex spectra. The Rydberg constant itself also reappeared in such empirical formulas. It became more and more clear that somehow the same physical mechanism had to be at work behind the variety of spectroscopic observations, although simultaneously the exciting search threatened to bog down under the sheer weight of data.

But as we know now, it might almost have been equally good for the progress in this field if for a generation after Balmer no line spectrum analysis had been attempted, because the key pieces for the construction of the necessary conceptual scheme had to come first from an entirely different direction. And we now turn to follow that path.

28.4 Niels Bohr and the Problem of Atomic Structure

The rest of this chapter is an account of the original work in 1913 of a young Danish physicist and its consequences for the physical sciences. Niels Bohr, born in 1885, the year of Balmer's publication concerning the hydrogen spectrum, had just received his Ph.D. from the University of Copenhagen when he joined J. J. Thomson at Cambridge in 1911 as a visiting researcher. After a few months he left to join the then-more-exciting group around Rutherford at Manchester. The work in that laboratory was just delivering proof of the theory of nuclear atoms; at the same time, the rest of the atom, the configuration of the electrons about the nucleus, was still a puzzle. There were only the more or less general suggestions that the electrons move about the nucleus in closed orbits, a scheme rather analogous to the planetary system. Indeed, Thomson recently had been able to show that the system would be stable if one assumed the electrons to be arranged in rings, with two electrons on the innermost ring, up to eleven in a second ring, and so on.

But those hypotheses had two fatal weaknesses. First, although they provided electrons to account for the *presence* of the phenomena of electric currents, ionization, photoelectricity, the neutrality of the ordinary atom, x-ray scattering, and so forth, they were unable to produce quantitative explanations for the details of any of these processes, nor could they explain even qualitatively the presence of line emission spectra.

Second, the assumption that the electrons revolved about the nucleus was necessary for explaining why the electrons did not follow the force of electric attraction by falling into the nucleus, but this immediately created a seemingly insuperable problem. According to the theory of light emission, any vibrating or revolving charge emits electromagnetic energy (Section 25.3), so that each revolving electron should continually spiral closer and closer to the nucleus with ever-increasing speed while emitting light of steadily increasing frequency.

As Rutherford said later: "I was perfectly aware when I put forward the theory of the nuclear atom that according to classical theory the electron ought to fall into the nucleus. . . ." Of course, both the emission of continuous spectra for ordinary (gaseous) elements and the annihilation of the electrons were predictions in catastrophic conflict with actual fact.

It was at that point that Bohr, age 28, entered the picture. In essence, he joined the new idea of the nuclear atom with another great recent conceptual scheme—the quantum theory—that had just begun to make its way into scientific thought in Britain. As Arthur Eddington wrote in 1936; "Let us go back to 1912. At that time quantum theory was a German invention which had scarcely penetrated to England at all. There were rumors that Jeans had gone to a conference on the continent and been converted; Lindemann, I believe, was expert on it; I cannot think of anyone else."

28.5 Energy Levels in Hydrogen Atoms

Bohr's original groundbreaking paper was published in the *Philosophical Magazine* in July 1913 under the title "On the Constitution of Atoms and Molecules"; other papers followed at short intervals. We cannot here follow Bohr's own line of thought exactly, partly because of its advanced mathematical level, partly because, as even Max Planck himself once exclaimed, Bohr's work is written "in a style that is by no means simple." Nevertheless, the main arguments are quite accessible to us.

Bohr begins by reviewing Rutherford's atomic model and the difficulties it raised concerning electronic orbits and light radiation. Then appears the new theme. He announces that classical electrodynamics, which had originally been confirmed by Hertz's experiments with large circuits creating radio waves, is inadequate to describe the behavior of systems of atomic size. Essentially this is the lesson to be drawn from the work of Planck and Einstein on the quantum theory of light, although they had not analyzed the detailed motion of the atomic "oscillator." The objection to Rutherford's theory that an electron revolving about the nucleus should continually lose energy had already been met by Planck, who had proved that the energy in the oscillator does *not* change except discontinuously, in steps of hf, where h is Planck's constant and f is the frequency of rotation or vibration.

In this light, consider an atom of hydrogen—going as usual for the simplest case first. It has a nucleus with one positive charge (a proton), circled by a single electron at a distance known from kinetic theory to be of the order of 10^{-10} m. Bohr's first postulate therefore is that while the electron is in a given orbit, which for the time being we may imagine to be circular, *it does not radiate energy*. To radiate a photon or to absorb energy, the electron must be either decreasing or increasing its orbit, that is, it must be in transition from one *stable* (that is, nonradiating) orbit to another. Thus to every level on the energy level diagram of the oscillator there corresponds a definite stable orbit. A sudden transition from one orbit to the next is accompanied by a change in energy level, that is, from some energy E'' to E', so that $E'' - E' = \Delta E$, which by Einstein's photon theory is equal to the energy of the photon emitted:

$$E'' - E' = \Delta E = h\nu \qquad (28.5)$$

Obviously there are now several questions, but above all: What are the actual energies in the different energy levels? Planck's theory only supplied this hint: The energy levels in any one oscillator are all equally spaced, and throughout any sample of radiating atoms there are distributed oscillators of every conceivable type of vibration. These assumptions, as we have seen, led Planck to the correct derivation of the emission curves for glowing solids and liquids, but they had proved powerless for the interpretation of the line emission from gases.

Now Bohr quite boldly denies that for radiating atoms in an ordinary gas the same process holds as in solids, liquids, or dense gases. He postulates that in this case *each atomic emitter of hydrogen has identically the same energy level scheme*. Furthermore, these energy levels cannot be evenly spaced; if they were, all hydrogen atoms would send out photons of only one frequency.

We seem to see a way of proceeding. We know the frequencies of light that hydrogen gas emits; therefore we know the energies of the photons. This gives us directly the spacing between energy levels, and so we may construct empirically the energy level diagram.

For example, to account for the first four lines in the visible spectrum ($\lambda = 656.3, 486.1, 434.0$, and 410.2 nm), we calculate the energy of these photons ($h\nu \approx 3.02 \times 10^{-19}, 4.07 \times 10^{-19}, 4.57 \times 10^{-19}$, and 4.84×10^{-19} J), then build up an energy "ladder" that incorporates these four intervals (Fig. 28.7).

Now the emission of the line H_γ ($\lambda = 434.0$ nm) corresponds to a change in energy level from E_D to E_C, and similarly for the others. In principle, we can extend such a ladder to include all observed emission lines.

There occurs a first doubt. Both Planck's and Einstein's work on the quantum theory of light

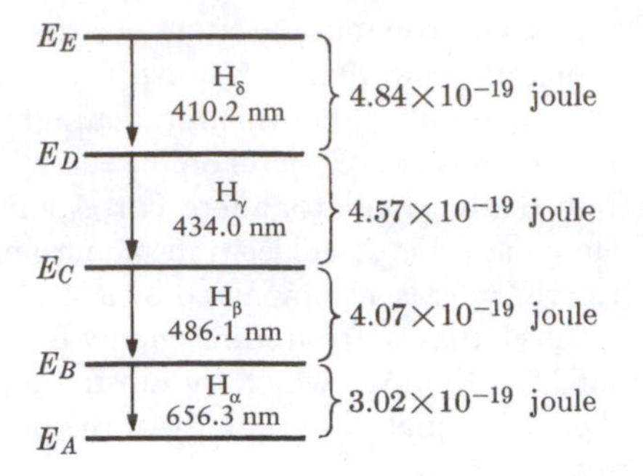

Fig. 28.7. A provisional energy level scheme to account for emission of first four Balmer lines.

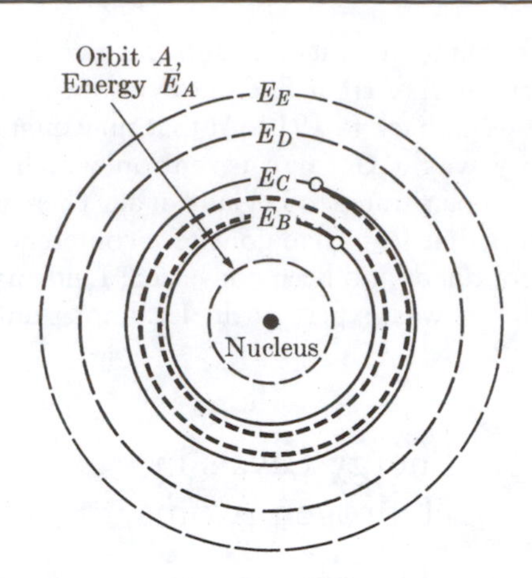

Fig. 28.8. Provisional model of five stable orbits for the electron around the hydrogen nucleus. The electron is spiraling from orbit *C* to orbit *B*.

assumed axiomatically that the frequency ν of the emitted light is equal to the frequency *f* of oscillation or rotation in the atom. It follows that an electron, while changing from one energy level to the next, must be rotating at a fixed orbital frequency $f = \nu$. This is at first sight not inconceivable, if an electron might suddenly spiral from one stable, nonradiating orbit into the next, say to one nearer the nucleus, while retaining its period and its frequency of revolution during the transition (Fig. 28.8). But, on the one hand, this raises the troublesome question of how the electron can thereafter change to a new orbital frequency when it may subsequently have to negotiate the interval from E_B to E_A to emit a photon of higher frequency, and on the other, one can prove quickly that the electron *cannot* keep a constant frequency or period while it spirals from one stable orbit to the next over any considerable distance.

The net force acting at any instant on the revolving electron, a centripetal Coulomb's law force of electric attraction (Section 24.7) directed to the central nucleus, as seen in Fig. 28.9, is simply given (numerically) by

$$F_c = m_e v^2/r = (k_e e \times e)/r^2, \qquad (28.6)$$

where m_e is the mass of electron, *e* the numerical charge on the electron and on the hydrogen nucleus, and *r* the distance between electron and nucleus.

But the speed of the electron is $v = 2\pi fr$, and although this was defined for circular motion, it should hold well enough for motion along portions of a close spiral. Substituting and rearranging terms in Eq. (28.6) yields

$$1/f^2 \equiv T^2 = (4\pi^2 m_e/k_e e^2)r^3. \qquad (28.7)$$

In short, $T^2 \propto r^3$—nothing less than Kepler's third law, now turning up even in an atom!

The conclusion is that the period (and frequency) of the electron's orbital revolution does indeed change with the distance from the nucleus; therefore radiation cannot be emitted at constant frequency in the transition between stable orbits. The whole scheme collapses.

There is, however, a daring way of saving the fundamental idea. In the discussion so far, it was assumed that ordinary laws of mechanics and electricity hold during the transition of the electron between energy levels. This led us to a contradiction with experience. Therefore Bohr states a second postulate: *Ordinary mechanics and electricity* [for example, Eqs. (28.6) and (28.7)] *might be used to discuss the motion of the electron in a stable and nonradiating orbit, but "the passing of the systems between different stationary states cannot be treated on that basis."* ("Stationary" is Bohr's term for what we have called "stable.")

Rather than try any further to visualize what happens during the transition, Bohr asks us to realize that the classical laws were not at all designed to accommodate such intra-atomic processes, that a different mode of thought is required for that region of phenomena, and that the most straightforward approach is to accept a third postulate: During the transition between energy levels, the electron, in a way that cannot

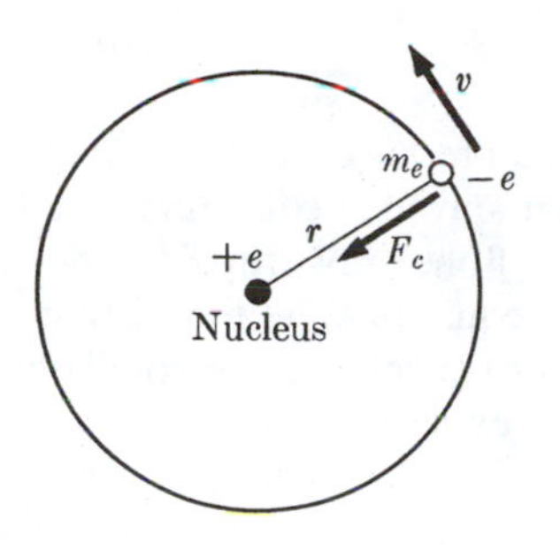

Figure 28.9.

be pictured, emits a photon whose frequency ν is given by the Planck-Einstein formula ($\Delta E = E'' - E' = h\nu$; therefore $\nu = \Delta E/h$), *regardless of the frequency f of orbital revolution of the electron.*

That f is not equal to ν was so completely foreign to all physical thinking that Bohr felt compelled to append this simple remark, "The [last] assumption is in obvious contrast to the ordinary ideas of electrodynamics, but appears to be necessary in order to account for experimental facts." In its impact on the contemporary scientist, it was a statement quite equivalent to Galileo's assertion some three centuries before that all bodies fall at equal rates of descent no matter what their weight.

At a later time, Bohr reemphasized the philosophy of sound scientific pragmatism: classical physics, when applied to the problem of electron stability and line emission, again and again has led to the wrong answers; evidently, the tool is not suited to the problem. "We stand here almost entirely on virgin ground, and upon introducing new assumptions we need only take care not to get into contradiction with experiment. Time will have to show to what extent this can be avoided, but the safest way is, of course, to make as few assumptions as possible."

To summarize: Bohr's three radical postulates do give us a scheme for explaining the emission of line spectra by a hydrogen atom (or for that matter, by any other atomic system). We need only examine all known emission lines for that element, calculate the photon energies, and construct an energy ladder with intervals corresponding to these energies. For each step on that ladder the electron is considered to move in a stable orbit. When the atom absorbs energy from the outside (by the collision with atoms in a hot gas, with particles in electric discharge, with photons, etc.), the electron moves to a new stable orbit with greater energy. When the electron suddenly returns to the previous orbit, the energy difference is given up as a photon.

Such a scheme, although it might be perfectly self-consistent and congruent with all known experimental data, would still be unsatisfactory, for it would leave the origin of the "energy ladder" itself completely unexplained. Before 1913 there had been several attempts to find relationships between the frequencies of spectral lines and the "allowed" frequencies of vibration of various mechanical models of the atom, corresponding to the fundamental and overtone frequencies of a string (Section 23.2). All such attempts had failed, in part because (as we have just seen) the frequency ν of the emitted light cannot be identified with the frequency f of the electron's motion in a particular orbit.

Nevertheless, Bohr suspected that there was probably some definite relation between the single frequency ν of the emitted light and the two different frequencies of the initial and final orbits, and that this relation involved the energies E' and E'' of these orbits. He saw that what was needed was, first of all, a formula expressing the frequency of rotation in an orbit, f, in terms of the energy of that orbit, E. Such a formula could easily be obtained from classical theory: The energy of the electron, orbiting as indicated in Fig. 28.9, is simply the sum of its kinetic energy and its potential energy (see Section 24.11):

$$E = \tfrac{1}{2}m_e v^2 - k_e e^2/r \qquad (28.8)$$

Since, according to Eq. (28.6),

$$\tfrac{1}{2}m_e v^2 = \tfrac{1}{2}k_e e^2/r$$

we can write the equation for the energy of the electron in its orbit of radius r as

$$E = -\tfrac{1}{2}k_e e^2/r. \qquad (28.9a)$$

Rearranging, we obtain

$$r = -\tfrac{1}{2}k_e e^2/E. \qquad (28.9b)$$

On substituting this formula for r into Eq. (28.7) and rearranging, we find the desired relation between orbital frequency and energy:

$$f = \frac{\sqrt{2}\,(-E)^{3/2}}{\pi k_e e^2 \sqrt{m}}\,. \qquad (28.10)$$

Note that the energy E is negative for electrons bound in orbits around nuclei, so $-E$ will be a positive quantity.

So far Bohr has not gone beyond classical physics, although in choosing to express frequency in terms of energy he has obviously been guided by Planck's hypothesis. But now he takes another bold step on the basis of little more than a hunch; he assumes that if the electron is initially at rest at a very large distance from the nucleus, so that [by Eqs. (28.9a) and (28.10)] its energy and frequency of revolution are both essentially zero, and if, when it is captured or bound by the atom, it drops into a final state with frequency and energy related by Eq. (28.10), *the frequency* ν *of the emitted light is simply the average of the initial and final frequencies of orbital revolution.* If the initial frequency is zero and the final frequency is, say, f, the average is just $\frac{1}{2}f$. In a paragraph that marks the crucial breakthrough in the development of the modern theory of atomic structure, Bohr writes:

> Let us now assume that, during the binding of the electron, a homogeneous radiation is emitted of a frequency ν, equal to half the frequency of revolution of the electron in its final orbit; then, from Planck's theory, we might expect that the amount of energy emitted by the process considered is equal to $nh\nu$, where h is Planck's constant and n an entire [integer] number. If we assume that the radiation emitted is homogeneous, the second assumption concerning the frequency suggests itself, since the frequency of revolution of the electron at the beginning of the emission is zero.

Note that here n means the *number of quanta* of energy $h\nu$ that are emitted, according to Planck's hypothesis, when an electron moves from rest at an infinite distance to an orbit with frequency f and energy E. Bohr proposes to set

$$E = \tfrac{1}{2}nhf$$

because he is asserting that the frequency of the emitted radiation is $\nu = \frac{1}{2}f$. Substituting this expression for E into Eq. (28.10), we can determine both ν and E in terms of the charge and mass of the electron, Planck's constant, and the integer n:

$$\nu = \frac{2\pi^2 m_e e^4 k_e^{\,2}}{n^3h^3}, \qquad (28.11)$$

$$E = -2\pi^2 e^4 m_e k_e^{\,2}/n^2h^2. \qquad (28.12)$$

Note that all terms except m in these equations are known universal constants.

When Bohr substituted numerical values into these equations, he found that—provided that n was set equal to 1—he could obtain almost exactly the experimental values for the "ionization energy" of the hydrogen atom and for the spectroscopic frequency corresponding to the capture of a free electron into the lowest energy level of a hydrogen atom. From Eq. (28.12) we find $E = -2.18 \times 10^{-18}$ J.

To put it another way, the electron must be given 2.18×10^{-18} J of energy to allow it to move (against the force of attraction to the nucleus) from the first orbit to infinity, or to a large enough distance from the nucleus so that the remaining energy is virtually zero. But this state of affairs, the loss of the electron, corresponds to the ionization of the hydrogen atom; and thus Bohr's theory predicts that one must supply 2.18×10^{-18} J to a hydrogen atom to cause its ionization.

The actual energy requirement for ionizing atoms can be measured experimentally by accelerating a beam of electrons into a tube filled with a little hydrogen gas. If the electrons in the beam are sufficiently energetic, during a collision they impart to the orbital electrons in hydrogen atoms enough energy to allow the orbital electrons to leave their parent atoms entirely, which in turn causes the gas to become suddenly ionized and highly conducting.

It turns out that it takes an accelerating voltage of 13.6 volts for the electron beam to induce a sudden change in conduction through or ionization in the gas tube. But with that accelerating voltage, each projectile, having the electronic charge e, carries an energy of $13.6 \times 1.6 \times 10^{-19} = 21.7 \times 10^{-19}$ J—in excellent agreement with the prediction from Bohr's theory.

But, even more remarkable, Bohr found that he could explain in a very simple way *all* the spectral series of hydrogen if he *changed the definition of n.* Instead of using n to represent the number of quanta emitted when an electron "falls from infinity" into a given state of energy E, Bohr proposed that n is the number of the allowed stable orbit the electron can be in, starting from the *orbit closest to* the nucleus ($n = 1$). The various possible orbits correspond to $n = 1, 2, 3, \ldots$ up to infinity; n *is* now called a *quantum number,* and is used to label the orbits, and Eq. (28.12) can be used to compute the energy of the electron in each of the allowed orbits.

To understand what these orbits are like, it is helpful to go back and substitute Eq. (28.12) for E into Eq. (28.9b) to get the radius r of the possible orbits. *It is then found that the radius of the orbit with quantum number n is proportional to* n^2:

$$r = n^2h^2/4\pi^2 m_e e^2 k_e. \qquad (28.13)$$

Orbits with small values of n are close to the nucleus and [by Eq. (28.12)] have high (negative) binding energies E; conversely, orbits with large values of n are far away from the nucleus and have low binding energies (since E is inversely proportional to n^2). When $n = 1$, r is smallest, and has by calculation the value

$$r = 0.53 \times 10^{-10}\ \text{m}.$$

But this is just the order of magnitude expected for the dimensions of the (normal, or unexcited, or ordinary) atom (see Loschmidt's estimate from kinetic theory in Section 22.3), and this success provides further evidence in favor of Bohr's model.

Bohr proposed that the frequency of the photon emitted when the electron jumps from a distant orbit with quantum number n'' to an orbit nearer the nucleus with quantum number n' can be computed from Planck's postulate [Eq. (28.5)], if one simply uses these quantum numbers to compute the energies E'' and E' from Eq. (28.12):

$$\nu = \frac{(E'' - E')}{h} = \left[\frac{-2\pi^2 e^4 m_e k_e^{\,2}}{(n'')^2 h^3}\right] - \left[\frac{-2\pi^2 e^4 m_e k_e^{\,2}}{(n')^2 h^3}\right]$$

$$= \frac{2\pi^2 e^4 m_e k_e^{\,2}}{h^3}\left[\frac{1}{(n')^2} - \frac{1}{(n'')^2}\right].$$

In terms of the more directly measurable wavelength λ, this becomes

$$\frac{1}{\lambda} = \frac{2\pi^2 e^4 m_e k_e^{\,2}}{h^3 c}\left[\frac{1}{(n')^2} - \frac{1}{(n'')^2}\right]. \qquad (28.14)$$

Here we recognize a most satisfying fact: This *derived* expression is indeed identical with the *empirical* formula for the hydrogen spectra [Eq. (28.3)],

$$1/\lambda = R[1/(n')^2 - 1/(n'')^2], \qquad (28.15)$$

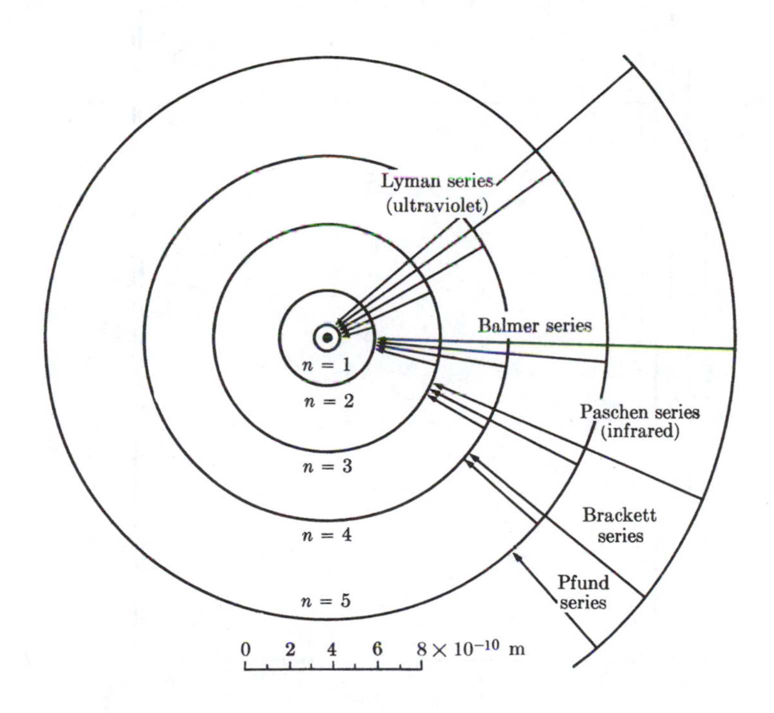

Fig. 28.10. Final scheme of transitions between stable Bohr orbits in hydrogen atom (not drawn to scale).

provided that: (a) the combination of constants $2\pi^2 m_e e^4 k_e^2/h^3 c$ is equal to 1.09678×10^7/m, Rydberg's empirical constant R (which indeed turns out, on calculation, to be the case within 0.01%), and (b) that we may interpret the quantum numbers n' and n'' according to the following scheme. (We have used modern values here; in early 1913, the discrepancy was 6%, yet the agreement was very impressive.)

When $n' = 2$ and $n'' = 3, 4, 5, \ldots$, which gives us the lines of the Balmer series from the empirical Eq. (28.15), the "ground state" (lowest-energy orbit) for all transitions is the second smallest orbit (see Fig. 28.10). When $n' = 3$ and $n' = 4, 5, 6, \ldots$ (the condition that is met for the Paschen series), the ground state is the third orbit; and so on for the other series.

For example, $n' = 1$, $n'' = 2, 3, 4, \ldots$ refers to the Lyman series, which was not fully identified until 1914, but which in 1913 was generally expected to exist in the ultraviolet. Figures 28.10 and 28.11 show where in the orbital scheme and on the energy level diagram the Lyman series fits in with the others.

The radii and energy levels for this series can be calculated, independent of spectroscopic data, from Eqs. (28.13) and (28.12) by assigning $n = 1$ (the ground state for the spectral line series is here the natural state for the *unexcited* atom). Note well that nothing of the kind was possible in our initial, unsuccessful attempt to construct energy level diagrams and orbital schemes from the available experimental data of line emission. Lacking lines from the Lyman series, we could not at that time avoid the (incorrect) assumption that the ground state for transitions connected with Balmer series emission was the lowest natural state; only with an independent postulate can one hope to fix the radii and energy levels before all the lines are known.

Thus we have come to the initial successes of Bohr's model. It predicts the size of the unexcited atom, and the prediction is of the correct order of magnitude. It yields an expression for

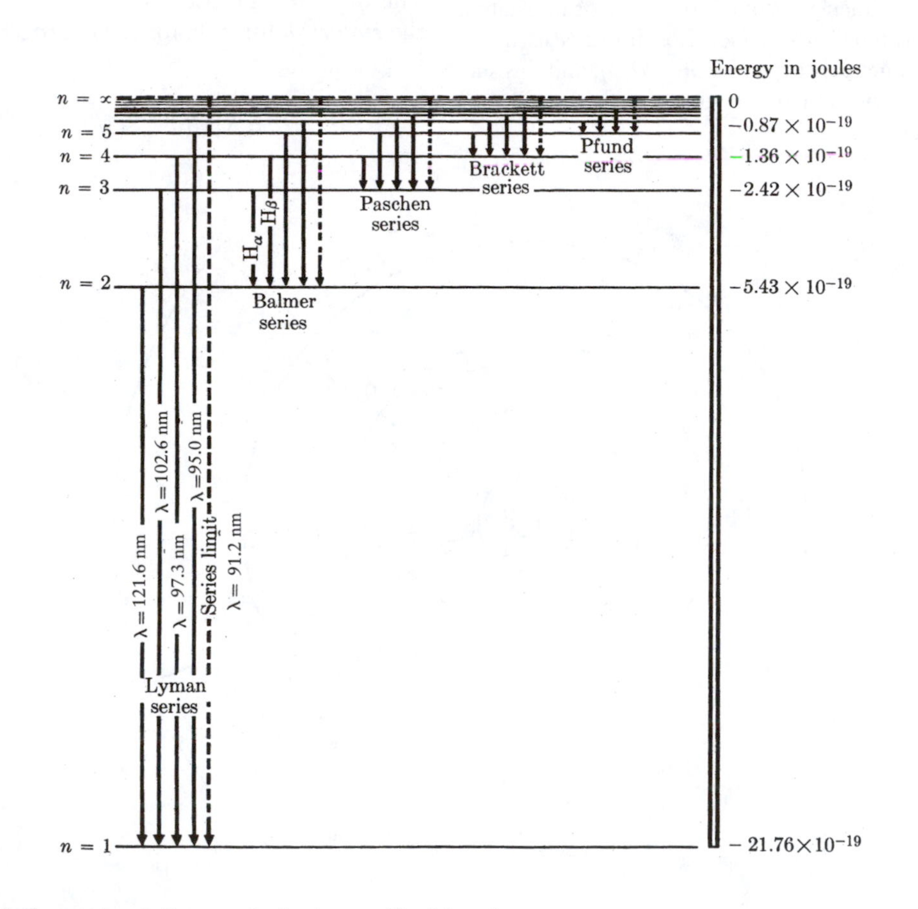

Fig. 28.11. Energy level diagram for hydrogen (final form).

the wavelengths of all lines that are known to come from radiating hydrogen atoms, and this expression coincides exactly with one that summarizes the experimental facts of line emission, including lines not known at the time the theory was formulated. It accounts for Rydberg's empirical constant in terms of known physical quantities. It provides us with a visualizable (although therefore perhaps dangerous) system, and establishes a physical order among the events accompanied by emission. The model introduces quantum theory into the atom and thereby on the one hand gives a physical basis for the idea that the energy in the atom is quantized, and on the other hand removes the problem of the stability of electron orbits from the classical theory, which could not provide the solution.

Bohr did not arrive at his theory of electron orbits by a process of logical reasoning from a few basic postulates, although the theory was later formulated in this way. Nor did he make his discovery by induction from experimental data, although he was guided by knowing the general type of phenomenon that he wanted to explain. The critical steps—assuming that the frequency of emitted radiation is the average of the frequencies of revolution in the two orbits, and then redefining the quantum number *n*—were leaps in the dark, which could be justified only by their subsequent success. Yet the discovery was not due to chance in the sense that anyone could have made it by guesswork; it was only after familiarizing himself with various aspects of the problem and trying out several approaches that, as Bohr reported in the paragraph quoted, the right assumption "suggested itself" to him.

Problem 28.7. Verify by calculation the agreement between Rydberg's constant and the expression in Eq. (28.14). (For data refer to Appendix.)

Problem 28.8. With proper techniques, each of these four lines may be found in the hydrogen spectrum: $\lambda = 121.57$ nm, $\lambda = 397.01$ nm, $\lambda = 950.0$ nm, $\lambda = 7400$ nm. Try to identify in each case to which series the line belongs, and fix the number of the orbits (or the value of the quantum numbers) involved in the particular transition.

28.6 Further Developments

Although the derivation of the empirical hydrogen spectrum formula was a superb feat in itself, Bohr also extended his argument from the very beginning to encompass a large variety of problems—rather like Newton, who followed his explanation of Kepler's empirical laws with a discussion of the inequalities in the moon's motions, the problem of comets, of the tides, and the like. In fact, it is the hallmark of great conceptual schemes that their applicability far exceeds the boundaries of initial intent, as has been amply illustrated in the preceding cases.

Bohr's original 1913 paper discussed a number of other applications. For example, as we have noted, E. C. Pickering in 1896–1897 had described a series of lines in stellar spectra that do not agree with the frequencies given by Eq. (28.15). In 1912, A. Fowler also found these lines in electric discharges through mixtures of hydrogen and helium. Their wavelength is given by [Eq. (28.4)]

$$1/\lambda = R[1/(n'/2)^2 - 1/(n''/2)^2],$$

However, Bohr continues, "we can account naturally for these lines if we ascribe them to helium [specifically, singly ionized helium, He^+]. A neutral atom of the latter element consists, according to Rutherford's theory, of a positive nucleus of charge 2*e* and two electrons." Now note that if one of the electrons is stripped off through the violence either of the electric discharge or of mutual atomic collisions in hot stars, the remaining helium ion is just like a hydrogen atom *except* for a doubly positive central charge. The light emitted by transitions of the single remaining electron can be calculated at once.

Problem 28.9. Starting with Eq. (28.6), revise the derivation leading to Eq. (28.14) for the case of an electron of charge *e* in an orbit around a nucleus with the numerical value of charge 2*e*. Show that the result may be written in the form

$$1/\lambda = (2\pi^2 m_e e^4 k_e^2/h^3 c)[1/(n'/2)^2 - 1/(n''/2)^2]. \quad (28.16)$$

But Eq. (28.16) coincides with the observations expressed in Eq. (28.4), since it has been proved that Rydberg's constant *R* is equivalent within small limits to the factor within the first bracket in Eq. (28.16)! Bohr says: "If we put $n' = 3$, and let n'' vary [that is, $n'' = 4, 5, \ldots$], we get a series which includes 2 of the [lines] observed by Fowler. . . . If we put $n' = 4$, we get the series observed by Pickering in the spectrum of ζ Puppis. Every second of the lines in this series is identical with a line in the Balmer series of the hydrogen spectrum." This is indeed an exciting interpretation of an old puzzle.

It is clear that Eq. (28.16) differs from Eq. (28.14) only by the factor $(2)^2$. If an atom with Z positive nuclear charges were stripped of all but one electron, Eq. (28.16) should apply to its spectrum except that $(2)^2$ is replaced by Z^2. And indeed, it is possible to produce such violent discharges that not only helium but each of the light elements up to the eighth, oxygen, is suitably ionized. The lines emitted from such gases are in full agreement with Bohr's theory.

Problem 28.10. The vapor of lithium ($Z = 3$) may be doubly ionized (Li^{++}). Derive the equation analogous to Eq. (28.16) for the line spectrum from Li^{++}, and draw an energy level diagram and a diagram of the first few orbits analogous to Figs. 28.10 and 28.11.

We must continue to test Bohr's model: Does it account for *absorption* of light as well as it did for emission? Two facts have been noted in the discussion of line absorption spectra: (i) that light with a continuous spectrum, on passing through a gas, may lose frequencies that correspond exactly to those the gas would itself emit in a line emission spectrum, and (ii) that the absorption spectrum of a gas is, however, usually less rich than its emission spectrum.

In this specific case, monatomic hydrogen gas generally absorbs only lines corresponding to the Lyman spectrum, as we now know; it has to be brought to an excited state if it is to absorb the lines corresponding to the Balmer spectrum, as Bohr pointed out. If we make the assumption, consistent with the whole model, that an orbital electron can absorb a photon only if the energy so absorbed brings it out exactly into another permitted orbit, the whole set of observations becomes clear.

Ordinarily, hydrogen atoms are in the most stable, the "natural," state, in the innermost orbit corresponding to $n = 1$. The energy differences between that level and all others correspond to lines in the Lyman spectrum; photons with less energy, as all those in the Balmer spectrum, cannot be absorbed because they are unable to make the electron go even from the first innermost orbit to the second one. Only if the gas is initially in an excited state (one with $n = 2$ or higher) will sufficient atoms be present in the particular ground state of the Balmer series ($n = 2$) to absorb Balmer series frequencies.

On the same model, another puzzle may be solved. It had been noted that above a certain frequency of incident light there is no longer *line* absorption, but *continuous* absorption of all frequencies. Evidently a photon that corresponds to the highest frequency line in any series is one that, if absorbed, produces complete ionization of the hydrogen atom by carrying the electron to the level $n = \infty$. That electron is then no longer bound to the atom, but is free; but for free electrons there is no energy quantization. They can accept and carry (as translational kinetic energy) any amount of energy whatsoever. Evidently, when a photon with more energy than is needed to free the electron is absorbed, part of the energy is used to ionize the atom, and the rest is given to the electron as kinetic energy.

We realize now that in the last paragraph we have actually been speaking about the photoelectric effect as applied to gases: Only photons above a threshold frequency ν_0 will cause photoelectric emission, and ν_0 is $\Delta E/h$, where ΔE is the difference in energy levels between the ground state of the illuminated gas and the state for which $n = \infty$. Bohr says, "Obviously, we get in this way the same expression for the kinetic energy of an electron ejected from an atom by photoelectric effect as that deduced by Einstein, that is KE $= h\nu - W$. . . ."

Problem 28.11. A sample of monatomic hydrogen gas is illuminated from a source emitting only light of wavelength $\lambda = 102.6$ nm (second Lyman line). If Bohr's picture of absorption and radiation is correct, and if the hydrogen atoms later reemit the energy that they absorb, what *two* wavelengths of light do we expect to be reradiated from the sample?

Problem 28.12. Calculate the work function W for the photoelectric effect in unexcited monatomic hydrogen gas.

Problem 28.13. As Bohr was aware, Rutherford had noted in 1912 that fast β-rays sent through a gas "lose energy in distinct finite quanta." Explain this observation in terms of Bohr's model.

Problem 28.14. According to the kinetic-molecular theory of gases, the average kinetic energy per gas molecule is given by $\langle KE_{trans}\rangle_{av}$ = {{3/2}}kT, where k is Boltzmann's constant, 1.3806×10^{-23} J/K, and T is the temperature in K [see Eq. (22.12)]. If the orbital electron can be thrown into a higher energy level by absorbing the kinetic energy during an inelastic collision between two atoms, calculate the temperature of the gas (for monatomic hydrogen) at which most gas atoms are in an energy level where $n = 2$ and consequently can contribute to the prominent absorption of lines with the Balmer series frequencies. [*Note:* Even at a lower temperature, some fraction of the total number of atoms is sufficiently excited to aid in this type of absorption. Why?]

Fig. 28.12. Wolfgang Pauli (1900-1958), Austrian physicist. A leader in the development of quantum theory, his Exclusion Principle provides a systematic procedure for explaining the electronic structure of atoms. His hypothesis of a new particle to explain β-decay led to the discovery of the neutrino (Section 32.1). He received the 1945 Nobel Prize in Physics.

To this point, Bohr's concern was only with single-electron systems, for example, hydrogen and ionized helium. After he had explained such physical properties of hydrogen atoms as dimensions, emission spectra, and ionization energies with brilliant success, the next problem was obviously atoms whose nuclei are surrounded by more than one electron. Bohr made a first attack on this problem in his 1913 paper, and returned to it again several times during the next decade. Other scientists, attracted by the initial success of Bohr's model for hydrogen, soon joined in this work. By 1920 there had been developed a *shell model* of atomic structure, which was able to account at least qualitatively for the chemical properties of the elements.

As the x-ray experiments of Moseley (Section 27.5) had shown, the atomic number of an element may be identified with the electric charge carried by the nucleus in Rutherford's model. It is also equal to the number of electrons circulating in orbits around the nucleus of the neutral atom. These orbits are (with slight modifications) the same ones available to the electron in a hydrogen atom; but it was found that certain apparently arbitrary restrictions must be imposed on the number of electrons that can occupy a given orbit. The innermost shell (corresponding to $n = 1$ in Fig. 28.10) is called the K shell, and has room for a maximum of two electrons. Helium's two electrons "fill" or "close" the shell, so that lithium's third electron must start a new one, the L shell. (Refer to Table 21.2 for atomic numbers of the elements.) The L shell is assumed to have room for eight electrons, so that neon, with a total of ten electrons, fills the K and L shells completely. The next element is sodium, whose eleventh electron places itself in the third or M shell; in this way the atomic structures of the rest of the elements can be built up.

The rules determining the maximum number of electrons that can occupy a shell were formulated in terms of three new quantum numbers in addition to Bohr's original quantum number n. Two of them, introduced by the German mathematical physicist Arnold Sommerfeld, specified the shape of the electron's orbit—which, by analogy with planets in the solar system, could be elliptical instead of circular—and its orientation in space. (Bohr's orbits were all in a plane, which was too simple an assumption.)

The fourth quantum number was introduced by the Austrian physicist Wolfgang Pauli (1900–1958), who called it merely a "nonclassical two-valuedness": some property of the electron that can have two values, $+\frac{1}{2}$ or $-\frac{1}{2}$. Pauli was then able to state the general rule for building up electronic shells of atoms: *No two electrons may have exactly the same set of four quantum numbers.* This is known as the *Pauli exclusion principle:* The presence of one electron in a quantum state, defined by giving the values of the four quantum numbers, "excludes" any other electron from occupying that state.

Shortly after Pauli announced his new quantum number, two Dutch graduate students, George Uhlenbeck and Samuel Goudsmit, proposed that this number could conceivably refer to a rotation or *spin* of the electron itself. The two values $+\frac{1}{2}$ and $-\frac{1}{2}$ correspond to clockwise or counterclockwise rotation with respect to some direction in space, defined for example by a magnetic field.

The chemical properties of the elements are explained by postulating that an atom tends to gain or lose electrons in such a way as to achieve a closed-shell structure. The rare gases (He, Ne, Ar, etc.), which already have such a structure, do not enter into chemical reactions at all, with few exceptions. The alkali metals (Li, Na, . . .) have one extra electron in the outermost shell; this electron is easily given up. The halogens (F, Cl, . . .), on the other hand, are lacking one electron, and thus tend to acquire one whenever possible. More generally the *valence* of an element (Section 20.11) can be identified with the number of electrons needed to be gained or lost in order to form a closed shell.

In summary, not only the main physical properties but also the chemical behavior of the elements seems to be governed in principle by the structure of the electron cloud around the nucleus. But the configuration itself seems determined largely by the atomic number, that is, the *number* of electrons present, and this recognition, a triumph of old Pythagorean and neo-Platonic tendencies still present in modern science, brought this remarkable reaction from Bohr himself:

> This interpretation of the atomic number may be said to signify an important step toward the resolution of a problem which for a long time has been one of the boldest dreams of natural science, namely, to build up an understanding of the regularities of nature upon the consideration of pure numbers.

The list of successes of Bohr's fundamental approach has by no means been exhausted in this recital. We might continue, as is done in the books recommended at the end of this chapter, to apply the theory and its subsequent modifications to solving such puzzles as the precise wavelengths of x-ray spectra (for example, Moseley's unexplained K_β line), the difference in the appearance of line spectra under different experimental conditions, the relative intensities of some of the observed lines, and much else besides.

But the fact is that Bohr's conceptual scheme, although today still indispensable and basic to an initial understanding of both physics and chemistry, has by now been transformed and overtaken by the swift progress of modern physical science. Several fundamental transformations became necessary to account for the wealth of phenomena not included even in such a sweeping synthesis as Bohr's. In essence, these transformations—which could not have happened except as consequences of Bohr's original model—centered on abandoning the last vestiges of visualizability of single events in the electron cloud. The rather comfortable and intuitively meaningful orbits, shells, and "jumps" of electrons had to be given up as essentially meaningless from an operational point of view, and, in fact, misleading for some purposes of prediction. The contemporary atom, which still retains the now familiar concept of energy levels, quite intentionally no longer presents any simple physical picture to guide our imagination. It is a mathematical model, a set of postulates and equations—the ultimate triumph of ascetic but superbly effective thought. As Bohr himself said at the very beginning of this phase of the development (1925):

> To the physicists it will at first seem deplorable that in atomic problems we have apparently met with such a limitation of our usual means of visualization. This regret will, however, have to give way to thankfulness that mathematics in this field, too, presents us with the tools to prepare the way for further progress.

Problem 28.15. Show that the kinetic energy of an electron in Bohr's model is (a) proportional to the total energy E, (b) inversely proportional to n^2. Hence show that (c) the velocity of the electron in the orbit model is inversely proportional to n. This result will be used in the next chapter.

RECOMMENDED FOR FURTHER READING

H. A. Boorse and L. Motz (editors), *World of the Atom,* extracts from the writings of Balmer and

Bohr, and biographical sketches; chapters 58 and 59 on electron spin and the exclusion principle.

A. P. French and T. B. Greenslade, Jr. (editors), *Physics History II,* articles by A. Leitner on Fraunhofer, pages 23–33, and by P. Jordan on Pauli, pages 79–82

C. C. Gillispie (editor), *Dictionary of Scientific Biography,* articles by L. Rosenfeld on Kirchhoff and Bohr, Vol. 7, pages 379–383 and Vol. 2, pages 239–254; by R. V. Jenkins on Fraunhofer, Vol. 5, pages 142–144; by S. G. Schacher on Bunsen, Vol. 2, pages 586–590; by A. E. Woodruff on Doppler, Vol. 4, pages 167–168; by C. L. Maier on Balmer, Vol. 1, pages 425–426; by M. Fierz on Pauli, Vol. 10, pages 422–425

J. L. Heilbron, "Lectures on the History of Atomic Physics 1900-1922," in *History of Twentieth Century Physics,* edited by C. Weiner, pp. 40–108, New York: Academic Press, 1977; "Bohr's first theories of the atom," *Physics Today,* Vol. 38, no. 10, pages 25–36 (1985)

Daniel Kleppner, "The Yin and Yang of hydrogen," *Physics Today,* Vol. 52, no. 4, pages 11, 13 (April 1999), on the role of hydrogen in the history of science, including the Bose-Einstein condensation.

Abraham Pais, *Niels Bohr's Times, in Physics, Philosophy and Polity,* New York: Oxford University Press, 1991

S. Rozental (editor), *Niels Bohr: His Life and Work as Seen by His Friends and Colleagues,* New York: Wiley/Interscience, 1967

C. J. Schneer, *Mind and Matter,* Chapter 15

Spencer Weart and Melba Phillips (editors), *History of Physics,* articles by S. A. Goudsmit and G. Uhlenbeck on "Fifty years of spin," pages 246–254, and by J. L. Heilbron on "J. J. Thomson and the Bohr atom," pages 303–309

CHAPTER 29

Quantum Mechanics

29.1 Recasting the Foundations of Physics Once More

With this chapter we come to the end of Newtonian mechanics as a fundamental basis for physical theory, even though almost all its practical applications remain valid at the present time. We have already seen the beginnings of one theory destined to use some of Newtonian mechanics but abandon the rest—the quantum hypothesis for light and Bohr's extension of it to a model of the atom. These results, part of which is termed quantum theory, were soon to be incorporated into a unified quantum mechanics. In the next chapter we shall discuss the other new theory, Einstein's theory of relativity. The complete amalgamation of these two theories into a consistent general system, which could play the same role as Newton's laws or Aristotle's natural philosophy, still lies in the future. It is partly for this reason and partly because of the considerable mathematical complexity of quantum mechanics and relativity that we have had to devote to them less space than their present importance might warrant.

To set the stage for the introduction of quantum mechanics, let us recall the situation in physics around 1920. First, the quantum hypothesis had been generally accepted as a basis for atomic theory, and the photon concept had been given strong support by Robert A. Millikan's experimental confirmation of Einstein's theory of the photoelectric effect. Even more definitive proof of the particle nature of electromagnetic radiation came from the demonstration of the Compton effect in 1923 (Section 16.7).

Yet, despite the early success of Bohr's theory of the hydrogen atom, no one had found a satisfactory way to extend the theory to multielectron atoms and molecules without introducing arbitrary additional hypotheses tailored to fit each particular case. Furthermore, it seemed necessary to use principles of Newtonian mechanics and Maxwellian electromagnetism in some parts of the theory, while rejecting them in others.

Although the photoelectric effect, as interpreted by Einstein, seemed to offer definite proof that light energy comes in discrete units, this did not by any means persuade physicists that the wave theory of light should be abandoned in favor of a particle theory. The experiments on interference and diffraction of light, and indeed all the properties of electromagnetic radiation discovered in the nineteenth century, seemed to require wave properties. Even the quantum theory, which assigned each "particle" of light an energy proportional to its "frequency" ($E = h\nu$), depended on a wave property: periodicity. Thus one was led to the paradoxical conclusion that light behaves in some respects like particles *and* in other respects like waves.

Another puzzling aspect of the physical theories developed in the first two decades of the twentieth century was their failure to prescribe definite *mechanisms* on the atomic level. Thus Bohr's theory specified the initial and final states of an atomic radiation process, but did not describe just how the electron got from one orbit to another. The theory of radioactive decay was even less definite: It simply predicted the *probability* that a nucleus would emit radiation, but did not state when or how any particular nucleus would decay. It appeared that the *determinism* or *causality* characteristic of the Newtonian world machine—the predictability, at least in principle, of the detailed course of events of any atomic process, given the initial masses, velocities, and forces—was slipping away. Just as Maxwell and Boltzmann had shown that the second law of thermodynamics was only statistically valid as a description of observable processes, so some physicists began to suspect that Newton's laws of motion and the laws of conservation of energy and momentum might, in the atomic world, be true only "on the average," but not necessarily at every instant of time.

But if neither the wave theory nor the particle theory of light could be proved exclusively correct and if the classical laws of mechanics and electromagnetism could no longer be relied on to describe atomic processes, clearly something new was needed, some fundamental law from which all these partially or statistically valid theories could be derived. And perhaps even the basic Newtonian presuppositions about the nature and purpose of theories would have to be modified. As they gradually recognized the seriousness of these problems, physicists early in the 1920s began to talk (much more urgently than they usually do) about a "crisis in the foundations of physics."

We recognize the old pattern that has characterized the advancement of physics from the beginning: Each breakthrough leads to the solution of old problems, but then becomes the basis for the necessary next stages as its deficiencies become evident. That is how humans have had to proceed—only the gods might know all from the start!

Fig. 29.1. Louis Victor, Prince de Broglie (1892–1987), French physicist, introduced the hypothesis that "particles" like electrons and protons have wave properties. This hypothesis was confirmed by experiments and became the basis for Schrödinger's version of quantum mechanics. De Broglie received the 1929 Nobel Prize in Physics.

29.2 The Wave Nature of Matter

It is a curious fact that radical conceptual changes in a science are often initiated by people who did not receive their initial professional training in that science. A person who first approaches the problems of a discipline with the mature perspective of an outsider, rather than having been indoctrinated previously with the established methods and attitudes of the discipline, can sometimes point to unorthodox, although (when seen in retrospect) remarkably simple solutions to those problems. As we saw in Chapter 17, for example, the generalized law of conservation of energy was introduced into physics from engineering, physiology, and philosophy, although its value was accepted eventually by physicists.

Louis de Broglie (1892–1987), formally titled "prince" as a member of a family related to the old French royalty, received his first degree in history at the University of Paris in 1910, intending to go into the civil service. But he became intrigued by scientific problems as a result of discussions with his brother, the physicist Maurice de Broglie, and by reading the popular scientific expositions of the mathematician Henri Poincaré. (These books, listed at the end of Chapter 14, are still delightful reading.) Although he managed to acquire sufficient competence in using the standard tools of theoretical physics to satisfy his professors, his real interest was in the more fundamental problems of the nature of space, time, matter, and energy.

In his doctoral dissertation of 1924 (based in part on papers published a year earlier), de Broglie proposed a sweeping symmetry for physics: Just as photons behave like particles as well as like waves, so electrons should behave like waves as well as like particles. In particular, an electron of mass m moving with velocity v will have an associated "wavelength" λ given by the simple formula

$$\lambda = h/mv \qquad (29.1)$$

where h is Planck's constant.

Although de Broglie's astonishing hypothesis was based originally on theoretical arguments,

Fig. 29.2. Clinton Joseph Davisson (1881–1958) and Lester Halbert Germer (1896–1971), American physicists whose experiments on electron diffraction confirmed Louis de Broglie's hypothesis on the wave nature of atomic particles. Davisson and G. P. Thomson (who independently performed a similar experiment) shared the 1937 Nobel Prize in Physics.

it was soon realized that it could be checked rather directly by experiments of a kind that had already been performed by the physicist Clinton J. Davisson (1881–1958) in the United States. In these experiments, electrons emitted from an electrode in a vacuum tube were allowed to strike a metal surface, and the scattering of the electron beam at various angles was measured. Walter Elsasser in Germany pointed out that Davisson's results could be interpreted as showing a *diffraction* of the beam of electrons as they passed through rows of atoms in the metal crystal and that those of other experiments on electron scattering by C. Ramsauer could be attributed to *interference* of waves that somehow determined the path of the electrons. In both cases the apparent wavelength associated with the electrons agreed with the values calculated from de Broglie's equation [Eq.(29.1)].

In 1926, Davisson together with another American physicist, Lester H. Germer (1896–1971), carried out further experiments, which confirmed de Broglie's hypothesis, and in 1927 the British physicist George Paget Thomson (1892–1975) demonstrated the phenomenon of electron diffraction by still another method. (Note again the international nature of the way science progresses.) Later experiments showed conclusively that not only electrons but protons, neutrons, heavier nuclei, and indeed all "material particles" share this wavelike behavior.

A small irony of history: G. P. Thomson shared the 1937 Nobel Prize in Physics with Davisson for demonstrating the *wave* nature of the electron, 31 years after his father J. J. Thomson received that prize for demonstrating the *particle* nature of the electron (Section 25.7).

Problem 29.1. From Eq. (29.1) calculate the de Broglie wavelength λ for an electron with kinetic energy equal to 50 eV. In order to show diffraction effects, a wave disturbance must go through an opening whose linear dimension is about the same order of magnitude as its wavelength, or smaller. Show that the space between rows of atoms in a crystal could be a suitable opening.

Problem 29.2. Estimate the de Broglie wavelength for a baseball moving at 100 m/sec and for a skier moving at 50 m/sec. On the basis of these results, suggest why the wave nature of matter was not discovered until the twentieth century.

In his 1924 memoir, de Broglie pointed out that the quantization of allowed orbits in the Bohr model of the hydrogen atom can also be deduced from his formula for the wavelength of an electron, if one makes the natural assumption that the circumference of the orbit must be just large enough to contain an integral number of wavelengths, so that a continuous waveform results, as in Fig. 29.3. This assumption, applied to a circular orbit of radius r, would mean that

$$2\pi r = n\lambda = nh/mv,$$

where n is an integer. According to Bohr's theory, the kinetic energy ($\frac{1}{2}mv^2$) of the electron is proportional to its total energy E, which varies inversely as the square of the quantum number n [Eq. (28.12)]. Therefore, if the mass m is fixed, the velocity is inversely proportional to n (see Problem 28.15), and the radius is proportional to n^2, in agreement with Eq. (28.13).

De Broglie's hypothesis suggests that the reason why atomic energy levels are quantized is the same as the reason why the frequencies of overtones of a vibrating string are "quantized" (Section 23.2): The waves must exactly fill up a certain space. Hence a theory of atomic properties might be constructed by analogy with the theory

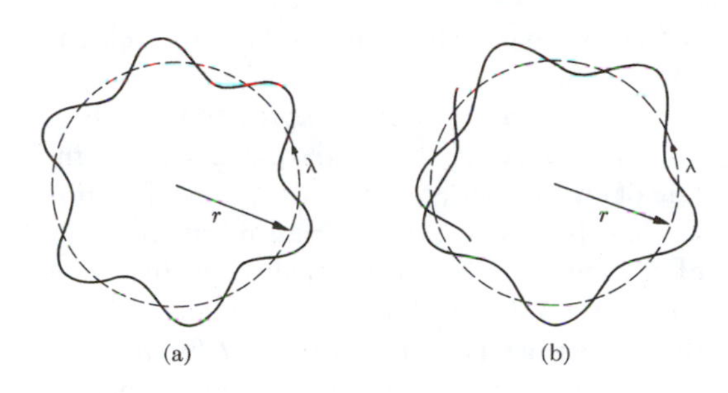

Fig. 29.3. Only certain wavelengths will "fit" around a circle, for example, in (a) not in (b).

of vibrations of mechanical systems—reminiscent of the old program of Pythagoras and Kepler, who based many of the laws of nature on the laws that govern the operation of musical instruments (Sections 1.1 and 4.4).

Another year passed before this clue was followed up by an Austrian physicist, Erwin Schrödinger (1887–1961), who happened to be quite familiar with the theory of sound. Stimulated by the favorable remarks of Einstein and Peter Debye on de Broglie's theory, Schrödinger developed a mathematical theory of atomic properties in which the quantization of energy levels was based on the allowed values, or *eigenvalues,* of the electron wavelengths. (The German prefix *eigen* means "own," "individual," "proper," or "characteristic." The complete German word *Eigenwert* has been half-translated into scientific English as "eigenvalue.")

Schrödinger's theory, known as "wave mechanics," was published early in 1926. It was soon recognized to be a satisfactory generalization of Bohr's theory; it provided a set of postulates that determined uniquely a fundamental equation governing the behavior of the electrons in any given atom or molecule [see below, Eq. (29.3)]. In the simple cases in which Schrödinger's equation can be solved exactly or almost exactly, it was found to predict the correct values of spectral frequencies and the intensities of spectral lines, as well as all other *observable* properties of the system. (For most atoms and molecules the equation has not been solved exactly because of its extreme complexity when many electrons are involved.)

The major defect (if it can be called that) of Schrödinger's theory, like de Broglie's theory from which it evolved, is that it fails to explain just what is vibrating or "waving"; it is definitely

Fig. 29.4. Erwin Schrödinger (1887–1961), Austrian physicist, whose wave equation, a convenient mathematical expression for the basic principles of quantum mechanics, governs the properties of any atomic or molecular system (except when relativistic effects are important). He shared the 1933 Nobel Prize in Physics with P. A. M. Dirac, who generalized the wave equation to make it consistent with special relativity theory (Section 30.9).

not the electron when conceived as a point particle analogous to a billiard ball, because that would still imply the classical conception of a particle localized at a particular spatial position and having a particular velocity, both of which change in a deterministic way from one instant to the next. Instead, the electron itself is in some (not visualizable) sense the wave.

If the electron is in a definite *state* or energy level in the atom (corresponding to one of the allowed orbits in Bohr's theory), then it would seem to occupy simultaneously the entire orbit and to have simultaneously all values of the velocity vector corresponding to the different possible directions of motion in a circular orbit at constant speed.

Even this statement is unduly restrictive, for the theory can be interpreted to mean that the electron is "spread out" over a region of space around the Bohr orbit and has a large range of different speeds (see below, Section 29.3). This is a natural description of a wave disturbance in a continuous medium; but it seems difficult to reconcile it with our earlier conception of the electron as a particle whose mass, charge, and velocity can be experimentally determined (Section 26.4). This is, indeed, just as difficult as thinking of an electromagnetic wave as a vibration of some ethereal medium (Section 23.3) that at the same time manifests itself as a particle having a definite momentum (Section 16.7) and energy (Section 26.5)!

In the case of electromagnetic waves, we could say that the vibrating or oscillating quantities were electric and magnetic field intensities (Section 25.3). For Schrödinger's electron waves, we can say that the vibrating quantity is a *wave function,* usually denoted by ψ (Greek letter psi). The word "function" is used because, like the electric and magnetic fields $\boldsymbol{E}$ and $\boldsymbol{H}$, ψ is a mathematical function of space and time variables, that is, its value depends on x, y, z, and t. (For an electron in a stationary state corresponding to a Bohr orbit, ψ does not depend on t.) However, *unlike* $\boldsymbol{E}$ and $\boldsymbol{H}$, the values of ψ may be *complex numbers,* numbers that depend on the imaginary unit $\sqrt{(-1)}$, and therefore it defies any simple physical interpretation.

Do not be alarmed if you don't understand the following paragraphs! As Richard Feynman, one of the world's experts on the subject, once remarked, no one *really* "understands" quantum mechanics. It is an extraordinarily powerful mathematical theory that gives accurate descriptions and predictions of the properties of matter and radiation in a wide variety of physical situations—from the way transistors work to modern cosmology—including many cases where Newtonian physics is inadequate. Yet physicists and philosophers still disagree about what it tells us about the ultimate nature of the world.

According to Schrödinger's theory, one must first construct the wave equation for a physical system by converting the expression for the kinetic energy of each particle into a mathematical operation on the wave function ψ. This operation is similar to finding the acceleration of a particle whose position s is expressed as a function of the time t (see Sections 6.4, 6.5, and 10.3), except that the independent variable t is replaced by each of the three space variables x, y, z in turn (corresponding to the three velocity components).

The operations can be expressed verbally as "find the rate of change with respect to x, of the rate of change with respect to x, of" the function ψ. (In other words, the operation "find the rate of change of . . ." is performed twice, first on ψ and then on the result of that operation.) A further difference is that ψ *is not yet known* as a function of x, y, z, so these operations cannot yet be performed, but are only written down symbolically as "kinetic energy operations on ψ," $(\mathrm{KE})_{\mathrm{op}}\psi$.

In order to determine the stationary states of an atomic or molecular system, Schrödinger's theory requires us to add this kinetic energy operation to a potential energy operation $(\mathrm{PE})_{\mathrm{op}}$, where $(\mathrm{PE})_{\mathrm{op}}$ is defined simply as multiplying ψ by the algebraic expression for the total potential energy of the system, PE. $(\mathrm{PE})_{\mathrm{op}}$ usually consists simply of a sum of terms of the form $e_i e_j / r_{ij}$, where e_i and e_j are the charges of two interacting electrons or nuclei and r_{ij} is the distance between them (see Section 24.12).

The total operation is called the *Hamiltonian operator, H:*

$$H = (\mathrm{KE})_{\mathrm{op}} + (\mathrm{PE})_{\mathrm{op}} \qquad (29.2)$$

(The name honors William Rowan Hamilton, the nineteenth-century Irish mathematician who first explored the analogies between optical and mechanical theories.)

Schrödinger's fundamental equation is then simply

$$H\psi = E\psi \qquad (29.3)$$

where E is a numerical constant, equal to the energy of the system.

The Schrödinger equation [Eq. (29.3)] must now be solved for the unknown wave function ψ and the unknown energy E. In general there are several possible ψ functions that are solutions of this equation; each has its own eigenvalue E. Thus *the fact that an atom has a discrete set of possible states and a discrete set of corresponding energy values emerges automatically as a mathematical property of the equation; the set of rules for constructing the Schrödinger equation replaces the previous postulates about allowed orbits in the Bohr theory.* These rules apply to all conceivable systems, whereas in Bohr's theory his postulates had to be constructed by trial and error for each case, with only his intuition and experience to guide him.

Recall that in Bohr's theory E is a negative quantity for closed orbits of an electron bound within an atom [Eq. (28.12)]. As noted in Section 28.5, if the electron is given sufficient energy it can escape the attraction of the nucleus and move freely in space; this process of "ionization" requires 13.6 eV if the electron is in the lowest orbit of a hydrogen atom, whose energy corresponds to what we now call the lowest eigenvalue of H. For free electrons *all positive values of E are allowed.* In other words, there is a *continuous spectrum* of positive energies in addition to a *discrete spectrum* of quantized negative energies. This remarkable fact is a direct mathematical consequence of Schrödinger's equation.

Equation (29.3) is called the *time-independent Schrödinger equation* because it does not contain a time variable; it describes only the stationary states of an atom, although the particle itself is by no means "stationary" but may have a rather high speed. Rather it is the *wave function* ψ that does not change with time.

However, in more general situations such as collisions and chemical reactions, we need to know how ψ itself changes with time. Schrödinger also proposed a time-dependent equation, similar to Eq. (29.3) except that the constant energy E on the right-hand side is replaced by an energy operator $(E)_{op}$, defined as "find the rate of change of time of":

$$H\psi = (E)_{op}\psi \tag{29.4}$$

and the wave function ψ is now a function of t as well as of x, y, and z.

Calculations based on Eq. (19.4) are valid for many physical situations, but we will see in Section 30.9 that a rather different way of including the time variable should really be used.

29.3 Knowledge and Reality in Quantum Mechanics

Although the wave function ψ appears to play a central role in Schrödinger's wave mechanics, it is according to some physicists only a superfluous auxiliary construct. The rationale for this viewpoint derives from the success of another theory published a few months before Schrödinger's: Werner Heisenberg's theory, now known as *matrix mechanics.* Heisenberg (1901–1976), a young German physicist working in Max Born's institute at Göttingen, constructed an alternative set of rules for calculating the frequencies and intensities of spectral lines, using only relations between observable quantities. With the collaboration of Born and Pascual Jordan, Heisenberg soon developed his method into a complete mathematical theory of atomic processes that could yield the same results as Schrödinger's theory.

Fig. 29.5. Werner Karl Heisenberg (1901–1976), German physicist, the first to discover quantum mechanics in the form known as matrix mechanics. He is also known for his uncertainty (or indeterminacy) principle and for his proton-neutron theory of the nucleus. He received the 1932 Nobel Prize in Physics.

As Max Jammer remarked in his comprehensive study of the history of this subject (*Conceptual Development of Quantum Mechanics*), "It is hard to find in the history of physics two theories designed to cover the same range of experience, which differ more radically than these two"—Schrödinger's wave mechanics and Heisenberg's matrix mechanics. The former emphasized the *continuity* of physical processes and the wavelike behavior of the electron—almost visualizable, despite the obscure meaning of the wave function. The latter proceeded from the *discontinuity* of physical processes, suggested by the observed discreteness of spectral lines, and regarded the electron as a particle, though without assigning it any definite space-time description in the classical sense. Nevertheless it was quickly discovered that the two theories are mathematically equivalent—the proof of this fact having been given by Schrödinger and independently by Carl Eckart (at the California Institute of Technology) in 1926.

The development of physics since 1926 has not undermined in any significant way the validity of the Heisenberg-Schrödinger theories, considered now as two alternative formulations of a single theory, *quantum mechanics*. The philosophical viewpoints about the nature of the physical world and our knowledge of it deriving from quantum mechanics are of considerable interest, the more so as this is still an area of vigorous research. For example, quantum mechanics has revived the old question: Do physical objects have an existence and properties entirely *independent* of the human observer, or at least of experimental apparatus?

The early success of quantum mechanics led physicists to try to interpret its concepts in terms of the classical categories applicable to particles. The first such attempt was made by Max Born in 1926. He proposed that the electron is really a particle and that the wave function simply represents the *probability* that it is located at a particular point in space.[1]

Fig. 29.6. Max Born (1882–1970), German physicist, who proposed the probabilistic interpretation of quantum mechanics, for which he received the 1954 Nobel Prize in Physics.

The use of the word "probability" in Born's interpretation is not the same as in classical physics, for example in the kinetic theory of gases (Chapter 22), in which it was assumed that the molecules have definite positions and velocities determined at each instant in time by Newton's laws and could in principle be calculated exactly. In the kinetic theory, statistical methods were introduced as a matter of convenience in doing calculations: It seemed impossible to measure all the positions and velocities of 10^{22} molecules or more, or to use this information even if it were available. "Probability" referred to the mode of description employed by the scientist and did not affect the idea that the properties of an individual molecule are considered exactly determinable.

But in Born's view, the positions and velocities of each subatomic particle are basically random; they cannot be said to have any definite values, even in principle, but have only certain probabilities of exhibiting particular values.

Born argued that the contrary assumption (adopted at first by Schrödinger), namely, that the electron is really a wave spread out in space with a charge density given by the wave function, is unacceptable. For if the electron is not bound in an atom, its wave function eventually will spread out over an infinite space; yet experiment indicates that all of the electron is actually present at some particular place. This result is still

[1]More precisely, since the value of the wave function is a complex number (and only the square of a complex number has entirely real components), it is the square of the absolute value of the wave function that is proportional to the probability that the electron has the position x, y, z. If the complex number ψ is expressed as a sum of real and imaginary parts, $\psi = a + ib$, where i is the square root of -1, then the square of its absolute value is defined as $|\psi|^2 = a^2 + b^2$, which is a real number. Since only the square of the wave function has a direct physical meaning, ψ itself is somewhat indeterminate: We can multiply it by a "phase factor" $e^{i\omega}$ without changing the value of $|\psi|^2$. This "invariance" property turns out to be connected, through Emmy Noether's theorem (Section 17.11), with the law of conservation of charge (Section 24.3).

consistent with the idea that the electron has a certain probability of being at any given place; performing the experiment to find its position converts this probability into a certainty—it either *is* or *is not* there, but it cannot be "partly there." It is already evident that any attempt to pin down a definite property of the electron involves an act of measurement.

Max Born was awarded the Nobel Prize in Physics in 1954, primarily for his statistical interpretation of quantum mechanics. The fact that the award came so long after his original publication (1926) indicates that this interpretation was not at once regarded as an indubitable, basic "discovery," but that it gradually came to be accepted as a fruitful viewpoint in physical theory. Insofar as many scientists do accept the notion that natural processes are fundamentally random rather than deterministic, we have to admit that the "Newtonian world machine," or the mechanistic viewpoint advocated by Descartes and Boyle in the seventeenth century, has been abandoned.

Another way of characterizing the consequences of quantum mechanics is through Heisenberg's indeterminacy principle, often called the "uncertainty principle." In 1927, Heisenberg proposed as a postulate that there are certain pairs of physical properties of a particle that cannot simultaneously be measured to an arbitrarily high degree of accuracy. The more accurately we try to measure one property in this pair, the less accurate will be our measurement of the other one.

For example, if we try to determine both the position (x) and momentum (p) of an electron, then the product of the average errors or uncertainties of measurement (symbolized by δ) must be at least as great as $h/2\pi$:

$$(\delta x) \cdot (\delta p) \geq h/2\pi \qquad (29.5)$$

Although the principle applies to the position and velocity of any object, it is a significant limitation only for atomic or subatomic particles, since Planck's constant h is so small.

Problem 29.3. (a) An electron is known to have a speed of 100 ± 1 m/sec. What will be the uncertainty in its measured position? (b) A rock of mass 100 g is known to have a speed of 100 ± 1 m/sec. What will be the uncertainty in its measured position?

Problem 29.4. For an electron in a Bohr orbit in a hydrogen atom, the uncertainty in position at any moment may be said to be equal to the radius of the orbit (its distance from the center being within $\pm r$), and the uncertainty in velocity at that moment may be said to equal the numerical value of the speed (the magnitude of the velocity component in any direction fluctuates between $+v$ and $-v$ as it goes around a circular orbit). Write down the condition that the de Broglie wavelength of the electron must be no greater than the circumference of the orbit; then write down the indeterminacy principle for position and momentum of this electron; then show that these two statements are equivalent.

The indeterminacy principle can be deduced from quantum mechanics, but its validity was also established by considering various possible experiments that might be designed to measure both x and p. Such experiments were analyzed in some detail in a famous discussion between Bohr and Einstein. The conclusion was that any measurement involves an *interaction* between the observers (or their apparatus) and the object being observed.

Heisenberg's principle also applies to the energy E of a particle and the time interval t during which it has that energy. It is impossible to determine both E and t simultaneously with unlimited accuracy, because of the restriction

$$(\delta E) \cdot (\delta t) \geq h/2\pi.$$

Heisenberg's principle might be interpreted as merely a restriction on how much we can *know* about the electron, taking into account the limitations of existing experimental methods, without thereby rejecting the belief that the electron really does have a definite position and momentum. The term "*uncertainty* principle" would then be appropriate, with its implication that the principle applies to the observer's knowledge rather than to nature itself.

There are two objections to that view. The first is the feeling of many physicists that science should be concerned only with concepts that have operational definitions (Section 12.4). The case of the ether (Sections 23.3 and 25.3) had shown that considerable time and effort can be wasted if one insists on constructing theories based on hypothetical entities whose properties cannot be measured experimentally. A second reason for being skeptical about the assumption that an electron really has a definite position and momentum is that no one has yet succeeded in constructing a satisfactory theory based on this assumption, despite numerous attempts.

Heisenberg did not suggest that the terms "position" and "momentum" be banished from physics, for they are needed to describe the results of experiments. Even the term "path" or "orbit" of an electron is still admissible as long as we take the attitude that "the path comes into existence only when we observe it"[2] (or, more cautiously, the path becomes for us a reality only as we observe it).

Heisenberg's remark appears to some philosophers to align quantum mechanics with the traditional philosophy of subjective idealism, according to which the real world consists only of the perceptions of an observer and physical objects have no objective existence or properties apart from such human observations. Taken together with Born's assertion that atomic quantities are inherently random, Heisenberg's viewpoint would seem to deny the possibility of finding out anything certain about a physical system taken by itself, since the observer or the measuring apparatus is an essential part of any physical system we may choose to study.

A somewhat similar situation is quite familiar in the behavioral sciences: In almost any psychological experiment, the subject knows that his or her behavior is being observed, and might not behave in the same way if no psychologist were present. But at least in that case one feels confident that there *is* such a thing as "behavior when the psychologist is not present," even if we do not observe it, and we probably believe that if psychologists were sufficiently clever they could carry out their experiments without the knowledge of the subject, if their scientific and moral code of ethics did not forbid it.

A better analogy might be the conventional intelligence test, which purports to measure an inherent property of the individual, but which is now recognized to be so strongly dependent on the circumstances of testing and on the group of other test-takers to whom the individual is being compared that it is doubtful whether "intelligence" really has an objective existence—except in the trivial sense that, as the psychological skeptics say, "intelligence is what the intelligence test tests."

Quantum mechanics is rescued from the quicksand of subjectivism by the fact that it makes perfectly definite predictions about the properties of physical systems, and these properties can be measured with the same results (including the measurement uncertainty) by all observers. If one prefers, one does not *have* to abandon the idea that there is a real world "out there;" and even the fact that the measuring apparatus is, strictly speaking, inseparable from the things we are measuring does not in practice make very much difference.

Moreover, in spite of the fact that our knowledge of the position and momentum of electrons, photons, and all other particles is restricted to probabilities, the wave function, which determines these probabilities, is itself completely determined by the theory, and changes with time in a very definite, "causal" way: If ψ is given at a particular time and the Hamiltonian operator of the entire system is known [Eq. (29.2)], then ψ can be computed at any other time from Eq. (29.4). (The "entire system" includes of course any measuring device that may interact with the system.)

Thus as Heisenberg himself pointed out, "probability" itself acquires a certain "intermediate reality," not unlike that of the *potentia* (possibility or tendency for an event to take place) in Aristotle's philosophy. An isolated system that is evolving in this way but not being observed is, according to Heisenberg, "potential but not actual in character"; its states are "objective, but not real" (in Pauli *et al.*, editors, *Niels Bohr*). Needless to say, Heisenberg's words do not represent the consensus among physicists.

Although quantum mechanics has stimulated much discussion by philosophers, it would be inaccurate to claim that the adoption of this theory in physics requires the acceptance of either of the traditional doctrines in philosophy, idealism or realism. The significance of quantum mechanics is rather that it forces us to take an attitude toward the descriptive categories when doing atomic or subatomic physics that is different from the attitude we employ in talking

[2]W. Heisenberg, *Zeitschrift für Physik*, Vol. 43, page 185 (1927), quotation translated by Jammer, *Conceptual Development.* According to this viewpoint, Schrödinger's wave function is also a somewhat dubious concept; it cannot be given an operational definition, although it may be quite useful in doing calculations with the theory.

about the physical world in other cases. (This attitude can itself be considered a philosophical doctrine, often called "instrumentalism.")

One expression of the new attitude is the *complementarity principle,* proposed by Niels Bohr in 1927. According to this principle, two modes of description of a given system may appear to be mutually exclusive, yet both are necessary for complete understanding of the system. Thus, on the one hand, we may wish to emphasize *causality* by pointing to the fact that there is a well-determined evolution of the state of the system (defined by its wave function), as mentioned in the last paragraph. Yet this description is meaningful only if we refrain from making any observations of space and time variables, since the process of making such observations would disturb the state in an unpredictable way and destroy the causality. On the other hand, we might prefer to emphasize the spatiotemporal description and sacrifice the causality. Bohr held that each description gives a partial glimpse of the total truth about the system taken as a whole.

Bohr accepted the indeterminacy principle as a useful quantitative limitation on the extent to which complementary (and therefore partial) descriptions may overlap—that is, an estimate of the price that must be paid for trying to use both descriptions at once. But he did not follow Born and Heisenberg in their insistence that the "wave nature of matter" meant nothing more than the inherently random behavior of entities that are really particles. Instead, Bohr argued that the wave and particle theories are examples of complementary modes of description, each valid by itself, although (in terms of Newtonian physics) incompatible with each other.

The complementarity principle is a major component of what is now called the *Copenhagen interpretation* of quantum mechanics. This interpretation, named for the city where Bohr did most of his work, assumes that the physical world has just those properties that are revealed by experiments, including the wave and particle aspects, and that the theory can only deal with the results of observations, not with a hypothetical "underlying reality" that may or may not lie beneath the appearances. Any attempt to go further than this, to specify more precisely the microscopic details of the structure and evolution of an atomic system, will inevitably encounter nothing but randomness and indeterminacy.

Einstein rejected Bohr's claim that quantum mechanics tells us all we can expect to know about nature. He tried to show that even though the theory is correct and useful as far as it goes, it is incomplete and fails to account for all aspects of the physical world. His objection was the inspiration for a paper published in 1935 under the title "Can Quantum Mechanical Description of Physical Reality Be Considered Complete?" by Einstein, Boris Podolsky, and Nathan Rosen. It presents the famous "EPR" objection or paradox, discussed extensively by philosophers of science and a few physicists during the past several decades.

Einstein, Podolsky, and Rosen began by defining reality—perhaps the first time this had ever been done in a scientific journal:

> If, without in any way disturbing a system, we can predict with certainty . . . the value of a physical quantity, then there exists an element of physical reality corresponding to this physical quantity.

They discussed a thought experiment with two particles, I and II, that interact for a short time and then remain completely separated so that neither one can possibly influence the other (imagine sending one of them off to a distant galaxy). For example, they might be two photons emitted by an atom. According to quantum mechanics the wave function of the system at any later time must still involve both particles—they are "entangled," as we would now say.

Now suppose we measure a property of photon I, for example its spin. If the decaying atom was initially in a state with no net angular momentum, we know that the two photons emitted must have equal and opposite spins. Moreover, according to quantum mechanics we can choose to measure the spin of one photon in any direction, and it will always be found to be +½ or −½ (clockwise or counterclockwise) with respect to that direction. Hence if photon I has spin +½ we can say that photon II has spin −½ *along the axis we have chosen for the measurement of photon I.*

Thus we have determined an "element of reality" (the value of the spin) for photon II "without in any way disturbing" it (since by hypothesis it is in a far-away galaxy when we make the measurement on photon I). Therefore, Einstein and his colleagues asserted, it must have had that property *before* we made our measurement.

(Relativity theory excludes the possibility that some influence caused by our measurement of I could have been transmitted to II by a signal going faster than the speed of light.) Yet Heisenberg's principle asserts that the spin of photon II does not have a real value until we have measured it; hence quantum mechanics, which is limited by his principle, gives an incomplete description of reality.

Schrödinger then entered the debate with his famous and rather macabre "cat paradox." In this thought experiment, a cat is placed in a closed box with a radioactive sample and a Geiger counter connected to an electrical device that will kill the cat if any radiation is detected. Based on the known half-life of the sample, one can set up the experiment so there is a 50% probability that such radiation will be detected in 1 hour. If quantum mechanics governs the world, then after 1 hour the wave function of the cat must be a mixture of two functions, ψ_A corresponding to the cat being alive and ψ_D to it being dead. According to the Copenhagen interpretation the cat is neither alive nor dead but in some intermediate state—until we open the chamber and look at it. At the instant when we *observe* the cat, its wave function "collapses" to either ψ_A or ψ_D and the cat *then* becomes either alive or dead.

Schrödinger's cat paradox makes, rather more vividly, the point Einstein had been pressing against Bohr: However accurate quantum mechanics may be in predicting the results of experiments, it fails to give an acceptable description of reality—that is, it fails to portray the world that is (we assume) independent of our observation of it.

Modern technology has made it possible to perform the EPR experiment. During the 1970s and 1980s it was shown by several physicists—most conclusively by Alain Aspect and his colleagues in Paris—that the measurement of a property of one particle *does* influence that of another particle that is entangled with it by a previous interaction. In the Aspect experiment it is impossible for any signal, even one traveling at the speed of light, to inform photon II that we have determined the spin of photon I to be +½ in a particular direction; yet the measurement of the spin of photon II reveals just the correlation with the spin of I that is required by quantum mechanics. Therefore one must conclude either that the spin of photon II did not exist until it was measured or that the universe simply cannot be divided into local regions, each of which is unaffected by what happens in other regions no matter how distant.

Thus the experiment that Einstein proposed in order to cast doubt on the validity of quantum mechanics turned out, when actually performed, to provide one of the most startling confirmations of that theory. The outcome is somewhat reminiscent of the hypothetical "bright spot" proposed by Poisson; he expected that the experimental proof of its nonexistence would refute Fresnel's wave theory of light, which predicted it—but instead it turned out to exist after all (Section 23.3).

The Copenhagen interpretation met with strong objections from a minority of leading physicists, including both de Broglie and Einstein. Others carry on the objection to this day. There is understandably a deep-seated resistance to abandoning the traditional conception of physical reality, and much appeal in Einstein's famous critique of inherent randomness: "God does not play dice" (letter to Max Born, December 4, 1926). But so far no satisfactory alternative interpretation of quantum mechanics (nor any experimental refutation of the theory itself) has been found.

29.4 Systems of Identical Particles

You may have heard the statement "every time you take a breath of air, you swallow one of the molecules that came from Julius Caesar's last breath." That seems rather remarkable, and is intended to impress you with the enormous number of molecules contained in a fairly small volume of gas (first estimated by Josef Loschmidt from the kinetic theory of gases).

But according to quantum mechanics, such a statement is not just wrong, it is meaningless: It assumes that you can put a label on a molecule and say "*this* particular molecule, which came out of Caesar's mouth a couple of millennia ago, is *the same* molecule that is now entering your mouth." On the contrary, all molecules of oxygen (for example) that contain the same isotopes are *identical* and *indistinguishable*. In particular, if two molecules, one coming from the north and one from the east, collide and scatter, one flying off to the west and the other to the south, it is meaningless to say that the molecule that goes toward the west after the collision is (or is not) the same one that came from the north before the collision.

The same principle applies to any set of identical particles, whether or not they have collided. For example, in describing the construction of an atom (Section 28.6), one sometimes carelessly says that an atom like lithium has two electrons in a closed shell and a third electron in the next shell that is "the valence electron." But in reality there is no particular electron that can be called "the valence electron" since all three are identical and indistinguishable. Moreover, when contemplating two lithium atoms, one cannot even say that three particular electrons belong to one atom and the others belong to the other atom.[3]

The principle of indistinguishability of identical particles in quantum mechanics has important consequences: It allows the theory to explain and predict in a very simple way phenomena that cannot be understood in Newtonian mechanics except by means of arbitrary extra hypotheses. For example, in 1927 the German physicists Walter Heitler (b. 1904) and Fritz London (1900–1954) calculated the force between two hydrogen atoms using quantum mechanics. As Heisenberg pointed out, one must use a single wave function for the system of two electrons rather than assuming that each electron has its own wave function. (Strictly speaking one has to have a single wave function that includes the two protons as well as the two electrons, but because of the much greater mass of the protons they can be treated for this purpose as fixed charges not influenced by quantum effects.)

One might try to approximate the wave function of the two electrons, which depends on their positions $(x_1, y_1, z_1, x_2, y_2, z_2)$ as a product of the wave function describing electron #1 bound to proton A, and the wave function describing electron #2 bound to proton B,

$$\psi(x_1 \ldots x_2) = [\psi_A(x_1 \ldots)\, \psi_B(x_2 \ldots)]. \qquad (29.6)$$

[3]According to some recent interpretations of quantum mechanics, it is possible under certain conditions to distunguish separate identical particles by "decoherence" (elimination of the entanglement produced by the Schrödinger equation). so that in some sense one might be able to justify the statement at the beginning of this section. This view, still controversial, does not affect the validity of the statements about the symmetry of wave functions, which lead to experimentally-confirmed predictions such as the Bose-Einstein condensation (page 460).

To take account of the indistinguishability of the identical electrons #1 and #2, one can simply add another term in which the labels are interchanged, giving equal weight to a state in which electron 2 is bound to proton A while electron 1 is bound to proton B. Because of the mathematics of complex numbers, the observable properties of the system (or the probability, in Born's interpretation) will be the same if one *subtracts* the extra term rather than adds it, so the approximate wave function will look something like this:

$$\psi(x_1 \ldots x_2) = [\psi_A(x_1 \ldots)\psi_B(x_2 \ldots) \pm \psi_A(x_2 \ldots)\psi_B(x_1 \ldots)]. \qquad (29.7)$$

Equation (29.7) describes two possible wave functions, called *symmetric* and *antisymmetric* depending on whether one takes the + or the − sign for the second term.

Heitler and London calculated the energy of the two hydrogen atoms using both symmetric and antisymmetric wave functions, taking a series of fixed values for the distance R between A and B. They found that the symmetric function gives a *minimum energy* at a particular distance $R = 0.08$ nm. This should correspond to stable equilibrium; if the atoms get any closer or any farther apart their energy will be higher, and a force will push them back toward this separation distance . For the antisymmetric function there is no minimum at any distance, and for the original "unsymmetrized" function [Eq. (29.7)] there is a very shallow minimum at a somewhat greater distance.

As it happens, the experimental value for the distance between the two atoms in a hydrogen molecule is 0.074 nm, so the new theory gave a rough approximation of the length of the bond between hydrogen atoms. More importantly, it showed that one could explain, at least qualitatively, *why atoms form molecules*. Previously one could explain this phenomenon only by postulating an attractive force between atoms, but this could not account for the peculiar fact that these forces seem to be "saturated": Once a hydrogen atom has bonded to one other atom, it will not bond with any more at the same time (Section 20.11). The quantum mechanical calculations, when applied to systems of three or more hydrogen atoms, did show that only two atoms can form a stable bond.

The quantum effect that explains chemical bonding is called *resonance* or *exchange*. The term *exchange* is more appropriate for the simple case of two hydrogen atoms studied by Heitler and London, since it can be visualized as a simple interchange of the two electrons between the two atoms. In more complicated cases like the benzene molecule, the wave function for the entire molecule can be approximated by combining two or more wave functions corresponding to particular arrangements of the electrons in the molecule that have equal energy; like a pair of resonant coupled oscillators in Newtonian mechanics, one imagines that energy shifts back and forth between these arrangements or "resonant structures."

As can be seen, quantum mechanics despite its strange features offers plenty of scope for the imagination of scientists, and indeed many of the "quantum effects" proposed in this way turn out to be misleading or nonexistent. The often-mentioned "exchange forces" and "resonance effects" are really only mathematical constructs, convenient for solving problems but entirely dependent on the particular approximation used to represent wave functions. For example, benzene and similar molecules can be described at least as well, and often better, by using "molecular orbitals" (wave functions for single electrons belonging to the entire benzene ring), rather than by the theory based on "resonance" (with pairs of electrons shared by adjacent atoms).

Bose-Einstein Statistics

But we have also seen a new *symmetry principle* enter physics in our discussion of symmetric and antisymmetric wave functions. The first hint of this came in 1924 when the Indian physicist Satyendranath Bose (1894–1974) sent Einstein a short paper (in English) on the derivation of Planck's blackbody radiation law, asking him to arrange for its publication in the leading German journal *Zeitschrift für Physik*. Einstein considered the paper so significant that he translated it into German himself and had it published with a note commending the author's important contribution.

In his derivation Bose assumed, in effect, that the quanta of radiation are identical and indistinguishable, so that when one computes the number of quantum states, one should not count states differing only in the interchange of quanta of the same frequency. This assumption allowed Bose to rederive Planck's formula without making use of Newtonian mechanics or Maxwellian electrodynamics (both of which seemed at that point to have dubious validity in the atomic domain).

Fig. 29.7. Satyendranath Bose (1894–1974), Indian physicist, who proposed a new derivation of Planck's radiation law, which was generalized by Einstein and became the basis for Bose-Einstein statistics.

Einstein extended Bose's method to a system of particles having (unlike photons) finite mass. He found that in this case there is a peculiar *condensation* phenomenon: If the number of particles in the system is increased beyond a certain critical value (which depends on the volume, temperature, Planck's constant, and Boltzmann's constant), some particles will fall into the lowest quantum state with zero kinetic energy. Yet the system includes no attractive forces that could hold the particles together in any particular region of space.

It was soon pointed out by the Irish-born mathematical physicist P. A. M. Dirac that the idealized system studied by Bose and Einstein corresponds to the choice of a symmetrical wave function: ψ remains unchanged when the labels of any two particles are interchanged. This became known as *Bose-Einstein statistics* because it prescribed a specific rule for doing statistical calculations on systems of identical particles.

Fermi-Dirac Statistics

Around the same time (in February 1926) the Italian physicist Enrico Fermi (1901–1954) proposed a quantum theory of an idealized quantum gas of identical particles, such as electrons, obeying the *Pauli exclusion principle* (Section 28.6). Like a giant atom but with no nucleus, this gas would have a huge number of quantum states, but only one electron could occupy each state.

Fig. 29.8. Enrico Fermi (1901–1954), Italian physicist, who formulated the quantum theory of systems of identical particles that satisfy the Pauli exclusion principle (such as electrons). This theory is known as Fermi-Dirac statistics (since Paul Dirac proposed a similar theory at about the same time). Fermi received the 1938 Nobel Prize in Physics for his experimental work on neutron bombardment of nuclei; he then moved to the United States, where he directed the project that achieved the first nuclear chain reaction, thus suggesting the possibility of a weapon based on nuclear fission (Section 32.3).

Dirac independently proposed the same theory a few months later (in August 1926), and noted that it corresponds to the antisymmetric wave function. In fact the Pauli principle is an automatic consequence of the antisymmetric property, now called *Fermi-Dirac statistics* in this case: If ψ_A and ψ_B in Eq. (29.7) are the *same function,* as they must be for similar states in the same atom or the same gas, then the second term just cancels the first and the overall wave function is zero.

It was quickly recognized that these two kinds of statistics must apply to different particles, which therefore became known as *bosons* and *fermions,* respectively. Electrons, protons, and neutrons are fermions; photons and certain other particles are bosons.

Fermi-Dirac statistics became the basis for a new quantum theory of electrons in metals, after Wolfgang Pauli showed that it could be used to resolve two of the discrepancies in the earlier theory of P. Drude and H. A. Lorentz. That theory treated the electrons as particles free to move throughout the entire metal, yet the electrons did not seem to increase their average speed when the metal got hotter as one would expect from the kinetic theory of gases—they made almost no contribution to the specific heat. Moreover, the theory could not explain the magnetic properties of metals, even if the electrons were assumed to behave like magnets (as might be expected from their spin). The Dutch physicist Johanna van Leeuwen had proved in 1919 that such a system could not respond to an external magnetic field: The electronic magnet did not tend to align either in the direction of the field (paramagnetism) or in the opposite direction (diamagnetism).

Pauli showed that if the electrons behaved like a Fermi-Dirac system, then at low or room temperature they would fill the lowest possible energy states up to a certain energy, now called the "Fermi energy"; very few electrons would be excited to higher states. This means that small increases in temperature have no effect at all on the speeds of most of the electrons; they cannot increase their speeds unless they have enough energy to jump to states above the Fermi energy. Only the electrons near or above the Fermi energy can absorb more energy and contribute to the specific heat. This peculiar property is known as *quantum degeneracy.* Moreover, an electron cannot change its spin direction in response to a magnetic field if the state with the same energy but opposite spin is already occupied. Only electrons that can go into vacant states with opposite spin can respond to the field. Pauli showed that the Fermi-Dirac theory gives the correct values for paramagnetic and diamagnetic properties of metals and that van Leeuwen's theorem does not apply to quantum systems.

Other physicists—especially Arnold Sommerfeld and Felix Bloch—followed Pauli's lead and quickly developed a new quantum theory of metals. Bloch used the Schrödinger equation to show that the electrons behave like a gas of free particles even within a regular lattice of atomic ions; the electrical conductivity of electrons in a *perfect* lattice would be infinite, and resistance arises only a result of irregularities or motions of the ions.

Another peculiar feature of this system, very convenient for the theorist, is that the strong electrical forces between electrons in a real metal have very little effect on the electrons stuck in states below the Fermi energy; interactions that would normally scatter them into different states cannot do so because there are no available empty states for them to go into. The properties of the system at low temperatures are therefore somewhat similar to those of an ideal Fermi-Dirac gas with no interactions at all.

Bose-Einstein Condensation

While physicists quickly found many systems—metals, stars, atomic nuclei—to which Fermi-Dirac statistics could be applied, it was not clear for several years how Bose-Einstein statistics would lead to any significant progress. In particular, the Bose-Einstein condensation seemed to have no counterpart in the real world until 1938, when Fritz London proposed that it might describe the so-called *λ-transition of liquid helium.*

The Dutch physicist Heike Kamerlingh Onnes (who also discovered superconductivity) was the first to liquefy helium in 1908, and in 1911 he noticed that as the liquid is cooled its density reaches a maximum at about 2.2 K. His colleague and successor as director of the low-temperature laboratory at Leiden, Willem Hendrik Keesom, found that this temperature corresponds to a peculiar kind of transition. The Greek letter lambda was used to name this transition because of the shape of the measured curve of specific heat as a function of temperature. The Russian physicist P. L. Kapitza found that below the λ-transition helium becomes a "superfluid"—for example, it flows through a narrow space with no resistance.

London pointed out that the nucleus of helium, isotope 4, has an even number of protons and neutrons, and may therefore obey Bose-Einstein statistics, at least approximately. On substituting the mass of the helium nucleus into Einstein's formula for the condensation temperature, he found that the transition should occur at 3.09 K, less than 1 degree above the measured temperature for liquid helium. Although superfluid helium seems to display some of the properties that would be expected for a Bose-Einstein system, it is not a very close match because of the strong forces between the helium atoms. Physicists continued to search for a better example of the predicted Bose-Einstein condensation.

In 1995 Einstein's prediction was fully confirmed by scientists at Boulder, Colorado, in a laboratory run by the National Institute of Standards and Technology and the University of Colorado. A group led by Eric A. Cornell and Carl E. Wieman was able to cool a cloud of about 2000 rubidium atoms to 20 billionths of a degree above absolute zero. The atoms were far enough apart (10^{-4} cm) that their mutual interactions were negligible, thus satisfying the criteria for an "ideal gas." Direct measurements of the distribution of atomic velocities showed the sudden appearance of a peak at zero, corresponding to the condensation of a substantial fraction of the atoms into a single ground state.

Fig. 29.9. Fritz London (1900–1954), German physicist, who was one of the first to show how quantum mechanics could explain the chemical bonds between atoms. He proposed that the superfluidity of liquid helium could be approximately described as a Bose-Einstein condensation.

According to Daniel Kleppner, a physicist at the Massachusetts Institute of Technology who was also experimenting to demonstrate the Bose-Einstein condensation, the significance of the Cornell-Wieman discovery is not that it was needed to confirm Einstein's and Bose's speculations. "There was no doubt the theory was correct," he told a reporter in July 1995. Rather, the discovery is important because it opens up new areas of physics. For example, by early 2000 the young Danish-born physicist Lene Vestergaard Hau of Harvard University demonstrated in her laboratory that in a Bose-Einstein condensate the speed of light was reduced to that of a slowly moving car.

Problem 29.5. The title of this book is *Physics, the Human Adventure: From Copernicus to Einstein and Beyond.* On the basis of your study so far, write a brief essay on the way physics has been evolving over the centuries through the cumulative contributions of human beings of the greatest variety (from many continents, from the most celebrated persons to the most obscure, etc.). Consider also in what sense it has been an adventure (errors leading to new theories; hunches, careful reasoning, sheer labor; good luck or unjust neglect, etc.).

RECOMMENDED FOR FURTHER READING

David Bohm, *Causality and Chance in Modern Physics,* Princeton, NJ: Van Nostrand, 1957. A lucid account of the development of mechanism in physical theory, ending with an argument that quantum mechanics should not be considered a complete theory with ultimate validity.

Niels Bohr, "Discussion with Einstein on epistemological problems in atomic physics," pages 199–241 in *Albert Einstein, Philosopher-Scientist* (edited by P. A. Schilpp), pages 199–241; see also Einstein's comments, pages 665–683 of the same book.

H. A. Boorse and L. Motz, *The World of the Atom,* extracts from the writings of Hamilton, Compton, de Broglie, Davisson and Germer, Schrödinger, Born, Heisenberg, Bose, and biographical sketches, Chapters 56, 61, 63–66, 68, 71

William H. Brock, *Norton History of Chemistry,* Chapter 13, "Nature of the Chemical Bond"

Richard Feynman, *The Character of Physical Law,* Chapters 6 and 7

C. C. Gillispie, *Dictionary of Scientific Biography,* articles by K. Koizumi on C. J. Davisson, Vol. 3, pages 597–598; by J. Hendry on G. P. Thomson, Vol. 18, pages 908–912; by A. Hermann on Schrödinger, Vol. 12, pages 217–223; by D. C. Cassidy on Heisenberg, Vol. 17, pages 394–403; by C. W. F. Everitt and W. M. Fairbank on London, Vol. 8, pages 473–479; by O. Darrigol on Dirac, Vol. 17, pages 224–233; by E. Segrè on Fermi, Vol. 4, pages 576–583

Werner Heisenberg, "The development of the interpretation of the quantum theory," pages 17–29 in *Niels Bohr and the Development of Physics* (edited by W. Pauli *et al.*), London: Pergamon Press, and New York: McGraw-Hill, 1965; *Physics and Philosophy,* New York: Harper and Row, 1962

Gerald Holton, "The roots of complementarity" and "On the hesitant rise of quantum physics in the United States," in *Thematic Origins,* pages 99–187

Daniel Kleppner, "The fuss about Bose-Einstein condensation," *Physics Today,* Vol. 49, no. 8, pages 11, 13 (1996). Explains why the Cornell-Wieman discovery is important.

David L. Mermin, "Is the moon there when nobody looks? Reality and the quantum theory," *Physics Today,* Vol. 38, no. 4, pp. 38–47 (1985).

Walter Moore, *A Life of Erwin Schrödinger,* New York: Cambridge University Press, 1994.

Mary Jo Nye, *Before Big Science,* Chapter 6

Melba Newell Phillips (editor), *Physics History,* articles by Compton on the Compton effect, D. M. Dennison on physics in the 1920s, Heisenberg on the history of quantum mechanics, and G. P. Thomson on electron diffraction, pages 105–108, 119–136

Michael Redhead, "Quantum theory," in *Companion* (edited by R. Olby), pages 458–478

Arthur L. Robinson, "Loophole closed in quantum mechanics test," *Science,* Vol. 219, pages 40–41 (1983), on the experiment by Aspect *et al.*

S. Sambursky, *Physical Thought,* extracts from writings of de Broglie, Heisenberg, Bohr, and Pauli, pages 510–540

Erwin Schrödinger, *Science, Theory, and Man,* New York: Dover, 1957

Abner Shimony, "The reality of the quantum world," *Scientific American* (Jan. 1988), pp. 46–53.

Stephen Toulmin (editor), *Physical Reality,* New York: Harper and Row, 1970. Essays by Planck, Mach, Einstein, Bohr, and others on the philosophical interpretation of physical theories.

Andrew Watson, "Quantum spookiness wins; Einstein loses in photon test," *Science,* Vol. 277, page 481 (1997). Report of the experiment in Switzerland by Nicolas Gisin's group on entangled particles.

S. R. Weart and M. Phillips (editors), *History of Physics,* articles by F. Bloch on "Heisenberg and the early days of quantum mechanics" and by R. K. Gehrenbeck on "Electron diffraction: Fifty years ago," pages 319–331

J. H. Weaver, *World of Physics,* excerpts from papers by Bohr, Schrödinger, Heisenberg, Born, Pauli, Dirac, and Pagels; Bohr's discussion with Einstein; Vol. II, pages 315–424, 470–483, and Vol. III, pages 801–834

Linda Wessels, "Schrödinger's route to quantum mechanics," in *History of Physics* (edited by S. G. Brush), pages 154–183

Bruce R. Wheaton, "Louis de Broglie and the origins of wave mechanics," in *Physics History II* (edited by French and Greenslade), pages 61–65

Arthur Zajonc, *Catching the Light,* Chapter 11

CHAPTER 30

Einstein's Theory of Relativity

30.1 Biographical Sketch of Albert Einstein

Turning to a brief account of the theory of relativity, let us begin by sketching the career of its chief developer. Albert Einstein was born on March 14, 1879, in Ulm, a middle-sized city in the Swabian part of Bavaria in southwestern Germany. A year later his family moved to Munich, the political and intellectual center of southern Germany. His father, Hermann Einstein, operated (not too successfully) a small electrochemical factory. His mother, born Pauline Koch, had artistic interests and instilled in her son a love for music. An uncle who lived nearby and ran the technical side of the business aroused Albert's first interest in mathematics by teaching him some algebra.

Einstein was not a child prodigy; on the contrary, he was late in learning to talk, and did not show much proficiency at learning his lessons in school. He disliked any kind of regimentation or pressure toward conformity, whether from his teachers or his classmates.

His family was Jewish, but "entirely irreligious," as Einstein later described it; young Einstein did not receive much religious instruction at home, and his parents sent him to a Catholic elementary school, in part because it was closer and more convenient. He became early, on his own, deeply interested in the Jewish religion. But, according to his "Autobiographical Notes," at age 12,

> . . . through the reading of popular scientific books I soon reached the conviction that much in the stories of the Bible could not be true. The consequence was a positively fanatic [orgy of] freethinking coupled with the impression that youth is intentionally being deceived by the state through lies; it was a crushing impression. Suspicion against every kind of authority grew out of this experience, a skeptical attitude towards the convictions which were alive in any specific social environment—an attitude which has never again left me, even though later on, because of a better insight into the causal connections, it lost some of its original poignancy.

When Albert was 15 his father's business failed. Hermann Einstein decided to leave Munich and try again in Italy. Albert was left behind to finish his secondary school education in Munich, while living in a boarding house. He was unhappy there, and soon left to join his parents. His exit apparently did not sadden the Munich school—in fact, one of his teachers had suggested he leave on the grounds that his mere presence in the class "destroyed the respect of the students." Apparently Einstein's obvious, honestly expressed distaste for the mechanical drill of lessons and for military discipline, combined with the fact that he was nevertheless exceedingly proficient in mathematics and physics, was too much for the teacher to take.

In Italy, his father's business did not go too well; it was now clear that young Einstein would soon have to support himself. He decided to study electrical engineering at the Swiss Federal Polytechnic University in Zurich. Though only 16, and two years younger than most of the candidates, he presented himself for the entrance examination—and failed (in biology and French). To prepare for another try, he went to a Swiss high school in Aarau, and for the first time found himself in a congenial, democratically run school. There he seems to have flourished, and he recalled later that it was at Aarau that he had his first ideas leading to the relativity theory. By the time he graduated from the Polytechnic, his interests had settled on theoretical physics (with the intention of becoming a high school teacher), but he had to teach himself such advanced subjects as Maxwell's electromagnetic theory, which had not yet become part of the standard curriculum.

After graduation, Einstein (still not a Swiss citizen) was unable to find a satisfactory teaching position in Switzerland. Finally the father of

Fig. 30.1. Albert Einstein at age 19. Although he received the Nobel prize for his work in quantum theory (Section 26.5), he is best known as the founder of the theory of relativity. His equation for mass-energy transformation, $E = mc^2$, is one of the most important in modern physics, since it governs nuclear fission and fusion (Chapter 32) as well the creation and destruction of particle-antiparticle pairs (Section 30.9). His general theory of relativity governs gravitational phenomena such as black holes, "Einstein rings" (see the cover of this book), and the expansion of the universe.

one of his friends and former classmates helped to get him a job at the patent office in Bern. Soon after taking this position, Einstein married Mileva Marić, a fellow student in Zurich who had come from Hungary. She shared Einstein's interest in physics. Two sons were born of the marriage, in addition to a daughter born earlier.

Although it might seem an unusual occupation for a theoretical physicist, the position as patent examiner did have certain advantages for Einstein at this stage in his career. The work was not dull; he was continually having to analyze and criticize ideas for new inventions. Moreover, unlike most teaching positions available to young physicists, the patent office job gave Einstein enough free time to do his own reading, thinking, and research. The complete absence of contact with professional physicists during this period was perhaps a blessing in disguise, for it permitted Einstein to develop his rather unorthodox approach to the problems of physics.

In 1905, Einstein published three papers, each of which was eventually recognized as the work of genius. The first, applying the kinetic-molecular theory to Brownian movement, was mentioned in Section 22.7. The second, on the photon hypothesis and the photoelectric effect, was discussed in Section 26.5. The third and most astonishing, on what was later called the special relativity theory, will be treated in the following sections. In a separate brief note, published shortly afterward, Einstein pointed out that as a consequence of his theory of relativity, the mass m of an object should be related to its energy content E through the formula $E = mc^2$ (see Section 30.8).

By 1909, these and other publications had brought him so much attention in the scientific world that he was appointed Professor Extraordinary (equivalent to Associate Professor) at the University of Zurich, and a year later he was called to the Chair of Theoretical Physics at the German University in Prague, Czechoslovakia. In 1912 he returned to Zurich as Professor of Theoretical Physics at the Polytechnic from which he had graduated, and began to develop his general theory of relativity (Section 30.10). The next year, he was offered what was at the time the highest position to which a European scientist could aspire: professor at the University of Berlin, research director at the newly founded Kaiser Wilhelm Institute there, and member of the Prussian Academy of Science.

The decision to go to Berlin was not an easy one for Einstein. As a pacifist and social democrat, he did not like the political and social atmosphere in Berlin. But in other respects the position was irresistible: Among Einstein's colleagues during his stay in Berlin (1913–1933) were many of Europe's leading physicists—Max Planck, Walther Nernst, Max von Laue, James Franck, Gustav Hertz, Lise Meitner, and Erwin Schrödinger.

The move to Berlin was also accompanied by a breakup of the marriage with Mileva, who stayed in Switzerland with their children. Einstein married his cousin Elsa Einstein, while assigning to Mileva, as part of the divorce settlement, the proceeds from the Nobel prize that he expected to receive.

One of Einstein's most spectacular triumphs came just after the end of World War I. He predicted, on the basis of his general theory of relativity, that light from a star should be deflected by the gravitational force of the sun, and calculated that the deflection should be a maximum of 1.75 sec of arc for a beam of light passing very close to the sun. This prediction could be tested accurately only during a solar eclipse, and the observations would have to be made from certain special places on the earth's surface where there would be a *total* eclipse.

Einstein's prediction became known in England before the end of World War I, and British astronomers decided to make an expedition to test it during the eclipse expected on March 29, 1919. It would be necessary to travel to northern Brazil and to an island off the coast of West Africa for optimum viewing conditions, and such expeditions would have been rather dangerous while England and Germany were at war. But after the war ended on November 11, 1918, the British had time to plan and carry out the project. The observed result coincided with the expected value within the estimated experimental error. Since there was no other known theory that predicted this result, it could be quickly celebrated as a victory for Einstein's theory.

One can readily imagine the impact on world opinion of a British-sponsored expedition confirming a "German" theory, just a few months after the British and their allies had defeated the Germans in a lengthy and costly war. After the bitterness of combat, the acclaim of this work was an example of international reconciliation through scientific cooperation. And of course the rapturous press reports about the scientific ideas being verified—the "weight of light" and the "curvature of space"—helped spread the idea that Einstein had wrought a revolution that overthrew the old order and opened a new and mysterious door to the deepest secrets of the universe.

Einstein was now, and for decades to come, the most famous scientist in the world, and inevitably became the target of appeals for support of worthy personal and international causes (as well as for attacks by crackpots). He decided to lend his prestige to the support of three undertakings outside science itself that have had, each in its own way, a major impact on twentieth-century history. The first was support of the Zionist movement, which led eventually to the establishment of the state of Israel; in this he was largely motivated by the tragic persistence of hidden and explicit antisemitism, which he saw all around him. The second was one letter he was persuaded by some of his colleagues to write to President Franklin D. Roosevelt, to suggest "watchfulness and, if necessary, quick action" concerning the possible use of atomic energy as a weapon, for there were already signs the Germans were beginning such research; although this letter was widely publicized in later years, it did not lead directly to the actual development of the atomic bomb (see Section 32.3). The third cause was peace, from his manifesto against German militarism in World War I to his support of the international "Pugwash" movement of scientists, started in the 1950s, on behalf of international arms control.

In the early 1920s, when Einstein began to identify himself publicly as a Jew, he attracted the criticism of the political movement that soon developed into the Nazi Party in Germany. This meant that from these quarters relativity theory itself was violently attacked on political and racial grounds; it was supposed to be an example of "Jewish" or "non-Nordic" physics, or even of "Bolshevik" physics. Some previously respectable German physicists joined in these attacks, in particular the Nobel prize winner Philipp Lenard, whose measurements on the photoelectric effect in 1902 had provided data for Einstein's early formulation of the quantum theory.

In early 1933, when Hitler came to power and the racial purges began in Germany, Einstein was just returning from one of his frequent visits to the United States. When he landed in Belgium, he decided that it would be impossible for him to go on to Berlin so long as the Nazis were in charge. (Indeed, they had already confiscated his property in Berlin.) He therefore accepted an offer to join the newly founded Institute for Advanced Study at Princeton, New Jersey, and this became his home until his death in 1955.

30.2 The FitzGerald-Lorentz Contraction

As will be seen in the next section, Einstein's primary concern in his 1905 paper on relativity was to show that Maxwell's electromagnetic theory could be understood in a way that would be both physically accurate and logically consistent for objects and observers moving at high speeds relative to each other. But this reformu-

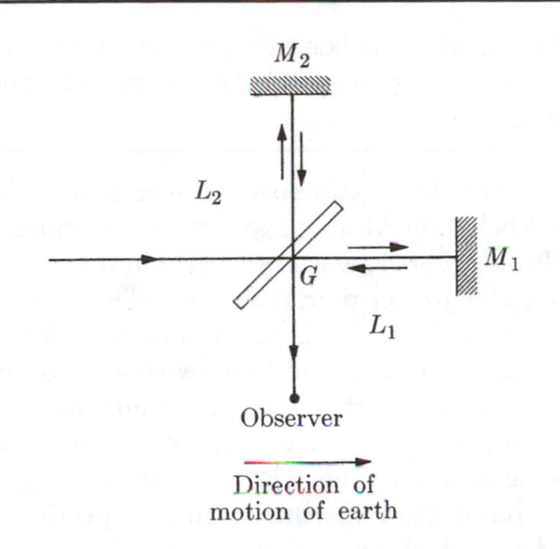

Fig. 30.2. The Michelson-Morley experiment (simplified sketch).

lation required him to make a fundamental critique of the traditional concepts of space and time and subsequently of mass and energy.[1]

At the heart of this critique was the adoption of a set of transformation equations (now known as the Lorentz transformation) by means of which the values of physical quantities for a phenomenon as observed or measured from one frame of reference could be translated into the values for the same phenomenon as observed or measured from another frame of reference that is moving at a constant speed with respect to the first. As it happens, the mathematical form of these equations had been deduced by other physicists before Einstein, but had been given a different physical interpretation by them, one that was more in line with traditional ideas.

Instead of introducing both the equations and their physical interpretation as Einstein presented them in 1905 right away, it will be pedagogically easier, and is more usual, to present first those earlier ideas (on which Einstein himself did not rely).

In Section 25.3 we mentioned Michelson's and Morley's beautiful experiment of 1887, designed to detect the motion of the earth relative to the ether. The "failure" of this experiment gave physicists such as H. A. Lorentz in the 1890s the first clue leading to the new transformation equations. In the experiment, shown schematically in Fig. 30.2, a beam of light is divided by a partly reflecting glass plate G into two beams at right angles, one of which travels along a path of length L_1 and the other along a path of length L_2. Each is reflected back by a mirror (M_1 or M_2) at the end of its path; the two beams are then recombined at G, and are seen by an observer. If the times required to travel the two separate routes are different, then the beams will be "out of phase" when they recombine, and destructive interference of the beam will result. Michelson had designed a very sensitive interferometer, which could detect extremely small phase differences in this way.

Suppose first that path L_1 is parallel to the direction of the earth's velocity in space, v. Michelson and Morley assumed that the speed of light traveling along this path would be compounded of the speed c with respect to the ether and the earth's speed v. As the light goes from G to M_1, M_1 is receding from it, so the trip should take *longer* than if the earth were at rest in the ether; but the return trip would take a shorter time since the plate G is moving forward to meet the light that is coming back from M_1. Thus the time required for the first part (from G to M_1) should be $L_1/(c-v)$, and the time for the second part (from M_1 to G) should be $L/(c+v)$. The total time required for the round trip along the path parallel to the earth's motion would therefore be

$$t_{\parallel} = L_1/(c-v) + L_1/(c+v)$$
$$= \frac{2L_1}{c}\,\frac{1}{(1-v^2/c^2)} \qquad (30.1)$$
$$= \frac{2L_1}{c(1-v^2/c^2)}\ .$$

The calculation of the time needed by the other beam, the one that goes along the path perpendicular to the earth's motion (GM_2G), is a little harder, and Michelson got it wrong the first time he tried it (in 1881). He thought that the time required to travel at right angles to the earth's motion would be unaffected by this motion, and would simply be equal to $2L_2/c$. However, one must see the situation as if one were fixed in the ether (Fig. 30.3): The central plate G has moved a distance equal to $vt_\perp$ during the time $t_\perp$ that it takes the beam of light to travel the total distance d (from G to M_2 and back to G). This distance d is not simply $2L_2$, but is related to it by the Pythagorean theorem (see Fig. 30.3):

[1]In May 1905, Einstein wrote to Conrad Habicht that his paper on the light quantum "is very revolutionary" but described his paper on relativity theory in much milder terms: It uses a "modification" of the concepts of space and time.

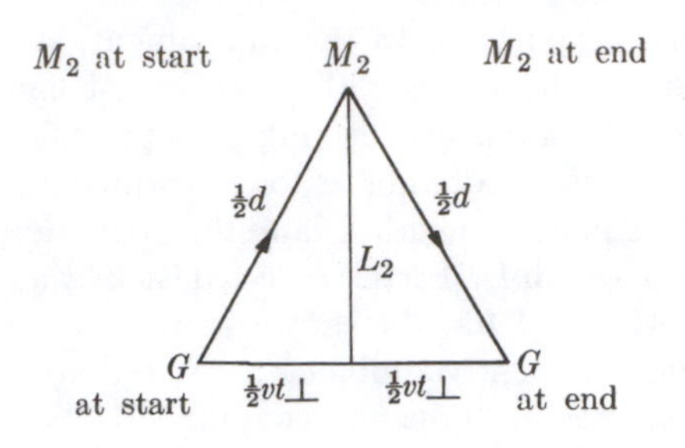

Fig. 30.3. Path of light from G to M_2 to G as seen by someone fixed in the ether.

$$(\tfrac{1}{2}d)^2 = (\tfrac{1}{2}vt_\perp)^2 + L_2^2$$

or

$$d = \sqrt{(v^2 t_\perp{}^2 + 4L_2^2)}.$$

Hence the time $t_\perp$ has to be found by solving the equation

$$t_\perp = d/c.$$

On substituting for d as above and solving, we find

$$t_\perp = \frac{2L_2}{c\sqrt{(1 - v^2/c^2)}} \quad (30.2)$$

Thus the time difference between the parallel and perpendicular paths would be

$$\tfrac{1}{2}\Delta t \equiv t_\parallel - t_\perp = \frac{2L_1}{c\,(1 - v^2/c^2)} - \frac{2L_2}{c\sqrt{(1 - v^2/c^2)}}. \quad (30.3)$$

The actual procedure was modified somewhat to take account of the possibility that the two lengths cannot in practice be made precisely equal, but the final result is the same. We have inserted the factor $\frac{1}{2}$ on the left-hand side of Eq. (30.3) to make the definition of Δt consistent with that in the modified procedure.

Problem 30.1. (a) Use commonly available data to show that the speed of the earth in its orbit is approximately 30 km/sec. (b) In Michelson's original apparatus, the lengths L_1 and L_2 were each about 120 cm. Assuming that at some time of the year the earth is moving at a speed of 30 km/sec relative to the ether, calculate Δt, the difference between time intervals [as defined in Eq. (30.3)] for the parallel and perpendicular paths, that one would expect to observe in this case. (c) If the wavelength of the light used in the experiment is 600 nm, what fraction of a cycle of a light wave would the value of Δt calculated in part (b) correspond to?

The value of Δt expected to be observed in the Michelson-Morley experiment, as estimated in Problem 30.1, was not large, but it was substantially greater than the expected experimental error. Thus, when Michelson and Morley did the experiment and found virtually no time difference at all, they were most surprised and uncomfortable. So were many distinguished physicists at the time, who urged them to repeat and refine the experiment, fully expecting to find a result showing that the ether exists.

One explanation that offered itself for the failure was that one of the basic assumptions employed in the derivation of Eq. (30.3) was incorrect. In 1889 the Irish physicist G. F. FitzGerald suggested that the negative result of the Michelson-Morley experiment might be explained by simply assuming that the arm of the interferometer contracts by a factor $\sqrt{(1 - v^2/c^2)}$ in the direction of motion through the ether at a speed v—somewhat as the length of a ship racing through the ocean is made slightly shorter by the resistance of the water. Thus by this assumption an arm of the interferometer measuring 1 m when perpendicular to v will really be only $\sqrt{(1 - v^2/c^2)}$ m long when it is placed parallel to v. Thus, by multiplying L_1 by this factor in the formula for $t_\parallel$ [Eq. (30.1)] we would obtain a corrected time interval,

$$t_\parallel{}^* = 2L_1\sqrt{(1 - v^2/c^2)}/c(1 - v^2/c^2) \quad (30.4)$$
$$= 2L_1/c\sqrt{(1 - v^2/c^2)}$$

Then

$$t_\parallel{}^* - t_\perp = (2L_1 - 2L_2)/c\sqrt{(1 - v^2/c^2)}$$

If the original lengths L_1 and L_2 were equal, then

$$t_\parallel{}^* - t_\perp = 0 \quad (30.5)$$

This would coincide with observation, and so tends to "explain" it.

FitzGerald tried to justify his contraction assumption (or hypothesis) by pointing out that intermolecular forces are probably electrical in nature. If (as in Maxwell's theory) electrical forces were propagated through the ether, then it would not be unreasonable to suppose that they would be affected by motion through it; so the contraction hypothesis might be made more plausible by suggesting the existence of an increase in intermolecular attractive forces that draw the molecules closer together.

The contraction hypothesis was also proposed a few years later by the Dutch physicist H. A. Lorentz (who probably thought of it independently, since FitzGerald's publication of 1889 was not widely known). Lorentz developed a general theory of electrodynamics, in which he assumed that the lengths of all objects in a system moving with respect to the ether shrink by a factor $\sqrt{(1 - v^2/c^2)}$ (the correction factor still applied only to the component in the direction of motion). Furthermore, Lorentz proposed that the values of *mass* and of *time* intervals in the moving system are *dilated* by a similar factor: They must be multiplied by a factor $1/\sqrt{(1 - v^2/c^2)}$, which means a significant *increase* unless v is very small compared to c. (The time dilation would not affect the interpretation of the Michelson-Morley experiment since, unlike the length contraction, it would be the same along both the perpendicular and the parallel path.)

Although the FitzGerald-Lorentz contraction hypothesis suggested that strange things happen when objects move at speeds close to the speed of light, it did not challenge traditional concepts of space and time. It was still assumed that certain phenomena depend on motion through an ether, and that one can distinguish in some absolute sense between a system that is moving and one that is not. Even the idea of time dilation did not contradict the presupposition that there is an absolute time scale for a system at rest in the ether, and that this time scale is more fundamental than any "local time" appropriate to a moving body and observed on clocks transported along with the moving body.

Problem 30.2. List *all* the assumptions made in deriving Eq. (30.5).

30.3 Einstein's Formulation (1905)

Einstein's first paper on relativity, published in 1905, appears to have emerged from a prolonged period of reflection on some aspects of Maxwell's electromagnetic theory rather than from any special concern on his part with the results of the Michelson-Morley experiment. At the age of 16, as he reported later in his "Autobiographical Notes," he had discovered the seminal paradox:

> If I pursue a beam of light with the velocity c (velocity of light in a vacuum), I should observe such a beam of light as a spatially oscillatory electromagnetic field at rest. However, there seems to be no such thing, whether on the basis of experience or according to Maxwell's equations. From the very beginning it appeared to me intuitively clear that, judged from the standpoint of such an observer, everything would have to happen according to the same laws as for an observer who, relative to the earth, was at rest. For how, otherwise, should the first observer know, that is, be able to determine, that he is in a state of uniform motion?

Thus the possibility of reducing the relative speed of light to zero for some observer seemed to be an absurdity; yet there was nothing in mechanics as then understood to exclude such a possibility.

In the opening paragraph of his 1905 paper, Einstein pointed out an inconsistency in the conventional interpretation of electromagnetic theory:

> It is known that Maxwell's electrodynamics—as usually understood at the present time—when applied to moving bodies, leads to asymmetries which do not appear to be inherent in the phenomena. Take, for example, the reciprocal electrodynamic action of a magnet and a conductor. The observable phenomenon [namely the direction and magnitude of the induced current] here depends only on the relative motion of the conductor and the magnet, whereas the customary view draws a sharp distinction between the two cases in which either the one or the other of these bodies is in motion.

For example, when the magnet is in motion and the conductor at rest, one explains the current in this way: There arises in the neighborhood of the magnet an electric field with a certain definite energy, producing a current at the places where parts of the conductor are situated. But in the second case, if the magnet is stationary and the conductor in motion, one does not say that an electric field arises in the neighborhood of the magnet. Rather, one says that in the conductor there appears an electromotive force; there is no energy corresponding to it, but the electromotive force gives rise—assuming equality of relative motion of conductor and magnet in the two cases discussed—to an electric current of the same path and intensity as that produced by the electric field in the first case. To put it differently: To calculate the *same* current in the two cases, one has to use two very different sets of ideas and sets of equations. Our physics seems to be more complicated than Nature. That is what Einstein meant by theoretical "assumptions that do not appear to be inherent in the phenomena."

Einstein then proceeds to locate the difficulty in the assumption of the existence of an absolute space. Influenced by his reading of the critiques of Newtonian science by the eighteenth-century Scottish philosopher David Hume and the nineteenth-century Austrian physicist Ernst Mach, he decided to reject this time-honored assumption, and to build physics on a new assumption:

> The same laws of electrodynamics and optics will be valid for all frames of reference for which the equations of mechanics hold good.

Those frames of reference are, of course, the "inertial" frames of reference—to put it simply, those systems that are moving at constant velocity with respect to the fixed stars,[2] and within which Newton's laws of motion apply.

We have already met this statement in Galilean relativity for mechanics; now Einstein says it applies to optics, electricity, and magnetism as well—a breathtaking example of the urge to simplify and unify the branches of physics.

Einstein continues briskly:

> We will raise this conjecture (the purport of which will hereafter be called the "Principle of Relativity") to the status of a postulate, and also introduce another postulate, which is only apparently irreconcilable with the former, namely, that light is always propagated in empty space with a definite velocity c which is independent of the state of motion of the emitting body. These two postulates suffice for the attainment of a simple and consistent theory of the electrodynamics of moving bodies based on Maxwell's theory for stationary bodies. The introduction of a "luminiferous ether" will prove to be superfluous inasmuch as the view here to be developed will not require an "absolute stationary space" provided with special properties, nor assign a velocity-vector to a point of the empty space in which electromagnetic processes take place.

The first postulate means that the same law of nature (for example Maxwell's equation) holds in an "inertial" frame of reference that is at rest with respect to a fixed star as in one that moves at, say, nearly the velocity of light. The second simply means that the failure of Michelson and others to find an effect on the velocity of light by motion is a God-given signal that there is a law of nature that makes c the same for all inertial systems, even though our intuition may rebel against it. No wonder that physicists like Michelson were outraged for a long time.

30.4 Galilean Transformation Equations

Einstein's first major goal (in what is now known as the "special theory of relativity," to distinguish it from the later, more general theory of relativity that applies to rotating and other accelerating frames of reference) was to determine the way in which the result of measurements of space and time intervals for the same phenomenon differ when made from two inertial frames of reference that are moving at constant velocity with respect to one another.

Einstein himself used the easily visualizable example of measurements made from a railroad train moving along an embankment and from the fixed station embankment.[3] (Note: The best way to become acquainted with and persuaded of the counterintuitive ideas we are now meeting is to work through the series of simple but mind-expanding problems that follow the discussion.)

---

[2]A terrestrial laboratory is not quite an inertial frame, since the earth is rotating around its axis. But for most purposes we can ignore or easily correct for such discrepancies. And once we have identified a single inertial frame, we can immediately identify an infinite number of others, namely all those frames moving at constant velocity with respect to the first one.

[3]In the following discussion we shall be using the approach of Einstein's book *Relativity: The Special and General Theory*, first published in German in 1917. This exposition, available in English translation, is one of the best for beginning students.

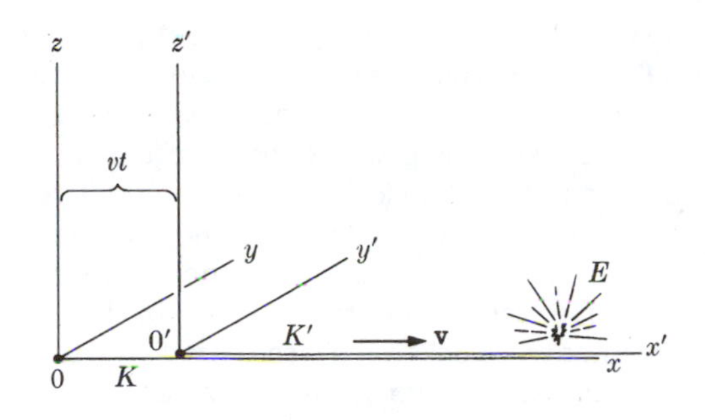

Fig. 30.4. Frame of reference K' is moving at constant velocity v in the x direction relative to K. At time $t = 0$ on the clock of O, when it is also $t' = 0$ on the clock of O', the origin of the two coordinate systems momentarily coincided. When these two origins had moved a distance $x = vt$ apart, an event E happened (for example, an explosion). O says it happened at time t and position x (with $y = 0$, $z = 0$); O' says it happened at time t' and position x' (with $y' = 0$, $z' = 0$). In classical physics, we naturally assume that $x' = x - vt$ and that $t' = t$, in accord with experience.

An observer O standing in the station on the embankment sees an explosion down along the track, and measures the distances to that spot and the time when it happened in his own frame of reference (let us call it K), using the variables x, y, z, and t. Another observer O', riding in the train (moving at a constant velocity v in the x direction relative to the embankment) and observing *the same event,* would measure distances and times with respect to his own frame of reference (let us call it K'), using variables that we may call x', y', z', and t'.

In the physics of Galileo and Newton, it would naturally be assumed that if at an initial time $t = t' = 0$ the origins of the two coordinate systems coincided (when the values of x, y, z and x', y', z' all equal zero for the *origins* of the coordinate systems), then at any later time an event that the observer on the embankment reports to occur at time t' and at position (x, y, z) would be reported to occur at time t' and at position (x', y', z') by the observer on the train (see Fig. 30.4). Moreover,

$$\left.\begin{aligned} x' &= x - vt \\ y' &= y \\ z' &= z \\ t' &= t \end{aligned}\right\} \qquad (30.6)$$

Einstein called this set of equations the "Galilean transformation equations." They are, so to speak, a dictionary by which an observer in K can translate into his own language a report coming to him from the observer in K', and vice versa.

Equation (30.6) is best understood by referring to Fig. 30.4: For any given point in space, the value assigned to its position coordinate by measuring in K' along the x' coordinate is *smaller* than the corresponding value measured in K along the x coordinate—and the difference is simply the distance that the frame K' has traveled relative to K, namely vt. The y and z coordinates are not affected by the relative motion of K and K' (since their motion is along the x and x' axes only). Of course, it is assumed also in all of classical physics that there is a single universal time variable t that is the same for all observers; hence $t = t'$.

Problem 30.3. Prove that it follows from Eq (30.6) that $x = x' + vt'$. This is called the inverse transformation equation, and it is useful for predicting what O will measure when all he has is the data taken by O'.

Problem 30.4. The length of a rigid stick resting in K and pointing along the x axis can be identified by the symbols L or Δx, where $\Delta x \equiv x_2 - x_1$. The symbols x_2 and x_1 refer of course to the coordinates of each end of the stick, measured along the x axis. Use Eq. (30.6) to show that when the length of the same stick is measured from K' (and is symbolized by $\Delta x'$), it is found that $\Delta x' = \Delta x$. This coincides with the intuitively obvious point that lengths in classical physics are not changed by measuring from moving frameworks, or to put it in technical language, "are invariant with respect to transformation." [*Note:* The length of an object at rest in one's own reference frame is called the "proper length." The above examples are not cases of measurement of "proper length"—but in classical physics the values obtained would be the same.)

Problem 30.5. Derive from the transformation equation [Eq.(30.6)] that when the speed of an object is u' as determined in K', the speed u for the same object as determined in K is given by the equation $u' + v$. [For example, a man walking toward the locomotive inside a train (K') is going at 3 mi/hr (u') with respect to the train, and the train itself is going at 80 mi/hr (v) with respect to the embankment. From the embankment the measured speed of the man (u) is 83 mi/hr.] Then use

Eq. (30.6), and show conversely that the equation $u' = u - v$ gives the speed u' (measured from K') of a man who is walking on the embankment with speed u with respect to the embankment. [*Note:* The equation $u = u' + v$ (or the equivalent equation $u' = u - v$) is called the classical addition theorem of velocities, and is in accord with ordinary experience and classical physics.]

Problem 30.6. Nothing is more important in problems of relativity than the proper use of words and concepts—for example, to distinguish between position (x) and length (which is a distance interval, Δx), instantaneous time (t) and time interval (Δt), measurement of events (made with instruments in one frame of reference) versus measurements of the same events but observed from another frame of reference. A related need is that of taking an operational view of measurements, for example, to be quite clear as to how a given measurement would be carried out. For this reason, do the following important problem: A newspaper editor wishes to know the exact length (in feet) at a given instant (say, exactly at noon) of a parade of people that has been moving down a straight road at an even pace. Describe exactly and in detail how you would arrange to measure this length (of a "moving object") without stopping the parade. [*Hints:* The best way will be to hire a good number of helpers, spaced out at known or measurable positions along the route. Give them synchronized watches, so that they can ascertain where the head or tail of the procession is at the same agreed-upon instant of time.]

Problem 30.7. In the above example, we discovered graphically that a measurement of the length of a moving object involves the use of watches, that is, the concept of time! (a) Why is this not the case in making a measurement of "proper length"? (b) In the above example, assume now that the helpers did not use watches that were well synchronized; the watches of those near the head of the parade showed 12:00 noon, while the watches of those placed near the tail still read 11:58 a.m. What effect would this lack of synchrony have on the measurement of the length of the parade?

Problem 30.8. In the light of the last problem, how would you arrange to ensure that a row of helpers, spaced along a line, did have synchronized clocks? Describe your procedure in detail. [*Hint:* Any two clocks at different places are synchronized in any inertial system if a light wave spreading out (preferably in vacuum) from a point exactly midway between the clocks reaches the two clock faces when both show the same time reading. To ensure that the two readings are the same not just by accident, and that the clocks do not have different rates, we might well do the experiment again a bit later to see if the clocks both show the same new time reading. An alternative way of checking whether two clocks are synchronized is by sending from each, at the same time, a beam of light *toward* a common point that is exactly midway between the clocks. An observer at the midpoint receives on his retina both light beams at the same instant ("simultaneously"), if and only if the two beams are each really sent out, say, at 12:00 noon as shown on both clocks, that is, if these were synchronized clocks. Again, to guard against an accidentally equal setting, one can repeat the experiment later. Use this alternative way to describe a method to double check that the editor's assistants all have synchronized clocks.]

Problem 30.9. Note that the definition of synchronization of clocks at different places involves light waves. Why could it not be done in just the same way, but using sound waves instead of light? How is it that we are allowed to use the method with light beams to synchronize clocks in *any* inertial system (even one moving at high speed past us)?

The Galilean transformation equations seem so obvious that they were rarely stated explicitly before 1905, least of all by Galileo himself; nevertheless they are an essential feature of Newtonian physics. But Einstein discovered that if we are to accept the two postulates: (1) that all laws of nature must be the same in K and K', and (2) that the speed of light is the same for an observer in K and in K', then we must discard the intuitively clear Galilean transformation equations and replace them by others that may go counter to our intuition. In particular, we shall have to accept the apparently strange notion that distance intervals Δx and time intervals Δt measured in K' are different from the values for $\Delta x'$ and $\Delta t'$ measured in K' for the same intervals.

30.5 The Relativity of Simultaneity

To demonstrate the need for this change, Einstein poses the following famous *gedanken* [thought]

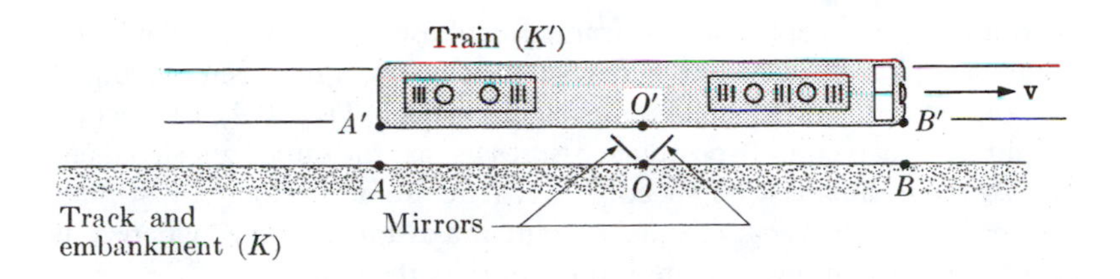

Fig. 30.5. The observer *O* on the embankment has before him two mirrors, tilted so that he can at the same time watch light coming from both *A* and *B*. Also *AO* = *OB*. Lightning bolts strike at *A* and *B* simultaneously, as judged by *O* later (when light from *A* and *B* comes to his eyes at the same time). But consider that when lightning struck, *O′* on the moving train was closest and opposite to *O*, *A* opposite *A′*, and *B* opposite *B′*. What will *O′* see, and when?

experiment: Suppose it is asserted that two events happen "simultaneously" at two different places—for example, that two bolts of lightning strike "simultaneously" at the two places marked *A* and *B* in Fig. 30.5. How can we verify this assertion? We might ask an observer O, placed on the fixed embankment, who happened to be at a point halfway between the charred spots *A* and *B* (that is, *AO* = *OB*). Let us assume that he had, in anticipation, set up before him an apparatus (for example, two mirrors at O) that enabled him to look at light from both *A* and *B* at the same time. If he perceived the arrival at O of two flashes, reaching his eye at the same time from two places that he knows to be equally distant (that is, from where lightning had struck at *A* and *B*), then he can say that the bolts that had struck there previously did so *simultaneously.*

Now suppose that there was also an observer O′, with similar mirrors, sitting on a moving train, and that O′ just happened to be abreast of (coincided with) O precisely when the bolts of lightning struck at *A* and *B*, as judged from the embankment. Since the train, with O′ on it, is moving relative to the embankment at a velocity *v*, the observer at O′ will have moved away from O by the time the two flashes reach O′. In fact O′ is moving toward *B* and away from *A*, so that he sees the flash coming from *B before* that flash reaches the "stationary" observer O on the embankment at *M*. But he sees the other flash, the one coming from *A*, *after* the observer at O sees it, since O′ is moving away from *A* and the beam of light from *A* takes a little extra time to catch up with him. Thus the observer on the train sees the flash from *B* before he sees the flash from *A*.

However, the lightning bolts will have burned or charred spots *A′* and *B′*, one at each end of the train as these ends were passing by and coincided with *A* and *B* at the instant that the bolts struck there. As O′ can check by measuring the distance along the train to *A′* and *B′* in each direction, *A′O′* = *B′O′*. In short, O′ will say that the two bolts struck at equally distant points, but that the light from these points reached him at different moments (first from *B*, then from *A*). Therefore O′ will say that the bolts *could not have struck A and B simultaneously.*

[*Note:* It is implied in this argument that O′ assumes that the speed of light is the same for light coming from *A* and from *B*. But this is precisely what the second principle of relativity allows, or rather commands, him to do: He too is in an inertial system, and by the second principle every observer in every inertial system can and must assume that the speed of light as measured in his own frame of reference is constant in all directions in space, and equal to *c*. (In fact, of course, we make this assumption safely all the time in our "fixed" laboratory—without worrying about the fact that our spaceship, earth, is in motion with respect to the sun and other stars.)]

In short, O′ judges that the two lightning bolts had not hit simultaneously. This is different from O, who judges that the same two events *were* simultaneous. From this example, Einstein concludes that *the concept of "simultaneity" is relative with respect to the coordinate system in which the determination is made.* The concept of simultaneity is frame-dependent and so has no "absolute" physical significance for two events occurring at different places. Two events may be simultaneous for an observer in one frame of reference, but not simultaneous for an observer in a frame that is moving relative to the first.

The discussion so far is not easy to accept at first reading. Somehow one has the intuitive feeling that if the observer O' on the train were to measure the speed of the light of a flash coming toward him from the point B on the embankment, he would find that the speed was greater than the speed of the light of a flash coming toward him from the point A.

Indeed this is what we would expect if we were thinking about *sound* waves from A and B. And the same would apply if the beams of light from A and B were ripples in an ether stationary with respect to the embankment; in such a case the speed of a beam as measured by O in his frame of reference K would be c, but the speed of the same beam relative to the observer O' on the train when the beam is coming to him from point B would be $c + v$—the beam's speed relative to him would be greater than c because they are approaching each other. Conversely, if light moved in a fixed ether, the beam of light from A would be expected to have a speed $c - v$ relative to O', since he is fleeing from A.

But all these conclusions contradict Einstein's second postulate, that the speed of light is the same (namely c) for all observers in all inertial frames. And they also contradict a great variety of experiments, none of which had found any evidence for different values of the speed of light from different directions. (The null result of the Michelson-Morley experiment was just one of the experiments of which Einstein may have known.)

As soon as we accept the second postulate and give up "absolute simultaneity," it follows, as Einstein next shows, that we must adopt a new set of transformation equations, *different* from the Galilean transformation equations [Eq. (30.6)]. To find the new equations, Einstein starts again from the assumption that for a given set of events the values of x' in K' must be related to the values of x as determined in K in such a way that the measured speed of light will come out to be equal to c in both frames. Thus, if a light signal starts from some point where $x = 0$ in frame K at a time $t = 0$ (as also determined in frame K), then at a later time t the light signal will have reached a position x, determined by the equation

$$x = ct$$

or

$$x - ct = 0. \qquad (30.7)$$

In frame K', whose origin initially coincides with K (when $t = t' = 0$) but has been moving at a relative velocity v (as in Fig. 30.4), the second postulate assures us that the same light signal has a speed $c' = c$. After some time t', it will have reached a point at a distance x' in K' where x' is related to c and t' by the equation

$$x' = ct'$$

or

$$x' - ct' = 0. \qquad (30.8)$$

Note that Eqs. (30.7) and (30.8) are *general relations* between the variables x and t, on the one hand, and x' and t', on the other; we have not yet said that any particular time t' and position x' as determined in one frame is to be associated with any particular position x and time t in the other.

But now we impose the additional requirement that *the appearance of a light signal at a particular time and place is an event that we must be able to describe consistently, by the appropriate equations, in either of the two frames of reference,* that is, by Eq. (30.7) as well as by Eq. (30.8).

30.6 The Relativistic (Lorentz) Transformation Equations

Einstein was able to show that x' and t' must be related to x and t by the equations

$$x' = \frac{x - vt}{\sqrt{1 - (v^2/c^2)}},$$
$$t' = \frac{t - (v/c^2)x}{\sqrt{1 - (v^2/c^2)}}. \qquad (30.9)$$

Equations (30.9), together with the equations

$$y' = y \quad \text{and} \quad z' = z$$

(which are valid for the case we have discussed, in which all motion is only in the x direction) give the "relativistic transformation equations" by which to translate information obtained in one inertial frame of reference into information valid for another. This, then, is the result of our two postulates, namely, that all inertial frames are equivalent and that the speed of light is equal to c in each.

At this point, to simplify our equations, we will adopt the Greek letter γ (gamma) to represent the "Lorentz factor":

$$\gamma \equiv 1/\sqrt{(1 - v^2/c^2)}$$

so Eq. (30.9) becomes

$$x' = \gamma\,(x - vt)$$

$$t' = \gamma[t - (v/c^2)x]$$

Problem 30.10. (a) By considering a signal that moves in frame *K* according to the equation $x = ct$, and applying Eqs. (30.9), verify that the speed of light is indeed equal to *c* in frame *K′*. (b) Compare Eqs. (30.9) with the Galilean transformation equations [Eqs. (30.6)], which are still found to hold well enough for problems in mechanics as long as the relative speeds involved are much smaller than *c*. For example, show that Eq. (30.9) reduces to (30.6) if $v << c$.

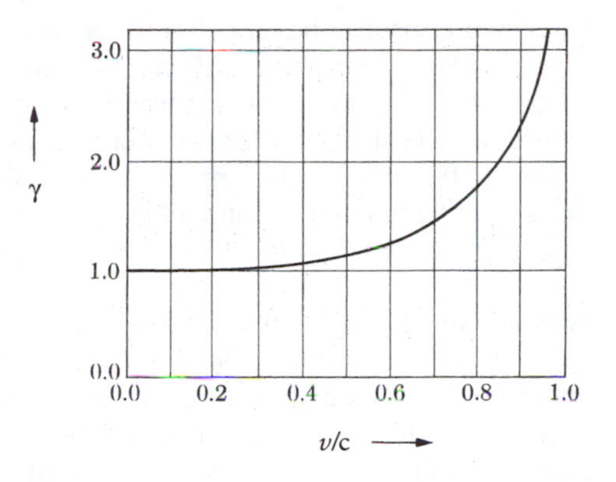

Fig. 30.6.

The transformation equations [Eqs. (30.9)], which Einstein derived in something like this abstract manner from his two postulates in 1905, had previously been proposed on the basis of different reasoning. As Einstein was aware, the equations correspond to the so-called "Lorentz transformation equations," which H. A. Lorentz had derived earlier in connection with his ether-based theory of electrons. The suggestion of an apparent length contraction of a moving object proportional to the factor $1/\gamma$, as in Eqs. (30.9), had been published by FitzGerald in 1889 as a possible explanation of the Michelson-Morley experiment (Section 30.2).

However, Lorentz and FitzGerald had used these equations with a different physical interpretation; they had assumed that there really is an absolute frame of reference, fixed in the ether, and that changes in the measured values of space and time intervals occur only in frames of reference that are moving with respect to absolute space.

For Einstein, on the contrary, all frames of reference moving at constant relative velocity with respect to a given inertial frame are equivalent, and none can be singled out as at rest in "absolute space." It was, so to speak, a complete democratization of all inertial frames, at the cost of removing absolute space and time from the throne they had held for centuries.

An observer in *each* frame would observe that lengths appear to be contracted in the other frames moving past him. Moreover, each observer would notice a "time dilation" of clocks carried in the other frames; if each observer carried a watch of identical physical construction (so that they would click off seconds at exactly the same rate if they were initially sitting side by side in the same frame of reference), then each observer would find that the intervals between clicks registered on the watches in other frames were longer when compared with his own. The amount of this dilation is given simply by the factor γ.

To summarize: *If an inertial frame of reference K′ is moving at a constant velocity v with respect to another inertial frame K, then an observer in either frame would find from his own measurements that length intervals of objects moving with the other's frame are contracted by a factor 1/γ, and time intervals between events are dilated (the moving clocks appear to run slow to the fixed observer) by a factor γ.*

Note that the effect of apparent contraction and dilation reduces to zero if $v << c$; and that it goes to extreme values as $v \to c$ (the length of a moving meter stick appears to contract to zero; the time interval between one click and the next on the moving clock appears to dilate to ∞). Note also that of course the moving observer does not see such effects when looking only at his own clocks and meter sticks—*they* are at rest with respect to him, and so are quite "normal." The apparent contraction of objects and dilation of time intervals is not something that "happens" to the objects and watches, but is the result of the measurement procedure.

Problem 30.11. The plot of γ vs. v/c in Fig. 30.6 shows that the relativistic form of the transformation equations is usually not needed if the relative speeds are less than about 20% of the speed of light. (a) Verify that the curve in Fig. 30.6 is correct by actual calculation for some one value of v/c above 0.2. (b) How fast must K' go for K to see a "contraction" by ½ and a "time dilation" by 2?

Problem 30.12. Imagine that a space probe could be made to go constantly at the speed of 0.8c, and that it is sent to a star located 5 light years from us. By our earthbound clocks, it would therefore take 6.25 years to reach the star. Show that the time elapsed and the distance covered as measured from within the space probe are, respectively, 3.75 years and 3 light years.

[*Note:* Additional problems are at the end of the chapter.]

30.7 Consequences and Examples

A well-known predicted consequence of the time-dilation effect is the *twin paradox* (sometimes called the *clock paradox*): If one twin goes off in a spaceship traveling at a speed close to the speed of light and then turns around and returns to earth, she should find that she is younger than her twin brother who stayed behind! Seen from the frame of reference of the earth, biological processes of aging, as well as the operation of clocks (including the interval between heartbeats), run more slowly in the spaceship than on earth.

Nevertheless the predicted outcome is difficult to believe intuitively; after all, why is the situation not symmetrical in *this* case? Why couldn't the spaceship be taken as the frame of reference, so that the earthbound twin would be younger from the viewpoint of the traveler? One has the feeling that when the traveler returns, all time-dilation effects on both sides should cancel out, since one cannot objectively be younger than the other when they meet again.

But that argument assumes incorrectly that the spaceship is an inertial frame of reference all the time; in fact, unlike the earth, it has to undergo large accelerations in order to start, turn around, and stop; hence the time-dilation effects are *not* the same for both twins. As discussed in detail in some of the readings cited at the end of this chapter, the "traveling twin" will in fact have aged less than the "stay-at-home twin"—although the actual differences for any realizable speeds of travel are completely insignificant for living beings.

It is, however, quite expected that such very small differences can be found when one compares the time shown on two initially identical, synchronized clocks after one of them has been sent through some long journey at high speed. And indeed the experiment has been done with atomic clocks carried by airplane.

Another consequence of time dilation, more easily testable, is the change in the half-life for decay of an unstable particle when it is moving relative to the observer. This effect was first measured in the 1930s for muons, particles intermediate in mass between protons and electrons (see Section 32.1). Werner Heisenberg recalled, many years later, that this measurement made relativity theory acceptable in a few places in Germany at a time when Einstein's theories were under attack by the Nazis and not taught by German professors.

30.8 The Equivalence of Mass and Energy

It had been suggested before 1905 that the mass or inertia of an electron must increase with increasing speed, so that the acceleration, by a given force, of an already fast-moving electron would be less than that of the same electron when starting from rest. This mass increase had been found experimentally by W. Kaufmann (1902) and others by deflecting in electric and magnetic fields the high-speed electrons (β-rays) emitted by radioactive nuclei.

Einstein deduced from his special theory of relativity in 1905 the conclusion that the "tangible," ponderable mass of any particle, whether charged like an electron or not, increases with speed according to the equation

$$m = \gamma m_0 \tag{30.10}$$

where m_0 is the mass of the particle at rest with respect to the observer, called the *rest mass,* and m is the mass of the particle measured while it moves at speed v relative to the observer, called the *observed* or *relativistic mass.* (This mass determination may be made by supplying a known centripetal force and measuring the radius of curvature of the path, that is, from $F = mv^2/r$.) Furthermore, c is the velocity of light, a constant, equal to about 3×10^8 m/sec.

Now it is evident from Eq. (30.10) that for ordinary speeds, where $v << c$, $m = m_0$ within the experimental error of measurement. But in the motion of atomic and subatomic particles this restriction no longer holds—some β-rays are ejected from disintegrating nuclei with speeds greater than 90% that of light.

At first glance this increase in mass with speed is deeply disturbing. Surely a moving particle has no more molecules, no more tangible, ponderable matter than before! If mass increases, mass is not simply a measure of the quantity of matter after all. And then the law of conservation of mass would appear to be no longer strictly true. Such are the speculations that we must consider next.

Problem 30.13. Evaluate the ratio m/m_0 for an object that has successively a speed v of 0.1, 0.3, 0.5, 0.7, 0.9, and 0.97 times the velocity of light c. (The experimental check, incidentally, has been found to be very good.) What should be the mass of any object if $v = c$?

Problem 30.14. How much would your mass increase if you were flying in a jet craft at the speed of sound (about 330 m/sec)?

The ratio v/c is usually replaced by the symbol β; thus the Lorentz factor $\gamma = (1 - \beta^2)^{-1/2}$. However, the binomial theorem from algebra teaches that $(1 - \beta^2)^{-1/2} = 1 + \beta^2/2 + 3\beta^4/16$ + other terms of still higher powers. Now if v is considerably smaller than c, β^2 is very small, and β^4, together with all higher terms, is negligible. Therefore we can write

$$m \approx m_0 \left(1 + \tfrac{1}{2}\beta^2\right) = m_0 \left(1 + v^2/2c^2\right)$$

$$= m_0 + \left(\tfrac{1}{2} m_0 v^2\right)/c^2.$$

This last result opens our eyes to an amazing physical interpretation of the increase in mass with speed, for it points out directly that the increase in mass is

$$\Delta m \doteq m - m_0 = \left(\tfrac{1}{2} m_0 v^2\right)/c^2$$

where we may at least tentatively identify the term $\frac{1}{2} m_0 v^2$ with kinetic energy. Thus

$$\Delta m = \mathrm{KE}/c^2$$

and in trying to understand the change in mass with speed, we were led to the concept that kinetic energy, when added to a ponderable object, also adds inertia in an amount KE/c^2. Whether we say that energy *has* mass or *is* mass or *appears as* mass is only playing with words and adds nothing to the last equation.

Although we have arrived at this result by an approximation, the equation $\mathrm{KE}/c^2 = \Delta m$ is true in general. Only we must realize that in such a purely mechanical situation *the kinetic energy of a fast-moving particle might be redefined as*

$$\mathrm{KE} = \Delta mc^2 = mc^2 - m_0 c^2$$

and is not separately evaluated by $\frac{1}{2} m_0 v^2$, for the mass increases from m_0 with speed, and so more kinetic energy is required for equal changes in speed at the latter stages of approaching final speed v than at the earlier stages.

Problem 30.15. Einstein's original approach was to postulate that no material body can be made to reach a velocity equal to or greater than the speed of light, with respect to the observer. Prove that the foregoing is in harmony with this postulate. [*Hint:* What kinetic energy would be needed if such speeds were desired?]

The idea that it is the increase in energy that changes the mass can now be extended further to include energies other than ordinary kinetic. Consider first a spring with a mass m_0 when limply extended. When we compress it and so give it elastic potential energy PE, its mass increases to m_0 + PE/c^2. A lump of metal likewise increases its observed mass when it is heated; here

$$\Delta m = (\text{heat energy supplied})/c^2.$$

In brief, the principle of mass equivalence of energy was soon extended to cover the increase (or decrease) of any type of energy; *for every unit* (1 joule) *of energy supplied* to a material object, whether it be kinetic energy given to a bullet, gravitational potential energy acquired by the change in relative distance of two stars, or any other form of energy, the *mass of the system thereby increases by*

$$(1\ \text{joule})/(3 \times 10^8\ \text{m/sec})^2 = 1.1 \times 10^{-24}\ \text{kg}.$$

To repeat, this does not mean there are now more molecules in the system than before; what has changed is the observable inertia of the energy-enriched material. Evidently the rate of exchange, that is, the factor c^2 in

$$\Delta m = (\text{energy})/c^2,$$

indicates that such changes in mass are beyond direct experience in ordinary mechanical experiments, and therefore there was no need to consider this academic point in the section on the conservation of mass. But in nuclear phenomena or in cyclotrons and other accelerators, where the masses of the particles involved are relatively small to start with and the energies are relatively large, the changes in mass may become very noticeable.

If you prefer, *mass* does not measure the amount of tangible or ponderable matter in a system; it measures the amount of matter *together with* the mechanical energies, electrical energies, etc. Therefore *we rewrite the law of conservation of mass* to read as follows: In a closed system,

$$\Sigma[m_0 + (\text{energy})/c^2] = \text{constant.} \qquad (30.11)$$

A typical instance of the use and range of these ideas may be found in comparing the mass of a nucleus with the mass of its component particles (nucleons) after separation. For example, a helium nucleus (or α-particle) has a rest mass of about 4.0028 amu; it is composed of four nucleons, namely, two protons and two neutrons. When the helium nucleus is broken up in a nuclear reaction, the rest mass of each proton measures about 1.0076 amu and that of each neutron about 1.0090 amu. But $(2 \times 1.0076 + 2 \times 1.0090)$ amu is more than 4.0028 amu by about 0.03 amu (or, since 1 amu = 1.66×10^{-27} kg, by 5×10^{-29} kg), and it would seem that the sum of the parts "weighs" more than the parent nucleus.

And so they do indeed, because energy had to be supplied to break up the helium nucleus, specifically

$$5 \times 10^{-29}\ \text{kg} \times (3 \times 10^8\ \text{m/sec})^2 = 4.5 \times 10^{-12}\ \text{J}.$$

The complete breakup of helium nuclei is therefore an energy-absorbing (*endothermic*) reaction, and the energy that must be supplied, called the *binding energy* of the nucleus, is stored as potential energy among the fragments. Indeed, it can be retrieved in the reverse reaction of building up helium nuclei from protons and neutrons by the emission of energy on synthesis (Section 32.2).

At last we can understand why Prout's hypothesis in its original form was incorrect (Section 21.1). Prout assumed that if all the elements were compounds of hydrogen, their atomic weights must necessarily be integer multiples of the atomic weight of hydrogen. Careful determinations of atomic weights disproved this hypothesis. According to Heisenberg's model of the nucleus one might expect that the total mass of the nucleus would be equal to the sum of the masses of the nucleons (protons and neutrons) of which it is composed. But that is not quite true either. The reason is that the nucleons are held together by a balance of forces (electrostatic repulsion of the protons, and a short-range "nuclear force" between all the nucleons, which is attractive at the normal internucleon distance). The work done by these forces in putting together the nucleus corresponds to the binding energy, and the mass-equivalent of this energy has to be added to the mass of the nucleons.

All that has been said so far demonstrates the *mass equivalence of energy.* Now nature generally exhibits a symmetry in its operations; that is, one may reasonably suspect that a particle of matter may lose some of its mass by giving up a corresponding amount of energy. This can, of course, be made to occur most directly by slowing down a high-speed particle. It will get "lighter" by an amount given as (loss of KE/c^2) although, of course, whatever agency served to slow the particle must then absorb that much energy and consequently must get that much "heavier" in turn.

But there are more striking ways to realize this exchange between mass and energy: It is possible for *all* of a particle's mass to be transformed into energy. But in order to explain how this can happen we must return to the development of quantum mechanics.

Problem 30.16. The combined installations for producing electric power in the United States have a capacity of about 10^{11} joules of energy per second. If instead of hydroelectric, steam, and internal combustion engines, one central nuclear power plant were to provide this energy by conversion of nuclear "fuel" into electrical energy, how much matter would be converted per day? How much material would be handled per year, if only 0.1% of the material supplied is converted by this process?

It stands to reason that the law of conservation of energy must be reformulated if energy-mass conversions are to be included. One simple way of doing this is to consider every object in the system as a potential source for complete annihilation, and therefore to assign to every rest mass m_0 a *rest energy* m_0c^2, a potential energy that in the course of events may be partly or completely converted into other forms of energy. Then we can safely say that in a closed system, the total amount of energy, namely, the rest energy (Σm_0c^2) plus all *other* forms of energy (ΣE) is constant, or

$$\Sigma(m_0c^2 + E) = \text{constant} \qquad (30.12)$$

Now this is really not a new law; Eq. (30.11), the extended law of conservation of mass, read $\Sigma(m_0 + E/c^2) = \text{constant}$, and if Eq. (30.12) is divided through by c^2 on both sides we obtain the very same expression. Either equation tells essentially the same story. As Einstein himself pointed out:

> Pre-relativity physics contains two conservation laws of fundamental importance, namely, the law of conservation of energy and the law of conservation of mass; these two appear there as completely independent of each other. Through relativity theory they melt together into *one* principle.

In this sense one may say that the concepts of mass and of energy also "melt together," as though they were two aspects of one physical reality. And once more we see Einstein's passion for unification at work.

Problem 30.17. If the laws of conservation of mass and of energy are jointly contained either in Eq. (30.11) or in Eq. (30.12), how shall we reformulate the law of conservation of linear momentum? Surely Eq. (30.10) shows that the old form, which assumed that mass is independent of velocity, is no longer strictly tenable.

30.9 Relativistic Quantum Mechanics

In addition to its specific consequences such as length-contraction, time-dilation, and mass-energy transformation, Einstein's theory of relativity imposes general rules on other physical theories. In particular, it requires that the space and time variables must appear in a symmetrical fashion in all fundamental equations; that is, those equations must be unchanged when one changes to a different coordinate system using the Lorentz transformation equations (Section 30.6).

Fig. 30.7. Paul Adrien Maurice Dirac (1902–1984), British physicist who found the most general way to formulate quantum mechanics, and used his relativistic wave equation to predict the existence of the positron. More generally, his theory suggested that all particles should have corresponding antiparticles with opposite electric charge. Dirac formulated, independently of Fermi, the quantum theory of systems of identical particles known as Fermi-Dirac statistics (Section 19.4). He received the Nobel Prize in Physics in 1933 (shared with Schrödinger), shortly after Anderson discovered the positron.

Schrödinger's wave equation [Eq. (29.3)] clearly violates this rule since it does not contain the time variable at all. His time-dependent equation [Eq. (29.4)] fails in a less obvious way: Setting aside the potential energy term $(\text{PE})_{op}$ in order to deal with the simplest case of a free particle, we see that the space variables (x, y, z) appear in $(\text{KE})_{op}$ as the "rate of change of the rate of change" (mathematically a "second derivative"), whereas the time variable appears in E_{op} as the "rate of change" (a first derivative). In other words, the Schrödinger equation implicitly singles out a particular reference frame as being at rest, and assumes an "absolute" time t.

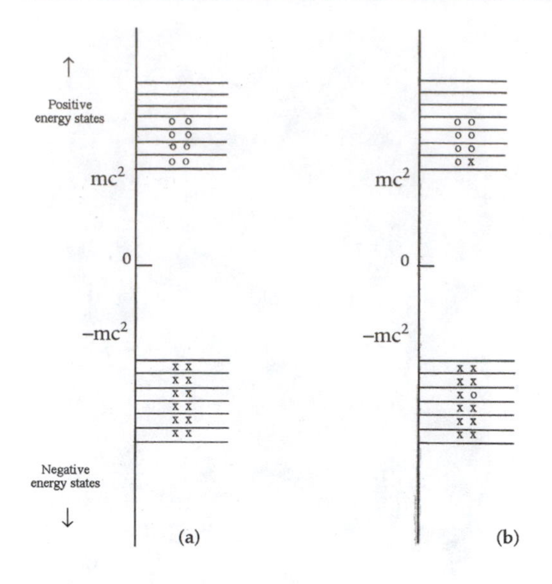

Fig. 30.8. Dirac's model for energy levels of particles governed by his relativistic wave equation. The positive energies (including rest-mass energy) extend from $+m_0c^2$ to $+\infty$; the negative energies range from $-m_0c^2$ to $-\infty$. (a) Normal situation with every negative-energy state filled, indicated by x, and every positive-energy state empty, indicated by o. (b) One particle has acquired enough energy to jump from a negative-energy state that is now empty (o) to a positive-energy state (x). The "hole" (o) in the sea of negative-energy states behaves like a particle with opposite charge. (The two states at each level correspond to spin $+\frac{1}{2}$ and $-\frac{1}{2}$.)

Schrödinger himself was familiar with relativity theory and realized that his equation was inconsistent with it. He tried to correct it by changing the time-dependent part to a second derivative. But the consequences of that equation did not agree with experiments for electrons, so he didn't publish it. (The equation was subsequently found to apply to some other particles and is now known as the "Klein-Gordon equation.")

The problem was solved in a different and very unusual way by P. A. M. Dirac (1902–1984). Dirac proposed an equation in which the space and time variables all enter as first derivatives (a single "rate of change of"). At the same time the equation, when squared, is consistent with the Schrödinger equation, so it should retain the correct empirical predictions of that equation.

Dirac found that the counterpart of the wave function governed by his equation was not one function but four: ψ_1, ψ_2, ψ_3, ψ_4.

Dirac's equation has a number of remarkable properties. First, ψ_1 and ψ_2 are easily interpreted as corresponding to particles with spins +½ and $-\frac{1}{2}$. Thus Wolfgang Pauli's postulate that the electron has a two-valued quantum number (interpreted by Uhlenbeck and Goudsmit as "spin"), which allowed him to explain the construction of multiple electron shells in atoms with the help of his exclusion principle, no longer had to be invoked as a mysterious *ad hoc* assumption; it was a direct consequence of a more general theory.

Second, the functions ψ_3 and ψ_4 correspond to particles with *negative energy*. This is puzzling, since no forces have been included that might lead to a negative potential energy as one would expect for an electron bound to a nucleus. It leads to a serious difficulty. Like the Schrödinger equation, Dirac's equation allows free-particle states of indefinitely large energy, but it also allows indefinitely large negative energies. It would seem that any state with a *finite* energy, positive or negative, would be unstable since the particle could always emit radiation and fall down to a lower energy, dropping toward $-\infty$.

Dirac devised the bold and imaginative postulate that *all the negative energy states are normally filled* and that the Pauli principle prevents more than one particle from occupying a state. This infinite "sea" of negative-energy particles is unobservable. But it is possible for a particle in a negative-energy state to acquire enough energy to jump to a positive-energy state. Note that in Fig. 30.8 the minimum positive energy is the rest-mass energy of a particle, m_0c^2, while the minimum negative energy is $-m_0c^2$. Thus the negative-energy particle must acquire at least $2m_0c^2$ of energy to jump to a positive-energy state.

Dirac argued that the empty state or "hole" left behind by the jumping particle would itself act like a particle of opposite charge. He interpreted the process as the transformation of two photons into two particles with opposite charge and a total energy of at least $2m_0c^2$. (In general there must be at least two photons, or a photon and some other particle, in order to conserve momentum as well as mass-energy; review the discussion of photon momentum in Section 16.7.)

Problem 30.18. Explain why momentum could not be conserved if a single photon were transformed into two oppositely charged particles with zero or negligibly small kinetic energy. If the particles are electrons, calculate the minimum frequency of the two photons (assumed to be equal) needed to create the two particles.

The inverse process can also occur: If there is a hole in the sea of negative-energy particles, a positive-energy particle can "fall into it," emitting radiation with total energy equal to at least twice the rest-mass energy of the particle.

Dirac initially tried to interpret the two particles described by his theory as the electron and the proton, but it soon became clear that this would not work: The mathematics required that the two particles have the same mass. So in 1931 he boldly predicted the existence of a new particle, a positively charged "antielectron."

In the heady days of the quantum revolution, no new idea was taken seriously unless it was at least half-outrageous. Dirac's theory was so outrageous it almost had to be right! And it was not long before it found a spectacular confirmation. In 1932 the American physicist Carl D. Anderson (1905–1991), who later said he had not at all been influenced by Dirac's theory, discovered in cosmic rays the tracks of particles that had the same mass as electrons but positive charge.

The existence of the *positron,* as it came to be called, was not only confirmed by many other investigators, but it also turned out to have exactly the strange new properties required by Dirac's theory. Positron-electron pairs could be created from radiation if the minimum energy $2m_0c^2$ was present. A positron, once produced, acted like Dirac's "hole": It was soon annihilated by colliding with an electron, producing very intense γ-rays (electromagnetic waves of wavelength shorter than x-rays). The positron's behavior thus provided the most extreme example of the mass-energy transformation predicted by Einstein: Mass can be *completely* converted into energy and back again.

In his biography of his father J. J. Thomson, George Thomson pointed out the curious similarity between Dirac's theory and Maxwell's concept of the electromagnetic field:

> Underlying [Maxwell's] idea of "displacement" is the idea that electricity is everywhere waiting to be moved, even in empty space where there are no charged bodies. He would have been delighted by Chadwick and Blackett's discovery that positive and negative electrons can be created by radiant energy out of empty space. Perhaps it was as well that this was not discovered in the nineteenth century, before the nature of charge was better, though it is still imperfectly, understood.

There was no reason to believe that Dirac's theory would apply only to electrons and positrons. It should be valid for any particle that obeys Fermi-Dirac statistics. Every such *particle* should have a corresponding *antiparticle.* Thus there should be (as Dirac himself predicted in his 1931 paper) an *antiproton,* a particle with the same mass as the proton but negatively charged. But they seem to be very rare (at least in our little corner of the universe), so physicists had to manufacture them (see Chapter 32, footnote 3). But now, with antiprotons, antineutrons, and positrons one should be able to make *antimatter*—for every atom such as hydrogen, carbon, or nitrogen, there could exist a similar atom in which every particle is replaced by its antiparticle. We have here a new kind of *symmetry* in the world.

Of course individual atoms of antimatter could not exist very long in our part of the universe because they would be annihilated by collisions with atoms of matter. But in the universe as a whole, it might seem that there should be just as much antimatter as matter—unless one can find a reason why more of the latter should exist. Again we see that the solution to one problem in physical science creates yet another problem for the next generation to solve—an adventure without forseeable end.

Although Dirac's original theory implied that electrons are "normal" while positrons are strange, ghostly particles made out of the absence of hypothetical negative-energy electrons, the theory can easily be made symmetrical so as to ascribe equal reality to both particles and antiparticles. Electrons could just as well be holes in an infinite sea of negative-energy positrons; quantum mechanics lets you have it both ways at the same time.

Alternatively one could adopt the intriguing proposal by the American physicist Richard P. Feynman (following earlier suggestions by John A. Wheeler and others): A positron is an electron traveling backward in time. This idea is just as "outrageous" as Dirac's original theory; but 1949 was not a revolutionary time for physics and it was not taken seriously by other physicists, although they were happy to use it for doing calculations.

Suppose we represent an electron by a line *AB* and a positron by a line *CB* in a plane where

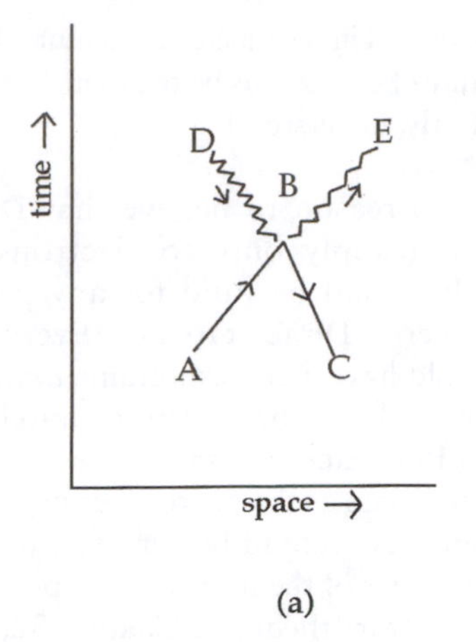

Fig. 30.9 (a). According to Feynman, the annihilation of an electron and a positron moving forward in time along paths *AB* and *CB* respectively, producing two γ-rays moving forward in time along paths *BD* and *BE,* can be more conveniently described as the scattering of an electron moving forward in time along *AB* by a γ-ray moving backward along *DB,* resulting in an electron moving backward in time along *BC* and a γ-ray moving forward along *BE.*

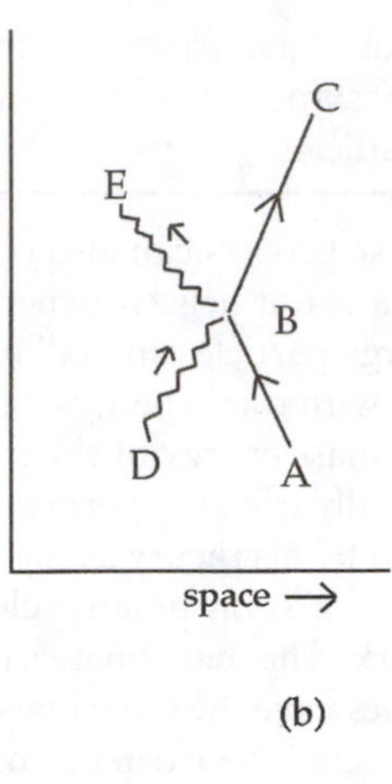

Fig. 30.9 (b). On interchanging the space and time axes, we find that an event previously described as electron-positron annihilation, creating a pair of γ-rays, is now the scattering of an electron (moving along *ABC*) by a γ-ray (moving along *DBE*).

the ordinate is time (Fig. 30.9a). They meet and annihilate each other at *B*, giving rise to two γ-rays *BD* and *BE*. Wouldn't it be more economical to say there is just *one* electron and *one* γ-ray? The electron starts at *A*, travels forward in time to *B* where it encounters a γ-ray moving backward in time from *D* to *B*. They collide and the electron bounces back in time along the path *BC*, while the γ-ray bounces forward in time along *BE*.

The advantage of this description is that one can rotate the axes, interchanging space and time, so that the same theoretical calculation applies to the scattering of an electron moving forward in time along the path *ABC* by a γ-ray moving forward in time along the path *DBE* (Fig. 30.9b). This is an example of the very popular "Feynman diagrams" approach, which makes use of the space-time symmetry inherent in relativity theory to facilitate calculations.

Even if you find "electrons moving backward in time" a little hard to swallow outside the realm of science fiction, you must admit that Feynman's idea has the aesthetic virtue that comes from its simplicity and symmetry. It was Dirac who advised scientists to apply aesthetic criteria in judging theories. Recalling how Schrödinger was discouraged from pursuing his relativistic wave equation because it seemed to disagree with experiment, Dirac wrote about the individual's creative phase of research at the frontiers of mathematical physics:

> It is more important to have beauty in one's equations than to have them fit experiment.... It seems that if one is working from the point of view of getting beauty in one's equations, and if one has really a good insight, one is on a sure line of progress. If there is not complete agreement between the results of one's work and experiment, one should not allow oneself to be too discouraged, because the discrepancy may well be due to minor features that are not properly taken into account and that will get cleared up with further developments of the theory.
>
> ("Evolution of the Physicist's Picture of Nature")

30.10 The General Theory of Relativity

The special theory of relativity can deal only with physical phenomena in which acceleration and gravitational forces are not involved. By 1916 Einstein had worked out the basic form of a more comprehensive theory, known as the general theory of relativity. Although some of the details of this theory are still under vigorous discussion and the details depend on considerable mathematical development, its basic principles and predictions are of such great interest and importance that at least a brief summary is required here.

In explaining his general theory in 1916, Einstein characteristically pointed out first a fact that had been well known, but never properly understood before: *The inertial mass of every object has the same value as its gravitational mass.* This fact was first illustrated by Galileo's discovery that all objects have the same gravitational acceleration at a given location, together with its interpretation by Newton's laws. Thus, by Newton's second law of motion, the net force acting on the object is proportional to its *inertial* mass,

$$\boldsymbol{F}_{\text{net}} = m_{\text{inert.}} \boldsymbol{a}.$$

But if the gravitational pull is the cause of the acceleration, a takes on the value equal to g and the gravitational force is proportional to the gravitational mass of the object:

$$\boldsymbol{F}_{\text{G}} = m_{\text{grav.}} \cdot \boldsymbol{g},$$

or, to put it into entirely equivalent terms,

$$\boldsymbol{F}_{\text{G}} = m_{\text{grav.}} \cdot f_{\text{G}},$$

where f_{G} is the gravitational field strength (see Section 24.8). When we equate these two expressions for the force, we have

$$\boldsymbol{a} = (m_{\text{grav.}}/m_{\text{inert.}}) \cdot f_{\text{G}}.$$

The fact that in free fall $\boldsymbol{a}$ is by experiment numerically the same for all bodies at a given location (where the field strength f_{G} must have a fixed value independent of the nature of the object it acts on) implies that the ratio $m_{\text{grav.}}/m_{\text{inert.}}$ is the same for all bodies. We can then adopt a system of units in which this ratio is 1 (as is usually done without explicit discussion). Hence the value of the gravitational mass of a body is equal to that of its inertial mass.

Einstein now argues that this equality must mean that "the *same* quality of a body manifests itself according to circumstances as 'inertia' or as 'weight.'" (Our quotations in this section are taken from Einstein's *Relativity: The Special and General Theory,* 1917.) The significance of this assertion may be illustrated by considering a large box (of negligible mass) in empty space far away from a planet or other gravitating objects. Inside the box is an observer, who is floating freely in space, since there is no gravitational field present.

Now suppose someone attaches a hook and rope to the outside of the top of the box and begins to pull on it with a constant force. (Of course up to this point the word "top" is inappropriate, since there is no physical way to determine up or down directions in the box. But as we shall see, the side to which the hook is attached becomes the top.) The box undergoes a uniform acceleration, which is transmitted to the observer inside the box by the floor when it scoops him up or hits him:

> He must therefore take up this pressure by means of his legs if he does not wish to be laid out full length on the floor. He is then standing in the box in exactly the same way as anyone stands in a room of a house on our earth. If he releases a body which he previously had in his hand, the acceleration of the box will no longer be transmitted to this body, and for this reason the body will approach the floor of the box with an accelerated relative motion. The observer will further convince himself that *the acceleration of the body toward the floor of the box is always of the same magnitude, whatever kind of body he may happen to use for the experiment.*
>
> . . . the man in the box will thus come to the conclusion that he and the box are in a gravitational field which is constant with regard to time. Of course he will be puzzled for a moment as to why the box does not fall in this gravitational field. Just then, however, he discovers the hook in the middle of the top of the box and the rope which is attached to it, and he consequently comes to the conclusion that the box is suspended at rest in the gravitational field.
>
> Ought we to smile at the man and say that he errs in his conclusion? I do not believe we ought to if we wish to remain consistent; we must rather admit that his mode of grasping the situation violates neither reason nor known mechanical laws.

Thus, even though to an *outside* observer the box may appear to be in accelerated motion in gravitation-free space, the observer *inside* the box may legitimately regard it as being at rest, but subject to a gravitational field. This discussion illustrates that physics can be formulated in such a way that frames of reference in uniformly accelerated motion relative to each other may also be considered equivalent *(principle of equivalence).* This extends the previous, more limited *special* theory of relativity, where only unaccelerated frameworks were equivalent.

Einstein's general theory has been accepted by most physicists, at least as a working hypothesis, primarily because of the successful confirmation of five predictions deduced from it.

a) *Gravitational action on light.* If the box discussed above has a little hole in one side, and a beam of light from a distant source comes in while the box is at rest, the beam will go straight across to the other side as long as the box is at rest. But when the box is accelerated rapidly enough, the beam will of course appear to the man inside to curve downward as it passes across the box. To put it differently, the observer in an accelerated frame of reference will notice that light does not travel in straight lines. But his frame of reference, by the principle of equivalence, can be equally well regarded as being fixed, though acted on by a gravitational field. By generalizing we may conclude that any "real" gravitational field should also have an effect on the gravitation of light.

Einstein thus predicted that the light from stars, passing near the sun, would be deflected toward the sun, so that the apparent positions of the stars themselves (as seen from the earth, say) would be shifted away from the sun. Of course this effect could only be observed during a total eclipse of the sun, and its magnitude is so small that it is not noticeable except by careful observation. As mentioned in Section 30.1, it was the successful test of this prediction by the British eclipse expedition of 1919 that provided one of the first convincing pieces of evidence for Einstein's general theory of relativity.

Einstein also showed in one of the first papers he published after his move to the United States that if a star emitting light happens to be behind a large mass that bends the path of the starlight on all sides, the deflecting mass could act as a lens, producing a circular image of the light source for an observer on the other side of the large mass.

He originally believed that, just as the bending of light by the sun could be observed only during a total eclipse since otherwise the sun's own light would "drown out" the light from the star, we would not actually be able to observe these theoretical circular or double images. But in the 1960s, astrophysicists analyzed this "gravitational lensing" effect in more detail and concluded that it might play a significant role in observations of distant galaxies, with galaxies at intermediate distances acting as lenses. In 1979 (around the time of the centennial of Einstein's birth) the English astronomer Dennis Walsh discovered two "quasars" (distant "quasi-stellar" objects with enormous energy output) that seemed to be almost identical, and proposed that they are in fact two images of the same quasar. Subsequently a number of examples of multiple images and even of complete circular "Einstein rings" have been found (part of such a ring's image is used on this book's cover). Gravitational lensing is now an important tool for detailed studies of the most distant (and therefore appearing to us to be the oldest) parts of the universe.

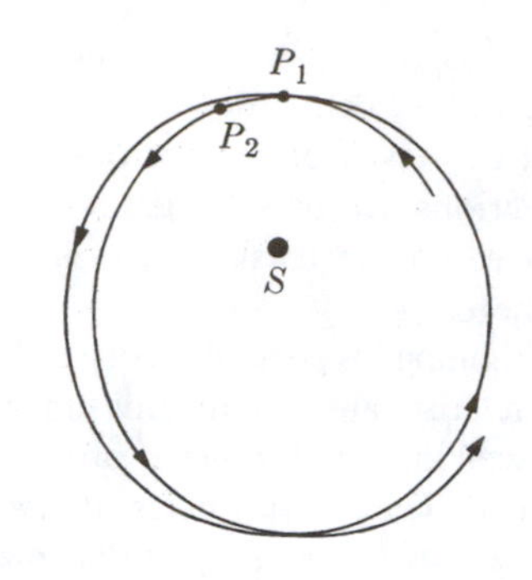

Fig. 30.10. Advance of the perihelion for the orbit of the planet Mercury from P_1 to P_2 (greatly exaggerated).

b) *Advance of the perihelion of Mercury.* Another success of the general theory, which Einstein developed in detail in 1915, was the calculation of a small correction to the theoretical Newtonian orbits of planets in the solar system. The observed fact that the orbit of the planet Mercury slowly rotates around the sun, so that the position of closest approach to the sun (perihelion) gradually shifts, had been a long-standing puzzle in celestial mechanics.

Einstein's explanation depends on the idea that energy is associated with mass, and that even energy in "empty" space, for example the energy represented in the sun's gravitational field, may be affected by and in turn act on other objects that exert gravitational forces. As a planet gets close to the sun, it experiences a gravitational force greater than would be expected from Newton's law of universal gravitation in its classical form. One way of putting it is that the effective mass of the sun, plus its gravitational field, increases. This effect is significant only for planets that come in very close to the sun, such as Mercury, and that have eccentric orbits, so that there is a considerable variation in their distance from the sun over which to get differences of the effect.

According to Einstein's theory, when a planet approaches the sun (neighborhood of the perihelion, P_1), it experiences an extra force due to the additional effective mass of the sun; the effect is to deflect the planet into a changed orbit so that the next perihelion is at P_2 (see Fig. 30.10). The result is that the planet still moves very nearly in an ellipse, but the ellipse itself rotates slowly around the sun. The shift from P_1 to P_2 between one pass and the next is called the "advance of the perihelion" of the orbit. The numerical value of the rate of advance, as calculated from the general theory of relativity, is almost exactly in agreement with the observed value.

Even though it was not, like the deflection of starlight, a new phenomenon, the advance of Mercury's perihelion actually provided stronger evidence for Einstein's theory. This was partly because it tested a "deeper" part of the theory (whereas deflection of starlight could be predicted qualitatively from the equivalence principle) and partly because the observational data were more accurate.

c) *Gravitational red-shift of light emitted by a massive star.* A further prediction of the general theory of relativity is that time intervals appear to be dilated (clocks are seen to run more slowly) when we look into regions of space in which there are strong gravitational fields. The rate at which a light wave is emitted from an atom also serves as a clock. Hence the frequency of electromagnetic radiation produced in such regions by a particular atomic process (say, in the emission of a Balmer line) is lower, as received and measured by an observer in another part of space. The result is an observed shift of spectral-line frequencies toward the red (low-frequency) end of the spectrum.

This gravitational "red-shift" of light that is coming to us from the sun or from more massive stars was observed in the 1920s by astronomers, who were motivated by Einstein's prediction. But the interpretation of their observations was controversial, and the validity of the effect remained in doubt until 1960. In that year Robert V. Pound and his student G. A. Rebka at Harvard succeeded in making a direct laboratory measurement of the red-shift due to the Earth's gravity.

Incidentally, this effect is quite different from another "red-shift," attributed to a Doppler effect that is caused by the relative motion of a star away from the earth (Section 28.2), rather than by its light going through a strong gravitational field.

d) *Black holes.* The gravitational red-shift is also involved in another relativistic phenomenon that has in recent decades aroused interest among astrophysicists: the "gravitational collapse" of extremely massive stars. It is believed that a star several times as massive as the sun would be unstable, because the gravitational attractive forces tending to make it contract would overcome the expansive tendency created by the random motions of its molecules, the pressure of the escaping radiation, and repulsive forces between nuclei. As the star collapses, its intense gravitational field would impose a greater and greater red-shift on all emitted radiation, until finally the frequency tends to zero and no radiation could escape at all; hence the star would no longer be seen. On the other hand, all radiation (and matter) approaching the vicinity of the star would be sucked into it by the strong gravitational field and could not escape. Such a star would thus appear to us as a "black hole" (the name was proposed by John Archibald Wheeler) in the sky, blocking our vision of anything directly on the other side of it.

The black hole concept first emerged in general relativity theory in 1916, when the German astronomer Karl Schwarzschild solved Einstein's equations for the special case of a point-mass alone in the universe. He found that the space-time curvature becomes infinite at a certain distance R_S from the point-mass M, given by the equation

$$R_S = 2GM/c^2$$

where G is the gravitational constant and c is the speed of light. This distance is now called the "Schwarzschild radius," and the theoretical infinity of the curvature is called the "Schwarzschild singularity." For a mass M equal to that of the sun the Schwarzschild radius is about 2.5 km.

One interpretation of the Schwarzschild singularity goes back to the eighteenth century, when the particle theory of light was accepted by many scientists. Laplace and other astronomers

Fig. 30.11.

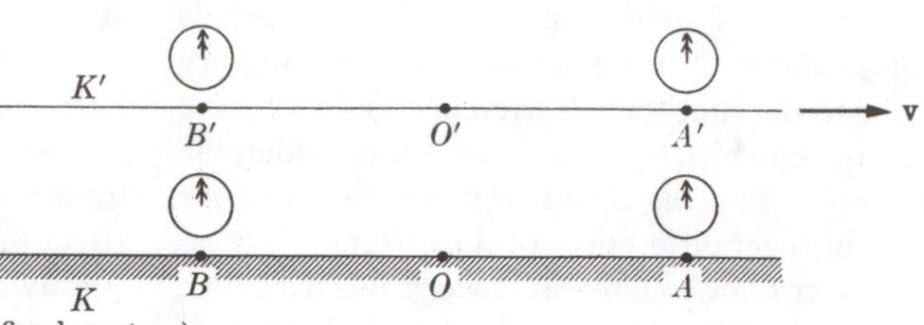

pointed out that if a star is sufficiently dense—more precisely, if it has a large enough ratio of mass to radius—the "escape velocity" required for any object to leave it would exceed the speed of light. Particles of light with any mass, no matter how small, could not escape the gravitational attraction of such a star and would eventually fall back into it.

Problem 30.19. Derive from Newtonian theory the formula for the velocity that a particle must have to escape from an object of mass M and radius R and show that it is consistent with the above formula for the Schwarzschild radius. [*Hint:* Apply to the gravitational force law the concept of "potential" introduced in Chapter 24; the kinetic energy of the particle must exceed its gravitational potential energy.]

Laplace thought that the universe might be filled with such dense stars, which would be completely unknown to us unless we happened to be close enough to feel their gravitational pull. But astrophysicists now believe that black holes can be observed by certain characteristic effects they produce. First, matter being sucked into a black hole would be strongly accelerated, and the radiation it emits would therefore show large Doppler shifts; at the high temperatures expected to be attained, there would also be strong x-ray emission. Second, the British astrophysicist Stephen Hawking proved mathematically that when quantum effects are included in the theory of black holes, one finds that black holes do radiate energy in the form of blackbody radiation. During the 1980s and 1990s, a number of individual radiation sources were identified as probable black holes, and astrophysicists also concluded that the mysterious quasars are enormous black holes at the centers of galaxies.

The possible fate of stars contracting to a small volume has attracted much scientific speculation in the past few decades, in part because of the discovery of "pulsars" in the 1960s. These are thought to be extremely dense, rapidly rotating stars composed mainly of neutrons and emitting radiation a bit like a rotating light beacon. A "neutron star" is believed to be the remnant of a "supernova" explosion that occurs when a star collapses.

e) *Gravitational waves.* Research on pulsars provided indirect confirmation of another phenomenon predicted by Einstein from his general relativity theory. He proved that his equations allowed the propagation of gravitational waves, just as Maxwell had shown that his electromagnetic equations allowed the propagation of electromagnetic waves. Just as the acceleration of a charged particle produces electromagnetic waves, the acceleration of a massive dense object should produce gravitational waves. But gravitational waves proved to be much harder to detect.

In 1974, Russell Hulse and Joseph Taylor discovered a binary pulsar system: two neutron stars orbiting around each other. They were able to show that the observed slowing down of the motions of these stars, which could be measured very precisely, is exactly what one would expect if their rotational energy is being converted into the energy of gravitational radiation, according to general relativity.

Further consequences of relativity are discussed in Chapter 32.

Additional Problems

Problem 30.20. This problem will reveal an important truth needed to understand special relativity theory better. Imagine four identically made clocks. Put two at equal distances from the midpoint O in one coordinate system K, and the two others at equal distances from the midpoint O' in another, relatively moving coordinate system K'. At some moment, all clocks are momentarily lined up in pairs, as indicated in Fig. 30.11. (Assume that all clocks happen to show 12:00 noon as seen from O, or are just then set to show 12:00 noon by someone positioned at each clock.) At that instant also, a light flash goes off at O (or at O' they coincide momentarily anyway). (a) Explain why O (in K) will find that the two clocks at A and B are synchronized. (b) Explain why O' will find that clocks A' and B' are not synchronized (for example, A' shows 12:04 p.m. on its face when B' shows only 12:02 p.m., if the distances involved are great lenough! (c) Explain why O' is justified in saying that there is something wrong with the setting of his clocks at A' and B', and why he is justified in changing the settings; for example, setting the clocks at A' and B' both to 12:03 p.m. (d) Explain why O, seeing this being done, may object to it, but to no avail. (e) Explain why from now on O and O' each have synchronized clocks, *but will never again get the same measurement for the length of any relatively moving "object"* (say for a parade marching in K, or simply a stick lying in K and measured both by O and by O', or for a stick lying in K' and measured both by O and O'). (f) Explain what has been lost and gained by the procedure of resetting the clocks.

Problem 30.21. From Eq. (30.9), show that there exists also a set of inverse transformation equations:

$$x = \gamma(x' + vt')$$

$$t = \gamma(t' + vx'/c^2)$$

[*Hint:* It is arbitrary which coordinate system is considered at rest, but by convention the relative velocity of K' with respect to K, shown to go to the right in Fig. 30.4, is considered positive, and the relative velocity between the systems is reversed and hence is changed to $-v$ if K' is considered stationary and K is considered moving.)

RECOMMENDED FOR FURTHER READING

Jeremy Bernstein, "The reluctant father of black holes," *Scientific American*, Vol. 274, no. 6, pages 80–85 (1996)

Michael Berry, "Paul Dirac: The purest soul in physics," *Physics World*, Vol. 22, no. 2, pages 36–40 (1998)

H. A. Boorse and L. Motz, *World of the Atom*, excerpts from papers by Dirac, Oppenheimer, and Anderson on the positron, Chapters 69, 70, and 72

P. A. M. Dirac, "The evolution of the physicist's picture of nature," *Scientific American*, Vol. 208, no. 5, pages 45–53 (1963)

Albert Einstein, "Autobiographical Notes"

Albert Einstein, *Relativity: The Special and General Theory*, translated by R. W. Lawson from the German book published in 1917.

Albert Einstein, "On the generalized theory of gravitation," *Scientific American*, Vol. 182, no. 4, pages 13–17 (April 1950).

Albert Einstein, *Ideas and Opinions*, New York: Crown Publishers, 1954

Phillipp Frank, *Einstein: His Life and Times*

Peter Galison, "Einstein's clocks: The place of time," in *The Best American Science Writing 2000* (edited by James Glieck), pages 213–238, New York: Harper-Collins/Ecco Press, 2000

Martin Gardner, *Relativity Simply Explained*, New York: Dover, 1976

C. C. Gillispie, *Dictionary of Scientific Biography*, articles by O. Darrigol on Dirac, Vol. 17, pages 224–233; by A. V. Douglas on Eddington, Vol. 4, pages 277–282; by L. S. Swenson, Jr., on Michelson, Vol. 9, pages 371–374; by E. G. Spittler on Morley, Vol. 9, pages 530–531; by A. M. Bork on FitzGerald, Vol. 5, pages 15–16; by R. McCormmach on Lorentz, Vol. 8, pages 487–500; by S. H. Dieke on Schwarzschild, Vol. 12, pages 247–253

Banesh Hoffman, *Relativity and Its Roots*, Mineola, NY: Dover, 1999

Gerald Holton, *The Advancement of Science*, Part I, Chapters 2–6

Gerald Holton, "Einstein and the cultural roots of modern science" and other essays in his book *Einstein, History, and Other Passions*, Cambridge, MA: Harvard University Press, 2000

Gerald Holton, *Thematic Origins*, Chapters 6, 7, and 9

Gerald Holton and Yehuda Elkana (editors), *Albert Einstein: Historical and Cultural Perspectives*, Princeton, NJ: Princeton University Press, 1982

Arthur I. Miller, "On Einstein's invention of special relativity," in *History of Physics* (edited by S. G. Brush), pages 210–235

Marshall Missner, "Why Einstein became famous in America," *Social Studies of Science*, Vol. 15, pages 267–291 (1985).

Paul J. Nahin, *Time Machines: Time Travel in Physics, Metaphysics, and Science Fiction*, second edition, New York: American Physical Society/Springer, 1999

R. C. Olby *et al.* (editors), *Companion*, articles by John Stachel on relativity, pages 442–457, and by J. J. Gray on geometry and space, pages 651–660

M. N. Phillips (editor), *Physics History*, articles by R. S. Shankland on the Michelson-Morley experiment, pages 55–74, and by S. Chandrasekhar on general relativity and astrophysics, pages 97–104

Lewis Pyenson and Susan Sheets-Pyenson, *Servants of Nature,* Chapter 16

Richard Alan Schwartz, "The F.B.I. and Dr. Einstein," *The Nation,* Vol. 237, no. 6, pages 168–173 (1983)

R. S. Shankland, "Michelson: America's first Nobel prize winner in science," in *Physics History II* (edited by French and Greenslade), pages 48–54

Irwin I. Shapiro, "A century of relativity," in *More Things in Heaven and Earth,* edited by B. Bederson, pages 69–88. New York: Springer/American Physical Society, 1999.

John Stachel (editor), *Einstein's Miraculous Year: Five Papers that Changed the Face of Physics,* with an introduction by the editor, Princeton, NJ: Princeton University Press, 1998

J. H. Weaver, *World of Physics,* excerpts from papers by Michelson and Morley, Einstein, Dirac, Anderson, Laplace, and Wheeler and Ruffini, Vol. II, pages 122–130, 213–234, 738–743, 746–751; Vol. III, pages 122–150, 212–217, 467–492

L. P. Williams, *Relativity Theory: Its Origins and Impact on Modern Thought,* New York: Wiley, 1968. An anthology including artists' responses to relativity as well as extracts from the writings of Einstein and other physicists.

Arthur Zajonc, *Catching the Light,* Chapter 10

CHAPTER 31

The Origin of the Solar System and the Expanding Universe

31.1 The Nebular Hypothesis

The origin of the solar system, and of the earth in particular, is one of the most fundamental problems in astronomy. Apart from the intrinsic fascination of understanding how and when our planet was formed, a plausible solution to that problem could suggest an answer to a more general question: Is the formation of a planetary system a normal feature of star formation or a rare accident? Are planetary systems likely to be so abundant throughout the universe that the chance of communicating with other civilizations is high, or is life on earth the lonely exception in a dying galaxy?

Let us trace the slow rise and fall of various ingenious theories over recent centuries, leading up to the current understanding—and once more see brilliant minds engaged in the unfolding adventure of understanding the cosmos through scientific thought.

Before the eighteenth century, European astronomers believed that God created the solar system in its present state only a few thousand years ago. In the 1700s, speculations proliferated about physical processes that might have formed the sun and planets over a longer period of time. The French naturalist Buffon (see Section 18.2) imagined a close encounter between the sun and a comet, which drew from the former hot gases that later condensed to form the planets. He estimated that the earth had cooled down over a period of about 75,000 years.

During the nineteenth century most astronomers accepted the "nebular hypothesis" proposed in 1796 by the French theorist Pierre Simon de Laplace and the German-British astronomer William Herschel (Section 16.8). Laplace stressed the uniformity of motions in the solar system: All planets go in the same direction around the sun in nearly the same plane—the "ecliptic" plane—and rotate in the same direction around their own axes, and all satellites go in the same direction around their primaries. Such uniformity would be extremely improbable if these bodies had been formed independently. For Newton that was a proof that the system had been designed by God; for Laplace it showed that the system had developed through a unitary physical process. (He was not worried by exceptions such as Uranus, which rotates and whose satellites move in orbits in a plane nearly perpendicular to the ecliptic plane.)

Laplace suggested that the atmosphere of the sun originally extended throughout the entire space now occupied by planetary orbits; it was a hot, luminous rotating cloud of gas, gradually losing its original heat to the cold surrounding space. As the cloud cooled it contracted, rotated faster, flattened into a disk, and broke up into rings, which then condensed into planets (Fig. 31.1). Satellites could be formed by a similar process from clouds surrounding the planets. The remaining central portion became the sun. (There is one obvious defect in the theory: The sun should rotate faster than the speed of revolution of Mercury around it, but in fact it rotates much more slowly.)

Herschel suggested that stars are formed by contraction from nebulae like those he observed through his telescope. Thus the Laplace-Herschel nebular hypothesis, a "monistic" process requiring no outside stimulus to start the development of the system, implied that planetary formation is a universal accompaniment of the birth of stars, and that with sufficiently powerful telescopes we should be able to observe the earlier stages of a process occurring in our galaxy right now. Also, it would seem likely that life has arisen or will arise on some of these many planets. A "dualistic" process like Buffon's, on the other hand, would be much rarer, especially when it is realized that the foreign body would have to be another star rather than a comet.

The German scientist Hermann von Helmholtz, in 1854, modified the Laplace-Herschel theory by assuming that the primordial nebula was cold rather than hot; gravitational contraction

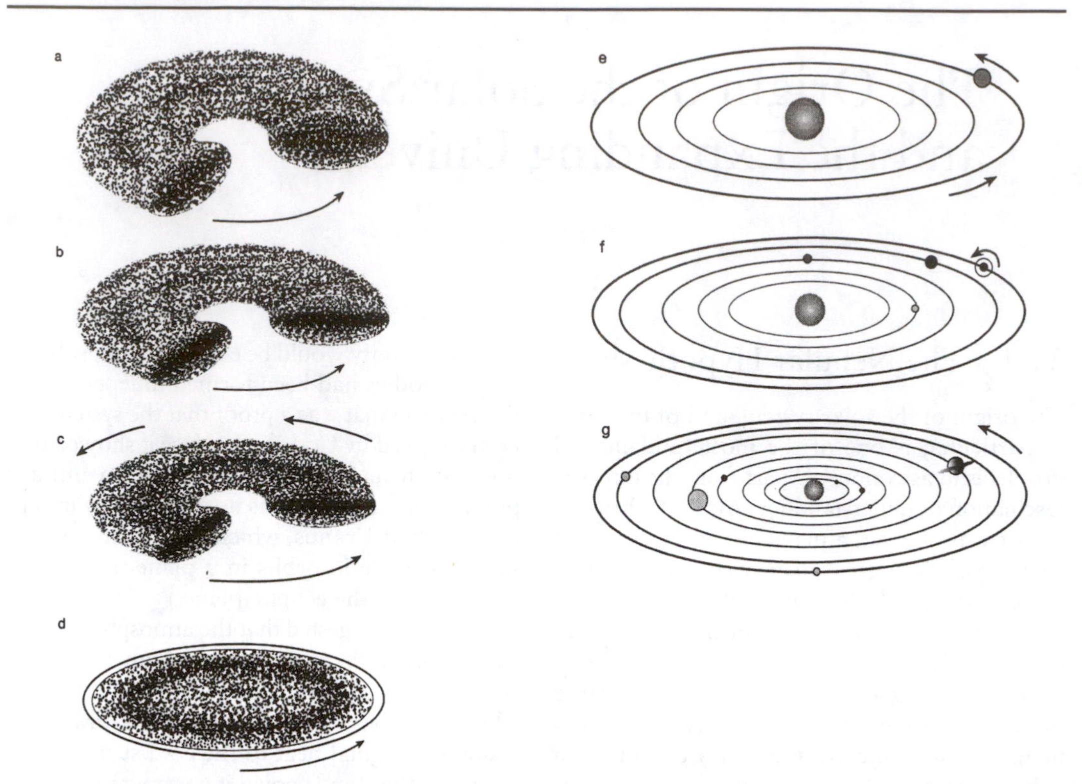

Fig. 31.1. Laplace's nebular hypothesis. (a) Rotating spherical cloud has started to flatten. (b) Most of the material has fallen to equatorial plane, forming a flat disk. (c) Disk cools and contracts, spins faster (conservation of angular momentum). (d) When the outside of the disk spins so fast that gravitational attraction can no longer counteract its inertial motion, it no longer contracts but is left as a separate ring. (e) This process is repeated, forming several rings; material in the outer ring condenses to a large blob, a "protoplanet" that may eventually become a planet. (f) Each ring forms a protoplanet; outer protoplanets repeat the original in miniature, spinning off rings, which form protosatellites. (g) At end of process, all rings have formed planets and most planets have satellites. One planet (Saturn) also has flat rings that did not condense into satellites. (The asteroid belt between Mars and Jupiter was subsequently interpreted as the remnant of a ring that could not collect into a single planet because of the interfering action of Jupiter, the most massive planet.)

would convert mechanical energy to heat (in accordance with the law of conservation of energy), and thus account for the present high temperature of the sun.

The nebular hypothesis was closely connected with nineteenth-century geological theories, which generally presumed that the earth had been formed as a hot fluid ball and then cooled down, solidifying on the outside first (Section 18.2).

The British physicist Kelvin used the cooling model to estimate the age of the earth and found it to be only 20 to 100 million years, compared with the several hundred million years assumed by geologists to have been available for slow processes like erosion to produce their observed effects. Adopting Helmholtz's hypothesis of heating by gravitational contraction, Kelvin also estimated that the sun could not be much older than 20 million years. Kelvin's estimates caused difficulties for Charles Darwin's theory of biological evolution, since Darwin had assumed, following the geologists, that long periods of time would be available for present-day species (including humans) to emerge by the slow process of natural selection (Section 18.3).

At the beginning of the twentieth century the credibility of the Laplace-Herschel-Kelvin scenario for the development and present state of the earth was undermined by the discovery of radioactivity. The earth's crust contained enough radium, the highly radioactive element isolated by Marie and Pierre Curie in Paris, to generate the heat needed to replace that lost from the earth's crust by conduction into space; the earth might actually be warming up rather than cooling down. Analysis of the amounts of uranium, radium, lead, and helium in rocks quickly led to estimates that these rocks are several thousand million years old (Section 27.2).

During the 1940s, scientists devised methods to estimate the ages of rocks by comparing their proportions of lead isotopes with the proportions in meteorites containing very little uranium. The latter could be regarded as the "primordial" proportions of lead isotopes at the time of formation of the solar system since none of their present lead isotopes was derived from radioactive decay of uranium. Arthur Holmes and F. G. Houtermans used this method to estimate that the earth was formed about 3 Ga (= 3 billion years) ago.

In 1953 the American geochemist Clair Cameron Patterson (1922–1995) used very accurate determinations of lead isotopes in rocks and meteorites to arrive at the presently accepted value for the age of the earth: about 4.5 Ga. Remarkably, this estimate has remained essentially unchanged for half-a-century, during a period when the "revolution in the earth sciences" (plate tectonics) has changed most of what was believed about the earth's past history.

Modern radioactive dating of the earth (part of the subject known as "geochronology") has settled the dispute between Kelvin and the nineteenth-century geologists who assumed that the earth is several hundred million years old, and has thereby demolished one of the strongest arguments against Charles Darwin's theory of evolution by natural selection. The geologists actually *under*estimated the age of the earth (insofar as they made any quantitative estimates); the antievolutionists, who relied on Kelvin's calculations to assert that there had not been enough time for Darwin's gradual process to work, were deprived of their scientific support. The shoe is on the other foot: It is now the "young-earth creationists," who claim that the earth and in fact the entire universe were created only a few thousand years ago, who find their views flatly contradicted by modern physical science.

31.2 Planetesimal and Tidal Theories

Even before radioactivity had cast doubts on the nineteenth-century cooling-earth model, the American geologist Thomas Chrowder Chamberlin (1843–1928) had proposed replacing that

Fig. 31.2. Thomas Chrowder Chamberlin (1843–1928), American geologist. Rejecting Laplace's nebular hypothesis, Chamberlin proposed that the planets formed by accretion of small solid particles, which he called planetesimals. This hypothesis is used in modern theories of the formation of earthlike planets, although Chamberlin's other hypothesis, the interaction of the sun with another star (Fig. 31.3), is no longer accepted.

model by a completely different hypothesis about the origin of the earth and other planets. His researches on past ice ages led him to doubt the doctrine that the earth had cooled down from an initial temperature of several thousand degrees. With the help of the American astronomer Forest Ray Moulton (1872–1952), he showed that the nebular hypothesis could not satisfactorily explain the physical properties of the solar system (for example, the fact the giant planets have most of the angular momentum in the solar system rather than the sun).

Chamberlin proposed in 1903 that the planets and satellites had been formed by the aggregation of small cold particles, which he called "planetesimals" (meaning "infinitesimal planets"). The formation process would generate enough heat to bring them up to their present temperatures.

Having rejected Laplace's theory of the formation of the solar system, Chamberlin still seemed to be attached to Herschel's old idea

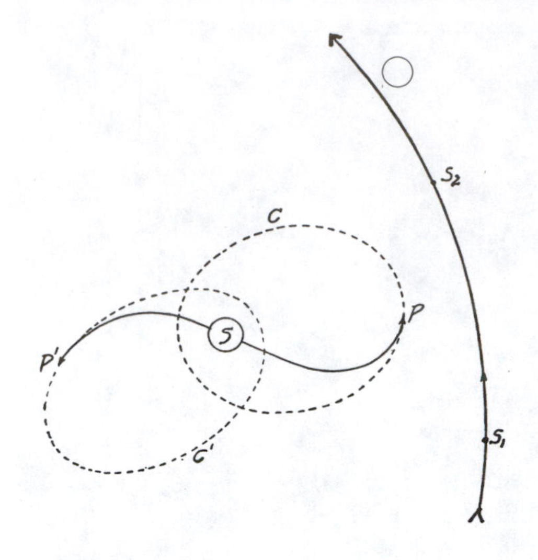

Fig. 31.3. Chamberlin's diagram showing development of orbits of material left near the sun *S*, under the gravitational influence of the receding intruder star at successive positions S_1, S_2.

that the nebulae one sees in the sky are associated with the early stages of stellar evolution. Looking at photographs of spiral nebulae taken by the American astronomer James Keeler, he speculated that the two prominent arms belonged to two previously distinct celestial objects. From this thought, and contemplation of solar prominences, he was led to the idea that a planetary system could have been generated when another star passed close to the sun. Just as the moon raises tides on the near and far sides of the earth, the gravitational force of the intruder would cancel the gravitational force holding the solar gases in on the near and far sides of the sun, allowing two filaments of material to flow out; the filaments would then be curved by the continued action of the intruder as it recedes (Fig. 31.3). Chamberlin assumed that the filaments would eventually condense into small solid particles, which would be captured into orbits around the sun.

When it became clear in the 1920s that spiral nebulae are galaxies rather than objects that could be as small as planetary systems, Chamberlin dropped this part of his theory but retained the assumption that two stars interacted in order to release into space the material from which planets formed.

As it happened, the British geophysicist Harold Jeffreys and the British physicist-astronomer James Hopwood Jeans independently proposed a similar assumption in 1916 and 1917, respectively. It became known as the "tidal" theory of the origin of the solar system. But Jeans and Jeffreys rejected Chamberlin's planetesimal hypothesis, retaining instead the older assumption that the planets formed from fluid balls.

Astronomers realized that any theory that required the close encounter of two stars in order to form planets would entail an extremely small number of planetary systems in the universe (and an even smaller number of planets that could support the evolution of life). This was consistent with the failure to find convincing evidence for planetary systems surrounding stars other than our own. Jeans seemed to take perverse pleasure in the idea that we are the result of a chance event that has happened only once in the history of the universe and (because the stars are decaying and moving farther apart) will probably never happen again.

The tidal theory, whether the Chamberlin-Moulton or Jeans-Jeffreys version, was generally accepted by astronomers until 1935, even though it was never worked out in sufficient detail to provide a convincing explanation of the quantitative properties of the solar system. Its supporters believed that the tidal theory could overcome the major defect of the nebular hypothesis by showing at least qualitatively how most of the original angular momentum could have been given to the major planets rather than to the sun.

But the tidal theory turned out to have serious defects as well. The American astronomer Henry Norris Russell found two major objections. First, theories of stellar structure developed by Arthur S. Eddington and others in the 1920s indicated that the gases drawn from the interior of the sun would be at such a high temperature—on the order of 1 million degrees—that they would dissipate into space before they could condense into planets. Second, a simple dynamical calculation showed that it would be impossible for the tidal encounter to leave enough material in orbits at distances from the sun corresponding to the giant planets. (In other words, like the nebular hypothesis it failed to explain the distribution of angular momentum in the solar system.)

There was now *no* satisfactory theory of the origin of the solar system, since the earlier objections to the nebular hypothesis still seemed valid.

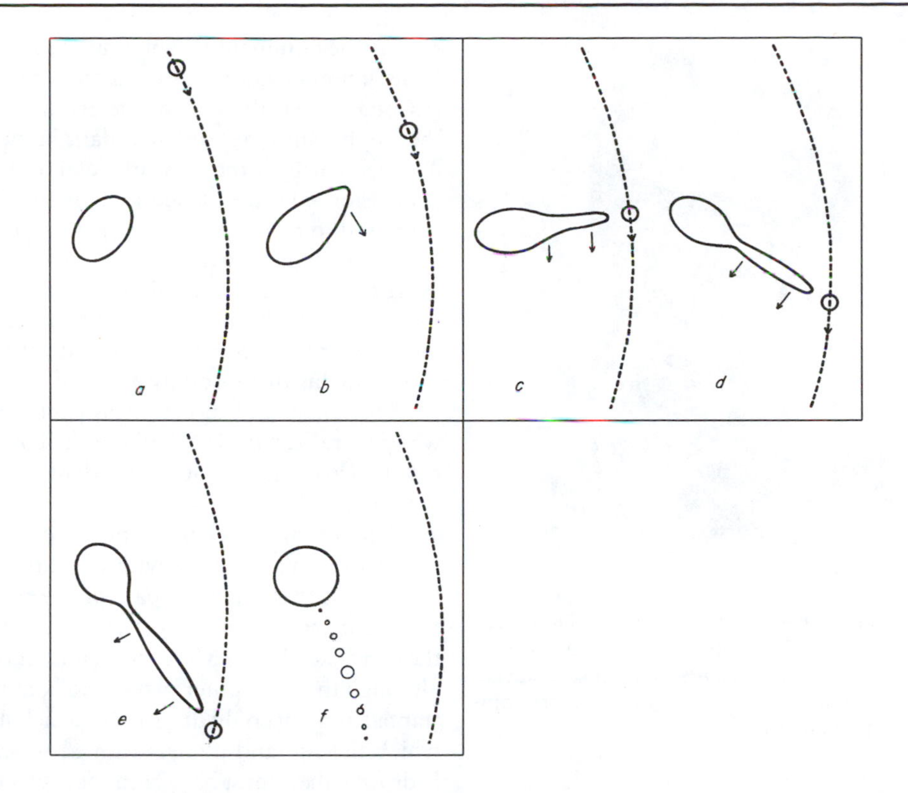

Fig. 31.4. Jeans' tidal theory. "This sequence of six diagrams (*a, b, c, d, e, f*) is intended to suggest the course of events in the sun, or other star, when its planets were coming into existence. For simplicity the second star is represented throughout by a small circle, although actually this would, of course, also experience deformation and possible break-up." ("The Evolution of the Solar System," 1943)

31.3 Revival of Monistic Theories after 1940

New ideas and observations provided a plausible basis for reviving a monistic theory after 1940. For example, the Swedish plasma physicist Hannes Alfvén proposed that the early sun had a strong magnetic field and was surrounded by an ionized gas. According to his reasoning from Maxwell's electromagnetic theory, lines of magnetic force, rotating with the sun, would be "trapped" in the ionized gas and transfer angular momentum to it. In this way one might explain why most of the angular momentum of the present solar system is carried by the giant planets rather than by the sun.

Alfvén's "magnetic braking" mechanism to slow the sun's rotation was adopted by several theorists in the 1950s and 1960s even though they rejected other aspects of his theory of planetary formation. For those familiar with early modern astronomy it is a curious echo of Kepler's theory, in which magnetic forces from the sun push the planets around (Section 4.3).

Another problem with the nebular hypothesis was the implausibility of the assumption that the nebula would start to condense by itself. As Jeans and the French mathematician Henri Poincaré had pointed out early in the twentieth century, if the total mass of the planets and the sun were uniformly distributed throughout the entire volume of solar system, the density would be so small that (unless the temperature is very low) the atoms would simply dissipate into space rather than condensing.

This difficulty could be avoided if the nebula had originally been much more massive, and then (after the condensation process started) had lost most of its mass. The clue to how this could have happened was furnished by a major discovery made by a British graduate student at Harvard, Cecilia Payne (later Payne-Gaposchkin, 1900–1979).

Up to 1925 it was generally believed that the sun and other stars are composed primarily of

Fig. 31.5. Cecilia Helena Payne-Gaposchkin (1900–1979), British-American astronomer. Her analysis of the spectra of the sun and stars led to the conclusion that they (and indeed the entire universe) are composed mostly of hydrogen.

elements like oxygen, silicon, and iron, because the spectral lines identified with these elements are quite prominent in solar and stellar spectra. Moreover, if the earth had once been part of the sun, as the dominant Chamberlin-Moulton and Jeans-Jeffreys theories assumed, then it would be reasonable that those elements are abundant in the sun because they are abundant in the earth. The nebular hypothesis would lead to the same conclusion since the planets and the sun were all supposed to be formed from the same nebula.

By a detailed analysis of the solar spectrum using quantum theory combined with the kinetic theory of gases, Payne found that *hydrogen is the most abundant element,* followed by helium. This conclusion, strongly resisted at first by other astronomers, was eventually confirmed, and was generalized in the 1930s to the conclusion that hydrogen is the most abundant element in the universe.

The earth's interior seems to consist primarily of iron and rocks, with very little hydrogen except that which is part of the oceans as H_2O. The same is probably true of the moon and the earthlike planets Mercury, Venus, and Mars, although the large outer planets do contain substantial amounts of hydrogen. If the nebula from which the sun and planets formed was mostly hydrogen, there must have been some process that removed most of the hydrogen from the inner part of the solar system. Therefore the original nebula must have been much more massive than the present solar system, and could have been dense enough to allow for the initial condensa-

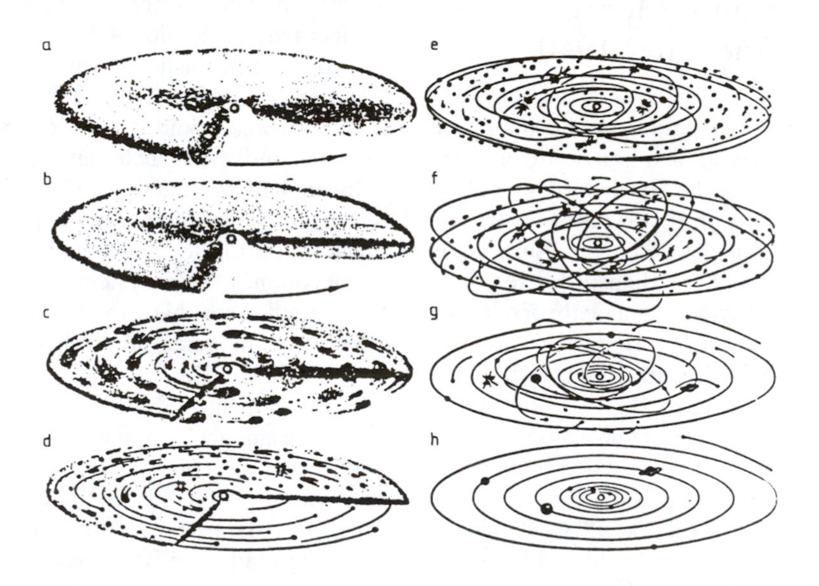

Figure 31.6. Safronov-Wetherill model for formation of planets. In contrast to the nebular hypothesis (Fig. 31.1), in which gaseous rings are formed and then each ring condenses to a planet, the Safronov-Wetherill model postulates direct condensation to solid planetesimals at an early stage.

Fig. 31.7. Viktor Sergeivitch Safronov (b. 1917), Russian astronomer whose theory of the formation of earthlike planets is widely accepted.

tion of solids from gas, thus meeting the objection of Jeans and Poincaré.

That idea provided the basis for monistic theories proposed in the 1940s and 1950s. The revival of monistic theories began with a 1944 paper by the German astrophysicist C. F. von Weizsäcker. At the same time Alfvén and the Russian astronomer Otto Schmidt developed theories that were dualistic, insofar as they postulated previously formed clouds of material captured by the sun, but their theories concentrated on the subsequent development of the solar system and eventually became monistic theories.

Schmidt's theory was developed by Victor Safronov and other Soviet scientists throughout the 1960s and 1970s. It became primarily a model for the accumulation of small solid particles (Chamberlin's planetesimals) from the protoplanetary cloud into planets. Safronov's model was adopted, with some modifications, by the American geophysicist George W. Wetherill, who explored its consequences with the help of computer calculations. The Safronov-Wetherill model (Fig. 31.6) is now considered the most plausible one for the formation of the earthlike planets, although it does not yet account quantitatively for their properties.

The Safronov-Wetherill model assumes that the process of planetary formation involves many collisions among solid particles of different sizes; some of these collisions will lead to sticking and buildup of larger objects, but some will cause the breakup of previously formed objects. The calculations predict that in each region (a doughnut-shaped ring around the sun) a single planet will probably become large enough to survive and sweep up the rest of the material in that region by gravitational attraction. (This may have failed to happen in the region of the asteroid belt between Mars and Jupiter because Jupiter's strong gravitational field interfered with the process.)

The Safronov-Wetherill model is also consistent with the current theory for the origin of earth's moon, proposed by several scientists in the 1970s: After the earth was formed it was struck by another planet, perhaps as large as Mars or Venus. The collision ejected and vaporized material from the earth's mantle (Fig. 31.8), and this material, together with fragments of the other planet, eventually condensed to form the moon.

The most rigorous test of the model began in the 1990s with the discovery of several planets orbiting other stars. This is one of the most exciting areas of science today, and many

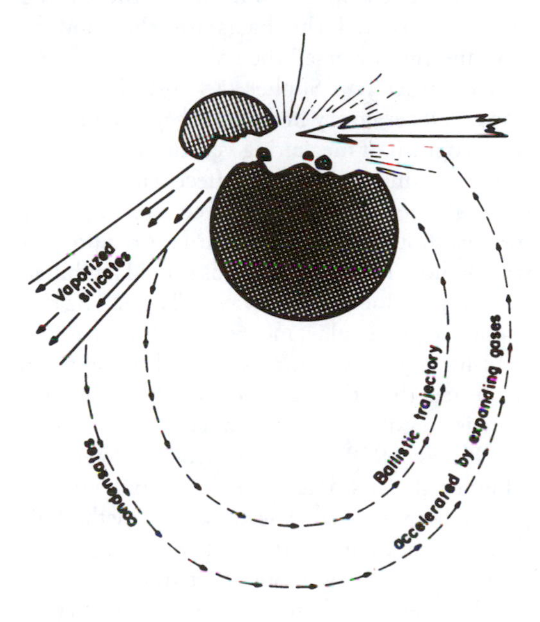

Fig. 31.8. Giant impact theory for the origin of the moon, proposed by William K. Hartmann and Donald R. Davis in 1975, and in a different form by A. G. W. Cameron and W. R. Ward in 1976. Impact of Mars-sized object (arrow to top) knocks off and vaporizes silicates from mantle of early earth. Portions of vapor then condense; expansion of remaining silicate vapor accelerates condensates into orbits around earth. (If solid particles were simply ejected from earth's surface they would go into orbits that return them to the surface.)

important questions remain to be answered by biologists as well as astronomers. In particular: What are the necessary conditions for the origin and evolution of life? Are those conditions likely to be met in a few or many planetary systems? Do most planetary systems (like some of the first ones discovered) have giant Jupiter-like planets moving in eccentric orbits close to the sun, a situation unfavorable to the stable existence of earth-like planets with moderate climates?

31.4 Nebulae and Galaxies

If the nebulae we see in the sky are not direct precursors of planetary systems, what are they? By the end of the nineteenth century, astronomers had decided that these fuzzy luminous patches, sometimes resolvable by high-powered telescopes into individual stars, are "island universes"—galaxies outside our own Milky Way and perhaps comparable to it in size and structure. Early in the twentieth century, observations with more powerful telescopes, primarily in the United States, provided the basis for the modern "expanding universe" theory.

Vesto Melvin Slipher (1875–1969) at the Lowell Observatory in Arizona began an extensive program of measuring velocities of nebulae in 1912, using the Doppler effect. Theoretically, spectral lines emitted by objects moving away from us should be shifted toward the red (longer-wave); those moving toward us should be shifted toward the blue (shorter-wave). By 1925 he had studied forty nebulae, most of which were found to be moving away from us. Several astronomers suggested that the velocities of those nebulae increase regularly with distance.

Before 1920, it was difficult to determine whether the nebulae were really outside our galaxy, because there was no reliable method for determining their distances. It was necessary to supplement the trigonometric parallax method (used by Bessel and others in the nineteenth century to measure the distances of the closest stars, see Problem 1.3) with some other method of estimating the distances of those stars too far away to have a measurable parallax. This was done by the Danish astronomer Ejnar Hertzsprung (1873–1967) and the American astronomer Harlow Shapley (1885–1972) on the basis of a discovery made by another American astronomer, Henrietta Swan Leavitt (1868–1921), who analyzed observations of Cepheid variable stars. These are a class of giant stars, first discovered in the constellation Cepheus, which brighten and dim at regular intervals.

Leavitt, working with observations made at the Harvard College Observatory, discovered in 1908 the general rule that brighter variables have longer periods. Since in one particular group all of the stars could be assumed to be at the same (known) distance, this meant that the intrinsic brightness (luminosity) was directly correlated with the period of variation. The observed (apparent) brightness of a star is inversely proportional to the square of its distance from us. It was reasonable to assume that a more distant Cepheid variable that had the same period of variation has the same *intrinsic* brightness, and by comparing that number with its *observed* brightness and using the general correlation formula, one could deduce its distance.

The island universe theory gained credibility after 1900 when several observations suggested that nebulae are not part of our galaxy. For example, Slipher's first measurement of the Doppler shift of Andromeda indicated that it is moving toward us at 300 km/sec, and soon afterward he found two other spiral nebulae receding from us at 1100 km/sec. Such high speeds were unknown for stars in our galaxy, and some astronomers concluded that the spiral nebulae were thus outside our galaxy.

A related problem was the size of our galaxy and our location within it. Statistical studies of the distribution of stars, by J. C. Kapteyn and others, led to a lens-shaped model of the galaxy with the sun near the center and a radius of about 30,000 LY (light years). But in 1918 Shapley announced that the galaxy has a diameter of 300,000 LY, with the sun located several tens of thousands LY away from the center. His model of the galaxy was based partly on his use of Leavitt's period-luminosity relation to estimate the distance of globular star clusters in our galaxy.

Shapley's "big galaxy" theory led him initially to reject the island universe theory because he thought it unlikely that the spiral nebulae could be as large as our galaxy. Heber D. Curtis at Lick Observatory in California was the leading supporter of the island universe theory.

On April 26, 1920, Curtis and Shapley presented their views at the National Academy of Sciences in Washington, DC. This has been called "astronomy's great debate." There were two points of contention: the validity of the island universe theory of nebulae, and the size and structure of our own Milky Way galaxy. Curtis, the advocate of the island universe theory, attacked the use of the period-luminosity relation to estimate distances of globular clusters, and argued that the galaxy is only about 30,000 LY in diameter (1/10 Shapley's value). This would be consistent with that theory if one assumes that nebulae are external galaxies comparable in size to ours, whereas if they were as big as Shapley's big galaxy they would have to be extremely distant and the novae we see in them would be impossibly bright. Shapley defended his big galaxy theory but didn't give much attention to the nebulae.

In 1923, the American astronomer Edwin P. Hubble (1889–1953) found a Cepheid variable in the Andromeda nebula, and in 1924 he detected several more there and in another spiral nebula. Using Leavitt's period-luminosity relation he estimated their distances to be about 800,000 LY. In retrospect this discovery is considered to have established the island universe theory once and for all, although acceptance of the result was somewhat delayed because the Cepheid yardstick was not generally adopted by astronomers. (In fact it was Shapley, rather than supporters of the theory, who placed confidence in the Cepheid method for measuring distances; but since he apparently didn't think that spirals contained stars he didn't look for Cepheids there.) It also appeared that the nebulae are much smaller than our galaxy, if Shapley's estimate is correct for the latter.

Thus while nebulae were shown to be outside our galaxy, it was not yet clear that they are really galaxies themselves; if they are, then our galaxy would seem to be exceptionally large compared to all the others. This problem was partly resolved when R. J. Trumpler in 1930 established the existence of interstellar absorption. If some of the light emitted by a star is partly absorbed before it gets to our telescopes, the star will appear farther away than it really is. Thus our galaxy is smaller (by a factor of 3) than Shapley thought, although other features of his big galaxy model were subsequently confirmed—the off-center position of the sun and the role of globular clusters in outlining the galaxy. The rest of the discrepancy was removed when Walter Baade revised the distance scale in 1952, making the nebulae farther away and hence larger than previously thought and closer in size to our galaxy.

The outcome of the "great debate" was a victory for each astronomer on the issue he considered most important: Curtis' island universe theory was adopted along with Shapley's model of the our galaxy.

31.5 The Expanding Universe

In the meantime astronomers, inspired in part by the Dutch cosmologist Willem de Sitter's 1917 solution of Einstein's general-relativistic field equations, had been attempting to find the average recession speeds of nebulae (v, assumed to be proportional to their Doppler red-shifts) and their distances (d). Hubble proposed such a relation in 1929:

$$v = Hd$$

with H, the Hubble constant, equal to the inverse of a *characteristic time* $1/H$. Hubble's original estimate for $1/H$ was about 1 billion y, later raised to 2 billion y (2 Ga or 2×10^9 y).

For example, if a galaxy is moving at a speed of 500 km/sec, we can estimate from Hubble's law, with $1/H = 2 \times 10^9$ y), that its distance is approximately

Fig. 31.9. Edwin Powell Hubble (1889–1953), American astronomer who found that the average recession speeds of galaxies are proportional to their distances. This result provided the basis for the expanding universe theory.

$$d = v/H = (500 \text{ km/sec}) \times (2 \times 10^9 \text{ y}).$$

Since 1 y = 3.16×10^7 sec, the distance is

$$d = 1000 \times 10^9 \times 3.16 \times 10^7 \text{ km}$$

$$\approx 3 \times 10^{19} \text{ km}.$$

This distance is approximately 1 megaparsec (1 million parsecs), a convenient unit for astronomical distances, equal to about 3 million LY (see Appendix IV). As we will see shortly, *H* was subsequently found to be much smaller than Hubble himself thought, so the speed of 500 km/sec corresponds to a much greater distance.

Hubble is often said to have "discovered the expanding universe" through the establishment of his velocity-distance relation. If one assumes that this relation can be used to extrapolate backward in time, with each galaxy keeping the same speed, then one could conclude that all the galaxies were concentrated in a very small region at a time $1/H$ years ago, thus arguing for $1/H$ being a measure of the "age of the universe." But that conclusion depends on several unproved assumptions, and even Hubble did not fully accept it. He was not even convinced that the observed red-shifts of distant galaxies were caused by their motions, and pointed out that the expanding-universe model was not consistent with other data on stars.

One reason for doubting the simple expanding-universe model was that, if one used astronomical data available up to 1950, one would obtain an unreasonably young universe: No more than 2 billion y (2 Ga) have elapsed since the beginning of the expansion. Since radiometric estimates of the age of the earth were in the range 3 to 5 Ga at that time (Section 31.1) there seemed to be a conflict between astronomy and geology: How could the earth be older than the universe?

The conflict was eventually resolved in favor of geology when the astronomers, following Walter Baade, reanalyzed their data and revised their estimates of distances and ages upward. Since the 1950s, estimates of $1/H$ have generally fallen within the range 10 to 20 Ga, while the age of the earth has been determined to be 4.5 Ga.

Fig. 31.10. Georges Lemaître (1894–1966), Belgian astronomer. His "primeval atom" theory offered an interpretation of the expanding universe consistent with relativity, quantum theory, and thermodynamics.

31.6 Lemaître's Primeval Atom

The modern concept of the expanding universe emerged from the mathematical calculations of Aleksandr Friedman, Willem de Sitter, and Georges Lemaître, based on Einstein's general relativity theory and the analysis of astronomical evidence on the distances and velocities of nebulae by Hubble.

Lemaître (1894–1966), a Belgian astronomer, proposed in 1927 that the universe originated as a single *primeval atom*. He arrived at this conception by considering how the second law of thermodynamics should be expressed in terms of quantum theory. The second law, as formulated by Rudolf Clausius and Ludwig Boltzmann in the nineteenth century, states that the *entropy* of a closed system or of the entire universe must always increase or stay constant; entropy is a measure of the randomness of the system on the atomic level (Section 22.9). The quantum theory (Chapter 26) states that light and other forms of electromagnetic radiation, although behaving like waves with definite frequencies of vibration, also behave like discrete particles whose energy E is proportional to their frequency ν. The quantum of energy is $E = h\nu$, where h is Planck's constant.

According to Lemaître, physical processes that correspond to increasing entropy should be interpreted as atomic processes in which the number of quanta increases; a high-energy quantum, by interacting with matter, is transformed into two or more quanta of lower energy. Moreover, because of Einstein's relation between mass and energy, $E = mc^2$, such a process is equivalent to the division of a single large mass into two or more smaller masses. Conversely, if we trace the history of the universe backward we should arrive at an initial state in which the total mass-energy is concentrated in a single superquantum or superatom.

Acceptance of such a theory would require us to abandon the assumption, implicit in Newtonian physics and articulated by Laplace, that the entire history of the world is determined by its state at any time. As Lemaître put it:

> Clearly the initial quantum could not conceal in itself the whole course of evolution; but according to the principle of indeterminacy, this is not necessary. Our world is now understood to be a world where something really happens; the whole story of the world need not have been written down in the first quantum like a song on the disc of a phonograph. The whole matter of the world must have been present at the beginning, but the story it has to tell may be written step by step.

This is a bold assertion that *randomness* supplies *novelty* in the world; something new and unpredictable is always coming into existence as the primeval atom subdivides into smaller and smaller portions of mass-energy.

Lemaître assumed that the entire mass of the universe, estimated at about 10^{54} grams, started as a single atomic nucleus that occupied a sphere of radius 10^{11} meters, i.e., about 1 astronomical unit (earth-sun distance) when the distance between particles was the same as that between protons in an ordinary nucleus, about 10^{-13} centimeters. This sphere is also identical with the entire space of the universe. It then begins to expand, the rate of expansion being governed by the force of "cosmic repulsion" (determined by the *cosmological constant* that Einstein introduced in his general relativity theory) and gravitational attraction. There will be three major phases in the expansion:

1. Fragmentation of the primeval atom, giving high-energy radiation and high-velocity particles.
2. A period of deceleration when gravitational attraction and cosmic repulsion are roughly balanced; in this period there will be condensations, forming galaxies and clusters of stars;
3. Renewed expansion as the repulsion dominates. Notice that the beginning of the expansion is later than the formation of stars in Lemaître's theory, thus allowing the possibility that the earth is older the characteristic time $1/H$ defined in Section 31.5. This feature was one reason for the popularity of Lemaître's cosmogony in the 1930s and 1940s when the age of the earth seemed to be greater than the "age of the universe."

The present radius of the universe, according to Lemaître, is about 10^{10} LY; up to that time, telescopes had seen out to a distance of about 5×10^8 LY, a not-insignificant part of the entire universe.

Lemaître himself gave a memorable summary of his cosmogony in 1931:

> The evolution of the world can be compared to a display of fireworks that has just ended: some few red wisps, ashes and smoke. Standing on a well-chilled cinder, we see the slow fading of the suns, and we try to recall the vanished brilliance of the origin of worlds.

Although some aspects of Lemaître's cosmogony are now considered incorrect, it is still admired as a remarkable precursor of the big bang theory established in the 1930s (Section 32.5).

RECOMMENDED FOR FURTHER READING

S. G. Brush, *History of Modern Planetary Physics,* Vol. 2, Chapters 1.1 and 2.2; Vol. 3, Chapter 1.1, 2.1, 3.1, and 4.1

Michael J. Crowe, *Modern Theories of the Universe from Herschel to Hubble,* New York: Dover Publications, 1994

Frank Durham and Robert D. Purrington, *The Frame of the Universe: A History of Physical Cosmology.* New York: Columbia University Press, 1983.

David E. Fisher, *The Birth of the Earth: A Wanderlied through Space, Time, and the Human Imagination,* New York: Columbia University Press, 1987.

N. S. Hetherington (editor), *Cosmology,* articles by Hetherington, "The great debate" and "Hubble's cosmology," pages 321–327, 347–369, and by Robert W. Smith, "Cosmology 1900–1931," pages 329–345

John North, *Norton History of Astronomy,* Chapters 16 and 17

Stuart Ross Taylor, *Destiny or Chance: Our Solar System and Its Place in the Cosmos,* New York: Cambridge University Press, 1998

J. H. Weaver, *World of Physics,* excerpts from Hubble, Lemaître, and Laplace, Vol. III, pages 332–254, 467–472

Charles Whitney, *The Discovery of Our Galaxy.* Ames, IA: Iowa State University Press, 1988

CHAPTER 32

Construction of the Elements and the Universe

32.1 Nuclear Physics in the 1920s and 1930s

At the end of Chapter 27 we had to postpone our account of nuclear structure until after we had explained the radical changes in physical theory brought by Einstein, Bohr, Heisenberg, Schrödinger, and Dirac. We can now return to this subject, and show how the physics of the smallest parts of matter may be linked to the structure and evolution of the largest objects in the universe: stars, galaxies, and the universe itself. This bringing together of initially two entirely different fields is one of the triumphs of recent science.

Let us first review the situation in the 1930s. *Prout's hypothesis* (Chapter 21) that all elements are compounds of hydrogen was revived in the twentieth century when it appeared that the deviations of atomic weights from integer values could be attributed to: (1) mixing of *isotopes* of an element with different atomic weights, and (2) deviation of isotopic weights from integers because of the *mass-energy conversion* involved in forces between parts of the *nucleus*. By the 1920s it was known that a hydrogen atom consists of a positively charged particle, the *proton,* which contains most of the mass, and a negatively charged particle, the *electron,* with mass about 1/1836 that of the proton.

Since the *mass number* A is defined as the integer closest to the atomic weight of an isotope (Section 27.1), one could assume at the time that the isotope has a nucleus containing A protons. The *nuclear charge* Z (also an integer) is generally only about one-half of A. Since it was known that electrons can be emitted in the kind of radioactive decay called *β-decay,* it seemed reasonable to suppose that the nucleus contains electrons, namely $A - Z$ electrons, which partly neutralize the positive charge of the protons and leave a net charge of $+Z$.

In 1919 Ernest Rutherford observed what was then a very strange occurrence. During the bombardment of nitrogen gas with α-particles from a radium preparation, some high-speed particles were produced during the collision that were far longer in range than the stream of incoming α-particles.

After detailed investigation, Rutherford proposed that the following was happening: The α-particles, known to come in with a kinetic energy $(KE)_\alpha = 12.3 \times 10^{-13}$ joule, collided (inelastically) with the nuclei of relatively stationary nitrogen atoms. An α-particle could knock out one of the protons in the nitrogen nucleus, causing the observed emission. But then the remaining nucleus lacks one positive charge, and is no longer an atom of nitrogen but one of carbon. It was a case of *transmutation* of the nitrogen nucleus into a carbon nucleus.

Subsequent research showed that Rutherford was right in his belief that he had achieved nuclear transmutation, but wrong in his interpretation of what the transmutation produced. Rather than yielding carbon, the reaction leads to a rare but well-known type (or isotope) of oxygen, observed to recoil with a kinetic energy $(KE)_{Ox} \approx 1 \times 10^{-13}$ joule; this transmutation is accompanied, as Rutherford suggested, by the ejection of a positively charged particle, a *proton* with $(KE)_p \approx 9.5 \times 10^{-13}$ joule.

Recognizing that α-particles and protons are respectively the same structures as the nuclei of ordinary helium (He) and hydrogen (H), we can write this nuclear reaction

$$ {}_7N^{14} + {}_2He^4 \rightarrow {}_8O^{17} + {}_1H^1 \quad (32.1)$$

	${}_7N^{14}$	${}_2He^4$	${}_8O^{17}$	${}_1H^1$
KE (joules):		(12.3×10^{-13})	(1×10^{-13})	(9.5×10^{-13})
Rest mass (amu):	(14.0075)	(4.0039)	(17.0045)	(1.0081)

Here we have indicated not only the mass numbers as superscripts after the element symbols, but the more accurate masses, as later separately and carefully determined for those particles *at rest,* in a separate line underneath. These rest masses refer to the inertia that the particles exhibit if they do not also have kinetic energy.[1]

Rutherford's suggestion was epochal. It was the fulfillment of the ancient alchemic dream in unexpected form—the first induced transmutation of matter, although, of course, in quantities too small to be directly recovered. The proposal initiated a whole new way of thinking in nuclear physics. Evidently, the total rest mass of $_8O^{17}$ and $_1H^1$ nuclei is larger than the rest mass of $_7N^{14}$ and $_2He^4$, by precisely 0.0012 amu; on the other hand, the kinetic energy of the incident particle is larger (by about 1.8×10^{-13} joule) than the sum of kinetic energies for the recoiling particles $_8O^{17}$ and $_1H^1$. We expect that the term $\Sigma(m_0 + \mathrm{KE}/c^2)$ will be constant before and after transmutation; and this is true. (Check this.) The observed decrease in kinetic energy is compensated for by the increase in rest mass.

For several years it was thought that the nucleus actually does consist of protons and electrons. But detailed calculations based on quantum mechanics showed that this assumption was inconsistent with other known properties of the nucleus. Chadwick's discovery of the *neutron,* discussed in Section 27.6, was a crucial breakthrough. He identified the neutron as a particle formed when beryllium is bombarded with α-particles (helium nuclei) and confirmed that its mass is close to that of the proton.

Problem 32.1. The nuclear reaction in which Chadwick discovered the neutron is a transmutation that may be written as follows:

$$_4Be^9 + {_2He^4} \rightarrow {_6C^{12}} + {_0n^1}.$$

The rest masses of neutral isotopes are $_4Be^9$ = 9.01219 amu, $_2He^4$ = 4.00260, $_6C^{12}$ = 12.00000 [by definition]; neutron mass ($_0n^1$) = 1.00866 amu; also, the energy of the bombarding $_2He^4$ nuclei, obtained from polonium, is known to be about 5.3×10^6 eV. Calculate the energy and speed of the emerging neutrons. (Recall that 1 amu $\approx 1.66 \times 10^{-27}$ kg, 1 eV $\approx 1.60 \times 10^{-19}$ joule. Clearly express any assumptions you make in your calculations.)

Werner Heisenberg then suggested that the nucleus of all atoms (except hydrogen) contains protons and neutrons rather than protons and electrons. Thus a nucleus of charge Z and mass number A is composed of Z protons and $A - Z$ neutrons (see Fig. 27.9). *Heisenberg's model of the nucleus* was attractive to other physicists since it explained the general properties of the nucleus as well as the proton-electron model, while eliminating the defects of the latter.[2] Yet Heisenberg himself was not completely consistent in his view of the neutron, and sometimes described it as a proton-electron combination. The neutron gradually established its identity as a single particle as a result of further research in the 1930s.

A crucial piece of evidence was an experiment done by Chadwick and Maurice Goldhaber in 1934, showing that the mass of the neutron, previously in dispute, is greater than the total mass of a proton plus an electron. This meant that the neutron could not be a combination of those two particles held together by an attractive force, whose negative potential energy would have made m_n smaller than $m_p + m_e$. Instead, it is a particle that is *unstable* because it can decay into a proton plus an electron, emitting a photon (exothermic reaction). The Chadwick-Goldhaber experiment also provided another confirmation of Einstein's mass-energy relation.

More New Particles

The year 1932 saw not only the discovery of the neutron but also of the *positron,* a particle similar to the electron except that it has positive electric charge. The existence of this particle was predicted by P. A. M. Dirac from his theory of relativistic quantum mechanics (Chapter 30). This was the first example of an *antiparticle,* and it led to speculation that other particles such as the proton should also have antiparticles.[3]

---

[1]Recall that 1 amu = 1.66×10^{-27} kg. The numerical values given here are recent and were not available to Rutherford with this accuracy. As is customary, the calculation is made on the basis of the mass of the *neutral atom,* that is, the nucleus plus the electrons around the nucleus. Why is this permissible?

[2]For example, according to Heisenberg's indeterminacy principle (Chapter 29), an electron confined to the very small space inside the nucleus would be expected to have a very high momentum, and would tend to escape by β-decay at a much higher rate than was observed.

[3]The antiproton was discovered, or perhaps one should say *manufactured,* with the help of the "Bevatron," an accelerator at the radiation laboratory in Berkeley, California, by a group of physicists led by Owen Chamberlain and Emilio Segrè in 1955. The Bevatron was expressly designed to produce a beam of protons (p^+) with enough energy (several billion electron volts) to create antiprotons (p^-) by

Deuterium ($_1H^2$) was also discovered in 1932, by a group led by the American chemist Harold C. Urey (1893–1981). It is a form of hydrogen with atomic weight 2, named from the Greek *deuteros,* meaning "second." Its nucleus, the *deuteron* (d), contains a neutron and a proton. It is called an *isotope* of hydrogen because, having the same atomic number, it occupies the same place in the periodic table (Chapter 27). According to Heisenberg's model of the nucleus, different isotopes of an element have the same number of protons in the nucleus but different numbers of neutrons.

Since the nucleus, according to Heisenberg's model, contains particles with positive electric charge but none with negative charge, the question arises: What holds it together? There must be a force, called the *nuclear force* or *strong force,* that can produce enough attraction among protons and neutrons to overcome the electrostatic repulsion of protons at very short distances but that disappears when the protons and neutrons are farther apart (outside the nucleus).

Heisenberg speculated that if the neutron is a compound particle ($n = p^+ + e^-$), there may be an *exchange force* between protons similar to the force used so successfully by Heitler and London to explain the force between hydrogen atoms (Section 29.4). Thus the electron would shuttle back and forth between two protons, so that each proton exists sometimes as a proton and sometimes as a neutron. In other words, proton and neutron are two states of the same particle.

Hideki Yukawa (1907–1981), a Japanese physicist, was inspired by Heisenberg's speculation to propose in 1935 a theory of nuclear forces involving the exchange of a hypothetical new particle. He showed that this exchange process would produce a force that can be described by a potential energy function in which the ordinary Coulomb potential e^2/r for electrostatic forces (Section 24.12) is multiplied by an exponential function, $e^{-r/R}$. The force would decrease rapidly to zero when the distance r is much greater than the parameter R, called the *range* of the force. The mass of the exchange particle would be inversely proportional to R. Assuming $R \approx 2 \times 10^{-15}$ m, a length about an order of magnitude greater than the size of the atomic nucleus (see Section 27.4), Yukawa estimated that the mass of his particle should be about 200 times the mass of the electron.

the reaction $p^+ + p^+ \rightarrow p^+ + p^+ + p^+ + p^-$. This experiment is often regarded as the beginning of the era of high-energy physics, in which new particles are discovered by being created.

Fig. 32.1. Hideki Yukawa (1907–1981), Japanese physicist, whose exchange theory of nuclear forces led him to predict the existence of a new particle, the meson. He received the 1949 Nobel Prize in Physics (the first Japanese scientist to receive the prize) after the π-meson was discovered in 1947.

In 1937, Seth Neddermeyer and Carl D. Anderson (and, at about the same time, J. C. Street and E. C. Stevenson) discovered a new cosmic-ray particle that turned out to have roughly the same mass as Yukawa's proposed particle. Although Neddermeyer and Anderson were not aware of Yukawa's theory prior to their discovery, the new particle was quickly identified as the one postulated by the previously unknown Japanese physicist. Initially known as the "mesotron" or "yukon," it was eventually named the *muon* or *μ-meson* (from Greek *mesos,* "middle," since it is intermediate in mass between the electron and proton).

Although it appeared at first that Yukuwa's theory had been spectacularly confirmed, and a number of physicists immediately devoted themselves to theoretical and experimental research on mesons, it soon became clear that the muon is probably *not* the particle that carries nuclear forces: Its lifetime is much longer and its interaction with neutrons and protons much weaker than the theory predicts for such a particle.

After World War II, cosmic-ray research was resumed with the help of new techniques such as the photographic emulsion for recording tracks, developed by the Viennese physicists Marietta Blau and Hertha Wambacher. In 1947 a group led by the British physicist Cecil F. Powell and the Italian physicist Giuseppe Occhialini observed tracks that they attributed to a new meson, soon named the *pion* or π-meson. This particle seemed to interact more strongly with protons and neutrons and had a much shorter lifetime than the muon, so it was generally accepted as the particle postulated by Yukawa as the carrier of nuclear forces. Yukawa was awarded the 1949 Nobel Prize in Physics for his far-seeing prediction of mesons, even though no one had yet developed a satisfactory quantitative theory of nuclear forces based on these particles.

Quantum Theory of Radioactive Decay

To understand the rapid development of the physics of elementary particles and the atomic nucleus in the 1930s, one must go back to the development of quantum mechanics in the previous decade (Chapter 29). One of the earliest applications of quantum mechanics to nuclear physics was made in 1928 by the Russian-American physicist George Gamow (1904–1968), and at the same time by the American physicists R. W. Gurney and E. U. Condon. They pointed out that if a particle is held inside a nucleus by attractive forces, one would not ordinarily expect it to escape unless it can somehow acquire enough speed so that its kinetic energy exceeds the potential energy of those forces. One can imagine the particle sitting at the bottom of a volcano; it would take some kind of violent upheaval to kick it up to the top and then out. However, according to quantum mechanics, the particle's wave function extends outside the nucleus; hence it has a small but finite chance of finding itself sooner or later outside the volcano even if it did not get up to the top. In effect it has "tunneled" through the side.

The *Gamow-Gurney-Condon theory* gave a very effective explanation of *radioactive decay*, a process that seems to be completely random and involves the escape of a particle that does not have enough energy to overcome the forces holding it in the nucleus. (If it *did* have enough energy it would be hard to explain why it does not escape right away instead of waiting thousands or millions of years.)

Other physicists, especially R. d'E. Atkinson and F. G. Houtermans, saw that tunneling could work in the other direction too: In particular, two protons, which ordinarily repel each other, would have a small but finite probability of getting close enough together for the short-range nuclear forces to fuse them. This opened up the possibility of explaining how elements are built up from hydrogen and how stars get enough energy to keep shining for billions of years (see the next section).

It was generally assumed that mass-energy is conserved in nuclear reactions; whenever there is a change in mass, the difference shows up as kinetic energy in accordance with Einstein's formula $E = mc^2$. This assumption, which is usually called just "energy conservation" in the general sense, was confirmed by Cockcroft and Walton in 1932, using their accelerator to bombard lithium atoms with protons to produce the reaction

$${}_3\text{Li}^7 + {}_1\text{H}^1 = 2\ {}_2\text{He}^4$$

Cockcroft and Walton measured the kinetic energies of the nuclei taking part in the reaction and found that the mass-energy equation balances within experimental error.

But late in the 1920s it was found that energy conservation does not seem to be valid for β-decay reactions. The emitted electron has on the average only about one-third of the energy it should have, and the rest of the energy seems to disappear. To cure the apparent violation of energy conservation, Wolfgang Pauli postulated that another particle is emitted and carries off the missing energy.

Enrico Fermi worked out a detailed theory of β-decay based on Pauli's proposed new particle, which he called the *neutrino* (Italian for "small neutral one"). The neutrino, symbol ν (Greek letter nu), was assumed to have zero mass and to travel at the speed of light; but it is different from a photon. Its interaction with matter is so extremely weak that it might pass all the way through the earth without being deflected. As a result it was not possible to detect neutrinos experimentally for more than 20 years after their existence was postulated. Yet physicists had so much faith in the law of conservation of energy that they preferred to believe in an apparently unobservable particle rather than abandon the law.

The neutrino was eventually detected in 1956 by Frederick Reines and Clyde Cowan, using a nuclear reactor to produce an extremely dense beam of neutrinos. Because they appear as by-products of nuclear reactions fueling the radiation of stars, they are present throughout the universe in immense numbers, even though so elusive. Recent experiments suggest that each neutrino may have a small mass; and that, coupled with their copious presence, suggests that neutrinos may account for some of the universe's total mass, so far "missing" according to calculations (Section 32.6).

The usual β-decay reaction amounts to changing a neutron into a proton and an electron. Comparison of the masses of the particles involved shows that this reaction goes with a significant liberation of energy. But it is possible to reverse the reaction, squeezing together a proton and an electron to form a neutron (and an antineutrino). This *inverse β-decay reaction* can occur only under conditions of very high pressure or temperature, but it does play an important role in stars.

Two versions of inverse β-decay were studied in the 1930s. The first was originally proposed in 1934 by two astronomers, Walter Baade and Fritz Zwicky: If a star collapses to high enough density, nearly all of its protons and electrons will become neutrons and we will have a *neutron star.* This idea was not taken seriously until about 30 years later, when astronomers discovered *pulsars* and decided that they are in fact rotating neutron stars.

The second version was studied by Gamow and the Hungarian-American physicist Edward Teller (b. 1908): the collision of two protons at high enough speeds can form a deuteron plus a positron. In 1938, the German-American physicist Hans Bethe (b. 1906) and C. L. Critchfield showed that this reaction could start a chain of nuclear reactions leading to the formation of helium in stars, generating enough energy to account for a large part of that emitted by the sun. C. F. von Weizsäcker also suggested about this time that sequences of nuclear reactions starting from hydrogen and using other light elements such as lithium and carbon could build up the elements and provide sufficient energy to keep the stars shining. Although Weizsäcker was the first to present a comprehensive qualitative discussion of the various fusion reactions, he did not succeed in establishing by detailed calculations just which reactions would be the most likely to take place.

Binding Energy, Fission, and Fusion

By 1935, fairly accurate atomic weights were known for the stable isotopes of all the elements. By comparing them with the masses of the proton and neutron one could calculate a *curve of binding energies* showing how much energy would be released or absorbed in nuclear transmutations. The curve, drawn as a function of mass number *A*, is U-shaped, starting with a high value for the light elements, going to a wide valley with the lowest point corresponding to iron, then slowly rising. Energy is released by any reaction that goes down the curve (just like rolling downhill) but energy must be supplied to go up.

In general, if the density and temperature are high enough to allow nuclear reactions to occur, light elements will combine to form medium-weight elements, a process called *nuclear fusion.* Also, heavy elements can split to form medium-weight elements, a process known as *nuclear fission.* Iron is the ultimate equilibrium state (except at very high densities, where like other elements it collapses into neutrons).

32.2 Formation of the Elements in Stars

Hydrogen Fusion

When the nuclei of hydrogen atoms are so agitated (as by the heat in stars or in "atomic" explosions) that they can collide with one another despite their mutual electric repulsion, then they may presumably join or fuse, and thereby two hydrogen nuclei (of the "heavy" variety, made of a proton and a neutron each) may become one nucleus of helium; the latter has, however, less rest mass than that of both building blocks together, and the difference is evidenced as an increase in the kinetic energy, available to the end product of the fusion process.

Problem 32.2. As first explained by Bethe, the sun obtains its energy by fusion processes that turn four hydrogen nuclei into one helium nucleus in cycles involving the emission of energy in different radiations. If each of the hydrogen atoms has a rest mass of 1.0081 and each helium atom a rest mass of 4.0039, calculate the energy available in each fusion cycle. [*Note:* The reaction happens so

Fig. 32.2. Hans Bethe (b. 1906), German-American physicist, founded the theory of energy generation and nuclear synthesis in stars. He won the 1967 Nobel Prize in Physics.

infrequently that only 1% of the sun's hydrogen is consumed in 10^9 y; still, many millions of tons of ponderable mass are lost each minute by radiation of energy into space.]

By 1938 the pieces were in place for a synthesis of ideas about stellar evolution and nuclear reactions. Briefly, the astronomers were ready to assume that hydrogen is the primordial material. Contradicting the earlier assumption that the sun and stars are composed of elements like iron and silicon, Cecilia Payne-Gaposchkin, as we noted in the previous chapter, had discovered in 1925 that the atmospheres of these bodies are mostly hydrogen and helium rather than heavy elements like iron. Her result was confirmed by Henry Norris Russell and Bengt Strömgren, so that in the 1930s it was generally believed that hydrogen is the most abundant element in the universe. So the most likely way to start the sequence of nuclear reactions is by fusing protons together to form deuterium (plus a positron and a neutrino), whether in stars or in the hypothetical hot compressed initial state of the universe known as the *big bang*.

$$2\,{}_1H^1 \rightarrow {}_1H^2 + e^+ + \nu \qquad (32.2a)$$

The probability that this reaction will occur when any two protons collide is so small that it has never been observed in the laboratory. But theoretical physicists were so sure of the validity of the β-decay theory that they seemed quite confident in the calculation indicating that the reaction will eventually occur at high densities and high temperatures.

The deuterium nucleus should then combine with another proton to form helium, emitting a γ-ray, in the two-stage process,

$${}_1H^2 + {}_1H^1 \rightarrow {}_2He^3 + \gamma \qquad (32.2b)$$

$$2\,{}_2He^3 \rightarrow {}_2He^4 + 2\,{}_1H^1 \qquad (32.2c)$$

The observed rate of energy generation by stars such as the sun is considerably greater than one would expect if hydrogen fusion provided the only source of energy. If one may assume that heavier elements are already present in a star (perhaps seeded by explosions of a previous generation of stars), a number of other reactions become possible. Bethe proposed the *carbon cycle* of nuclear reactions. It begins with the absorption of a proton by a carbon nucleus:

$${}_6C^{12} + p \rightarrow {}_7N^{13} + \gamma. \qquad (32.3a)$$

To understand (or rationalize) what follows we may invoke an empirical rule: A nucleus will tend to be unstable if it contains more protons than neutrons, because there are not enough neutrons to shield the protons from their mutual electrostatic repulsion. It will try to stabilize itself by converting one of its protons into a neutron and expelling a positron, e^+ (another example of inverse β-decay). Thus the nitrogen isotope ${}_7N^{13}$, having 7 protons and 6 neutrons turns back into a heavier carbon isotope:

$${}_7N^{13} \rightarrow {}_6C^{13} + e^+ + \nu \qquad (32.3b)$$

This carbon isotope absorbs yet another proton, forming a heavier isotope of nitrogen (${}_7N^{14}$), which is stable and can absorb yet another proton, turning into an oxygen isotope, ${}_8O^{15}$:

$$_6C^{13} + p \rightarrow {}_7N^{14} + \gamma \qquad (32.3c)$$

$$_7N^{14} + p \rightarrow {}_8O^{15} + \gamma \qquad (32.3d)$$

This oxygen isotope is unstable (more protons than neutrons) so it expels a positron and a neutrino, decaying back to nitrogen ($_7N^{15}$). The nitrogen absorbs a proton, but the resulting nucleus $_8O^{16}$, which would be stable under ordinary conditions, is unstable in the hot dense environment; it emits an α-particle ($_2He^4$), leaving the original $_6C^{12}$ carbon nucleus from which we started:

$$_8O^{15} \rightarrow {}_7N^{15} + e^+ + \nu \qquad (32.3e)$$

$$_7N^{15} + p \rightarrow {}_2He^4 + {}_6C^{12} \qquad (32.3f)$$

The net result of the complex carbon cycle is simply to convert 4 protons into a helium nucleus; a chemist might say that the carbon nucleus acts as a "catalyst" to facilitate the process, remaining unchanged itself in the long run.

Bethe showed that the carbon cycle would produce about the right amount of energy to explain the present rate of radiation from the sun; subsequent work using more accurate data on nuclear reactions indicated that the carbon cycle should be the main source of energy for stars brighter than the sun, while the proton-proton reaction should be the main source for stars dimmer than the sun. The sun itself could use either or both; this question does not seem to be completely settled yet.

Bethe did not, in his 1939 article on stellar energy, explain the origin of the carbon atoms that seemed to be needed as catalysts for his reactions. He concluded that no elements heavier than helium can be built up in "ordinary" stars; the heavier elements found in stars must already exist when the star is formed. The goal of synthesizing all the elements from hydrogen seemed to be unattainable because there are no stable isotopes with mass numbers 5 and 8 that could serve as intermediate stages in building carbon from hydrogen. The only hope seemed to be a reaction in which three α-particles simultaneously come together to form $_6C^{12}$; this might happen occasionally at the very high densities and temperatures inside stars, but calculations indicated that this "triple-α" reaction would not produce enough carbon to explain the synthesis of heavier elements.

Fig. 32.3. Fred Hoyle (b. 1915), astrophysicist and cosmologist, one of the advocates of the steady state theory of the universe. He found the mechanism by which carbon could be synthesized from lighter elements.

The problem was solved in 1953 when the British astrophysicist Fred Hoyle proposed the existence of an excited state of the $_6C^{12}$ nucleus, with an energy of about 7.7 MeV. Such a state would have the same energy as the transient three-α-particle combination, producing a "resonance" (see Section 29.4) between the two states. This resonance should, theoretically, make the reaction about 10 million times more likely to occur. The predicted resonant state, previously unknown, was quickly confirmed by laboratory experiments.

Problem 32.3. Discuss the question, "Why does the $_6C^{12}$ nucleus have a resonance that allows it to be easily formed by fusion of three α-particles?," without using any information beyond what is in this book. Is it a scientific question? Would it undermine our current scientific view of the world to consider it a scientific question?

By 1957, a comprehensive scheme for synthesizing most of the elements in stars from hydrogen and helium had been worked out by William A. Fowler, Hoyle, Margaret and Geoffrey Burbidge, and (independently) by A. G. W. Cameron in Canada. This scheme did not, however, provide a satisfactory explanation for the cosmic abundance of helium itself. That puzzle was left to be solved by Gamow's big bang cosmology (Section 32.4).

32.3 Fission and the Atomic Bomb

Artificial radioactivity was discovered in 1934 by the French physicists Irène Joliot-Curie (1897–1956), daughter of Marie Curie, with her husband Frédéric Joliot-Curie (1900–1958). They found that when certain light elements (boron, magnesium, aluminum) were bombarded with α-particles from polonium, positrons as well as protons and neutrons were emitted, and the source continued to emit positrons *after the α-particle source was removed.* It appeared that an initially stable nucleus had been changed into a radioactive one.

Enrico Fermi and his colleagues in Italy immediately undertook a systematic study of nuclear reactions induced by neutrons (rather than α-particles), looking for new elements that might be produced in this way. In particular, Fermi thought that by bombarding uranium, the heaviest known element, an even heavier (transuranium) element might result. He did produce a radioactive element, but in such a small amount that he could not determine what it was. In fact, although he did not know it, he had caused the nuclei to be broken into large, radioactive fragments.

The discovery of nuclear fission was accomplished by a group in Berlin. The Berlin group originally consisted of Otto Hahn (1879–1968), Lise Meitner (1878–1968), and Fritz Strassmann (1902–1980). It was Meitner, a physicist, who recognized the need for working with chemists—Hahn was an organic and nuclear chemist, and Strassmann was an analytical and physical chemist. (The Rome and Paris groups were composed almost exclusively of physicists.)

In 1938 Hahn and Strassmann showed that one of the substances that Fermi thought might be a transuranium element was actually an isotope of barium (atomic number 56); they also found that isotopes of lanthanum (57), strontium (38), yttrium (39), krypton (36), and xenon (54) were produced. But, as indicated by their statement quoted in Section 3.4, they were reluctant to accept what now seems the obvious conclusion: *The nucleus of uranium (92) had split into two small nuclei of unequal size.* This conclusion was soon proposed by Meitner and her nephew Otto Frisch, who introduced the term *fission* and noted that enormous amounts of energy could be released by the transformation of a small amount of the mass of the uranium nucleus.

Fig. 32.4. Otto Hahn (1879–1968), German chemist, whose experiments with Lise Meitner and Fritz Strassmann led to the discovery of nuclear fission, for which he received the 1944 Nobel Prize in Chemistry.

Uranium Fission

An isotope of uranium, ${}_{92}U^{235}$, can absorb a neutron and break up into isotopes of barium and krypton, plus three neutrons:

$$ {}_{92}U^{235} + {}_{0}n^{1} \rightarrow {}_{56}Ba^{140} + {}_{36}Kr^{93} + 3\,{}_{0}n^{1} \quad (32.4) $$

(This is only one of several possible reactions.) The emission of several neutrons in this *fission* reaction makes it possible for one such nuclear process to trigger off several like it in the immediate neighborhood, and as each reaction generates more of the same, an explosive *chain reaction* is created.

Problem 32.4. Following are the data for Eq. (32.4). The rest masses of U^{235} and the neutron are 235.04392 and 1.00867 amu, respectively. The energy of the incident neutron is negligible. After fission, the fragments have a total of about

Fig. 32.5. Lise Meitner (1878–1968), Austrian physicist who proposed the concept of nuclear fission to explain the results obtained by Hahn and Strassmann.

3.2×10^{-11} joules of kinetic energy. (a) What is the total rest mass of the fragments? (b) What is the total inertia of the fragments as it would be measured right after fission, before they have handed on their kinetic energy by collision with other atoms?

Scientists in several countries immediately started to study nuclear fission. Niels Bohr and the American physicist John A. Wheeler developed a theory of fission that could explain why some nuclei undergo fission only when hit by a very fast neutron, while others split when hit by a slow neutron.

The Hungarian-American physicist Leo Szilard (1898–1964), while living in England in 1934, recognized the possibility of weapons based on fission. He thought this research should not be made public; he applied for a patent on the chain reaction, but through an agency that allowed patents to be kept secret. After hearing of the discovery of fission in 1939 he did an experiment that showed that more than one neutron is liberated in the disintegration of uranium, hence that a chain reaction is in fact possible. Fermi confirmed this independently.

In September 1939, the Germans unleashed a fierce war that soon engulfed most of Europe. Their government, led by Adolf Hitler, vowed to replace Western civilization by a tyrannical totalitarianism, and for several years the advance of their troops seemed unstoppable. The story of the "atom bomb" development must be understood against this background (and the later entry of Japan into the war on the side of the Germans).

Thus, in September 1939, the British magazine *Discovery* carried a piece by its editor, C. P. Snow,[4] which sums up remarkably well the knowledge and attitudes of physicists at that time:

> Some physicists think that, within a few months, science will have produced for military use an explosive a million times more violent than dynamite. It is no secret; laboratories in the United States, Germany, France and England have been working on it feverishly since the Spring . . . The principle is fairly simple: . . . a slow neutron knocks a uranium nucleus into two approximately equal pieces, and two or more *faster* neutrons are discharged at the same time. These faster neutrons go on to disintegrate other uranium nuclei, and the process is self-accelerating . . . [The weapon] must be made, if it really is a physical possibility. If it is not made in America this year, it may be next year in Germany.

Leo Szilard and other European scientists who had emigrated to the United States alerted the American government to the need to see if it was indeed possible to develop an atomic bomb. In fact, the Germans had started on such research before the others, and had stopped exporting uranium to other countries. It is well known that Albert Einstein was persuaded to sign a letter to President Franklin D. Roosevelt endorsing this proposal (see Section 30.1); but while this did lead eventually to a small-scale project in the United States—for example, Einstein's letter resulted in a mere $6000 to be spent—the prospects for actually producing a weapon in time for use in the war seemed so remote that little effort was devoted by the U. S. government to the project until news of progress in Britain was received. Indeed, nuclear energy was then considered to have such low military priority that emigré

[4]Snow, a physicist-turned-novelist, later became famous for his 1959 lecture "The Two Cultures and the Scientific Revolution," which described the widening gap between the worlds of scientists and literary intellectuals; he also gave an insider's perspective on the debate between scientists advising the British government about the effectiveness of bombing in World War II, in his book *Science and Government.*

physicists in the United States and Britain were left free to work on it while being excluded (because of their legally "alien" status) from participating in projects such as radar.

Two weeks after the outbreak of World War II, the British government received a report of a speech by Adolf Hitler in which he threatened the use of a "weapon which is not yet known." Though Hitler often warned of resorting to "secret weapons," this particular speech later turned out to have been badly translated—at that point he was really talking about the German Air Force—but it worried one British official, Lord Hankey, who recalled that several years earlier Ernest Rutherford had told him that experiments on nuclear transmutation might someday be of great importance to the defense of Britain. He asked James Chadwick, discoverer of the neutron, to look into the question.

It turned out that two German physicists familiar with nuclear theory, Otto Frisch and Rudolph Peierls, were thinking about fission in Birmingham, England, and their calculations showed the feasibility of an atomic bomb project, which was launched first in Britain and subsequently in the United States.

Vannevar Bush (director of the U.S. government's Office of Scientific Research and Development) learned about the British project in the summer of 1941. He urged President Roosevelt to accelerate the American work, which up to that time had been stalled because there seemed little chance of success. In 1942 Brig. Gen. Leslie R. Groves of the Army Corps of Engineers became officer-in-charge of the *Manhattan Project*, which was set up to develop the atomic bomb. The British then terminated their own project and sent as many of their scientists as possible to help the American project produce the bomb—and also so that the British would acquire knowledge and experience that would be useful after the war.

Fermi, who moved to the United States, led a group that tried to demonstrate experimentally the feasibility of a nuclear chain reaction. They succeeded in obtaining a controlled chain reaction on December 2, 1942. This historic event took place at the University of Chicago in a laboratory set up under the football field bleachers.

A few weeks earlier, the American physicist J. Robert Oppenheimer (1904–1967) and others suggested setting up a new laboratory to design the atomic bomb. To preserve secrecy scientists should be brought together at a remote location. A site at Los Alamos, New Mexico, was selected.

Fig. 32.6. J. Robert Oppenheimer (1904–1967), American theoretical physicist, director of the Manhattan Project that designed the first atomic bombs.

The first staff assembled there in March 1943, with Oppenheimer as director.

In order to make a fission bomb—if the laws of physics really allowed it—it was necessary either to enrich the 235 isotope of uranium or to use the 238 isotope to produce the new element 94, now called *plutonium.*[5] Gaseous diffusion, one of the methods used to enrich U^{235}, was discussed in Section 22.5.

The Los Alamos group considered two possible methods for assembling a *critical mass* of fissionable material. (If more than the critical mass is present, the neutrons produced by the initial fission reactions will cause further reactions in a self-sustaining chain reaction; if less than the critical mass is present, enough neutrons will escape from the system so that the fission reactions will not continue.) The simpler method was

[5]One of the scientific byproducts of the Manhattan Project was the artificial creation and study of the *transuranic elements,* those with atomic numbers higher than that of uranium. Edwin M. McMillan and Philip H. Abelson had already produced element 93 (neptunium) at Berkeley in 1940. Under the leadership of Glenn Seaborg, the Berkeley group created plutonium, followed by elements 95 (americium), 96 (curium), 97 (berkelium), 98 (californium) and others (see the periodic table in Section 21.4).

a "gun" that would fire one subcritical mass into another, so that the combination would be critical. The more complex was implosion, suggested by Seth Neddermeyer: surrounding a subcritical mass with high explosives, deployed in such a way as to produce a converging shock wave that would compress the material into a small volume; this would be equivalent to exceeding the critical mass. George B. Kistiakoswki, an expert on explosives, was brought in to develop this method. It was discovered that, for technical reasons, only implosion could be used with plutonium.

To be sure that at least one workable bomb could be produced, the Los Alamos group developed two kinds of bombs: a gun weapon, using U^{235}, and an implosion weapon, using plutonium. Since they were not sure the implosion method would work, it was thought necessary to test it first, and this was done at Alamogordo, New Mexico, on July16, 1945.

The bomb dropped on Hiroshima on August 6, 1945, was a uranium weapon using the gun method; the bomb dropped on Nagasaki on August 9, 1945, was a plutonium weapon using the implosion method.

The use of the atomic bomb against Japan in 1945 shocked the world and raised questions that still ignite strong passions many decades later. Would the Japanese have surrendered anyway, without being atomic-bombed, or would they have put up a bitter if suicidal resistance, forcing the further sacrifice of many American and Japanese lives? Conversely, how did the atomic bomb influence the outcome of the war and the history of international relations after the war? (It is correct to say that the bomb *ended* the war, but radar, cryptography, and such developments as synthetic rubber helped to *win* it.)

And finally, why did the government not listen to two groups of distinguished physicists—one led by Szilard and the other by James Franck—who, before the work on the bombs was completed, urged the U. S. government not to use the bombs on civilian targets? It is worth noting that most of the major physicists after the war were leaders in urging universal arms control (through organizations they founded, such as the Federation of American Scientists, the Union of Concerned Scientists, and the Pugwash movement, all of which still exist and continue to be active).

A major motivation for scientists who were asked by the government to join the Manhattan Project was the reasonable fear that the Germans, triumphant so far in their brutal war, would resort to the use of atomic weapons. After all, that country had the expertise, with such nuclear scientists as Otto Hahn and Werner Heisenberg. The Anglo-American effort was thus justified as being primarily defensive. It was not until November 1944 that a scientific intelligence mission, "Alsos," discovered that the German atomic bomb program was far behind the American effort; thus there was no longer a strong basis for concern that the Germans might be able to use the bomb.

For a while after 1945 it was believed that Heisenberg and other German physicists intentionally pushed the German atomic bomb project into a dead end because they did not want Hitler to have the bomb, and Heisenberg did not publicly deny this suggestion. But recently released evidence (transcripts of secret recordings of conversations between Heisenberg and his colleagues made while they were detained in England at the end of the war) shows that Heisenberg did not have a clear understanding of some technical details of chain reactions, and consequently overestimated the critical mass of uranium that would be needed to make a bomb. Based on this mistake (and several others made by his co-workers), rather than on moral considerations, he did not make a serious effort to develop the atomic bomb.

There was also a Japanese nuclear project, which did not come close to developing a workable bomb; but it is hard for Americans who lived through World War II and experienced the brutality of it to believe that the Japanese military would have refrained from using the atomic bomb if they had had it.

32.4 Big Bang or Steady State?

George Gamow first established his reputation in physics with a 1928 paper explaining how radioactive decay works (Section 32.1). Gamow quickly became one of the experts in the new subject of nuclear physics. Later he emigrated to the United States; he taught at George Washington University from 1934 to 1956, and then moved to the University of Colorado. His interests turned in the 1930s to the astrophysical and cosmological aspects of nuclear reactions, and in particular to the problem of explaining how the elements had originally been formed.

In the 1930s, Hans Bethe showed how helium could be synthesized by fusion of hydrogen in stars and how heavier elements could be formed by adding protons and neutrons to the carbon nucleus (Section 32.2). But Gamow suggested that the elements could have been synthesized at the beginning of the universe rather than waiting for stars to form. He postulated an original high-density, high-temperature gas of neutrons, which quickly started to break down into protons and electrons. In 1948 he assigned the task of developing the theory to Ralph Alpher, a graduate student at George Washington University; Alpher was later joined by Robert Herman from the Johns Hopkins Applied Physics Laboratory.

According to Gamow's theory as worked out by Alpher, all the elements were formed by the successive capture of neutrons. The process stopped when the supply of neutrons was exhausted, the temperature dropped (thus reducing the reaction rates), and particles dispersed as the universe expanded. It came to be known as the *big bang* theory—originally intended as a pejorative phrase but eventually adopted by the theory's advocates.

Fig. 32.7. George Gamow (1904–1968), Russian-American nuclear physicist and founder of the modern big bang cosmology.

Alpher and Herman soon realized that the radiation pervading the universe in their model would maintain the spectrum characteristic of "blackbody radiation" (Chapter 26) as it cooled. Moreover, they could calculate the changing density and temperature of this radiation during the expansion and cooling of the universe; they could estimate its present temperature if the present density of matter were known. They found that the radiation would now be about 5 K above absolute zero.

The original big bang theory had two major drawbacks. First, it failed to explain the formation of the elements beyond helium. Helium's mass number (total number of protons and neutrons) is 4; there are no stable isotopes with mass numbers 5 and 8, so it is difficult to build larger nuclei by addition of neutrons one at a time. This problem could be solved only by invoking the rival theory of nucleosynthesis in stars, and indeed the modern big bang theory relies on the research results of Hoyle, Fowler, and their collaborators to construct the elements beyond helium.

The second objection to the big bang theory was, as we noted, that the age of the universe, as estimated from Hubble's law using astronomical data available in the 1940s, was only about 2 Ga. This was significantly less than the age of the earth and its oldest rocks, as determined by radiometric dating. Estimates of the age of the earth in the 1940s were around 3 to 5 Ga, and by 1953 the American geochemist C. C. Patterson had arrived at the currently accepted value, 4.5 Ga. How could the earth be older than the universe?

One night in 1946 three young scientists in Cambridge, England—Hermann Bondi, Thomas Gold, and Fred Hoyle—went to see a ghost-story film. As Hoyle later recalled it, the film had four separate parts linked ingeniously together in such a way that it became circular, its end the same as its beginning. Gold asked his friends: What if the universe is constructed like that?

The result of the ensuing discussion was not a cyclic model for the universe, but one that always looks the same even though it is always changing. This became known as the "perfect cosmological principle": There is nothing special about the particular place or time in which we live; on a large enough scale, all times and places are equivalent. Anyone, anywhere, anytime, ought to observe the same large-scale features of the universe.

Bondi, Gold, and Hoyle proposed that the universe had no beginning in time but has always existed. It appears to us to be expanding because we see galaxies rushing away from us, but that does not necessarily mean that our own galaxy will eventually be left alone, when all the others are so far away that their light never gets back to us. Instead, they suggested that matter is continuously being created as hydrogen atoms, at a rate just sufficient to compensate for the matter that is disappearing from that part of the universe visible to us. This newly created matter will eventually form into stars and galaxies so that the universe will always look about the same to any observer at any time.

The Bondi-Gold-Hoyle theory immediately resolved the timescale problem; the inverse Hubble constant $1/H$ was no longer interpreted as the "age" of the universe but simply as a measure of the rate of its expansion.

An obvious objection to this *steady state* cosmology is that the continuous creation of matter—not out of energy, but out of *nothing*—violates the well-established law of conservation of mass-energy. As Lucretius asserted two millennia ago, "nothing can come from nothing." The obvious answer is that the big bang also violates this law—but does it all at once at the beginning of time, where it is beyond the reach of scientific study. The steady state cosmology is more scientific, according to its proponents, because it postulates a process that might actually be observed, even though the rate of creation is so small as to be undetectable with present-day instruments.

Bondi several times declared himself a follower of Karl Popper's philosophy of falsificationism; he argued that the steady state theory is a "good scientific hypothesis" in Popper's sense precisely because it makes very definite statements that should be vulnerable to disproof (see Sections 3.1, 12.5, and 14.2). He described it as being very "inflexible" compared to evolutionary models (including the big bang) that use adjustable parameters (for example, the cosmological constant)—inflexibility being desirable because it allows an "all-or-nothing" test.

The steady state theory at first seemed less comprehensive than the big bang theory because it did not explain the formation of the chemical elements. Gamow, Alpher, and Herman had shown that helium and other elements could be built up from hydrogen under conditions of extremely high pressure and temperature such as would exist for a brief time after the big bang, although they had some difficulty in finding a plausible theoretical scheme for getting past mass numbers 4 and 7. This problem was partly solved by Hoyle's reaction (three helium nuclei form a carbon nucleus); in 1957 Margaret and Geoffrey Burbidge, William Fowler, and Fred Hoyle showed how heavy elements could be synthesized by neutron reactions in stars, especially supernovae. Thus a big bang was not needed to account for the formation of the elements. Nevertheless the big bang theory is still considered to give a more satisfactory explanation for the fact that 25 to 30% of the mass of the universe is in the form of helium.

One would expect that cosmological theories could best be tested by looking at the most distant objects whose radiation was emitted several billion years ago. According to the big bang theory, the first galaxies and stars were being formed at that time, so we might well expect to see things that look different from what we find in the closer (that is, newer) part of the universe.

The rapid growth of radio astronomy in the 1950s promised to allow just this kind of cosmological test to be made. Martin Ryle and his colleagues at Cambridge University counted the number of radio sources of various intensities and compared the results with those predicted by the steady state and other theories. The first results, in 1955, were said to provide conclusive evidence against the steady state theory but were later criticized as inaccurate. Subsequent results, although subject to some disagreement, were still unfavorable. According to a 1973 summary by Ryle:

> As we proceed outwards from the most intense—and presumably nearest—sources, we find a great excess of fainter ones. Now this suggests that in the past either the power or the space density of the sources was greater than it is now. Whichever way it is, the universe must have changed radically within the time-span accessible to our radio telescopes.
>
> But at still smaller intensities we find a sudden reversal of this trend—a dramatic reduction in the number of the faintest sources. This convergence is so abrupt that we must suppose that before a certain epoch in the past, there were no radio sources. Both these observations, therefore, seem to indicate that we are living in an evolving universe—which has not always looked the same . . .

> So the picture presented by the radio source observations supports the idea of an expanding universe which evolves with time, from an initial state of very high temperature and high density.
>
> [In L. John (editor), *Cosmology Now*]

The radio-source evidence was somewhat confused by the recognition in the 1960s of a separate category of peculiar sources, the quasi-stellar objects or *quasars*. All quasars have very large red-shifts, and if one converts the red-shifts to distance by the usual Doppler formula, one must assume that quasars are very far away and generate exceptionally large amounts of energy. If that is the correct interpretation of their red-shifts, then quasars are objects that existed only long ago and thus their presence contradicts the steady state theory.

Some theorists argued, however, that the red-shifts of quasars are due to some other cause and that they are really much closer (that is, more recent) and less energetic objects. Some such explanation is needed to save the steady state theory; if quasars are nearby rather than distant objects, their existence would be compatible with the perfect cosmological principle. Moreover, by subtracting them from Ryle's radio source counts of distant objects, one would get results that do not disagree with the predictions of the steady state theory.

32.5 Discovery of the Cosmic Microwave Radiation

The American physicist Steven Weinberg, in his book *The First Three Minutes* (1977), speculated about why no one made a systematic search for the background radiation before 1965 despite the prediction by Alpher and Herman in 1948. He suggested, first, that the credibility of the big bang theory was diminished by its failure to explain the origin of elements heavier than helium, so it did not seem important to test its other predictions. By contrast the Burbidge-Burbidge-Fowler-Hoyle theory of nucleosynthesis in stars—a theory that was associated with the steady state cosmology—did seem to provide a satisfactory explanation for the construction of the heavy elements from hydrogen and helium, even though it did not provide enough helium to start with.

Second, Weinberg pointed to a breakdown of communication between theorists and experimentalists: The theorists did not realize that the radiation could be observed with equipment already available, and the experimentalists did not realize the theoretical significance of what they had observed. From this perspective it is significant that Robert Dicke, who is both a theorist and an experimentalist, played a major role in the discovery.

The most remarkable missed opportunity occurred in a discussion between the two theorists, Gamow and Hoyle. While each criticized the other's theory, they could still have friendly discussions. In the summer of 1956 Gamow told Hoyle that the universe must be filled with microwave radiation at a temperature of about 50 K. As it happened Hoyle was familiar with Andrew McKellar's estimate (from the cyanogen spectrum) that the temperature of space is about 3 K. So Hoyle argued that the temperature could not be as high as Gamow claimed. But neither of them realized that if the 3 K value could be confirmed by a direct measurement, and if the radiation had the Planckian blackbody spectrum, it would refute the steady state theory, which predicted *zero* temperature for space, while giving at least qualitative support for the big bang theory.

A different kind of communication problem did lead to the actual discovery of the cosmic microwave background. To achieve optimum accuracy in transmitting information by microwaves using artificial satellites, it was necessary to design an antenna that could minimize the noise from all sources. The Bell Telephone Laboratories had been involved in microwave technology since World War II, because of the need to develop radar equipment. In 1942, Bell engineers Harold T. Friis and A. C. Beck designed a horn-shaped reflector antenna that was widely used as a microwave radio relay. Another Bell engineer, Arthur B. Crawford, built a 20-ft horn-reflector at Bell's Crawford Hill facility near Holmdel, New Jersey, in 1960, originally to receive signals bounced from a plastic balloon placed high in the atmosphere. After it had served its purpose in Bell's communication satellite project, the antenna was available for research in radio astronomy—just in time for the arrival of Arno Penzias and Robert Wilson.

Penzias and Wilson wanted to use the Crawford horn-reflector antenna for radio astronomy, but first they had to get rid of some excess noise that had been found in the antenna. They failed to identify the source of this noise despite several attempts. Finally, in January 1965, Penzias happened to be using the means of communication that paid his salary: the telephone. In a telephone conversation with another radio astronomer at the Massachusetts Institute of Technology, Bernard Burke, he learned about a theory proposed by P. J. E. Peebles, that might explain the origin of the microwave noise.

Peebles was working with Robert Dicke at Princeton, a few miles from the Holmdel lab. Dicke did not accept the hypothesis that the universe began with a big bang; he did not believe that all the matter and energy could have suddenly appeared at an instant of time. He thought it more likely that the universe went through phases of expansion and contraction. But at the end of each contraction phase, all matter went through a fiery furnace, hot enough and dense enough to break down the heavier nuclei into protons and neutrons.

Thus, although Dicke's universe not did *start* with a big bang, each cycle of it must begin and end in conditions that are very similar to those of the big bang theory. Moreover, Dicke's cosmology implied a "primeval fireball" of high-temperature radiation that retains its Planckian blackbody character as it cools down, and he estimated that the present temperature of the radiation would be 45 K. (He had forgotten his own 1946 measurement that suggested the existence of background radiation at a temperature less than 20 K.) Peebles made further calculations from Dicke's theory and obtained an estimate of about 10 K.

Dicke and Peebles, together with two other scientists, P. G. Roll and D. T. Wilkinson, then started to construct an antenna at Princeton to measure the cosmic background radiation. Before they had a chance to get any results, Dicke received the call from Penzias, suggesting that they get together to discuss the noise in the Crawford antenna, which could be interpreted as a signal from a residual microwave background having the temperature of 3.5 ± 1.0 K.

It was soon apparent that Penzias and Wilson, in trying to reduce "noise," had already detected the radiation predicted by Dicke and Peebles. The reports of the Bell Labs and Princeton groups were sent to the *Astrophysical Journal* in May 1965 and published together in the July 1 issue.

The mass media as well as the scientific journals were soon filled with articles about cosmic microwaves and statements noting the triumph of the big bang over the steady state cosmology. Even Hoyle admitted that the steady state theory, at least in its original form, "will now have to be discarded," although he tried to hang on to a modified version that could explain the microwave radiation. But Bondi's emphasis on the testability of the steady state theory had come back to haunt its proponents; any attempt to twist the theory to explain the new discoveries risked the stigma of Popper's "pseudoscience" label.

Although the press was quick to conclude that the prediction from the big bang theory had been confirmed by the Penzias-Wilson measurement and others, as reported in 1965, scientists realized that these results were limited to only a few wavelengths, all of them on one side of the Planck curve (Chapter 26). Other explanations of the background radiation, such as a combination of radio sources, could explain those data points but would have difficulty accounting for a spectrum that agreed with Planck's law over a wide range of frequencies. It was not until the mid-1970s that enough measurements at different frequencies had been made to convince the skeptics that the background radiation really follows Planck's law.

By 1980, nearly all of those who had supported the steady state theory had explicitly abandoned it or simply stopped publishing on the subject. A survey of American astronomers by Carol M. Copp in the late 1970s found that most considered it very probable that the microwave background is a relic of the big bang, while very few expected that it would eventually be accounted for in a steady state cosmology.

The rapid conversion of many cosmologists from the steady state to the big bang (or at least to some kind of evolutionary model) provides a counterexample to Planck's principle of scientific progress (Section 3.4). The discovery of the cosmic microwave radiation, combined with arguments about helium abundance and observations of distant radio sources and quasars, was responsible for this conversion. No need to wait for the older generation of scientists to retire from the field!

32.6 Beyond the Big Bang

The cosmic background radiation created a serious theoretical problem because it seemed to be *too* uniform. After the first fraction of a second, the different parts of the universe could no longer be causally connected with each other; this is because light signals could not pass from one part to another in time to have any influence. Thus the fact that we observe radiation to have the same temperature in different directions becomes a new mystery.

One proposed cosmology that may deal with this problem is the "new inflationary universe" proposed by the American physicist Alan Guth. With the help of the "grand unified theory" of elementary particles, one may imagine a phase transition (breaking the symmetry between the fundamental forces) that suddenly releases enough energy to produce a rapid expansion (exponential with time). Regions small enough to have been causally connected before the phase transition now become so inflated in size that they may encompass all of the universe that we now observe after the inflationary phase is over.

If galaxies were eventually going to form, the universe could not continue to be a uniform entity, but some nonuniformities in the cosmic radiation must eventually develop. These "wrinkles" were in fact observed in 1992 by a team led by the American astronomer George Smoot; they can be used to test different theories of galaxy formation.

The major question that still remains—if we accept the big bang cosmology in some version—is whether the universe will continue to expand indefinitely or eventually start contracting. The answer depends primarily on two factors: first, the density of matter in the universe; if it is great enough, the gravitational attraction can slow down the expansion and reverse it, just as the earth's gravity forces an object thrown upward to fall down again. We need about one atom per million cubic centimeters to brake the expansion; but the amount of material we can see in galaxies is only about one-thirtieth of this. Hence the evidence seems to suggest indefinite expansion.

On the other hand, the fact that rotating galaxies do not fly apart under their own centrifugal force shows that they must contain much more matter than is accounted for by the stars we can see in them. Hence the problem for cosmologists in the 1990s was: What and where is this "dark matter"? Is it interstellar dust, black holes, or neutrinos (if, as now seems likely, the last have finite mass)? Must some new kind of matter be imagined and discovered?

The second factor governing the fate of the universe is the possible existence of a repulsive force, such as the one described by Einstein's once-despised *cosmological constant*. Observations of distant galaxies in 1998 suggested that the expansion of the universe is actually accelerating (implying a repulsive force), rather than slowing down as one would expect if the only effective long-range force is gravitational attraction.

If the universe does contract at some time in the future, it would presumably collapse into a black hole. It might then be reborn in a new big bang. The American physicist John Archibald Wheeler has suggested that every time this happens, the various dimensionless physical constants such as the proton-electron mass ratio may acquire new values. (A "dimensionless" constant is one that has no units such as meters or grams—it is just a number like 10 or 1836.)

So far, theoretical physicists have not been able to explain why these dimensionless constants have the values they do. Perhaps they are random, not determined by any law of nature. But Wheeler and other cosmologists have shown that if these constants were very much different from their actual values in our universe, the formation of planetary systems and the evolution of higher forms of life might be impossible. This observation has given rise to the so-called *anthropic principle* suggested by Brandon Carter. One way to put it is that the reason why the physical constants have the values they do is that if they did not, we would not be here to measure them! While most physicists do not believe it is worth their attention, the principle fascinates a few—and makes a suitable conclusion for a text, by hinting at the deep human passions behind so many of the advances we have covered.

Fred Hoyle, in his autobiography, claims that his discovery of the nuclear reaction forming carbon from three helium nuclei was an application of the general philosophy of the anthropic principle, although that principle had not been formulated explicitly at the time. He believed that there must be an explanation for his own existence as a carbon-based living organism; hence there must be a physical process that produces carbon from lighter elements; he just had to figure out what it was, then get a chance to check it in the laboratory.

Here are two versions of the anthropic principle:

1. Many universes have been and will be created with different values of the physical con-

stants, but life can exist in only a few of them, including of course our own. This seems rather wasteful; it might be called a "weak version."

2. According to a "strong" version, there is a coupling between the creation of the universe and the future evolution of life in that universe, because (according to one interpretation of quantum theory) nothing can have a real existence unless it is observed directly or indirectly by a conscious intelligence. As in the old "extramission" theory of vision (Section 3.3), humans do not passively receive information about the universe but actively probe their environment—with the new twist that this probing touches the past (because of the finite speed of light). So the universe must come into existence with those properties that will allow the evolution of intelligent life, which in turn will confer reality on the universe by observing it. This is rather hard to believe.

But here, just as in our study of the earliest beginnings of science, we are on very debatable grounds, with the scientific elements still shrouded in philosophical and personal preferences.

RECOMMENDED FOR FURTHER READING

Lawrence Badash, *Scientists and the Development of Nuclear Weapons,* Amherst, NY: Humanity Books, 1998

Marcia Bartusiak, "The genesis of the inflationary universe hypothesis," *Mercury,* Vol. 16, pages 34–45 (1987).

Bernadette Bensaude-Vincent, "Star scientists in a Nobelist family: Irène and Frédéric Joliot-Curie," in *Creative Couples in the Sciences* (edited by H. M. Pycior *et al.*), pages 57–71, New Brunswick, NJ: Rutgers University Press, 1996

Jeremy Bernstein, "The birth of modern cosmology," *American Scholar,* Vol. 55, no. 1, pages 7–18 (1985/1986)

H. Bondi and T. Gold, "The steady state theory of the expanding universe," *Monthly Notices of the Royal Astronomical Society,* Vol. 108, pages 252–270 (1948)

H. A. Boorse and L. Motz (editors), *World of the Atom,* on Rutherford, Gamow, Chadwick, Cockcroft and Walton, Yukawa, Heisenberg, Bethe, Meitner and Frisch, Chapters 49, 67, 74, 76, 80, 89–91

Andrew Brown, *The Neutron and the Bomb: A Biography of Sir James Chadwick,* New York: Oxford University Press, 1997

Stephen G. Brush, "How cosmology became a science." *Scientific American,* Vol. 247, no. 2, pages 62–70 (1992). On the big bang-steady state controversy and its resolution.

David C. Cassidy, "A historical perspective on 'Copenhagen'," *Physics Today,* Vol. 53, no. 7, pages 28–32 (2000). On reasons for the Germans' failure to build a bomb or even a functioning reactor by 1945.

Dan Cooper, *Enrico Fermi and the Revolution in Modern Physics,* New York: Oxford University Press, 1999

Elisabeth Crawford, Ruth Lewin Sime, and Mark Walker, "A Nobel tale of postwar injustice," *Physics Today,* Vol. 50, no. 9, pages 26–32 (September 1997). Why Meitner didn't get the Nobel prize.

Enrico Fermi, "Experimental production of a divergent chain reaction" (1942), in *Physics History II* (edited by French and Greenslade), pages 153–158

Timothy Ferris, *The Whole Shebang: A State-of-the-Universe Report,* New York: Simon and Schuster, 1997

George Gamow, "The origin and evolution of the universe," *American Scientist,* Vol. 39, pages 393–407 (1951); "The evolutionary universe," *Scientific American,* Vol. 195, no. 3, pages 136–154 (September 1956) (see also following article by Hoyle); George Gamow, *My World Line: An Informal Autobiography,* New York: Viking, 1970.

Kurt Gottfried, "Physicists in politics," *Physics Today,* Vol. 52, no. 3, pages 42–48 (March 1999)

Stephen Hawking, *A Brief History of Time,* updated and expanded edition, New York: Bantam Books, 1998

N. S. Hetherington (editor), *Cosmology,* articles by H. Kragh, A. H. Guth, D. N. Schramm, A. Vilenkin, J. J. Halliwell, P. A. Wilson, S. J. Dick, G. Gale, and J. R. Urani, pages 371–568

Gerald Holton, "'Success sanctifies the means': Heisenberg, Oppenheimer, and the transition to modern physics," in his book *The Advancement of Science,* chapter 7

Fred Hoyle, "The steady-state universe," *Scientific American,* Vol. 195, no. 3, pages 157–166 (September 1956); see also preceding article by Gamow.

Kenji Kaneyuki and Kate Scholberg, "Neutrino oscillations," *American Scientist,* Vol. 87, pages 222–231 (1999)

Helge Kragh, "Particle science," in *Companion* (edited by R. C. Olby), pages 661–676

John C. Mather and John Boslough, *The Very First Light: The True Inside Story of the Scientific Journey back to the Dawn of the Universe,* New York: Basic Books, 1996

John North, *Norton History of Astronomy,* Chapter 17

Mary Jo Nye, *Before Big Science,* Chapter 7

Roger Penrose, "The modern physicist's view of nature," in *The Concept of Nature* (edited by J. Torrance), pages 117–166

Melba Newell Phillips (editor), *Physics History,* articles by T. H. Osgood and H. S. Hirst on Rutherford's experiments, by C. D. Anderson on the positron and muon, by E. U. Condon on tunneling, and by E. B. Sparberg on fission

Marelene Rayner-Canham and Geoffrey W. Rayner-Canham (editors), *A Devotion to Their Science: Pioneer Women of Radioactivity*

Bertram Schwarzschild, "Very distant supernovae suggest that the cosmic expansion is speeding up," *Physics Today,* Vol. 51, no. 6, pages 17–19 (June 1998). Report on the observations by Saul Perlmutter and others.

Ruth Lewin Sime, "Lise Meitner and the discovery of nuclear fission," *Scientific American,* Vol. 278, no. 1, pages 80–85 (1998)

S. A. Watkins, "The making of a physicist [Lise Meitner]," in *Physics History II* (edited by A. P. French and T. B. Greenslade), pages 75–78

Spencer Weart and Melba Phillips (editors), *History of Physics,* articles by F. G. Brickwedde, C. Weiner, L. M. Brown, and L. H. Hoddeson on new particles in the 1930s, pages 208–213, 332–353; articles by E. Fermi on the genesis of the nuclear energy project, and by O. R. Frisch and J. A. Wheeler on the discovery of fission, pages 272–286

J. H. Weaver (editor), *World of Physics,* excerpts from papers by Gamow, Weinberg, Bondi, Weisskopf, Guth and Steinhardt, Bethe, Davies, Oppenheimer, and Wheeler, Vol. III, pages 257–348, 392–404, 655–694, 1032–1044

Steven Weinberg, "A designer universe?" in *The Best of American Science Writing 2000* (edited by James Glieck), pages 239–248, New York: Harper-Collins/Ecco Press, 2000

Charles Weiner, "1932—moving into the new physics," *Physics Today,* Vol. 25, no. 5, pages 40–49 (May 1972)

J. A. Wheeler, "Our universe: The known and the unknown," *American Scientist,* Vol. 56, pages 1–20 (1968); *Geons, Black Holes, and Quantum Foam: A Life in Physics* (with Kenneth Ford), New York: Norton, 1998

CHAPTER 33

Thematic Elements and Styles in Science

The introduction of the "anthropic principle" to explain the values of the dimensionless constants of physics (Section 32.6) makes many physicists uncomfortable, because it seems so out of harmony with the accepted ways in which we are now accustomed to understand the physical world. Yet it would not be at all out of place for a scientist following Aristotelian principles, one of which explicitly authorizes the use of *final causes* to explain why things happen in the world. Indeed, this kind of explanation, under the name of *design* or *natural theology,* was commonplace in biology before the publication of Charles Darwin's *Origin of Species*. Nevertheless, it seems like a regression to an obsolete, discredited kind of scientific reasoning.

In this final chapter we look at some aspects of science, such as the acceptance or rejection of final causes, that may add to our understanding of the overall development of science and its place in our civilization.

33.1 The Thematic Element in Science

In our discussion of the discovery of laws (Chapter 14), we stressed heavily the role of "fact" and of calculation. And it is true that regardless of what scientific statements they believe to be "meaningless," all philosophies of science agree that two types of propositions are *not* meaningless and are important, namely statements concerning empirical matters of "fact" (which ultimately boil down to meter readings) and statements concerning the calculus of logic and mathematics. Let us call them, respectively, *empirical* and *analytical* statements, and think of them as if they were arrayed, respectively, on orthogonal x and y axes; thereby we can represent these two "dimensions" of usual scientific discourse by a frank analogy, and generate terminology that will be useful as long as we do not forget that all analogy has its limits.

Now we may use the xy plane to analyze the concepts of science (such as force) and the propositions of science, for example, an hypothesis (such as "x-rays are made of high-energy photons") or a general scientific law (such as the law of universal gravitation). The *concepts* are analogous to points in the xy plane, having x and y coordinates. The *propositions* are analogous to line elements in the same plane, having projected components along x and y axes.

To illustrate, consider a concept such as force. It has empirical, x-dimension meaning because forces can be qualitatively discovered and, indeed, quantitatively measured by, say, the observable deflection of solid bodies. And it has analytical, y-dimension meaning because forces obey the mathematics of vector calculus (namely, the parallelogram law of composition of forces), rather than, for example, the mathematics of scalar quantities.

Now consider a proposition (an hypothesis or a law): The law of universal gravitation has an empirical dimension or x component—for example, the observation in the Cavendish experiment (Section 11.3) in which massive objects are "seen" to "attract" and in which this mutual effect is measured. And the law of universal gravitation has an analytical or y component, the vector algebra rules for the manipulation of forces in Euclidean space. Whether they are arbitrary or not, the xy axes have, since the seventeenth and eighteenth centuries, more and more defined the total allowable content of science and even (rightly or not) of sound scholarship generally. The eighteenth-century Scottish philosopher David Hume, in a famous passage, expressed eloquently that only what can be resolved along x and y axes is worthy of discussion:

> If we take in our hands any volume, of divinity, or school metaphysics, for instance: Let us ask, Does it contain any abstract reasoning

> concerning quantity or number? No. Does it contain any experimental reasoning concerning matter of fact or criteria? No. Commit it then to the flames. For it can contain nothing but sophistry and illusion.

Yet there remains a third element in science, at which we have hinted only briefly. Following our analogy we could assign it to a *z* axis, perpendicular to *x* and *y*. But if we now leave the *xy* plane, we are going off into an undeniably dangerous direction. For it must be confessed at once that the tough-minded thinkers who try to live entirely in the *xy* plane are more often than not quite justified in their doubts about the claims of the more tender-minded people (to use a characterization made by the American psychologist William James). The region below or above this plane, if it exists at all, might well be a muddy or maudlin realm, even if the names of those who have sometimes gone in this direction are distinguished. As the Dutch historian of science, E. J. Dijksterhuis, has said:

> Intuitive apprehensions of the inner workings of nature, though fascinating indeed, tend to be unfruitful. Whether they actually contain a germ of truth can only be found out by empirical verification; imagination, which constitutes an indispensable element of science, can never even so be viewed without suspicion.
>
> (*Mechanization of the World Picture*)

And yet, the need for going beyond the *xy* plane in understanding science and, indeed, in doing science, has been consistently voiced since long before Copernicus, who said that the ultimate restriction on the choice of scientific hypotheses is not only that they must agree with observation but also "that they must be consistent with certain preconceptions called 'axioms of physics,' such that every celestial motion is circular, every celestial motion is uniform, and so forth." (Rosen, *Three Copernican Treatises*). And if we look carefully, we can find even among the most hardheaded modern philosophers and scientists a tendency to admit the necessity and existence of this additional dimension in scientific work. Thus Bertrand Russell spoke of cases in which "the premises turn out to be a set of presuppositions neither empirical nor logically necessary." (*Human Knowledge*)

One could cite and analyze similar opinions by a number of other scientists and philosophers. In general, however, there has been no systematic development of the point until recently. What is needed may be called *thematic analyses* of science (by analogy with thematic analyses that have for so long been used to great advantage in scholarship outside science), which considers the role of the elements on the *z* axis in scientific thought. This third dimension is the dimension of fundamental presuppositions, notions, terms, methodological judgments and decisions—in short, of *themata* (the Greek plural of *thema,* a word meaning that which is given or adopted)—which are themselves neither directly evolved from, nor resolvable into, objective observation on the one hand or logical, mathematical, and other formal analytic ratiocination on the other hand.

With the addition of the thematic dimension, we generalize the plane in which concepts and statements were previously analyzed. It is now a three-dimensional "space"—using the terms always in full awareness of the limits of analogy—which may be called proposition space. A concept (such as force) or a proposition (such as the law of universal gravitation) is to be considered, respectively, as a point or as a configuration (line) in this threefold space. Its resolution and projection is in principle possible on each of the three axes.

To illustrate: The *x* and *y* components of the physical concept "force" (its projections in the *xy* plane) have been mentioned. We now look at the thematic component, and see that throughout history there has existed in science a "principle of potency." It is not difficult to trace this from Aristotle's *energeia* through the neo-Platonic *anima motrix,* and the active *vis* that still is to be found in Newton's *Principia,* to the mid-nineteenth century when *Kraft* is still used in the sense of energy (Mayer, Helmholtz). Each of the four terms could be translated "force" or "energy" if we set aside the important distinction modern physics makes between those two English words. In view of the obstinate preoccupation of the human mind with the theme of the potent, active principle, before and quite apart from any science of dynamics (and also with its opposite, the passive, persisting principle on which it acts), it is difficult to imagine any science in which there would not exist a conception of force (and of its opposite, inertia).

It would also be difficult to understand certain conflicts. Scholastic physics defined "force" by a projection in the empirical dimension that concentrated on the observation of continuing terrestrial motions against a constantly acting obstacle; Galilean-Newtonian physics defined "force" quite differently, namely, by a projection in the empirical dimension that concentrated on a thought experiment such as that of an object being accelerated on a friction-free horizontal plane. The projections onto the analytic dimension differed also in the two forms of physics (that is, attention to magnitudes vs. vector properties of forces). On these two (x, y) axes, the concepts of force are entirely different. Nevertheless, the natural philosophers in the two camps in the early seventeenth century thought they were speaking about the same thing, and the reason was that they shared the need or desire to incorporate into their physics the same thematic conception of *anima,* or *vis,* or *Kraft*—in short, of force.

A second example of thematic analysis might be the way one would consider not a concept but a general scientific proposition. Consider the clear thematic element in the powerful laws of conservation in physics, for example the law of conservation of momentum, as formulated for the first time in useful form by Descartes. In Descartes' physics, as Dijksterhuis wrote:

> All changes taking place in nature consist in motions of . . . three kinds of particles. The primary cause of these motions resides in God's *Concursus ordinarius,* the continuous act of conservation. He so directs the motion that the total *quantitas motus* (momentum), that is, the sum of all the products of mass and velocity, remain constant.
>
> (*Mechanization of the World Picture*)

In Descartes' view this relation (Σmv = constant) constitutes the supreme natural law. This law, Descartes believes, springs from the invariability of God, by virtue of which, now that He has wished the world to be in motion, the variation must be as invariable as possible.

Since the seventeenth century, we have learned to change the *analytic* content of the conservation law—again, from a scalar to a more complex version—and we have extended the empirical applicability of this law from impact between palpable bodies to other events. But we have always tried to cling to this and to other conservation laws, even at a time when the observations seem to make it very difficult to do so. The French mathematician Henri Poincaré clearly saw this role of themata in his book *Science and Hypothesis:*

> The principle of the conservation of energy simply signifies that there is a *something* which remains constant. Whatever fresh notions of the world may be given us by future experiments, we are certain beforehand that there is something which remains constant, and which may be called energy.

(to which we now add: even when we used to call it only mass). The thema of conservation has remained a guide, even when the language has had to change. We now do not say that the law springs from the "invariability of God," but with that curious mixture of arrogance and humility that scientists have learned to put in place of theological terminology, we say instead that the law of conservation is the physical expression of the elements of constancy by which Nature makes herself understood by us.

In physical science, themata often come in pairs of opposites. The opposite of the active *force* theme might be the inert *atom* theme of the seventeenth-century mechanical philosophers (Gassendi and Boyle), reaching perhaps its highest level of effectiveness and popularity in the "billiard-ball" kinetic theory of gases developed by Clausius, Maxwell, and Boltzmann in the nineteenth century; it emerged again in the late twentieth century in the version of quantum mechanics that explains forces (electromagnetic, nuclear, gravitational) as the exchange of particles (photons, mesons, gravitons). Or there may be three competing *themata: conservation laws* are tempered on one hand by *dissipation* and the *increase of disorder* (entropy and the second law of thermodynamics), and enhanced on the other by *progress* and the *growth of complexity* (evolution in an expanding universe).

The strong hold that certain themes have on the mind helps to explain the stubborn faith with which some scientists cling to one explanation or another when the "facts" cannot decide which of the two is right. The passionate motivation comes out clearly when two scientists respond to each other's antithetical constructs. For example, Werner Heisenberg, whose physics was built on the theme of the discrete, atomistic, quantized, *discontinuous,* wrote about Erwin Schrödinger's physics: "The more I ponder about [his] theory, the more disgusting it appears to me." And Schrödinger, in turn, whose physics was based on the idea of *continuity,* wrote that Heisenberg's approach left him "discouraged if not repelled."

The fact that many scientists have thematic preferences usually also explains their initial commitment to some point of view that may in fact run exactly counter to all accepted doctrine and to the clear evidence of the senses. Of this no one has spoken more eloquently and memorably than Galileo when he commented on the fact that to accept the idea of a moving earth one must overcome the strong impression that one can "see" that the sun is really moving:

> Nor can I ever sufficiently admire the outstanding acumen of those [Galileo's *Salviati* says in the Third Day of the *Dialogue Concerning the Two Chief World Systems*] who have taken hold of this opinion [the Copernican system] and accepted it as true; they have through sheer force of intellect done such violence to their own senses as to prefer what reason told them over that which sensible experience plainly showed them to the contrary. . . . there is no limit to my astonishment when I reflect that Aristarchus and Copernicus were able to make reason so conquer sense that, in defiance of the latter, the former became mistress of their belief.

Among the themata that permeate Galileo's work and which helped reason to do "such violence" to the plain evidence of sense experience, we can readily discern the then widely current thema of the once-given real world that God supervises from the center of His temple; the thema of mathematical nature; and the thema that the behavior of things is the consequence of their geometrical shapes (for which reason Copernicus said the earth rotates "because" it is spherical, and William Gilbert, following the lead, is said to have gone so far as to find experimentally, at least to his own satisfaction, that a carefully mounted magnetized sphere keeps up a constant rotation).

33.2 Themata in the History of Science

While developing the position that themata have as legitimate and necessary a place in the pursuit and understanding of science as have observational experience and logical construction, we should make it clear that we need not decide now on the *source* of themata. Our first aim is simply to see their role in science, and to describe some of them, as folklorists might when they catalog the traditions and practices of a people. It is not necessary to go further and to make an association of themata with any of the following conceptions: Platonic, Keplerian, or Jungian archetypes or images; myths (in the nonderogatory sense, so rarely used in the English language); synthetic *a priori* knowledge; intuitive apprehension or Galileo's "reason"; a realistic or absolutistic or, for that matter, any other philosophy of science. To show whether any such associations do or do not exist is a task not yet fulfilled.

We also do not want to imply that the occurrence of themata is characteristic only of science in the last centuries. On the contrary, we see the thematic component at work from the very beginning, in the sources of cosmogonic ideas later found in Hesiod's *Theogony* and in *Genesis*. Indeed, nowhere can one see the persistence of great questions and the obstinacy of certain preselected patterns for defining and solving problems better than in cosmologic speculations. The ancient Greek cosmologic assumptions presented a three-step scheme: At the beginning, in F. M. Cornford's words, there was

> . . . a primal Unit, a state of indistinction or fusion in which factors that will later become distinct are merged together. (2) Out of this Unity there emerge, by separation, parts of opposite things. . . . This separating out finally leads to the disposition of the great elemental masses constituting the world-order, and the formation of the heavenly bodies.(3) The Opposites interact or reunite, in meteoric phenomena, or in the production of individual living things. . . .
>
> (*Principium Sapientiae*)

Now the significant thing to notice is that when we move these conceptions from the animistic to the physical level, this formula of ancient cosmogony recurs point for point, in our day, in the evolutionist camps of modern cosmology. That theory of the way the world started, the big bang, proposes a progression of the universe from a mixture of radiation and particles (or a "primeval fireball" of energy) at time $t = 0$; through the subsequent stages of differentiation by expansion and transformation of radiation to matter; and finally to the building up of heavier elements by thermonuclear fusion processes, preparing the ground for the later formation of atoms and molecules, eventually resulting in galaxies, solar systems, and living beings. And even the ancient main *opposition* to the evolutionary cosmology itself, namely, the tradition of Parmenides, had its equivalent in the steady state theory of cosmology (Section 32.4).

So the questions persist (for example, concerning the possibility of some "fundamental stuff," of evolution, of structure, of spatial and temporal infinities). And the choices among alternative problem solutions also persist. These thematic continuities indicate the obverse side of the iconoclastic role of science; for science, since its dawn, has also had its more general themata-creating and themata-using function. James Clerk Maxwell expressed this well over a century ago in an address on the subject of molecular mechanics:

> The mind of man has perplexed itself with many hard questions. Is space infinite, and in what sense? Is the material world infinite in extent, and are all places within that extent equally full of matter? Do atoms exist, or is matter infinitely divisible?
>
> The discussion of questions of this kind has been going on ever since man began to reason, and to each of us, as soon as we obtain the use of our faculties, the same old questions arise as fresh as ever. They form as essential a part of science of the nineteenth century of our era, as of that of the fifth century before it.
>
> ("Molecules")

We may add that thematic questions do not get solved and disposed of. Cartesian vortices in a world packed with matter are replaced by Newton's gravitating masses in empty space; two centuries later, Kelvin reduces atoms to vortex motions in a continuous fluid. Then discrete atoms triumph over the ether vortices, but field theories arise, which deal with matter particles again as singularities, now in a twentieth-century-type continuum. (Bohr's "complementarity" is a rare, perhaps unstable attempt to compromise between continuous and discrete themata.) The modern version of the cosmological theory based on the thema of a life cycle (beginning, evolution, and end) triumphed on experimental grounds over the rival theory based on a thema of continuous existence, and threw it out the window—but this thema has already come in again through the back door in attempts to explain the big bang itself. For contrary to the physical theories in which they find embodiment in *xy* terms, themata are not proved or disproved. Rather, they rise and fall and rise again with the tides of contemporaneous usefulness or intellectual fashion. And occasionally a great thema disappears from view, or a new one develops and struggles to establish itself—at least for a time.

Maxwell's is an unusual concession—he had a remarkably deep comprehension of the history and philosophy of science—but it is not difficult to understand why scientists speak only rarely in such terms. One must not lose sight of the obvious fact that science itself has grown strong because its practitioners have seen how to project their discourse onto the *xy* plane. This is the plane of public science (S_2, see Section 13.2), of fairly clear conscious formulations. Here a measure of public agreement is in principle easy to obtain, so that scientists can fruitfully cooperate or disagree with one another, can build on the work of their predecessors, and can teach more or less unambiguously the current content and problems of the field. All fields that claim or pretend to be scientific try similarly to project their concepts, statements, and problems onto the *xy* plane, to emphasize the empirical and analytic aspects.

But it is clear that while there can be automatic factories run by means of programmed computers and the feedback from sensing elements, there can be no automatic laboratory. The essence of the automaton is its success in the *xy* plane at the expense of the *z* direction (hence automata do not make qualitatively new findings). And the essence of the scientific genius is often exactly the opposite—sensitivity in the *z* direction even at the expense of success in the *xy* plane. For while the *z* dimension is never absent even in the most exact of the sciences as pursued by actual persons; it is a direction in which most of us must move *without* explicit or conscious formulation and without training; it is the direction in which the subject matter and the media for communication are entirely different from those invented specifically for discussion of matters in the *xy* plane with which the scientist after long training can feel at home.

Therefore it is difficult to find people who are bilingual in this sense. We are not surprised that for most contemporary scientists any discussion that tries to move self-consciously away from the *xy* plane is out of bounds. However, it is significant that even in the twentieth century persons of genius—such as Einstein, Bohr, Pauli, Born, Schrödinger, Heisenberg—felt it to be necessary and important to try just that. For the others, for the main body of scientists, the plane of discourse has been progressively tilted or projected from *xyz* space onto the *xy* plane. (Perhaps prompted by this example, the same thing is happening more and more in other fields of scholarship.) The themata actually used in science are now largely left implicit rather than made explicit. But they are no less important. To understand fully the role an hypothesis or a law has in the development of science we need to see it also as an exemplification of persistent motifs, for example the thema of "constancy" or of "conservation," of quantification, of atomistic discreteness, of inherently probabilistic behavior, or—to return to Newton—of the interpenetration of the worlds of physics and of theology.

We have spoken mostly of the physical sciences. We might, with equal or greater advantage, have dealt with sciences that are still coming to their maturity, which do not have a highly developed corpus either of phenomena or of logical calculi and rational structures. In those cases, the *z* elements are not only still relatively more prominent but also are discussed with much greater freedom—possibly because at its early stage a field of study still bears the overwhelming imprint of one or a few geniuses. It is they who are particularly "themataprone," and who have the necessary courage to make decisions on thematic grounds.

It should be clear that the most significant fact, implied in the example given, is that most and perhaps all of these themata are not restricted merely to uses in scientific context, but seem to come from the less specialized ground of our general imaginative capacity. This view leads us at once beyond the usual antithetical juxtaposition between science and the humanities. For the laments on the separation between science and the other components of our culture depend on the oversimplification that science has only *x* and *y* dimensions, whereas scholarly or artistic work involves basic decisions of a different kind, with predominantly aesthetic, qualitative, mythic elements. In our view this dichotomy is much attenuated, if not eliminated, if we see that in science, too, the *xy* plane is not enough, and never has been.

It is surely unnecessary to give warning that despite the appearance and reappearance of the same thematic elements in science and outside, we shall not make the mistake of thinking that science and nonscience are at bottom somehow the same activity. There are differences that we should treasure. As Alfred North Whitehead once said about the necessity to tolerate—no, to welcome—national differences: "Men require of their neighbors something sufficiently akin to be understood, something sufficiently *different* to provoke attention, and something great enough to command admiration." It is in the same sense that we should be prepared to understand the separateness that gives identity to the study of each field as well as the kinship that exists between them.

33.3 Styles of Thought in Science and Culture

Since science is, and always has been, one of the central parts of a civilization's culture, one many now ask: To what extent are changes in scientific theories related to the social, political, and cultural environment of science and to prevailing philosophical views?

During the 1970s and 1980s, there was a fashion among some historians and sociologists of science to claim that scientific knowledge is not chiefly a collection of facts about nature discovered by scientists, but on the contrary is largely or entirely "socially constructed." This phrase, whose precise meaning was always in dispute, suggested that scientific theories and concepts did not refer to an objective reality acknowledged by all scientists, but were developed by an elite group of influential scientists to suit their own interests—economic, political, racial, or gender-based. (Strangely enough, the social constructionists did not count intellectual curiosity as an "interest"!) Other scientists, according to this view, had to accept these socially constructed theories and concepts in order to gain access to the resources (positions, funding, publications) controlled by the elite group. According to social constructionists, a statement is (by their definition) true because it is accepted by scientists, rather than being accepted because it is true.[1]

[1]Apparently already in Galileo's time there were people who thought scientific facts were socially constructed. In his *Dialogue* (Second Day) he wrote about people who "so firmly believe the earth to be motionless . . ." that when they hear "that someone grants it to have motion" they "imagine it to have been set in motion when Pythagoras (or whoever it was) first said that it moved, and not before." They "believe it first to have been stable, from its creation up to the time of Pythagoras, and then made movable only after Pythagoras deemed it to be so."

Despite much effort, little evidence has been found to support the claims of social constructionists. It is true that in a few extreme cases such as the adoption of Lysenko's biological fantasies in the Soviet Union, scientists there were constrained to accept socially constructed theories instead of following facts and logic, at least for a short time until reality, in the form of agricultural failures, reared its ugly head. But even there, the fulminations of Communist Party functionaries against quantum mechanics and relativity had little effect on their popularity or the mode of presentation; a multivolume treatise on theoretical physics by the leading Russian physicists L. D. Landau and E. M. Lifshitz was widely used in both capitalist and socialist countries.

Leaving aside the unproven assertions about social construction of scientific knowledge, which refer primarily to what we have called the *xy* plane, we may still ask whether there is a connection between cultural or political ideas and the *z* axis of science.

Indeed, historians of nineteenth-century science have extensively discussed the influence of one cultural movement, romanticism, through German "nature philosophy." As we noted in Section 25.2, the Danish physicist Hans Christian Oersted said that he was convinced of the unity of all physical forces, in particular electricity and magnetism, through his study of nature philosophy, and this inspired his successful search for the action of an electric current on a magnetic compass needle. More generally, nature philosophy is said to have encouraged the formulation of the law of conservation of energy through the doctrine of the unity of forces of nature.

The romantic movement in the arts emphasized the individual and the unique, freedom from the restrictions of classical rules of form and structure, direct expression of the emotions, and an insistence that the whole is greater than its parts (now called "holism") because it is pervaded by a spirit that cannot be rationally or "reductionistically" explained but can only be intuitively felt. Accordingly, nature philosophy was directly opposed to the mathematical-empirical tradition of Galileo, Newton, and their eighteenth-century followers. Instead of conceiving matter as composed of separate atoms moving through empty space, the nature philosophers insisted that atoms, if they exist at all, can affect other atoms only through attractive and repulsive forces; thus *force* is a more fundamental reality than *matter.* This viewpoint or thema was sometimes called "dynamism"; it was opposed to atomism and to "materialism" (the view that matter is the most fundamental entity). Faraday's qualitative concepts of electric and magnetic fields—forces that can be imagined to exist in space apart from their material sources—showed the fertility of this approach.

Nevertheless, atomism, an outgrowth of quantitative laboratory work in chemistry around 1800, became enormously popular and successful in physical science in the middle of the nineteenth century. The conservation thema, no longer just a qualitative principle asserting the unity of different forces without supposing any one of them to be more fundamental than the others, was transformed into a quantitative law, widely regarded as authorizing the *reduction* of heat, electricity, magnetism and perhaps even gravity to *mechanical phenomena.* Thus, to oversimplify somewhat, in gases heat is *nothing but the energy of atomic motion,* according to the kinetic theory.

Moreover, the atom itself could now be given quantitative mechanical properties—size, mass, speed, etc.—to an extent never imagined before the nineteenth century. Chemists, following August Kekulé, postulated quantitative rules ("valence") governing the combination of atoms into molecules, and proposed detailed two- and even three-dimensional structures for these molecules. Meanwhile Maxwell, with the help of a very different mechanical model, together with a mathematical description of Oersted's electromagnetic interaction and Faraday's fields, discovered the nature of light and established a quantitative relation between its speed and the empirical data of electromagnetism.

Kelvin best articulated this approach in the famous statements we have already quoted in Sections 3.1 and 13.1: "I never satisfy myself until I can make a mechanical model of a thing" and "when you can measure what you are speaking about and express something in numbers you know something about it." Note that mechanical modeling and quantification are two distinct themata that in this case go together with each other and with a third thema, atomism, but need not always do so.

Historians and philosophers of science sometimes use the phrase *style of thought* to characterize the particular combination of themata epitomized by one scientist (such as Galileo, Descartes, Newton, Faraday, Kelvin, Curie,

Einstein, Meitner) and shared to a greater or lesser extent by other scientists working at the same time.

There is no evidence that the mid-nineteenth-century flowering of atomistic/mechanistic/quantitative physical science was in any way influenced by the contemporaneous movement called realism by cultural historians. In literature, European and American writers, scorning romantic idealizations and fantasies, attempted to portray life as it really is, with all its sordid and trivial aspects: Flaubert, Zola, Dostoevsky, Gogol, Hardy, Dickens, and Whitman. Similarly, Goya, Daumier, and Courbet were realist painters. In politics, romanticism, despite its reputation for rebelling against convention, had been profoundly conservative, looking backward to a simpler society with its hierarchical structure in which each person was satisfied to occupy a predetermined role; the excesses of the French Revolution had highlighted the dangers of overthrowing an established order. But realism fostered democracy: It became easier in the late nineteenth century to advocate the extension of civil rights and freedoms to poor people, women, and (in America) blacks, and to work toward the transfer of power from authoritarian rulers to representative legislatures—although the realization of these goals was still mostly in the future.

Romantic poets—notably Byron, Keats and Wordsworth—had denounced Newtonian science, and Goethe even proposed an alternative to Newton's theory of color. Realists tended to approve of science, seeing it as a source of economic progress and intellectual enlightenment; some, like Zola, thought literature could provide a scientific description of human behavior. But, aside from Darwin's theory of evolution, mid-nineteenth-century science had little direct influence on social philosophy and cultural activity.

A brief consideration of late-nineteenth-century science and culture will show why, as we noted at the beginning of Section 33.2, it is risky to make any simplistic correlation between themata and more general philosophies or styles of thought. Many historians identify the rise, at the time, of a neoromantic movement in literature, art, and music that seems to echo the romanticism of the early part of the century. Political conservatism returned, amplified by nationalism and imperialism. Intellectuals, especially in France, discussed the alleged "bankruptcy of science" and articulated a "reaction against materialism."

But within the scientific community, the critique of atomism and mechanism took a direction very different from romantic nature philosophy. The critics—including some eminent scientists such as Mach, Poincaré, Ostwald, Hertz, and Duhem—argued that science, rather than speculating about atoms, forces, and ethers, should exclude *all* speculation about unseen entities. Instead, they insisted, the purpose of science is to discover empirical facts and regularities; theories are useful only as an efficient way to summarize these facts and regularities—to achieve "economy of thought" in Mach's phrase. The writings of these men exemplify an *empirical style of thought* that is clearly different from both nature philosophy and mechanistic atomism, yet shares some themata with each. Like the nature philosophers they thought of energy as a fundamental entity rather than something reducible to the motion of atoms; like the mechanists they preferred quantitative experiment and mathematical reasoning rather than qualitative observation and intuition.

From a twentieth-century perspective the mechanists came closer to the truth about the atomic world, but the empiricists played a more important role in facilitating the quantum and relativity revolution by breaking down misplaced confidence in Newtonian physics; Bohr, Heisenberg, and Bridgman were their philosophical heirs.

Modernism is a word now often applied to styles in architecture, music, literature, and painting, yet it was used earlier to characterize theological movements, starting in the late nineteenth century, aimed at harmonizing science and religion. One feature of cultural modernism in the early twentieth century was the use of technology and geometrical forms, exemplified by the German *Bauhaus* school. The similarity between the conceptions of space and time in cubist paintings (early works of Picasso, Braque, and Duchamp) and in special relativity theory has often been noted, although their simultaneous appearance around 1905, and the lack of documentary evidence of any cause-and-effect relation, precludes either having directly influenced the other.

Although Einstein credited Mach with having the courage to question Newtonian ideas of absolute space, time, and mass, and his formulation of special relativity is couched in terms of what would be measured by different "observers," Einstein's style of thought cannot be called empiricist. It was certainly atomistic—his theory of Brownian movement enabled Jean Perrin to establish once and for all the grainy structure of matter, and Einstein's quantum theory proposed to do the same for light and other electromagnetic radiation. Einstein's work was highly quantitative, and relied on advanced mathematics for the general theory of relativity. In his later debates with Bohr, Einstein defended *philosophical* realism (the atom has definite properties whether or not you choose to measure them), yet his discoveries swept away many of the mechanisms of the physical science that flourished in the mid-nineteenth-century realist period.

Themata recur in the history of science, but each time in different combinations and contexts. In the 1970s, physicists stopped talking about "elementary" particles after they had found more than a hundred of them, renamed their field "high-energy physics," and reformulated their theories in terms of just four fundamental forces (strong, weak, electromagnetic, and gravitational), hoping that those four could eventually by unified into just one. (Some important progress toward that goal has been achieved.) Superficially this looks like the attempt of nature philosophers to replace atomism by dynamism in the early nineteenth century, but a glance at some representative writings of each group would quickly convince you that the shared themata are greatly outnumbered by their opposites.

At the end of the twentieth century cultural modernism in Europe and America had been replaced, in the opinion of many social commentators, by "postmodernism," a thought-style that (like romanticism and neoromanticism) rejected the claim of mathematical-experimental science to uncover the real nature of the world. Some postmodernists argued that modern science had been constructed only to represent the interests of Western "imperialism," and that the knowledge constructed by other groups is equally valid, including aboriginal ones. If so, we might have expected postmodern science to fragment into many styles of thought, one for each group. But on the contrary, at the beginning of the twenty-first century we see in fact increased *participation* in an international scientific community by men and women of many ethnic, national, and religious groups; most of these scientists would vigorously deny that their style of thought and work is determined by their gender or cultural identity. Increasingly they share a common heritage, which includes the concepts and theories in physical science presented in this book.

33.4 Epilogue

Our aim throughout the book has been to present both the fact and meaning of physical science within the context of its historical development, not abstractly but through the theories and experiments of actual innovators in their astonishing variety. Science is not one equation after another; rather, it is a grand continuing adventure, with its spectacular advances, periods of failure, and pauses for sober reflection. It is appropriate to conclude with an overview and reminder of high points of the story of achievements, looking back to its origins in Greek antiquity and coming up to today's advances, mentioning not only its successes but some of the problems that have not yet been solved.

For this purpose it is convenient to employ some of the terms used by the eminent British historian of science A. C. Crombie, in his book *Styles of Scientific Thinking in the European Tradition.* Crombie, who wrote extensively on medieval and early modern science, discerned six styles: postulation, experimental argument, hypothetical modeling, taxonomy, probabilistic and statistical analysis, and historical derivation. Several of these styles will already be familiar from our discussions of physical science.

The legacy of Greek science and mathematics is dominated by postulation, most explicitly in Euclid's work on geometry but also in the mathematical theories of Pythagoras and Plato, Aristotle's comprehensive philosophical system, Ptolemy's astronomy, and the very different writings of the atomists. In each case there is an intense effort to find and justify fundamental principles, and then to show by logical reasoning that these principles account for the observed phenomena.

In astronomy, one principle was so self-evidently true that it scarcely needed justification: The earth is fixed at the center of the universe. *Therefore* the apparent motions of the sun, planets, and stars must be explained by some kind of actual motion around the motionless earth. To construct such explanations, Plato proposed that the motions of celestial bodies must be represented by *circles;* this postulate was adopted not only by Aristotle and Ptolemy, but also by Copernicus and Galileo, even as they rejected the geocentric postulate. Only Kepler, after a great struggle, was able to break free of the Platonic commandment and substitute ellipses for circles.

Experimental argument as a style of thought developed in the Middle Ages; Crombie calls Alhazen's research in optics, which finally established the intromission theory of vision, a model for the combination of experimental and mathematical reasoning. This style also promoted the use of *perspective* in painting, and in turn Galileo's training in perspective helped him to "see" patterns of light and shade on the moon as bulges and concavities.

Galileo became famous as the foremost practitioner of experimental argument in seventeenth-century physics, even though—as can be seen from the extensive quotations from his writings presented in Chapters 5 and 7—he often resorted to the postulational style of Plato and Aristotle in order to persuade his readers. Galileo brilliantly combined the two styles in overturning the geocentric system while providing a firm foundation for the heliocentric system. To establish the crucial concept of uniformly accelerated motion, Galileo invented an experiment—rolling a ball down an inclined plane—that is still regarded as a classic example of how to do good physics.

Newton, too, is known as much for his postulates—the laws of motion and gravity, the concepts of absolute space and time—as for the experiments and calculations he performed to support them. Nevertheless, through the work of Kepler, Galileo, Boyle, Newton, and others, "quantitative precision," Crombie reminds us, "came to replace logical certainty in the seventeenth century as the primary demand of the theoretical sciences, as of the practical arts." Scientific research might be motivated by the search for truth, but scientists came to realize that what they could actually find was not necessarily an infallible truth about the world but a theory that provided accurate explanations and predictions of observed facts. Newton's theory of universal gravity, which did not say what gravity "really is" but explained Kepler's laws of planetary motion and successfully predicted many phenomena such as small deviations from those laws, the shape of the earth, and the date of the next return of Halley's comet, was an outstanding example.

The style of hypothetical modeling was new in the seventeenth century, even though one suspects that Ptolemaic astronomers did not really believe in the objective existence of all their epicycles, eccentrics, and equants. Kepler used the "camera obscura" (see Fig. 4.6) as a hypothetical model for the formation of optical images in the eye. But it was Descartes who explicitly stated that his theoretical statements about the world were not to be considered postulates representing his own beliefs, but as useful models from which one could deduce consequences in agreement with observation. This was quite obvious in his theories of light, where he proposed two contradictory models or analogies. First, light is an *impulse* transmitted through a medium, so that a person "sees" just as a blind man perceives the world by poking with his cane and interpreting the impulses that come back to his hand; second, light is a *stream of particles,* and the refraction of a ray at an interface is caused by attractive or repulsive forces exerted on those particles.

Boyle and Newton apparently thought Descartes had gone too far in using implausible hypotheses. Boyle proposed several requirements for a good hypothesis, especially that it must be "conceivable" (not absurd) and self-consistent, as well as being able to explain the facts. Newton declined to "feign" hypotheses for the nature of gravity when he could not find a satisfactory one; he said we must simply accept gravity as a fact.

The most popular hypothetical model in the seventeenth and eighteenth centuries was the "clockwork universe," originally proposed by Boyle and others by analogy with an actual clock in Strasbourg Cathedral in the 1570s. This clock had a number of well-designed mechanical animals that moved and made noises at different times of day. Boyle argued that if mere humans could design a mechanism that simulated animals so effectively, certainly God could design a clockwork mechanism like the universe. Moreover, since God must be a perfect workman, His machine can last forever: It never runs down and its parts never wear out. Translated into postulates supported by experimental argument, the clockwork universe model led to the laws of conservation of mass, momentum, and energy.

In the nineteenth century, hypothetical modeling guided the construction of atomic theories in chemistry and physics. Dalton's and Avogadro's atoms and molecules were not only unseen, there was at first no way to determine their sizes, shapes, and motions; only their relative weights and valences could be estimated. Yet they were remarkably successful in explaining and predicting chemical reactions. The atoms and molecules of Clausius, Maxwell, and Boltzmann were still somewhat hypothetical; for example, Maxwell used, as convenient, two different models, the billiard ball and the point center of force, not considering it necessary to decide which one better represented "real" atoms. But these three physicists did something else equally important: they established probability and statistical analysis (previously introduced by Pascal, Jacob and Daniel Bernoulli, and Laplace) as a productive new style of thinking. At about the same time taxonomy, a classifying style of thought usually more appropriate for biology or geology, briefly entered chemistry in the search for a "natural" periodic system of elements.

Like certain other concepts that used to be considered hypothetical models, such as the heliocentric solar system and biological evolution, the atom eventually became such a well-defined entity and so essential to chemistry and physics that it had to be considered an established fact. Although many of its properties were quantitatively determined by the 1880s with the help of the kinetic theory of gases, the empirical skepticism of the late nineteenth century delayed the general acceptance of "molecular reality" until around 1910. The actual *discovery* of the discrete atomic structure of matter—that is, the accomplishment of the task of persuading the scientific community that atoms really exist—may thus be credited to Jean Perrin and his experiments confirming Einstein's theory of Brownian movement.

The first person to "see" an atom was Erwin Mueller, in 1955, with the field-ion microscope that he perfected. Like Bessel's observation of stellar parallax in 1838, this was an anticlimax from the viewpoint of a historical controversy that could have been settled if it had been done two centuries earlier; its real value was to create a new field of research that could answer different kinds of questions. In the meantime, hypothetical modeling was being used in other areas of nineteenth-century physics—first optics and then electromagnetism. Young and Fresnel revived and improved Huygens' impulse model of light by creating the wave theory, but at the cost of creating a hypothetical ether that failed to satisfy our criteria for a good theory. Their ether, in order to account for the propagation of transverse waves, had to be an ephemeral elastic solid filling all space—an absurd idea.

The solution of that puzzle came later in the century, through the development of electromagnetism. Oersted, having been led by nature philosophy to *postulate* a connection between electricity and magnetism, vindicated his speculative approach by actually discovering such a connection in the laboratory: An electric current exerts a force on a magnetic compass needle. Faraday established that the inverse effect also occurs (a magnet can push a current-carrying wire), and then went on to make an even more remarkable discovery: The electromagnetic interaction can be used to *generate* electric current, transforming mechanical energy into electricity. These three discoveries, translated into practical devices (the telegraph, the electric motor, and the generator) launched modern electrical technology.

They also became the basis for Maxwell's electromagnetic theory, which started as an incredible hypothetical mechanical model combining vortices and ball bearings, representing magnetism and electricity, respectively. In its final form the theory was a set of equations governing the variations of electrical and magnetic fields—now taken as primary entities, not represented by or reduced to anything else—which propagate through space at the speed of light. Since the theoretical waves behaved just like light in all respects, they must *be* light—without the need for any awkward elastic solid to transmit them. The ether still existed but it had been dematerialized, dissolved into a collection of varying force fields. Moreover, light was no longer a unique phenomenon; rather it was just one instance (a narrow range of wavelengths perceptible by the human eye) of a general phenomenon: electromagnetic radiation, which can theoretically have any wavelength, and includes radio waves, microwaves, infrared radiation, x-rays and γ-rays.

Maxwell, quiet and studious, was practically unknown to his British compatriots, compared to the romantic hero Faraday and the flamboyant Kelvin, honored for his success in making the Atlantic Cable work and notorious for his controversial calculations on the age of the earth, which cast doubt on Darwin's theory of evolution. But it was Maxwell whom the British and their allies should have thanked in 1940, when *radar* (an acronym for *ra*dio *d*etection *a*nd *r*anging), using his electromagnetic waves, saved them from defeat by the German Air Force in the Battle of Britain.

Around the turn of the twentieth century the pace of discovery and theorizing became so rapid that the story of physics can no longer be described as a linear sequence. Within a decade (1895–1905), experimental argument had established x-rays, radioactivity, the electron, the quantum theory of blackbody radiators, and the transmutation of elements. (The same period saw the birth of the motion picture, the radio, the automobile, and the airplane.) Postulation, combined with a good dose of hypothetical modeling, created the quantum theory and special relativity. Probabilistic and statistical analysis played an essential role in the formulation of quantum theory, Brownian movement theory, and the theory of radioactive decay. Light (electromagnetic radiation) was given a special status in many of these theories, behaving like a stream of particles as well as like a wave phenomenon in the quantum theory, and providing an absolute standard against which to describe the motion of physical objects.

Having finally proved the existence of the atom, physicists immediately proceeded to show that it was not an atom at all in the classical sense, but could be divided into parts, both theoretically and experimentally. In Rutherford's model of 1911, the atom consisted of a tiny nucleus containing most of the mass, surrounded by light electrons. The electrons could be knocked out and the nucleus could eject α-, β-, or γ-rays, leaving behind a nucleus of a different element. Bohr's model of 1913, using quantum theory to calculate the possible orbits and jumps between orbits of the electrons, was somewhat more hypothetical than Rutherford's and combined contradictory classical and quantum assumptions, but was nevertheless a major advance in describing the atom.

In the 1920s the quantum revolution started to move toward its conclusion when Arthur Compton confirmed the particle nature of electromagnetic radiation by his experiment on the scattering of x-rays by electrons; while Louis de Broglie proposed that "particles" like electrons also have a wave nature, a suggestion soon confirmed by the electron-diffraction experiments of Davisson and Germer, and of G. P. Thomson. Heisenberg and Schrödinger now postulated, in different but equivalent equations, the theory known as quantum mechanics. This theory allows one to calculate with great accuracy the value of any observable property of any atom, molecule, or many-particle system (such as a metal or a star), provided that: (a) relativistic effects are unimportant, (b) one knows the nature of the constituent particles and the forces between them, and (c) one has available a computer of unlimited speed and memory. One must also observe the restrictions imposed by the Pauli exclusion principle (no more than one electron can have the same set of quantum numbers), or more generally the symmetry rules that define Fermi-Dirac and Bose-Einstein statistics.

The first proviso could not be ignored, because in the meantime relativity had become an essential part of all fundamental theories in physics. Einstein's general theory was initially comprehensible to only a few experts, but the sensational confirmation of its light-bending prediction by Eddington's 1919 eclipse observations was known to all. So P. A. M. Dirac undertook the task of making quantum mechanics compatible with relativity—or at least with the special theory of relativity. The result was a new equation that had two remarkable features. First, it explained the spin of the electron, a property that had previously been arbitrarily assumed. Second, it led, in one interpretation, to the prediction that "antielectrons" (electrons with positive charge) should exist—a prediction quickly confirmed by the discovery of the positron in 1932.

This early success of Dirac's relativistic quantum mechanics was followed, 15 years later, by the development of a greatly refined theory of electrons, positrons, and photons called "quantum electrodynamics," whose predictions have been confirmed with extremely high accuracy. (One of the practical applications of the theory was the development of the transistor, making high-speed computers possible.) Moreover, "antiprotons" (protons with negative charge) were also found to exist. But new questions arose to challenge theorists: Why are there so few antiparticles compared to the number of particles in the universe (or at least in our part of it)? How can one construct an even more general synthesis of relativity and quantum mechanics—one that would unify gravity and electromagnetic forces?

Yet another line of research, interacting with both relativity and quantum theory, led to new ideas about the structure and properties of the nucleus and to the prediction and discovery of new elementary particles. Before 1932 the nucleus was assumed to consist of protons and electrons, but as early as 1920 Rutherford had suggested the probable existence of a neutral combination of a proton and an electron. As soon as this particle, now called the neutron, was discovered by Chadwick in 1932, Heisenberg postulated that the nucleus consists of protons and neutrons. (Although electrons can be *emitted* from the nucleus, in β-decay, that does not mean they existed there as distinct particles.) Although more satisfactory in many ways than the older proton-electron model, the proton-neutron model exacerbated one difficulty: Since protons repel each other, what holds the nucleus together if there is not even any electrostatic attraction between oppositely charged particles?

One answer came from Hideki Yukawa, who suggested that strong short-range attractive forces between protons and neutrons are caused by the continual exchange of a new particle, subsequently called the *meson*. Like quantum electrodynamics, which postulates that electromagnetic forces are caused by the exchange of photons between charged particles, Yukawa's theory was another shot in the continuing thematic contest between particles and forces. Which is more fundamental: Are particles reducible to forces (as Boscovich and the dynamists claimed) or are forces reducible to moving particles (as Descartes and the kinetic theorists assumed)?

That scientific battle was temporarily suspended while scientists participated in a real world war. Nuclear fission was discovered just before the beginning of World War II, and several physicists realized that it could provide the basis for a powerful new weapon. Fearing on good grounds that German physicists would develop such a weapon for the Nazis, physicists from Britain and America cooperated to create the atomic bomb at the Los Alamos laboratory in New Mexico. As it happened, the Germans were not able to develop a workable bomb before the end of the European war in 1945, but by the decision of President Harry Truman the American bombs were used anyway against the other enemy, Japan. With dismay, mixed with relief that the war had been ended without further loss of American and Japanese lives, the atomic scientists saw their invention emerge as a new danger to civilization and a major factor in the cold war hostility between the West and the Soviet Union.

For a couple of decades after World War II, physicists enjoyed greatly increased funding because of the obvious military value of some of their past discoveries such as nuclear fission and radar. The money helped to bring many new discoveries and theories in the expensive field of elementary particle physics. But other sciences also shared the general support for science and began to capture more public attention. Space science, stimulated by the competition between the United States and the Soviet Union to develop intercontinental rockets for military purposes, led many physicists, chemists, and engineers to turn their attention to the moon and the planets. Samples returned by lunar missions, beginning in 1969, were carefully analyzed for clues to the origin of the moon and perhaps the earth itself. Oceanographic research, sponsored by the U.S. Navy, made significant contributions toward the "revolution in the earth sciences" of the 1960s, leading to the establishment of the theory of continental drift and plate tectonics.

These advances in planetary science, which relied heavily on physical methods and instruments, embodied a significant thematic switch. For more than a century before 1960, Anglo-American geology had been dominated by the *uniformitarian* principle: The earth's past history should be explained in terms of the slow operation of causes similar in nature and magnitude to those presently observed in operation, such as erosion and vulcanism. Even the origin of the moon, a problem that might seem to lie outside the proper domain of uniformitarianism, was generally attributed to continuous processes like gravitational capture or accretion of particles. But new evidence forced planetary scientists to admit that discontinuous changes and catastrophic collisions had shaped the earth and solar system. Plate tectonics implied not only the abrupt separation of continents but also sudden reversals of the earth's magnetic field. The lunar samples were interpreted to mean that the moon had been formed as the result of a giant impact of another planet on the earth. And, most dramatic in its effect on the public imagination, the theory of physicist Luis Alvarez and his colleagues, which postulated the impact of an asteroid about 65 million years ago, seemed to account for the extinction of the dinosaurs, which perhaps cleared the way for the evolution of other animal species, including humans.

Astronomy and cosmology also fascinated the public as well as physicists in the second half of the twentieth century. In the 1930s, Hans Bethe had applied his expertise in nuclear physics to constructing a sequence of reactions that could fuse hydrogen or carbon into heavier elements and explain how energy is generated in the sun and other stars. In the 1940s another nuclear physicist, George Gamow, developed the big bang cosmology to explain not only the expanding universe but also the synthesis of elements from hydrogen or neutrons. The debate between Gamow and proponents of the rival steady state cosmology, led by Fred Hoyle, attracted much attention—partly because of the profundity of the questions being asked, like "did the universe exist forever or was it created at a particular time?"—and partly because both Gamow and Hoyle were excellent expositors of science.

The renewed interest in cosmology brought with it a revival of research on general relativity, which had languished since the 1920s. One result of this research was the *black hole* concept, based on the old idea that light could not escape the gravitational attraction of a sufficiently massive star, making it invisible; the black hole was assigned more interesting properties by Stephen Hawking. Not only are black holes now considered detectable, because of the radiation emitted by matter being sucked into them, but little ones may be all around us and big ones may be eating up entire galaxies or providing gateways to other universes. Science fiction writers must hustle to keep up with the fantastic ideas generated by scientists.

At the end of the twentieth century, as at the end of the nineteenth, a few pessimists proclaimed that all the basic principles of physical science had been discovered and there was nothing left to do except to apply them to more phenomena and determine the constants of nature to a few more decimal places. And the pessimists of the 1990s were almost certainly wrong, as were their predecessors in the 1890s. The *unsolved* problems of the physical world now seem even more formidable than those solved in the twentieth century.

Though in application it works splendidly, we do not even understand the physical meaning of quantum mechanics, much less how it might be united with general relativity. We don't know why the dimensionless constants (ratios of masses of elementary particles, ratios of strength of gravitational to electric forces, fine structure constant, etc.) have the values they do, unless we appeal to the implausible anthropic principle, which seems like a regression to Aristotelian teleology. We don't know what, if anything came before the big bang, or whether there are other universes besides our own, as some cosmologists suggest. Some of the new planetary systems discovered around other stars are so unlike our own solar system as to make us realize that we have not yet discerned the basic principles governing the formation of planetary systems in general.

Without even mentioning the numerous physical problems to be solved in earth science, chemistry, biology, and psychology, there are enough unsolved problems within physics and astronomy to keep you, our reader, fully occupied with fascinating work during your lifetime.

RECOMMENDED FOR FURTHER READING

A. C. Crombie, *Styles of Scientific Thinking,* Synopsis, Chapters 1 and 2

Gerald Holton, "'Themata' in scientific thought," in his book *The Scientific Imagination,* pages 3–24, Cambridge, MA: Harvard University Press, 1998

Gerald Holton, "Thematic presuppositions and the direction of scientific advance," in his book *The Advancement of Science and Its Burdens,* pages 3–27, Cambridge, MA: Harvard University Press, 1998

Marjorie Malley, "The discovery of atomic transmutation: Scientific styles and philosophies in France and Britain," in *History of Physics* (edited by S. G. Brush), pages 184–194

Current issues of *Scientific American, The American Scientist, Science News,* and similar publications

Refer to the website for this book, www.ipst.umd.edu/Faculty/brush/physicsbibliography.htm, for Sources, Interpretations, and Reference Works and for Answers to Selected Numerical Problems.

APPENDIX I

Abbreviations and Symbols

(See Index to locate definitions)

a	annum (year)
a	atto (metric prefix)
a	acceleration
$\mathbf{a}$	acceleration vector
a_c	centripetal acceleration
a	half major axis of ellipse
A	[see amp or Ar as appropriate]
A	area
A	mass number
Å	angstrom (unit of length, no longer used)
Ag	silver
Al	aluminum
Am	Americium (element)
amp	ampere (unit of electric current)
amu	atomic mass unit
Ar	argon
As	arsenic
At	astatine (element)
atm	average pressure of earth's atmosphere
Au	gold
AU	astronomical unit (earth-sun distance)
b	half-minor axis of ellipse
B	boron
Ba	barium
Be	beryllium
Bh	bohrium (element)
Bi	bismuth
Bk	berkelium (element)
Br	bromine
c	centi (metric prefix)
c	speed of light
c	rms speed of gas molecules
c	specific heat
c_p	specific heat at constant pressure
c_v	specific heat at constant volume
C	[see coul]
C	Celsius (temperature)
C	carbon
C	molar specific heat
cal	calorie
Cd	cadmium (element)
Ce	cerium (element)
Cf	californium (element)
Cl	chlorine
cm	centimeter
Cm	curium (element)
Co	cobalt
cos	cosine
coul	coulomb (unit of electrical charge)
Cr	chromium
Cs	cesium
csc	cosecant
Cu	copper
d	day
d	deco (metric prefix)
d	deuterium
da	deka (metric prefix)
Db	dubnium (element)
Dy	dysprosium (element)
dyn	[dyne, unit of force = 10^{-5} N, no longer used]
e	electron
e	elementary electric charge
e	base of natural logarithms
e	eccentricity of ellipse
E	East
E	exa (metric prefix)
E	energy
Er	erbium (element)
erg	[unit of energy = 10^{-7} J, no longer used]
Es	einsteinum (element)
Eu	europium (element)
eV	electron volt (unit of energy)
f	frequency
f	femto (metric prefix)
F	Fahrenheit (temperature)
F	fluorine
F	Faraday constant (in electrolysis)
F	force
$\mathbf{F}$	force vector
Fe	iron
fm	[fermi, unit of length = 10^{-15} m, no longer used]
Fm	fermium (element)
Fr	francium (element)
ft	foot or feet
g	gram
g	gravitational acceleration at earth's surface
G	giga (metric prefix)
G	universal gravitational constant
Ga	gallium (element)
Ga	Giga annum
Gd	gadolinium (element)
Ge	germanium (element)
h	hecto (metric prefix)
h	Planck's constant
H	hydrogen
H	heat
H	Hubble constant
He	helium
Hf	hafnium (element)
Hg	mercury
Ho	holmium (element)
hr	hour

Hs	hassium (element)
i	running index for quantities listed or added
i	imaginary unit $\sqrt{(-1)}$
I	iodine
I	rotational inertia
in.	inch
In	indium (element)
Ir	iridium
J	joule (unit of energy)
J	mechanical equivalent of heat
k	kilo (metric prefix)
k	force constant of spring
k	Boltzmann constant
k_e	electrostatic force constant
K	kelvin (unit of temperature)
K	potassium
KE	kinetic energy
kg	kilogram
km	kilometer
Kr	krypton
kwh	kilowatt-hour
l	length
l	angular momentum
L	liter (unit of volume)
La	lanthanum (element)
lb	pound
Li	lithium
log	(natural) logarithm
Lr	lawrencium (element)
Lu	lutetium (element)
LY	light year
m	meter (unit of length) or mass
m	milli (metric prefix)
m	mass
m_e	mass of electron
m_e	mass of earth
m_n	mass of neutron
m_p	mass of planet
m_p	mass of proton
m_0	rest mass
M	mega (metric prefix)
M	(relative) molecular weight
Ma	million years
Md	mendelevium (element)
Mg	magnesium
mi	mile
min	minute
Mn	manganese
Mo	molybdenum (element)
mol	mole (gram-molecular-weight)
ms	millisecond
Mt	meitnerium (element)
n	nano (metric prefix)
n	neutron
N	north
N	newton (unit of force)
N	nitrogen
N_A	Avogadro's number
N_L	Loschmidt's number
Na	sodium
Nb	niobium (element)
Ne	neon
Ni	nickel
nm	nanometer
No	nobelium (element)
Np	neptunium (element)
O	oxygen
Os	osmium (element)
p	pico (metric prefix)
p	proton
P	peta (metric prefix)
P	pressure
P	phosphorus
Pa	pascal (unit of pressure)
parsec	parallax second (unit of distance)
Pb	lead
Pd	palladium (element)
pdl	poundal
PE	potential energy
Po	polonium
Pr	praseodymium (element)
Pt	platinum
Pu	plutonium
Q	heat
Q	equant point
Q.E.D.	quod erat demonstrandum (that which was to be proved)
r	radius (or distance from a point)
R	gas constant
R_H	Rydberg constant
R_∞	Rydberg constant for infinitely heavy nucleus
R	radius of planetary orbit
Ra	radium
rad	radian (unit of angle)
Rb	rubidium
Re	rhenium (element)
Rf	rutherfordium (element)
Rn	radon
Ru	ruthenium
s	distance
S	sulfur
S	south
S	entropy
Sb	antimony
Sc	scandium (element)
Se	selenium
sec	second
sec	secant
Sg	seaborgium (element)
Si	silicon
sin	sine
Sm	samarium (element)
Sn	tin
Sr	strontium
t	time
T	tera (metric prefix)
T	half-life (of radioactive decay)
T	absolute temperature
T	period (of planetary or oscillating motion)
Ta	tantalum (element)
tan	tangent
Tb	terbium (element)
Tc	technetium (element)
Te	tellurium (element)
Th	thorium
Ti	titanium
Tl	thallium (element)
Tm	thulium (element)
torr	Torricelli (unit of pressure, no longer used)
u	[see amu]
U	uranium
Uun	ununnilum (element)
Uuu	unununium (element)
v	velocity or speed
$\boldsymbol{v}$	velocity vector
V	volume
W	tungsten

W	west
W	work
W	number of molecular arrangements
V	vanadium
Xe	xenon
Y	yttrium (element)
y	year
y	yocto (metric prefix)
Y	yotta (metric prefix)
Yb	ytterbium (element)
z	zepto (metric prefix)
Z	zetta (metric prefix)
Z	atomic number
Zn	zinc
Zr	zirconium

Greek Letters Used as Scientific Terms

α	alpha particle (helium nucleus)
α	separation factor (in diffusion)
α	fine structure constant
β	beta particle (electron)
γ	gamma ray (high-energy photon)
γ	gamma (Lorentz factor)
δ	delta (number of active degrees of freedom)
Δ	delta (change of . . .)
η	eta (efficiency of machine)
θ	theta (angle)
λ	lambda (wavelength)
λ	lambda (transition in liquid helium)
μ	mu meson (muon)
μ	micro (metric prefix)
ν	neutrino
ν	nu (frequency)
π	pi (circumference/diameter of circle)
π	pi meson (pion)
ρ	rho (density)
Σ	sigma (sum over . . .)
Σ_i	sigma subscript *i* (sum over values of *i*)
τ	tau
ψ	psi (wave function)
ω	omega (angular frequency)

Mathematical Symbols Other than Greek Letters

$\propto$	proportional to
$\approx$	approximately equal to
$\equiv$	defined as equal to
$\sqrt{}$	square root
$\langle \ldots \rangle_{av}$	average value of . . .
$'$	minute
$''$	second

Metric System Prefixes, Greek Alphabet, Numerals

Metric System Prefixes

yotta (Y)	10^{24}	septillion
zetta (Z)	10^{21}	sextillion
exa (E)	10^{18}	quintillion
peta (P)	10^{15}	quadrillion
tera (T)	10^{12}	trillion
giga (G)	10^{9}	billion
mega (M)	10^{6}	million
kilo (k)	10^{3}	thousand
hecto (h)	10^{2}	hundred
deka (da)	10	ten
deco (d)	10^{-1}	tenth
centi (c)	10^{-2}	hundredth
milli (m)	10^{-3}	thousandth
micro (μ)	10^{-6}	millionth
nano (n)	10^{-9}	
pico (p)	10^{-12}	
femto (f)	10^{-15}	
atto (a)	10^{-18}	
zepto (z)	10^{-21}	
yocto (y)	10^{-24}	

Greek Alphabet

Α	α	Alpha	Ν	ν	Nu
Β	β	Beta	Ξ	ξ	Xi
Γ	γ	Gamma	Ο	ο	Omicron
Δ	δ	Delta	Π	π	Pi
Ε	ε	Epsilon	Ρ	ρ	Rho
Ζ	ζ	Zeta	Σ	σ	Sigma
Η	η	Eta	Τ	τ	Tau
Θ	θ	Theta	Υ	υ	Upsilon
Ι	ι	Iota	Φ	ϕ, φ	Phi
Κ	κ	Kappa	Χ	χ	Chi
Λ	λ	Lambda	Ψ	ψ	Psi
Μ	μ	Mu	Ω	ω	Omega

Roman Numerals, Arabic Equivalents, and IUPAC Numerical Roots

Roman	*Arabic*	*IUPAC*[1]
	0	nil
I	1	un [pronounced as in "t*une*"]
II	2	bi
III	3	tri
IV	4	quad
V	5	pent
VI	6	hex
VII	7	sept
VIII	8	oct
IX	9	enn
X	10	unnil
XI	11	unun
XIX	19	unen
XX	20	binil
XXX	30	trinil
XL	40	quadnil
L	50	pentnil
LX	60	hexnil
XC	90	ennil
C	100	unnilnil
CI	101	unnilun
CX	110	ununun
CXII	112	ununbi
CC	200	binilnil
CD	400	quadnilnil
D	500	pentnil
CM	900	ennnil
M	1000	
MM	2000	

[1]Recommended by the International Union of Pure and Applied Chemistry, Commission on Nomenclature of Inorganic Chemistry (1978, published 1979) for constructing temporary names of hypothetical elements of atomic numbers greater than 100. The suffix "ium" would be added except that the final "i" of "bi" or "tri" is elided; thus the name for the element of atomic number 112 would be *ununbium*. Also, the final "n" of "enn" is elided when it occurs before "nil" so the name of the element of atomic number 190 would be *unennilium*. The element will have a three-letter symbol made from the first letters of the roots, thus *Uuu* means unununium. When the element is discovered, IUPAC will eventually approve a permanent name following the usual procedure.

Defined Values, Fundamental Constants, and Astronomical Data

Defined Values

Name of Quantity	*Symbol*	*Value*
Meter	m	Length of the path traveled by light in vacuum during a time interval of 1/299,792,458 of a second
Second	sec	9,192,631,770 periods of the radiation corresponding to the transition between the two hyperfine levels of the ground state of the Cs^{133} atom
Kilogram	kg	Mass of the standard kilogram at International Bureau of Weights and Measures, Sèvres, France
Degree	K	1/273.16 of absolute temperature of triple point of water
Atomic mass unit	amu	1/12 of the mass of an atom of C^{12} isotope
Gravitational acceleration at surface of earth at sea level	g	9.80665 m/sec^2
Normal atmosphere	atm	101325 N/m^2 = 1.01325 × 10^5 Pa
Mole	mol	Amount of substance containing same number of molecules as 12 g of pure C^{12}
Liter	L	1,000.028 cm^3
Ampere	amp	The current that, when flowing in each of two long parallel wires 1 m apart, causes them to exert on each other a force of 2×10^{-7} N
Electrostatic force constant	k_e	8.9874×10^9 N·m^2/coul^2

Astronomical Data

Name of quantity	*Value*
Mass of earth	5.98×10^{24} kg
Mean radius of earth	6.37×10^6 m
Mean distance from earth to sun	1.496×10^{11} m
Eccentricity of earth's orbit	0.0167
Mean distance from earth to moon	3.82×10^8 m
Sun's radius	6.96×10^8 m
Sun's mass	1.99×10^{30} kg

Mathematical Constants

$\pi = 3.14159$;	$1/\pi = 0.31831$;
$\pi^2 = 9.8690$;	$\sqrt{\pi} = 1.77245\ldots$
$e = 2.71828$;	$1/e = 0.36788\ldots$
$\sqrt{2} = 1.41421\ldots$	
$\sqrt{3} = 1.73205\ldots$	

Fundamental Constants[1]

Name of Quantity	*Symbol*	*Value*
Velocity of light	c	2.9979×10^8 m/sec (defined)
Planck's constant	h	6.626×10^{-34} J·sec
Charge on electron	e	(–)1.602×10^{-19} coul
Fine structure constant	$\alpha = e^2/hc$	1/137.036
Universal gravitational constant	G	6.674×10^{-11} N·m^2/kg^2
1 atomic mass unit	amu	1.6605×10^{-27} kg
Mass of electron	m_e	9.109×10^{-31} kg = 5.486×10^{-4} amu
Electron charge to mass ratio	e/m_e	1.7588×10^{11} coul/kg
Mass of neutral hydrogen	(H^1)	1.007825 amu
Mass of proton (p)	m_p	1.00728 amu
Mass of neutron (n)	m_n	1.008665 amu
Mass of α particle	m_α	1.001506 amu
Ratio of proton to electron mass	m_p/m_e	1836.15
Rydberg constant for infinitely heavy nucleus	R_∞	1.09737×10^7/m
Rydberg constant for hydrogen	R_H	1.09678×10^7/m
Boltzmann constant	k	1.3806×10^{-23} J/K
Avogadro's number	N_A	6.022×10^{23} /mol
Loschmidt's number	N_L	2.69×10^{19}/cm^3
Universal gas constant	R	8.314 J/mol·K
Volume of 1 mole of gas at standard temperature and pressure		2.24×10^{-2} m^3
Faraday constant	F	96,485 coul/mol

[1]In each case, the figures are known to higher accuracy than implied here, but are given in the form most useful for the purposes of this book. See Appendix V for a general discussion of units and constants. For further data, consult a collection such as the *Handbook of Chemistry and Physics* (Chemical Rubber Publishing Co., Cleveland; issued yearly).

Conversion Factors

Length

1 meter (m) = 39.37 inches
1 inch (in.) = 0.0254 meter
1 meter = 3.281 feet
1 foot (ft) = 0.3048 meter
1 kilometer (km) = 0.6214 mile (mi)
1 mile = 1.609 kilometers
1 cubit ≈ 0.5 m
1 angstrom (Å, no longer used) = 10^{-10} m
1 astronomical unit (AU) = 14.96×10^7 km = 9.29×10^7 mi
1 light year (LY) = 9.46×10^{12} km = 6.32×10^4 AU
1 parsec (distance at which 1 AU subtends 1 sec of arc) = 206,265 AU = 3.084×10^{16} meters = 3.26 LY
1 Mpc = 10^6 parsec = 3.26×10^6 LY

Angular Measure

1 degree (°) = 1/360 of revolution around circle = (π/180) radian = 60 minutes (min or ′) = 3600 seconds (sec or ″)
1 radian (rad) = 57.3° = 0.159 revolutions
π rad = 180° = ½ revolution

Area

1 m^2 = 10.76 ft^2
1 ft^2 = 9.29×10^{-2} m^2
1 m^2 = 1.55×10^3 $in.^2$
1 $in.^2$ = 6.452×10^{-4} m^2
1 cm^2 = 0.1550 $in.^2$
1 $in.^2$ = 6.452 cm^2
1 ft^2 = 144 $in.^2$
1 $in.^2$ = 6.944×10^{-3} ft^2

Volume

1 m^3 = 35.31 ft^3; 1 ft^3 = 2.832×10^{-2} m^3
1 liter (L) = 10^{-3} m^3
1 m^3 = 10^3 L
1 cm^3 = 6.102×10^{-2} $in.^3$
1 $in.^3$ = 16.39 cm^3

Time

1 minute (min) = 60 seconds (sec)
1 hour (hr) = 60 min = 3600 sec
1 day (d) = 24 hr = 86,400 sec
1 year (y or a) ≈ 365¼ d = 3.16×10^7 sec
1 millennium (ka) = 10^3 y
1 million years (Ma) = 10^6 y
1000 million years (Ga) = 1 billion years = 10^9 y

Velocity

1 m/sec = 3.281 ft/sec; 1 ft/sec = 0.3048 m/sec
1 m/sec = 3.6 km/hr; 1 km/hr = 0.2778 m/sec
1 m/sec = 2.237 mi/hr; 1 mi/hr = 0.4470 m/sec
1 mi/hr = 1.467 ft/sec = 0.4470 m/sec

Acceleration

1 m/sec^2 = 3.281 ft/sec^2
1 ft/sec^2 = 0.3048 m/sec^2

Force

1 newton (N) = 7.233 poundals (pdl)
1 pdl = 0.1383 N
1 pound [force][1] (lbf) = 4.448 N
1 N = 0.2248 lbf

Mass

1 kilogram (kg) $\leftrightarrow$ 2.205 pound[1] (lb)
1 lb $\leftrightarrow$ 0.4536 kg
1 gram (g) $\leftrightarrow$ 2.205×10^{-3} lb
1 lb $\leftrightarrow$ 453.6 g

Pressure

1 pascal (Pa) = 1 N/m^2 = 1.45×10^{-4} lbf/in^2
1 atmosphere (atm) = 1.013×10^5 N/m^2
1 atm = 14.7 $lbf/in.^2$
1 bar = 10^5 N/m^2

Energy

1 joule (J) = 0.239 cal; 1 cal = 4.184 J
1 J = 0.2778×10^{-6} kilowatt-hour (kwh)
1 kwh = 3.6×10^6 J
1 J = 0.625×10^{19} electron volts (eV)
1 eV = 1.60×10^{-19} J

[1]The double arrow "$\leftrightarrow$" means "corresponds to," not "equal to" since the pound is a unit of weight (i.e., force) and cannot be equal to a unit of mass. Common usage in the United States does not observe the scientific distinction between "pound" as a unit of *weight* (force) and (incorrectly) as a unit of *mass*. In this book we designate the former by the abbreviation "lbf" (pound-force) and the latter by "lb" (pound). Thus the statement "1 kg weighs 2.205 pounds" means "a 1 kg *mass* is an object that *weighs* 2.205 lbf at a location where g has the standard value 9.80665 m/sec^2, i.e. mg = 9.8 kgxm/sec^2 = 2.205 lbf." The pound-force should not be confused with the "poundal," which is no longer used either in science or in daily life (see Appendix V).

Systems of Units

The system of units we have been using is usually called "SI," the initials of the French term *Système Internationale* for the system universally accepted by the scientific community. It is a generalization of the earlier "MKS" system, which refers to the meter, the kilogram, and the second, as the three fundamental standards of measurement; it includes four other units, the ampere (current), kelvin (temperature), mole (amount of substance), and candela (luminous intensity). These are defined in Appendix III, except for the candela, which is used primarily in engineering rather than in physics.

The standards of length and mass go back to the time of the French Revolution, when a committee of scientists (led by J. L. LaGrange) appointed by the new government developed what became known as the *metric system.* The meter was originally defined as one ten-millionth of the quadrant of the earth's meridian, and the kilogram was defined as the mass of one cubic decimeter of water at 4°C. Advances in the technology of precision measurements made these definitions unsatisfactory, and we eventually arrived at the present SI units in which, for example the meter is defined in terms of the speed of light, and second is defined in terms of a particular type of radiation from the Cs^{133} atom.

But the laws of nature do not, of course, depend on the magnitude of the arbitrarily chosen standards. It is evident that it would make no fundamental difference if one used (as is sometimes done in the text of this book) the centimeter instead of the meter or the gram instead of the kilogram; those choices correspond to the cgs (centimeter-gram-second) system found in a number of older books. The whole subject of units is of interest only to a few specialists, though it sometimes seems to loom large to introductory students. Do not feel that the following comments, intended for the curious, are of deep significance or have to be memorized!

We might, for example, have referred throughout the book to the units used for daily transactions in the United States and Great Britain, that is, the *foot* instead of the meter, the *pound avoirdupois* (lb) instead of the kilogram. The latter corresponds to the mass of 27.692 $in.^3$ of water at 4°C and ordinary pressure—a quantity of matter having an inertia about 0.4536 times that of an object of l-kg mass. As in the case of the kilogram, the pound is expressed by the equivalent inertia of a cylinder of precious metal, this one being kept by the Bureau of Standards for the United Kingdom in London.[1] The relation 1 lb = 0.4536 kg or 1 kg = 2.2 lb is a choice that had no deep profound reason, and is historically rather accidental. (It does explain the fact, puzzling to Americans who have refused to accept the metric system in everyday life, that the weight limit for personal baggage on some international airlines is 44 lb rather than a "round number" like 40 or 50 lb.) At any rate, together with the equivalence between 1 ft and 0.3048 m, it comprises the relation between the SI system of units and what is called the British Absolute System, or foot-pound-second (fps) system. In a more general sense, the one is called a metric system and the other a British system.

Certainly, spring balances might be calibrated by accelerating a 1-lb standard mass, so that $F_{net} = m$ (in lb) $\times\, a$ (in ft/sec^2). The unit of force so obtained, and used in connection with the fps system, is the lb-ft/sec^2, called the *poundal* (abbreviated pdl). It is quite analogous to the name *newton* (N) given to 1 kg-m/sec^2, the unit of force used in the SI system, or to the name *dyne* given to 1 g-cm/sec^2, the unit of force used in the cgs system. As you can confirm by direct

[1]In the U.S., the relation between the pound and the gram (like that between the foot and the meter) is fixed not by a comparison between two metal pieces at the National Institute of Standards and Technology in Gaithersburg, Maryland, but by act of Congress. However, there is no conflict here, for the legal decision was designed to agree with the results of direct comparison of standard masses within the limits of ordinary measurement.

calculation, 1 pdl = 0.1383 N = 1.383×10^4 dyn, or 1 N = 7.23 pdl (1 N = 105 dyn).

The introduction of electrical quantities further complicates the situation. For example, the SI and cgs systems are rather similar as long as one considers only combinations of mass, length, and time, since the basic units are related to each other by powers of 10. (See the conversion factors given in Appendix IV.) But it used to be customary to associate the cgs system with so-called *electrostatic units* (esu) of charge. In this system, Coulomb's law was written in the form

$$F = q_1 q_2 / r^2$$

with the constant k_e in Eq. (24.1) being set equal to 1. The unit of electric charge (1 esu, also called 1 statcoulomb) was defined as the quantity of a point charge that, 1 cm away from an equally large point charge, exerts on it 1 dyn of force. One could also define electric current in *electromagnetic units* (emu), using the *abampere* as the amount of current that, when flowing in two parallel straight wires at a distance of 1 cm, gives rise to a force of 2 dyn on each cm length of one wire. (Compare the definition of the ampere given in Section 25.3.) The *abcoulomb* was then defined as the amount of charge that passes through a wire when a current of 1 abamp is flowing for 1 sec. It can then be shown that 1 abcoul = c statcoul, where c is a conversion factor numerically equal to the speed of light expressed in cm/sec; furthermore one can convert to the SI system by using the relation: 1 coul = 1/10 abcoul.

Although the cgs/esu/emu system does have some advantages, the majority of physicists have decided that the SI system is preferable, and this system will probably be used in nearly all modern textbooks.

The choice of *primary standards* is still under discussion in the scientific community, and several of the definitions given in Appendix III are based on fairly recent but tentative decisions taken at international conferences. The general tendency has been to abandon the use of macroscopic physical objects kept in a particular place (such as the "standard meter") in favor of atomic properties that can easily be reproduced with great accuracy anywhere in the universe. Thus one no longer has to rely on a comparison with the standard objects kept in France to determine the magnitude of the meter or the second; one now measures the wavelength or frequency of a specified electronic transition of a particular atom (Cs^{133}). Similarly, a Martian scientist could adopt the Kelvin absolute temperature scale without having to know the melting and boiling points of water at the pressure corresponding to the earth's atmosphere; she would only have to locate the triple point of water in terms of her own common units of temperature and pressure. (See the last paragraph of Section 19.3.) And she could use our scale of atomic masses without knowing the precise isotopic composition of terrestrial oxygen, as long as she had available some pure C^{12}. Only one of the fundamental units, the kilogram, has so far not been given a definition in terms of universal atomic properties.

RECOMMENDED FOR FURTHER READING

J. L. Heilbron, "The politics of the meter stick," in *Physics History II* (edited by A. P. French and T. B. Greenslade), pages 219–223

Alphabetic List of the Elements

Element	*Symbol*	*Atomic Number Z*	*Element*	*Symbol*	*Atomic Number Z*	*Element*	*Symbol*	*Atomic Number Z*
Actinium	Ac	89	Holmium	Ho	67	Rhodium	Rh	45
Aluminum	Al	13	Hydrogen	H	1	Rubidium	Rb	37
Americium	Am	95	Indium	In	49	Ruthernium	Ru	44
Antimony	Sb	51	Iodine	I	53	Rutherfordium	Rf	104
Argon	Ar	18	Iridium	Ir	77	Samarium	Sm	62
Arsenic	As	33	Iron	Fe	26	Scandium	Sc	21
Astatine	At	85	Krypton	Kr	36	Seaborgium	Sg	106
Barium	Ba	56	Lanthanum	La	57	Selenium	Se	34
Berkelium	Bk	97	Lawrencium	Lr	103	Silicon	Si	14
Beryllium	Be	4	Lead	Pb	82	Silver	Ag	47
Bismuth	Bi	83	Lithium	Li	3	Sodium	Na	11
Bohrium	Bh	107	Lutetium	Lu	71	Strontium	Sr	38
Boron	B	5	Magnesium	Mg	12	Sulfur	S	16
Bromine	Br	35	Manganese	Mn	25	Tantalum	Ta	73
Cadmium	Cd	48	Meitnerium	Mt	109	Technetium	Tc	43
Calcium	Ca	20	Mendelevium	Md	101	Tellurium	Te	52
Californium	Cf	98	Mercury	Hg	80	Terbium	Tb	65
Carbon	C	6	Molybdenum	Mo	42	Thallium	Tl	81
Cerium	Ce	58	Neodymium	Nd	60	Thorium	Th	90
Cesium	Cs	55	Neon	Ne	10	Thulium	Tm	69
Chlorine	Cl	17	Neptunium	Np	93	Tin	Sn	50
Chromium	Cr	24	Nickel	Ni	28	Titanium	Ti	22
Cobalt	Co	27	Niobium	Nb	41	Tungsten	W	74
Copper	Cu	29	Nitrogen	N	7	Ununbium[1]	Uub	112
Curium	Cm	96	Nobelium	No	102	Ununhexium	Uuh	116
Dysprosium	Dy	66	Osmium	Os	76	Ununnilium	Uun	110
Einsteinium	Es	99	Oxygen	O	8	Ununoctium	Uuo	118
Erbium	Er	68	Palladium	Pd	46	Ununpentium	Uup	115
Europium	Eu	63	Phosphorus	P	15	Ununseptium	Uus	117
Fermium	Fm	100	Platinum	Pt	78	Unununium	Uuu	111
Fluorine	F	9	Plutonium	Pu	94	Uranium	U	92
Francium	Fr	87	Polonium	Po	84	Vanadium	V	23
Gadolinium	Gd	64	Potassium	K	19	Xenon	Xe	54
Gallium	Ga	31	Praseodymium	Pr	59	Ytterbium	Yb	70
Germanium	Ge	32	Promethium	Pm	61	Yttrium	Y	39
Gold	Au	79	Protactinium	Pa	91	Zinc	Zn	30
Hafnium	Hf	72	Radium	Ra	88	Zirconium	Zr	40
Hassium	Hs	108	Radon	Rn	86			
Helium	He	2	Rhenium	Re	75			

[1]See Appendix II for the system for naming Ununbium, etc.

Periodic Table of Elements

Atomic weights are the most recent (1997) adopted by the International Union of Pure and Applied Chemistry. The numbers are all relative to the atomic weight of the principal isotope of carbon, C^{12}, defined as 12.00000. For some of the artificially produced elements, the approximate atomic weight of the most stable isotope is given in parentheses. The full names of the elements are given in Appendix V, and the system for constructing names of hypothetical heavy elements is given in Appendix II.

Period	Series	I	II	III	IV	V	VI	VII	VIII			0
1	1	**1 H** 1.01										**2 He** 4.00
2	2	**3 Li** 6.94	**4 Be** 9.01	**5 B** 10.81	**6 C** 12.01	**7 N** 14.01	**8 O** 16.00	**9 F** 19.00				**10 Ne** 20.18
3	3	**11 Na** 22.99	**12 Mg** 24.31	**13 Al** 26.98	**14 Si** 28.09	**15 P** 30.97	**16 S** 32.07	**17 Cl** 35.45				**18 Ar** 39.95
4	4	**19 K** 39.10	**20 Ca** 40.08	**21 Sc** 44.96	**22 Ti** 47.87	**23 V** 50.94	**24 Cr** 52.00	**25 Mn** 54.94	**26 Fe** 55.8	**27 Co** 58.9	**28 Ni** 58.7	
	5	**29 Cu** 63.55	**30 Zn** 65.39	**31 Ga** 69.72	**32 Ge** 72.61	**33 As** 74.92	**34 Se** 78.96	**35 Br** 79.90				**36 Kr** 83.80
5	6	**37 Rb** 85.47	**38 Sr** 87.62	**39 Y** 88.91	**40 Zr** 91.22	**41 Nb** 92.91	**42 Mo** 95.94	**43 Tc** (98)	**44 Ru** 101.1	**45 Rh** 102.9	**46 Pd** 106.4	
	7	**47 Ag** 107.9	**48 Cd** 112.4	**49 In** 114.8	**50 Sn** 118.7	**51 Sb** 121.8	**52 Te** 127.6	**53 I** 126.9				**54 Xe** 131.3
6	8	**55 Cs** 132.9	**56 Ba** 137.3	**57–71***	**72 Hf** 178.5	**73 Ta** 180.9	**74 W** 183.8	**75 Re** 186.2	**76 Os** 190.2	**77 Ir** 192.2	**78 Pt** 195.1	
	9	**79 Au** 197.0	**80 Hg** 200.6	**81 Tl** 204.4	**82 Pb** 207.2	**83 Bi** 209.0	**84 Po** 209.0	**85 At** 210.0				**86 Rn** (222)
7	10	**87 Fr** (223.0)	**88 Ra** (226)	**89–103†**	**104 Rf** (261)	**105 Db** (262)	**106 Sg** (266)	**107 Bh** (264)	**108 Hs** (269)	**109 Mt** (268)	**110**	
	11	**111**	**112**	**113**	**114**	**115**	**116**	**117**				**118**

*Lanthanide series	**57 La** 138.9	**58 Ce** 140.1	**59 Pr** 140.9	**60 Nd** 144.2	**61 Pm** (145)	**62 Sm** 150.4	**63 Eu** 152.0	**64 Gd** 157.3	**65 Tb** 158.9	**66 Dy** 162.5	**67 Ho** 164.9	**68 Er** 167.3	**69 Tm** 168.9	**70 Yb** 173.0	**71 Lu** 175.0
†Actinide series	**89 Ac** (227)	**90 Th** 232.0	**91 Pa** 231.0	**92 U** 238.0	**93 Np** (237)	**94 Pu** (244)	**95 Am** (243)	**96 Cm** (247)	**97 Bk** (247)	**98 Cf** (251)	**99 Es** (252)	**100 Fm** (257)	**101 Md** (258)	**102 No** (259)	**103 Lr** (262)

Summary of Some Trigonometric Relations

You will need a calculator with trigonometric functions to do the examples in this appendix.

1) *Right triangles.* (a) The sum of the angles in any plane triangle being 180°, and angle γ in Fig. A.1 being 90°, it follows that $\alpha + \beta = 90°$. (b) By the Pythagorean theorem, $c^2 = a^2 + b^2$.

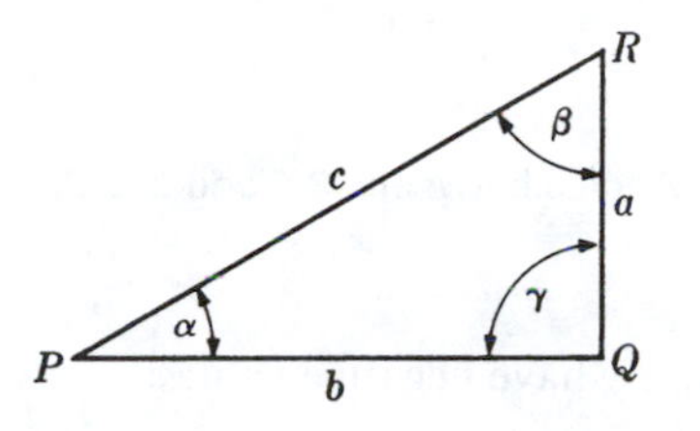

Fig. A.1. Definition of sides and angles in a right triangle.

Definitions for right triangles. We can *define* six functions of any one angle in a right triangle, namely sine (sin), cosine (cos), tangent (tan), cotangent (cot), secant (sec), and cosecant (csc). Specifically,

$$\sin \alpha \equiv \text{(opposite side)/(hypotenuse)} = a/c$$

$$\cos \alpha \equiv \text{(adjacent side)/(hypotenuse)} = b/c$$

$$\tan \alpha \equiv \text{(opposite side)/(adjacent side)} = a/b$$

$$\cot \alpha \equiv b/a$$

$$\sec \alpha \equiv c/b$$

$$\csc \alpha \equiv c/a$$

It follows directly from these definitions that:

(a) $$\tan \alpha = (\sin \alpha)/(\cos \alpha)$$

$$\cot \alpha = (\cos \alpha)/(\sin \alpha) = 1/(\tan \alpha)$$

$$\sec \alpha = 1/(\cos \alpha)$$

$$\csc \alpha = 1/(\sin \alpha).$$

Therefore only the sine and cosine functions of an angle need be discussed in detail.

(b) $$\sin^2 \alpha \equiv (\sin \alpha)^2 = a^2/c^2$$

$$\cos^2 \alpha \equiv (\cos \alpha)^2 = b^2/c^2$$

Thus

$$\sin^2 \alpha + \cos^2 \alpha = (a^2 + b^2)/c^2 = 1$$
(by the Pythagorean theorem).

(c) Because $\sin \beta = b/c$, $\cos \beta = a/c$ and so on, therefore

$$\sin \alpha = \cos \beta,$$

$$\cos \alpha = \sin \beta,$$

$$\sin \alpha = \cos (90 - \alpha),$$

$$\cos \alpha = \sin (90 - \alpha).$$

EXAMPLE 1
In a right triangle ($\gamma = 90°$), $\alpha = 35°$, length b = 4 cm; what is length a?

$$\tan \alpha = a/b,$$

therefore $a = b \tan \alpha = 4 \times 0.700 = 2.8$ cm.

EXAMPLE 2
In a right triangle, a = 8.3 cm, c = 21 cm; what are α and β?

$$\sin \alpha = a/c = 8.3/21 = 0.395.$$

Depending on what kind of calculator you have, you may need to interpolate here. If you know the values of the sine function for only integer values of the angle, you would find sin 23° = 0.391 and sin 24° = 0.407. But

$$(0.395 - 0.391)/(0.407 - 0.391) = (0.004/0.016) = 1/4.$$

Thus angle α is (roughly) ¼ of 1° higher than 23°, or 23°15′. And $\beta = 90° - \alpha = 66° 45'$.

3) *General definitions.* One can define the same trigonometric functions for any angle, even for an angle not in a triangle but enclosed between an inclined line and the horizontal

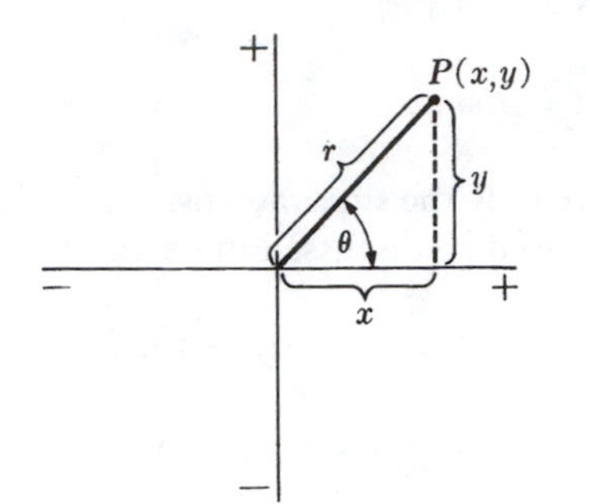

Fig. A.2. General definition of sides and angles.

If r is the length of the line and (x, y) the x and y coordinates of its end-point P, the other end being at the origin of the coordinate system, then

$$\sin\theta = y/r, \quad \cos\theta = x/r, \quad \tan\theta = y/x.$$

By convention, θ is measured counterclockwise from the right (positive) section of the abscissa.

4) *Larger angles.* If θ is larger than 90°, its trigonometric functions are defined as before, e.g.:

$$\sin\theta = y/r$$

(a) $\theta = 90°$ to 180° (Fig. A.3). Note that x has a negative value, being on the left side of the xy coordinate system, or in the second quadrant.

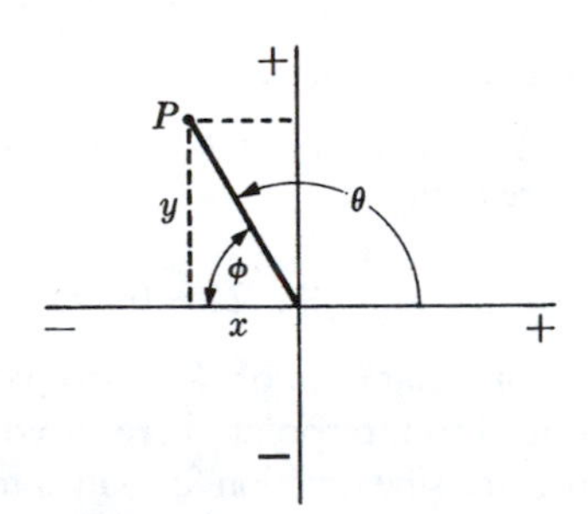

Fig. A.3. Angles between 90° and 180°.

Thus

$$\cos\theta = x/r = -\cos\phi = \cos(180° - \theta),$$

$$\sin\theta = y/r = \sin\phi = \sin(180° - \theta).$$

(b) $\theta = 180°$ to 270° (Fig. A.4); $\phi = \theta\ 180°$.

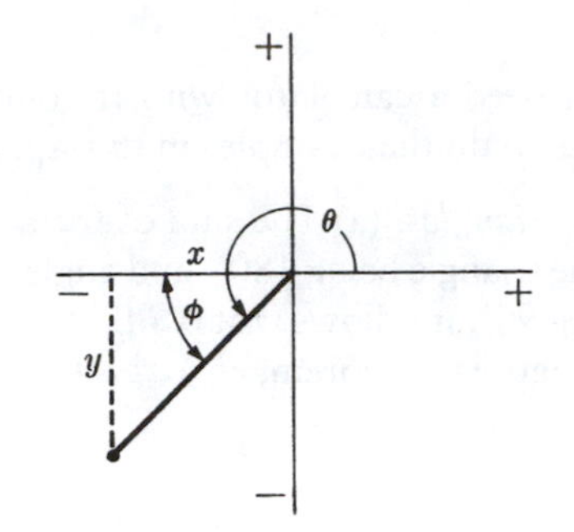

Fig. A.4. Angles between 180° and 270°.

Both x and y have negative values:

$$\sin\theta = -\sin\phi = -\sin(\theta - 180°),$$

$$\cos\theta = -\cos\phi = -\cos(\theta - 180°),$$

(c) $\theta = 270°$ to 360° (Fig. A.5); $\phi = \theta - 360°$.

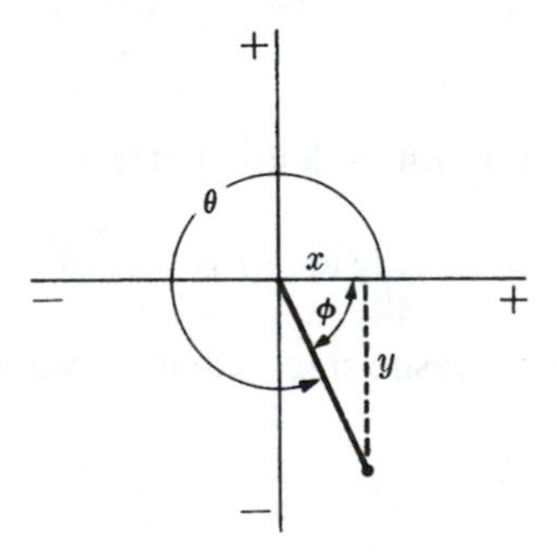

Fig. A.5. Angles between 270° and 360°.

Now y has a negative value, x a positive one:

$$\sin\theta = -\sin\phi = -\sin(360° - \theta),$$

$$\cos\theta = \cos\phi = \cos(360° - \theta).$$

These relations show that values of trigonometric functions computed for the range 0° to 90° serve for all angles.

EXAMPLE 3
Find sin θ and tan θ if θ = 295°. Note that θ is between 270° and 360°;

$$\sin \theta = -\sin (360° - \theta) = -\sin 65° = -0.906.$$

And

$$\tan \theta = (\sin \theta)/(\cos \theta) = -0.906/0.423 = -2.15.$$

5) *General triangles.* Even when a triangle does not have a right angle, as in Fig. A.6, there still exist several simple relations between the sides and the angles.

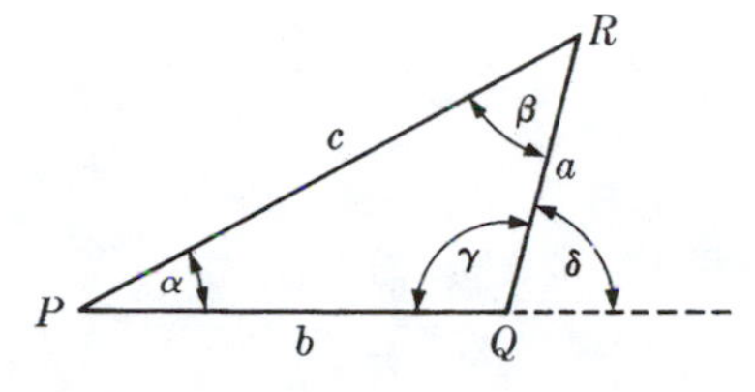

Fig. A.6. Definition of sides and angles for general triangles.

(a) Law of cosines: The square of any one side is equal to the sum of the squares of the other two sides minus twice the product of those two, multiplied by the cosine of their included angle. For example,

$$c^2 = a^2 + b^2 - 2ab \cos \gamma.$$

Thus one may find one side in terms of two others and one angle, or one angle in terms of three sides. Because the complementary angle $\delta = 180 - \gamma$, and $\cos \delta = \cos (180 - \gamma) = \cos \gamma$, we may write

$$c^2 = a^2 + b^2 + 2ab \cos \delta$$

(b)Law of sines: In any triangle, the ratio of any side and the sine of the opposite angle is constant, or

$$(a/\sin \alpha) = (b/\sin \beta) = (c/\sin \gamma)$$

Thus two sides and one angle determine the remaining angles and side, and so forth.

6) *Other useful trigonometric relations.*

$$\sin (\alpha + \beta) = \sin \alpha \cos \beta + \cos \alpha \sin \beta$$

$$\cos (\alpha + \beta) = \cos \alpha \cos \beta + \sin \alpha \sin \beta$$

$$2 \sin \alpha \cos \alpha = \sin 2\alpha$$

$$\cos 2\alpha = \cos^2 \alpha - \sin^2 \alpha.$$

Vector Algebra

We noted in Section 10.1 that motion with constant speed around a circle implies that the velocity vector is continually changing in direction though not in magnitude. A seemingly paradoxical situation arises at this point and demands a brief mathematical digression: In order to change the direction of a vector quantity at constant magnitude, one must, in general, add to it another vector with far from negligible magnitude.

Consider as an example the case of a disabled ship, adrift in wind and tide at some velocity $\boldsymbol{v}_1$, perhaps 2 m/sec due N. Now a rescue boat throws a line and pulls to give the ship a new velocity $\boldsymbol{v}_2$, again 2 m/sec, but now due W. What is the velocity $\boldsymbol{v}_3$ that the rescue vessel superposes on $\boldsymbol{v}_1$ to modify it to $\boldsymbol{v}_2$?

Figure A.7 gives the solution in familiar graphic form: v_2 is to be the resultant of v_1, and v_3, but you notice that the case is a little different from the one in Fig.8.8, where we were given v_3 and v_1 (or v_a and v_b) and asked to find v_2 (or v). Now the unknown is not the resultant, but one side of the vector parallelogram.

Accordingly, we enter first v_1, then v_2—both are known in direction and magnitude—and then the connecting line from the head of the arrow representing v_1 to the head of v_2. Thus the direction and magnitude of the missing side are found, and by measurement from a scale drawing we should find that v_3 in this example is about 2.8 knots to SW.

Comparison of parts (c) and (d) of Fig. A.7 indicates that the last stage (d) is really superfluous; the connecting line in (c) adequately represented v_3 even before the parallelogram was completed (Fig. A.8). From now on we may find the difference of any two vector quantities A and B by this abbreviated graphical procedure.

Furthermore, we shall occasionally follow the convention of using boldface italic letters to refer to vector quantities ($\boldsymbol{F}$, $\boldsymbol{v}$, $\boldsymbol{a}$). The addition of two vector quantities $\boldsymbol{P}$ and $\boldsymbol{Q}$ may then be written $\boldsymbol{P} + \boldsymbol{Q} = \boldsymbol{R}$. On reading such equations,

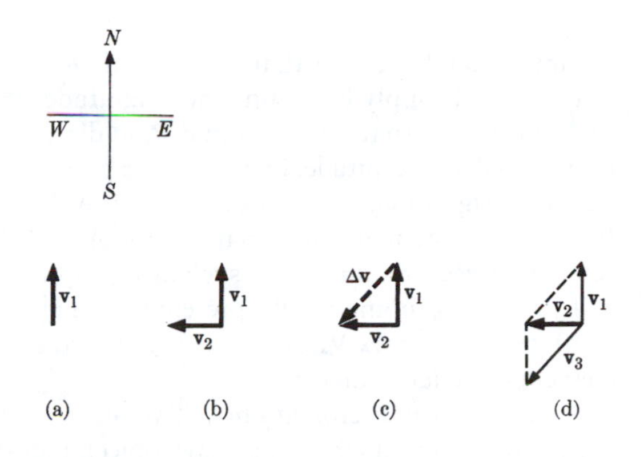

Fig. A.7. Vector addition $\boldsymbol{v}_2 = \boldsymbol{v}_1 + \boldsymbol{v}_3$.

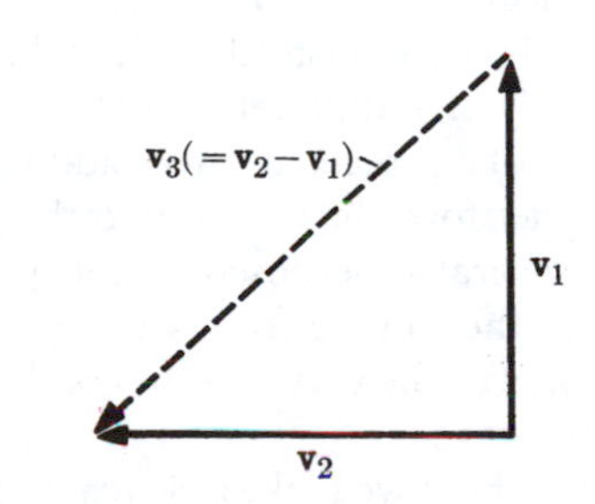

Fig. A.8. Vector subtraction $\boldsymbol{v}_3 = \boldsymbol{v}_2 - \boldsymbol{v}_1$.

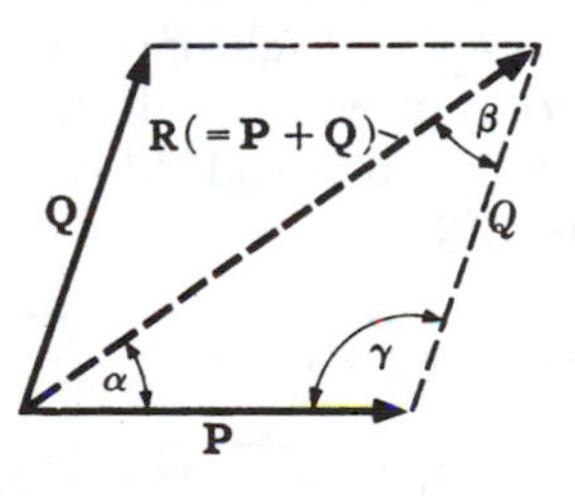

Fig. A.9. Vector addition, graphically and trigonometrically.

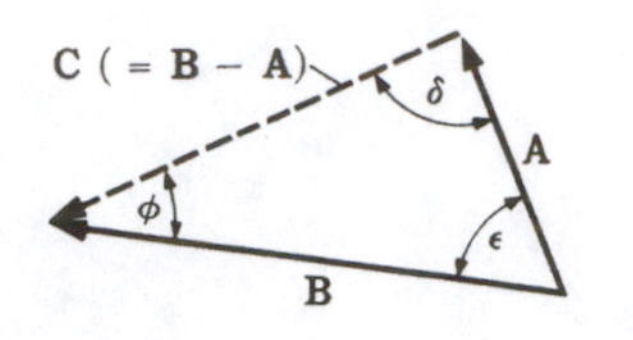

Fig. A.10. Vector subtraction, graphically and trigonometrically.

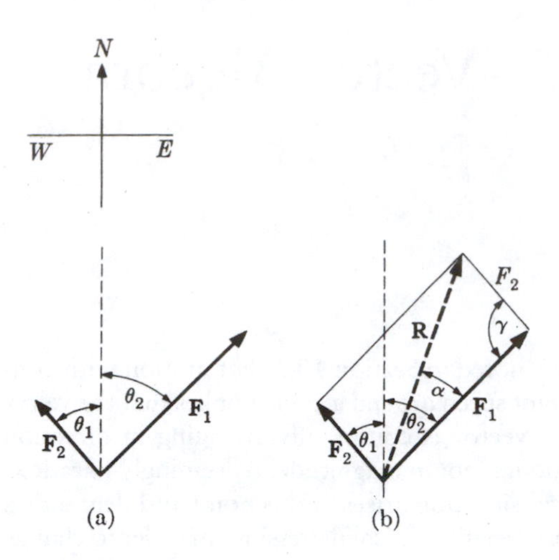

Fig. A.11. Vector addition.

we immediately realize that, in general, $\boldsymbol{R}$ is not obtained simply by adding the magnitudes of $\boldsymbol{P}$ and $\boldsymbol{Q}$, but that one must find $\boldsymbol{R}$, in direction as well as magnitude, by means of a graphical or a trigonometric solution (see Fig. A.9). In the same manner, the equation $\boldsymbol{B} - \boldsymbol{A} = \boldsymbol{C}$ refers to a vector subtraction such as is graphically and trigonometrically presented in Fig. A.10. (See Appendix VIII for a summary of the elements of trigonometry.)

a) *Example of vector addition.* Two men are pulling on ropes attached to a heavy object; the first pulls with force $\boldsymbol{F}_1$ = 90 newtons at 40° E of N, the second with force $\boldsymbol{F}_2$ = 45 newtons in a direction 30° W of N. Find the resultant of these two forces, that is, the vector sum, the single equivalent force $\boldsymbol{R}$ that might just as well be applied to this object instead of $\boldsymbol{F}_1$, and $\boldsymbol{F}_2$. (Our previous study of adding velocities to obtain the true velocity of a projectile and of adding forces to find the net force has illustrated the general truth that several superposed vector quantities provide just the same effect as a *single* vector quantity equal to the vectorial sum of the superposed quantities.)

Solution: First we make a sketch in terms of the chosen arrow symbolism of the relative magnitude and placement of vectors $\boldsymbol{F}_1$ and $\boldsymbol{F}_2$ (Fig. A.11a). Then we complete the parallelogram; the diagonal corresponds to $\boldsymbol{R}$, the vectorial sum of $\boldsymbol{F}_1$ and $\boldsymbol{F}_2$. The right half of the parallelogram is a triangle composed of $\boldsymbol{R}$, $\boldsymbol{F}_1$, and a side corresponding to $\boldsymbol{F}_2$; the angle opposite $\boldsymbol{R}$, indicated as γ, is directly given by $\gamma = 180° - (\theta_1 + \theta_2)$, in this case 180° – 70°, or 110°.[1] Now we find the magnitude of $\boldsymbol{R}$ by

$$R = \sqrt{(F_1{}^2 + F_2{}^2 - 2F_1F_2 \cos \gamma)}$$

$$= \sqrt{[8100 + 2025 - 2 \times 45 \times 90 \times (-0.34)]}$$

$$= 114 \text{ newtons.}$$

The direction of $\boldsymbol{R}$ may be fixed in several equivalent ways, but here it would seem most consistent to determine α, then $(\theta_2 - \alpha)$, the angle of $\boldsymbol{R}$ with respect to the north direction:

$$\sin \alpha = F_2 (\sin \gamma)/R = 45 \times (0.94/114) = 0.37;$$

$$\alpha \approx 22°$$

and

$$(\theta_2 - \alpha) = 18°.$$

The answer to the original question, then, is that the resultant force is 114 newtons along a direction 18° E of N. And incidentally, the same procedure is applicable for calculating the sum *any* two vectors, no matter at what angle to each other and no matter what their physical nature. Even the specific cases in which a right angle exists between the two vectors are, in substance, the same problem simplified by the convenient facts that there the factor corresponding to $2F_1F_2 \cos \gamma$ under the square root is zero (since cos 90° = 0), and that $\sin \alpha = F_2 \sin \gamma/R = F_2/R$ (since sin 90° = 1).

Just as examination of Fig. A.7 leads us to a simplification of the parallelogram method of vector subtraction, namely to Fig. A.8, so will a moment's thought prove that one may, if one so desires, similarly abbreviate the process of vector addition. In both Fig. 5.5 and Fig. A.11b, only the right half of the parallelogram is really needed; the left half is a congruent triangle, and therefore superfluous for our construction. Therefore, to add two vector quantities, we draw their

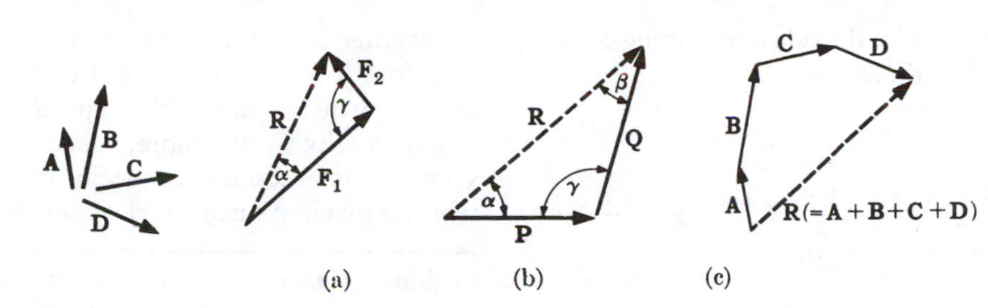

Fig. A.12. Vector addition.

arrowlike representation and join them head to tail (Fig. A.12a, b), the resultant R then being represented by the arrow running from the tail of the first drawn to the head of the last drawn arrow.

When there are more than two vectors to be added, we may do it either by adding two of them, then adding a third one to their resultant, then a fourth to *that* resultant, and so forth; or more conveniently, though less accurately, we may do it graphically by adding the corresponding arrows head to tail, as in Fig. A.12c. This is generally referred to as the *polygon method* of vector addition; it will soon prove to be an important tool in our argument.

Problem A.1. Note, describe, and justify carefully the difference between abbreviated vector subtraction (Fig. A.8) and vector addition (Fig. A.12a).

Problem A.2. (a) Prove, by graphical construction, that the polygon method of vector addition must give the same result as adding several vectors one at a time. (b) Prove by graphical construction that in Fig. A.12c the same resultant is obtained regardless of the *order* in which the individual arrows are joined together.

Problem A.3. (a) $\boldsymbol{F}_1$ = 5 newtons due W, $\boldsymbol{F}_2$ = 8 newtons at 30° S of E. Find the resultant force, in direction and magnitude, by the parallelogram method (graphically), by trigonometric calculation, and by the abbreviated graphical method (polygon method, which here becomes simply a triangle construction). (b) To the two forces in part (a), a third is added: $\boldsymbol{F}_3$ = 4 newtons 10° W of N. Find the resultant of all three forces by the graphical polygon method (in two different sequences of joining the arrows), and then by adding $\boldsymbol{F}_3$ to the resultant of $\boldsymbol{F}_1$ and $\boldsymbol{F}_2$ in a trigonometric calculation.

b)*Example of vector subtraction.* A bullet is shot in a direction 40° above the horizontal plane and with an initial velocity $v_1 = 3.5 \times 10^4$ cm/sec. After 46 sec the bullet returns to the same plane with a velocity v_2 having the same magnitude as v_1, namely, 3.5×10^4 cm/sec, but directed 40° below the horizontal. What was the change of velocity Δv during that time interval (magnitude and direction)?

Solution: Again, first make a sketch of the physical situation (Fig. A.13a), and then a sketch of the scheme previously discussed for the subtraction of vector quantities (Fig. A.13b). Here

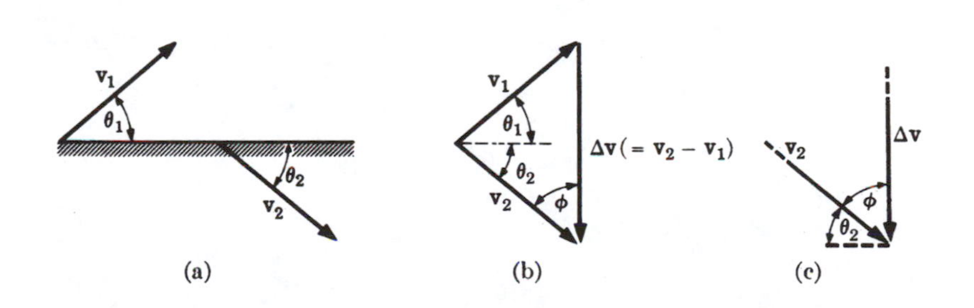

Fig. A.13. $\mathbf{v} = \mathbf{v}_2 - \mathbf{v}_1$.

the angle opposite the side representing Δv is $(\theta_1 + \theta_2)$, or 80°; therefore

$$\Delta v = \sqrt{[v_1^{\,2} + v_2^{\,2} - 2v_1v_2 \cos(\theta_1 + \theta_2)]}$$

$$= \{(12.25 \times 10^8 + 12.25 \times 10^8 - 2 \times 3.5 \times 10^4 \times 3.5 \times 10^4 \times 0.174)\}^{1/2}$$

$$= 4.5 \times 10^4 \text{ cm/sec.}$$

The direction of $\Delta \boldsymbol{v}$ is "obviously" vertically downward, but it is important that this be confirmed rigorously. In Fig. A.13c, angle $(\theta_2 + \phi)$ gives the inclination of Δv with the vertical; θ_2 is known, and ϕ can be easily computed from

$$\sin \phi = v_1 \sin(\theta_1 + \theta_2)/\Delta v$$

and the sum $\theta_2 + \phi$ is thereby confirmed to be 90°; that is, Δv is perpendicular to the horizontal plane.

This problem, which incidentally was designed to illustrate again the result of a change in direction of a vector even at constant magnitude, can be carried one important step further. Knowing now that during 46 sec the change in velocity was 4.5×10^4 cm/sec, we at once can find the average acceleration. By definition, $a = \Delta v/t$; therefore $a = [(4.5 \times 10^4)/46]$ cm/sec^2 = 980 cm/sec^2, which is in fact the expected acceleration of gravity. Furthermore, as acceleration is a vector, it is necessary to specify its direction, which is that of $\Delta \boldsymbol{v}$, namely, vertically downward.

Problem A.4. A sailboat cruises on a river that flows in a direction 15° N of E. If there were no wind, the river current would impart to the boat a speed of 5 mi/hr in that same direction. On the other hand, without current, the wind alone would move the boat at 9 mi/hr due SE. Find (a) the resultant of these two vectors, and (b) the difference between these two vectors. The result of (a) represents the total velocity of the boat referred to some fixed point on shore. What is the physical significance of the other result?

Problem A.5. Two forces, $\boldsymbol{F}_1$ = 10 newtons, $\boldsymbol{F}_2$ = 6 newtons, act on the same object at a 30° angle with each other. What is the resultant net force? What third force must be applied to establish equilibrium? (The latter force is called the *equilibrant*.)

General Bibliography

We list here books (and a few articles) that have been cited in more than one chapter (Recommended for Further Reading) or are of general interest for students of the history of physical science. A good small college library should have almost all of them, even if not in the latest edition.

Alioto, Anthony M., *A History of Western Science,* second edition, Englewood Cliffs, NJ: Prentice-Hall, 1993

Aristotle, *On the Heavens* [*De Caelo*], translated by W. K. C. Guthrie, Cambridge, MA: Harvard University Press, 1960

Bederson, Benjamin (editor), *More Things in Heaven and Earth: A Celebration of Physics at the Millennium,* New York: Springer-Verlag, 1999. A collection of review articles, some historical.

Ben-David, Joseph, *The Scientist's Role in Society: A Comparative Study,* Chicago: University of Chicago Press, 1984

Bensaude-Vincent, Bernadette, and Stengers, Isabelle, *A History of Chemistry,* Cambridge, MA: Harvard University Press, 1996

Blay, Michel, *Reasoning with the Infinite: From the Closed World to the Mathematical Universe.* Chicago: University of Chicago Press, 1998

Boas, Marie, *see* Hall, Marie Boas

Boorse, H. A., and Motz, L. (editors), *The World of the Atom.* New York: Basic Books, 1966

Bridgman, P. W., *The Logic of Modern Physics,* New York: Macmillan, 1960

Brock, William H., *The Norton History of Chemistry,* New York: Norton, 1993

Brown, Laurie M., Pais, Abraham, and Pippard, Brian (editors), *Twentieth Century Physics,* Philadelphia: Institute of Physics Publishing, 1995

Brush, Stephen G., *A History of Modern Planetary Physics,* 3 vols., New York: Cambridge University Press, 1996

Brush, Stephen G. (editor), *History of Physics: Selected Reprints,* College Park, MD: American Association of Physics Teachers, 1988

Brush, Stephen G. (editor), *Kinetic Theory,* Vol. 1, *The Nature of Gases and of Heat;* Vol. 2, *Irreversible Processes;* Vol. 3, *The Chapman-Enskog Solution of the Transport Equation for Moderately Dense Gases,* New York: Pergamon Press, 1965–1972

Brush, Stephen G. (editor), *Resources for the History of Physics,* Hanover, NH: University Press of New England, 1972

Brush, Stephen G., *Statistical Physics and the Atomic Theory of Matter, from Boyle and Newton to Landau and Onsager,* Princeton, NJ: Princeton University Press, 1983

Brush, Stephen G., and Belloni, L., *The History of Modern Physics: An International Bibliography,* New York: Garland, 1983

Buchwald, Jed Z. (editor), *Scientific Practice: Theories and Stories of Doing Physics,* Chicago: University of Chicago Press, 1995

Butterfield, Herbert, *Origins of Modern Science, 1300–1800.* Revised edition. New York: Free Press, 1997

Cohen, I. Bernard, *The Birth of a New Physics,* revised and updated edition, New York: Norton, 1985

Cohen, I. Bernard, and Westfall, Richard S. (editors), *Newton: Texts, Backgrounds, Commentaries,* New York: Norton, 1995

Collingwood, R. G., *The Idea of Nature,* Oxford: Clarendon Press, 1964

Conant, J. B., *On Understanding Science,* New York: New American Library, 1951

Crease, Robert P., and Mann, Charles C., *The Second Creation: Makers of the Revolution in 20th-Century Physics,* revised edition, New Brunswick, NJ: Rutgers University Press, 1996

Crombie, A. C., *Medieval and Early Modern Science,* Cambridge, MA: Harvard University Press, 1967

Crombie, A. C., *Science, Art and Nature in Medieval and Modern Thought,* London: Hambledon Press, 1996

Crowe, Michael J., *Modern Theories of the Universe from Herschel to Hubble,* New York: Dover, 1994

Crowe, Michael J., *Theories of the World from Antiquity to the Copernican Revolution,* New York: Dover, 1990

Cushing, J. T., *Philosophical Concepts in Physics: The Historical Relation between Philosophy and Scientific Theories,* New York: Cambridge University Press, 1998

Davis, E. A. (editor), *Science in the Making: Scientific Development as Chronicled by Historic*

Papers in the Philosophical Magazine, Levittown, PA: Taylor & Francis, 1995–1999. Four volumes covering 1798–1998

Dear, Peter (editor), *The Scientific Enterprise in Early Modern Europe: Readings from Isis,* Chicago: University of Chicago Press, 1997

Densmore, Dana, *Newton's Principia: The Central Argument,* translation, notes, and expanded proofs by Dana Densmore; translation and illustrations by W. H. Donahue, Santa Fe, NM: Green Lion Press, 1995

DeVorkin, David, *The History of Modern Astronomy and Astrophysics: A Selected, Annotated Bibliography,* New York: Garland, 1982

Drake, Stillman, *Galileo,* New York: Hill and Wang, 1980

Dugas, René, *History of Mechanics,* New York: Dover, 1988

Dugas, René, *Mechanics in the 17th Century,* New York: Central Book Co., 1958

Duhem, Pierre, *Aim and Structure of Physical Theory,* translated by P. P. Wiener, New York: Atheneum, 1962

Einstein, Albert, "Autobiographical Notes," in *Albert Einstein Philosopher-Scientist* (edited by P. A. Schilpp), pages 1–95, New York: Library of Living Philosophers, 1949; also reprinted as a separate book, *Autobiographical Notes: A Centennial Edition,* Chicago: Open Court, 1991.

Einstein, Albert, *Collected Papers,* edited by John Stachel *et al.,* Princeton, NJ: Princeton University Press, 1987–.

Einstein, Albert, *Ideas and Opinions,* translated by Sonja Bargmann, New York: Modern Library, 1994

Einstein, Albert, *Relativity: The Special and the General Theory,* translated by R. W. Lawson, second edition, New York: Crown, 1995

Einstein, Albert, and Infeld, Leopold, *The Evolution of Physics, from Early Concepts to Relativity and Quanta,* New York: Simon and Schuster, 1966

Everdell, William R., *The First Moderns: Profiles in the Origins of 20th-Century Thought,* Chicago: University of Chicago Press, 1997

Everitt, C. W. F., *James Clerk Maxwell, Physicist and Natural Philosopher,* New York: Scribner, 1975

Feynman, Richard P., *The Character of Physical Law,* Cambridge, MA: MIT Press, 1973

Frank, Phillipp, *Einstein, His Life and Times,* New York: Da Capo Press, 1979

French, A. P., and Greenslade, T. B., Jr. (editors), *Physics History from AAPT Journals, II,* College Park, MD: American Association of Physics Teachers, 1995

Galileo Galilei, *Dialogue Concerning the Two Chief World Systems,* translated with revised notes by S. Drake, Berkeley, CA: University of California Press, 1967

Galileo Galilei, *Discoveries and Opinions of Galileo,* translated with notes by S. Drake, Garden City, NY: Doubleday Anchor Books, 1957

Galileo Galilei, *Two New Sciences, Including Centers of Gravity & Force of Percussion,* translated by S. Drake, Madison, WI: University of Wisconsin Press, 1974

Galison, Peter, *How Experiments End,* Chicago: University of Chicago Press, 1987

Galison, Peter, *Image and Logic: A Historical Culture of Microphysics,* Chicago: University of Chicago Press, 1997.

Garber, Elizabeth, *The Language of Physics: The Calculus and the Development of Theoretical Physics in Europe, 1750–1914,* Boston: Birkhäuser, 1999

Garber, Elizabeth, Brush, Stephen G., and Everitt, C. W. F. (editors), *Maxwell on Molecules and Gases,* Cambridge, MA: MIT Press, 1986

Geymonat, Ludovico, *Galileo Galilei: A Biography and Inquiry into His Philosophy of Science,* translated by S. Drake, New York: McGraw-Hill, 1965

Gillispie, C. C. (editor), *Dictionary of Scientific Biography,* Vols. 1–16, New York: Scribner, 1970–1980; supplements, Vols. 17–18 (1990) edited by F. L. Holmes

Gjertsen, D., *The Newton Handbook,* London: Routledge and Kegan Paul, 1986

Grant, Edward, *The Foundations of Modern Science in the Middle Ages: Their Religious, Institutional, and Intellectual Contexts,* New York: Cambridge University Press, 1996

Hall, A. Rupert, *From Galileo to Newton, 1630–1720,* New York: Dover, 1981

Hall, A. Rupert, *The Revolution in Science 1500–1750,* New York: Longman, 1983

Hall, A. Rupert, and Hall, Marie Boas, *A Brief History of Science,* Ames: Iowa State University Press, 1988

Hall, Marie Boas, *The Scientific Renaissance 1450–1630,* New York: Harper, 1966

Hanson, Norwood Russell, *Constellations and Conjectures,* Boston: Reidel, 1973

Harman, P. M., *Energy, Force, and Matter: The Conceptual Development of Nineteenth-Century Physics,* New York: Cambridge University Press, 1982

Heilbron, J. L., *Elements of Early Modern Physics,* Berkeley: University of California Press, 1982

Heilbron, J. L., and Wheaton, B. R., *Literature on the History of Physics in the 20th Century,* Berkeley, CA: University of California, Office for the History of Science and Technology, 1981

Henry, John, *The Scientific Revolution and the Origins of Modern Science,* New York: St. Martin's Press, 1997

Hetherington, Norriss S. (editor), *Cosmology: Historical, Literary, Philosophical, Religious, and*

Scientific Perspectives, New York: Garland, 1993

Hetherington, Norriss S. (editor), *Encyclopedia of Cosmology: Historical, Philosophical, and Scientific Foundations of Modern Cosmology,* New York: Garland, 1993

Hoffmann, Dieter, *et al.* (editors), *The Emergence of Modern Physics,* Pavia, Italy: Universita degli Studi di Pavia, 1996

Holmes, F. L. (editor), *Dictionary of Scientific Biography,* Supplement II (Vols. 17 & 18), New York: Scribner, 1990, includes biographies of scientists who died since 1970.

Holton, Gerald, *The Advancement of Science and Its Burdens,* Cambridge, MA: Harvard University Press, 1998

Holton, Gerald, *The Scientific Imagination,* Cambridge, MA: Harvard University Press, 1998

Holton, Gerald, *Thematic Origins of Scientific Thought: Kepler to Einstein,* revised edition, Cambridge, MA: Harvard University Press, 1988

Hoskin, Michael (editor), *The Cambridge Illustrated History of Astronomy,* New York: Cambridge University Press, 1997

Ihde, Aaron J., *The Development of Modern Chemistry,* New York: Harper & Row, 1964.

Jacob, James R., *The Scientific Revolution: Aspirations and Achievements, 1500–1700,* Amherst, NY: Humanity Books, 1999

Jungnickel, Christa, and McCormmach, Russell, *Intellectual Mastery of Nature,* Vol. I, *The Torch of Mathematics 1800–1870;* Vol. 2, *The Now Mighty Theoretical Physics 1870–1925,* Chicago: University of Chicago Press, 1986

Kevles, Daniel J., *The Physicists: The History of a Scientific Community in Modern America,* Cambridge, MA: Harvard University Press, 1995

Kitcher, Philip, *The Advancement of Science: Science without Legend, Objectivity without Illusions,* New York: Oxford University Press, 1993

Knight, David, *Ideas in Chemistry,* New Brunswick, NJ: Rutgers University Press, 1992

Koyré, Alexandre, *From the Closed World to the Infinite Universe,* Baltimore, MD: Johns Hopkins University Press, 1968

Koyré, Alexandre, *Metaphysics and Measurement,* New York: Gordon & Breach, 1992

Koyré, Alexandre, *Newtonian Studies,* Cambridge, MA: Harvard University Press, 1965

Kragh, Helge, *Quantum Generations: A History of Physics in the Twentieth Century,* Princeton, NJ: Princeton University Press, 1999.

Krige, John, and Pestre, Domninique (editors), *Science in the 20th Century,* Amsterdam: Harwood Academic, 1997

Kuhn, Thomas S., *The Copernican Revolution,* New York: Fine Communications, 1997

Kuhn, Thomas S., *The Structure of Scientific Revolutions,* third edition, Chicago: University of Chicago Press, 1996

Lang, Kenneth R., and Gingerich, Owen (editors), *Source Book in Astronomy and Astrophysics, 1900–1975,* Cambridge, MA: Harvard University Press, 1979.

Lankford, John (editor), *History of Astronomy: An Encyclopedia,* New York: Garland, 1997

Leverington, David, *A History of Astronomy, from 1890 to the Present,* New York: Springer-Verlag, 1995

Lindberg, David, *The Beginnings of Western Science,* Chicago: University of Chicago Press, 1992

Lucretius, *On the Nature of the Universe,* translated by Ronald Melville from *De Rerum Natura,* with introduction and notes by Don and Peta Fowler, Oxford: Clarendon Press, 1997

Mach, Ernst, *Popular Scientific Lectures,* Chicago: Open Court, 1986

Mach, Ernst, *The Science of Mechanics,* sixth edition, LaSalle, IL: Open Court, 1960

Machamer, Peter (editor), *The Cambridge Companion to Galileo,* New York: Cambridge University Press, 1998

Mauskopf, Seymour H. (editor), *Chemical Sciences in the Modern World,* Philadelphia: University of Pennsylvania Press, 1993

Mason, Stephen F., *A History of the Sciences,* new revised edition, New York: Collier Books, 1962

Maxwell, James Clerk, *Maxwell on Molecules and Gases* (edited by E. Garber *et al.*), Cambridge, MA: MIT Press, 1986

McClellan, James E., III, and Dorn, Harold, *Science and Technology in World History: An Introduction,* Baltimore: Johns Hopkins University Press, 1999

McKenzie, A. E. E., *The Major Achievements of Science,* Ames: Iowa State University Press, 1988

Merton, Robert, *The Sociology of Science,* Chicago: University of Chicago Press, 1973

Moyer, Albert E., *American Physics in Transition,* Los Angeles: Tomash, 1983

Moyer, Albert E., "History of Physics" [in 20th-Century America], *Osiris,* new series, Vol. 2, pages 163–182 (1985)

Newton, Isaac, *Newton's Philosophy of Nature,* selections from his writings, edited and arranged with notes by H. S. Thayer, New York: Hafner, 1953

Newton, Isaac, *The Principia: Mathematical Principles of Natural Philosophy,* translated by I. Bernard Cohen and Anne Whitman, assisted by Julia Budnez, preceded by "A Guide to Newton's Principia" by I. Bernard Cohen. Berkeley, CA: University of California Press, 1999

Newton, Isaac, *Opticks: Or, A Treatise of the Reflections, Refractions, Inflections & Colours of Light,* based on the fourth edition (1730), New York: Dover, 1979

North, John, *Norton History of Astronomy and Cosmology,* New York: Norton, 1995

Nye, Mary Jo, *Before Big Science: The Pursuit of Modern Chemistry and Physics 1800–1940,* Cambridge, MA: Harvard University Press, 1996

Nye, Mary Jo, *From Chemical Philosophy to Theoretical Chemistry: Dynamics of Matter and Dynamics of Disciplines, 1800–1950,* Berkeley, CA: University of California Press, 1993

Olby, R. C., Cantor, G. N., Christie, J. R. R., and Hodge, M. J. S. (editors), *Companion to the History of Modern Science,* New York: Routledge, 1990

Pauli, Wolfgang, *et al.* (editors), *Niels Bohr and the Development of Physics,* New York: McGraw-Hill, 1965

Planck, Max, *Scientific Autobiography and Other Papers,* New York: Philosophical Library, 1949

Plato, *Phaedon,* translated by D. Gallop, New York: Oxford University Press, 1993

Poincaré, Henri, *Science and Method,* translated by F. Maitland, New York: Dover, 1952

Popper, Karl, *Conjectures and Refutations: The Growth of Scientific Knowledge,* second edition, New York: Basic Books, 1965

Popper, Karl, *Logic of Scientific Discovery,* New York: Basic Books, 1959

Purrington, Robert D., *Physics in the Nineteenth Century,* New Brunswick, NJ: Rutgers University Press, 1997

Pyenson, Lewis, and Sheets-Pyenson, Susan, *Servants of Nature: A History of Scientific Institutions, Enterprises and Sensibilities.* New York: Norton, 1999

Rayner-Canham, Marelene, and Rayner-Canham, Geoffrey W. (editors), *A Devotion to Their Science: Pioneer Women of Radioactivity,* Philadelphia: Chemical Heritage Foundation, 1997

Rigden, John S. (editor), *Macmillan Encyclopedia of Physics,* 4 volumes, New York: Macmillan Reference USA/Simon & Schuster/Macmillan, 1996

Ronan, Colin A., *Science: Its History and Development among the World's Cultures,* New York: Facts on File, 1982

Sambursky, S., *Physical Thought from the Presocratics to the Quantum Physicists: An Anthology,* New York: Pica, 1975

Sarton, George, *The Study of the History of Science,* New York: Dover, 1957

Schilpp, P. A. (editor), *Albert Einstein Philosopher-Scientist,* New York: Library of Living Philosophers, 1949

Schirmacher, Wolfgang (editor), *German Essays on Science in the 20th Century,* New York: Continuum, 1996

Schlagel, Richard H., *From Myth to Modern Mind: A Study of the Origins and Growth of Scientific Thought,* Vol. 1, *Theogony through Ptolemy,* Vol. 2, *Copernicus through Quantum Mechanics,* New York: Peter Lang, 1996

Schneer, Cecil J., *The Evolution of Physical Science,* Lanham, MD: University Press of America, 1984

Schneer, Cecil J., *Mind and Matter,* Ames, IA: Iowa State University Press, 1988

Schorn, Ronald A., *Planetary Astronomy: From Ancient Times to the Third Millennium,* College Station, TX: Texas A & M University Press, 1998

Segrè, Emilio, *From X-Rays to Quarks,* San Francisco: Freeman, 1980

Serres, M. (editor), *A History of Scientific Thought,* Oxford: Blackwell, 1995

Solomon, Joan, *The Structure of Matter: The Growth of Man's Ideas on the Nature of Matter,* New York: Wiley/Halsted, 1974

Stroke, H. Henry (editor), *The Physical Review: The First Hundred Years. A Selection of Seminal Papers and Commentaries,* Woodbury, NY: AIP Press, 1995

Suplee, Curt, *Physics in the 20th Century,* New York: Abrams, 1999

Tomonaga, Sin-itiro, *The Story of Spin,* translated by T. Oka., Chicago: University of Chicago Press, 1997

Torrance, John (editor), *The Concept of Nature,* New York: Oxford University Press, 1992

Toulmin, Stephen, and Goodfield, June, *The Architecture of Matter,* New York: Harper, 1977

Toulmin, Stephen, and Goodfield, June, *The Fabric of the Heavens,* New York: Harper, 1965

Van Helden, Albert, *Measuring the Universe: Cosmic Dimensions from Aristarchus to Halley,* Chicago: University of Chicago Press

Weart, Spencer R., and Phillips, Melba N. (editors), *History of Physics,* New York: American Institute of Physics, 1985. Articles reprinted from *Physics Today.*

Weaver, Jefferson Hane (editor), *The World of Physics: A Small Library of the Literature of Physics from Antiquity to the Present,* 3 volumes, New York: Simon and Schuster, 1987

Westfall, Richard S., *The Life of Isaac Newton,* New York: Cambridge University Press, 1993

Wheaton, Bruce R., *The Tiger and the Shark: Empirical Roots of Wave-Particle Dualism,* New York: Cambridge University Press, 1983

Whitehead, Alfred North, *Science and the Modern World,* New York: Free Press, 1967

Whittaker, E. T., *A History of the Theories of Aether and Electricity,* Volume I, *The Classical Theories;* Vol. II, *The Modern Theories,* New York: Tomash Publishers and American Institute of Physics, 1987

Zajonc, Arthur, *Catching the Light: The Entwined History of Light and Mind,* New York: Oxford University Press, 1993

Credits

AIP Emilio Segrè Visual Archives: Figs. 6.1, 18.2, 25.14, 29.2 (Photo by Francis Simon), 29.6, 29.7, 29.9 (Photo by Francis Simon), 30.7 (Photo by A. Börtzells Tryckeri) 32.2, 32.3 (Photo by Ramsey and Muspratt), 32.4, 32.5, 32.7.

AIP Emilio Segrè Visual Archives, E. Scott Barr Collection: Fig. 32.1.

AIP Emilio Segrè Visual Archives, Goudsmit Collection, Fig. 28.12.

AIP Emilio Segrè Visual Archives, Landé Collection: Fig. 22.1.

AIP Emilio Segrè Visual Archives, Dorothy Davis Locanthi Collection: Fig. 31.10.

AIP Emilio Segrè Visual Archives, W. F. Meggers Collection: Figs. 25.14, 26.4, 27.1

AIP Emilio Segrè Visual Archives, *Physics Today* Collection: Figs. 18.1, 31.5, part E (unnumbered figure).

AIP Emilio Segrè Visual Archives, T. J. J. See Collection: Fig. 11.5.

AIP Emilio Segrè Visual Archives, Zeleny Collection: Fig. 18.3.

Margrethe Bohr Collection, courtesy AIP Emilio Segrè Visual Archives: Part H (unnumbered figure).

Bryn Mawr College Archives: Fig. 17.26.

Bulletin of the Atomic Scientists, courtesy AIP Emilio Segrè Visual Archives: Fig. 32.6.

Carnegie Institution of Washington: Fig. 31.2. Image courtesy of Carnegie Institution of Washington.

Cavendish Laboratory, Department of Physics, Cambridge University: Fig. 25.15.

Einstein Archives, Hebrew University of Jerusalem by permission of the Albert Einstein Estate: Fig. 30.1.

Elsevier Science: Fig. 31.4. Reprinted from Endeavour, Vol. 2. "The Evolution of the Solar System," (1943) pp. 3–11, with permission from Elsevier Science.

Hale Observatories, courtesy AIP Emilio Segrè Visual Archives: Fig. 31.9.

Harvard News Office: Part D (unnumbered figure).

Lunar and Planetary Science Institute, Houston, Texas (U.S. Government royalty-free license under NASA Contracts Numbers NASW-3389 and NASW-4066 with Universities Space Research Association): Fig. 31.8.

Max-Planck-Institut für Physik, courtesy AIP Emilio Segrè Visual Archives: Fig. 29.5.

Mendeleéff Museum & Archives, courtesy I. S. Dmitriev: Fig. 21.1.

National Institute of Standards and Technology: Fig. 9.3.

National Portrait Gallery, London, courtesy AIP Emilio Segrè Visual Archives: Fig. 25.2.

National Aeronautics and Space Administration: cover (Einstein ring).

Palais de la Decouverte: Fig. 29.1. Copyright Palais de la Decouverte. BROGLIE (Louis de), physicien français (1892–1987). Prix Nobel 1929.

Princeton University Press: Fig. 23.13. Park, David, *The Fire within the Eye,* Copyright 1997 by Princeton University Press. Reprinted by permission of Princeton University Press.

Springer-Verlag New York: Fig. 31.7.

U.K. Atomic Energy Authority, courtesy AIP Emilio Segrè Visual Archives: Fig. 27.2.

University of California Press: excerpts from translation of Isaac Newton, *The Principia: Mathematical Principles of Natural Philosophy* by I. Bernard Cohen and Ann Whitman, copyright 1999 by The Regents of the University of California; excerpts from Stillman Drake's translation of Galileo Galilei, *Dialogue Concerning the Two Chief World Systems,* copyright 1953 and 1962 by The Regents of the University of California.

University of Wisconsin Press: excerpts from Stillman Drake's translation of Galileo Galilei, *Two New Sciences,* copyright 1974 by University of Wisconsin Press.

University of Chicago, courtesy AIP Emilio Segrè Visual Archives: Fig. 29.8.

Index

Page numbers for definitions are in italics.

a (acceleration), *70*
a (half major axis of ellipse), *41,* 43
A (mass number), *412*
Å (angstrom; unit of length), *388*
abbreviations, 531–533
Abelson, Philip H., 508
absorption line spectra, 428–432
acceleration, *70–75*, 83, *84*–86
  centripetal, *125–126*
  conversion factors, 539
accidental discoveries, 145, 187–188
actinium, radioactive decay series, 415, *416*
action at a distance, *151,* 277
ad hoc hypothesis, 147, 194
Adams, Henry (1838–1914, American historian), 260–261
Adams, John Couch (1819–1892, English astronomer), 145, 146
adiabatic process, *327*
Aepinus, Franz (1724–1802, German physicist)
  electrostatics, 235
aesthetic criteria for theories, 480
Aether. *See* Ether
Agassiz, Louis (1807–1873, Swiss-American naturalist), 256
age
  of the earth, xiii, 414
  of rocks, xiii
agriculture, 186
Air (element), 7
air
  compressibility, 270
  homogeneity of mixture, 276, 278
  pressure, 267–270
Alexander the Great (356–323 B.C.), 8
Alexandria, 8–9
Alfvén, Hannes (1908–1995, Swedish plasma physicist), 491
Alhazen (Abu `Ali al-Hasan ibn al-Hasan ibn al'Haytham, 965–1039, Arab mathematician and physicist), 34–35, 46, 526
alkali metals, 297
al-Kindi, Abu Yusuf Ya'qub ibn (c. 801–c. 866, Arabic scientist), 33–34
α (alpha; separation factor), *321*
α Centauri (star), 11
α-rays or particles, 168, 384, 410, 418
Alpher, Ralph (b. 1921, American physicist), 510–511
Alvarez, Luis (1911–1988, American physicist), 529
americium (element), 508
ammonia, atomic composition of molecule, 287
Amontons, Guillaune (1663–1705, French physicist), 270
amp (ampere; unit of electric current), *358, 371,* 375–376, *537*
Ampère, André Marie (1775–1836, French physicist), 387
  force between two currents, 370–371, *375*
amu (atomic mass unit), *412,* 500
analytic dimension in science, 517–518
Anderson, Carl David (1905–1991, American physicist)
  discovery of muon, 501
  discovery of positron, 479
Andromeda galaxy
  distance, 495
  Doppler shift, 494
angstrom (unit of length), *388*
Ångström, Anders Jonas (1814–1874, Swedish physicist), 388
angular measure, conversion factors, 539
angular momentum, 217–*218,* 231
angular velocity, *124,* 218
anode, *366*
anthropic principle, *514–515*
anthropology, 157
antiparticles, antimatter, 479
  paucity, 528
antiproton, 479, 500
Apollonius, 9, 11
appearances. *See* "save the appearances"
Aquarius (constellation), 4
  coming of Age of, 4
Arab science, 15, 33–35, 187
Archimedes (c. 287–212 B.C., Greek mathematician-physicist), 51, 233
area, conversion factors, 539
argon
  atomic weight, 305
  discovery, 305
  masses of isotopes, 412
  molecular weight, 293
Aries (constellation), 4
Aristarchus of Samos (c. 310 B.C.–c. 230 B.C., Greek astronomer), 9, 22

Aristarchus of Samos (*continued*)
  heliocentric theory, *10*–11, 18, 19
Aristotle (427 B.C.–347 B.C. Greek philosopher), 6
  cosmology, 7–8
  elements, *7*, 234
  falling bodies, 51, 78, 81
  followers of, 51, *54*, 77–80
  force, 108
  philosophy, xiii, *5*, 7, 43, 63, 78, 80, 168
  physics, 7, 22, 51
  potential being, 31–32
  projectile motion, 99
  vacuum, 82
  vision, 31–32, 34
arms control movement, 464
arrow (Zeno's paradox), 30
Aspect, Alain, 456
asteroids
  discovery, 146
  impact on earth, 144
  origin, 147, 488
astrology, 4, 14, 23
astronomy, 1
  data, 537
  "great debate," 494–495
  Greek, 3–16
  history, 15–16
  radio, 511–512
atheism, 80
Athens, 3, 9
Atkinson, Robert d'Escourt (1898–1982, Welsh-American physicist), 502
Atlantic Cable, 257, 383
atmosphere
  density variation with height, 328–329
  escape from planets and satellites, 328
  homogeneity, 276, 328
  standard, *269, 537*
atom
  in chemistry, 275–295
  diameter, 328
  "discovered" by Perrin, 337
  existence, 164, 172, 328, 337
  forces between, 266–267, 288, 457
  origins of theory in physics and chemistry, 263–337
  made "real" by measurement (Kelvin), 328
  "seen" by Mueller, 527
  thematic dimension (opposed to force), 519, 523
atomic bomb, 321, 464, 506–509, 515
atomic mass unit, *412*
atomic number, 297, 304, 412
  change in β decay, 416
  and nuclear charge, 420
  and number of electrons in atom, 444
  and x-ray spectrum, 423
atomic weight, *283–284, 292*
atomists, 30, 80, 234, 273, 274. *See also* Democritus; Lucretius; Gassendi, Pierre
attitude, scientific, 190–191
Atwood's machine, 116–117
AU (astronomical unit), *11*
autumnal equinox, *4*, 13. *See also* equinox
Avogadro, Amedeo (1776–1856, Italian physicist and chemist), 286
  hypothesis, 187, *286*–289, 297, 312
  number, *289*, 328, 367

b (half minor axis of ellipse), *41*, 43
Baade, Walter (1893–1960, German-American astronomer), 496, 503
Bacon, Francis (1561–1626, English philosopher), 24, 40, 43, 63, 165, 167, 177, 179, 189, 234
Bailly, Jean-Sylvain (1736–1793, French astronomer), 254
balance
  equal-arm, 114–115
  spring, 108, 113, 116
Balmer, Johann Jakob (1825–1898, Swiss schoolteacher)
  formula for spectral lines, 148, 432–434
  series, 432, 443, 440, 441
bar (unit of pressure), *269*
Barberini, Cardinal Maffeo (1568–1644, Pope Urban VIII from 1623 to 1644), 58
barometer, Torricelli, 267
baseball, 97
battery, 365
Beck, A. C., 512
Becquerel, [Antoine] Henri (1852–1908, French physicist)
  radioactivity, 184–185, 384, 409
Being, *30*, 31
bell-shaped curve, 323
Bell Telephone Laboratories (New Jersey), 512–513
Benedetti, Giovanni Battista (1530–1590, Italian mathematician-physicist), 78
Bentley, Richard (1662–1742, English scholar-theologian), 152, 175
benzene
  Kekulé's theory, 292
  quantum mechanics, 458
  structural formula, 291
Berkeley, George (1685–1753, Anglo-Irish philosopher and clergyman), 2, 46–47
Berkeley (California), radiation laboratory, 500
berkelium (element), 508
Berlin University, 463
Bernoulli, Daniel (1700–1782, Swiss mathematician), 273–274
Bernoulli, James (1654–1705, Swiss mathematician), 217
Berry, Michael, 485
Berthollet, Claude Louis (1748–1822, French chemist), 280–281
berthollides, 281
beryllium, 298–299
Berzelius, Jöns Jacob (1779–1848, Swedish chemist), 280
Bessel, Friedrich Wilhelm (1874–1846, German astronomer-mathematician), 11, 38
β-decay, 384, 410, 414, 416–417, 502–504
  inverse, 503, 504
β-rays, 384, 410. *See also* electron
Bethe, Hans (b. 1906, German-American physicist), 504
  formation of elements, xiii, 503–505

BeV (billion electron volts), *414*
Bevatron, 500
Beveridge, W. I. B., 188
Big Bang cosmology, 395, 504, 510–513, 515
  blackbody radiation, 510
  synthesis of elements, 510
  thematic dimension, 520
biology and physics, 184
Black, Joseph (1728–1799, Scottish chemist), 235, 236, 273
blackbody radiation, 389–395
black-box problem, *118,* 233
black hole, *483–484,* 485
Blau, Marietta (1894–1969, Austrian physicist), 502
Bloch, Felix (1905–1983, Swiss physicist), 459
block and tackle, 233
Bode, Johann Elbert (1747–1826, German astronomer), 146
  law, *146*–148
Bohr, Niels (1885–1962, Danish physicist), 17, 38, 407, 434
  atomic structure theory, 192, 434–445
  complementarity principle, 455
  Copenhagen interpretation of quantum mechanics, 455
  and Einstein, 455
  nuclear fission, theory, 507
Boisbaudran. *See* LeCoq de Boisbaudran
Boltzmann, Ludwig (1844–1906, Austrian physicist), 335, 337
  constant, *319*
  disordered states, 335
  entropy and probability, 259, 335
  kinetic theory, 310, 392
  music and science, 159–160
  recurrence paradox, 336–337
  reversibility paradox, 335
  thermodynamics, second law, 335–336
  time reversal, 337
bond (chemical), double or triple, 291
Bondi, Hermann (b. 1919, Austrian-English mathematician-cosmologist)
  Popper's falsificationism, 511, 513
  steady state cosmology, 510–511
Born, Max (1882–1970, German physicist), 401, 452
  probability and quantum mechanics, 452–453
Bose, Satyendranath (1894–1974, Indian physicist), 458
  quantum statistics, 458
Bose-Einstein condensation, 445, 460
boson, *459*
Boyle, Robert (1627–1691, British chemist-physicist), 103, 269, 274
  atomic models, 272
  clockwork universe, 144, 253, 266, 270, 526
  gas pressure law, 148, 270
  heat, 234
braccio (unit of length), *81*
Brackett series (in hydrogen spectrum), 433, 440, 441
Bradwardine, Thomas (c. 1290–1349, English mathematician), 79
Bragg, William Henry (1862–1942, Australian-British physicist, 385
Bragg, [William] Lawrence (1890–1971, British physicist), 385
Brahe, Tycho (1546–1601, Danish astronomer)
  observations, 40, 45
  Tychonic system, 25
Brauner, Bohuslav (1855–1935, Czechoslovakian chemist), 299
Bridgman, Percy Williams (1882– 1961, American physicist), 38, 162–164, 188
British Absolute System of units, 541
Broglie, Louis de (1892–1987, French physicist), 447, 456
  wavelength of moving particle, 447–449
Brown, Robert (1773–1858, Scottish botanist), 311
Brownian movement, *310–311,* 321
Bruno, Giordano (1548–1600, Italian philosopher), 40, 149
Buffon, Georges Louis Leclerc, Comte de (1707–1788, French naturalist), 253–254, 487
Bunsen, Robert Wilhelm (1811–1899, German chemist), 427
Burbidge, Geoffrey (b. 1925, British-American astronomer), 511
Burbidge, Margaret (b. 1919, British-American astronomer), 511
Bureau Internationale des Poids et Mesures, 111
Buridan, Jean (c. 1295–c.1385, French philosopher), 209
Bush, Vannevar (1890–1974, American engineer and science administrator), 508
Butterfield, Herbert (1900–1979, British historian) 23–24, 25, 26
Buys-Ballot, Christoph Hendrik Diederik (1817–1890, Dutch meteorologist), 319, 430

c (circumference of earth), 9
c (distance from origin to focus of ellipse), *41*
c (specific heat), *235*
c (speed of light), *344*
  relation to electric and magnetic force constants, 376
c_p (specific heat at constant pressure), 241
c_v (specific heat at constant volume), 241
C (Celsius scale), 241
cal (calorie; unit of heat), *240*
calcium, valence, 290
calculus, 44
calendar, 4
  Gregorian, 20
  Julian, 20
californium (element), 508
caloric theory of heat, 28, 166, 235–236, 273

calorie (unit of heat), *240*
calorimetry, 235
Cambridge University, 104–105
camera obscura, 46
Cameron, A. G. W., 493
canal rays, 384, 387
Cannizzaro, Stanislao (1826–1910, Italian chemist), 289, 297
capacitor, *359*
carbon
  atom, 279
  atomic weight, 284, 293
  masses of isotopes, 412
carbon 12 isotope
  as atomic weight standard, 304, 412
  synthesis by triple-α reaction, 505
carbon cycle, *504–505*
carbon dioxide, 280, 290
  molecular weight, 293
carbon monoxide, 290
  molecular weight, 293
Carnap, Rudolph (1891–1970, German-American philosopher), 167
Carnot, Nicolas Léonard Sadi (1796–1832, French physicist-engineer), 244, 256–257
Carter, Brandon, 514
Cassini, Giovanni Domenico (1625–1712, Italian-French astronomer), 21, 128–129, 142
Cassini, Jacques II (1677–1756, Italian-French astronomer), 142
cathode, *366*
cathode rays, 379–382, 385–386
  deflection by electric fields, 386
cause and effect, 166, 174, 240, 395
Cavendish, Henry (1731–1810, English physicist-chemist)
  experiment on gravitational force, 137
  explosion of hydrogen and oxygen, 284
celestial sphere(s), 1, *2,* 6–8, 22
  harmony, 5
  diurnal motion, 7–8, *55*
Celsius, Anders (1701–1744, Swedish astronomer)
  temperature scale, 240–*241*
centigrade, 241
centrifugal force. *See* centripetal force
centripetal acceleration, *125–126*
  due to earth's rotation, 128, 142
centripetal force, 114
Cepheid variable (star), 494
Ceres (asteroid), *146*
cgs units, 542
Chadwick, James (1891–1974, British physicist), 425, 500, 508
Chamberlain, Owen (b. 1920, American physicist), 500
Chamberlin, Thomas Chrowder (1843–1928, American geologist), *489*
  planetesimals, 489
  solar system, origin, 489–490
change, possibility of, 30–31, 36
chaos, 144
Charles, Jacques Alexandre César (1746–1823, French physicist), 270
Châtelet, Gabrielle-Emilie du (1706–1749, French physicist), 141
Chemical Revolution, *206*
chemistry
  atomic-molecular theory, 28
  bond, quantum theory, 457
  mass and weight, 204–207
  oxygen theory, 204–206
  phlogiston theory, 28, 204–206
  relation to physics, xiii
China, science in, 149, 187
chlorine
  ion, 367
  molecular weight, 293
cinematograph, 382–383
circle as limiting case of ellipse, 41
Clairaut, Alexis Claude (1713–1765, French mathematician), 141, 142
Clarke, Samuel (1675–1729, English theologian), 253
Clausius, Rudolf (1822–1888, German physicist), 309
  entropy, 259
  kinetic theory of gases, 308–310
  mean free path, 319
  thermodynamics, second law, 259
  specific heats of diatomic molecules, 330
clock (twin) paradox, 474
clock synchronization, 484–485
clockwork universe, 48, 144, 211, 251–253, 453, *526*
cm (centimeter), *65*
cobalt, wavelength of K line (x-ray spectrum), 423
Cockcroft, John (1897–1967, British physicist), 502
Cohen, I. Bernard (b. 1914, American historian of science), 82
Colding, Ludvig August (1815–1888, Danish engineer-physicist), 244
Cole, Jonathan R., 179
Cole, Stephen, 179
collisions, 210–212, 213–215
  elastic, *211,* 221, 249
  inelastic, *211,* 228
color, 350–351
Columbus, Christopher (1451–1506, Italian explorer), 10, 17, 24–25
comets, *141, 153*
  direction of tail, 217
Commentariolus. See Copernicus, Nicolaus
common sense, 188–189
communication in science, 180
complementarity principle, 455
  attempt to compromise opposing themata, 521
Compton, Arthur Holly (1892–1962, American physicist), 217
  effect, *217,* 385, 400, 461
Conant, James Bryant (1893–1978, chemist, educator and diplomat), 60, 167
concepts, physical, 161–164, 172–174, 197
  meaningless, 163–164
condenser, parallel plate, *359*

Condon, Edward Uhler (1902–1974, American physicist), 502
conservation laws, 201–261
angular momentum, 217–218
energy, 148, 166, 219–250
mass, 166, 203–207
mass-energy, *476–477,* 502
momentum, 64, 148, 166, 209–218
motion, 64
and symmetry, 247–248
thematic element, 519
violation by cosmological theories, 511
constants, fundamental, 538
continuity/discontinuity as opposing themata, 519
conversion factors, 539–540
coordinate system, 98
Copenhagen interpretation of quantum mechanics, 455–456
Copernicus, Nicolaus (1473–1543, Polish astronomer), 17–18, 19–23, 26, 36, 40, 58
Commentariolus, 18
earth-sun distance, 129
heliocentric system, 1, 11, 18–26, 28, 98
model, 194
opposition to system, 23–25
Revolutions of the Heavenly Spheres, 17, 20
thematic element, 518
vindication by Newton, 139
Copp, Carol M., 513
copper, wavelength of K line (x-ray spectrum), 423
Cornell, Eric Allin (American physicist), 460
Cornford, F. M., 520
cos (cosine), *547*
cosmic microwave radiation, 512–513
cosmogony, thematic dimension, 520
cosmological principle, perfect, *510*
cosmology, 1–26, 40–46, 47–49, 494–497, *513–515*
coul (coulomb), *358*
coulomb (unit of electric charge), *358, 371*
Coulomb, Charles Augustin (1736–1806, French physicist-engineer), 358
electric force law, *358,* 421
Cowan, Clyde Lorrain, Jr. (1919–1974, American physicist), 503
Crawford, Arthur B., 512
creationism, 489
creativity, 170–171
Crick, Francis Harry Crompton (b. 1916, English physicist-biologist), 385
Crombie, A. C., 525
Crookes, William (1832–1919, English chemist-physicist), 381
crystal structure, 385
Curie, Marie Sklodowska (1867–1934, Polish-French physicist-chemist), xiii, 185, 258, 409–410
Curie, Pierre (1859–1906, French physicist), 185, 258, 409
curium (element), 508
Curtis, Heber Doust (1872–1942, American astronomer), 494–495

d (diameter of molecule), 327–328
Dalton, John (1766–1844, English chemist), xiii, 242, 263
atmosphere's homogeneity, 276, 328
atomic-molecular theory, 28, 276–285, 308
atomic weights, 283–284
definite proportions, law, 281
and Gay-Lussac, 285
multiple proportions, law, 282
partial pressures, law, 319
symbols for elementary and compound atoms, 279
Darwin, Charles Robert (1809–1882, English naturalist)
Origin of Species, 17, 105
theory of evolution, xiii, 27, 160, 164, 258, 414, 488–489
Darwin, George Howard (1845–1912, English astronomer-mathematician, son of Charles Robert), 142
Daub, E. E., 337
Davis, Donald R., 493
Davisson, Clinton Joseph (1881–1958, American physicist), 448, 461
electron diffraction, 448
Davy, Humphry (1778–1829, English chemist), 276
heat, 237
discovery of elements by electrolysis, 365
day, *3*
Debye, Peter (1884–1966, Dutch physicist-chemist), 449
decoherence, 457
deferent, *12*
degree of freedom of motion, *331*
De Morgan, Augustus (1806–1871, English mathematician), 253
Democritus (c. 460–c. 370 B.C., Greek philosopher), 31, 39
density, *106*
Descartes, René (1596–1650, French mathematician-philosopher), 46, 47, 63–64, 101, 104
action at a distance, 266
analytic geometry, 90, 103
conservation, thematic element of law, 519
conservation of motion, 210–211
cosmology and vortices, 107, 133, 144, 152, 234
dualism, 64
light, 341–342
mechanistic philosophy, 266
motion, 64–65, *210*–211
and Newton, 133, 140, 152
vacuum, 83
vortex-fluid (kinetic) model of gas, 272
determinism, 153, 322–323, 446
deuterium, *425,* 501
deuteron, *225, 501*
Dewey, John (1859–1952, American philosopher-educator), 192
dichotomy (Zeno's paradox), 30

Dicke, Robert (1916–1997, American physicist), 512–513
Dijksterhuis, E. J., 518, 519
dimensionless constants, 514, 538
Dirac, Paul Adrien Maurice (1902–1984, English physicist), 477
  beauty in equations, 480
  Fermi-Dirac statistics, 458–460
  hole theory, 478–479
  positron, 479, 500
  relativistic wave equation, 478
  relativity, 176
discontinuity, 403–404, 451
  thematic element, 519
discoveries, 160
  accidental, 145, 159
  simultaneous, 244
disorder, 335–336
disk, race with hoop, 231
displacement, 89, 94–95
displacement law, 389–390
dissipative system, *227*
DNA, 385
Dobereiner, Johann Wolfgang (1780–1849, German chemist), 297
Donne, John (1572–1631, English poet), 80
Doppler, Johann Christian (1803–1853, Austrian physicist), 430
  effect, 430
Drude, Paul Karl Ludwig (1863–1906, German physicist), 459
duality of science, 171–174, 178
Duhem, Pierre (1861–1916, French physical chemist, historian of science), 193
  Duhem-Quine thesis, *194*
Dulong, Pierre Louis (1785–1838, French chemist-physicist), 238
dynamics, *77, 99*
dynamism, thema opposed to atomism, 523
dynamo, 374

e (base of natural logarithms), *168*
e (charge of electron), 367
e (eccentricity of ellipse), *41*
Earth (element), 7
  natural motion, 7
  natural place, 7, *10*
earth (planet)
  age, 254, 258, 414, 488–489
  atmosphere, effect of rotation, 23, 24
  cooling, 253–256, 488
  core, 256
  eccentricity of orbit, 249
  falling bodies, effect of earth's rotation, *56–58*, 98
  inertia, 120
  location, 6
  mass, 121, 249
  orbit, 21, 42, 44, 129, 466
  origin. *See* solar system, origin
  precession of axis, 20
  radius, 249
  revolution around sun, 10, 19, 22, 29, 125
  rotation, 10, 19, 22, 55–56, 125, 249
  rotation, effect on shape, 129, 142
  rotation speed needed to cancel gravity, 128
  shape, 10, 129, 142
  size, 9–10, 125
earth sciences revolution, 489
Easter, *20*
eccentric motion, *12*–13, 42
Eckart, Carl (1902–1973, American physicist), 452
eclipse expedition to test relativity, 464
ecliptic, *3*, 144
Eddington, Arthur Stanley (1882– 1944, English astronomer)
  mysticism, 176
  quantum theory in England, 435
  relativity, 176
  stellar structure, 490
Edison, Thomas Alva (1847–1931, American inventor)
  electric light, 383
education, general, xiv
eigenvalue, *449*
Einstein, Albert (1879–1955, German physicist), xiii, xiv, 17, 192, 398, 408, 462–464
  and Bohr, 455–456
  Bose-Einstein statistics and condensation, 458–460
  on de Broglie theory, 449
  Brownian movement, 337
  cosmological constant, 497, 514
  criteria for judging theories, 38
  $E = mc^2$, 475–477, 497, 502
  EPR objection, 455
  equivalence principle, 481
  gravitational deflection of light, 464, 482
  gravitational waves, 484
  and Jews, 464
  mass, inertial and gravitational, 115
  mass-energy transformation, *475–477*, 497, 500
  and Mileva Marić, 463
  Maxwell's electromagnetic theory, 464, 467
  Mercury perihelion, advance of, 482–483
  philosophies of scientists, 171
  photon, 399
  and Planck, 399
  and quantum mechanics, 455–456
  quantum theory, 398–401
  randomness, attitude toward, 456
  relativity, general theory, 35, 115, 189, 247, 480–484
  relativity, special theory, 163, 247, 377, 467–477
  Relativity: The Special and General Theory, 468, 481, 485
  ring, 482
  Roosevelt, letter to, 507
  style of thought, 524–525
  thought experiments, 57
  Zionism, support for, 464
elastic potential energy, *229–230*
Electrical Age, 371, 374
electricity, electrical charge, *352–353*
  conductor, 354, 467–468
  conservation, 353, 452
  current, 354
  field, *359–360*
  ground, 355, 362

induction, *355*
insulator, 354
magnetism, interaction with, 369–371
negative, 353
positive, 353
potential difference, 362–*365*
social consequences, 374
electrical force
on test charge between plates, 365
Coulomb's law, 358–362, 421
electrical power produced in United States, 476
electrochemistry, 365–366
electrode, 366
electromagnetism, electromagnetic fields, 369–377
electromagnetic radiation spectrum, 402
electromagnetic units, 542
electromagnetic worldview, 384
electron, 170, 172–173, 353–354, 387
in Bohr model of atom, 128
charge, 173, 367, 380, 387
mass, 173, 418
in metals, 459–460
in nucleus, 424–425
radiation of energy, 216
spin, 444
electron-positron pair, creation and annihilation, 479
electron volt, *414*
electroscope, *356–357*
electrostatics, 352–368
field, 359
elements
alphabetic list, 543
Aristotle's, 7
inert gases, 305
Lavoisier's, *275–276*
names for new, 305, 307, 536
periodic table, 304, *545*
lanthanide, 304
rare earth, 304
transuranic, 305
elevator, 116
ellipse, 11, *41–44*
eccentricity, *41*
Elsasser, Walter Maurice (1904–1991, German-American geophysicist), 448
emission spectra
continuous, 388–395
line, 427–441
empirical dimension of science, *517–518*
empirical or experimental style of thought, 524, 526, 528
emu (electromagnetic units), 542
endothermic reaction, *476*
energy, 238
dissipation, 251–261
equipartition, 332
kinetic, 212, 219–221
kinetic energy operator, 450
levels, 393
mass-energy transformation, 206–207, *475–476*
operator, 451
potential, 225
potential energy operator, 450
rest mass energy, *415*, 477
in space, 369
English science, 159
entropy, *259*
Epicurus (341 B.C.–279 B.C., Greek philosopher), 31, 80, 203
epicycle, *12*, 23, 24, 42
myth of "epicycles on epicycles," 18
EPR objection, 455–456
ε (epsilon; elasticity), 349
ε_0 (permittivity constant), *358*
Equal Areas Law, 43–45, 131–132, 218
equant, *13*, 18, 40, 45
equilibrium, *108*, 116, *172*, 233
thermal, 320, 395
equinox
autumnal, *4*, 13
precession, 3, *4*, 20
vernal, *4*, 13, 20
Eratosthenes (c. 275 B.C.–c. 195 B.C., Greek astronomer and geographer), 9–10, 22
error, experimental, probable, *162*, 207
ether, *7*, 454
elastic solid model, 194, 349
elimination, 377
natural motion, 7
transmitter of electromagnetic fields, 172, 376
transmitter of gravity, 151–152
transmitter of heat, 153, 376
transmitter of light, 349, 376
Euclid (fl. 300 B.C., Greek mathematician), 9, 51, 105
ray theory of vision, 32, 34, 35
Eudoxus of Cnidus (fl. 130 B.C., Greek astronomer-mathematician), 6, 7
Euler, Leonhard (1707–1783, Swiss mathematician), 109, 217, 253
eV (electron volt), *414*
evaporation, 310
evolution of science, 37, 178–179
exchange forces, 457–458, 501
Exclusion Principle. *See* Pauli, Wolfgang
experiment, 189, 195. *See also* thought experiment
explanation, 167–168
explosions, 215
extramission theory. *See* vision
extrapolation, *65*
eye, 2. *See also* vision
retina, 46–47

F_{grav} (gravitational force), 112
fall, free, 7, 86
faraday (unit of electric charge), *366*
Faraday, Michael (1791–1867, English physicist-chemist), 192, 371
electrochemistry, 366
electromagnetic induction, *373*
electromagnetic rotator (motor), *372*
field, 359
impact of his electrical discoveries, 182
interconnections of natural activities, energy, 238, 244
research that led to discovery of electron, 379–380
Federation of Concerned Scientists, 509
Fermi, Enrico (1901–1954, Italian-American physicist), 459
chain reaction, 508

Fermi, Enrico (*continued*)
impact of experiments on radioactivity, 182
neutrino, 502
quantum (Fermi-Dirac) statistics, 458–460, 479
synthesis of transuranic elements or fission, 506
fermion, *459*
Feynman, Richard Phillips (1918–1988, American physicist), 450
positron as electron going backwards in time, 479–480
field, 369
electric, 359–361
gravitational, 225
fingo, 108
Fire (element), 7, 234
fission, nuclear, 36, 182, 506
FitzGerald, George Francis (1851–1901, Irish physicist), 408
contraction hypothesis, 466–467
Fizeau, Armand Hippolyte Louis (1819–1896, French physicist)
Doppler effect for stars, 430
light, speed of, 193, 348
Flammarion, Camille (1842–1925, French astronomer), 260, 261
Flamsteed, John (1646–1719, English astronomer), 129
fluorescence, 402
football, 97
Forbes, James David (1809–1868, Scottish physicist), 238
force, *108,* 162, 172. *See also* action at a distance; dynamism
central, 370
conversion factors, 539
inverse square, 43
lines of electric, 361, 372
and motion, 77, 103, 108
nuclear, 501–502
perpendicular, 370
pushing planets around orbits, 43, 48
reality of, 240
thematic dimension, 518
Foucault, Jean Bernard Léon (1819–1868, French physicist)
light, speed of, 193, 348
pendulum experiment, 25, 106
Fourier, Jean Baptiste Joseph (1768–1830, French mathematician), 255
cooling of earth, heat conduction theory, 255, 258
Fowler, William (1911–1995, American physicist)
formation of elements, xiii, 511
frame of reference, 120, 468–473
accelerated, 481
inertial, 468
Franck, James (1882–1964, German physicist), 509
Franklin, Benjamin (1706–1790, American scientist and statesman)
electrostatics, 235, 352–353
Franklin, Rosalind Elsie (1920–1958, British crystallographer), 385
Fraunhofer, Joseph von (1787–1826, German optician-physicist), 428
frequency
rotation, *123*
vibration, *345*
French science, 159, 190
Fresnel, Augustin (1788–1827, French physicist), 238
wave theory of light, 348
friction, 215
coefficient of, 226
heat generated by, 237
work done to overcome, 224, 226
Friedrich, Walther (1883–1968, German physicist), 385
Friis, Harold T., 512
Frisch, Otto Robert (1904–1979, German-British physicist), 37, 506, 508
fundamental vibration, *346*
funiculus, 269–270

g (acceleration due to gravity), 86, 89, *94,* 113–114, 115, 142, *537*
value at various locations, 113
g (gram), *111*
G (gravitational constant), *137*
Ga (giga annum), *414*
galaxies, 151
distances, 494
island universe theory, 494
galaxy (Milky Way), 150
Galilei, Galileo (1564–1642, Italian physicist-astronomer), xiii, xiv, 6, 22, 36, 38, 40, 50–52, 61, 63, 87, 101, 103–104, 163, 177
acceleration, 70, 85, 109, 166, 173, 192–193
Copernican system, 52–58
and Descartes, 140
Dialogue Concerning the Two Chief World Systems, 54–59, 80, 98
experiments, 82, 98, 190
falling bodies, 51, 55–56, 77–87, 113, 137, 189
heat, 234
inclined-plane experiment, 85–86, 166, 189, 231
Leaning Tower of Pisa, experiment, 51, 74
light, speed of, 343
and Newton, 139, 140
mathematics, 165, 168
model, 194
pendulum, 50, 188, 228
pile driver, 222
primary and secondary qualities, 164–165
projectile motion, 27, 28, 97–98, 148, 167, 195–196
relativity principle, *98–99*
social construction of scientific knowledge, 522
telescope and telescopic observations, 52–54, 63
themata, 520
tides, 143
transformation equations, *469–470*
trial, condemnation and vindication, 59–60
Two New Sciences, 80–87, 97
Galilei, Vincenzio (c. 1520–1591, Italian musician, father of Galileo), 50
Galle, Johann Gottfried (1812–1910, German astronomer), 145

- gallium, 302–303
- γ-rays, 384–385, 410
  - interaction with electrons and positrons, 480
  - range of wavelengths, 402
- Gamow, George (1904–1968, Russian-American physicist), 509–510
  - big bang theory, 510–512
  - quantum theory of radioactive decay, 502
- gas, 265–274. *See also* molecule
  - cooling by expansion, 327
  - diffusion, 311, 320–322
  - effusion, 320
  - equation of state, 312
  - free expansion, 241
  - heat conduction, 310
  - heating by compression, 310, 327
  - ideal, *271*
  - inert or rare, 305, 444
  - kinetic theory, 38, 272, 308–337
  - specific heats, *241,* 310, 329–333
  - static model, 272–273
- Gassendi, Pierre (1592–1655, French philosopher), 57, 58
- gauge pressure, *269*
- Gay-Lussac, Joseph Louis (1778–1850, French chemist-physicist)
  - combining volumes law, *284–285*
  - free-expansion experiment, 241, 242, 313–314
  - thermal expansion law, 148, *270–271,* 309
- Geiger, Hans (1882–1945, German physicist), 417–420
- Geissler, Johann Heinrich Wilhelm (1815–1879, German glassmaker-technician), 380
  - tube for study of cathode rays, 380–381
- generator, electrical, 374
- geocentric system, 2, 36, 54–55
- geology and physics, 258
- geometry
  - analytic, 63, 65–66, 90
  - used in astronomy, 5, 28, 41
- geophysics and nuclear arms control, 186
- germanium, 302
  - prediction and discovery, 302
- Germany, Nazi, science and technology in, 180, 182
- Germer, Lester Halbert (1896–1971, American physicist), 448
  - electron diffraction, 448
- GeV (giga electron volt), *414*
- Gibbs, Josiah Willard (1839–1903, American physicist), 310
- Gilbert, William (1544–1603, English physician-physicist), 40, 48, 520
- Gingerich, Owen, 18
- gnomon, *3*
- gold
  - density, 137
  - scattering of α particles, 421
  - transmutation into, 246
- Gold, Thomas (b. 1920, Austrian-American astronomer) 510–511
- Goldhaber, Maurice (b. 1911, Austrian-American physicist), 500
- Goldstein, Eugene (1850–1930, Polish physicist), 381, 382
- Göttingen University, 248
- Goudsmit, Samuel (1902–1978, Dutch-American physicist), 444
- Gouy, Léon [Louis-Georges] (1845–1926, French physicist), 351
- Graham, Thomas (1805–1869, British chemist), 320
- gram, *111*
- Gramme, Zénobe-Théophile (1826–1901, Belgian engineer)
  - electrical generator, 383
- gram-molecular-weight, *294*
- gravity. *See also* Newton, Isaac
  - acceleration, 86, 89, *94,* 113–114
  - at center of earth, 143, 154
  - force, 112
  - law of universal, 86, 136, 148
  - variation in space, 113–114, 128, 142–143
- gravitational lensing, 482
- gravitational mass, 115
- gravitational waves, 484
- Greek alphabet, 535
- Greek letters used as scientific terms, 533
- Greek science and mathematics, 3–15, 30–33, 190
  - thematic dimension of cosmogony, 520
- Greenwich Observatory, 103
- Gregory XIII, Pope (1502–1585), 20
  - calendar, 20
- Grove, William Robert (1811–1896, British physicist), 244
- Groves, Leslie R., 508
- Guericke, Otto von (1602–1686, German physicist), 103, 268, 269
- Gurney, R. W., 502
- Guth, Alan (b. 1947, American physicist-cosmologist), 514

- *H* (Hubble constant), *495–496,* 511
- *h* (Planck constant), *394*
- Hahn, Otto (1879–1968, German physical chemist), 36, 506
- half-life for radioactive decay, *413*
- Halley, Edmund (1656–1742, British astronomer), 105, 133
  - comet, 141
  - secular acceleration of moon, 143
- Hamilton, William Rowan (1805–1865, Irish mathematician), 450
  - operator, 450
- Hankey, Lord, 508
- Harmonic Law, 45–46, 132–133, 144, 151, 436
- Hartmann, William K., 493
- Harvard University
  - general education report, xiv
- Hau, Lene Vestergaard, 460
- Hawking, Stephen (b. 1942, English physicist), xiii, 484
- heat, 234–247
  - caloric theory, 28, 235–236
  - death, 254, 256, *259–261,* 336
  - as energy, 238

heat (*continued*)
  fluid theory, 234
  latent, 235, 273
  mechanical equivalent, 242, *243*, 245, 329
  motion theory, 234, 237
  radiant, 238
  sensible, 235
  specific, *235*, 241
  wave theory, 153, *238*, 351
Heisenberg, Werner (1901–1976, German physicist), 425, 451
  atomic bomb, 509
  discontinuity thema, 519
  indeterminacy principle, 453–454
  matrix mechanics, 451
  nucleus, model, 476, *500*
Heitler, Walter (1904–1981, German-Swiss physicist), 457
heliocentric theory, 10, 18–19, 38. *See also* Copernicus
helium
  electrons, number of extranuclear, 421
  formation in big bang, 511
  isotope masses, 412
  molecular weight, 293
  nucleus, 424
  spectrum, 441–442
Hellenism, 15
Helmholtz, Hermann von (1821–1894, German physiologist and physicist)
  energy conservation, 244
  energy source of sun, 247
  heat death, 259
  nebular hypothesis, 487–488
Henry, Joseph (1797–1878, American physicist), 372
Heraclides of Pontus (c. 390 B.C.– c. 315 B.C., Greek astronomer), 10
Herapath, John (1790–1868, English physicist), 308–309, 318
Herman, Robert, 510
Herschel, John (1792–1871, British astronomer), 149, 244
Herschel, William (1738–1822, German-British astronomer)
  double stars, 149
  infrared radiation, radiant heat, 238
  nebular hypothesis and star formation, 487, 489–490
  Uranus, 144, 146
Hertz, Heinrich (1857–1894, German physicist), 378
  cathode rays, 382
  electromagnetic waves, discovery, 378–379
  force, elimination, 384
  photoelectric effect, 396
Hertzsprung, Ejnar (1873–1967, Danish astronomer), 494
heuristic, *398*
Hilbert, David (1862–1943, German mathematician), 247–248
Hipparchus (c. 190 B.C.–125 B.C., Greek astronomer), 9, 11, 21–22
Hirn, Gustav Adolphe (1815–1890, French engineer-physicist), 244
history of science, xiii–xv, 15, 16, 161
Hitler, Adolf (1889–1945, German dictator), 464, 507–508
Hittorf, Johann Wilhelm (1824–1914, German chemist-physicist), 381
Hodgkin, Dorothy Crowfoot (1910–1994, British crystallographer), 385
Holden, N. E., 412, 414
holism, 523
Holtzmann, Karl (1811–1865, German physicist), 244
Hooke, Robert (1635–1703, English physicist), 103, 105, 211, 269
  law, 230
  light, 343
hoop, race with disk, 231
Hopkins, William (1793–1866, English geophysicist, mathematics teacher), 255–256
Houtermans, Friedrich Georg (1903–1966, German-Swiss physicist), 502
Hoyle, Fred (b. 1915, English astronomer)
  anthropic principle, 514
  formation of elements, xiii, 505
  steady state cosmology, 510–511, 513
Hubble, Edwin Powell (1889–1953, American astronomer), 495
  constant, 495–496, 511
  distances of galaxies, 495
  expanding universe, doubts about, 496
  velocity-distance equation for galaxies, 495–496
Huggins, William (1824–1910, English astronomer), 430
Hulse, Russell Alan (b. 1950, American astronomer), 484
human body as an engine, 246
Hume, David (1711–1776, Scottish philosopher), 2, 468, 517–518
Hutton, James (1726–1797, Scottish geologist), 254
Huygens, Christiaan (1629–1695, Dutch physicist), 103
  centripetal acceleration, 126, 128
  collisions, 211–212
  g, 86, 101
  kinetic energy (*vis viva*), 212, 219–221
  light, speed of, 129, 344
  pendulum, 228
  wave theory of light, 343
hydrogen, 298
  abundance in universe, xiii, 492
  atom, 278, 279
  diatomic molecule, 331
  electrons, number of extranuclear, 421
  escape from atmosphere, 329
  heavy isotope, 225, 412, 501
  masses of isotopes, 412
  molecular weight, 293
  specific heat, variation with temperature, 332–333
  spectrum, 432–441
hypotheses, 107, 152, 195
  ad hoc, 147
hypothetical modeling (style of thought). *See* model

I (rotational inertia), *230*
Ice Age, 256

ideal gas law, *271,* 312–313
idealism, *171,* 454
identical particles, 456
I-Hsing (682–727, Buddhist monk and astronomer), 149
impotency, postulates of, 246
incidence, angle of, *33*
inclined plane, 85–86, 166, 189, 226, 231
indeterminacy principle, *453,* 500
Index of forbidden books, 23, 58
indistinguishability, 457
inequalities (in astronomy), *143*–144
inert (rare) gases, 305, 444
inertia, *109. See also* mass
  in circular motion, 126
  law of, 57, 210
  rotational, 230
  work done to overcome, 224
Infeld, Leopold (1898–1968, Polish physicist), 115
inflationary cosmology, 514
infrared waves, 238, 351, 402
Ingen-Housz, Jan (1730–1799, Dutch physician-biologist), 205
Inquisition, 40, 58, 80
intelligence, use of, 188, 190
intelligence test, 454
intelligibility, principle of, 30, 36
International Union of Pure and Applied Chemistry. *See* IUPAC
interpolation, *65*
intromission theory. *See* vision
intuition, 190
ion, 354, 367
ionization, 357, 438, 451
iron
  atomic weight, 293
  combustion, 205
  valence, 291
  wavelength of K line (x-ray spectrum), 423
irreversibility, 251, 256, 258
Islam, science in, 33–35, 187
island universes, 151
isolation, isolated system, 117, 203–204, 245
isomer, 291–292
isotopes, 410–412
  abundances, 411–412
  masses, 411–412, 499
IUPAC (International Union of Pure and Applied Chemistry)
  numerical roots for naming transuranic elements, *536*

J (joule, unit of energy), *224,* *243*
J (Joule's equivalent between work and heat), *243*
James, William (1842–1910, American psychologist-philosopher), 518
Jammer, Max, 451
Jeans, James Hopwood (1887–1946, English physicist-astronomer), 435, 490, 491
Jeffreys, Harold (1891–1989, British geophysicist), 490
Joliot-Curie, Frédéric (1900–1958, French physicist), 425, 506
Joliot-Curie, Irène (1897–1956, French physicist), 425, 506
joule (unit of energy), *224*
Joule, James Prescott (1818–1889, British physicist), 201, 242
  energy conservation, 242–244, 249
  free-expansion experiment, 241
  priority dispute with Mayer, 244
  specific heats of gases, 330
Julian calendar, 20
Jupiter (planet)
  destruction of earth by, 144
  mass, 139
  motion, 20–21
  orbit, 21, 42
  period of revolution, 29
  radio noise, 37
  satellites, 52–53, *55*
  and Saturn, long inequality, 143
  shape, 142

k (Boltzmann constant), *319*
k_e (constant in Coulomb's law), *358*
k_m (electromagnetic force constant), *375*
K (kelvin, unit of temperature), *271, 537*
Kandinsky, Wassily (1866–1944, Russian artist), 418
Kapteyn, J. C., 494
Kaufmann, Walther (1871–1947, German physicist), 373
Keeler, James Edward (1857–1900, American astronomer), 490
Kekulé, Friedrich August (1829–1896, German chemist), 160, 292, 523
kelvin (unit of temperature), *271*
Kelvin, (Lord) William Thomson (1824–1907, Scottish physicist), 243, 257, 369
  age of earth, 258, 414, 488
  age of sun, 488
  atom, size and reality of, 328
  dissipation principle, *257,* 260
  free-expansion experiment, 241
  on mechanical models, 28
  on quantitative measurement, 170
  reversibility paradox, 251
  temperature scale, *271*
Kepler, Johannes (1571–1630, German astronomer), 6, 40–41, 42–49, 148, 158, 216, 449
  alchemy, 23
  clockwork universe, 48, 144
  first law, 41–43
  mathematical laws, 165
  second law, 43–45, 131–132, 218
  third law, 45–46, 132–133, 144, 151, 436
  vision theory, 35, 46–47
  vindication by Newton, 139
kilogram (kg), *110–111, 537*
kinematics, *77*
kinetic energy, 212, 219–221. *See also* vis viva
  operator, 450
kinetic theory of gases, 38, 308–337
  statistical methods, 322–325

Kirchhoff, Gustav Robert (1824–1887, German physicist), 427–428
Kistiakowski, George Bogdan (1900–1982, American physical chemist), 509
Klein-Gordon equation, 478
Kleppner, Daniel, 460
Knipping, Paul (1883–1935, German physicist), 385
knowledge, 452–456
Kölliker, Albert von, 383
Koyré, Alexandre (1892–1964, French philosopher and historian of science), 82, 122, 152
Krönig, August Karl (1822–1879, German chemist-physicist), 309

l (angular momentum), *218*
Lagrange, Joseph Louis (1736–1813, Italian-French mathematician), 141, 144, 227, 254, 541
λ (lambda; wavelength), *345*
λ-transition of liquid helium, 460
  Bose-Einstein condensation, 460
Landau, Lev Davidovich (1908–1968, Russian physicist), 523
Landolt, Hans (1831–1910, Swiss-German chemist), 206–207
Laplace, Pierre Simon de (1749–1827, French astronomer-physicist), 141, 144
  black hole, 483–484
  clockwork universe, 253, 254
  determinism, 153, 322–323
  God "hypothesis" rejected, 253
  nebular hypothesis, 218, 253, 487–*8*
  statistical theory, 323
Laue, Max von (1879–1960, German physicist), 385
Lavoisier, Antoine Laurent (1743–1794, French chemist), xiii, 7, 204, 205–208, 235, 275–276
laws (in science)
  definitional, *148*
  derivative, *148*
  discovery of, 187–199
  empirical, *148*, 517
  errors, 323
  formulation, 191–195
  limitations, 195–196
  mathematical, 165–167
  of nature, 187
lb (pound), *111–112*
lead (element), in finding age of earth, 489
Leavitt, Henrietta Swan (1868–1921, American astronomer), 454
  period-luminosity relation, *494*
Lecoq de Boisbaudran, François Paul Émile (1838–1912, French chemist), 302–303
Lehmann, Inge (1888–1993, Danish seismologist), 256
length
  contraction, 466–467, 472–473
  conversion factors, 539
  measurement of, 161–162
Leibniz, Gottfried Wilhelm (1646–1716, German philosopher-mathematician), 44, 73, 103, 104, 219
  cooling of earth, 253
  critique of Newton's views on God, 252–253
Lemaître, Georges (1894–1966, Belgian astronomer), 496, 498
  cosmogony, 496–497
  primeval atom, 496
Lenard, Philipp (1862–1947, German physicist), 397, 417
Leo X, Pope (1475–1521), 20
level of reference (for potential energy), *225*
lever, *232–233*
Leverrier, Urbain Jean Joseph (1811–1877, French astronomer), 145, 146
Lexell, Anders Johan (1740–1784, Swedish astronomer), 144
Liebig, Justus (1803–1873, German chemist), 244
Lifshitz, E. M., 523
light
  bending by sun, 35, 464
  diffraction, 348
  as energy propagated through vacuum, 28
  as ether vibration, 28, 376
  gravitational action on, 35, 464, 482, 483
  instantaneous transmission, 32
  interference, *347–348*
  momentum, 216–217
  particle theory, 238, 342–343
  polarization, 342, *349*, *376*
  polarization rotated by magnetic field, 374
  pulse theory, 343, 351
  refraction, *33*, 35, 38, 341–343
  speed, 129, 193–194, 341–342, 344, 375–376, 467
  theories, 148, 163, 193–194, 196
  transverse waves, 376
  visible, range of wavelengths, 389, 402
  wave front, 343
  wave theory, 194
light year, *150*
Lindemann, Frederick Alexander (1886–1957, British physicist), 435
Linus, Franciscus [Francis Line] (1595–1675, English scientist), 269–270
liquid
  cooling by evaporation, 327
  motion of molecules in, 310
Lodge, Oliver Joseph (1851–1940, British physicist), 144
London (city), 9
London, Fritz (German physicist, 1900–1954), 460
  Bose-Einstein condensation in helium, 460
  force between hydrogen atoms, 457
long inequality of Jupiter and Saturn, *143*
Lonsdale, Kathleen Yardley (1903–1971, British crystallographer), 385

Lorentz, Hendrik Antoon (1853–1928, Dutch physicist), 408, 485
absolute motion and time, 467
contraction hypothesis, 467
electrodynamics, 467
factor, *473*
mass increase with speed, 467
time dilation, 467
transformation, 465, *472–473*
Los Alamos (New Mexico) laboratory, 508
Loschmidt, Josef (1821–1895, Austrian physicist)
diameter of molecule, 328
number, *289,* 295, 328
reversibility paradox, 333
Lowell, Percival (1855–1916, American astronomer), *145*
Observatory, 494
Lucretius Carus, Titus (c. 99 B.C.– c. 55 B.C., Roman poet-philosopher), 31, 80, 203, 265
Lumière, Auguste (1862–1954) and Louis Jean (1864–1948) (French inventors), 382–383, 384
luminosity, 494
LY (light year), *150*
Lyell, Charles (1797–1875, British geologist), 254–255, 256, 261
Lyman series (in hydrogen spectrum), 433, 440, 441
Lysenko, Trofim Denisovich (1898–1976, Russian agronomist), 523

m (mass), *110*
m (meter), *65*
M (molecular weight), 294
Mach, Ernst (1838–1916, Austrian physicist-philosopher), 195, 468
atom, 164, 172
economy of thought, 199
empiricist view of science, 384
masses, relative, 121
Newtonian absolute space, time, mass, 106, 468, 524
machines, 232–234
Magdeburg experiment, 269
magic numbers, 175
magnetism, 48. *See also* electromagnetism
interaction with electricity, 369–371, 467–468
monopoles, 369
quantum theory, 459
sun's interaction with planets, 43, 48, 491
Manhattan Project, 508–509
Marci, Marcus (1595–1667, Bohemian physician), 213
Marconi, Guglielmo (1874–1937, Italian inventor), 379
Marić, Mileva, 463
Mariotte, Edme (c. 1620–1684, French physicist), 103
Mars (planet)
motion, 14, 19, 20–21
orbit, 21, 42, 129
period of revolution, 29
triangulation to find distance from earth, 129
Marsden, Ernest (1889–1970, British physicist), 417–420
mass, *106, 111*
conversion factors, 540
critical (for nuclear chain reaction), 508–509
earth, sun, and planets, 138–139
energy-mass transformation, 206–207, 415, 426, *475–476,* 499
gravitational, *115,* 481
inertial, *115,* 481
kilogram (standard unit), *110–111*
measured by reaction-car experiment, 121
missing (in Milky Way), 151
rest mass energy, *415*
increase with speed, 474–475
and weight, 540
mass number (of isotope), *412*
change in α decay, 416
materialism, *240,* 251
mathematical constants, 537
mathematical symbols, 533
mathematics and science, 28, 47, 63–76, 90, 99, 105, 199. *See also* geometry
matter, wave nature, 447–450
Maxwell, James Clerk (1831–1879, Scottish physicist), xiii, 324, 374
Demon, *334*
displacement current, 375, 479
distribution of molecular velocities, *322–323,* 324–326
electromagnetic fields and waves, 172, 374–377, 431, 467
electron rejected, 380
equipartition of energy, 332
ether model, 194, 375
forces between molecules, 315
kinetic (dynamical) theory of gases, 159–160, 310, 392
light as electromagnetic waves, 194, 340, 395, 523
quantitative style, 523
thematic questions, 521
viscosity of gas, 327
Mayer, Julius Robert (1814–1878, German physician), 239–240, 242
energy conservation, 239–241, 244, 308
energy source of sun, 247
priority dispute with Joule, 244
McKellar, Andrew, 512
McMillan, Edwin Matteson (1907–1991, American physicist), 508
mean-speed theorem, 72–73
measurement, 161
mechanics, 61
mechanistic science, 28, 48, 63–64, 153, 251, 369, 446
medicine and physics, 184
megaparsec, *496*
Meitner, Lise (1878–1968, Austrian physicist), 34, 507
nuclear fission, 506
meitnerium (element), radioactive decay of isotope, 414
Melloni, Macedonio (1798–1854, Italian physicist), 238
Melville, Ronald, 203

Mendeléeff (Mendeleev), Dmitri Ivanovitch (1834–1907, Russian chemist), 297–298, 299–303
  mass-energy transformation, 206
  periodic law, 148
mercury (element)
  molecular weight, 293
  monatomic molecule, 331
Mercury (planet)
  advance of perihelion, 35, 153, *482–483*
  mass, 139
  motion, 14, 20, 21
  orbit, 21, 42, 146
  period of revolution, 21, 29
Merton, Robert King (b. 1910, American sociologist), 140, 181, 244
Merton theorem, 72
meson. *See* muon, pion
mesotron, 501
metals, quantum theory of, 459
metaphysics, *5*
meter, *65, 537*
metric system, 65, *535*, 541
  politics, 542
MeV (million electron volts), *414*
Meyer, Lothar (1830–1895, German chemist), 206, 297, 301
Michelson, Albert Abraham (1852–1931, American physicist)
  completeness of science, 369, 466
Michelson-Morley experiment, 376, 408, *465*
microwaves, 402, *512–513*
Milky Way (galaxy), 53, *150*, 494
millibar (unit of pressure), *269*
Millikan, Robert Andrews (1868–1953, American physicist)
  oil-drop experiment, 367, 386–387
  photoelectric effect, 400, 446
model, 526. *See also* clockwork universe
  atomic, 160, 276–292
  electromagnetic ether, 374–375, 527
  eye as camera obscura, 46, 527
  gas, 276–278, 311–326
  light, 341–349
  mechanical, 7, *28*
  theoretical, 194
modernism, 524
Mohr, Karl Friedrich (1806–1879, German chemist), 244
mole, *294*
molecular orbital, 458
molecular weight, *292–293*
molecule, 279
  collisions, 314–315, 327
  degree of freedom of motion, 331
  force between, 313–315
  mean free path, 313
  number of atoms in, 288–289
  rotation, 309, 330
  size, 313, 327–328
  specific heats, 330, 331
  speed, 308, 318, 322–325
  unsaturated, 291
  vibration, 309
moment of momentum. *See* angular momentum
momentum, *210–211*
  angular, 217–218, 231
  of light, 216–217
  and Newton's laws, 212
moon
  distance from earth, 134
  earth's force on, 133–134
  earth's tides, effect on, *143*
  g at surface, 113–114
  mass, 138, 139
  motion, 14, 20
  origin, 142, 493
  period of revolution around earth, 134
  perturbations, 141
  radius, 134
  secular acceleration, 143
  surface appearance, 53
Morley, Edward William (1838–1923, American chemist). *See* Michelson-Morley experiment
Moseley, Henry Gwyn Jeffrey (1887–1915, English physicist), 422
  characteristic x-ray spectrum, 422–423
motion
  chaotic, 144
  circular, 5, 7
  natural, 7
  perpetual, 246
  possibility of, 30–31, 36
  relativity of, 64
  "uniform and ordered," 5–6
motion pictures, 382–383
motivation of scientists, 174–176
Mueller, Erwin Wilhelm (1911–1977, German-American physicist), 527
μ (mu; coefficient of friction), 226
μ-meson, 501
multiplicity of effort, 179–181
muon, 501
music, 45, 449
mysticism, 176

n (frequency of rotation), *123*
n (frequency of vibration), 345
N (newton; unit of force), *110*
N (or N_A; Avogadro's number), *328*, 367
Nagaoka, Hantaro (1865–1950, Japanese physicist), 422
nanometer, 388
Nature Philosophy, 184, 369
  and romantic movement, 523
nebula, 494
  spiral, 151, 490
nebular hypothesis, 218, 487–488
Neddermeyer, Seth Henry (1907–1988, American physicist)
  discovery of muon, 501
  implosion method in atomic bomb, 509
neon
  masses of isotopes, 412
neoPlatonism, 40
Neptune (planet)
  discovery, 145, 148
  mass, 139
  orbit, 42, 147
neptunium (element), 508
Nernst, Walther (1864–1941, German chemist), 271
neutrino
  β-decay, 502–503
  discovered, 503

mass, *503*
postulated, 502
neutron, 36, 225, *425*
discovery, 425, 500
mass, 426, 500
in nucleus, 425, 500
proton-electron combination, 425, 500
Newlands, John (1838–1898, English chemist), 297, 299
newton (unit of force), *110,* 224
Newton, Isaac (1642–1727, English physicist-mathematician), xiii, xiv, 28, 38, 80, 101–103, 104–105, 139–140
action at a distance, 152, 266
active principles, 252
apple, 105, 136, 154, 188
astronomy, 104, 107, 108
atoms, 265–266
bucket experiment, 106
calculus, 73
centripetal acceleration and force, 126–128, 130, 132–133
color, 350–351
cosmology, 63, 139
dissipative forces, 252
equant, 134
ether, 153
and Flamsteed, 140
forces between atoms, 266
and Galileo, 139, 140
gas model (repulsive particles), 272, 276
God's role in universe, 252–253, 487
gravity, 25, 48, 86, 104, 131–154, 168
and Halley, 133, 140
and Hooke, 134, 351
hypotheses, 107–108
"hypotheses non fingo," *152*
influences on, 139–140
influence of, on Europe and America, 182
Kepler's laws, 131–134, 136, 140
light, 238, 342–343
mass, 106, 204
mechanics as foundation of physics, end of, 446
model, 194
momentum and laws of motion, 212
moon, 104, 127, 141, 144
motion, first law 57, *108*–109
motion, laws, 103–122, 196
motion, second law, *109*–111, 116–118, 132–133, 148, 212, 222
motion, third law, *118*–121, 135, 371
"Newtonian world machine," 251–253, 261, 453
optics, 105, 107, 351
precession of equinoxes, 149
Principia, 60, 103, 105–108, 131, 134, 187
religion, 174–175
revolution, 2
rules of reasoning, *107*
scientific method, 155–156
space, 106, 153, 163
synthesis, *139*
tides, 143
time, 106, 163
nickel (element), wavelength of K line (x-ray spectrum), 423
Nilson, Lars-Frederik (1840–1899, Swedish chemist), 299, 303
nitrogen
atomic weight, 284
atoms, 277, 279
molecular mass, 322
molecular weight, relative, 293
nucleus, transmutation to oxygen, 499
valence, 291
Noether, Emmy (1882–1935, German mathematician), 202, 247–248
theorem, 247–248, 452
normal distribution, 323
North Celestial Pole, *3,* 4
ν (neutrino), 502
ν (nu; frequency), 394
ν_0 (threshold frequency), 397–400
nuclear fission, 36, 182
discovery, 506
nuclear force, 501–502
nuclear fusion, 503–505
nucleus, 216, 499–500
binding energy, 476, 503
charge, 420, 424
Heisenberg model, 476, *500*
stability, 425
structure, 424–426
numerology, 146, 175, 432

objectivity, 176–177
oblate spheroid, *142*
obliquity of ecliptic, 143–144
observations, 189
and frame of reference, 468–473
influence on object observed, 454
uncertainties, 8, 453
Occhialini, Giuseppe (1907–1993, Italian physicist), 502
Ockham, William (William of Ockham; c. 1285–c. 1349, English philosopher), 209
Oersted, Hans Christian (1777–1851, Danish physicist), 369–370, 373
Oldenberg, O., 400–401
ω (omega; angular velocity), *124*
operational definitions, *161–162,* 454
Oppenheimer, J. Robert (1904–1967, American physicist), 508
optics, 46, 63
Oresme, Nicolas (1320–1382, French mathematician), 72
organismic science, 28, 48
Ortega y Gasset, José (1883–1955, Spanish philosopher), 179
hypothesis, 179
oscillator, 392–394, 430
overtone, *346*
Oxford University, 72, 79
oxygen
atom, 277–278
atomic weight, 284, 293
masses of isotopes, 412
molecular weight, 293

p (momentum), *211*–213
Pa (pascal; unit of pressure), *268*
Pappus of Alexandria (fl. 300–350, Greek mathematician), 105
parabola, 11, 90

Paracelsus, Philippus Aureolus (c. 1493–1541, Swiss physician and alchemist), 23
parallax
  Mars, 129–130
  stellar, 11, 14, 22, 24, 38, 150
Paris
  Observatory, 128
  University of, 78
Parmenides of Elea (c. 515 B.C.–c. 450 B.C., Greek philosopher), 30, 36, 521
parsec (unit of distance), *150*
pascal (unit of pressure), *268*
Pascal, Blaise (1623–1662, French scientist and religious philosopher), 103, 268
Paschen, Friedrich (1865–1940, German physicist), 433
  series (of spectral lines), 433, 440, 441
Pasteur, Louis (1822–1895, French chemist and biologist), 17, 176–177
Patterson, Clair Cameron (1922–1995, American geochemist)
  age of earth, 489
Paul III, Pope (1468–1549), 20
Pauli, Wolfgang (1900–1958, German physicist), 443
  electrons in metals, 459
  exclusion principle, 443–444, 458
  neutrino, 502
  on Sommerfeld, 175
Payne-Gaposchkin, Cecilia (1900–1979, British-American astronomer), xiii, 491–492, 504
Peebles, P. J. E., 513
Peierls, Rudolph Ernst (1907–1995, British physicist), 508
pendulum, 50, 128, 214, 227–228
  ballistic, 228
  motion, law, 148
  seconds, *128*
Penzias, Arno A. (b. 1933, American astrophysicist), 512–513
perihelion, *43*, 482
period
  of pendulum, 50, *128*, 137
  of rotation, *123*
  sidereal, *45*
periodicity of waves, 344–345
periodic table of elements, 296–307
  inversions, 299, 305–306, 423–424
  Mendeléeff's, 300
  modern, 304
perpetuum mobile (perpetual mover), 246, 247
Perrier, Florin (brother-in-law of Pascal), 268
perturbations of celestial bodies, 138, 141, 145, 252
Pettersson, (Sven) Otto (Swedish chemist), 299
Pfund series (in hydrogen spectrum), 433, 440, 441
"phenomena, save the," 5, 6, 13, 20
philosophy
  influence on science, 5
  mechanical, 266, 268
  natural, 105
  "new" (17th century), 80, 103
  and quantum mechanics, 454–455
  of science, xiii–xv, 161, 198
phonograph, *124–125*
phosphorus
  atom, 279
  atomic weight, 284
  molecular weight and formula, 288, 293
  valence, 290
photoelectric effect, 396, 442, 446
photon, 399–401, 435
  spin, 455–456
photon-wave duality, 400–402
photosynthesis, 246
phlogiston theory, 28, *204–206*
Piazzi, Giuseppe (1746–1826, Italian astronomer), 146
Pickering, Edward Charles (1846–1919, American astronomer), 434
  series (of spectral lines), 434, 441–442
pile driver, 222, 225
pinhole camera, 46
π (pi; circumference/diameter of circle), 123
π meson, 502
pion, 502
place, natural, 5, 7
Planck, Max (1858–1947, German physicist), 17, 188, 339, 391
  black-body radiation distribution law, *394–395*, 458
  on Bohr, 435
  constant, *394*
  and Einstein's quantum theory, 401
  relativity, 176
  principle (concerning acceptance of theories), 37, 198, 348, 513
  quantum hypothesis, 392–396, 407
planets, *5*
  discovery of new, 144–145
  extrasolar systems, 494
  masses, 138–139
  motions, 7, 12–14, 28, 47, 109
  orbits, 21–22, 40
  periods of revolution, 28–29
  shapes, 141–142
  variation of speeds in orbit, 44
Plato (c. 428 B.C.–c. 347 B.C., Greek philosopher), 5–6, 15, 22, 31, 43, 103, 190
Platonists, 28, 171. *See also* neoPlatonism
plausibility of theories, 36
Playfair, John (1748–1819, Scottish mathematician-geologist), 254
Plücker, Julius (1801–1868, German mathematician-physicist), 380–381
Pluto (planet?)
  discovery, 145
  mass, 139
  orbit, 42, 147
plutonium (element)
  atomic bomb, 508–509
  radioactive decay of isotope, 414, 416
Podolsky, Boris (1896–1966, American physicist), 455
Poincaré, Henri (1854–1912, French mathematician),

162, 175–176, 189, 199, 384, 403, 491
conservation thema, *519*
recurrence theorem, 336
Poisson, Simon Denis (1781–1840, French physicist), 348
bright spot, *348*, 456
Pole Star (Polaris), 3
Popper, Karl Raimund (1902–1994, Austrian-British philosopher of science), 29, 199
falsifiability criterion, *164*, 382, 511, 513
principle, *29*, 34, 37, 179, 303, 348, 350
positivism, *171*
positron, *479*, *500*
electron going backwards in time, 479–480
post hoc, ergo hoc fallacy, *193*
postmodernism, *525*
postulation (style of thought), 190, *525*, *527*, *528*
potential difference, electric, 362
potential being, *31*, 32, 454
pound (unit of mass), 541
Pound, Robert Vivian (b. 1919, American physicist), 483
poundal (unit of force), 541
Powell, Cecil Frank (1903–1969, British physicist), 502
Power, Henry (1623–1668, English physician), 270
precession of equinoxes, 3, 11, 55
prediction, predictability, 5, 24, 29, *35*, 37, 38, 144, 146, 164, 167, 299, 302–303, 348, 384, 423, 425, 454, 464, 482–484, 500
pressure, *268*
atmospheric, 268, *269*, 270
conversion factors, 540
gauge, *269*
molecular impacts as cause, 309, 315–318
Priestley, Joseph (1733–1804, English chemist), 159, 187–188, 191
prime mover (*primum mobile*), 7, 24
priority disputes, 244
probabilistic style of thought (statistical view of nature), 322, 527–528
probability, *326*
probable error, *162*
projectile motion, 27, 88–99
prolate spheroid, *142*
proportions (chemical compounds)
definite, law, *280–281*
multiple, law, *282*
proton, 225, 384, 387
mass, 426
in nucleus, 424, 499
Proust, Joseph Louis (1755–1826, French chemist), 280–281
Prout, William (1785–1850, English chemist), 295, 296
hypothesis, 295–296, 476, 499
prouton, 387
psychology, 454
pseudoscience, 23, 513
Ptolemy, Clausius (c. 100–c. 170, Alexandrian astronomer), 9, 15
geocentric system, 11–15, 21, 22, 23, 28
model, 194
ray theory of vision, 32–34
Pugwash movement, 464, 509
pulley, 116–117, 233, 245
pulsar, 484
Puritanism, 104
purpose. *See* teleology
Puy de Dôme experiment, 268
Pythagoras of Samos (c. 560 B.C.–c. 480 B.C., Greek mathematician-philosopher) and the Pythagoreans, 5, 15, 31, 40, 45, 171, 449, 522

q (electric charge), 357–360
qualities, 107
primary and secondary, *164*, 177
quantization in chemistry and biology, 403
quantum degeneracy, 459
quantum mechanics, 449–461
Quantum Revolution, 392, 528
quantum theory, 38, 437–445
quasar, 512
Quine, Willard Van Orman (b. 1908, American philosopher), 194
quintessence, 7. *See also* ether

r (radius), 123
R (Rydberg constant), *433*
R (universal gas constant), *312*
rad (radian; measure of angle), *123*
radio astronomy, 511–512
radio waves, 379, 402
radioactive decay, xiii, 168
half-life, 413
quantum theory, 502
random process, 413, 502
series, 415–417
radioactivity, 184–185, 384, 409–417
age of earth, 258, 414
artificial, 506
radium, 168
and age of earth, 488–489
radioactive decay of isotopes, 414–416
radon, 168
randomness, 334–335, 453, 497
range
of force, 501
of projectile, *89*, 97
ray, light, 32–35, 46–47
Rayleigh, (Lord) John William Strutt (1842–1919, English physicist), 321, 390
reaction. *See* Newton, Isaac, third law of motion
reaction-car experiment, 120–121, 215
realism, *171*, 524
reality, 452–456, 461
Rebka, G. A., 483
record-player, *124*–125
Redhead, Michael, 461
red-shift of light
Doppler, *430*
gravitational, *483*
reducing agent (in chemistry), 290
reductionist explanation, 523
refraction, *33*, 35, 63
angle of, *33*, 38–39
Reines, Frederick (1918–1998, American physicist), 503
relativism, 24

relativity, 176, 467–496. *See also* Einstein, Albert
  general theory, xiii, 35
  Galilean principle, *98–99*
  special theory, 163, 247, 377, 467–477
religion and science, 22, 23, 28, 31, 59, 64, 104, 153, 174–175, 252–253, 266
Renaissance, 17
resonance in quantum mechanics, 457–458
retina, 46–47
retrograde motion, *12–13,* 19, 23
reversibility paradox, 251, 260, 333
Revolution, Scientific, 25
Revolutions of the Heavenly Spheres. See Copernicus, Nicolaus
ρ (rho; density), 349
Richer, Jean (1630–1696, French astronomer), 21, *128,* 129, 130, 142
Robins, Benjamin (1707–1751, English engineer-mathematician), 228
Roll, P. G., 513
Roman Catholic Church, 20, 23
  Inquisition, 40, 58, 80
  trial of Galileo, 59–60
Roman numerals with Arabic equivalents, 536
romantic movement, 523
Rømer, Ole (1644–1710, Danish astronomer), 344
Röntgen, Wilhelm Conrad (1845–1923, German physicist), 383
  discovery of x-rays, 192, 382
Roosevelt, Franklin Delano (1882–1945, President of the United States 1933–1945), 374, 464
Rosen, Nathan, 455
rotation, 123–130
  acceleration due to, 125–126
  circular motion, 123–128
  diurnal, *3*
  frequency, *123*
  of infinite sphere, 8
  kinematics, 123–125
  kinetic energy, 230
  period, *123*
rotational inertia, *230*
Royal Society of London, 103, 211, 308
Rumford, (Count) Benjamin Thompson (1753–1914, American physicist), 236
  heat, 236–238, 243
  homogeneity of air, 276
Russell, Bertrand (1872–1970, English philosopher-mathematician), 105, 176, 193, 354, 518
Russell, Henry Norris (1877–1957, American astrophysicist), 490, 504
Rutherford, Ernest (1871–1937, New Zealand-British physicist), xiii, 17, 410, 508
  age of earth, 258, 414
  α-rays, 384
  α-particle scattering experiment, 417–420
  β-rays, 384
  neutron, 425
  nuclear model of atom, 417–421, 434
  proton, 387
  radioactivity, 409–410
  transmutation of nitrogen to oxygen, 499
Rydberg, Johannes Robert (1854–1919, Swedish physicist), 175
  constant, *433,* 434
Ryle, Martin (1918–1984, British astronomer), 511–512

s (distance), *65*
s (length of arc), 123
s_x (range of projectile), *89*
S_1 (private science), *171–172,* 173–174, 176–180
S_2 (public science), *171–172,* 173–174, 176–181, 521
Safronov, Viktor Sergeivitch (b. 1917, Russian astronomer), 493
Sagredo, Giovanni Francesco, 54
Salviati, Filippo, 54
Sarton, George (1884–1956, Belgian-American historian of science), 8–9, 160
satellites (of planets), 141
saturation of interatomic forces, 291, 457
Saturn (planet)
  destruction of outer planets by, 144
  and Jupiter, long inequality, 143
  mass, 138, 139
  motion, 20–21
  orbit, 21, 42
  period of revolution, 29
  satellites, 138
"save the appearances" (or "save the phenomena"), *5,* 6, 13, 20
scalar, *125,* 364
scandium (element), prediction and discovery, 303
Schelling, Friedrich (1775–1854, German philosopher), 369
Schmidt, Otto (1891–1956, Russian astronomer and Arctic researcher), 493
scholastics, 79
Schrödinger, Erwin (1887–1961, Austrian physicist), 449
  cat paradox, 456
  continuity thema, 519
  relativistic wave equation, 478
  wave mechanics, 449–452, 477
Schuster, Arthur (1851–1934, British physicist), 382
Schwarzschild, Karl (1873–1916, German astronomer), 483
  singularity, 483–484
science
  distinguished from non-science, 158
  private, 171–172, 173–174, 176–180
  public, 171–172, 173–174, 176–181, 521
scientific method or procedure, 57, 147, 155–156, 159, 187–191
Scientific Revolution, 25, 187
Seaborg, Glenn (1912–1999, American nuclear chemist), 302, 508
seaborgium (element), 302
seasons, length of, 4, 13, 44

sec (secant), *547*
sec (second), *537*
Second Scientific Revolution, 407
seconds pendulum, *128*
secular acceleration of moon, *143*
secular inequalities, *144*
Segrè, Emilio (1905–1989, Italian-American physicist), 387, 500
Seguin, Marc (1786–1875, French engineer), 244
selection of ideas, 179
Settle, Thomas, 82, 85
Shapley, Harlow (1885–1972, American astronomer), 494–495
shell model
  atomic structure, 443–444
SI (Système Internationale) units, xiv, 224, 541
Σ (sigma; summation), *210, 213*
simplicity, 24, 35, 38, 45, 85, 281, 284
Simplicius (6th century A.D. Greek philosopher), 54
simultaneity, 471–472
sin (sine), *547*
Singer, Charles, 204
Sizzi, Francesco, 52–53
Slipher, Vesto Melvin (1875–1969, American astronomer), 494
Smoot, George, 514
Smyth, Henry DeWolf (1898–1986, American physicist), 171, 320–321
Snell, Willebrord (1591–1626, Dutch mathematician)
  law of refraction, 63, 341–342
Snow, Charles Percy (1905–1980, English physicist and author), 261, 507
social construction of scientific knowledge, *522, 525*
societies, scientific, 103
sociology of science, 104
Socratic dialogue, *54–55, 57*
Soddy, Frederick (1877–1956, British physicist), 415
  age of earth, 258
  radioactivity, 410
  transmutation, 410
sodium
  ion, 367
  spectrum, 430
  valence, 290
sodium hydroxide, atomic weight, 284
solar system, 21, 42
  motion through space, 99, 149
  position in Milky Way galaxy, 150
  stability, 144
solar system origin, 247, 487–494, 497. *See also* nebular hypothesis
  Chamberlin-Moulton theory, 489–490
  Jeans-Jeffreys theory, 490
  Safronov-Wetherill theory, 492–493
  tidal theory, 490
solid, motion of molecules in, 310
solstice
  summer, *3, 9*
  winter, *3*
Sommerfeld, Arnold (1868–1951, German mathematician-physicist), 175, 443, 459
soul, atomic theory of, 31
sound, speed of, 318
Soviet Union, science in, 180
space, 485
  absolute, *106,* 153, 467, 468
  divisibility of, 30
  relative, *106*
span (unit of length), *81*
specific heat, *235*
  at constant pressure, 241
  at constant volume, 241
spectrometer, 429
spectrum, 350–351
  absorption line, 428–430
  continuous, 388–395
  convergence limit, 432
  electromagnetic radiation, 402
  emission line, 427–428
  x-ray, 422–423
speed, *65–66*
  average, *68*
  constant, *65–66*
  infinite, 8
  instantaneous, *69, 85*
  planet, variation with distance from sun, 44
sphere. *See* celestial sphere
spheroid, *142*
spring
  balance, 108, 113, 116
  elastic potential energy of compression, 230
stadium (unit of distance), *9*
stars
  chemical composition, 492
  diurnal motion, 3, 25
  element formation, *503–505,* 511
  gravitational collapse, 483
  proper motion (relative to other stars), *149*
  size, 53
stationary state, 436, 451
statistical view of nature (style of thought), 322, *527–528*
steady state cosmology, 510–512
  thematic dimension, 521
steam engine, 256–257
Stern, Otto (1888–1969, German physicist), 324
Stevenson, E. C., 501
Stevinus (or Stevin), Simon (1548–1620, Dutch engineer and mathematician), 78
Stoney, George Johnstone (1826–1911, Irish physicist), 367, 380
Strassmann, Fritz (1902–1980, German chemist), 506
Stratton, George M. (American psychologist), 47
Street, J. C., 501
Strömgren, Bengt, 504
Strutt, Robert John (1875–1947, British physicist), 414
Stukely, William (biographer of Newton), 105
styles of thought in science and culture, *522–530*
subjectivism, 454
suction, 267, 269–270
sulfur
  atom, 279
  atomic weight, 284

sun
  age, 488
  energy source, 247, *505*
  force on comet's tail, 217
  apparent motion, 3–4, 14
  distance from earth, 21, 129
  g at surface, 113
  mass, 138, 139
  radiation from surface, 389
  spectrum, 429
  tides, effect on earth's, 143
supernova, 484
symbols, *531–533*
symmetry
  and conservation laws, 247
  particle-antiparticle, 479
  of wave function, 457–458
system
  dissipative, 227
  isolated, 203–204, *245*
Système Internationale (SI), 224
Szilard, Leo (1898–1964, Hungarian-American physicist), 507, 509

T (half-life for radioactive decay), *413*
T (Kelvin temperature), *271*
T (period), 123, 128
T (tension), *117*
t (time), *65*
T_C (Celsius temperature), *241*
tan (tangent), 69, *547*
Tartaglia, Niccolò (c. 1500–1577, Italian engineer and mathematician), 78
Taylor, F. S., 79–80
Taylor, Joseph (American astronomer), 484
technology and science, 185
teleology, 43, 79, 174, 267
telescope, *52–54*, 63, 104
tellurium (element), atomic weight, 299, 302
temperature
  absolute zero, 241, 327
  Celsius scale, *241*
  centigrade scale, *241*
  Kelvin scale, 241, *271*
  and molecular kinetic energy, 319–320
  seasonal variations, 4, 144
tension, *117*
thematic elements, themata, 517–522
  pairs of opposing, *519*
  three competing, *519*
theories, *27*
  criteria for good, 35–38, 480
  plausibility, 36
  purpose, 27–29
  role of abandoned, 28
thermionic tube, 397
thermodynamics
  first law, *245*
  second law, 251, 256, *257–261*, 335–336, 496
  third law, *271*
θ (theta; angular measure for circular motion), *123–124*
θ_i (angle of incidence), *33*
θ_r (angle of refraction), *33*
Thomas Aquinas, Saint (1225–1274, Italian philosopher-theologian), 23
Thomson, George Paget (1892–1975, British physicist)
  electron diffraction, 448
  Maxwell and Dirac theories, 479
Thomson, Joseph John (1856–1940, English physicist), 379, 385–386
  atomic model, 417
  discovery of electron, 379, 385–386, 396
thorium (element)
  radioactive decay series, 415, 416–417
thought experiment, *57*, 82, 110, 190, 470–471
tidal theory of origin of solar system, 490–491
tides, *143*
  neap, *143*
  in solid earth, 143
  spring, *143*
time
  absolute, *106*, 467
  conversion factors, *539*
  dilation, 467, 474
  direction ("arrow"), 260
  divisibility of, 30
  relative, *106*
  simultaneity, 471–472
Titius, Johann Daniel (1729–1796, German astronomer), 146
Titius-Bode law, *146*
Tombaugh, Clyde William (1906–1997, American astronomer), 145
Torricelli, Evangelista (1608–1647, Italian physicist and mathematician), 103, 267–268
  barometer, 267–268
Towneley, Richard (1628– 1707, English scientist), 270
tradition vs. innovation in science, 36
trajectory, *88*, 97
transmutation, 419, 499
transparency, 32
transuranic elements, 305, 508
trigonometry, *547–549*, *552*, *553*
Tropic of Cancer, *9*
twin paradox, 474
Tychonic system, 25
Tyndall, John (1820–1893, British physicist), 242

Uhlenbeck, George Eugene (1900–1988, Dutch-American physicist), 444
ultraviolet catastrophe, 390
ultraviolet light, 402
uncertainty principle (indeterminacy principle), *453*, 500
Uniformitarianism, *254–255*, 258
uniformity, 6, 18, 45
  of circular motion, *124*
Union of Concerned Scientists, 509
units, systems of, 541–542
universe
  age, 496, 510
  cyclic, 336–337, 513
  expanding, 151, 495–496, 514
  island, 494
  mass, 497
  origin, 515. *See also* Big Bang cosmology
uranium, 36
  atomic weight, 299
  decay series, *415–416*
  fission, 506–509
  isotope separation, 320–322, 508
  masses of isotopes, 412

radioactive decay of isotopes, 414
uranium hexafluoride, 321
molecular weight, 293
Uranus (planet)
discovery, 144–145, 146
mass, 139
orbit, 42
Urey, Harold C. (1893–1981, American nuclear chemist), 501

v (speed or velocity), *68–69*
V_s (stopping potential), 397
v_w (speed of wave), *345*
vacuum, 82
motion in, 79
"Nature abhors," 267
pump, 104, 267–269
valence, *290*–292, 306, 444
shell of electrons, 290
values of physical quantities (defined), 537
Van de Graaff, Robert Jemison (1901–1967, American physicist)
generator, 365
van den Broek, Antonius (1870–1926, Dutch lawyer-physicist), 420
van der Waals, Johannes Diderik (1837–1923, Dutch physicist), 310
van Leeuwen, Johanna (Dutch physicist), 459
Varley, Cromwell Fleetwood (1828–1883, English engineer), 381
Varmus, Harold, 184
vector, *93,* 108, 125, 211, 551–554
Velikovsky, Immanuel, 37, 249
velocity, 85, 89–91. *See also* speed
circular motion, *124*
conversion factors, 539
superposition of components, *91,* 165
vector, 91–95, 125
Venus (planet)
mass, 139
motion, 14, 20–21, 29
orbit, 21
period of revolution, 21, 29
phases, 29, 54
source of illumination, 29
vernal equinox, *4,* 13, 20. *See also* equinox
Villard, Paul, 384
viscosity of gas, 327
vision, 31
extramission theory, 2, *31,* 32–35
intromission theory, 2, 34–35, 46–47
visual fire, *31,* 34
visualizability, 444
vis viva (living force), *212,* 219, 309. *See also* energy, kinetic
Viviani, Vincenzo (1622–1703, Italian scientist), 51
volt (unit of potential difference), *362, 371*
Volta, Alessandro (1745–1827, Italian physicist), 365
volume, conversion factors, 539

W (weight), *112*
Wallis, John (1616–1703, English mathematician), 103, 211
Walton, Ernest Thomas Sinton (b. 1903. Irish-British physicist), 502
Wambacher, Hertha, 502
Ward, W. R., 493
water (compound)
molecule formed from hydrogen and oxygen, 276, 280, 284, 286–287, 290
triple point, 241, *272*
Water (element), 7
Watt, James (1736–1819, Scottish inventor), 182
wave
frequency, *346*
longitudinal, *345*
node, 346
periodicity, 344–35
phase, *345*
speed, 345
stationary (standing), 346
transverse, *345*
wavelength, *345*
wave function, *450*–452, 454
collapse, 456
complex-valued, 452
determinism and, 454
entangled, 455, 456–457
phase invariance and charge conservation, 452
wave mechanics, 449–452
wave-photon duality, 400–402, 446
weight, *112*–113. *See also* atomic weight
weightlessness, *114*
Weinberg, Steven (b. 1933, American physicist), 512
Weizsäcker, Carl Friedrich von (b. 1912, German physicist), 493, 503
Wetherill, George West (b. 1925, American geophysicist), 493
Wheeler, John Archibald (b. 1911, American physicist)
anthropic principle, 514
black hole, 483
nuclear fission, theory, 507
Whipple, F. L., 138
Whitehead, Alfred North (1861–1947, British philosopher), 36, 101, 522
Whittaker, Edmund Taylor (1873–1956, British physicist and historian of science), 209
Wieman, Carl E., 460
Wien, Wilhelm (1864–1928, German physicist), 384, 389
Wilkinson, D. T., 513
William of Ockham (c. 1285–c. 1349, English philosopher), 209
Wilson, Robert Woodrow (b. 1936, American astrophysicist), 512–513
Winckler, Clemens (1838–1904, German chemist), 302
women in science and mathematics, 248, 256, 385, 409, 425, 459, 460, 491–492, 494, 502, 504, 506, 507, 511
work, *223–229*
done by expanding gas, 241
world machine. *See* clockwork universe
Wren, Christopher (1632–1723, English architect), 103, 105, 211

x-rays, 192, 217, 382–385
apparatus for producing, 403
diffraction, 385
electromagnetic radiation pulses, 385
electromagnetic waves, 385
longitudinal ether waves, 385
photons, 402–403
range of wavelengths, 402

Young, Thomas (1773–1829, English physician and physicist)
double-slit interference experiment, *347–348*
light, 238, 347–348, 401
Yukawa, Hideki (1907–1981, Japanese physicist), 501–502
yukon, 501

Z (atomic number), 420
Z (nuclear charge), 499
Zeeman, Pieter (1865–1943, Dutch physicist), 379
Zeno of Elea (c. 490 B.C.–c. 430 B.C., Greek philosopher), 30
paradoxes, 30, 36, 67
Zermelo, Ernst (1871–1953, German mathematician), 336–337
Zodiac, 3, 4
Zwicky, Fritz (1898–1974, Swiss-American astrophysicist), 503